ESSENTIAL ARMY MANUAL

CONTAINS
SURVIVAL ★ LEADERSHIP ★ FITNESS ★ CONCEALMENT
Contains Hundreds More on DVD!

MW00680353

TABLE OF CONTENTS

© 2011 by WaterMark, Inc.

Questions and concerns please contact us at info@armymanualsonline.com

ISBN-978-1-882077-24-3

Book Design: TaylorDesign

Printed in China

FM 21-76   US ARMY SURVIVAL MANUAL

TABLE OF CONTENTS

CHAPTER 1 - INTRODUCTION

This manual is based entirely on the keyword SURVIVAL. The letters in this word can help guide you in your actions in any survival situation. Whenever faced with a survival situation, remember the word SURVIVAL.

SURVIVAL ACTIONS

The following paragraphs expand on the meaning of each letter of the word survival. Study and remember what each letter signifies because you may some day have to make it work for you.

S -Size Up the Situation

If you are in a combat situation, find a place where you can conceal yourself from the enemy. Remember, security takes priority. Use your senses of hearing, smell, and sight to get a feel for the battlefield. What is the enemy doing? Advancing? Holding in place? Retreating? You will have to consider what is developing on the battlefield when you make your survival plan.

Size Up Your Surroundings
Determine the pattern of the area. Get a feel for what is going on around you. Every environment, whether forest, jungle, or desert, has a rhythm or pattern. This rhythm or pattern includes animal and bird noises and movements and insect sounds. It may also include enemy traffic and civilian movements.

Size Up Your Physical Condition
The pressure of the battle you were in or the trauma of being in a survival situation may have caused you to overlook wounds you received. Check your wounds and give yourself first aid. Take care to prevent further bodily harm. For instance, in any climate, drink plenty of water to prevent dehydration. If you are in a cold or wet climate, put on additional clothing to prevent hypothermia.

Size Up Your Equipment
Perhaps in the heat of battle, you lost or damaged some of your equipment. Check to see what equipment you have and what condition it is in.

Now that you have sized up your situation, surroundings, physical condition, and equipment, you are ready to make your survival plan. In doing so, keep in mind your basic physical needs--water, food, and shelter.

U -Use All Your Senses, Undue Haste Makes Waste

You may make a wrong move when you react quickly without thinking or planning. That move may result in your capture or death. Don't move just for the sake of taking action. Consider all aspects of your situation (size up your situation) before you make a decision and a move. If you act in haste, you may forget or lose some of your equipment. In your haste you may also become disoriented so that you don't know which way to go. Plan your moves. Be ready to move out quickly without endangering yourself if the enemy is near you. Use all your senses to evaluate the situation. Note sounds and smells. Be sensitive to temperature changes. Be observant.

R -Remember Where You Are

Spot your location on your map and relate it to the surrounding terrain. This is a basic principle that you must always follow. If there are other persons with you, make sure they also know their location. Always know who in your group, vehicle, or aircraft has a map and compass. If that person is killed, you will have to get the map and compass from him. Pay close attention to where you are and to where you are going. Do not rely on others in the group to keep track of the route. Constantly orient yourself. Always try to determine, as a minimum, how your location relates to--

- The location of enemy units and controlled areas.
- The location of friendly units and controlled areas.
- The location of local water sources (especially important in the desert).
- Areas that will provide good cover and concealment.

This information will allow you to make intelligent decisions when you are in a survival and evasion situation.

V -Vanquish Fear and Panic

The greatest enemies in a combat survival and evasion situation are fear and panic. If uncontrolled, they can destroy your ability to make an intelligent decision. They may cause you to react to your feelings and imagination rather than to your situation. They can drain your energy and thereby cause other negative emotions. Previous survival and evasion training and self-confidence will enable you to vanquish fear and panic.

I -Improvise

In the United States, we have items available for all our needs. Many of these items are cheap to replace when damaged. Our easy come, easy go, easy-to-replace culture makes it unnecessary for us to improvise. This inexperience in improvisation can be an enemy in a survival situation. Learn to improvise. Take a tool designed for a specific purpose and see how many other uses you can make of it.
Learn to use natural objects around you for different needs. An example is using a rock for a hammer. No matter how complete a survival kit you have with you, it will run out or wear out after a while. Your imagination must take over when your kit wears out.

V -Value Living

All of us were born kicking and fighting to live, but we have become used to the soft life. We have become creatures of comfort. We dislike inconveniences and discomforts. What happens when we are faced with a survival situation with its stresses, inconveniences, and discomforts? This is when the will to live- placing a high value on living-is vital. The experience and knowledge you have gained through life and your Army training will have a bearing on your will to live. Stubbornness, a refusal to give in to problems and obstacles that face you, will give you the mental and physical strength to endure.

A -Act Like the Natives

The natives and animals of a region have adapted to their environment. To get a feel of the area, watch how the people go about their daily routine. When and what do they eat? When, where, and how do they get their food? When and where do they go for water? What time do they usually go to bed and get up? These actions are important to you when you are trying to avoid capture.
Animal life in the area can also give you clues on how to survive. Animals also require food, water, and shelter. By watching them, you can find sources of water and food.

Keep in mind that the reaction of animals can reveal your presence to the enemy.

If in a friendly area, one way you can gain rapport with the natives is to show interest in their tools and how they get food and water. By studying the people, you learn to respect them, you often make valuable friends, and, most important, you learn how to adapt to their environment and increase your chances of survival.

L -Live by Your Wits, *But for Now,* Learn Basic Skills

Without training in basic skills for surviving and evading on the battlefield, your chances of living through a combat survival and evasion situation are slight.

Learn these basic skills **now**--not when you are headed for or are in the battle. How you decide to equip yourself before deployment will impact on whether or not you survive. You need to know about the environment to which you are going, and you must practice basic skills geared to that environment. For instance, if you are going to a desert, you need to know how to get water in the desert.

Practice basic survival skills during all training programs and exercises. Survival training reduces fear of the unknown and gives you self-confidence. It teaches you to *live by your wits.*

S   **Size Up the Situation**
(Surroundings, Physical Condition, Equipment)

U   **Use All Your Senses,**
Undue Haste Makes Waste

R   **Remember Where You Are**

V   **Vanquish Fear and Panic**

I   **Improvise**

V   **Value Living**

A   **Act Like the Natives**

L   **Live by Your Wits,** *But for Now,* **Learn Basic Skills**

PATTERN FOR SURVIVAL

Develop a survival pattern that lets you beat the enemies of survival. This survival pattern must include food, water, shelter, fire, first aid, and signals placed in order of importance. For example, in a cold environment, you would need a *fire* to get warm; a *shelter* to protect you from the cold, wind, and rain or snow; traps or snares to get *food;* a means to *signal* friendly aircraft; and *first aid* to maintain health. *If injured, first aid has top priority* no matter what climate you are in.

Change your survival pattern to meet your immediate physical needs as the environment changes.

As you read the rest of this manual, keep in mind the keyword SURVIVAL and the need for a survival pattern.

CHAPTER 2 - PSYCHOLOGY OF SURVIVAL

It takes much more than the knowledge and skills to build shelters, get food, make fires, and travel without the aid of standard navigational devices to live successfully through a survival situation. Some people with little or no survival training have managed to survive life-threatening circumstances. Some people with survival training have not used their skills and died. A key ingredient in any survival situation is the mental attitude of the individual(s) involved. Having survival skills is important; having the will to survive is essential. Without a desk to survive, acquired skills serve little purpose and invaluable knowledge goes to waste.

*There is a psychology to survival. The soldier in a survival environment faces many stresses that ultimately impact on his mind. These stresses can produce thoughts and emotions that, if poorly understood, can transform a confident, well-trained soldier into an indecisive, ineffective individual with questionable ability to survive. Thus, every soldier must be aware of and be able to recognize those stresses commonly associated with survival. Additionally, it is imperative that soldiers be aware of their reactions to the wide variety of stresses associated with survival. This chapter will identify and explain the nature of stress, the stresses of survival, and those internal reactions soldiers will naturally experience when faced with the stresses of a real-world survival situation. The knowledge you, the soldier, gain from this chapter and other chapters in this manual, will prepare you to come through the toughest times **alive.***

A LOOK AT STRESS

Before we can understand our psychological reactions in a survival setting, it is helpful to first know a little bit about stress.
Stress is not a disease that you cure and eliminate. Instead, it is a condition we all experience. Stress can be described as our reaction to pressure. It is the name given to the experience we have as we physically, mentally, emotionally, and spiritually respond to life's tensions.

Need for Stress

We need stress because it has many positive benefits. Stress provides us with challenges; it gives us chances to learn about our values and strengths. Stress can show our ability to handle pressure without breaking; it tests our adaptability and flexibility; it can stimulate us to do our best. Because we usually do not consider unimportant events stressful, stress can also be an excellent indicator of the significance we attach to an event--in other words, it highlights what is important to us.
We need to have some stress in our lives, but too much of anything can be bad. The goal is to have stress, but not an excess of it. Too much stress can take its toll on people and organizations. Too much stress leads to distress. Distress causes an uncomfortable tension that we try to escape and, preferably,

avoid. Listed below are a few of the common signs of distress you may find in your fellow soldiers or yourself when faced with too much stress:

- Difficulty making decisions.
- Angry outbursts.
- Forgetfulness.
- Low energy level.
- Constant worrying.
- Propensity for mistakes.
- Thoughts about death or suicide.
- Trouble getting along with others.
- Withdrawing from others.
- Hiding from responsibilities.
- Carelessness.

As you can see, stress can be constructive or destructive. It can encourage or discourage, move us along or stop us dead in our tracks, and make life meaningful or seemingly meaningless. Stress can inspire you to operate successfully and perform at your maximum efficiency in a survival situation. It can also cause you to panic and forget all your training. Key to your survival is your ability to manage the inevitable stresses you will encounter. The survivor is the soldier who works with his stresses instead of letting his stresses work on him.

Survival Stressors

Any event can lead to stress and, as everyone has experienced, events don't always come one at a time. Often, stressful events occur simultaneously. These events are not stress, but they produce it and are called "stressors." Stressors are the obvious cause while stress is the response. Once the body recognizes the presence of a stressor, it then begins to act to protect itself.
In response to a stressor, the body prepares either to "fight or flee." This preparation involves an internal SOS sent throughout the body. As the body responds to this SOS, several actions take place. The body releases stored fuels (sugar and fats) to provide quick energy; breathing rate increases to supply more oxygen to the blood; muscle tension increases to prepare for action; blood clotting mechanisms are activated to reduce bleeding from cuts; senses become more acute (hearing becomes more sensitive, eyes become big, smell becomes sharper) so that you are more aware of your surrounding and heart rate and blood pressure rise to provide more blood to the muscles. This protective posture lets a person cope with potential dangers; however, a person cannot maintain such a level of alertness indefinitely. Stressors are not courteous; one stressor does not leave because another one arrives. Stressors add up. The cumulative effect of minor stressors can be a major distress if they all happen too close together. As the body's resistance to stress wears down and the sources of stress continue (or increase), eventually a state of exhaustion arrives. At this point, the ability to resist stress or use it in a positive way gives out and signs of distress appear. Anticipating stressors and developing strategies to cope with them are two ingredients in the effective management of stress. It is therefore essential that the soldier in a survival setting be aware of the types of stressors he will encounter. Let's take a look at a few of these.
Injury, Illness, or Death
Injury, illness, and death are real possibilities a survivor has to face. Perhaps nothing is more stressful than being alone in an unfamiliar environment where you could die from hostile action, an accident, or from eating something lethal. Illness and injury can also add to stress by limiting your ability to maneuver, get food and drink, find shelter, and defend yourself. Even if illness and injury don't lead to death, they add to stress through the pain and discomfort they generate. It is only by con-trolling the stress associated with the vulnerability to injury, illness, and death that a soldier can have the courage to take the risks associated with survival tasks.
Uncertainly and Lack of Control
Some people have trouble operating in settings where everything is not clear-cut. The only guarantee in a survival situation is that nothing is guaranteed. It can be extremely stressful operating on limited

information in a setting where you have limited control of your surroundings. This uncertainty and lack of control also add to the stress of being ill, injured, or killed.

Environment

Even under the most ideal circumstances, nature is quite formidable. In survival, a soldier will have to contend with the stressors of weather, terrain, and the variety of creatures inhabiting an area. Heat, cold, rain, winds, mountains, swamps, deserts, insects, dangerous reptiles, and other animals are just a few of the challenges awaiting the soldier working to survive. Depending on how a soldier handles the stress of his environment, his surroundings can be either a source of food and protection or can be a cause of extreme discomfort leading to injury, illness, or death.

Hunger and Thirst

Without food and water a person will weaken and eventually die. Thus, getting and preserving food and water takes on increasing importance as the length of time in a survival setting increases. For a soldier used to having his provisions issued, foraging can be a big source of stress.

Fatigue

Forcing yourself to continue surviving is not easy as you grow more tired. It is possible to become so fatigued that the act of just staying awake is stressful in itself.

Isolation

There are some advantages to facing adversity with others. As soldiers we learn individual skills, but we train to function as part of a team. Although we, as soldiers, complain about higher headquarters, we become used to the information and guidance it provides, especially during times of confusion. Being in contact with others also provides a greater sense of security and a feeling someone is available to help if problems occur. A significant stressor in survival situations is that often a person or team has to rely solely on its own resources.

The survival stressors mentioned in this section are by no means the only ones you may face. Remember, what is stressful to one person may not be stressful to another. Your experiences, training, personal outlook on life, physical and mental conditioning, and level of self-confidence contribute to what you will find stressful in a survival environment. The object is not to avoid stress, but rather to manage the stressors of survival and make them work for you.

We now have a general knowledge of stress and the stressors common to survival; the next step is to examine our reactions to the stressors we may face.

NATURAL REACTIONS

Man has been able to survive many shifts in his environment throughout the centuries. His ability to adapt physically and mentally to a changing world kept him alive while other species around him gradually died off. The same survival mechanisms that kept our forefathers alive can help keep us alive as well! However, these survival mechanisms that can help us can also work against us if we don't understand and anticipate their presence.

It is not surprising that the average person will have some psychological reactions in a survival situation. We will now examine some of the major internal reactions you and anyone with you might experience with the survival stressors addressed in the earlier paragraphs. Let's begin.

Fear

Fear is our emotional response to dangerous circumstances that we believe have the potential to cause death, injury, or illness. This harm is not just limited to physical damage; the threat to one's emotional and mental well-being can generate fear as well. For the soldier trying to survive, fear can have a positive function if it encourages him to be cautious in situations where recklessness could result in injury. Unfortunately, fear can also immobilize a person. It can cause him to become so frightened that he fails to perform activities essential for survival. Most soldiers will have some degree of fear when placed in unfamiliar surroundings under adverse conditions. There is no shame in this! Each soldier must train himself not to be overcome by his fears. Ideally, through realistic training, we can acquire the knowledge and skills needed to increase our confidence and thereby manage our fears.

Anxiety

Associated with fear is anxiety. Because it is natural for us to be afraid, it is also natural for us to experience anxiety. Anxiety can be an uneasy, apprehensive feeling we get when faced with dangerous situations (physical, mental, and emotional). When used in a healthy way, anxiety urges us to act to end, or at least master, the dangers that threaten our existence. If we were never anxious, there would be little motivation to make changes in our lives. The soldier in a survival setting reduces his anxiety by performing those tasks that will ensure his coming through the ordeal alive. As he reduces his anxiety, the soldier is also bringing under control the source of that anxiety--his fears. In this form, anxiety is good; however, anxiety can also have a devastating impact. Anxiety can overwhelm a soldier to the point where he becomes easily confused and has difficulty thinking. Once this happens, it becomes more and more difficult for him to make good judgments and sound decisions. To survive, the soldier must learn techniques to calm his anxieties and keep them in the range where they help, not hurt.

Anger and Frustration

Frustration arises when a person is continually thwarted in his attempts to reach a goal. The goal of survival is to stay alive until you can reach help or until help can reach you. To achieve this goal, the soldier must complete some tasks with minimal resources. It is inevitable, in trying to do these tasks, that something will go wrong; that something will happen beyond the soldier's control; and that with one's life at stake, every mistake is magnified in terms of its importance. Thus, sooner or later, soldiers will have to cope with frustration when a few of their plans run into trouble. One outgrowth of this frustration is anger. There are many events in a survival situation that can frustrate or anger a soldier. Getting lost, damaged or forgotten equipment, the weather, inhospitable terrain, enemy patrols, and physical limitations are just a few sources of frustration and anger. Frustration and anger encourage impulsive reactions, irrational behavior, poorly thought-out decisions, and, in some insta nces, an "I quit" attitude (people sometimes avoid doing something they can't master). If the soldier can harness and properly channel the emotional intensity associated with anger and frustration, he can productively act as he answers the challenges of survival. If the soldier does not properly focus his angry feelings, he can waste much energy in activities that do little to further either his chances of survival or the chances of those around him.

Depression

It would be a rare person indeed who would not get sad, at least momentarily, when faced with the privations of survival. As this sadness deepens, we label the feeling "depression." Depression is closely linked with frustration and anger. The frustrated person becomes more and more angry as he fails to reach his goals. If the anger does not help the person to succeed, then the frustration level goes even higher. A destructive cycle between anger and frustration continues until the person becomes worn down-physically, emotionally, and mentally. When a person reaches this point, he starts to give up, and his focus shifts from "What can I do" to "There is nothing I can do." Depression is an expression of this hopeless, helpless feeling. There is nothing wrong with being sad as you temporarily think about your loved ones and remember what life is like back in "civilization" or "the world." Such thoughts, in fact, can give you the desire to try harder and live one more day. On the other hand, if you allow yours elf to sink into a depressed state, then it can sap all your energy and, more important, your will to survive. It is imperative that each soldier resist succumbing to depression.

Loneliness and Boredom

Man is a social animal. This means we, as human beings, enjoy the company of others. Very few people want to be alone *all the time!* As you are aware, there is a distinct chance of isolation in a survival setting. This is not bad. Loneliness and boredom can bring to the surface qualities you thought only others had. The extent of your imagination and creativity may surprise you. When required to do so, you may discover some hidden talents and abilities. Most of all, you may tap into a reservoir of inner strength and fortitude you never knew you had. Conversely, loneliness and boredom can be another source of depression. As a soldier surviving alone, or with others, you must find ways to keep your mind productively occupied. Additionally, you must develop a degree of self-sufficiency. You must have faith in your capability to "go it alone."

Guilt

The circumstances leading to your being in a survival setting are sometimes dramatic and tragic. It may be the result of an accident or military mission where there was a loss of life. Perhaps you were the only, or one of a few, survivors. While naturally relieved to be alive, you simultaneously may be mourning the deaths of others who were less fortunate. It is not uncommon for survivors to feel guilty about being spared from death while others were not. This feeling, when used in a positive way, has encouraged people to try harder to survive with the belief they were allowed to live for some greater purpose in life. Sometimes, survivors tried to stay alive so that they could carry on the work of those killed. Whatever reason you give yourself, do not let guilt feelings prevent you from living. The living who abandon their chance to survive accomplish nothing. Such an act would be the greatest tragedy.

PREPARING YOURSELF

Your mission as a soldier in a survival situation is to stay alive. As you can see, you are going to experience an assortment of thoughts and emotions. These can work for you, or they can work to your downfall. Fear, anxiety, anger, frustration, guilt, depression, and loneliness are all possible reactions to the many stresses common to survival. These reactions, when controlled in a healthy way, help to increase a soldier's likelihood of surviving. They prompt the soldier to pay more attention in training, to fight back when scared, to take actions that ensure sustenance and security, to keep faith with his fellow soldiers, and to strive against large odds. When the survivor cannot control these reactions in a healthy way, they can bring him to a standstill. Instead of rallying his internal resources, the soldier listens to his internal fears. This soldier experiences psychological defeat long before he physically succumbs. Remember, survival is natural to everyone; being unexpectedly thrust into the life and death struggle of survival is not. Don't be afraid of your "natural reactions to this unnatural situation." Prepare yourself to rule over these reactions so they serve your ultimate interest--staying alive with the honor and dignity associated with being an American soldier.

It involves preparation to ensure that your reactions in a survival setting are productive, not destructive. The challenge of survival has produced countless examples of heroism, courage, and self-sacrifice. These are the qualities it can bring out in you if you have prepared yourself. Below are a few tips to help prepare yourself psychologically for survival. Through studying this manual and attending survival training you can develop the *survival attitude.*

Know Yourself

Through training, family, and friends take the time to discover who you are on the inside. Strengthen your stronger qualities and develop the areas that you know are necessary to survive.

Anticipate Fears

Don't pretend that you will have no fears. Begin thinking about what would frighten you the most if forced to survive alone. Train in those areas of concern to you. The goal is not to eliminate the fear, but to build confidence in your ability to function despite your fears.

Be Realistic

Don't be afraid to make an honest appraisal of situations. See circumstances as they are, not as you want them to be. Keep your hopes and expectations within the estimate of the situation. When you go into a survival setting with unrealistic expectations, you may be laying the groundwork for bitter disappointment. Follow the adage, "Hope for the best, prepare for the worst." It is much easier to adjust to pleasant surprises about one's unexpected good fortunes than to be upset by one's unexpected harsh circumstances.

Adopt a Positive Attitude

Learn to see the potential good in everything. Looking for the good not only boosts morale, it also is excellent for exercising your imagination and creativity.

Remind Yourself What Is at Stake

Remember, failure to prepare yourself psychologically to cope with survival leads to reactions such as depression, carelessness, inattention, loss of confidence, poor decision-making, and giving up before the body gives in. At stake is your life and the lives of others who are depending on you to do your share.

Train

Through military training and life experiences, begin today to prepare yourself to cope with the rigors of survival. Demonstrating your skills in training will give you the confidence to call upon them should the need arise. Remember, the more realistic the training, the less overwhelming an actual survival setting will be.

Learn Stress Management Techniques

People under stress have a potential to panic if they are not well-trained and not prepared psychologically to face whatever the circumstances may be. While we often cannot control the survival circumstances in which we find ourselves, it is within our ability to control our response to those circumstances. Learning stress management techniques can enhance significantly your capability to remain calm and focused as you work to keep yourself and others alive. A few good techniques to develop include relaxation skills, time management skills, assertiveness skills, and cognitive restructuring skills (the ability to control how you view a situation).
Remember, "the will to survive" can also be considered to be "the refusal to give up."

CHAPTER 3 - SURVIVAL PLANNING AND SURVIVAL KITS

Survival planning is nothing more than realizing something could happen that would put you in a survival situation and, with that in mind, taking steps to increase your chances of survival. Thus, survival planning means preparation. Preparation means having survival items and knowing how to use them People who live in snow regions prepare their vehicles for poor road conditions. They put snow tires on their vehicles, add extra weight in the back for traction, and they carry a shovel, salt, and a blanket. Another example of preparation is finding the emergency exits on an aircraft when you board it for a flight. Preparation could also mean knowing your intended route of travel and familiarizing yourself with the area. Finally, emergency planning is essential.

IMPORTANCE OF PLANNING

Detailed prior planning is essential in potential survival situations. Including survival considerations in mission planning will enhance your chances of survival if an emergency occurs. For example, if your job re-quires that you work in a small, enclosed area that limits what you can carry on your person, plan where you can put your rucksack or your load-bearing equipment. Put it where it will not prevent you from getting out of the area quickly, yet where it is readily accessible.

One important aspect of prior planning is preventive medicine. Ensuring that you have no dental problems and that your immunizations are current will help you avoid potential dental or health problems. A dental problem in a survival situation will reduce your ability to cope with other problems that you face. Failure to keep your shots current may mean your body is not immune to diseases that are prevalent in the area.

Preparing and carrying a survival kit is as important as the considerations mentioned above. All Army aircraft normally have survival kits on board for the type area(s) over which they will fly. There are kits for over-water survival, for hot climate survival, and an aviator survival vest (see Appendix A for a description of these survival kits and their contents). If you are not an aviator, you will probably not have access to the survival vests or survival kits. However, if you know what these kits contain, it will help you to plan and to prepare your own survival kit.

Even the smallest survival kit, if properly prepared, is invaluable when faced with a survival problem. Before making your survival kit, however, consider your unit's mission, the operational environment, and the equipment and vehicles assigned to your unit.

SURVIVAL KITS

The environment is the key to the types of items you will need in your survival kit. How much equipment you put in your kit depends on how you will carry the kit. A kit carried on your body will have to be smaller than one carried in a vehicle. Always layer your survival kit, keeping the most important items on your body. For example, your map and compass should always be on your body. Carry less important items on your load-bearing equipment. Place bulky items in the rucksack.

In preparing your survival kit, select items you can use for more than one purpose. If you have two items that will serve the same function, pick the one you can use for another function. Do not duplicate items, as this increases your kit's size and weight.

Your survival kit need not be elaborate. You need only functional items that will meet your needs and a case to hold the items. For the case, you might want to use a Band-Aid box, a first aid case, an ammunition pouch, or another suitable case. This case should be--

- Water repellent or waterproof.
- Easy to carry or attach to your body.
- Suitable to accept varisized components.
- Durable.

In your survival kit, you should have--

- First aid items.
- Water purification tablets or drops.
- Fire starting equipment.
- Signaling items.
- Food procurement items.
- Shelter items.

Some examples of these items are--

- Lighter, metal match, waterproof matches.
- Snare wire.
- Signaling mirror.
- Wrist compass.
- Fish and snare line.
- Fishhooks.
- Candle.
- Small hand lens.
- Oxytetracycline tablets (diarrhea or infection).
- Water purification tablets.
- Solar blanket.
- Surgical blades.
- Butterfly sutures.
- Condoms for water storage.
- Chap Stick.
- Needle and thread.
- Knife.

Include a weapon only if the situation so dictates. Read about and practice the survival techniques in this manual. Consider your unit's mission and the environment in which your unit will operate. Then prepare your survival kit.

CHAPTER 4 - BASIC SURVIVAL MEDICINE

Foremost among the many problems that can compromise a survivor's ability to return to safety are medical problems resulting from parachute descent and landing, extreme climates, ground combat, evasion, and illnesses contracted in captivity.

Many evaders and survivors have reported difficulty in treating injuries and illness due to the lack of training and medical supplies. For some, this led to capture or surrender.

Survivors have related feeling of apathy and helplessness because they could not treat themselves in this environment. The ability to treat themselves increased their morale and cohesion and aided in their survival and eventual return to friendly forces.

One man with a fair amount of basic medical knowledge can make a difference in the lives of many. Without qualified medical personnel available, it is you who must know what to do to stay alive.

REQUIREMENTS FOR MAINTENANCE OF HEALTH

To survive, you need water and food. You must also have and apply high personal hygiene standards.

Water

Your body loses water through normal body processes (sweating, urinating, and defecating). During average daily exertion when the atmospheric temperature is 20 degrees Celsius (C) (68 degrees Fahrenheit), the average adult loses and therefore requires 2 to 3 liters of water daily. Other factors, such as heat exposure, cold exposure, intense activity, high altitude, burns, or illness, can cause your body to lose more water. You must replace this water.

Dehydration results from inadequate replacement of lost body fluids. It decreases your efficiency and, if injured, increases your susceptibility to severe shock. Consider the following results of body fluid loss:

- A 5 percent loss of body fluids results in thirst, irritability, nausea, and weakness.
- A 10 percent loss results in dizziness, headache, inability to walk, and a tingling sensation in the limbs.
- A 15 percent loss results in dim vision, painful urination, swollen tongue, deafness, and a numb feeling in the skin.
- A loss greater than 15 percent of body fluids may result in death.

The most common signs and symptoms of dehydration are--

- Dark urine with a very strong odor.

- Low urine output.
- Dark, sunken eyes.
- Fatigue.
- Emotional instability.
- Loss of skin elasticity.
- Delayed capillary refill in fingernail beds.
- Trench line down center of tongue.
- Thirst. Last on the list because you are already 2 percent dehydrated by the time you crave fluids.

You replace the water as you lose it. Trying to make up a deficit is difficult in a survival situation, and thirst is not a sign of how much water you need.

Most people cannot comfortably drink more than 1 liter of water at a time. So, even when not thirsty, drink small amounts of water at regular intervals each hour to prevent dehydration.

If you are under physical and mental stress or subject to severe conditions, increase your water intake. Drink enough liquids to maintain a urine output of at least 0.5 liter every 24 hours.

In any situation where food intake is low, drink 6 to 8 liters of water per day. In an extreme climate, especially an arid one, the average person can lose 2.5 to 3.5 liters of water *per hour.* In this type of climate, you should drink 14 to 30 liters of water per day.

With the loss of water there is also a loss of electrolytes (body salts). The average diet can usually keep up with these losses but in an extreme situation or illness, additional sources need to be provided. A mixture of 0.25 teaspoon of salt to 1 liter of water will provide a concentration that the body tissues can readily absorb.

Of all the physical problems encountered in a survival situation, the loss of water is the most preventable. The following are basic guidelines for the prevention of dehydration:

- *Always drink water when eating.* Water is used and consumed as a part of the digestion process and can lead to dehydration.
- *Acclimatize.* The body performs more efficiently in extreme conditions when acclimatized.
- *Conserve sweat not water.* Limit sweat-producing activities but drink water.
- *Ration water.* Until you find a suitable source, ration your water sensibly. A daily intake of 500 cubic centimeter (0.5 liter) of a sugar-water mixture (2 teaspoons per liter) will suffice to prevent severe dehydration for at least a week, provided you keep water losses to a minimum by limiting activity and heat gain or loss.

You can estimate fluid loss by several means. A standard field dressing holds about 0.25 liter (one-fourth canteen) of blood. A soaked T-shirt holds 0.5 to 0.75 liter.

You can also use the pulse and breathing rate to estimate fluid loss. Use the following as a guide:

- With a 0.75 liter loss the wrist pulse rate will be under 100 beats per minute and the breathing rate 12 to 20 breaths per minute.
- With a 0.75 to 1.5 liter loss the pulse rate will be 100 to 120 beats per minute and 20 to 30 breaths per minute.
- With a 1.5 to 2 liter loss the pulse rate will be 120 to 140 beats per minute and 30 to 40 breaths per minute. Vital signs above these rates require more advanced care.

Food

Although you can live several weeks without food, you need an adequate amount to stay healthy. Without food your mental and physical capabilities will deteriorate rapidly, and you will become weak. Food replenishes the substances that your body burns and provides energy. It provides vitamins, minerals, salts, and other elements essential to good health. Possibly more important, it helps morale. The two basic sources of food are plants and animals (including fish). In varying degrees both provide the calories, carbohydrates, fats, and proteins needed for normal daily body functions.

Calories are a measure of heat and potential energy. The average person needs 2,000 calories per day to function at a minimum level. An adequate amount of carbohydrates, fats, and proteins without an adequate caloric intake will lead to starvation and cannibalism of the body's own tissue for energy.

Plant Foods

These foods provide carbohydrates--the main source of energy. Many plants provide enough protein to keep the body at normal efficiency. Although plants may not provide a balanced diet, they will sustain you even in the arctic, where meat's heat-producing qualities are normally essential. Many plant foods such as nuts and seeds will give you enough protein and oils for normal efficiency. Roots, green vegetables, and plant food containing natural sugar will provide calories and carbohydrates that give the body natural energy.
The food value of plants becomes more and more important if you are eluding the enemy or if you are in an area where wildlife is scarce. For instance--

- You can dry plants by wind, air, sun, or fire. This retards spoilage so that you can store or carry the plant food with you to use when needed.
- You can obtain plants more easily and more quietly than meat. This is extremely important when the enemy is near.

Animal Foods

Meat is more nourishing than plant food. In fact, it may even be more readily available in some places. However, to get meat, you need to know the habits of, and how to capture, the various wildlife.
To satisfy your immediate food needs, first seek the more abundant and more easily obtained wildlife, such as insects, crustaceans, mollusks, fish, and reptiles. These can satisfy your immediate hunger while you are preparing traps and snares for larger game.

Personal Hygiene

In any situation, cleanliness is an important factor in preventing infection and disease. It becomes even more important in a survival situation. Poor hygiene can reduce your chances of survival.
A daily shower with hot water and soap is ideal, but you can stay clean without this luxury. Use a cloth and soapy water to wash yourself. Pay special attention to the feet, armpits, crotch, hands, and hair as these are prime areas for infestation and infection. If water is scarce, take an "air" bath. Remove as much of your clothing as practical and expose your body to the sun and air for at least 1 hour. Be careful not to sunburn.
If you don't have soap, use ashes or sand, or make soap from animal fat and wood ashes, if your situation allows. To make soap--

- Extract grease from animal fat by cutting the fat into small pieces and cooking them in a pot.
- Add enough water to the pot to keep the fat from sticking as it cooks.
- Cook the fat slowly, stirring frequently.
- After the fat is rendered, pour the grease into a container to harden.
- Place ashes in a container with a spout near the bottom.
- Pour water over the ashes and collect the liquid that drips out of the spout in a separate container. This liquid is the potash or lye. Another way to get the lye is to pour the slurry (the mixture of ashes and water) through a straining cloth.
- In a cooking pot, mix two parts grease to one part potash.
- Place this mixture over a fire and boil it until it thickens.

After the mixture--the soap--cools, you can use it in the semiliquid state directly from the pot. You can also pour it into a pan, allow it to harden, and cut it into bars for later use.

ESSENTIAL ARMY MANUAL
[21] ARMY SURVIVAL MANUAL

Keep Your Hands Clean

Germs on your hands can infect food and wounds. Wash your hands after handling any material that is likely to carry germs, after visiting the latrine, after caring for the sick, and before handling any food, food utensils, or drinking water. Keep your fingernails closely trimmed and clean, and keep your fingers out of your mouth.

Keep Your Hair Clean

Your hair can become a haven for bacteria or fleas, lice, and other parasites. Keeping your hair clean, combed, and trimmed helps you avoid this danger.

Keep Your Clothing Clean

Keep your clothing and bedding as clean as possible to reduce the chance of skin infection as well as to decrease the danger of parasitic infestation. Clean your outer clothing whenever it becomes soiled. Wear clean underclothing and socks each day. If water is scarce, "air" clean your clothing by shaking, airing, and sunning it for 2 hours. If you are using a sleeping bag, turn it inside out after each use, fluff it, and air it.

Keep Your Teeth Clean

Thoroughly clean your mouth and teeth with a toothbrush at least once each day. If you don't have a toothbrush, make a chewing stick. Find a twig about 20 centimeters long and 1 centimeter wide. Chew one end of the stick to separate the fibers. Now brush your teeth thoroughly. Another way is to wrap a clean strip of cloth around your fingers and rub your teeth with it to wipe away food particles. You can also brush your teeth with small amounts of sand, baking soda, salt, or soap. Then rinse your mouth with water, salt water, or willow bark tea. Also, flossing your teeth with string or fiber helps oral hygiene.
If you have cavities, you can make temporary fillings by placing candle wax, tobacco, aspirin, hot pepper, tooth paste or powder, or portions of a ginger root into the cavity. Make sure you clean the cavity by rinsing or picking the particles out of the cavity before placing a filling in the cavity.

Take Care of Your Feet

To prevent serious foot problems, break in your shoes before wearing them on any mission. Wash and massage your feet daily. Trim your toenails straight across. Wear an insole and the proper size of dry socks. Powder and check your feet daily for blisters.
If you get a small blister, do not open it. An intact blister is safe from infection. Apply a padding material around the blister to relieve pressure and reduce friction. If the blister bursts, treat it as an open wound. Clean and dress it daily and pad around it. Leave large blisters intact. To avoid having the blister burst or tear under pressure and cause a painful and open sore, do the following:

- Obtain a sewing-type needle and a clean or sterilized thread.
- Run the needle and thread through the blister after cleaning the blister.
- Detach the needle and leave both ends of the thread hanging out of the blister. The thread will absorb the liquid inside. This reduces the size of the hole and ensures that the hole does not close up.
- Pad around the blister.

Get Sufficient Rest

You need a certain amount of rest to keep going. Plan for regular rest periods of at least 10 minutes per hour during your daily activities. Learn to make yourself comfortable under less than ideal conditions. A change from mental to physical activity or vice versa can be refreshing when time or situation does not permit total relaxation.

Keep Camp Site Clean

Do not soil the ground in the camp site area with urine or feces. Use latrines, if available. When latrines are not available, dig "cat holes" and cover the waste. Collect drinking water upstream from the camp site. Purify all water.

MEDICAL EMERGENCIES

Medical problems and emergencies you may be faced with include breathing problems, severe bleeding, and shock.

Breathing Problems

Any one of the following can cause airway obstruction, resulting in stopped breathing:

- Foreign matter in mouth of throat that obstructs the opening to the trachea.
- Face or neck injuries.
- Inflammation and swelling of mouth and throat caused by inhaling smoke, flames, and irritating vapors or by an allergic reaction.
- "Kink" in the throat (caused by the neck bent forward so that the chin rests upon the chest) may block the passage of air.
- Tongue blocks passage of air to the lungs upon unconsciousness. When an individual is unconscious, the muscles of the lower jaw and tongue relax as the neck drops forward, causing the lower jaw to sag and the tongue to drop back and block the passage of air.

Severe Bleeding

Severe bleeding from any major blood vessel in the body is extremely dangerous. The loss of 1 liter of blood will produce moderate symptoms of shock. The loss of 2 liters will produce a severe state of shock that places the body in extreme danger. The loss of 3 liters is usually fatal.

Shock

Shock (acute stress reaction) is not a disease in itself. It is a clinical condition characterized by symptoms that arise when cardiac output is insufficient to fill the arteries with blood under enough pressure to provide an adequate blood supply to the organs and tissues.

LIFESAVING STEPS

Control panic, both your own and the victim's. Reassure him and try to keep him quiet.
Perform a rapid physical exam. Look for the cause of the injury and follow the ABCs of first aid, starting with the airway and breathing, but be discerning. A person may die from arterial bleeding more quickly than from an airway obstruction in some cases.

Open Airway and Maintain

You can open an airway and maintain it by using the following steps.
Step 1. Check if the victim has a partial or complete airway obstruction. If he can cough or speak, allow him to clear the obstruction naturally. Stand by, reassure the victim, and be ready to clear his airway and

perform mouth-to-mouth resuscitation should he become unconscious. If his airway is completely obstructed, administer abdominal thrusts until the obstruction is cleared.

Step 2. Using a finger, quickly sweep the victim's mouth clear of any foreign objects, broken teeth, dentures, sand.

Step 3. Using the jaw thrust method, grasp the angles of the victim's lower jaw and lift with both hands, one on each side, moving the jaw forward. For stability, rest your elbows on the surface on which the victim is lying. If his lips are closed, gently open the lower lip with your thumb (Figure 4-1).

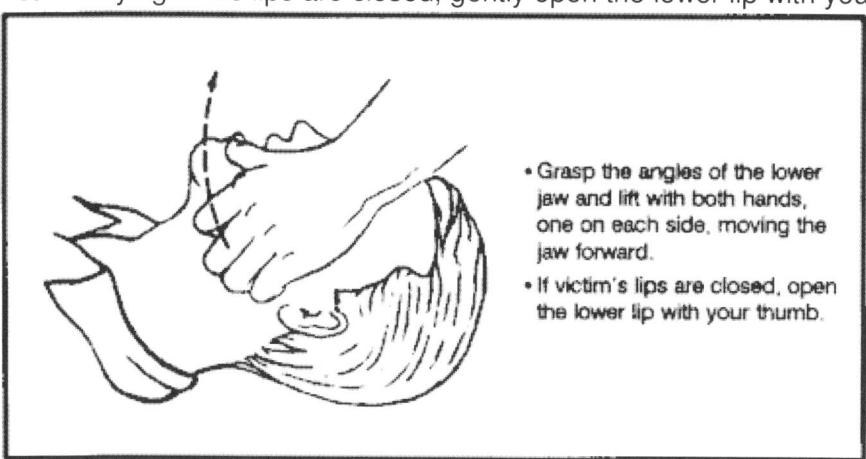

Figure 4-1. Jaw thrust method.

Step 4. With the victim's airway open, pinch his nose closed with your thumb and forefinger and blow two complete breaths into his lungs. Allow the lungs to deflate after the second inflation and perform the following:

- *Look* for his chest to rise and fall.
- *Listen* for escaping air during exhalation.
- *Feel* for flow of air on your cheek.

Step 5. If the forced breaths do not stimulate spontaneous breathing, maintain the victim's breathing by performing mouth-to-mouth resuscitation.

Step 6. There is danger of the victim vomiting during mouth-to-mouth resuscitation. Check the victim's mouth periodically for vomit and clear as needed.

> *Note: Cardiopulmonary resuscitation (CPR) may be necessary after cleaning the airway, but only after major bleeding is under control. See FM 21-20, the American Heart Association manual, the Red Cross manual, or most other first aid books for detailed instructions on CPR.*

Control Bleeding

In a survival situation, you must control serious bleeding immediately because replacement fluids normally are not available and the victim can die within a matter of minutes. External bleeding falls into the following classifications (according to its source):

- *Arterial.* Blood vessels called arteries carry blood away from the heart and through the body. A cut artery issues *bright red* blood from the wound in *distinct spurts* or pulses that correspond to the rhythm of the heartbeat. Because the blood in the arteries is under high pressure, an individual can lose a large volume of blood in a short period when damage to an artery of significant size occurs. Therefore, arterial bleeding is the most serious type of bleeding. If not controlled promptly, it can be fatal.

- *Venous.* Venous blood is blood that is returning to the heart through blood vessels called veins. A steady flow of *dark red, maroon, or bluish blood* characterizes bleeding from a vein. You can usually control venous bleeding more easily than arterial bleeding.
- *Capillary.* The capillaries are the extremely small vessels that connect the arteries with the veins. Capillary bleeding most commonly occurs in minor cuts and scrapes. This type of bleeding is not difficult to control.

You can control external bleeding by direct pressure, indirect (pressure points) pressure, elevation, digital ligation, or tourniquet.

Direct Pressure

The most effective way to control external bleeding is by applying pressure directly over the wound. This pressure must not only be firm enough to stop the bleeding, but it must also be maintained long enough to "seal off" the damaged surface.

If bleeding continues after having applied direct pressure for 30 minutes, apply a pressure dressing. This dressing consists of a thick dressing of gauze or other suitable material applied directly over the wound and held in place with a tightly wrapped bandage (Figure 4-2). It should be tighter than an ordinary compression bandage but not so tight that it impairs circulation to the rest of the limb. Once you apply the dressing, *do not remove it,* even when the dressing becomes blood soaked.

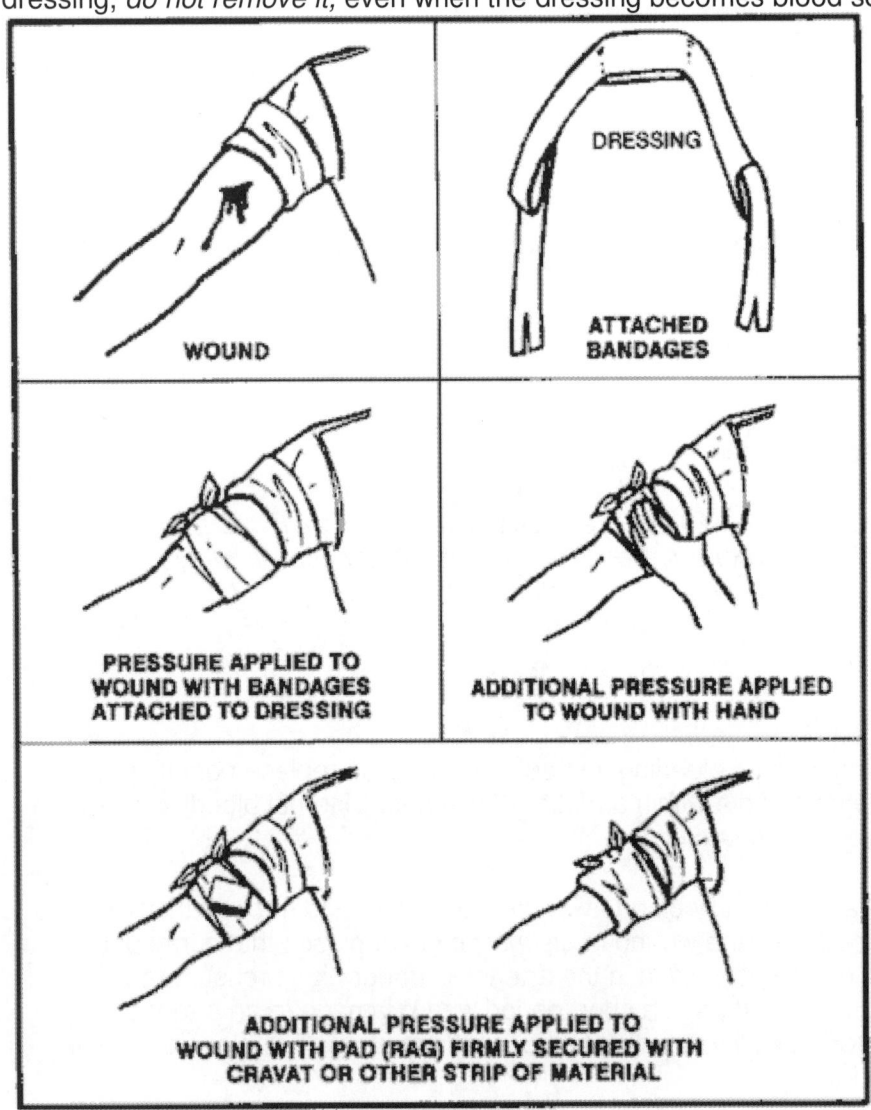

Figure 4-2. Application of a pressure dressing.

Leave the pressure dressing in place for 1 or 2 days, after which you can remove and replace it with a smaller dressing.
In the long-term survival environment, make fresh, daily dressing changes and inspect for signs of infection.

Elevation

Raising an injured extremity as high as possible above the heart's level slows blood loss by aiding the return of blood to the heart and lowering the blood pressure at the wound. However, elevation alone will not control bleeding entirely; you must also apply direct pressure over the wound. When treating a snakebite, however, keep the extremity lower than the heart.

Pressure Points

A pressure point is a location where the main artery to the wound lies near the surface of the skin or where the artery passes directly over a bony prominence (Figure 4-3). You can use digital pressure on a pressure point to slow arterial bleeding until the application of a pressure dressing. Pressure point control is not as effective for controlling bleeding as direct pressure exerted on the wound. It is rare when a single major compressible artery supplies a damaged vessel.

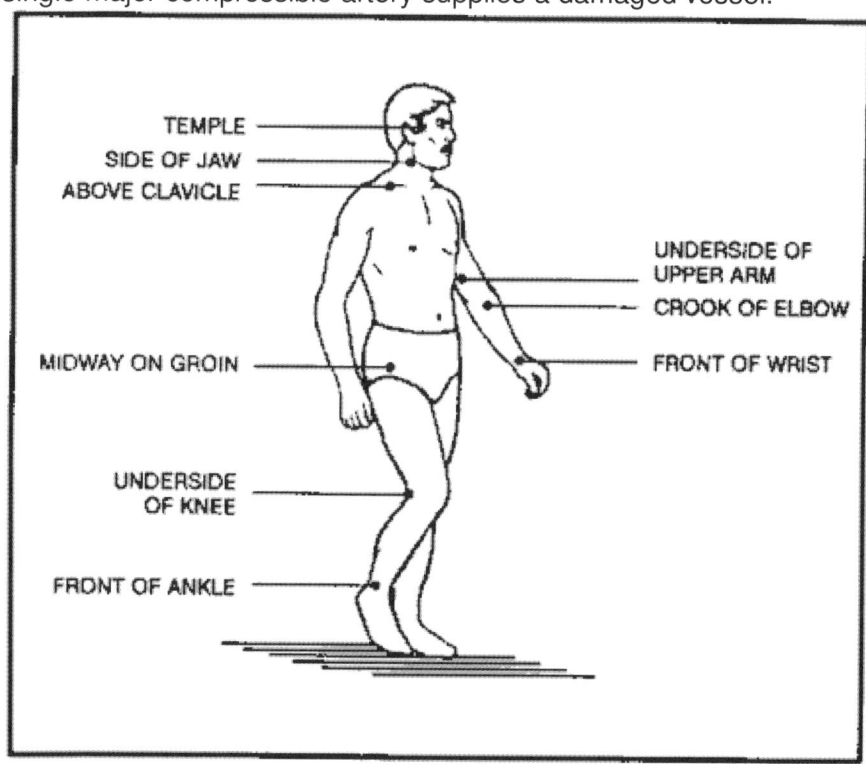

Figure 4-3. Pressure points.

If you cannot remember the exact location of the pressure points, follow this rule: Apply pressure at the end of the joint just above the injured area. On hands, feet, and head, this will be the wrist, ankle, and neck, respectively.

WARNING
Use caution when applying pressure to the neck. Too much pressure for too long may cause unconsciousness or death. Never place a tourniquet around the neck.

Maintain pressure points by placing a round stick in the joint, bending the joint over the stick, and then keeping it tightly bent by lashing. By using this method to maintain pressure, it frees your hands to work in other areas.

Digital Ligation

You can stop major bleeding immediately or slow it down by applying pressure with a finger or two on the bleeding end of the vein or artery. Maintain the pressure until the bleeding stops or slows down enough to apply a pressure bandage, elevation, and so forth.

Tourniquet

Use a tourniquet only when direct pressure over the bleeding point and all other methods did not control the bleeding. If you leave a tourniquet in place too long, the damage to the tissues can progress to gangrene, with a loss of the limb later. An improperly applied tourniquet can also cause permanent damage to nerves and other tissues at the site of the constriction.
If you must use a tourniquet, place it around the extremity, between the wound and the heart, 5 to 10 centimeters above the wound site (Figure 4-4). Never place it directly over the wound or a fracture. Use a stick as a handle to tighten the tourniquet and tighten it only enough to stop blood flow. When you have tightened the tourniquet, bind the free end of the stick to the limb to prevent unwinding.

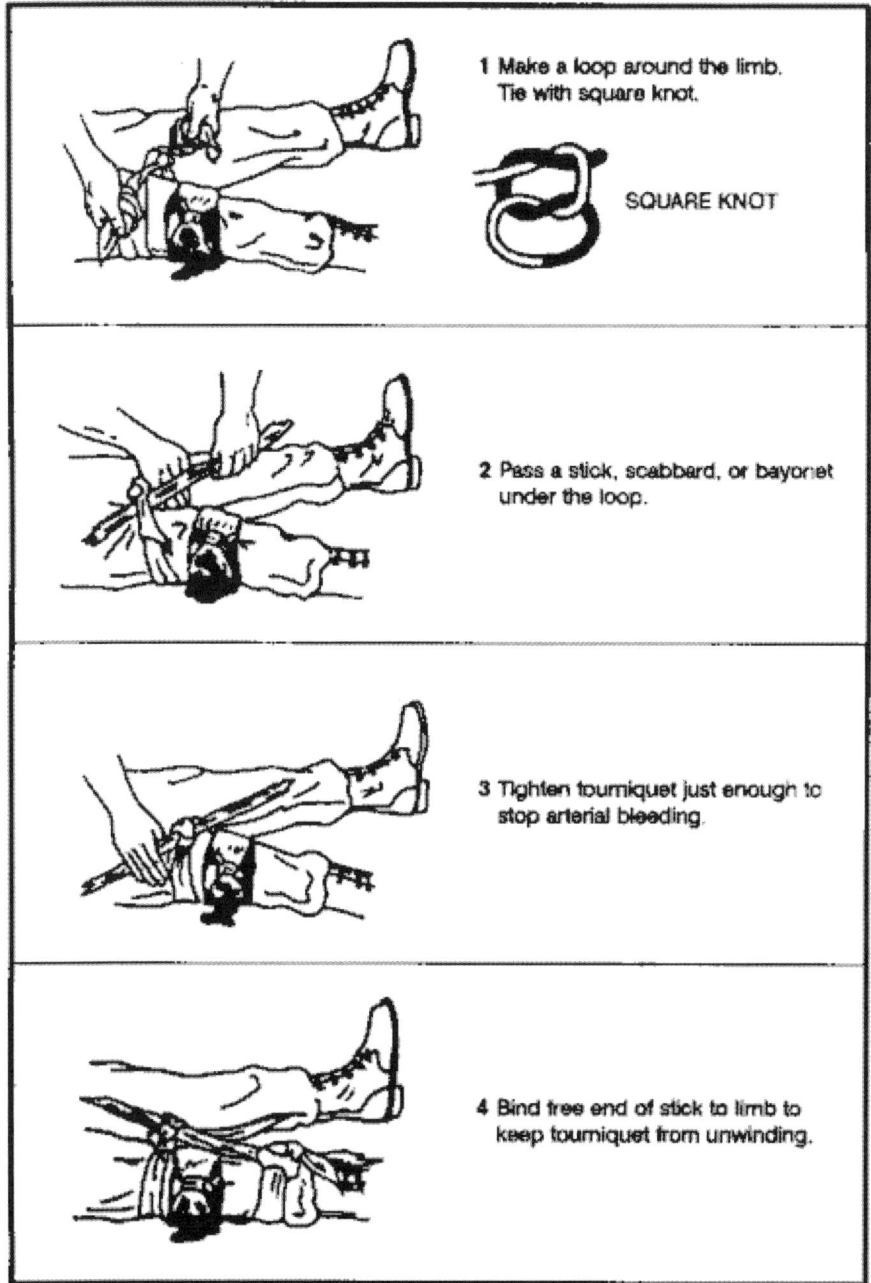

Figure 4-4. Application of tourniquet.

After you secure the tourniquet, clean and bandage the wound. A lone survivor **does not** remove or release an applied tourniquet. In a buddy system, however, the buddy can release the tourniquet pressure every 10 to 15 minutes for 1 or 2 minutes to let blood flow to the rest of the extremity to prevent limb loss.

Prevent and Treat Shock

Anticipate shock in all injured personnel. Treat all injured persons as follows, regardless of what symptoms appear (Figure 4-5):

- If the victim is conscious, place him on a level surface with the lower extremities elevated 15 to 20 centimeters.
- If the victim is unconscious, place him on his side or abdomen with his head turned to one side to prevent choking on vomit, blood, or other fluids.

- If you are unsure of the best position, place the victim perfectly flat. Once the victim is in a shock position, do not move him.
- Maintain body heat by insulating the victim from the surroundings and, in some instances, applying external heat.
- If wet, remove all the victim's wet clothing as soon as possible and replace with dry clothing.
- Improvise a shelter to insulate the victim from the weather.
- Use warm liquids or foods, a prewarmed sleeping bag, another person, warmed water in canteens, hot rocks wrapped in clothing, or fires on either side of the victim to provide external warmth.
- If the victim is conscious, slowly administer small doses of a warm salt or sugar solution, if available.
- If the victim is unconscious or has abdominal wounds, do not give fluids by mouth.
- Have the victim rest for at least 24 hours.
- If you are a lone survivor, lie in a depression in the ground, behind a tree, or any other place out of the weather, with your head lower than your feet.
- If you are with a buddy, reassess your patient constantly.

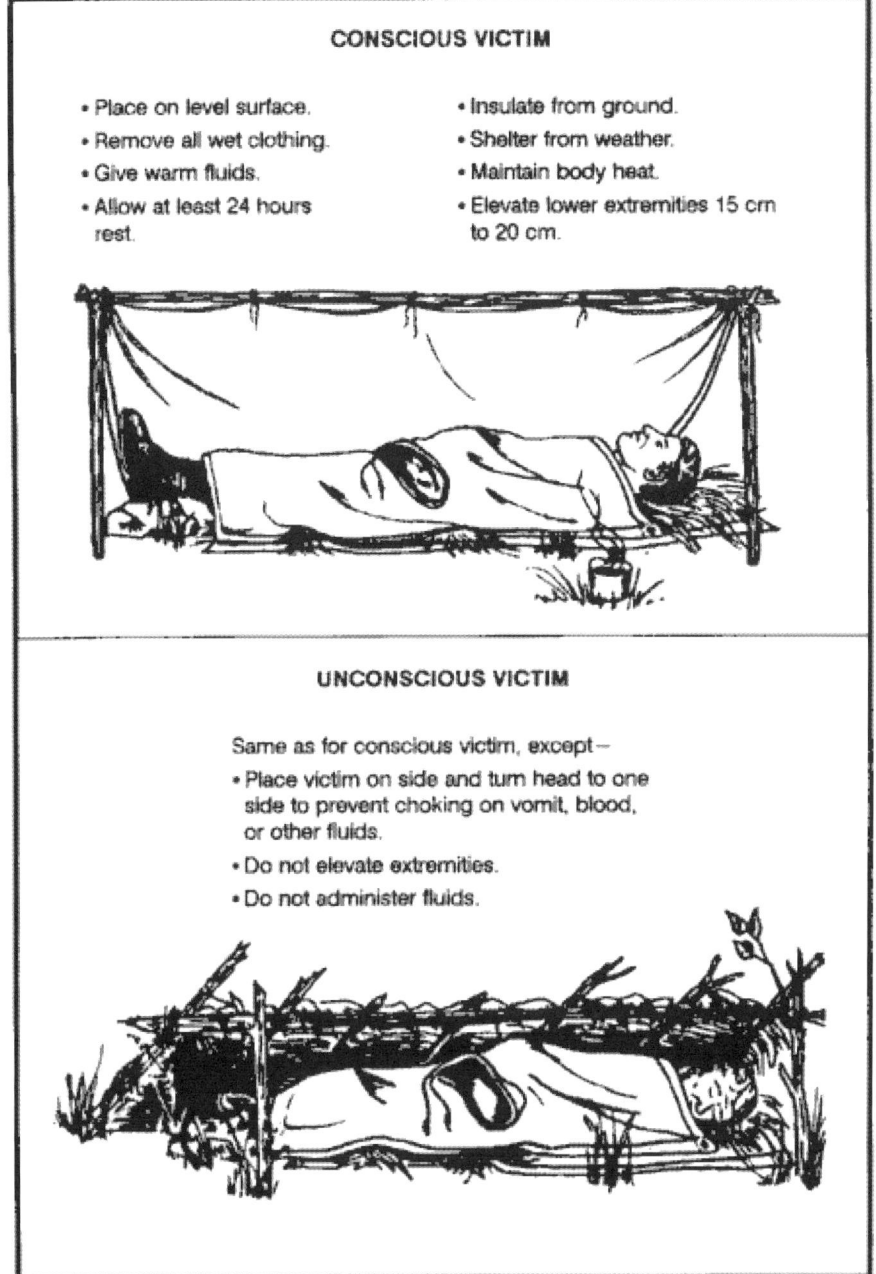

CONSCIOUS VICTIM

- Place on level surface.
- Remove all wet clothing.
- Give warm fluids.
- Allow at least 24 hours rest.
- Insulate from ground.
- Shelter from weather.
- Maintain body heat.
- Elevate lower extremities 15 cm to 20 cm.

UNCONSCIOUS VICTIM

Same as for conscious victim, except—

- Place victim on side and turn head to one side to prevent choking on vomit, blood, or other fluids.
- Do not elevate extremities.
- Do not administer fluids.

Figure 4-5. Treatment for shock.

BONE AND JOINT INJURY

You could face bone and joint injuries that include fractures, dislocations, and sprains.

Fractures

There are basically two types of fractures: open and closed. With an open (or compound) fracture, the bone protrudes through the skin and complicates the actual fracture with an open wound. After setting the fracture, treat the wound as any other open wound.

The closed fracture has no open wounds. Follow the guidelines for immobilization, and set and splint the fracture.

The signs and symptoms of a fracture are pain, tenderness, discoloration, swelling deformity, loss of function, and grating (a sound or feeling that occurs when broken bone ends rub together).

The dangers with a fracture are the severing or the compression of a nerve or blood vessel at the site of fracture. For this reason minimum manipulation should be done, and only very cautiously. If you notice the area below the break becoming numb, swollen, cool to the touch, or turning pale, and the victim shows signs of shock, a major vessel may have been severed. You must control this internal bleeding. Rest the victim for shock, and replace lost fluids.

Often you must maintain traction during the splinting and healing process. You can effectively pull smaller bones such as the arm or lower leg by hand. You can create traction by wedging a hand or foot in the V-notch of a tree and pushing against the tree with the other extremity. You can then splint the break.

Very strong muscles hold a broken thighbone (femur) in place making it difficult to maintain traction during healing. You can make an improvised traction splint using natural material (Figure 4-6) as follows:

- Get two forked branches or saplings at least 5 centimeters in diameter. Measure one from the patient's armpit to 20 to 30 centimeters past his unbroken leg. Measure the other from the groin to 20 to 30 centimeters past the unbroken leg. Ensure that both extend an equal distance beyond the end of the leg.
- Pad the two splints. Notch the ends without forks and lash a 20- to 30-centimeter cross member made from a 5-centimeter diameter branch between them.

Using available material (vines, cloth, rawhide), tie the splint around the upper portion of the body and down the length of the broken leg. Follow the splinting guidelines.

- With available material, fashion a wrap that will extend around the ankle, with the two free ends tied to the cross member.
- Place a 10- by 2.5-centimeter stick in the middle of the free ends of the ankle wrap between the cross member and the foot. Using the stick, twist the material to make the traction easier.
- Continue twisting until the broken leg is as long or slightly longer than the unbroken leg.
- Lash the stick to maintain traction.

  Note: Over time you may lose traction because the material weakened. Check the traction periodically. If you must change or repair the splint, maintain the traction manually for a short time.

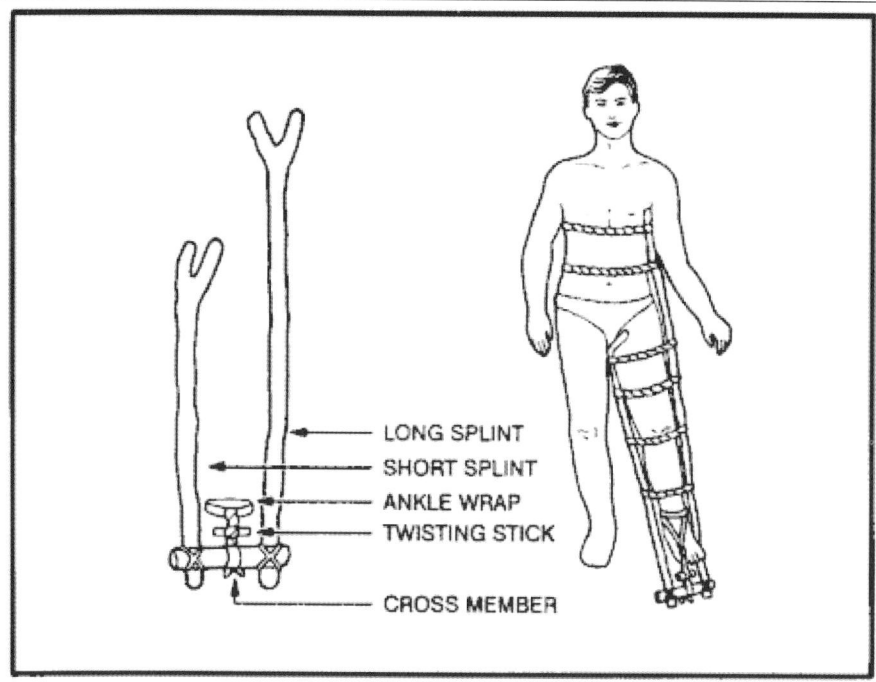

Figure 4-6. Improvised traction splint.

Dislocations

Dislocations are the separations of bone joints causing the bones to go out of proper alignment. These misalignments can be extremely painful and can cause an impairment of nerve or circulatory function below the area affected. You must place these joints back into alignment as quickly as possible.
Signs and symptoms of dislocations are joint pain, tenderness, swelling, discoloration, limited range of motion, and deformity of the joint. You treat dislocations by reduction, immobilization, and rehabilitation. Reduction or "setting" is placing the bones back into their proper alignment. You can use several methods, but manual traction or the use of weights to pull the bones are the safest and easiest. Once performed, reduction decreases the victim's pain and allows for normal function and circulation. Without an X ray, you can judge proper alignment by the look and feel of the joint and by comparing it to the joint on the opposite side.
Immobilization is nothing more than splinting the dislocation after reduction. You can use any field-expedient material for a splint or you can splint an extremity to the body. The basic guidelines for splinting are--

- Splint above and below the fracture site.
- Pad splints to reduce discomfort.
- Check circulation below the fracture after making each tie on the splint.

To rehabilitate the dislocation, remove the splints after 7 to 14 days. Gradually use the injured joint until fully healed.

Sprains

The accidental overstretching of a tendon or ligament causes sprains. The signs and symptoms are pain, swelling, tenderness, and discoloration (black and blue).
When treating sprains, think RICE--

    R   Rest injured area.
    -

I   Ice for 24 hours, then heat after that.
-

C   Compression-wrapping and/or splinting to help stabilize. If possible, leave the boot on a
-   sprained ankle unless circulation is compromised.

E   Elevation of the affected area.
-

BITES AND STINGS

Insects and related pests are hazards in a survival situation. They not only cause irritations, but they are often carriers of diseases that cause severe allergic reactions in some individuals. In many parts of the world you will be exposed to serious, even fatal, diseases not encountered in the United States.
Ticks can carry and transmit diseases, such as Rocky Mountain spotted fever common in many parts of the United States. Ticks also transmit the Lyme disease.
Mosquitoes may carry malaria, dengue, and many other diseases.
Flies can spread disease from contact with infectious sources. They are causes of sleeping sickness, typhoid, cholera, and dysentery.
Fleas can transmit plague.
Lice can transmit typhus and relapsing fever.
The best way to avoid the complications of insect bites and stings is to keep immunizations (including booster shots) up-to-date, avoid insect-infested areas, use netting and insect repellent, and wear all clothing properly.
If you get bitten or stung, do not scratch the bite or sting, it might become infected. Inspect your body at least once a day to ensure there are no insects attached to you. If you find ticks attached to your body, cover them with a substance, such as Vaseline, heavy oil, or tree sap, that will cut off their air supply. Without air, the tick releases its hold, and you can remove it. Take care to remove the whole tick. Use tweezers if you have them. Grasp the tick where the mouth parts are attached to the skin. Do not squeeze the tick's body. Wash your hands after touching the tick. Clean the tick wound daily until healed.

Treatment

It is impossible to list the treatment of all the different types of bites and stings. Treat bites and stings as follows:

- If antibiotics are available for your use, become familiar with them before deployment and use them.
- Predeployment immunizations can prevent most of the common diseases carried by mosquitoes and some carried by flies.
- The common fly-borne diseases are usually treatable with penicillins or erythromycin.
- Most tick-, flea-, louse-, and mite-borne diseases are treatable with tetracycline.
- Most antibiotics come in 250 milligram (mg) or 500 mg tablets. If you cannot remember the exact dose rate to treat a disease, 2 tablets, 4 times a day for 10 to 14 days will usually kill any bacteria.

Bee and Wasp Stings

If stung by a bee, immediately remove the stinger and venom sac, if attached, by scraping with a fingernail or a knife blade. Do not squeeze or grasp the stinger or venom sac, as squeezing will force more venom into the wound. Wash the sting site thoroughly with soap and water to lessen the chance of a secondary infection.
If you know or suspect that you are allergic to insect stings, always carry an insect sting kit with you. Relieve the itching and discomfort caused by insect bites by applying--

- Cold compresses.
- A cooling paste of mud and ashes.
- Sap from dandelions.
- Coconut meat.
- Crushed cloves of garlic.
- Onion.

Spider Bites and Scorpion Stings

The black widow spider is identified by a red hourglass on its abdomen. Only the female bites, and it has a neurotoxic venom. The initial pain is not severe, but severe local pain rapidly develops. The pain gradually spreads over the entire body and settles in the abdomen and legs. Abdominal cramps and progressive nausea, vomiting, and a rash may occur. Weakness, tremors, sweating, and salivation may occur. Anaphylactic reactions can occur. Symptoms begin to regress after several hours and are usually gone in a few days. Threat for shock. Be ready to perform CPR. Clean and dress the bite area to reduce the risk of infection. An antivenin is available.

The funnelweb spider is a large brown or gray spider found in Australia. The symptoms and the treatment for its bite are as for the black widow spider.

The brown house spider or brown recluse spider is a small, light brown spider identified by a dark brown violin on its back. There is no pain, or so little pain, that usually a victim is not aware of the bite. Within a few hours a painful red area with a mottled cyanotic center appears. Necrosis does not occur in all bites, but usually in 3 to 4 days, a star-shaped, firm area of deep purple discoloration appears at the bite site. The area turns dark and mummified in a week or two. The margins separate and the scab falls off, leaving an open ulcer. Secondary infection and regional swollen lymph glands usually become visible at this stage. The outstanding characteristic of the brown recluse bite is an ulcer that does not heal but persists for weeks or months. In addition to the ulcer, there is often a systemic reaction that is serious and may lead to death. Reactions (fever, chills, joint pain, vomiting, and a generalized rash) occur chiefly in children or debilitated persons.

Tarantulas are large, hairy spiders found mainly in the tropics. Most do not inject venom, but some South American species do. They have large fangs. If bitten, pain and bleeding are certain, and infection is likely. Treat a tarantula bite as for any open wound, and try to prevent infection. If symptoms of poisoning appear, treat as for the bite of the black widow spider.

Scorpions are all poisonous to a greater or lesser degree. There are two different reactions, depending on the species:

- Severe local reaction only, with pain and swelling around the area of the sting. Possible prickly sensation around the mouth and a thick-feeling tongue.
- Severe systemic reaction, with little or no visible local reaction. Local pain may be present. Systemic reaction includes respiratory difficulties, thick-feeling tongue, body spasms, drooling, gastric distention, double vision, blindness, involuntary rapid movement of the eyeballs, involuntary urination and defecation, and heart failure. Death is rare, occurring mainly in children and adults with high blood pressure or illnesses.

Treat scorpion stings as you would a black widow bite.

Snakebites

The chance of a snakebite in a survival situation is rather small, if you are familiar with the various types of snakes and their habitats. However, it could happen and you should know how to treat a snakebite. Deaths from snakebites are rare. More than one-half of the snakebite victims have little or no poisoning, and only about one-quarter develop serious systemic poisoning. However, the chance of a snakebite in a survival situation can affect morale, and failure to take preventive measures or failure to treat a snakebite properly can result in needless tragedy.

The primary concern in the treatment of snakebite is to limit the amount of eventual tissue destruction around the bite area.

A bite wound, regardless of the type of animal that inflicted it, can become infected from bacteria in the animal's mouth. With nonpoisonous as well as poisonous snakebites, this local infection is responsible for a large part of the residual damage that results.

Snake venoms not only contain poisons that attack the victim's central nervous system (neurotoxins) and blood circulation (hemotoxins), but also digestive enzymes (cytotoxins) to aid in digesting their prey. These poisons can cause a very large area of tissue death, leaving a large open wound. This condition could lead to the need for eventual amputation if not treated.

Shock and panic in a person bitten by a snake can also affect the person's recovery. Excitement, hysteria, and panic can speed up the circulation, causing the body to absorb the toxin quickly. Signs of shock occur within the first 30 minutes after the bite.

Before you start treating a snakebite, determine whether the snake was poisonous or nonpoisonous. Bites from a nonpoisonous snake will show rows of teeth. Bites from a poisonous snake may have rows of teeth showing, but will have one or more distinctive puncture marks caused by fang penetration. Symptoms of a poisonous bite may be spontaneous bleeding from the nose and anus, blood in the urine, pain at the site of the bite, and swelling at the site of the bite within a few minutes or up to 2 hours later. Breathing difficulty, paralysis, weakness, twitching, and numbness are also signs of neurotoxic venoms. These signs usually appear 1.5 to 2 hours after the bite.

If you determine that a poisonous snake bit an individual, take the following steps:

- Reassure the victim and keep him still.
- Set up for shock and force fluids or give an intravenous (IV).
- Remove watches, rings, bracelets, or other constricting items.
- Clean the bite area.
- Maintain an airway (especially if bitten near the face or neck) and be prepared to administer mouth-to-mouth resuscitation or CPR.
- Use a constricting band between the wound and the heart.
- Immobilize the site.
- Remove the poison as soon as possible by using a mechanical suction device or by squeezing.

Do not--

- Give the victim alcoholic beverages or tobacco products.
- Give morphine or other central nervous system (CNS) depressors.
- Make any deep cuts at the bite site. Cutting opens capillaries that in turn open a direct route into the blood stream for venom and infection.

    *Note: If medical treatment is over one hour away, make an incision (no longer than 6 millimeters and no deeper than 3 millimeter) over each puncture, cutting just deep enough to enlarge the fang opening, but only through the first or second layer of skin. Place a suction cup over the bite so that you have a good vacuum seal. Suction the bite site 3 to 4 times. Use mouth suction **only as a last resort and only if you do not have open sores in your mouth.** Spit the envenomed blood out and rinse your mouth with water. This method will draw out 25 to 30 percent of the venom.*

- Put your hands on your face or rub your eyes, as venom may be on your hands. Venom may cause blindness.
- Break open the large blisters that form around the bite site.

After caring for the victim as described above, take the following actions to minimize local effects:

- If infection appears, keep the wound open and clean.
- Use heat after 24 to 48 hours to help prevent the spread of local infection. Heat also helps to draw out an infection.
- Keep the wound covered with a dry, sterile dressing.
- Have the victim drink large amounts of fluids until the infection is gone.

WOUNDS

An interruption of the skin's integrity characterizes wounds. These wounds could be open wounds, skin diseases, frostbite, trench foot, and burns.

Open Wounds

Open wounds are serious in a survival situation, not only because of tissue damage and blood loss, but also because they may become infected. Bacteria on the object that made the wound, on the individual's skin and clothing, or on other foreign material or dirt that touches the wound may cause infection. By taking proper care of the wound you can reduce further contamination and promote healing. Clean the wound as soon as possible after it occurs by--

- Removing or cutting clothing away from the wound.
- Always looking for an exit wound if a sharp object, gun shot, or projectile caused a wound.
- Thoroughly cleaning the skin around the wound.
- Rinsing (not scrubbing) the wound with large amounts of water under pressure. You can use fresh urine if water is not available.

The "open treatment" method is the safest way to manage wounds in survival situations. Do not try to close any wound by suturing or similar procedures. Leave the wound open to allow the drainage of any pus resulting from infection. As long as the wound can drain, it generally will not become life-threatening, regardless of how unpleasant it looks or smells.

Cover the wound with a clean dressing. Place a bandage on the dressing to hold it in place. Change the dressing daily to check for infection.

If a wound is gaping, you can bring the edges together with adhesive tape cut in the form of a "butterfly" or "dumbbell" (Figure 4-7).

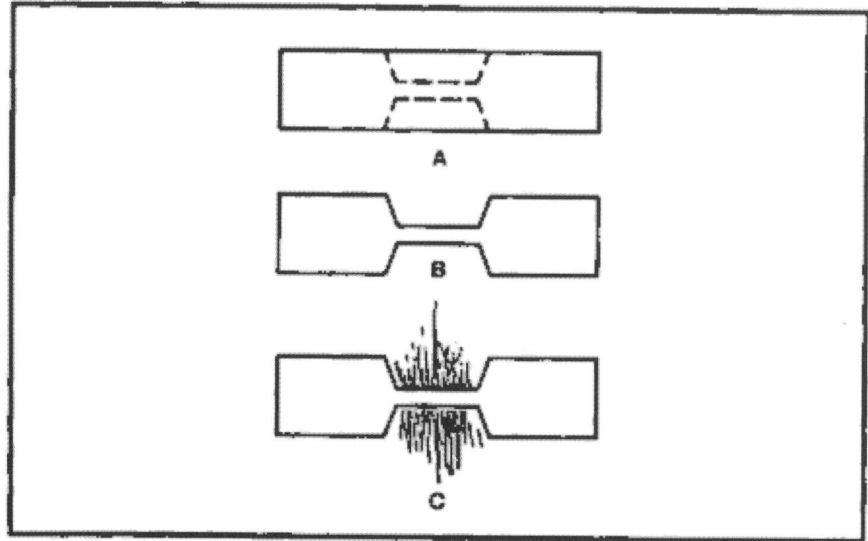

Figure 4-7. Butterfly closure.

In a survival situation, some degree of wound infection is almost inevitable. Pain, swelling, and redness around the wound, increased temperature, and pus in the wound or on the dressing indicate infection is present.

To treat an infected wound--

- Place a warm, moist compress directly on the infected wound. Change the compress when it cools, keeping a warm compress on the wound for a total of 30 minutes. Apply the compresses three or four times daily.
- Drain the wound. Open and gently probe the infected wound with a sterile instrument.
- Dress and bandage the wound.
- Drink a lot of water.

Continue this treatment daily until all signs of infection have disappeared.
If you do not have antibiotics and the wound has become severely infected, does not heal, and ordinary debridement is impossible, consider maggot therapy, despite its hazards:

- Expose the wound to flies for one day and then cover it.
- Check daily for maggots.
- Once maggots develop, keep wound covered but check daily.
- Remove all maggots when they have cleaned out all dead tissue and before they start on healthy tissue. Increased pain and bright red blood in the wound indicate that the maggots have reached healthy tissue.
- Flush the wound repeatedly with sterile water or fresh urine to remove the maggots.
- Check the wound every four hours for several days to ensure all maggots have been removed.
- Bandage the wound and treat it as any other wound. It should heal normally.

Skin Diseases and Ailments

Although boils, fungal infections, and rashes rarely develop into a serious health problem, they cause discomfort and you should treat them.

Boils

Apply warm compresses to bring the boil to a head. Then open the boil using a sterile knife, wire, needle, or similar item. Thoroughly clean out the pus using soap and water. Cover the boil site, checking it periodically to ensure no further infection develops.

Fungal Infections

Keep the skin clean and dry, and expose the infected area to as much sunlight as possible. *Do not scratch* the affected area. During the Southeast Asian conflict, soldiers used antifungal powders, lye soap, chlorine bleach, alcohol, vinegar, concentrated salt water, and iodine to treat fungal infections with varying degrees of success. *As with any "unorthodox" method of treatment, use it with caution.*

Rashes

To treat a skin rash effectively, first determine what is causing it. This determination may be difficult even in the best of situations. Observe the following rules to treat rashes:

- If it is moist, keep it dry.
- If it is dry, keep it moist.
- Do not scratch it.

Use a compress of vinegar or tannic acid derived from tea or from boiling acorns or the bark of a hardwood tree to dry weeping rashes. Keep dry rashes moist by rubbing a small amount of rendered animal fat or grease on the affected area.

Remember, treat rashes as open wounds and clean and dress them daily. There are many substances available to survivors in the wild or in captivity for use as antiseptics to treat wound:

- *Iodine tablets.* Use 5 to 15 tablets in a liter of water to produce a good rinse for wounds during healing.
- *Garlic.* Rub it on a wound or boil it to extract the oils and use the water to rinse the affected area.
- *Salt water.* Use 2 to 3 tablespoons per liter of water to kill bacteria.
- *Bee honey.* Use it straight or dissolved in water.
- *Sphagnum moss.* Found in boggy areas worldwide, it is a natural source of iodine. Use as a dressing.

Again, use noncommercially prepared materials with caution.

Frostbite

This injury results from frozen tissues. Light frostbite involves only the skin that takes on a dull, whitish pallor. Deep frostbite extends to a depth below the skin. The tissues become solid and immovable. Your feet, hands, and exposed facial areas are particularly vulnerable to frostbite.
When with others, prevent frostbite by using the buddy system. Check your buddy's face often and make sure that he checks yours. If you are alone, periodically cover your nose and lower part of your face with your mittens.
Do not try to thaw the affected areas by placing them close to an open flame. Gently rub them in lukewarm water. Dry the part and place it next to your skin to warm it at body temperature.

Trench Foot

This condition results from many hours or days of exposure to wet or damp conditions at a temperature just above freezing. The nerves and muscles sustain the main damage, but gangrene can occur. In extreme cases the flesh dies and it may become necessary to have the foot or leg amputated. The best prevention is to keep your feet dry. Carry extra socks with you in a waterproof packet. Dry wet socks against your body. Wash your feet daily and put on dry socks.

Burns

The following field treatment for burns relieves the pain somewhat, seems to help speed healing, and offers some protection against infection:

- First, stop the burning process. Put out the fire by removing clothing, dousing with water or sand, or by rolling on the ground. Cool the burning skin with ice or water. For burns caused by white phosphorous, pick out the white phosphorous with tweezers; do not douse with water.
- Soak dressings or clean rags for 10 minutes in a boiling tannic acid solution (obtained from tea, inner bark of hardwood trees, or acorns boiled in water).
- Cool the dressings or clean rags and apply over burns.
- Treat as an open wound.
- Replace fluid loss.
- Maintain airway.
- Treat for shock.
- Consider using morphine, unless the burns are near the face.

ENVIRONMENTAL INJURIES

Heatstroke, hypothermia, diarrhea, and intestinal parasites are environmental injuries you could face.

Heatstroke

The breakdown of the body's heat regulatory system (body temperature more than 40.5 degrees C [105 degrees F]) causes a heatstroke. Other heat injuries, such as cramps or dehydration, do not always precede a heatstroke. Signs and symptoms of heatstroke are--

- Swollen, beet-red face.
- Reddened whites of eyes.
- Victim not sweating.
- Unconsciousness or delirium, which can cause pallor, a bluish color to lips and nail beds (cyanosis), and cool skin.

   Note: By this time the victim is in severe shock. Cool the victim as rapidly as possible. Cool him by dipping him in a cool stream. If one is not available, douse the victim with urine, water, or at the very least, apply cool wet com-presses to all the joints, especially the neck, armpits, and crotch. Be sure to wet the victim's head. Heat loss through the scalp is great. Administer IVs and provide drinking fluids. You may fan the individual.

Expect, during cooling--

- Vomiting.
- Diarrhea.
- Struggling.
- Shivering.
- Shouting.
- Prolonged unconsciousness.
- Rebound heatstroke within 48 hours.
- Cardiac arrest; be ready to perform CPR.

   Note: Treat for dehydration with lightly salted water.

Hypothermia

Defined as the body's failure to maintain a temperature of 36 degrees C (97 degrees F). Exposure to cool or cold temperature over a short or long time can cause hypothermia. Dehydration and lack of food and rest predispose the survivor to hypothermia.
Unlike heatstroke, you must gradually warm the hypothermia victim. Get the victim into dry clothing. Replace lost fluids, and warm him.

Diarrhea

A common, debilitating ailment caused by a change of water and food, drinking contaminated water, eating spoiled food, becoming fatigued, and using dirty dishes. You can avoid most of these causes by practicing preventive medicine. If you get diarrhea, however, and do not have antidiarrheal medicine, one of the following treatments may be effective:

- Limit your intake of fluids for 24 hours.
- Drink one cup of a strong tea solution every 2 hours until the diarrhea slows or stops. The tannic acid in the tea helps to control the diarrhea. Boil the inner bark of a hardwood tree for 2 hours or more to release the tannic acid.
- Make a solution of one handful of ground chalk, charcoal, or dried bones and treated water. If you have some apple pomace or the rinds of citrus fruit, add an equal portion to the mixture to make it more effective. Take 2 tablespoons of the solution every 2 hours until the diarrhea slows or stops.

Intestinal Parasites

You can usually avoid worm infestations and other intestinal parasites if you take preventive measures. For example, never go barefoot. The most effective way to prevent intestinal parasites is to avoid uncooked meat and raw vegetables contaminated by raw sewage or human waste used as a fertilizer. However, should you become infested and lack proper medicine, you can use home remedies. Keep in mind that these home remedies work on the principle of changing the environment of the gastrointestinal tract. The following are home remedies you could use:

- *Salt water.* Dissolve 4 tablespoons of salt in 1 liter of water and drink. Do not repeat this treatment.
- *Tobacco.* Eat 1 to 1.5 cigarettes. The nicotine in the cigarette will kill or stun the worms long enough for your system to pass them. If the infestation is severe, repeat the treatment in 24 to 48 hours, *but no sooner.*
- *Kerosene.* Drink 2 tablespoons of kerosene *but no more.* If necessary, you can repeat this treatment in 24 to 48 hours. Be careful not to inhale the fumes. They may cause lung irritation.
- *Hot peppers.* Peppers are effective only if they are a steady part of your diet. You can eat them raw or put them in soups or rice and meat dishes. They create an environment that is prohibitive to parasitic attachment.

HERBAL MEDICINES

Our modern wonder drugs, laboratories, and equipment have obscured more primitive types of medicine involving determination, common sense, and a few simple treatments. In many areas of the world, however, the people still depend on local "witch doctors" or healers to cure their ailments. Many of the herbs (plants) and treatments they use are as effective as the most modern medications available. In fact, many modern medications come from refined herbs.

WARNING

Use herbal medicines with extreme care, however, and only when you lack or have limited medical supplies. Some herbal medicines are dangerous and may cause further damage or even death. See Chapter 9, Survival Use of Plants, for some basic herbal medicine treatments.

CHAPTER 5 - SHELTERS

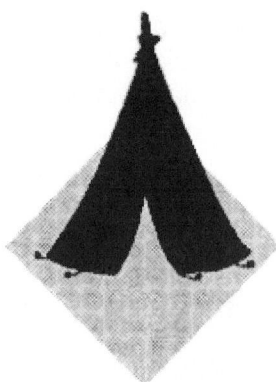

A shelter can protect you from the sun, insects, wind, rain, snow, hot or cold temperatures, and enemy observation. It can give you a feeling of well-being. It can help you maintain your will to survive.

In some areas, your need for shelter may take precedence over your need for food and possibly even your need for water. For example, prolonged exposure to cold can cause excessive fatigue and weakness (exhaustion). An exhausted person may develop a "passive" outlook, thereby losing the will to survive.

The most common error in making a shelter is to make it too large. A shelter must be large enough to protect you. It must also be small enough to contain your body heat, especially in cold climates.

SHELTER SITE SELECTION

When you are in a survival situation and realize that shelter is a high priority, start looking for shelter as soon as possible. As you do so, remember what you will need at the site. Two requisites are--

- It must contain material to make the type of shelter you need.
- It must be large enough and level enough for you to lie down comfortably.

When you consider these requisites, however, you cannot ignore your tactical situation or your safety. You must also consider whether the site--

- Provides concealment from enemy observation.
- Has camouflaged escape routes.
- Is suitable for signaling, if necessary.
- Provides protection against wild animals and rocks and dead trees that might fall.
- Is free from insects, reptiles, and poisonous plants.

You must also remember the problems that could arise in your environment. For instance--

- Avoid flash flood areas in foothills.
- Avoid avalanche or rockslide areas in mountainous terrain.
- Avoid sites near bodies of water that are below the high water mark.

In some areas, the season of the year has a strong bearing on the site you select. Ideal sites for a shelter differ in winter and summer. During cold winter months you will want a site that will protect you from the cold and wind, but will have a source of fuel and water. During summer months in the same area you will want a source of water, but you will want the site to be almost insect free.

When considering shelter site selection, use the word BLISS as a guide.

B - Blend in with the surroundings.
L - Low silhouette.
I - Irregular shape.
S - Small.
S - Secluded location.

TYPES OF SHELTERS

When looking for a shelter site, keep in mind the type of shelter (protection) you need. However, you must also consider--

- How much time and effort you need to build the shelter.
- If the shelter will adequately protect you from the elements (sun, wind, rain, snow).
- If you have the tools to build it. If not, can you make improvised tools?
- If you have the type and amount of materials needed to build it.

To answer these questions, you need to know how to make various types of shelters and what materials you need to make them.

Poncho Lean-To

It takes only a short time and minimal equipment to build this lean-to (Figure 5-1). You need a poncho, 2 to 3 meters of rope or parachute suspension line, three stakes about 30 centimeters long, and two trees or two poles 2 to 3 meters apart. Before selecting the trees you will use or the location of your poles, check the wind direction. Ensure that the back of your lean-to will be into the wind.

Figure 5-1. Poncho lean-to.

To make the lean-to--

- Tie off the hood of the poncho. Pull the drawstring tight, roll the hood longways, fold it into thirds, and tie it off with the drawstring.
- Cut the rope in half. On one long side of the poncho, tie half of the rope to the corner grommet. Tie the other half to the other corner grommet.

- Attach a drip stick (about a 10-centimeter stick) to each rope about 2.5 centimeters from the grommet. These drip sticks will keep rainwater from running down the ropes into the lean-to. Tying strings (about 10 centimeters long) to each grommet along the poncho's top edge will allow the water to run to and down the line without dripping into the shelter.
- Tie the ropes about waist high on the trees (uprights). Use a round turn and two half hitches with a quick-release knot.
- Spread the poncho and anchor it to the ground, putting sharpened sticks through the grommets and into the ground.

If you plan to use the lean-to for more than one night, or you expect rain, make a center support for the lean-to. Make this support with a line. Attach one end of the line to the poncho hood and the other end to an overhanging branch. Make sure there is no slack in the line.
Another method is to place a stick upright under the center of the lean-to. This method, however, will restrict your space and movements in the shelter.
For additional protection from wind and rain, place some brush, your rucksack, or other equipment at the sides of the lean-to.
To reduce heat loss to the ground, place some type of insulating material, such as leaves or pine needles, inside your lean-to.

Note: When at rest, you lose as much as 80 percent of your body heat to the ground.

To increase your security from enemy observation, lower the lean-to's silhouette by making two changes. First, secure the support lines to the trees at knee height (not at waist height) using two knee-high sticks in the two center grommets (sides of lean-to). Second, angle the poncho to the ground, securing it with sharpened sticks, as above.

Poncho Tent

This tent (Figure 5-2) provides a low silhouette. It also protects you from the elements on two sides. It has, however, less usable space and observation area than a lean-to, decreasing your reaction time to enemy detection. To make this tent, you need a poncho, two 1.5- to 2.5-meter ropes, six sharpened sticks about 30 centimeters long, and two trees 2 to 3 meters apart.

Figure 5-2. Poncho tent using overhanging branch.

To make the tent--

- Tie off the poncho hood in the same way as the poncho lean-to.
- Tie a 1.5- to 2.5-meter rope to the center grommet on each side of the poncho.

- Tie the other ends of these ropes at about knee height to two trees 2 to 3 meters apart and stretch the poncho tight.
- Draw one side of the poncho tight and secure it to the ground pushing sharpened sticks through the grommets.
- Follow the same procedure on the other side.

If you need a center support, use the same methods as for the poncho lean-to. Another center support is an A-frame set outside but over the center of the tent (Figure 5-3). Use two 90- to 120-centimeter-long sticks, one with a forked end, to form the A-frame. Tie the hood's drawstring to the A-frame to support the center of the tent.

Figure 5-3. Poncho tent with A-frame.

Three-Pole Parachute Tepee

If you have a parachute and three poles and the tactical situation allows, make a parachute tepee. It is easy and takes very little time to make this tepee. It provides protection from the elements and can act as a signaling device by enhancing a small amount of light from a fire or candle. It is large enough to hold several people and their equipment and to allow sleeping, cooking, and storing firewood.
You can make this tepee using parts of or a whole personnel main or reserve parachute canopy. If using a standard personnel parachute, you need three poles 3.5 to 4.5 meters long and about 5 centimeters in diameter.
To make this tepee (Figure 5-4)--

- Lay the poles on the ground and lash them together at one end.
- Stand the framework up and spread the poles to form a tripod.
- For more support, place additional poles against the tripod. Five or six additional poles work best, but do not lash them to the tripod.
- Determine the wind direction and locate the entrance 90 degrees or more from the mean wind direction.
- Lay out the parachute on the "backside" of the tripod and locate the bridle loop (nylon web loop) at the top (apex) of the canopy.
- Place the bridle loop over the top of a free-standing pole. Then place the pole back up against the tripod so that the canopy's apex is at the same height as the lashing on the three poles.
- Wrap the canopy around one side of the tripod. The canopy should be of double thickness, as you are wrapping an entire parachute. You need only wrap half of the tripod, as the remainder of the canopy will encircle the tripod in the opposite direction.

- Construct the entrance by wrapping the folded edges of the canopy around two free-standing poles. You can then place the poles side by side to close the tepee's entrance.
- Place all extra canopy underneath the tepee poles and inside to create a floor for the shelter.
- Leave a 30- to 50-centimeter opening at the top for ventilation if you intend to have a fire inside the tepee.

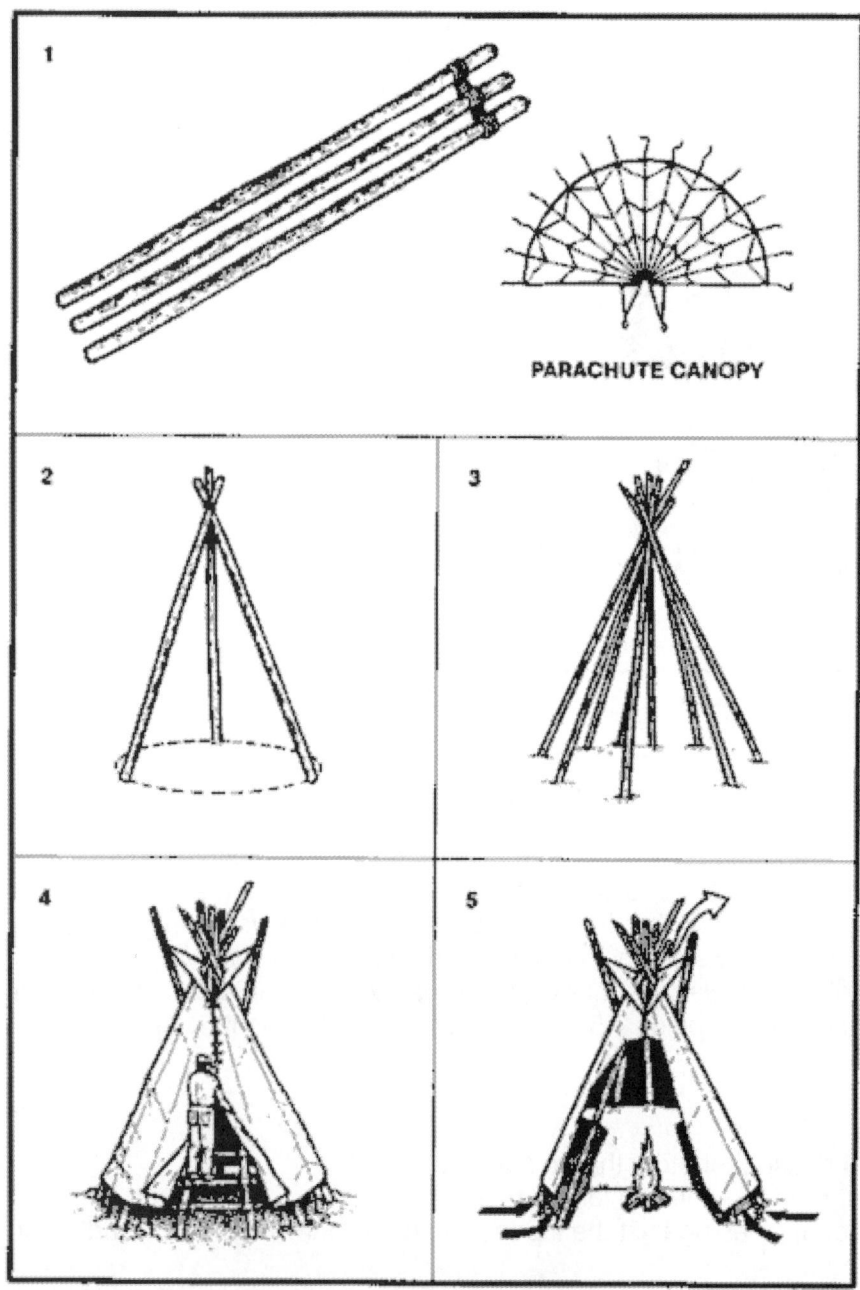

PARACHUTE CANOPY

Figure 5-4. Three-pole parachute tepee.

One-Pole Parachute Tepee

You need a 14-gore section (normally) of canopy, stakes, a stout center pole, and inner core and needle to construct this tepee. You cut the suspension lines except for 40- to 45-centimeter lengths at the canopy's lower lateral band.
To make this tepee (Figure 5-5)--

- Select a shelter site and scribe a circle about 4 meters in diameter on the ground.
- Stake the parachute material to the ground using the lines remaining at the lower lateral band.
- After deciding where to place the shelter door, emplace a stake and tie the first line (from the lower lateral band) securely to it.
- Stretch the parachute material taut to the next line, emplace a stake on the scribed line, and tie the line to it.
- Continue the staking process until you have tied all the lines.
- Loosely attach the top of the parachute material to the center pole with a suspension line you previously cut and, through trial and error, determine the point at which the parachute material will be pulled tight once the center pole is upright.
- Then securely attach the material to the pole.
- Using a suspension line (or inner core), sew the end gores together leaving 1 or 1.2 meters for a door.

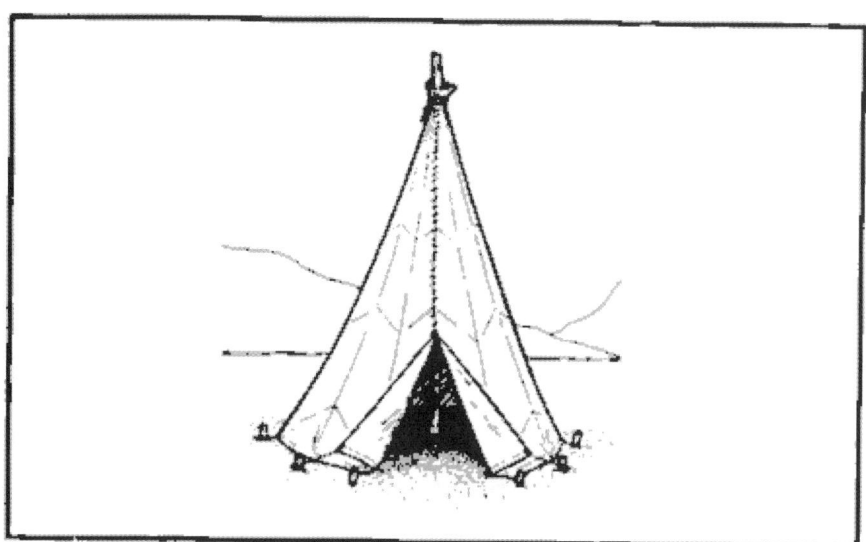

Figure 5-5. One-pole parachute tepee.

No-Pole Parachute Tepee

You use the same materials, except for the center pole, as for the one-pole parachute tepee.
To make this tepee (Figure 5-6)--

- Tie a line to the top of parachute material with a previously cut suspension line.
- Throw the line over a tree limb, and tie it to the tree trunk.
- Starting at the opposite side from the door, emplace a stake on the scribed 3.5- to 4.3-meter circle.
- Tie the first line on the lower lateral band.
- Continue emplacing the stakes and tying the lines to them.
- After staking down the material, unfasten the line tied to the tree trunk, tighten the tepee material by pulling on this line, and tie it securely to the tree trunk.

Figure 5-6. No-pole parachute tepee.

One-Man Shelter

A one-man shelter you can easily make using a parachute requires a tree and three poles. One pole should be about 4.5 meters long and the other two about 3 meters long.
To make this shelter (Figure 5-7)--

- Secure the 4.5-meter pole to the tree at about waist height.
- Lay the two 3-meter poles on the ground on either side of and in the same direction as the 4.5-meter pole.
- Lay the folded canopy over the 4.5 meter pole so that about the same amount of material hangs on both sides.
- Tuck the excess material under the 3-meter poles, and spread it on the ground inside to serve as a floor.
- Stake down or put a spreader between the two 3-meter poles at the shelter's entrance so they will not slide inward.
- Use any excess material to cover the entrance.

Figure 5-7. One-man shelter.

The parachute cloth makes this shelter wind resistant, and the shelter is small enough that it is easily warmed. A candle, used carefully, can keep the inside temperature comfortable. This shelter is unsatisfactory, however, when snow is falling as even a light snowfall will cave it in.

Parachute Hammock

You can make a hammock using 6 to 8 gores of parachute canopy and two trees about 4.5 meters apart (Figure 5-8).

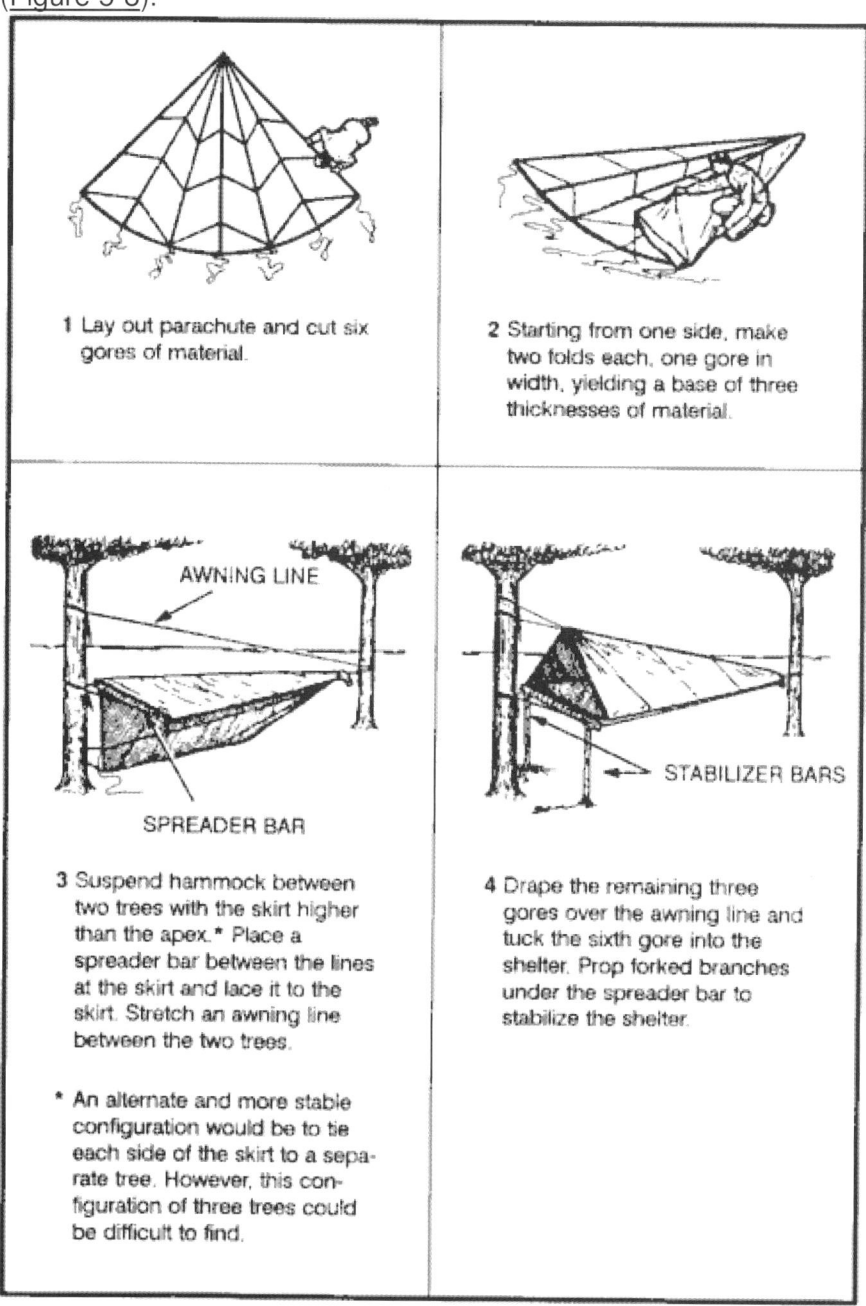

1 Lay out parachute and cut six gores of material.

2 Starting from one side, make two folds each, one gore in width, yielding a base of three thicknesses of material.

AWNING LINE

SPREADER BAR

3 Suspend hammock between two trees with the skirt higher than the apex.* Place a spreader bar between the lines at the skirt and lace it to the skirt. Stretch an awning line between the two trees.

* An alternate and more stable configuration would be to tie each side of the skirt to a separate tree. However, this configuration of three trees could be difficult to find.

STABILIZER BARS

4 Drape the remaining three gores over the awning line and tuck the sixth gore into the shelter. Prop forked branches under the spreader bar to stabilize the shelter.

Figure 5-8. Parachute hammock.

Field-Expedient Lean-To

If you are in a wooded area and have enough natural materials, you can make a field-expedient lean-to (Figure 5-9) without the aid of tools or with only a knife. It takes longer to make this type of shelter than it does to make other types, but it will protect you from the elements.

Figure 5-9. Field-expedient lean-to and fire reflector.

You will need two trees (or upright poles) about 2 meters apart; one pole about 2 meters long and 2.5 centimeters in diameter; five to eight poles about 3 meters long and 2.5 centimeters in diameter for beams; cord or vines for securing the horizontal support to the trees; and other poles, saplings, or vines to crisscross the beams.

To make this lean-to--

- Tie the 2-meter pole to the two trees at waist to chest height. This is the horizontal support. If a standing tree is not available, construct a biped using Y-shaped sticks or two tripods.
- Place one end of the beams (3-meter poles) on one side of the horizontal support. As with all lean-to type shelters, be sure to place the lean-to's backside into the wind.
- Crisscross saplings or vines on the beams.
- Cover the framework with brush, leaves, pine needles, or grass, starting at the bottom and working your way up like shingling.
- Place straw, leaves, pine needles, or grass inside the shelter for bedding.

In cold weather, add to your lean-to's comfort by building a fire reflector wall (Figure 5-9). Drive four 1.5-meter-long stakes into the ground to support the wall. Stack green logs on top of one another between the support stakes. Form two rows of stacked logs to create an inner space within the wall that you can fill with dirt. This action not only strengthens the wall but makes it more heat reflective. Bind the top of the support stakes so that the green logs and dirt will stay in place.

With just a little more effort you can have a drying rack. Cut a few 2-centimeter-diameter poles (length depends on the distance between the lean-to's horizontal support and the top of the fire reflector wall). Lay one end of the poles on the lean-to support and the other end on top of the reflector wall. Place and tie into place smaller sticks across these poles. You now have a place to dry clothes, meat, or fish.

Swamp Bed

In a marsh or swamp, or any area with standing water or continually wet ground, the swamp bed (Figure 5-10) keeps you out of the water. When selecting such a site, consider the weather, wind, tides, and available materials.

Figure 5-10. Swamp bed.

To make a swamp bed--

- Look for four trees clustered in a rectangle, or cut four poles (bamboo is ideal) and drive them firmly into the ground so they form a rectangle. They should be far enough apart and strong enough to support your height and weight, to include equipment.
- Cut two poles that span the width of the rectangle. They, too, must be strong enough to support your weight.
- Secure these two poles to the trees (or poles). Be sure they are high enough above the ground or water to allow for tides and high water.
- Cut additional poles that span the rectangle's length. Lay them across the two side poles, and secure them.
- Cover the top of the bed frame with broad leaves or grass to form a soft sleeping surface.
- Build a fire pad by laying clay, silt, or mud on one comer of the swamp bed and allow it to dry.

Another shelter designed to get you above and out of the water or wet ground uses the same rectangular configuration as the swamp bed. You very simply lay sticks and branches lengthwise on the inside of the trees (or poles) until there is enough material to raise the sleeping surface above the water level.

Natural Shelters

Do not overlook natural formations that provide shelter. Examples are caves, rocky crevices, clumps of bushes, small depressions, large rocks on leeward sides of hills, large trees with low-hanging limbs, and fallen trees with thick branches. However, when selecting a natural formation--

- Stay away from low ground such as ravines, narrow valleys, or creek beds. Low areas collect the heavy cold air at night and are therefore colder than the surrounding high ground. Thick, brushy, low ground also harbors more insects.
- Check for poisonous snakes, ticks, mites, scorpions, and stinging ants.
- Look for loose rocks, dead limbs, coconuts, or other natural growth than could fall on your shelter.

Debris Hut

For warmth and ease of construction, this shelter is one of the best. When shelter is essential to survival, build this shelter.
To make a debris hut (<u>Figure 5-11</u>)--

- Build it by making a tripod with two short stakes and a long ridgepole or by placing one end of a long ridgepole on top of a sturdy base.
- Secure the ridgepole (pole running the length of the shelter) using the tripod method or by anchoring it to a tree at about waist height.
- Prop large sticks along both sides of the ridgepole to create a wedge-shaped ribbing effect. Ensure the ribbing is wide enough to accommodate your body and steep enough to shed moisture.
- Place finer sticks and brush crosswise on the ribbing. These form a latticework that will keep the insulating material (grass, pine needles, leaves) from falling through the ribbing into the sleeping area.
- Add light, dry, if possible, soft debris over the ribbing until the insulating material is at least 1 meter thick--the thicker the better.
- Place a 30-centimeter layer of insulating material inside the shelter.
- At the entrance, pile insulating material that you can drag to you once inside the shelter to close the entrance or build a door.
- As a final step in constructing this shelter, add shingling material or branches on top of the debris layer to prevent the insulating material from blowing away in a storm.

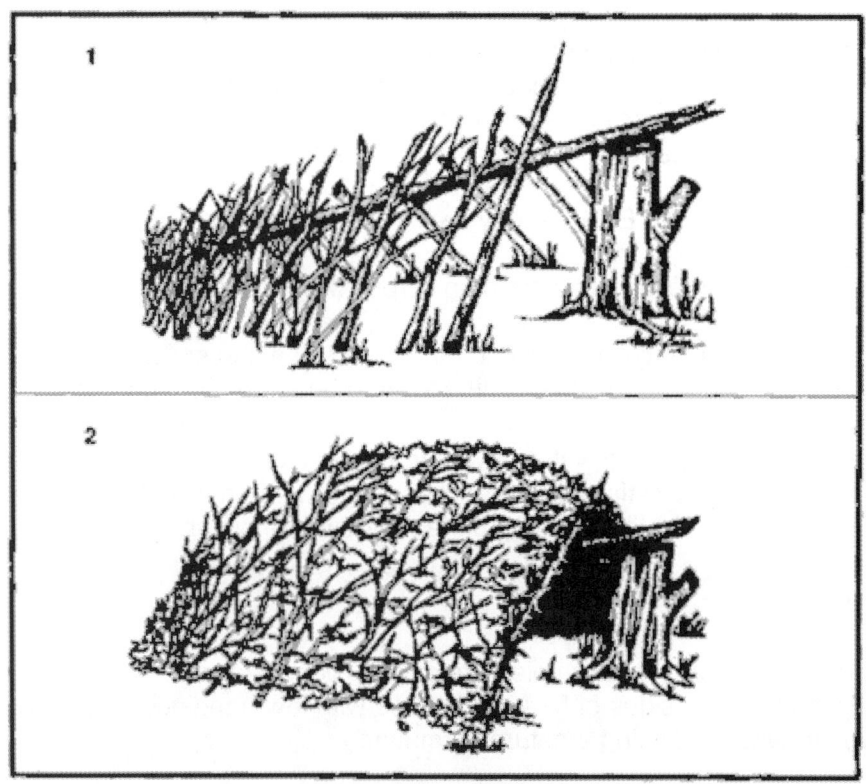

Figure 5-11. Debris hut.

Tree-Pit Snow Shelter

If you are in a cold, snow-covered area where evergreen trees grow and you have a digging tool, you can make a tree-pit shelter (Figure 5-12).

Figure 5-12. Tree-pit snow shelter.

To make this shelter--

- Find a tree with bushy branches that provides overhead cover.
- Dig out the snow around the tree trunk until you reach the depth and diameter you desire, or until you reach the ground.
- Pack the snow around the top and the inside of the hole to provide support.
- Find and cut other evergreen boughs. Place them over the top of the pit to give you additional overhead cover. Place evergreen boughs in the bottom of the pit for insulation.

See Chapter 15 for other arctic or cold weather shelters.

Beach Shade Shelter

This shelter protects you from the sun, wind, rain, and heat. It is easy to make using natural materials. To make this shelter (Figure 5-13)--

- Find and collect driftwood or other natural material to use as support beams and as a digging tool.
- Select a site that is above the high water mark.
- Scrape or dig out a trench running north to south so that it receives the least amount of sunlight. Make the trench long and wide enough for you to lie down comfortably.
- Mound soil on three sides of the trench. The higher the mound, the more space inside the shelter.
- Lay support beams (driftwood or other natural material) that span the trench on top of the mound to form the framework for a roof.
- Enlarge the shelter's entrance by digging out more sand in front of it.
- Use natural materials such as grass or leaves to form a bed inside the shelter.

Figure 5-13. Beach shade shelter.

Desert Shelters

In an arid environment, consider the time, effort, and material needed to make a shelter. If you have material such as a poncho, canvas, or a parachute, use it along with such terrain features as rock outcropping, mounds of sand, or a depression between dunes or rocks to make your shelter. Using rock outcroppings--

- Anchor one end of your poncho (canvas, parachute, or other material) on the edge of the outcrop using rocks or other weights.
- Extend and anchor the other end of the poncho so it provides the best possible shade.

In a sandy area--

- Build a mound of sand or use the side of a sand dune for one side of the shelter.
- Anchor one end of the material on top of the mound using sand or other weights.
- Extend and anchor the other end of the material so it provides the best possible shade.

    Note: If you have enough material, fold it in half and form a 30-centimeter to 45-centimeter airspace between the two halves. This airspace will reduce the temperature under the shelter.

A belowground shelter (Figure 5-14) can reduce the midday heat as much as 16 to 22 degrees C (30 to 40 degrees F). Building it, however, requires more time and effort than for other shelters. Since your physical effort will make you sweat more and increase dehydration, construct it before the heat of the day.

ESSENTIAL ARMY MANUAL
(53) ARMY SURVIVAL MANUAL

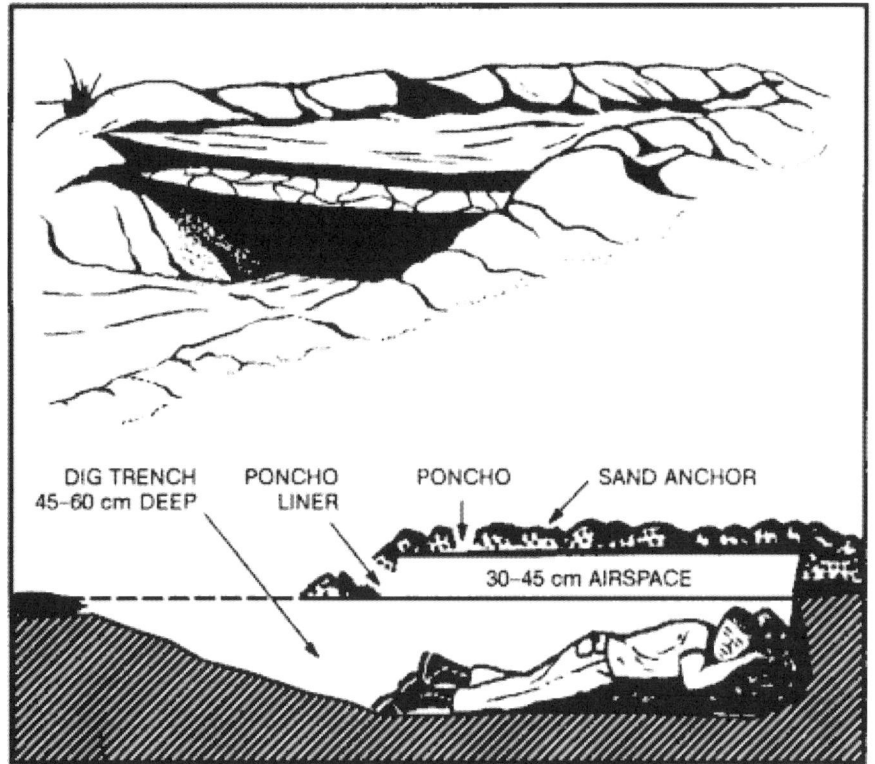

DIG TRENCH
45–60 cm DEEP

PONCHO
LINER

PONCHO

SAND ANCHOR

30–45 cm AIRSPACE

Figure 5-14. Belowground desert shelter.

To make this shelter--

- Find a low spot or depression between dunes or rocks. If necessary, dig a trench 45 to 60 centimeters deep and long and wide enough for you to lie in comfortably.
- Pile the sand you take from the trench to form a mound around three sides.
- On the open end of the trench, dig out more sand so you can get in and out of your shelter easily.
- Cover the trench with your material.
- Secure the material in place using sand, rocks, or other weights.

If you have extra material, you can further decrease the midday temperature in the trench by securing the material 30 to 45 centimeters above the other cover. This layering of the material will reduce the inside temperature 11 to 22 degrees C (20 to 40 degrees F).
Another type of belowground shade shelter is of similar construction, except all sides are open to air currents and circulation. For maximum protection, you need a minimum of two layers of parachute material (Figure 5-15). White is the best color to reflect heat; the innermost layer should be of darker material.

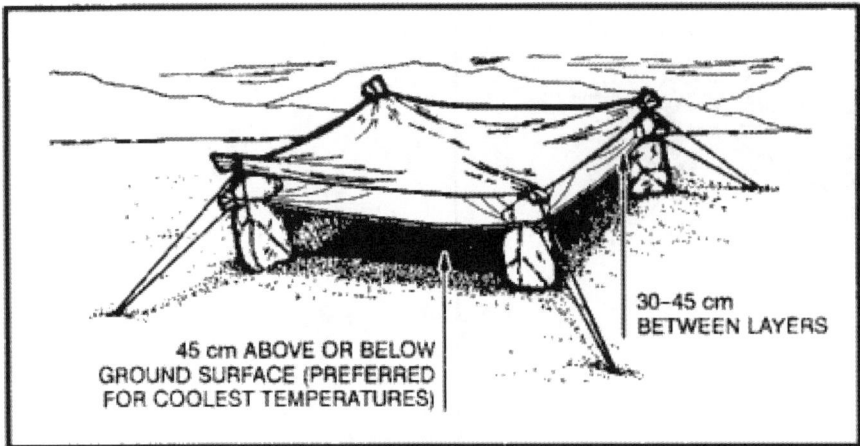

Figure 5-15. Open desert shelter.

CHAPTER 6 - WATER PROCUREMENT

Water is one of your most urgent needs in a survival situation. You can' t live long without it, especially in hot areas where you lose water rapidly through perspiration. Even in cold areas, you need a minimum of 2 liters of water each day to maintain efficiency.

More than three-fourths of your body is composed of fluids. Your body loses fluid as a result of heat, cold, stress, and exertion. To function effectively, you must replace the fluid your body loses. So, one of your first goals is to obtain an adequate supply of water.

WATER SOURCES

Almost any environment has water present to some degree. Figure 6-1 lists possible sources of water in various environments. It also provides information on how to make the water potable.

Environment	Source of Water	Means of Obtaining and/or Making Potable	Remarks
Frigid areas	Snow and ice	Melt and purify.	**Do not eat** without melting! Eating snow and ice can reduce body temperature and will lead to more dehydration. Snow and ice are no purer than the water from which they come. Sea ice that is gray in color or opaque is salty. Do not use it without desalting it. Sea ice that is crystalline with a bluish cast has little salt in it.
At sea	Sea	Use desalter kit.	**Do not** drink seawater without desalting.
	Rain	Catch rain in tarps or in other water-holding material or containers.	If tarp or water-holding material has become encrusted with salt, wash it in the sea before using (very little salt will remain on it).
	Sea ice		See remarks above for frigid areas.

Figure 6-1. Water sources in different environments.

Environment	Source of Water	Means of Obtaining and/or Making Potable	Remarks
Beach	Ground	Dig hole deep enough to allow water to seep in; obtain rocks, build fire, and heat rocks; drop hot rocks in water; hold cloth over hole to absorb steam; wring water from cloth.	Alternate method if a container or bark pot is available: Fill container or pot with seawater; build fire and boil water to produce steam; hold cloth over container to absorb steam; wring water from cloth.
Desert	Ground • in valleys and low areas • at foot of concave banks of dry river beads • at foot of cliffs or rock outcrops • at first depression behind first sand dune of dry desert lakes • wherever you find damp surface sand • wherever you find green vegetation	Dig holes deep enough to allow water to seep in.	In a sand dune belt, any available water will be found beneath the original valley floor at the edge of dunes.
	Cacti	Cut off the top of a barrel cactus and mash or squeeze the pulp. **CAUTION: Do not eat pulp. Place pulp in mouth, suck out juice, and discard pulp.**	Without a machete, cutting into a cactus is difficult and takes time since you must get past the long, strong spines and cut through the tough rind.

Figure 6-1. Water sources in different environments (continued).

Environment	Source of Water	Means of Obtaining and/or Making Potable	Remarks
Desert (continued)	Depressions or holes in rocks		Periodic rainfall may collect in pools, seep into fissures, or collect in holes in rocks.
	Fissures in rock	Insert flexible tubing and siphon water. If fissure is large enough, you can lower a container into it.	
	Porous rock	Insert flexible tubing and siphon water.	
	Condensation on metal	Use cloth to absorb water, then wring water from cloth.	Extreme temperature variations between night and day may cause condensation on metal surfaces.
			Following are signs to watch for in the desert to help you find water:
			• All trails lead to water. You should follow in the direction in which the trails converge. Signs of camps, campfire ashes, animal droppings, and trampled terrain may mark trails.
			• Flocks of birds will circle over water holes. Some birds fly to water holes at dawn and sunset. Their flight at these times is generally fast and close to the ground. Bird tracks or chirping sounds in the evening or early morning sometimes indicate that water is nearby.

Figure 6-1. Water sources in different environments (continued).

Note: If you do not have a canteen, a cup, a can, or other type of container, improvise one from plastic or water-resistant cloth. Shape the plastic or cloth into a bowl by pleating it. Use pins or other suitable items--even your hands--to hold the pleats.

If you do not have a reliable source to replenish your water supply, stay alert for ways in which your environment can help you.

CAUTION

Do not substitute the fluids listed in Figure 6-2 for water.

Fluid		Remarks
Alcoholic beverages		Dehydrate the body and cloud judgment.
Urine		Contains harmful body wastes. Is about 2 percent salt.
Blood		Is salty and considered a food; therefore, requires additional body fluids to digest. May transmit disease.
Seawater		Is about 4 percent salt. It takes about 2 liters of body fluids to rid the body of waste from 1 liter of seawater. Therefore, by drinking seawater you deplete your body's water supply, which can cause death.

Figure 6-2. The effects of substitute fluids.

Heavy dew can provide water. Tie rags or tufts of fine grass around your ankles and walk through dew-covered grass before sunrise. As the rags or grass tufts absorb the dew, wring the water into a container. Repeat the process until you have a supply of water or until the dew is gone. Australian natives sometimes mop up as much as a liter an hour this way.

Bees or ants going into a hole in a tree may point to a water-filled hole. Siphon the water with plastic tubing or scoop it up with an improvised dipper. You can also stuff cloth in the hole to absorb the water and then wring it from the cloth.

Water sometimes gathers in tree crotches or rock crevices. Use the above procedures to get the water. In arid areas, bird droppings around a crack in the rocks may indicate water in or near the crack.

Green bamboo thickets are an excellent source of fresh water. Water from green bamboo is clear and odorless. To get the water, bend a green bamboo stalk, tie it down, and cut off the top (Figure 6-3). The water will drip freely during the night. Old, cracked bamboo may contain water.

Figure 6-3. Water from green bamboo.

CAUTION
Purify the water before drinking it.

Wherever you find banana or plantain trees, you can get water. Cut down the tree, leaving about a 30-centimeter stump, and scoop out the center of the stump so that the hollow is bowl-shaped. Water from the roots will immediately start to fill the hollow. The first three fillings of water will be bitter, but succeeding fillings will be palatable. The stump (Figure 6-4) will supply water for up to four days. Be sure to cover it to keep out insects.

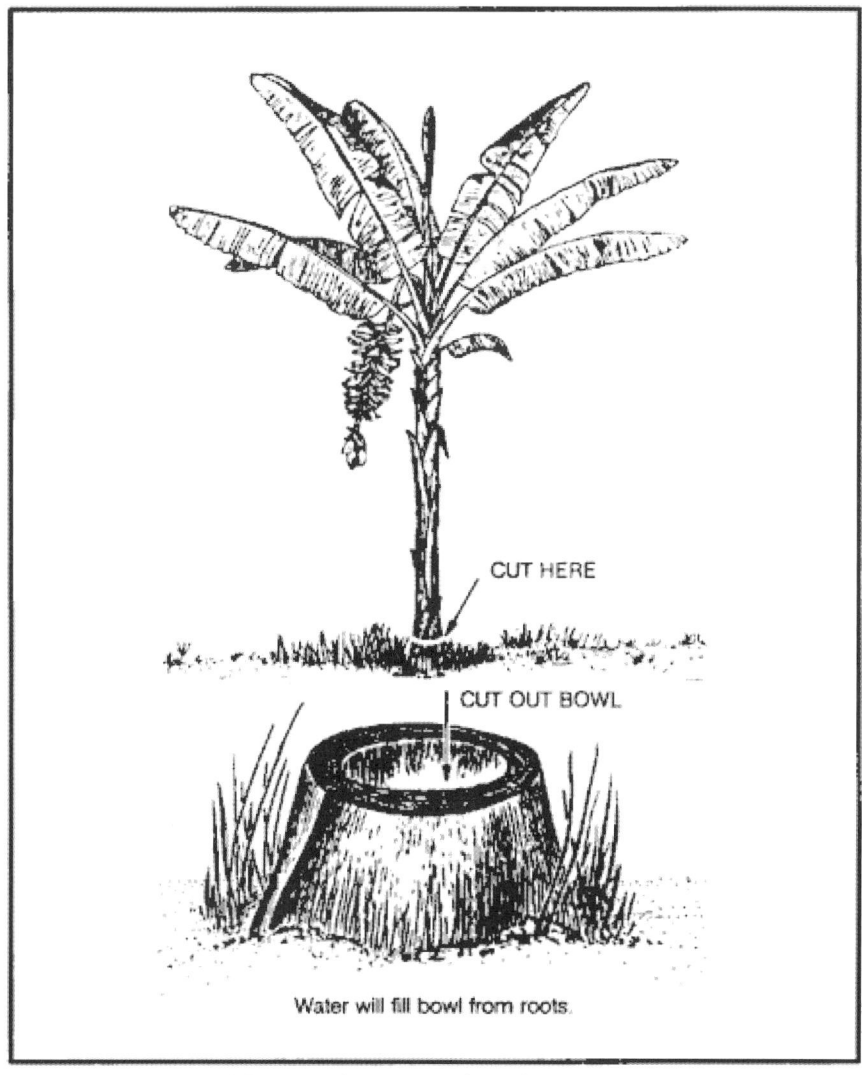

Figure 6-4. Water from plantain or banana tree stump.

Some tropical vines can give you water. Cut a notch in the vine as high as you can reach, then cut the vine off close to the ground. Catch the dropping liquid in a container or in your mouth (Figure 6-5).

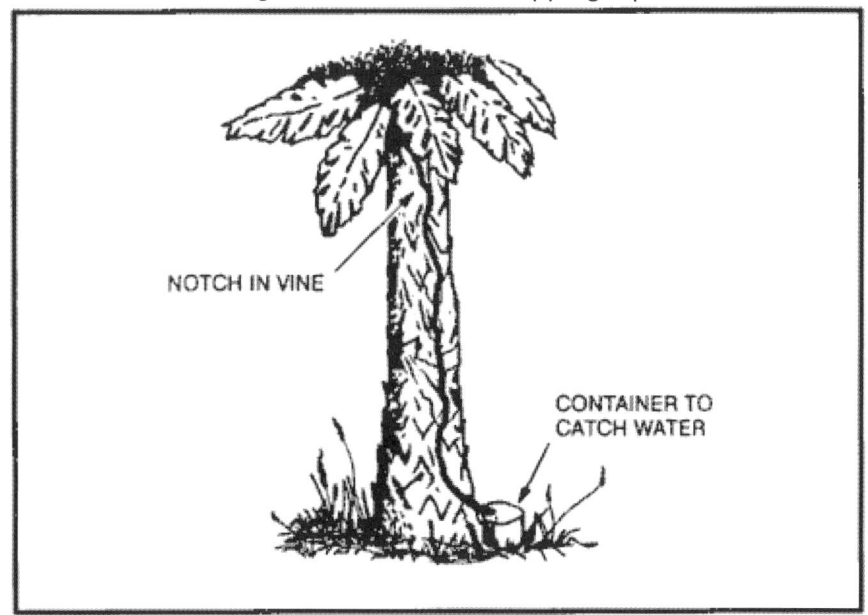

Figure 6-5. Water from a vine.

CAUTION

Do not drink the liquid if it is sticky, milky, or bitter tasting.

The milk from green (unripe) coconuts is a good thirst quencher. However, the milk from mature coconuts contains an oil that acts as a laxative. Drink in moderation only.

In the American tropics you may find large trees whose branches support air plants. These air plants may hold a considerable amount of rainwater in their overlapping, thickly growing leaves. Strain the water through a cloth to remove insects and debris.

You can get water from plants with moist pulpy centers. Cut off a section of the plant and squeeze or smash the pulp so that the moisture runs out. Catch the liquid in a container.

Plant roots may provide water. Dig or pry the roots out of the ground, cut them into short pieces, and smash the pulp so that the moisture runs out. Catch the liquid in a container.

Fleshy leaves, stems, or stalks, such as bamboo, contain water. Cut or notch the stalks at the base of a joint to drain out the liquid.

The following trees can also provide water:

- *Palms.* Palms, such as the buri, coconut, sugar, rattan, and nips, contain liquid. Bruise a lower frond and pull it down so the tree will "bleed" at the injury.
- *Traveler's tree.* Found in Madagascar, this tree has a cuplike sheath at the base of its leaves in which water collects.
- *Umbrella tree.* The leaf bases and roots of this tree of western tropical Africa can provide water.
- *Baobab tree.* This tree of the sandy plains of northern Australia and Africa collects water in its bottlelike trunk during the wet season. Frequently, you can find clear, fresh water in these trees after weeks of dry weather.

CAUTION

Do not keep the sap from plants longer than 24 hours. It begins fermenting, becoming dangerous as a water source.

STILL CONSTRUCTION

You can use stills in various areas of the world. They draw moisture from the ground and from plant material. You need certain materials to build a still, and you need time to let it collect the water. It takes about 24 hours to get 0.5 to 1 liter of water.

Aboveground Still

To make the aboveground still, you need a sunny slope on which to place the still, a clear plastic bag, green leafy vegetation, and a small rock (Figure 6-6).

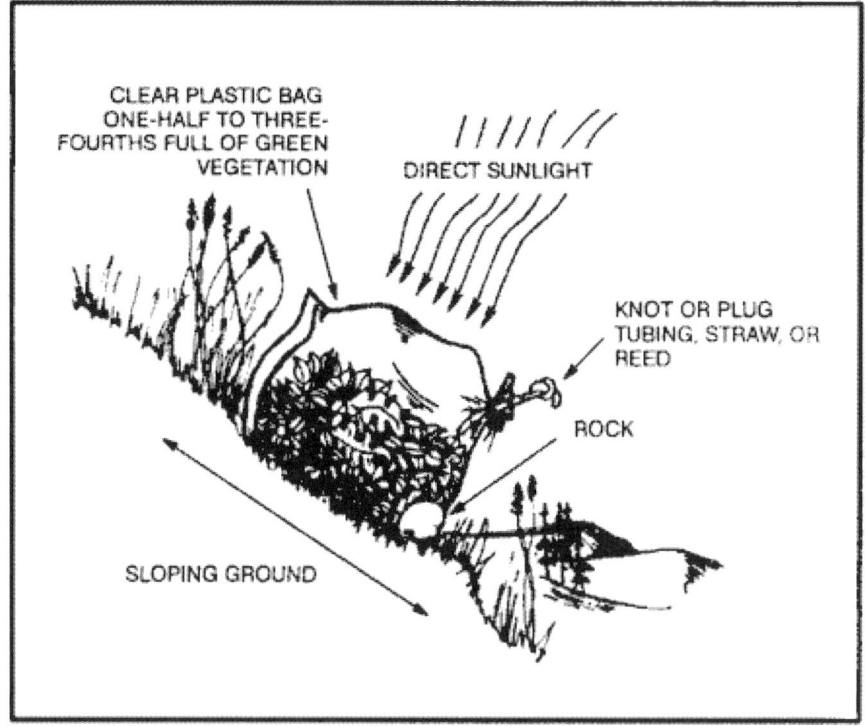

Figure 6-6. Aboveground solar water still.

To make the still--

- Fill the bag with air by turning the opening into the breeze or by "scooping" air into the bag.
- Fill the plastic bag half to three-fourths full of green leafy vegetation. Be sure to remove all hard sticks or sharp spines that might puncture the bag.

CAUTION
Do not use poisonous vegetation. It will provide poisonous liquid.

- Place a small rock or similar item in the bag.
- Close the bag and tie the mouth securely as close to the end of the bag as possible to keep the maximum amount of air space. If you have a piece of tubing, a small straw, or a hollow reed, insert one end in the mouth of the bag before you tie it securely. Then tie off or plug the tubing so that air will not escape. This tubing will allow you to drain out condensed water without untying the bag.
- Place the bag, mouth downhill, on a slope in full sunlight. Position the mouth of the bag slightly higher than the low point in the bag.
- Settle the bag in place so that the rock works itself into the low point in the bag.

To get the condensed water from the still, loosen the tie around the bag's mouth and tip the bag so that the water collected around the rock will drain out. Then retie the mouth securely and reposition the still to allow further condensation.

Change the vegetation in the bag after extracting most of the water from it. This will ensure maximum output of water.

Belowground Still

To make a belowground still, you need a digging tool, a container, a clear plastic sheet, a drinking tube, and a rock (Figure 6-7).

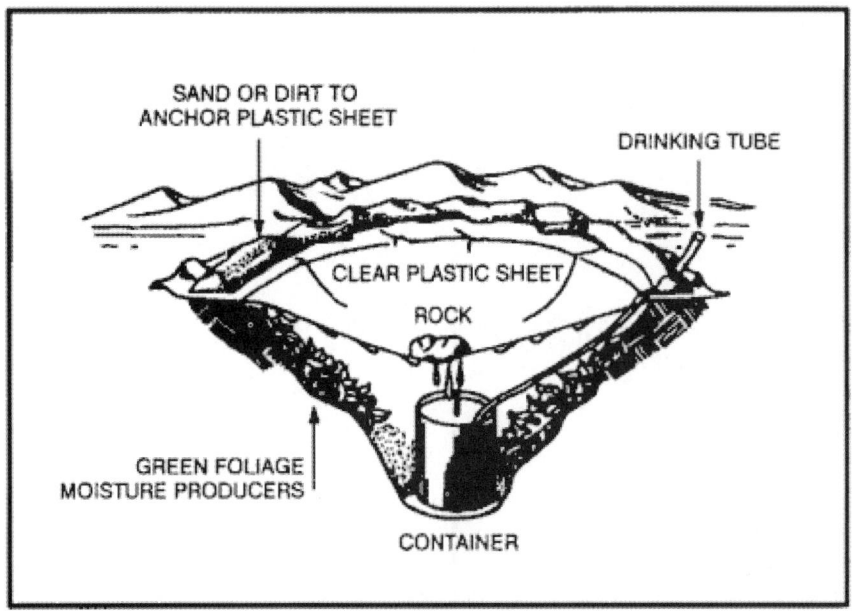

Figure 6-7. Belowground still.

Select a site where you believe the soil will contain moisture (such as a dry stream bed or a low spot where rainwater has collected). The soil at this site should be easy to dig, and sunlight must hit the site most of the day.

To construct the still--

- Dig a bowl-shaped hole about 1 meter across and 60 centimeters deep.
- Dig a sump in the center of the hole. The sump's depth and perimeter will depend on the size of the container that you have to place in it. The bottom of the sump should allow the container to stand upright.
- Anchor the tubing to the container's bottom by forming a loose overhand knot in the tubing.
- Place the container upright in the sump.
- Extend the unanchored end of the tubing up, over, and beyond the lip of the hole.
- Place the plastic sheet over the hole, covering its edges with soil to hold it in place.
- Place a rock in the center of the plastic sheet.
- Lower the plastic sheet into the hole until it is about 40 centimeters below ground level. It now forms an inverted cone with the rock at its apex. Make sure that the cone's apex is directly over your container. Also make sure the plastic cone does not touch the sides of the hole because the earth will absorb the condensed water.
- Put more soil on the edges of the plastic to hold it securely in place and to prevent the loss of moisture.
- Plug the tube when not in use so that the moisture will not evaporate.

You can drink water without disturbing the still by using the tube as a straw.

You may want to use plants in the hole as a moisture source. If so, dig out additional soil from the sides of the hole to form a slope on which to place the plants. Then proceed as above.

If polluted water is your only moisture source, dig a small trough outside the hole about 25 centimeters from the still's lip (Figure 6-8). Dig the trough about 25 centimeters deep and 8 centimeters wide. Pour the polluted water in the trough. Be sure you do not spill any polluted water around the rim of the hole where the plastic sheet touches the soil. The trough holds the polluted water and the soil filters it as the still draws it. The water then condenses on the plastic and drains into the container. This process works extremely well when your only water source is salt water.

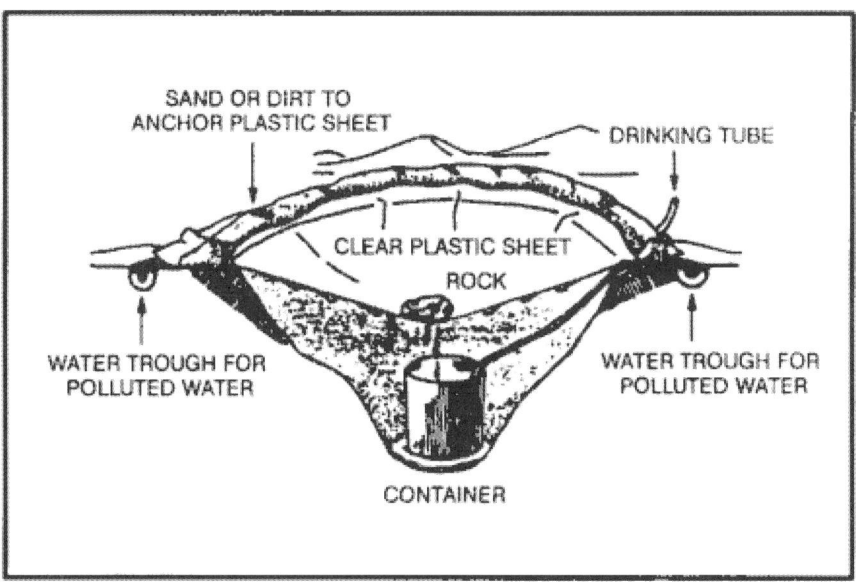

Figure 6-8. Belowground still to get potable water from polluted water.

You will need at least three stills to meet your individual daily water intake needs.

WATER PURIFICATION

Rainwater collected in clean containers or in plants is usually safe for drinking. However, purify water from lakes, ponds, swamps, springs, or streams, especially the water near human settlements or in the tropics.

When possible, purify all water you got from vegetation or from the ground by using iodine or chlorine, or by boiling.

Purify water by--

- Using water purification tablets. (Follow the directions provided.)
- Placing 5 drops of 2 percent tincture of iodine in a canteen full of clear water. If the canteen is full of cloudy or cold water, use 10 drops. (Let the canteen of water stand for 30 minutes before drinking.)
- Boiling water for 1 minute at sea level, adding 1 minute for each additional 300 meters above sea level, or boil for 10 minutes no matter where you are.

By drinking nonpotable water you may contract diseases or swallow organisms that can harm you. Examples of such diseases or organisms are--

- *Dysentery.* Severe, prolonged diarrhea with bloody stools, fever, and weakness.
- *Cholera and typhoid.* You may be susceptible to these diseases regardless of inoculations.
- *Flukes.* Stagnant, polluted water--especially in tropical areas--often contains blood flukes. If you swallow flukes, they will bore into the bloodstream, live as parasites, and cause disease.
- *Leeches.* If you swallow a leech, it can hook onto the throat passage or inside the nose. It will suck blood, create a wound, and move to another area. Each bleeding wound may become infected.

WATER FILTRATION DEVICES

If the water you find is also muddy, stagnant, and foul smelling, you can clear the water--

- By placing it in a container and letting it stand for 12 hours.
- By pouring it through a filtering system.

Note: These procedures only clear the water and make it more palatable. You will have to purify it.

To make a filtering system, place several centimeters or layers of filtering material such as sand, crushed rock, charcoal, or cloth in bamboo, a hollow log, or an article of clothing (<u>Figure 6-9</u>).

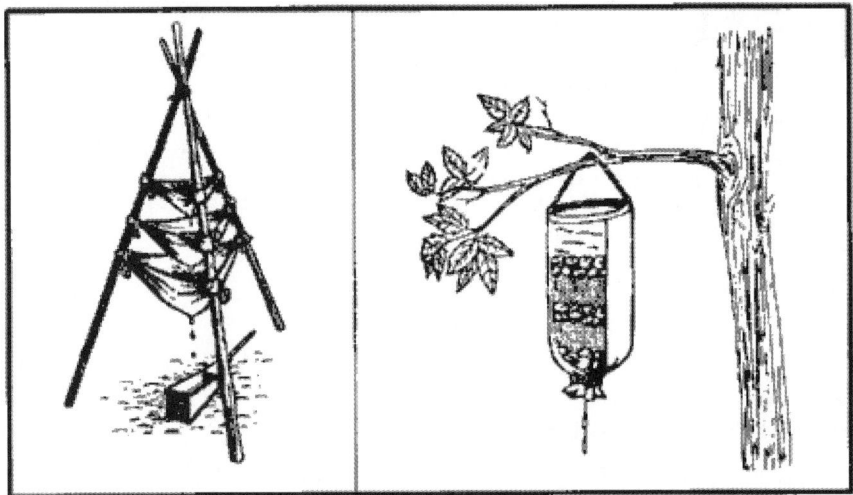

Figure 6-9. Water filtering systems.

Remove the odor from water by adding charcoal from your fire. Let the water stand for 45 minutes before drinking it.

CHAPTER 7 - FIRECRAFT

In many survival situations, the ability to start a fire can make the difference between living and dying. Fire can fulfill many needs. It can provide warmth and comfort. It not only cooks and preserves food, it also provides warmth in the form of heated food that saves calories our body normally uses to produce body heat. You can use fire to purify water, sterilize bandages, signal for rescue, and provide protection from animals. It can be a psychological boost by providing peace of mind and companionship. You can also use fire to produce tools and weapons.

Fire can cause problems, as well. The enemy can detect the smoke and light it produces. It can cause forest fires or destroy essential equipment. Fire can also cause burns carbon monoxide poisoning when used in shelters.

Remember weigh your need for fire against your need to avoid enemy detection.

BASIC FIRE PRINCIPLES

To build a fire, it helps to understand the basic principles of a fire. Fuel (in a nongaseous state) does not burn directly. When you apply heat to a fuel, it produces a gas. This gas, combined with oxygen in the air, burns.
Understanding the concept of the fire triangle is very important in correctly constructing and maintaining a fire. The three sides of the triangle represent *air, heat,* and *fuel.* If you remove any of these, the fire will go out. The correct ratio of these components is very important for a fire to burn at its greatest capability. The only way to learn this ratio is to practice.

SITE SELECTION AND PREPARATION

You will have to decide what site and arrangement to use. Before building a fire consider--

- The area (terrain and climate) in which you are operating.
- The materials and tools available.
- Time: how much time you have?
- Need: why you need a fire?
- Security: how close is the enemy?

Look for a dry spot that--

- Is protected from the wind.
- Is suitably placed in relation to your shelter (if any).
- Will concentrate the heat in the direction you desire.

Has a supply of wood or other fuel available. (See Figure 7-4 for types of material you can use.)

Tinder	Kindling	Fuel
• Birch bark	• Small twigs	• Dry, standing wood and dry, dead branches
• Shredded inner bark from cedar, chestnut, red elm trees	• Small strips of wood	• Dry inside (heart) of fallen tree trunks and large branches
• Fine wood shavings	• Split wood	
• Dead grass, ferns, moss, fungi	• Heavy cardboard	• Green wood that is finely split
• Straw	• Pieces of wood removed from the inside of larger pieces	• Dry grasses twisted into bunches
• Sawdust	• Wood that has been doused with highly flammable materials, such as gasoline, oil, or wax	• Peat dry enough to burn (this may be found at the top of undercut banks)
• Very fine pitchwood scrapings		• Dried animal dung
• Dead evergreen needles		• Animal fats
• Punk (the completely rotted portions of dead logs or trees)		• Coal, oil shale, or oil lying on the surface
• Evergreen tree knots		
• Bird down (fine feathers)		
• Down seed heads (milkweed, dry cattails, bulrush, or thistle)		
• Fine, dried vegetable fibers		
• Spongy threads of dead puffball		
• Dead palm leaves		
• Skinlike membrane lining bamboo		
• Lint from pocket and seams		
• Charred cloth		
• Waxed paper		
• Outer bamboo shavings		
• Gunpowder		
• Cotton		
• Lint		

Figure 7-4. Materials for building fires.

If you are in a wooded or brush-covered area, clear the brush and scrape the surface soil from the spot you have selected. Clear a circle at least 1 meter in diameter so there is little chance of the fire spreading.

If time allows, construct a fire wall using logs or rocks. This wall will help to reflector direct the heat where you want it (Figure 7-1). It will also reduce flying sparks and cut down on the amount of wind blowing into the fire. However, you will need enough wind to keep the fire burning.

CAUTION
Do not use wet or porous rocks as they may explode when heated.

Figure 7-1. Types of fire walls.

In some situations, you may find that an underground fireplace will best meet your needs. It conceals the fire and serves well for cooking food. To make an underground fireplace or Dakota fire hole (Figure 7-2)--

* Dig a hole in the ground.
* On the upwind side of this hole, poke or dig a large connecting hole for ventilation.
* Build your fire in the hole as illustrated.

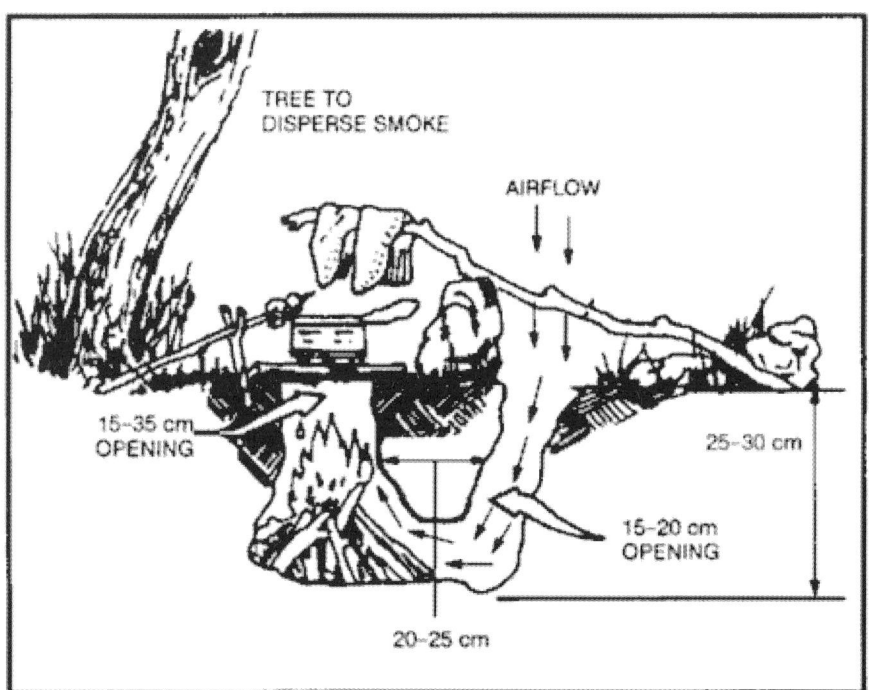

Figure 7-2. Dakota fire hole.

If you are in a snow-covered area, use green logs to make a dry base for your fire (Figure 7-3). Trees with wrist-sized trunks are easily broken in extreme cold. Cut or break several green logs and lay them side by side on top of the snow. Add one or two more layers. Lay the top layer of logs opposite those below it.

Figure 7-3. Base for fire in snow-covered area.

FIRE MATERIAL SELECTION

You need three types of materials (Figure 7-4) to build a fire--tinder, kindling, and fuel.

Tinder is dry material that ignites with little heat--a spark starts a fire. The tinder must be absolutely dry to be sure just a spark will ignite it. If you only have a device that generates sparks, charred cloth will be almost essential. It holds a spark for long periods, allowing you to put tinder on the hot area to generate a small flame. You can make charred cloth by heating cotton cloth until it turns black, but does not burn. Once it is black, you must keep it in an airtight container to keep it dry. Prepare this cloth well in advance of any survival situation. Add it to your individual survival kit.

Kindling is readily combustible material that you add to the burning tinder. Again, this material should be absolutely dry to ensure rapid burning. Kindling increases the fire's temperature so that it will ignite less combustible material.

Fuel is less combustible material that burns slowly and steadily once ignited.

HOW TO BUILD A FIRE

There are several methods for laying a fire, each of which has advantages. The situation you find yourself in will determine which fire to use.

Tepee

To make this fire (Figure 7-5), arrange the tinder and a few sticks of kindling in the shape of a tepee or cone. Light the center. As the tepee burns, the outside logs will fall inward, feeding the fire. This type of fire burns well even with wet wood.

ESSENTIAL ARMY MANUAL
(69) ARMY SURVIVAL MANUAL

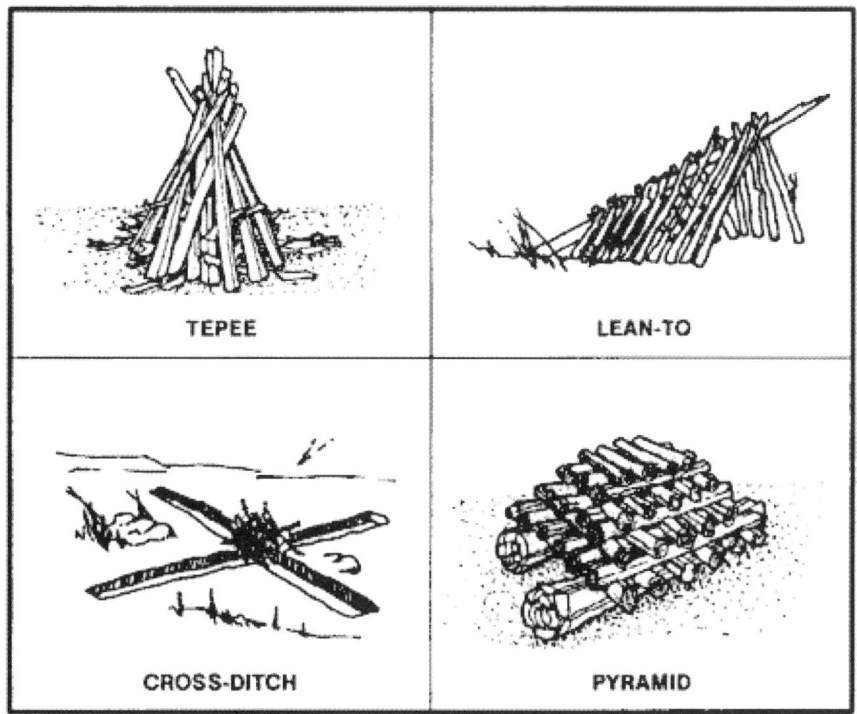

Figure 7-5. Methods for laying fires.

Lean-To

To lay this fire (Figure 7-5), push a green stick into the ground at a 30-degree angle. Point the end of the stick in the direction of the wind. Place some tinder deep under this lean-to stick. Lean pieces of kindling against the lean-to stick. Light the tinder. As the kindling catches fire from the tinder, add more kindling.

Cross-Ditch

To use this method (Figure 7-5), scratch a cross about 30 centimeters in size in the ground. Dig the cross 7.5 centimeters deep. Put a large wad of tinder in the middle of the cross. Build a kindling pyramid above the tinder. The shallow ditch allows air to sweep under the tinder to provide a draft.

Pyramid

To lay this fire (Figure 7-5), place two small logs or branches parallel on the ground. Place a solid layer of small logs across the parallel logs. Add three or four more layers of logs or branches, each layer smaller than and at a right angle to the layer below it. Make a starter fire on top of the pyramid. As the starter fire burns, it will ignite the logs below it. This gives you a fire that burns downward, requiring no attention during the night.

There are several other ways to lay a fire that are quite effective. Your situation and the material available in the area may make another method more suitable.

HOW TO LIGHT A FIRE

Always light your fire from the upwind side. Make sure to lay your tinder, kindling, and fuel so that your fire will burn as long as you need it. Igniters provide the initial heat required to start the tinder burning. They fall into two categories: modern methods and primitive methods.

Modern Methods

Modem igniters use modem devices--items we normally think of to start a fire.

Matches

Make sure these matches are waterproof. Also, store them in a waterproof container along with a dependable striker pad.

Convex Lens

Use this method (Figure 7-6) only on bright, sunny days. The lens can come from binoculars, camera, telescopic sights, or magnifying glasses. Angle the lens to concentrate the sun's rays on the tinder. Hold the lens over the same spot until the tinder begins to smolder. Gently blow or fan the tinder into flame, and apply it to the fire lay.

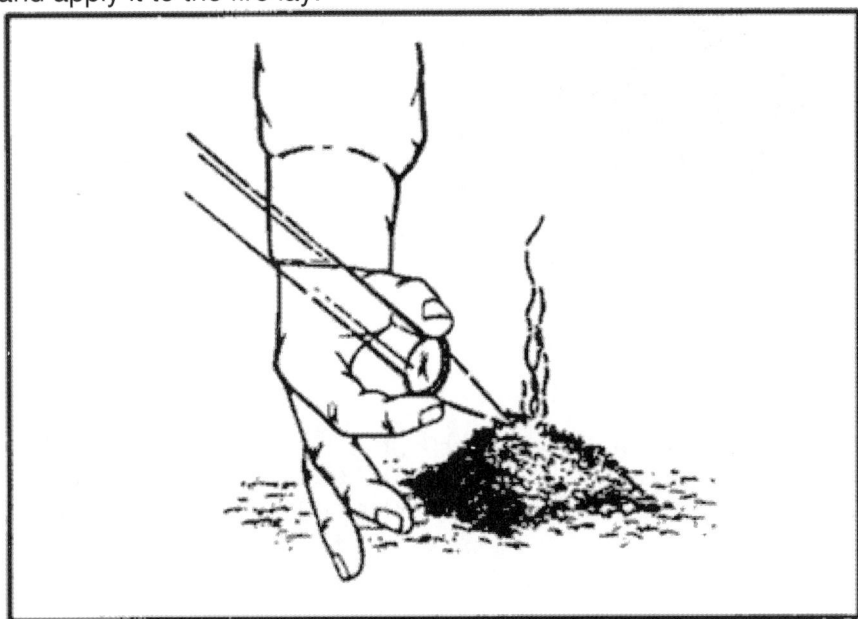

Figure 7-6. Lens method.

Metal Match

Place a flat, dry leaf under your tinder with a portion exposed. Place the tip of the metal match on the dry leaf, holding the metal match in one hand and a knife in the other. Scrape your knife against the metal match to produce sparks. The sparks will hit the tinder. When the tinder starts to smolder, proceed as above.

Battery

Use a battery to generate a spark. Use of this method depends on the type of battery available. Attach a wire to each terminal. Touch the ends of the bare wires together next to the tinder so the sparks will ignite it.

Gunpowder

Often, you will have ammunition with your equipment. If so, carefully extract the bullet from the shell casing, and use the gunpowder as tinder. A spark will ignite the powder. Be extremely careful when extracting the bullet from the case.

Primitive Methods

Primitive igniters are those attributed to our early ancestors.

Flint and Steel

The direct spark method is the easiest of the primitive methods to use. The flint and steel method is the most reliable of the direct spark methods. Strike a flint or other hard, sharp-edged rock edge with a piece of carbon steel (stainless steel will not produce a good spark). This method requires a loose-jointed wrist and practice. When a spark has caught in the tinder, blow on it. The spark will spread and burst into flames.

Fire-Plow

The fire-plow (Figure 7-7) is a friction method of ignition. You rub a hardwood shaft against a softer wood base. To use this method, cut a straight groove in the base and plow the blunt tip of the shaft up and down the groove. The plowing action of the shaft pushes out small particles of wood fibers. Then, as you apply more pressure on each stroke, the friction ignites the wood particles.

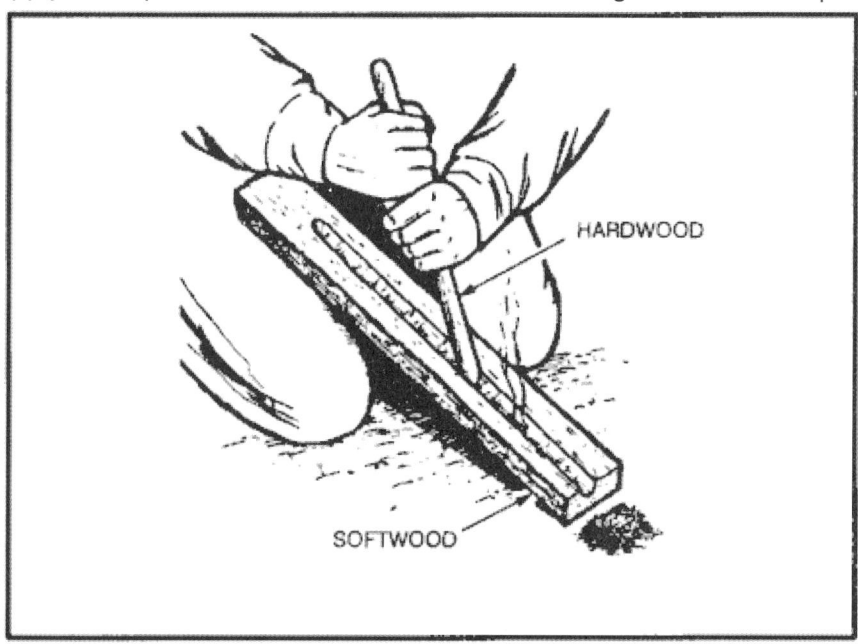

Figure 7-7. Fire-plow.

Bow and Drill

The technique of starting a fire with a bow and drill (Figure 7-8) is simple, but you must exert much effort and be persistent to produce a fire. You need the following items to use this method:

- *Socket.* The socket is an easily grasped stone or piece of hardwood or bone with a slight depression in one side. Use it to hold the drill in place and to apply downward pressure.
- *Drill.* The drill should be a straight, seasoned hardwood stick about 2 centimeters in diameter and 25 centimeters long. The top end is round and the low end blunt (to produce more friction).
- *Fire board.* Its size is up to you. A seasoned softwood board about 2.5 centimeters thick and 10 centimeters wide is preferable. Cut a depression about 2 centimeters from the edge on one side of the board. On the underside, make a V-shaped cut from the edge of the board to the depression.

- *Bow.* The bow is a resilient, green stick about 2.5 centimeters in diameter and a string. The type of wood is not important. The bowstring can be any type of cordage. You tie the bowstring from one end of the bow to the other, without any slack.

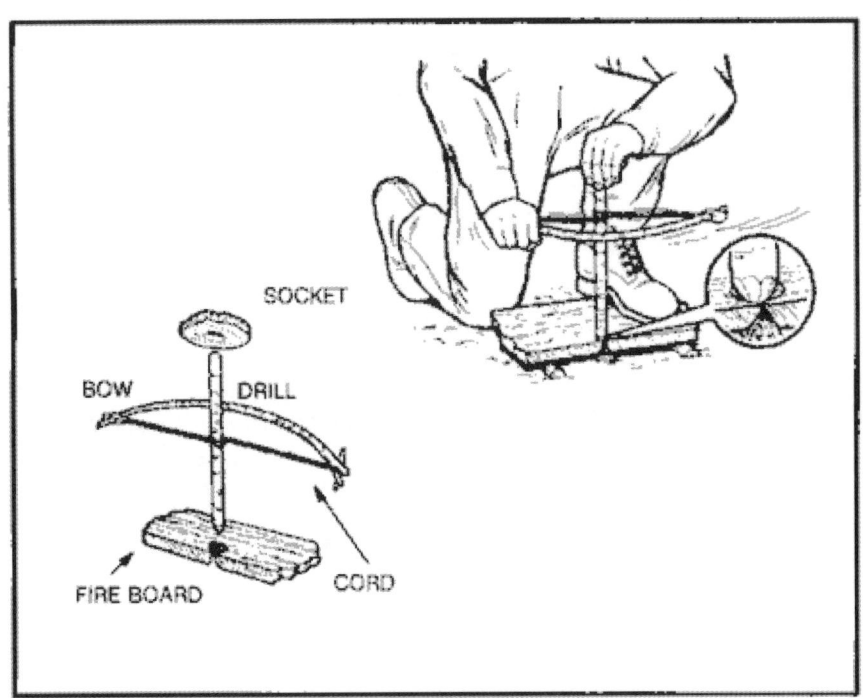

Figure 7-8. Bow and drill.

To use the bow and drill, first prepare the fire lay. Then place a bundle of tinder under the V-shaped cut in the fire board. Place one foot on the fire board. Loop the bowstring over the drill and place the drill in the precut depression on the fire board. Place the socket, held in one hand, on the top of the drill to hold it in position. Press down on the drill and saw the bow back and forth to twirl the drill (Figure 7-8). Once you have established a smooth motion, apply more downward pressure and work the bow faster. This action will grind hot black powder into the tinder, causing a spark to catch. Blow on the tinder until it ignites.

> *Note: Primitive fire-building methods are exhaustive and require practice to ensure success.*

HELPFUL HINTS

Use nonaromatic seasoned hardwood for fuel, if possible.

Collect kindling and tinder along the trail.

Add insect repellent to the tinder.

Keep the firewood dry.

Dry damp firewood near the fire.

Bank the fire to keep the coals alive overnight.

Carry lighted punk, when possible.

Be sure the fire is out before leaving camp.

Do not select wood lying on the ground. It may appear to be dry but generally doesn't provide enough friction.

CHAPTER 8 - FOOD PROCUREMENT

After water, man's most urgent requirement is food. In contemplating virtually any hypothetical survival situation, the mind immediately turns to thoughts of food. Unless the situation occurs in an arid environment, even water, which is more important to maintaining body functions, will almost always follow food in our initial thoughts. The survivor must remember that the three essentials of survival--water, food, and shelter--are prioritized according to the estimate of the actual situation. This estimate must not only be timely but accurate as well. Some situations may well dictate that shelter precede both food and water.

ANIMALS FOR FOOD

Unless you have the chance to take large game, concentrate your efforts on the smaller animals, due to their abundance. The smaller animal species are also easier to prepare. You must not know all the animal species that are suitable as food. Relatively few are poisonous, and they make a smaller list to remember. What is important is to learn the habits and behavioral patterns of classes of animals. For example, animals that are excellent choices for trapping, those that inhabit a particular range and occupy a den or nest, those that have somewhat fixed feeding areas, and those that have trails leading from one area to another. Larger, herding animals, such as elk or caribou, roam vast areas and are somewhat more difficult to trap. Also, you must understand the food choices of a particular species.
You can, with relatively few exceptions, eat anything that crawls, swims, walks, or flies. The first obstacle is overcoming your natural aversion to a particular food source. Historically, people in starvation situations have resorted to eating everything imaginable for nourishment. A person who ignores an otherwise healthy food source due to a personal bias, or because he feels it is unappetizing, is risking his own survival. Although it may prove difficult at first, a survivor must eat what is available to maintain his health.

Insects

The most abundant life-form on earth, insects are easily caught. Insects provide 65 to 80 percent protein compared to 20 percent for beef. This fact makes insects an important, if not overly appetizing, food source. Insects to avoid include all adults that sting or bite, hairy or brightly colored insects, and caterpillars and insects that have a pungent odor. Also avoid spiders and common disease carriers such as ticks, flies, and mosquitoes.
Rotting logs lying on the ground are excellent places to look for a variety of insects including ants, termites, beetles, and grubs, which are beetle larvae. Do not overlook insect nests on or in the ground. Grassy areas, such as fields, are good areas to search because the insects are easily seen. Stones, boards, or other materials lying on the ground provide the insects with good nesting sites. Check these sites. Insect larvae are also edible. Insects such as beetles and grasshoppers that have a hard outer

shell will have parasites. Cook them before eating. Remove any wings and barbed legs also. You can eat most insects raw. The taste varies from one species to another. Wood grubs are bland, while some species of ants store honey in their bodies, giving them a sweet taste. You can grind a collection of insects into a paste. You can mix them with edible vegetation. You can cook them to improve their taste.

Worms

Worms (*Annelidea*) are an excellent protein source. Dig for them in damp humus soil or watch for them on the ground after a rain. After capturing them, drop them into clean, potable water for a few minutes. The worms will naturally purge or wash themselves out, after which you can eat them raw.

Crustaceans

Freshwater shrimp range in size from 0.25 centimeter up to 2.5 centimeters. They can form rather large colonies in mats of floating algae or in mud bottoms of ponds and lakes.
Crayfish are akin to marine lobsters and crabs. You can distinguish them by their hard exoskeleton and five pairs of legs, the front pair having oversized pincers. Crayfish are active at night, but you can locate them in the daytime by looking under and around stones in streams. You can also find them by looking in the soft mud near the chimneylike breathing holes of their nests. You can catch crayfish by tying bits of offal or internal organs to a string. When the crayfish grabs the bait, pull it to shore before it has a chance to release the bait.
You find saltwater lobsters, crabs, and shrimp from the surf's edge out to water 10 meters deep. Shrimp may come to a light at night where you can scoop them up with a net. You can catch lobsters and crabs with a baited trap or a baited hook. Crabs will come to bait placed at the edge of the surf, where you can trap or net them. Lobsters and crabs are nocturnal and caught best at night.

Mollusks

This class includes octopuses and freshwater and saltwater shellfish such as snails, clams, mussels, bivalves, barnacles, periwinkles, chitons, and sea urchins (Figure 8-1). You find bivalves similar to our freshwater mussel and terrestrial and aquatic snails worldwide under all water conditions.

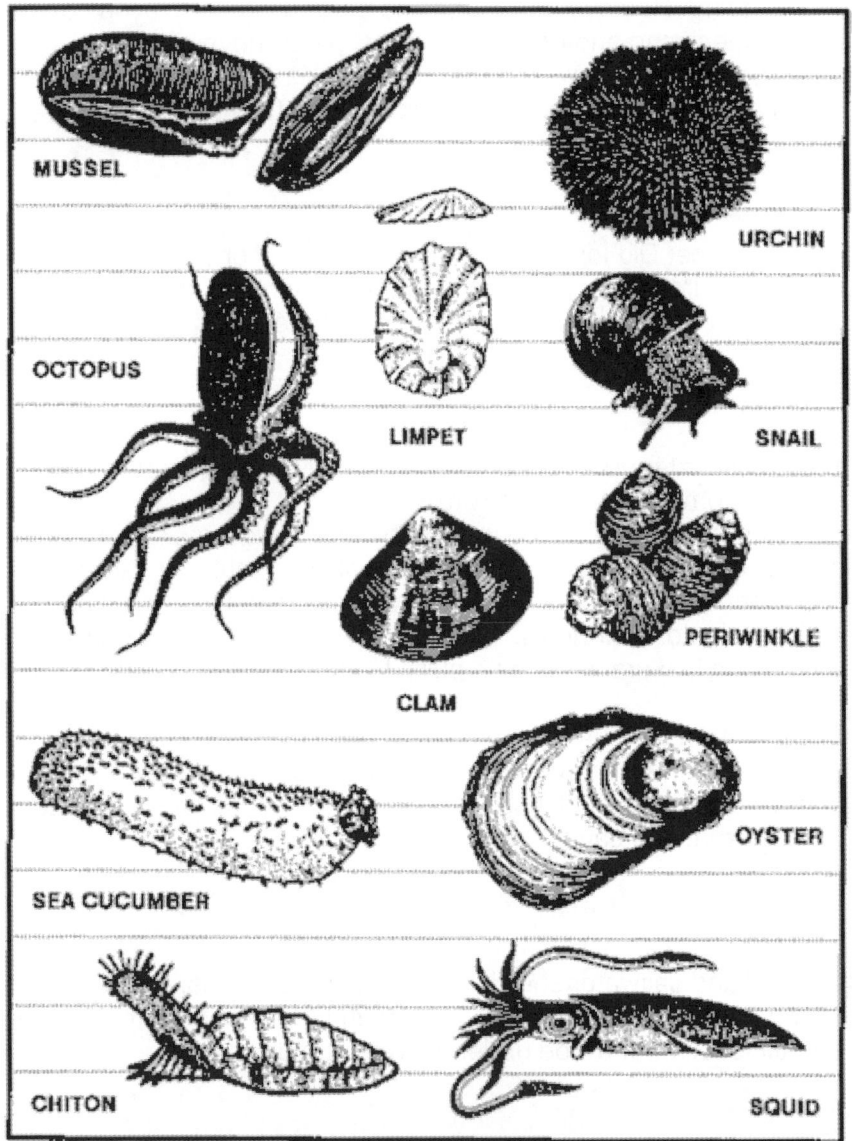

Figure 8-1. Edible mollusks.

River snails or freshwater periwinkles are plentiful in rivers, streams, and lakes of northern coniferous forests. These snails may be pencil point or globular in shape.

In fresh water, look for mollusks in the shallows, especially in water with a sandy or muddy bottom. Look for the narrow trails they leave in the mud or for the dark elliptical slit of their open valves.

Near the sea, look in the tidal pools and the wet sand. Rocks along beaches or extending as reefs into deeper water often bear clinging shellfish. Snails and limpets cling to rocks and seaweed from the low water mark upward. Large snails, called chitons, adhere tightly to rocks above the surf line.

Mussels usually form dense colonies in rock pools, on logs, or at the base of boulders.

CAUTION
Mussels may be poisonous in tropical zones during the summer!

Steam, boil, or bake mollusks in the shell. They make excellent stews in combination with greens and tubers.

CAUTION
Do not eat shellfish that are not covered by water at high tide!

Fish

Fish represent a good source of protein and fat. They offer some distinct advantages to the survivor or evader. They are usually more abundant than mammal wildlife, and the ways to get them are silent. To be successful at catching fish, you must know their habits. For instance, fish tend to feed heavily before a storm. Fish are not likely to feed after a storm when the water is muddy and swollen. Light often attracts fish at night. When there is a heavy current, fish will rest in places where there is an eddy, such as near rocks. Fish will also gather where there are deep pools, under overhanging brush, and in and around submerged foliage, logs, or other objects that offer them shelter.

There are no poisonous freshwater fish. However, the catfish species has sharp, needlelike protrusions on its dorsal fins and barbels. These can inflict painful puncture wounds that quickly become infected. Cook all freshwater fish to kill parasites. Also cook saltwater fish caught within a reef or within the influence of a freshwater source as a precaution. Any marine life obtained farther out in the sea will not contain parasites because of the saltwater environment. You can eat these raw.

Certain saltwater species of fish have poisonous flesh. In some species the poison occurs seasonally in others, it is permanent. Examples of poisonous saltwater fish are the porcupine fish, triggerfish, cowfish, thorn fish, oilfish, red snapper, jack, and puffer (Figure 8-2). The barracuda, while not actually poisonous itself, may transmit ciguatera (fish poisoning) if eaten raw.

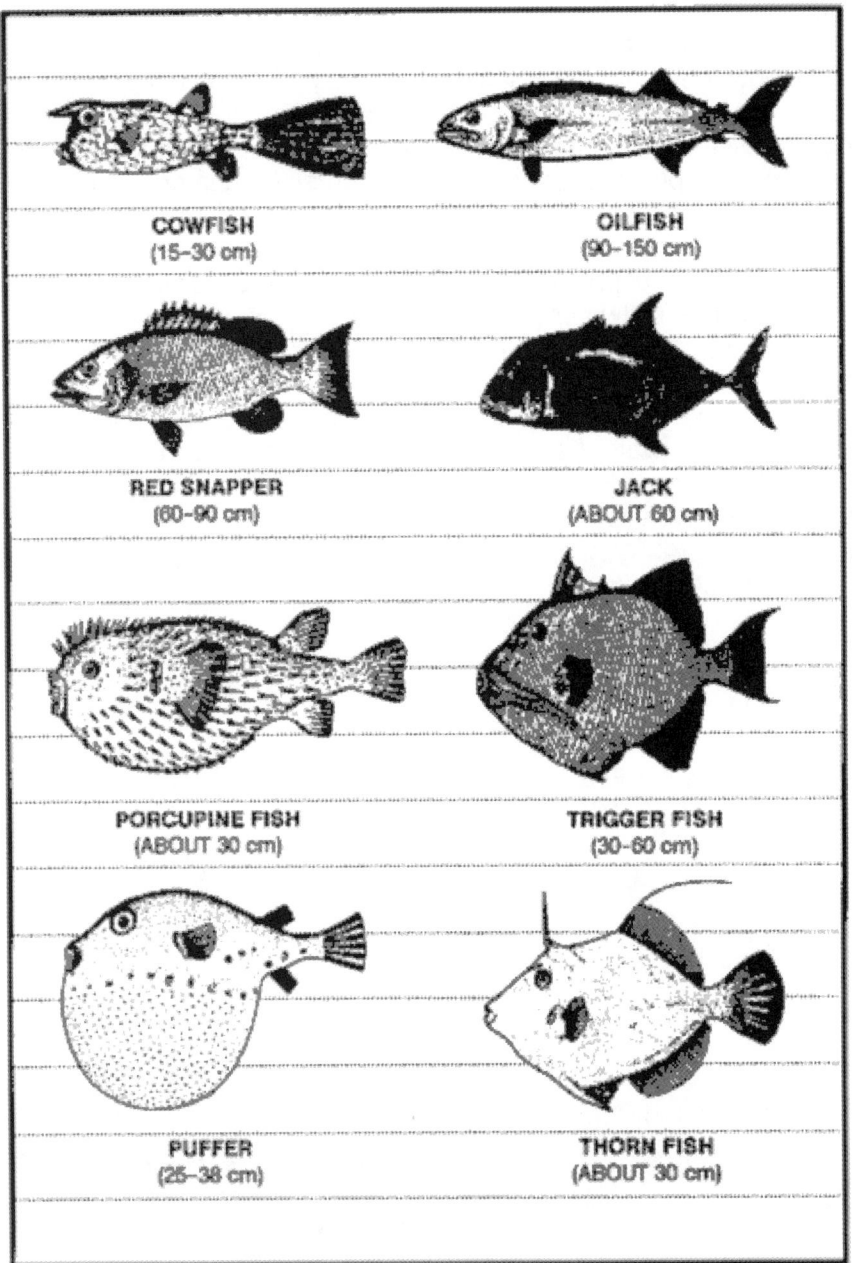

Figure 8-2. Fish with poisonous flesh.

Amphibians

Frogs and salamanders are easily found around bodies of fresh water. Frogs seldom move from the safety of the water's edge. At the first sign of danger, they plunge into the water and bury themselves in the mud and debris. There are few poisonous species of frogs. Avoid any brightly colored frog or one that has a distinct "X" mark on it's back. Do not confuse toads with frogs. You normally find toads in drier environments. Several species of toads secrete a poisonous substance through their skin as a defense against attack. Therefore, to avoid poisoning, do not handle or eat toads.

Salamanders are nocturnal. The best time to catch them is at night using a light. They can range in size from a few centimeters to well over 60 centimeters in length. Look in water around rocks and mud banks for salamanders.

Reptiles

Reptiles are a good protein source and relatively easy to catch. You should cook them, but in an emergency, you can eat them raw. Their raw flesh may transmit parasites, but because reptiles are cold-blooded, they do not carry the blood diseases of the warm-blooded animals.

The box turtle is a commonly encountered turtle that you should not eat. It feeds on poisonous mushrooms and may build up a highly toxic poison in its flesh. Cooking does not destroy this toxin. Avoid the hawksbill turtle, found in the Atlantic Ocean, because of its poisonous thorax gland. Poisonous snakes, alligators, crocodiles, and large sea turtles present obvious hazards to the survivor.

Birds

All species of birds are edible, although the flavor will vary considerably. You may skin fish-eating birds to improve their taste. As with any wild animal, you must understand birds' common habits to have a realistic chance of capturing them. You can take pigeons, as well as some other species, from their roost at night by hand. During the nesting season, some species will not leave the nest even when approached. Knowing where and when the birds nest makes catching them easier (Figure 8-3). Birds tend to have regular flyways going from the roost to a feeding area, to water, and so forth. Careful observation should reveal where these flyways are and indicate good areas for catching birds in nets stretched across the flyways (Figure 8-4). Roosting sites and waterholes are some of the most promising areas for trapping or snaring.

Types of Birds	Frequent Nesting Places	Nesting Periods
Inland birds	Trees, woods, or fields	Spring and early summer in temperate and arctic regions; year round in the tropics
Cranes and herons	Mangrove swamps or high trees near water	Spring and early summer
Some species of owls	High trees	Late December through March
Ducks, geese, and swans	Tundra areas near ponds, rivers, or lakes	Spring and early summer in arctic regions
Some sea birds	Sandbars or low sand islands	Spring and early summer in temperate and arctic regions
Gulls, auks, murres, and cormorants	Steep rocky coasts	Spring and early summer in temperate and arctic regions

Figure 8-3. Bird nesting places.

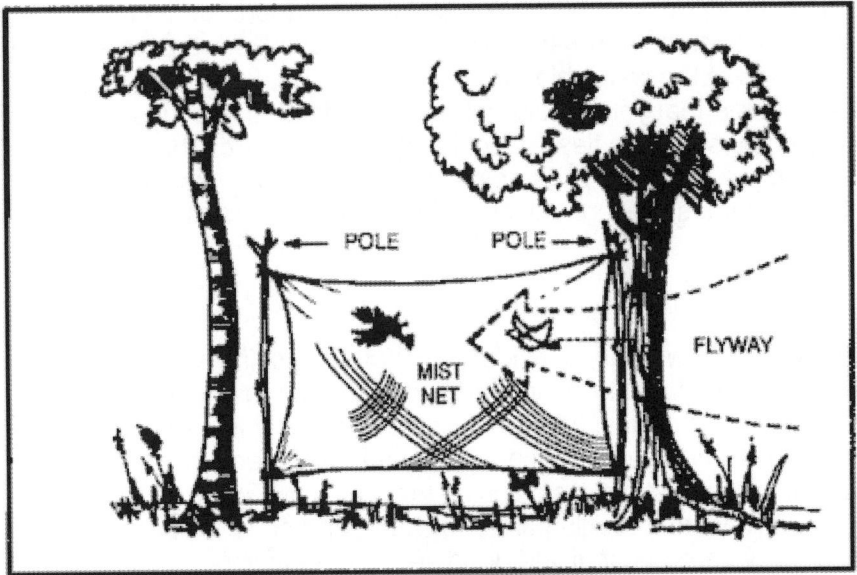

Figure 8-4. Catching birds in a net.

Nesting birds present another food source--eggs. Remove all but two or three eggs from the clutch, marking the ones that you leave. The bird will continue to lay more eggs to fill the clutch. Continue removing the fresh eggs, leaving the ones you marked.

Mammals

Mammals are excellent protein sources and, for Americans, the most tasty food source. There are some drawbacks to obtaining mammals. In a hostile environment, the enemy may detect any traps or snares placed on land. The amount of injury an animal can inflict is in direct proportion to its size. All mammals have teeth and nearly all will bite in self-defense. Even a squirrel can inflict a serious wound and any bite presents a serious risk of infection. Also, a mother can be extremely aggressive in defense of her young. Any animal with no route of escape will fight when cornered.

All mammals are edible; however, the polar bear and bearded seal have toxic levels of vitamin A in their livers. The platypus, native to Australia and Tasmania, is an egg-laying, semiaquatic mammal that has poisonous glands. Scavenging mammals, such as the opossum, may carry diseases.

TRAPS AND SNARES

For an unarmed survivor or evader, or when the sound of a rifle shot could be a problem, trapping or snaring wild game is a good alternative. Several well-placed traps have the potential to catch much more game than a man with a rifle is likely to shoot. To be effective with any type of trap or snare, you must--

- Be familiar with the species of animal you intend to catch.
- Be capable of constructing a proper trap.
- Not alarm the prey by leaving signs of your presence.

There are no catchall traps you can set for all animals. You must determine what species are in a given area and set your traps specifically with those animals in mind. Look for the following:

- Runs and trails.
- Tracks.
- Droppings.
- Chewed or rubbed vegetation.
- Nesting or roosting sites.
- Feeding and watering areas.

Position your traps and snares where there is proof that animals pass through. You must determine if it is a "run" or a "trail." A trail will show signs of use by several species and will be rather distinct. A run is usually smaller and less distinct and will only contain signs of one species. You may construct a perfect snare, but it will not catch anything if haphazardly placed in the woods. Animals have bedding areas, waterholes, and feeding areas with trails leading from one to another. You must place snares and traps around these areas to be effective.

For an evader in a hostile environment, trap and snare concealment is important. It is equally important, however, not to create a disturbance that will alarm the animal and cause it to avoid the trap. Therefore, if you must dig, remove all fresh dirt from the area. Most animals will instinctively avoid a pitfall-type trap. Prepare the various parts of a trap or snare away from the site, carry them in, and set them up. Such actions make it easier to avoid disturbing the local vegetation, thereby alerting the prey. Do not use freshly cut, live vegetation to construct a trap or snare. Freshly cut vegetation will "bleed" sap that has an odor the prey will be able to smell. It is an alarm signal to the animal.

You must remove or mask the human scent on and around the trap you set. Although birds do not have a developed sense of smell, nearly all mammals depend on smell even more than on sight. Even the slightest human scent on a trap will alarm the prey and cause it to avoid the area. Actually removing the scent from a trap is difficult but masking it is relatively easy. Use the fluid from the gall and urine bladders of previous kills. Do not use human urine. Mud, particularly from an area with plenty of rotting vegetation, is also good. Use it to coat your hands when handling the trap and to coat the trap when setting it. In nearly all parts of the world, animals know the smell of burned vegetation and smoke. It is only when a fire is actually burning that they become alarmed. Therefore, smoking the trap parts is an effective means to mask your scent. If one of the above techniques is not practical, and if time permits, allow a trap to weather for a few days and then set it. Do not handle a trap while it is weathering. When you position the trap, camouflage it as naturally as possible to prevent detection by the enemy and to avoid alarming the prey.

Traps or snares placed on a trail or run should use channelization. To build a channel, construct a funnel-shaped barrier extending from the sides of the trail toward the trap, with the narrowest part nearest the trap. Channelization should be inconspicuous to avoid alerting the prey. As the animal gets to the trap, it cannot turn left or right and continues into the trap. Few wild animals will back up, preferring to face the direction of travel. Channelization does not have to be an impassable barrier. You only have to make it inconvenient for the animal to go over or through the barrier. For best effect, the channelization should reduce the trail's width to just slightly wider than the targeted animal's body. Maintain this constriction at least as far back from the trap as the animal's body length, then begin the widening toward the mouth of the funnel.

Use of Bait

Baiting a trap or snare increases your chances of catching an animal. When catching fish, you must bait nearly all the devices. Success with an unbaited trap depends on its placement in a good location. A baited trap can actually draw animals to it. The bait should be something the animal knows. This bait, however, should not be so readily available in the immediate area that the animal can get it close by. For example, baiting a trap with corn in the middle of a corn field would not be likely to work. Likewise, if corn is not grown in the region, a corn-baited trap may arouse an animal's curiosity and keep it alerted while it ponders the strange food. Under such circumstances it may not go for the bait. One bait that works well on small mammals is the peanut butter from a meal, ready-to-eat (MRE) ration. Salt is also a good bait. When using such baits, scatter bits of it around the trap to give the prey a chance to sample it and develop a craving for it. The animal will then overcome some of its caution before it gets to the trap. If you set and bait a trap for one species but another species takes the bait without being caught, try to determine what the animal was. Then set a proper trap for that animal, using the same bait.

Note: Once you have successfully trapped an animal, you will not only gain confidence in your ability, you also will have resupplied yourself with bait for several more traps.

Trap and Snare Construction

Traps and snares *crush, choke, hang,* or *entangle* the prey. A single trap or snare will commonly incorporate two or more of these principles. The mechanisms that provide power to the trap are almost always very simple. The struggling victim, the force of gravity, or a bent sapling's tension provides the power.

The heart of any trap or snare is the trigger. When planning a trap or snare, ask yourself how it should affect the prey, what is the source of power, and what will be the most efficient trigger. Your answers will help you devise a specific trap for a specific species. Traps are designed to catch and hold or to catch and kill. Snares are traps that incorporate a noose to accomplish either function.

Simple Snare

A simple snare (Figure 8-5) consists of a noose placed over a trail or den hole and attached to a firmly planted stake. If the noose is some type of cordage placed upright on a game trail, use small twigs or blades of grass to hold it up. Filaments from spider webs are excellent for holding nooses open. Make sure the noose is large enough to pass freely over the animal's head. As the animal continues to move, the noose tightens around its neck. The more the animal struggles, the tighter the noose gets. This type of snare usually does not kill the animal. If you use cordage, it may loosen enough to slip off the animal's neck. Wire is therefore the best choice for a simple snare.

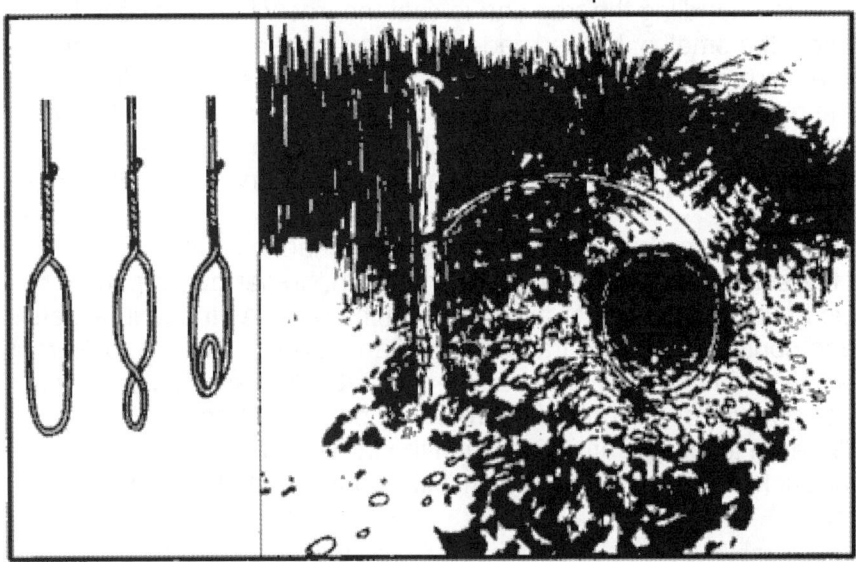

Figure 8-5. Simple snare.

Drag Noose

Use a drag noose on an animal run (Figure 8-6). Place forked sticks on either side of the run and lay a sturdy crossmember across them. Tie the noose to the crossmember and hang it at a height above the animal's head. (Nooses designed to catch by the head should never be low enough for the prey to step into with a foot.) As the noose tightens around the animal's neck, the animal pulls the crossmember from the forked sticks and drags it along. The surrounding vegetation quickly catches the crossmember and the animal becomes entangled.

ESSENTIAL ARMY MANUAL
(83) ARMY SURVIVAL MANUAL

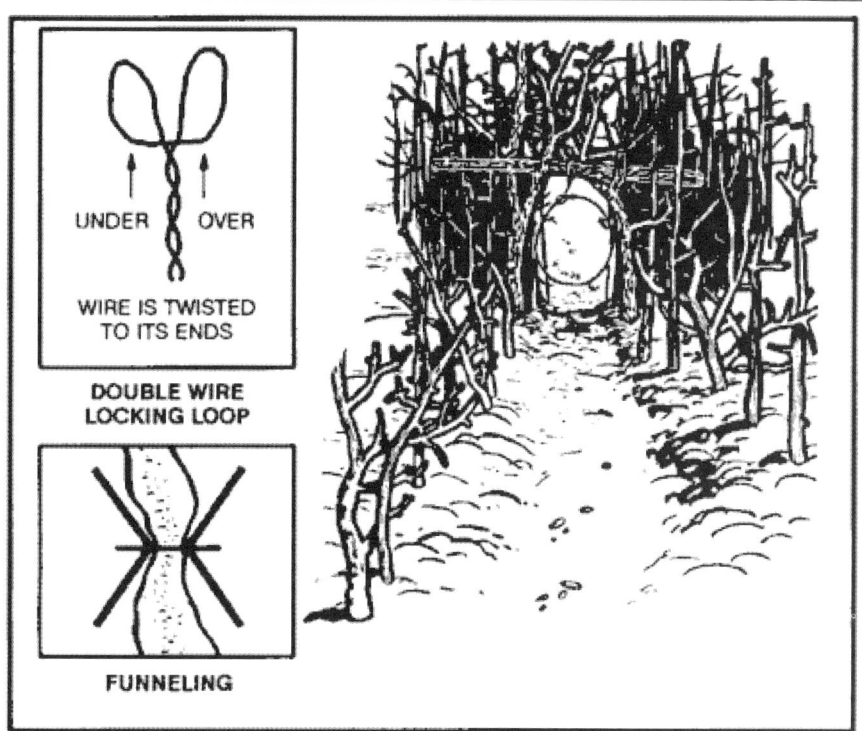

UNDER   OVER

WIRE IS TWISTED
TO ITS ENDS

**DOUBLE WIRE
LOCKING LOOP**

FUNNELING

Figure 8-6. Drag noose.

Twitch-Up

A twitch-up is a supple sapling, which, when bent over and secured with a triggering device, will provide power to a variety of snares. Select a hardwood sapling along the trail. A twitch-up will work much faster and with more force if you remove all the branches and foliage.

Twitch-Up Snare

A simple twitch-up snare uses two forked sticks, each with a long and short leg (Figure 8-7). Bend the twitch-up and mark the trail below it. Drive the long leg of one forked stick firmly into the ground at that point. Ensure the cut on the short leg of this stick is parallel to the ground. Tie the long leg of the remaining forked stick to a piece of cordage secured to the twitch-up. Cut the short leg so that it catches on the short leg of the other forked stick. Extend a noose over the trail. Set the trap by bending the twitch-up and engaging the short legs of the forked sticks. When an animal catches its head in the noose, it pulls the forked sticks apart, allowing the twitch-up to spring up and hang the prey.

> Note: Do not use green sticks for the trigger. The sap that oozes out could glue them together.

Figure 8-7. Twitch-up snare.

Squirrel Pole

A squirrel pole is a long pole placed against a tree in an area showing a lot of squirrel activity (Figure 8-8). Place several wire nooses along the top and sides of the pole so that a squirrel trying to go up or down the pole will have to pass through one or more of them. Position the nooses (5 to 6 centimeters in diameter) about 2.5 centimeters off the pole. Place the top and bottom wire nooses 45 centimeters from the top and bottom of the pole to prevent the squirrel from getting its feet on a solid surface. If this happens, the squirrel will chew through the wire. Squirrels are naturally curious. After an initial period of caution, they will try to go up or down the pole and will get caught in a noose. The struggling animal will soon fall from the pole and strangle. Other squirrels will soon follow and, in this way, you can catch several squirrels. You can emplace multiple poles to increase the catch.

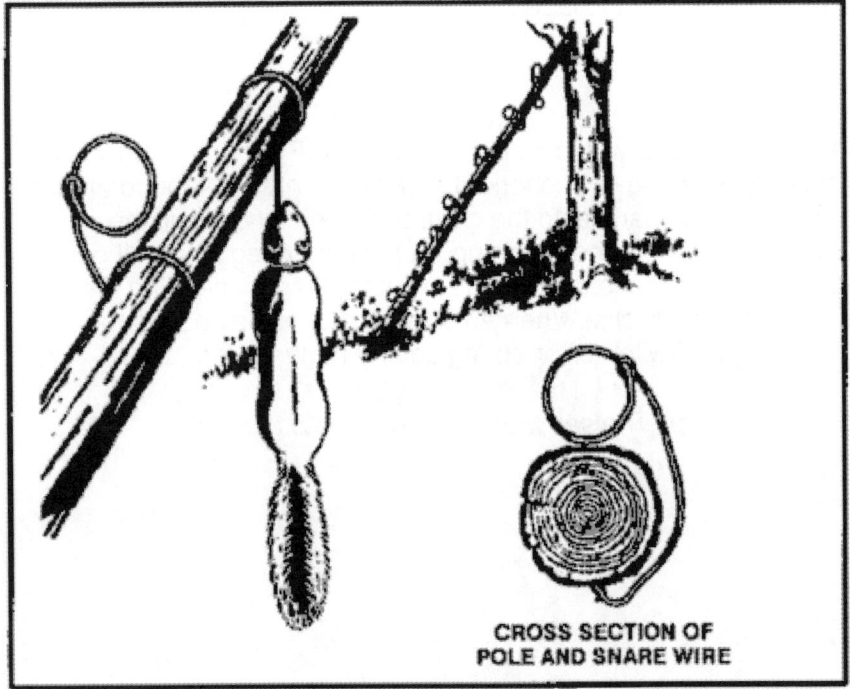

CROSS SECTION OF
POLE AND SNARE WIRE

Figure 8-8. Squirrel pole.

Ojibwa Bird Pole

ESSENTIAL ARMY MANUAL
(85) ARMY SURVIVAL MANUAL

An Ojibwa bird pole is a snare used by native Americans for centuries (Figure 8-9). To be effective, place it in a relatively open area away from tall trees. For best results, pick a spot near feeding areas, dusting areas, or watering holes. Cut a pole 1.8 to 2.1 meters long and trim away all limbs and foliage. Do not use resinous wood such as pine. Sharpen the upper end to a point, then drill a small diameter hole 5 to 7.5 centimeters down from the top. Cut a small stick 10 to 15 centimeters long and shape one end so that it will almost fit into the hole. This is the perch. Plant the long pole in the ground with the pointed end up. Tie a small weight, about equal to the weight of the targeted species, to a length of cordage. Pass the free end of the cordage through the hole, and tie a slip noose that covers the perch. Tie a single overhand knot in the cordage and place the perch against the hole. Allow the cordage to slip through the hole until the overhand knot rests against the pole and the top of the perch. The tension of the overhand knot against the pole and perch will hold the perch in position. Spread the noose over the perch, ensuring it covers the perch and drapes over on both sides. Most birds prefer to rest on something above ground and will land on the perch. As soon as the bird lands, the perch will fall, releasing the over-hand knot and allowing the weight to drop. The noose will tighten around the bird's feet, capturing it. If the weight is too heavy, it will cut the bird's feet off, allowing it to escape.

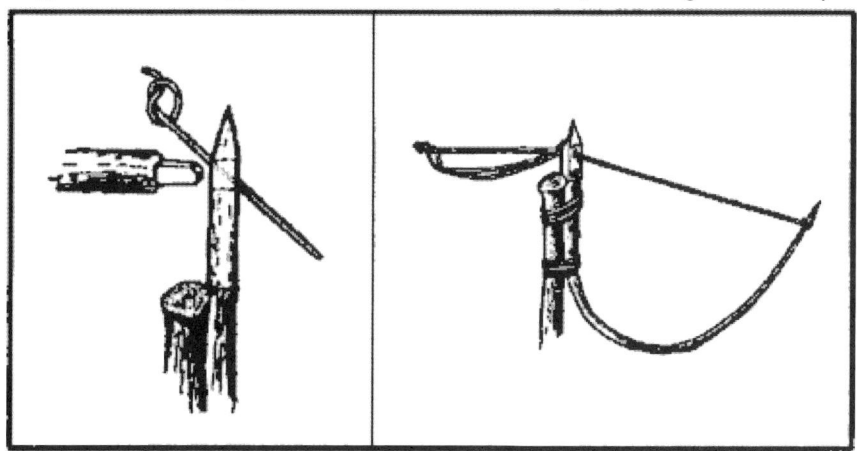

Figure 8-9. Ojibwa bird pole.

Noosing Wand

A noose stick or "noosing wand" is useful for capturing roosting birds or small mammals (Figure 8-10). It requires a patient operator. This wand is more a weapon than a trap. It consists of a pole (as long as you can effectively handle) with a slip noose of wire or stiff cordage at the small end. To catch an animal, you slip the noose over the neck of a roosting bird and pull it tight. You can also place it over a den hole and hide in a nearby blind. When the animal emerges from the den, you jerk the pole to tighten the noose and thus capture the animal. Carry a stout club to kill the prey.

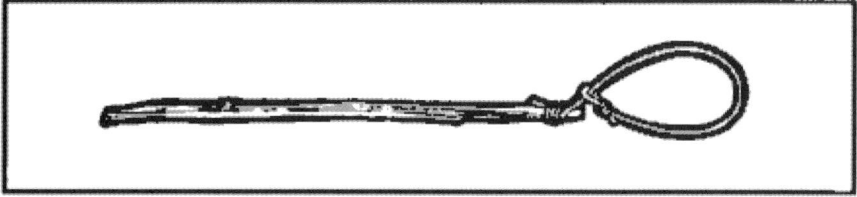

Figure 8-10. Noosing wand.

Treadle Spring Snare

Use a treadle snare against small game on a trail (Figure 8-11). Dig a shallow hole in the trail. Then drive a forked stick (fork down) into the ground on each side of the hole on the same side of the trail. Select two fairly straight sticks that span the two forks. Position these two sticks so that their ends engage the forks. Place several sticks over the hole in the trail by positioning one end over the lower horizontal stick and the other on the ground on the other side of the hole. Cover the hole with enough sticks so that the prey must step on at least one of them to set off the snare. Tie one end of a piece of cordage to a twitch-up or to a weight suspended over a tree limb. Bend the twitch-up or raise the suspended weight to determine where You will tie a 5 centimeter or so long trigger. Form a noose with the other end of the cordage. Route and spread the noose over the top of the sticks over the hole. Place the trigger stick against the horizontal sticks and route the cordage behind the sticks so that the tension of the power source will hold it in place. Adjust the bottom horizontal stick so that it will barely hold against the trigger. A the animal places its foot on a stick across the hole, the bottom horizontal stick moves down, releasing the trigger and allowing the noose to catch the animal by the foot. Because of the disturbance on the trail, an animal will be wary. You must therefore use channelization.

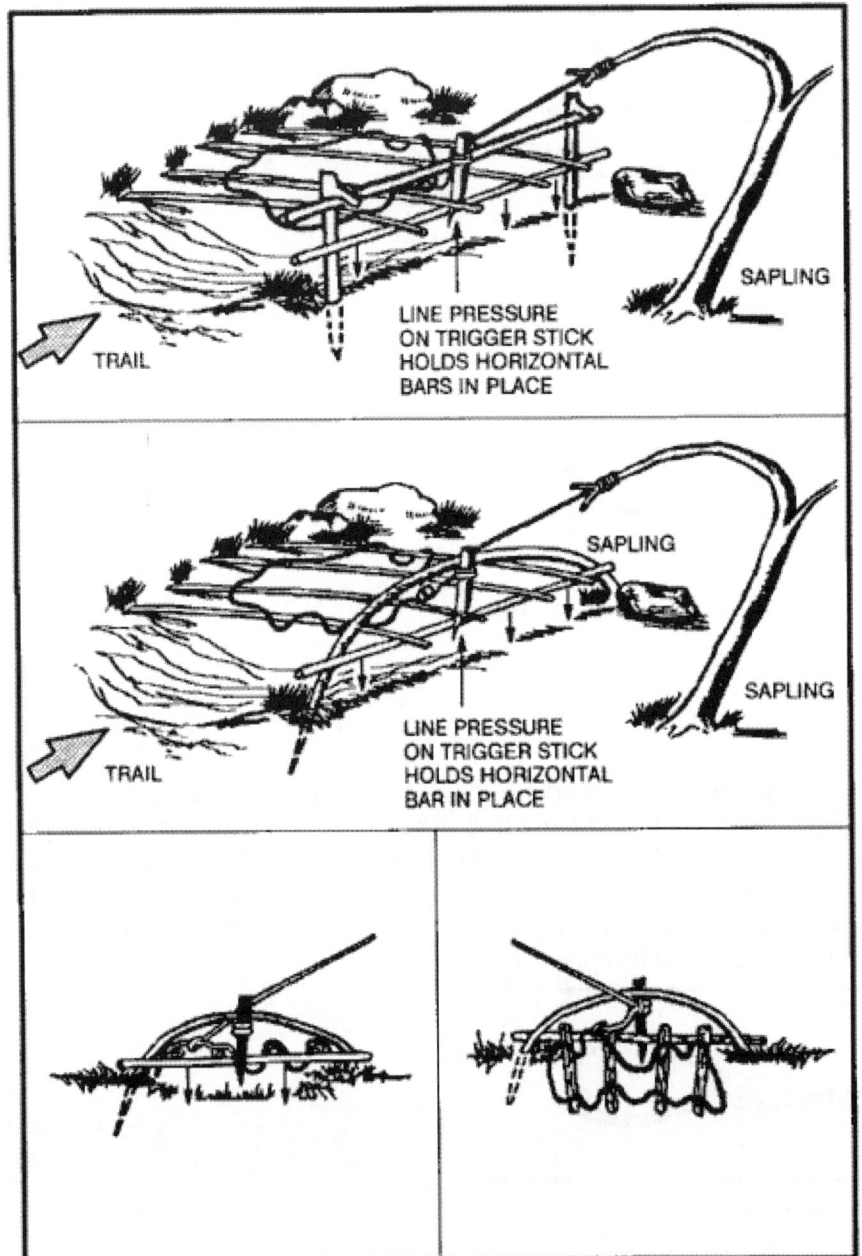

Figure 8-11. Treadle spring snare.

Figure 4 Deadfall

The figure 4 is a trigger used to drop a weight onto a prey and crush it (Figure 8-12). The type of weight used may vary, but it should be heavy enough to kill or incapacitate the prey immediately. Construct the figure 4 using three notched sticks. These notches hold the sticks together in a figure 4 pattern when under tension. Practice making this trigger before-hand; it requires close tolerances and precise angles in its construction.

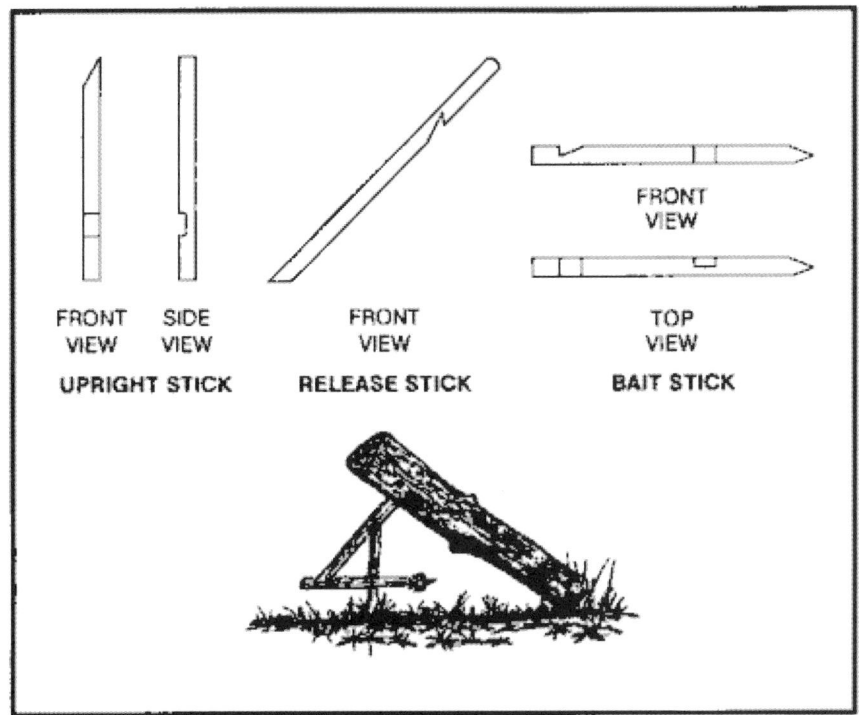

Figure 8-12. Figure 4 deadfall.

Paiute Deadfall

The Paiute deadfall is similar to the figure 4 but uses a piece of cordage and a catch stick (Figure 8-13). It has the advantage of being easier to set than the figure 4. Tie one end of a piece of cordage to the lower end of the diagonal stick. Tie the other end of the cordage to another stick about 5 centimeters long. This 5-centimeter stick is the catch stick. Bring the cord halfway around the vertical stick with the catch stick at a 90-degree angle. Place the bait stick with one end against the drop weight, or a peg driven into the ground, and the other against the catch stick. When a prey disturbs the bait stick, it falls free, releasing the catch stick. As the diagonal stick flies up, the weight falls, crushing the prey.

Figure 8-13. Paiute deadfall.

Bow Trap

A bow trap is one of the deadliest traps. It is dangerous to man as well as animals (Figure 8-14). To construct this trap, build a bow and anchor it to the ground with pegs. Adjust the aiming point as you anchor the bow. Lash a toggle stick to the trigger stick. Two upright sticks driven into the ground hold the trigger stick in place at a point where the toggle stick will engage the pulled bow string. Place a catch stick between the toggle stick and a stake driven into the ground. Tie a trip wire or cordage to the catch stick and route it around stakes and across the game trail where you tie it off (as in Figure 8-14). When the prey trips the trip wire, the bow looses an arrow into it. A notch in the bow serves to help aim the arrow.

WARNING
This is a lethal trap. Approach it with caution and from the rear only!

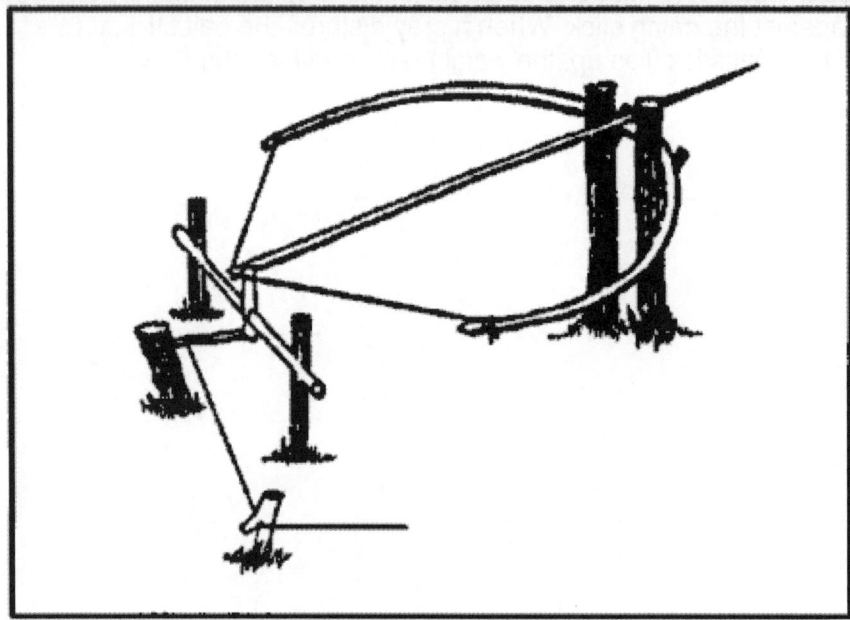

Figure 8-14. Bow trap.

ESSENTIAL ARMY MANUAL

(89) ARMY SURVIVAL MANUAL

Pig Spear Shaft

To construct the pig spear shaft, select a stout pole about 2.5 meters long (Figure 8-15). At the smaller end, firmly lash several small stakes. Lash the large end tightly to a tree along the game trail. Tie a length of cordage to another tree across the trail. Tie a sturdy, smooth stick to the other end of the cord. From the first tree, tie a trip wire or cord low to the ground, stretch it across the trail, and tie it to a catch stick. Make a slip ring from vines or other suitable material. Encircle the trip wire and the smooth stick with the slip ring. Emplace one end of another smooth stick within the slip ring and its other end against the second tree. Pull the smaller end of the spear shaft across the trail and position it between the short cord and the smooth stick. As the animal trips the trip wire, the catch stick pulls the slip ring off the smooth sticks, releasing the spear shaft that springs across the trail and impales the prey against the tree.

WARNING
This is a lethal trap. Approach it with caution!

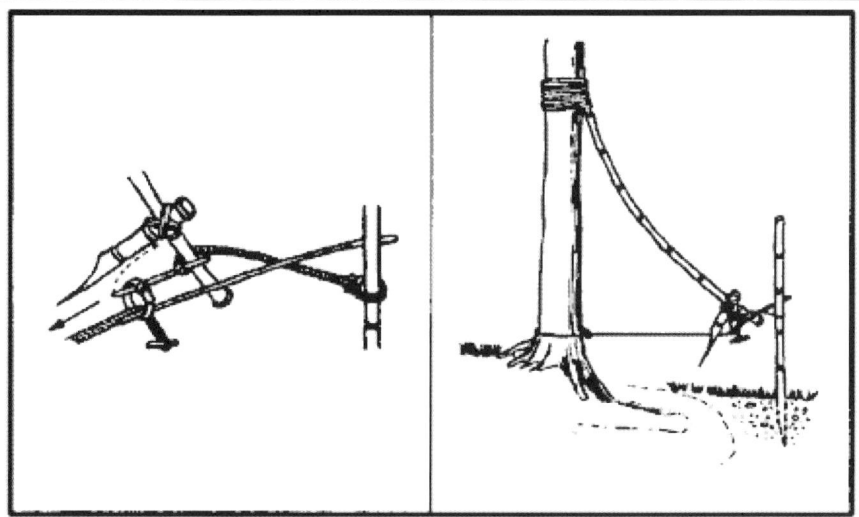

Figure 8-15. Pig spear shaft.

Bottle Trap

A bottle trap is a simple trap for mice and voles (Figure 8-16). Dig a hole 30 to 45 centimeters deep that is wider at the bottom than at the top. Make the top of the hole as small as possible. Place a piece of bark or wood over the hole with small stones under it to hold it up 2.5 to 5 centimeters off the ground. Mice or voles will hide under the cover to escape danger and fall into the hole. They cannot climb out because of the wall's backward slope. Use caution when checking this trap; it is an excellent hiding place for snakes.

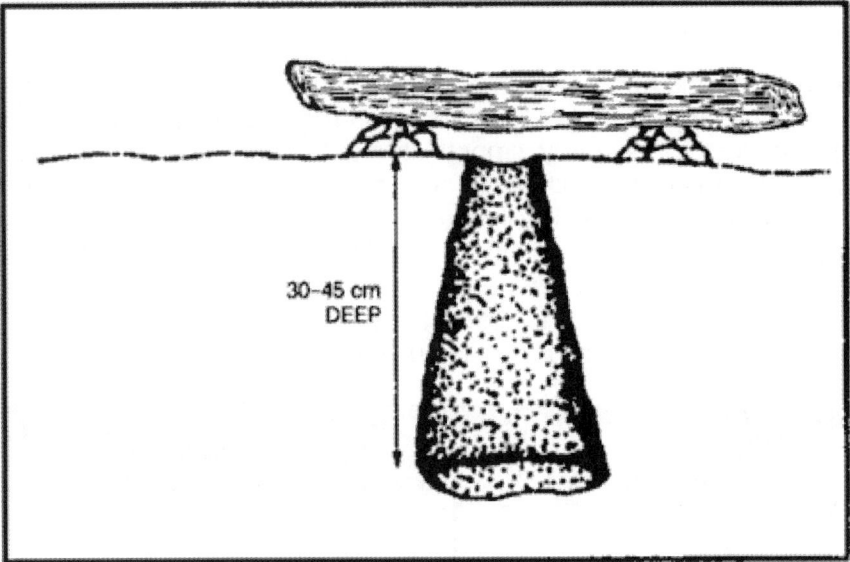

Figure 8-16. Bottle trap.

KILLING DEVICES

There are several killing devices that you can construct to help you obtain small game to help you survive. The rabbit stick, the spear, the bow and arrow, and the sling are such devices.

Rabbit Stick

One of the simplest and most effective killing devices is a stout stick as long as your arm, from fingertip to shoulder, called a "rabbit stick." You can throw it either overhand or sidearm and with considerable force. It is very effective against small game that stops and freezes as a defense.

Spear

You can make a spear to kill small game and to fish. Jab with the spear, do not throw it. See spearfishing below.

Bow and Arrow

A good bow is the result of many hours of work. You can construct a suitable short-term bow fairly easily. When it loses its spring or breaks, you can replace it. Select a hardwood stick about one meter long that is free of knots or limbs. Carefully scrape the large end down until it has the same pull as the small end. Careful examination will show the natural curve of the stick. Always scrape from the side that faces you, or the bow will break the first time you pull it. Dead, dry wood is preferable to green wood. To increase the pull, lash a second bow to the first, front to front, forming an "X" when viewed from the side. Attach the tips of the bows with cordage and only use a bowstring on one bow.
Select arrows from the straightest dry sticks available. The arrows should be about half as long as the bow. Scrape each shaft smooth all around. You will probably have to straighten the shaft. You can bend an arrow straight by heating the shaft over hot coals. Do not allow the shaft to scorch or bum. Hold the shaft straight until it cools.
You can make arrowheads from bone, glass, metal, or pieces of rock. You can also sharpen and fire harden the end of the shaft. To fire harden wood, hold it over hot coals, being careful not to bum or scorch the wood.
You must notch the ends of the arrows for the bowstring. Cut or file the notch; do not split it. Fletching (adding feathers to the notched end of an arrow) improves the arrow's flight characteristics, but is not necessary on a field-expedient arrow.

Sling

You can make a sling by tying two pieces of cordage, about sixty centimeters long, at opposite ends of a palm-sized piece of leather or cloth. Place a rock in the cloth and wrap one cord around the middle finger and hold in your palm. Hold the other cord between the forefinger and thumb. To throw the rock, spin the sling several times in a circle and release the cord between the thumb and forefinger. Practice to gain proficiency. The sling is very effective against small game.

FISHING DEVICES

You can make your own fishhooks, nets and traps and use several methods to obtain fish in a survival situation.

Improvised Fishhooks

You can make field-expedient fishhooks from pins, needles, wire, small nails, or any piece of metal. You can also use wood, bone, coconut shell, thorns, flint, seashell, or tortoise shell. You can also make fishhooks from any combination of these items (Figure 8-17).

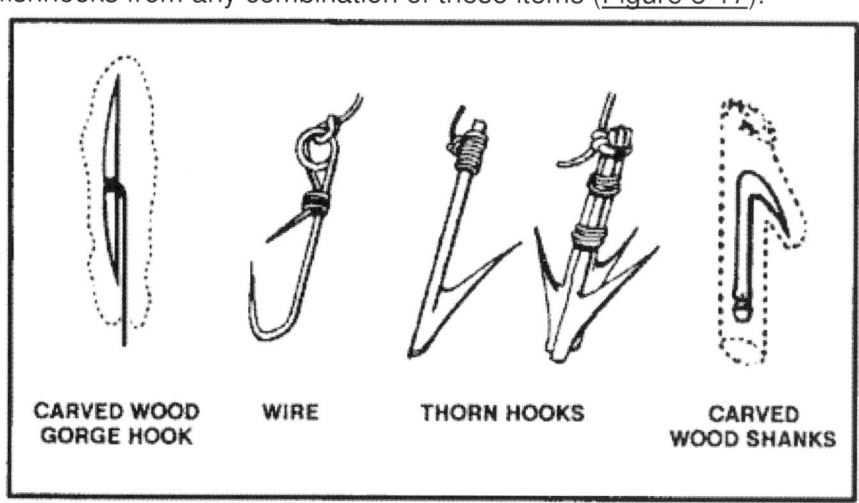

Figure 8-17. Improvised fishhooks.

To make a wooden hook, cut a piece of hardwood about 2.5 centimeters long and about 6 millimeters in diameter to form the shank. Cut a notch in one end in which to place the point. Place the point (piece of bone, wire, nail) in the notch. Hold the point in the notch and tie securely so that it does not move out of position. This is a fairly large hook. To make smaller hooks, use smaller material.

A gorge is a small shaft of wood, bone, metal, or other material. It is sharp on both ends and notched in the middle where you tie cordage. Bait the gorge by placing a piece of bait on it lengthwise. When the fish swallows the bait, it also swallows the gorge.

Stakeout

A stakeout is a fishing device you can use in a hostile environment (Figure 8-18). To construct a stakeout, drive two supple saplings into the bottom of the lake, pond, or stream with their tops just below the water surface. Tie a cord between them and slightly below the surface. Tie two short cords with hooks or gorges to this cord, ensuring that they cannot wrap around the poles or each other. They should also not slip along the long cord. Bait the hooks or gorges.

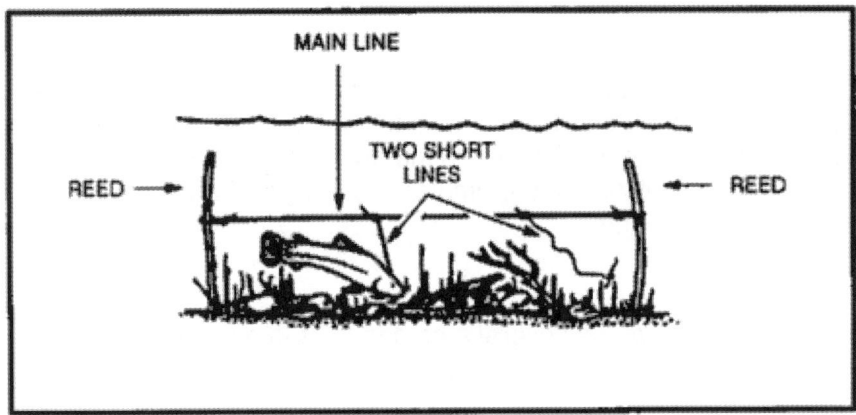

Figure 8-18. Stakeout.

Gill Net

If a gill net is not available, you can make one using parachute suspension line or similar material (Figure 8-19). Remove the core lines from the suspension line and tie the easing between two trees. Attach several core lines to the easing by doubling them over and tying them with prusik knots or girth hitches. The length of the desired net and the size of the mesh determine the number of core lines used and the space between them. Starting at one end of the easing, tie the second and the third core lines together using an overhand knot. Then tie the fourth and fifth, sixth and seventh, and so on, until you reach the last core line. You should now have all core lines tied in pairs with a single core line hanging at each end. Start the second row with the first core line, tie it to the second, the third to the fourth, and so on.

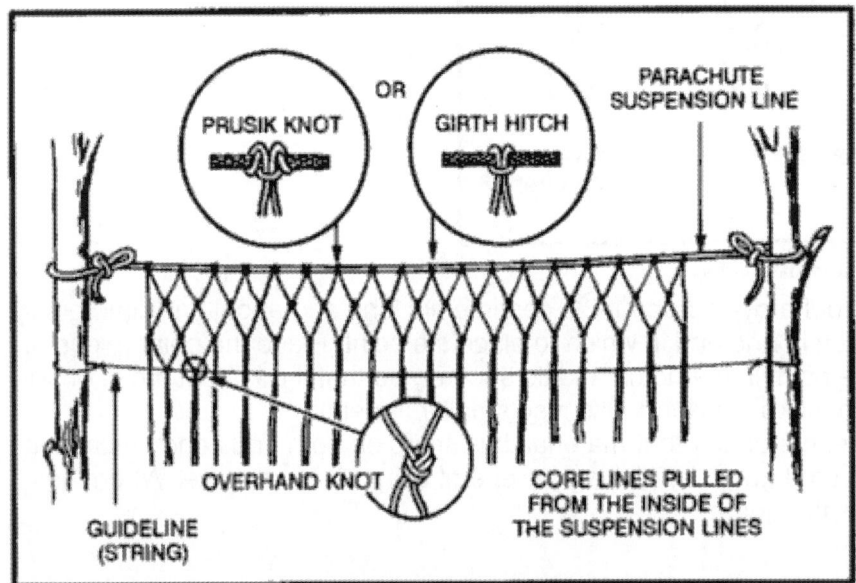

Figure 8-19. Making a gill net.

To keep the rows even and to regulate the size of the mesh, tie a guideline to the trees. Position the guideline on the opposite side of the net you are working on. Move the guideline down after completing each row. The lines will always hang in pairs and you always tie a cord from one pair to a cord from an adjoining pair. Continue tying rows until the net is the desired width. Thread a suspension line easing along the bottom of the net to strengthen it. Use the gill net as shown in Figure 8-20.

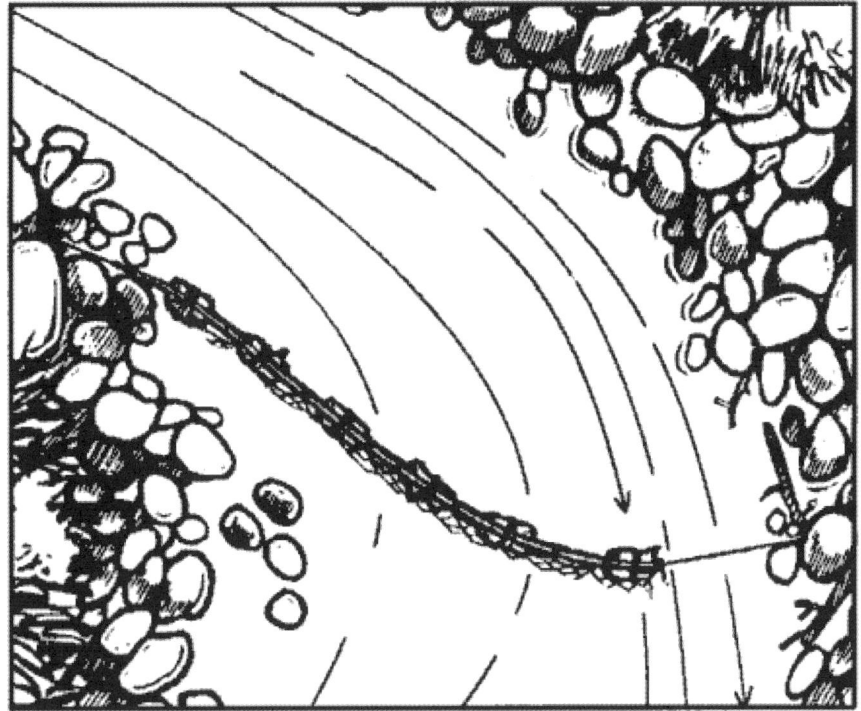

Figure 8-20. Setting a gill net in the stream.

Fish Traps

You may trap fish using several methods (Figure 8-21). Fish baskets are one method. You construct them by lashing several sticks together with vines into a funnel shape. You close the top, leaving a hole large enough for the fish to swim through.

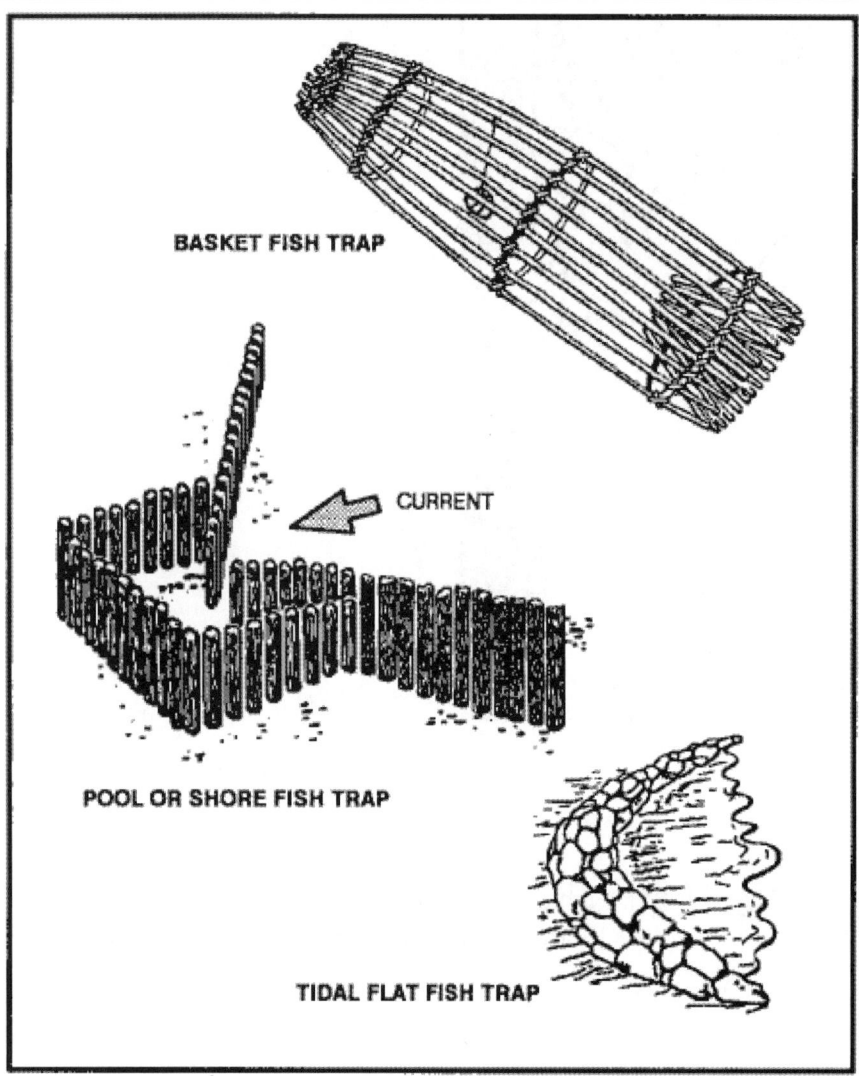

Figure 8-21. Various types of fish traps.

You can also use traps to catch saltwater fish, as schools regularly approach the shore with the incoming tide and often move parallel to the shore. Pick a location at high tide and build the trap at low tide. On rocky shores, use natural rock pools. On coral islands, use natural pools on the surface of reefs by blocking the openings as the tide recedes. On sandy shores, use sandbars and the ditches they enclose. Build the trap as a low stone wall extending outward into the water and forming an angle with the shore.

Spearfishing

If you are near shallow water (about waist deep) where the fish are large and plentiful, you can spear them. To make a spear, cut a long, straight sapling (Figure 8-22). Sharpen the end to a point or attach a knife, jagged piece of bone, or sharpened metal. You can also make a spear by splitting the shaft a few inches down from the end and inserting a piece of wood to act as a spreader. You then sharpen the two separated halves to points. To spear fish, find an area where fish either gather or where there is a fish run. Place the spear point into the water and slowly move it toward the fish. Then, with a sudden push, impale the fish on the stream bottom. Do not try to lift the fish with the spear, as it with probably slip off and you will lose it; hold the spear with one hand and grab and hold the fish with the other. Do not throw the spear, especially if the point is a knife. You cannot afford to lose a knife in a survival situation. Be alert to the problems caused by light refraction when looking at objects in the water.

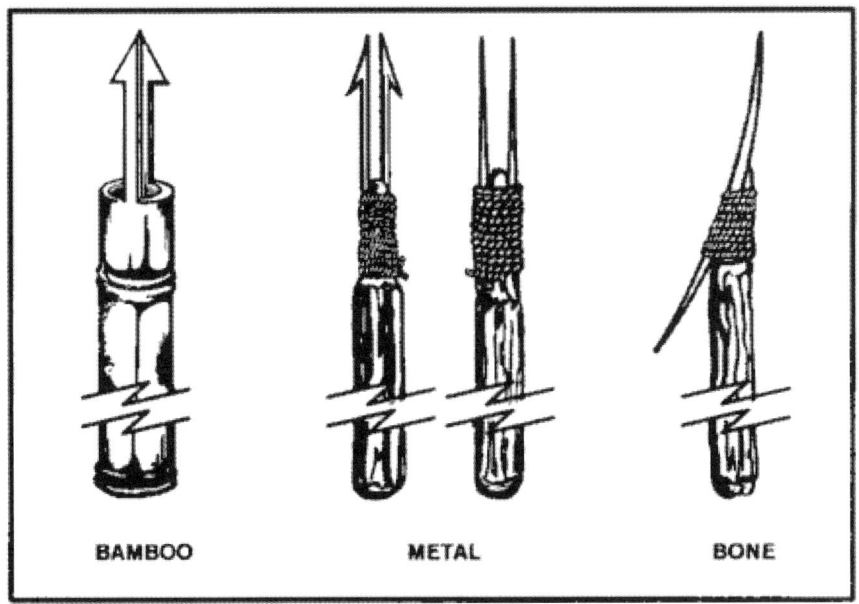

Figure 8-22. Types of spear points.

Chop Fishing

At night, in an area with a good fish density, you can use a light to attract fish. Then, armed with a machete or similar weapon, you can gather fish using the back side of the blade to strike them. Do not use the sharp side as you will cut them in two pieces and end up losing some of the fish.

Fish Poison

Another way to catch fish is by using poison. Poison works quickly. It allows you to remain concealed while it takes effect. It also enables you to catch several fish at one time. When using fish poison, be sure to gather all of the affected fish, because many dead fish floating downstream could arouse suspicion. Some plants that grow in warm regions of the world contain rotenone, a substance that stuns or kills cold-blooded animals but does not harm persons who eat the animals. The best place to use rotenone, or rotenone-producing plants, is in ponds or the headwaters of small streams containing fish. Rotenone works quickly on fish in water 21 degrees C (70 degrees F) or above. The fish rise helplessly to the surface. It works slowly in water 10 to 21 degrees C (50 to 70 degrees F) and is ineffective in water below 10 degrees C (50 degrees F). The following plants, used as indicated, will stun or kill fish:
Anamirta cocculus (Figure 8-23). This woody vine grows in southern Asia and on islands of the South Pacific. Crush the bean-shaped seeds and throw them in the water.
Croton tiglium (Figure 8-23). This shrub or small tree grows in waste areas on islands of the South Pacific. It bears seeds in three angled capsules. Crush the seeds and throw them into the water.
Barringtonia (Figure 8-23). These large trees grow near the sea in Malaya and parts of Polynesia. They bear a fleshy one-seeded fruit. Crush the seeds and bark and throw into the water.
Derris eliptica (Figure 8-23). This large genus of tropical shrubs and woody vines is the main source of commercially produced rotenone. Grind the roots into a powder and mix with water. Throw a large quantity of the mixture into the water.
Duboisia (Figure 8-23). This shrub grows in Australia and bears white clusters of flowers and berrylike fruit. Crush the plants and throw them into the water.
Tephrosia (Figure 8-23). This species of small shrubs, which bears beanlike pods, grows throughout the tropics. Crush or bruise bundles of leaves and stems and throw them into the water.

- *Lime.* You can get lime from commercial sources and in agricultural areas that use large quantities of it. You may produce your own by burning coral or seashells. Throw the lime into the water.
- *Nut husks.* Crush green husks from butternuts or black walnuts. Throw the husks into the water.

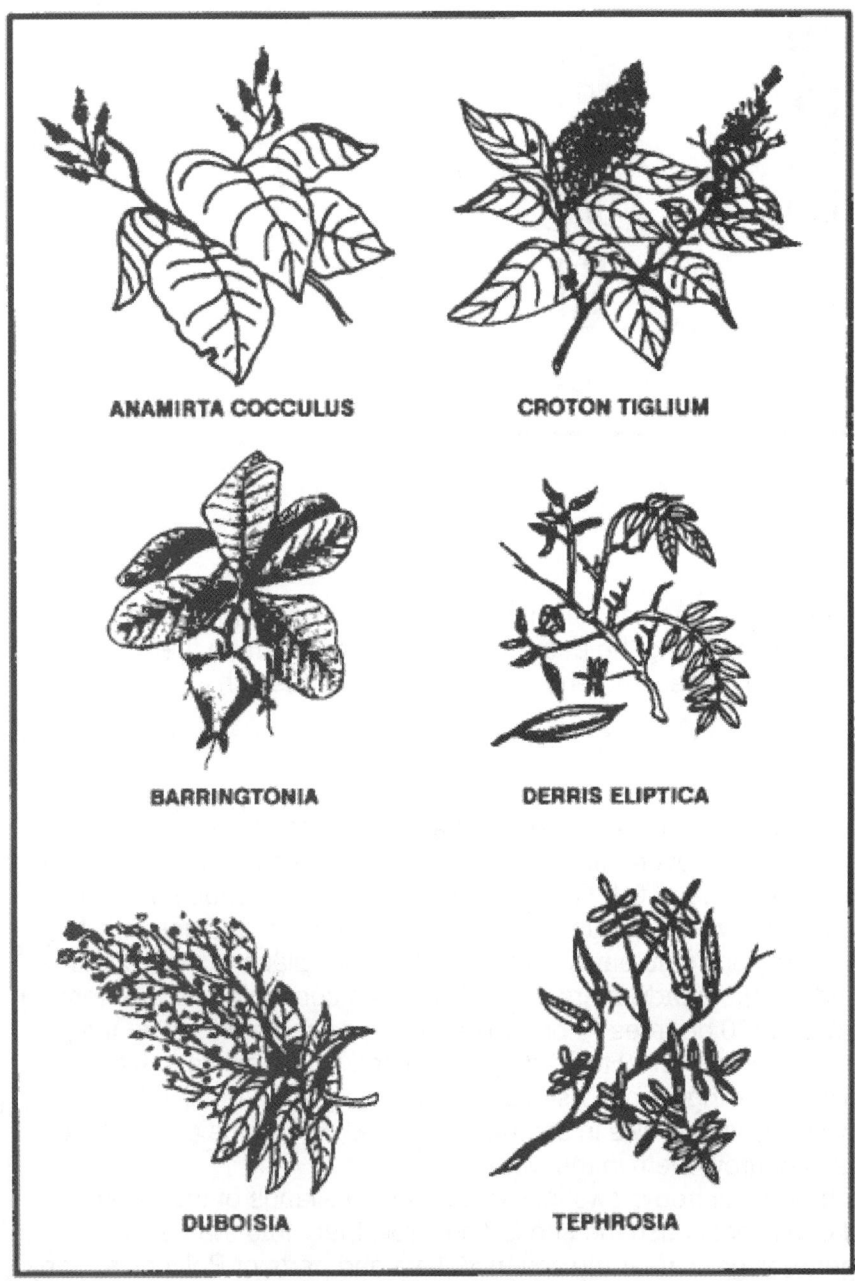

Figure 8-23. Fish-poisoning plants.

PREPARATION OF FISH AND GAME FOR COOKING AND STORAGE

You must know how to prepare fish and game for cooking and storage in a survival situation. Improper cleaning or storage can result in inedible fish or game.

Fish

Do not eat fish that appears spoiled. Cooking does not ensure that spoiled fish will be edible. Signs of spoilage are--

- Sunken eyes.
- Peculiar odor.
- Suspicious color. (Gills should be red to pink. Scales should be a pronounced shade of gray, not faded.)
- Dents stay in the fish's flesh after pressing it with your thumb.
- Slimy, rather than moist or wet body.
- Sharp or peppery taste.

Eating spoiled or rotten fish may cause diarrhea, nausea, cramps, vomiting, itching, paralysis, or a metallic taste in the mouth. These symptoms appear suddenly, one to six hours after eating. Induce vomiting if symptoms appear.

Fish spoils quickly after death, especially on a hot day. Prepare fish for eating as soon as possible after catching it. Cut out the gills and large blood vessels that lie near the spine. Gut fish that is more than 10 centimeters long. Scale or skin the fish.

You can impale a whole fish on a stick and cook it over an open fire. However, boiling the fish with the skin on is the best way to get the most food value. The fats and oil are under the skin and, by boiling, you can save the juices for broth. You can use any of the methods used to cook plant food to cook fish. Pack fish into a ball of clay and bury it in the coals of a fire until the clay hardens. Break open the clay ball to get to the cooked fish. Fish is done when the meat flakes off. If you plan to keep the fish for later, smoke or fry it. To prepare fish for smoking, cut off the head and remove the backbone.

Snakes

To skin a snake, first cut off its head and bury it. Then cut the skin down the body 15 to 20 centimeters (Figure 8-24). Peel the skin back, then grasp the skin in one hand and the body in the other and pull apart. On large, bulky snakes it may be necessary to slit the belly skin. Cook snakes in the same manner as small game. Remove the entrails and discard. Cut the snake into small sections and boil or roast it.

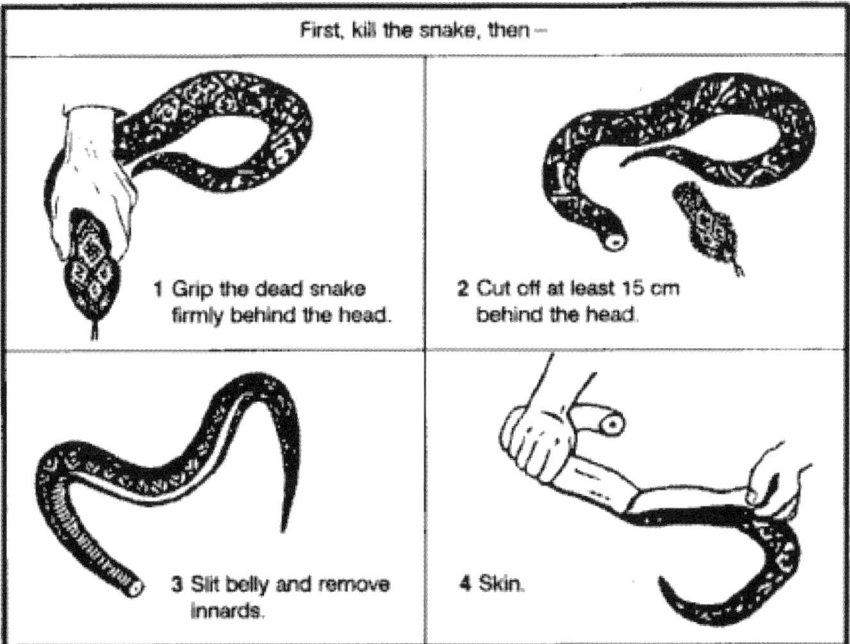

First, kill the snake, then --

1 Grip the dead snake firmly behind the head.

2 Cut off at least 15 cm behind the head.

3 Slit belly and remove innards.

4 Skin.

Figure 8-24. Cleaning a snake.

Birds

After killing the bird, remove its feathers by either plucking or skinning. Remember, skinning removes some of the food value. Open up the body cavity and remove its entrails, saving the craw (in seed-eating birds), heart, and liver. Cut off the feet. Cook by boiling or roasting over a spit. Before cooking scavenger birds, boil them at least 20 minutes to kill parasites.

Skinning and Butchering Game

Bleed the animal by cutting its throat. If possible, clean the carcass near a stream. Place the carcass belly up and split the hide from throat to tail, cutting around all sexual organs (Figure 8-25). Remove the musk glands at points A and B to avoid tainting the meat. For smaller mammals, cut the hide around the body and insert two fingers under the hide on both sides of the cut and pull both pieces off (Figure 8-26).

Note: When cutting the hide, insert the knife blade under the skin and turn the blade up so that only the hide gets cut. This will also prevent cutting hair and getting it on the meat.

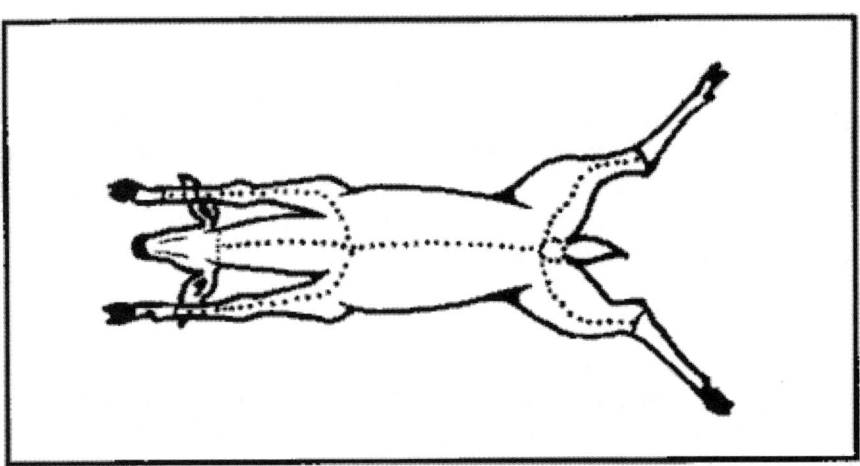

Figure 8-25. Skinning and butchering large game.

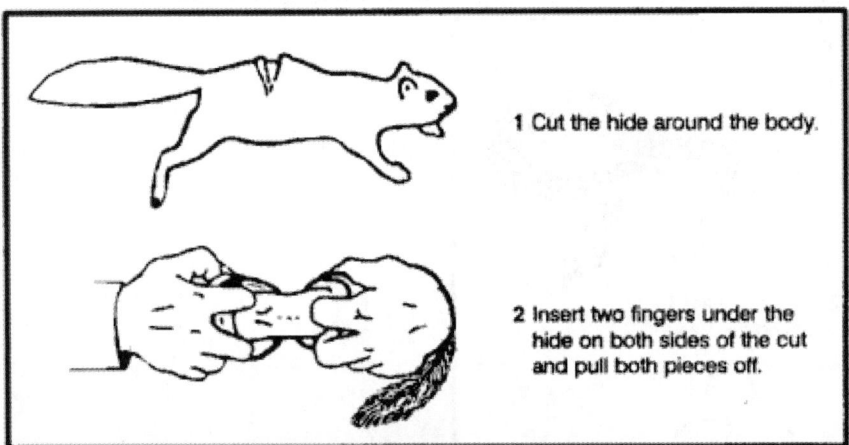

1 Cut the hide around the body.

2 Insert two fingers under the hide on both sides of the cut and pull both pieces off.

Figure 8-26. Skinning small game.

Remove the entrails from smaller game by splitting the body open and pulling them out with the fingers. Do not forget the chest cavity. For larger game, cut the gullet away from the diaphragm. Roll the entrails out of the body. Cut around the anus, then reach into the lower abdominal cavity, grasp the lower intestine, and pull to remove. Remove the urine bladder by pinching it off and cutting it below the fingers. If you spill urine on the meat, wash it to avoid tainting the meat. Save the heart and liver. Cut these open and inspect for signs of worms or other parasites. Also inspect the liver's color; it could indicate a diseased animal. The liver's surface should be smooth and wet and its color deep red or purple. If the liver appears diseased, discard it. However, a diseased liver does not indicate you cannot eat the muscle tissue.

Cut along each leg from above the foot to the previously made body cut. Remove the hide by pulling it away from the carcass, cutting the connective tissue where necessary. Cut off the head and feet.
Cut larger game into manageable pieces. First, slice the muscle tissue connecting the front legs to the body. There are no bones or joints connecting the front legs to the body on four-legged animals. Cut the hindquarters off where they join the body. You must cut around a large bone at the top of the leg and cut to the ball and socket hip joint. Cut the ligaments around the joint and bend it back to separate it. Remove the large muscles (the tenderloin) that lie on either side of the spine. Separate the ribs from the backbone. There is less work and less wear on your knife if you break the ribs first, then cut through the breaks.
Cook large meat pieces over a spit or boil them. You can stew or boil smaller pieces, particularly those that remain attached to bone after the initial butchering, as soup or broth. You can cook body organs such as the heart, liver, pancreas, spleen, and kidneys using the same methods as for muscle meat. You can also cook and eat the brain. Cut the tongue out, skin it, boil it until tender, and eat it.

Smoking Meat

To smoke meat, prepare an enclosure around a fire (Figure 8-27). Two ponchos snapped together will work. The fire does not need to be big or hot. The intent is to produce smoke, not heat. Do not use resinous wood in the fire because its smoke will ruin the meat. Use hardwoods to produce good smoke. The wood should be somewhat green. If it is too dry, soak it. Cut the meat into thin slices, no more than 6 centimeters thick, and drape them over a framework. Make sure none of the meat touches another piece. Keep the poncho enclosure around the meat to hold the smoke and keep a close watch on the fire. Do not let the fire get too hot. Meat smoked overnight in this manner will last about 1 week. Two days of continuous smoking will preserve the meat for 2 to 4 weeks. Properly smoked meat will look like a dark, curled, brittle stick and you can eat it without further cooking. You can also use a pit to smoke meat (Figure 8-28).

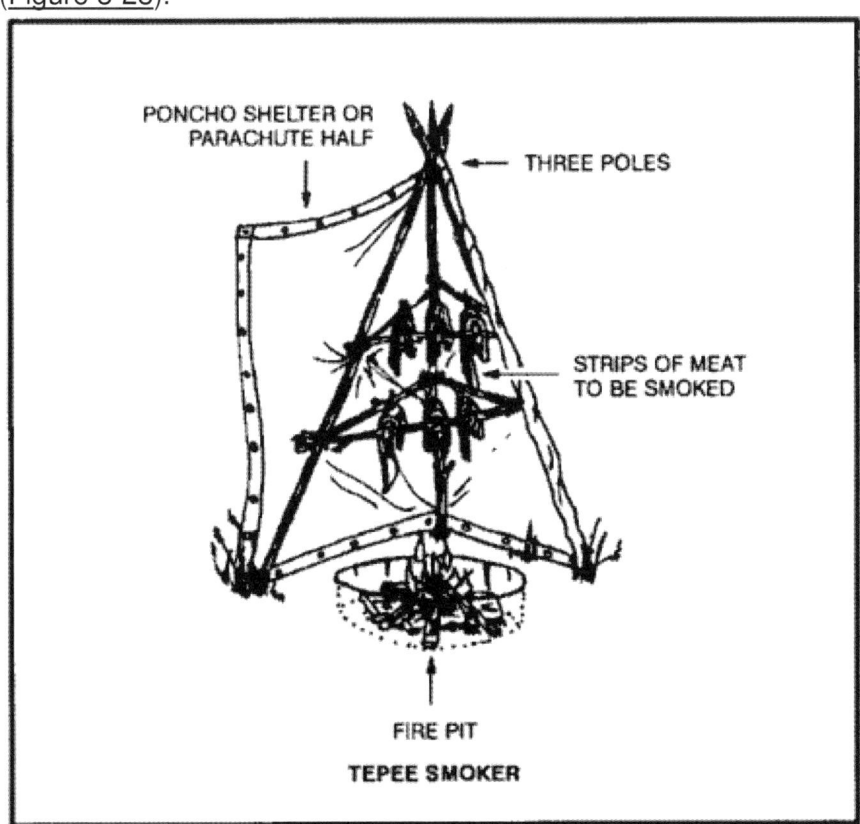

Figure 8-27. Smoking meat.

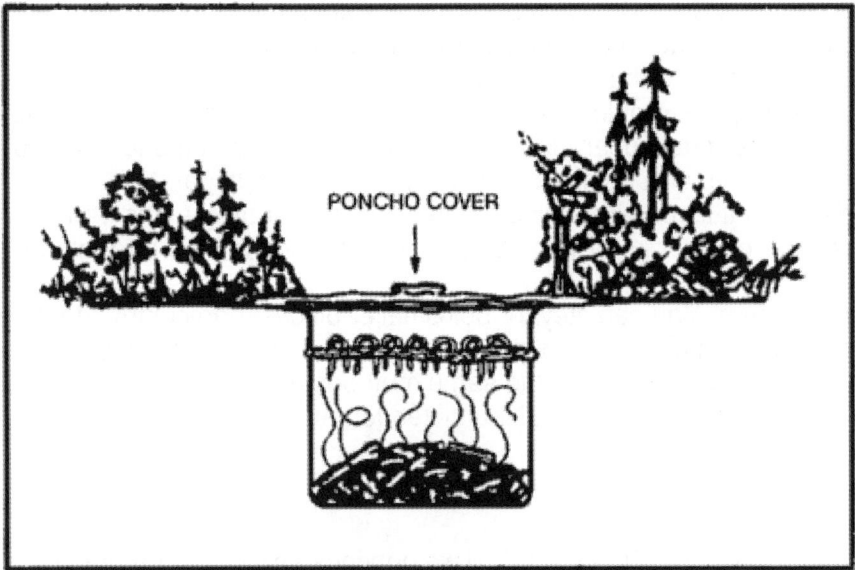

Figure 8-28. Smoking meat over a pit.

Drying Meat

To preserve meat by drying, cut it into 6-millimeter strips with the grain. Hang the meat strips on a rack in a sunny location with good air flow. Keep the strips out of the reach of animals and cover them to keep blowflies off. Allow the meat to dry thoroughly before eating. Properly dried meat will have a dry, crisp texture and will not feel cool to the touch.

Other Preservation Methods

You can also preserve meats using the freezing or brine and salt methods.

Freezing

In cold climates, you can freeze and keep meat indefinitely. Freezing is not a means of preparing meat. You must still cook it before eating.

Brine and Salt

You can preserve meat by soaking it thoroughly in a saltwater solution. The solution must cover the meat. You can also use salt by itself. Wash off the salt before cooking.

CHAPTER 9 - SURVIVAL USE OF PLANTS

After having solved the problems of finding water, shelter, and animal food, you will have to consider the use of plants you can eat. In a survival situation you should always be on the lookout for familiar wild foods and live off the land whenever possible.

You must not count on being able to go for days without food as some sources would suggest. Even in the most static survival situation, maintaining health through a complete and nutritious diet is essential to maintaining strength and peace of mind.

Nature can provide you with food that will let you survive any ordeal, if you don't eat the wrong plant. You must therefore learn as much as possible beforehand about the flora of the region where you will be operating. Plants can provide you with medicines in a survival situation. Plants can supply you with weapons and raw materials to construct shelters and build fires. Plants can even provide you with chemicals for poisoning fish, preserving animal hides, and for camouflaging yourself and your equipment.

Note: You will find illustrations of the plants described in this chapter in Appendixes B and C.

EDIBILITY OF PLANTS

Plants are valuable sources of food because they are widely available, easily procured, and, in the proper combinations, can meet all your nutritional needs.

> **WARNING**
> The critical factor in using plants for food is to avoid accidental poisoning. Eat only those plants you can positively identify and you know are safe to eat.

Absolutely identify plants before using them as food. Poison hemlock has killed people who mistook it for its relatives, wild carrots and wild parsnips.
At times you may find yourself in a situation for which you could not plan. In this instance you may not have had the chance to learn the plant life of the region in which you must survive. In this case you can use the Universal Edibility Test to determine which plants you can eat and those to avoid.
It is important to be able to recognize both cultivated and wild edible plants in a survival situation. Most of the information in this chapter is directed towards identifying wild plants because information relating to cultivated plants is more readily available.
Remember the following when collecting wild plants for food:

- Plants growing near homes and occupied buildings or along roadsides may have been sprayed with pesticides. Wash them thoroughly. In more highly developed countries with many automobiles, avoid roadside plants, if possible, due to contamination from exhaust emissions.
- Plants growing in contaminated water or in water containing *Giardia lamblia* and other parasites are contaminated themselves. Boil or disinfect them.
- Some plants develop extremely dangerous fungal toxins. To lessen the chance of accidental poisoning, do not eat any fruit that is starting to spoil or showing signs of mildew or fungus.
- Plants of the same species may differ in their toxic or subtoxic compounds content because of genetic or environmental factors. One example of this is the foliage of the common chokecherry. Some chokecherry plants have high concentrations of deadly cyanide compounds while others have low concentrations or none. Horses have died from eating wilted wild cherry leaves. Avoid any weed, leaves, or seeds with an almondlike scent, a characteristic of the cyanide compounds.
- Some people are more susceptible to gastric distress (from plants) than others. If you are sensitive in this way, avoid unknown wild plants. If you are extremely sensitive to poison ivy, avoid products from this family, including any parts from sumacs, mangoes, and cashews.
- Some edible wild plants, such as acorns and water lily rhizomes, are bitter. These bitter substances, usually tannin compounds, make them unpalatable. Boiling them in several changes of water will usually remove these bitter properties.
- Many valuable wild plants have high concentrations of oxalate compounds, also known as oxalic acid. Oxalates produce a sharp burning sensation in your mouth and throat and damage the kidneys. Baking, roasting, or drying usually destroys these oxalate crystals. The corm (bulb) of the jack-in-the-pulpit is known as the "Indian turnip," but you can eat it only after removing these crystals by slow baking or by drying.

WARNING
Do not eat mushrooms in a survival situation! The only way to tell if a mushroom is edible is by positive identification. There is no room for experimentation. Symptoms of the most dangerous mushrooms affecting the central nervous system may show up after several days have passed when it is too late to reverse their effects.

Plant Identification

You identify plants, other than by memorizing particular varieties through familiarity, by using such factors as leaf shape and margin, leaf arrangements, and root structure.
The basic leaf margins (Figure 9-1) are toothed, lobed, and toothless or smooth.

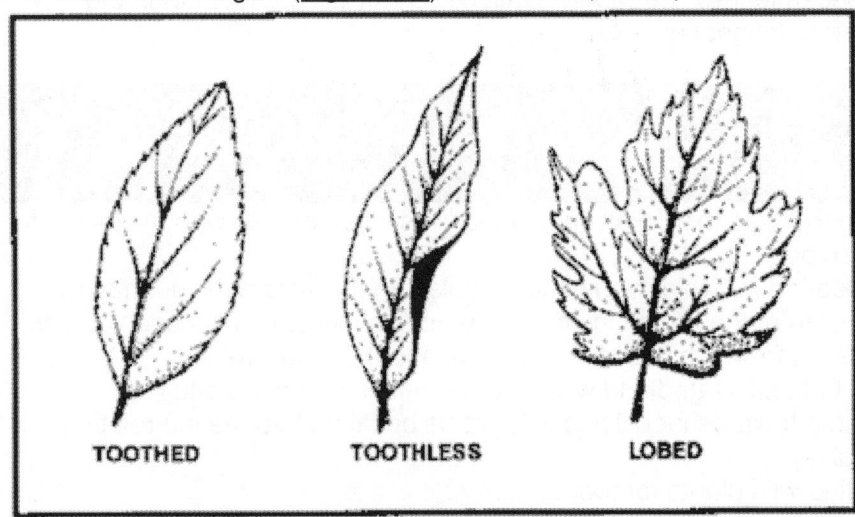

TOOTHED   TOOTHLESS   LOBED

Figure 9-1. Leaf margins.

These leaves may be lance-shaped, elliptical, egg-shaped, oblong, wedge-shaped, triangular, long-pointed, or top-shaped (Figure 9-2).

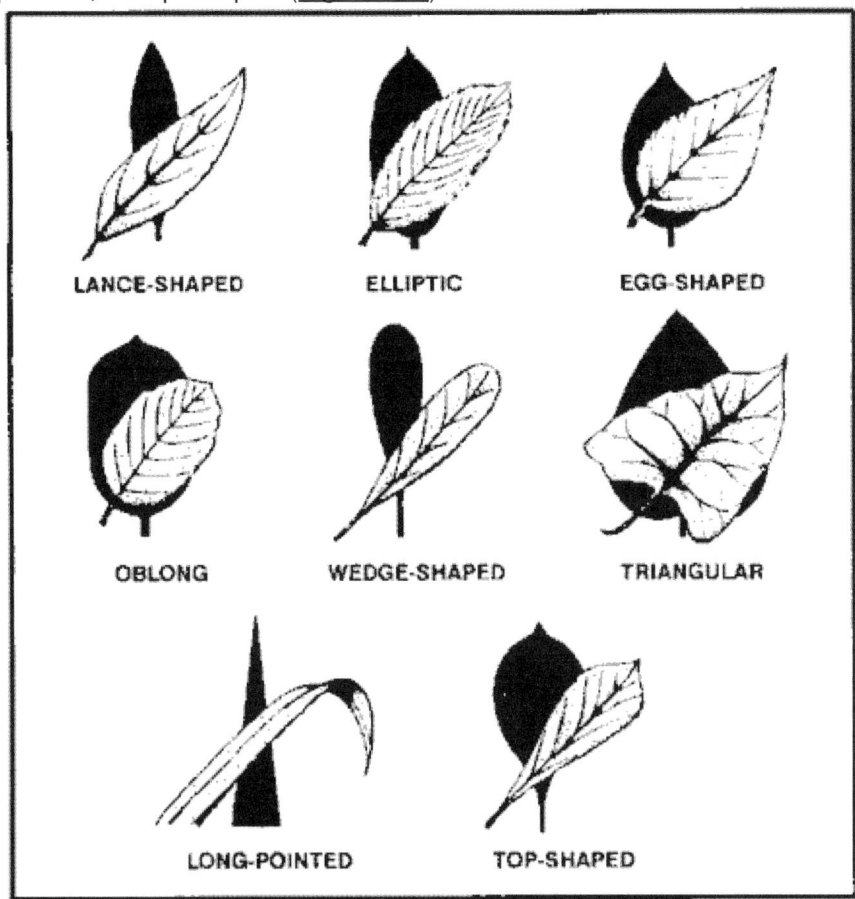

Figure 9-2. Leaf shapes.

The basic types of leaf arrangements (Figure 9-3) are opposite, alternate, compound, simple, and basal rosette.

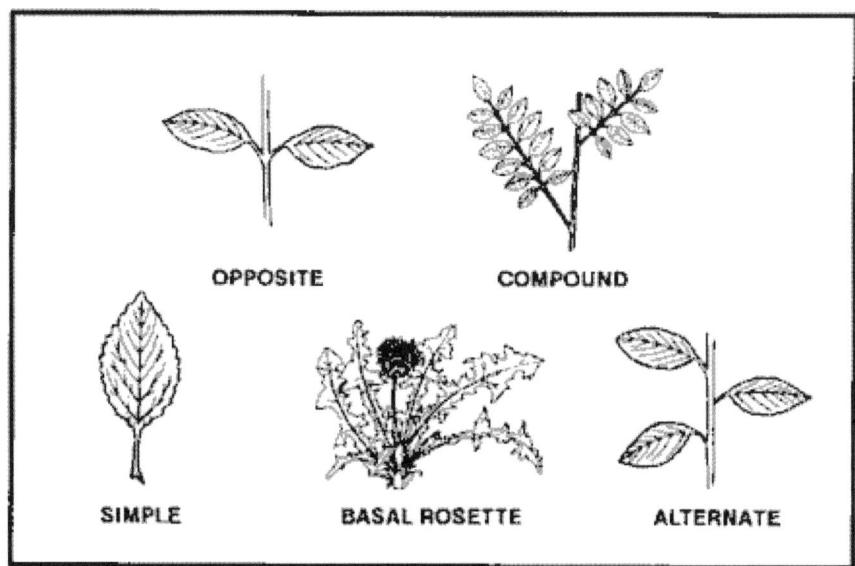

Figure 9-3. Leaf arrangements.

The basic types of root structures (Figure 9-4) are the bulb, clove, taproot, tuber, rhizome, corm, and crown. Bulbs are familiar to us as onions and, when sliced in half, will show concentric rings. Cloves are those bulblike structures that remind us of garlic and will separate into small pieces when broken apart. This characteristic separates wild onions from wild garlic. Taproots resemble carrots and may be single-rooted or branched, but usually only one plant stalk arises from each root. Tubers are like potatoes and daylilies and you will find these structures either on strings or in clusters underneath the parent plants. Rhizomes are large creeping rootstock or underground stems and many plants arise from the "eyes" of these roots. Corms are similar to bulbs but are solid when cut rather than possessing rings. A crown is the type of root structure found on plants such as asparagus and looks much like a mophead under the soil's surface.

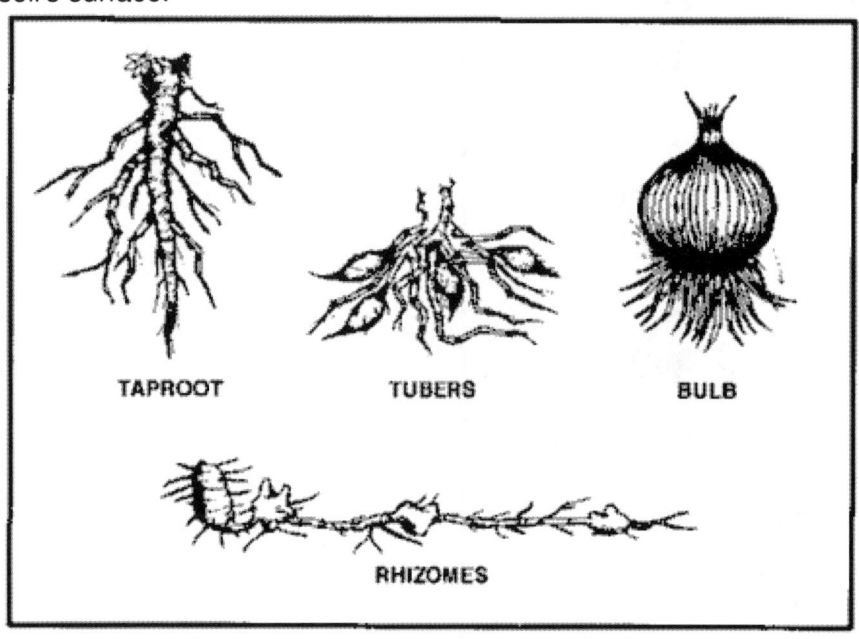

Figure 9-4. Root structures.

Learn as much as possible about plants you intend to use for food and their unique characteristics. Some plants have both edible and poisonous parts. Many are edible only at certain times of the year. Others may have poisonous relatives that look very similar to the ones you can eat or use for medicine.

Universal Edibility Test

There are many plants throughout the world. Tasting or swallowing even a small portion of some can cause severe discomfort, extreme internal disorders, and even death. Therefore, if you have the slightest doubt about a plant's edibility, apply the Universal Edibility Test (Figure 9-5) before eating any portion of it.

1	Test only one part of a potential food plant at a time.
2	Separate the plant into its basic components — leaves, stems, roots, buds, and flowers.
3	Smell the food for strong or acid odors. Remember, smell alone does not indicate a plant is edible or inedible.
4	Do not eat for 8 hours before starting the test.
5	During the 8 hours you abstain from eating, test for contact poisoning by placing a piece of the plant part you are testing on the inside of your elbow or wrist. Usually 15 minutes is enough time to allow for a reaction.
6	During the test period, take nothing by mouth except purified water and the plant part you are testing.
7	Select a small portion of a single part and prepare it the way you plan to eat it.
8	Before placing the prepared plant part in your mouth, touch a small portion (a pinch) to the outer surface of your lip to test for burning or itching.
9	If after 3 minutes there is no reaction on your lip, place the plant part on your tongue, holding it there for 15 minutes.
10	If there is no reaction, thoroughly chew a pinch and hold it in your mouth for 15 minutes. **Do not swallow.**
11	If no burning, itching, numbing, stinging, or other irritation occurs during the 15 minutes, swallow the food.
12	Wait 8 hours. If any ill effects occur during this period, induce vomiting and drink a lot of water.
13	If no ill effects occur, eat 0.25 cup of the same plant part prepared the same way. Wait another 8 hours. If no ill effects occur, the plant part as prepared is safe for eating.

CAUTION

Test all parts of the plant for edibility, as some plants have both edible and inedible parts. Do not assume that a part that proved edible when cooked is also edible when raw. Test the part raw to ensure edibility before eating raw. The same part or plant may produce varying reactions in different individuals.

Figure 9-5. Universal Edibility Test.

Before testing a plant for edibility, make sure there are enough plants to make the testing worth your time and effort. Each part of a plant (roots, leaves, flowers, and so on) requires more than 24 hours to test. Do not waste time testing a plant that is not relatively abundant in the area.

Remember, eating large portions of plant food on an empty stomach may cause diarrhea, nausea, or cramps. Two good examples of this are such familiar foods as green apples and wild onions. Even after testing plant food and finding it safe, eat it in moderation.

You can see from the steps and time involved in testing for edibility just how important it is to be able to identify edible plants.

To avoid potentially poisonous plants, stay away from any wild or unknown plants that have--

- Milky or discolored sap.
- Beans, bulbs, or seeds inside pods.
- Bitter or soapy taste.
- Spines, fine hairs, or thorns.
- Dill, carrot, parsnip, or parsleylike foliage.

- "Almond" scent in woody parts and leaves.
- Grain heads with pink, purplish, or black spurs.
- Three-leaved growth pattern.

Using the above <u>criteria</u> as eliminators when choosing plants for the Universal Edibility Test will cause you to avoid some edible plants. More important, these criteria will often help you avoid plants that are potentially toxic to eat or touch.

An entire encyclopedia of edible wild plants could be written, but space limits the number of plants presented here. Learn as much as possible about the plant life of the areas where you train regularly and where you expect to be traveling or working. Listed <u>below</u> and later in this chapter are some of the most common edible and <u>medicinal plants</u>. Detailed descriptions and photographs of these and other common plants are at Appendix B.

TEMPERATE ZONE FOOD PLANTS

- Amaranth (*Amaranthus retroflexus* and other species)
- Arrowroot (*Sagittaria* species)
- Asparagus *(Asparagus officinalis)*
- Beechnut (*Fagus* species)
- Blackberries (*Rubus* species)
- Blueberries (*Vaccinium* species)
- Burdock *(Arctium lappa)*
- Cattail (*Typha* species)
- Chestnut (*Castanea* species)
- Chicory *(Cichorium intybus)*
- Chufa *(Cyperus esculentus)*
- Dandelion *(Taraxacum officinale)*
- Daylily *(Hemerocallis fulva)*
- Nettle (*Urtica* species)
- Oaks (*Quercus* species)
- Persimmon *(Diospyros virginiana)*
- Plantain (*Plantago* species)
- Pokeweed *(Phytolacca americana)*
- Prickly pear cactus (*Opuntia* species)
- Purslane *(Portulaca oleracea)*
- Sassafras *(Sassafras albidum)*
- Sheep sorrel *(Rumex acetosella)*
- Strawberries (*Fragaria* species)
- Thistle (*Cirsium* species)
- Water lily and lotus (*Nuphar, Nelumbo,* and other species)
- Wild onion and garlic (*Allium* species)
- Wild rose (*Rosa* species)
- Wood sorrel (*Oxalis* species)

TROPICAL ZONE FOOD PLANTS

- Bamboo (*Bambusa* and other species)
- Bananas (*Musa* species)
- Breadfruit *(Artocarpus incisa)*
- Cashew nut *(Anacardium occidental)*
- Coconut *(Cocos nucifera)*
- Mango *(Mangifera indica)*
- Palms (various species)

- Papaya (*Carica* species)
- Sugarcane *(Saccharum officinarum)*
- Taro (*Colocasia* species)

DESERT ZONE FOOD PLANTS

- Acacia *(Acacia farnesiana)*
- Agave (*Agave* species)
- Cactus (various species)
- Date palm *(Phoenix dactylifera)*
- Desert amaranth *(Amaranths palmeri)*

Seaweeds

One plant you should never overlook is seaweed. It is a form of marine algae found on or near ocean shores. There are also some edible freshwater varieties. Seaweed is a valuable source of iodine, other minerals, and vitamin C. Large quantities of seaweed in an unaccustomed stomach can produce a severe laxative effect.

When gathering seaweeds for food, find living plants attached to rocks or floating free. Seaweed washed onshore any length of time may be spoiled or decayed. You can dry freshly harvested seaweeds for later use.

Its preparation for eating depends on the type of seaweed. You can dry thin and tender varieties in the sun or over a fire until crisp. Crush and add these to soups or broths. Boil thick, leathery seaweeds for a short time to soften them. Eat them as a vegetable or with other foods. You can eat some varieties raw after testing for edibility.

SEAWEEDS

- Dulse *(Rhodymenia palmata)*
- Green seaweed *(Ulva lactuca)*
- Irish moss *(Chondrus crispus)*
- Kelp *(Alaria esculenta)*
- Laver *(Porphyra* species)
- Mojaban *(Sargassum fulvellum)*
- Sugar wrack *(Laminaria saccharina)*

Preparation of Plant Food

Although some plants or plant parts are edible raw, you must cook others to be edible or palatable. Edible means that a plant or food will provide you with necessary nutrients, while palatable means that it actually is pleasing to eat. Many wild plants are edible but barely palatable. It is a good idea to learn to identify, prepare, and eat wild foods.

Methods used to improve the taste of plant food include soaking, boiling, cooking, or leaching. Leaching is done by crushing the food (for example, acorns), placing it in a strainer, and pouring boiling water through it or immersing it in running water.

Boil leaves, stems, and buds until tender, changing the water, if necessary, to remove any bitterness.

Boil, bake, or roast tubers and roots. Drying helps to remove caustic oxalates from some roots like those in the *Arum* family.

Leach acorns in water, if necessary, to remove the bitterness. Some nuts, such as chestnuts, are good raw, but taste better roasted.

You can eat many grains and seeds raw until they mature. When hard or dry, you may have to boil or grind them into meal or flour.

The sap from many trees, such as maples, birches, walnuts, and sycamores, contains sugar. You may boil these saps down to a syrup for sweetening. It takes about 35 liters of maple sap to make one liter of maple syrup!

PLANTS FOR MEDICINE

In a survival situation you will have to use what is available. In using plants and other natural remedies, positive identification of the plants involved is as critical as in using them for food. Proper use of these plants is equally important.

Terms and Definitions

The following terms, and their definitions, are associated with medicinal plant use:

- *Poultice.* The name given to crushed leaves or other plant parts, possibly heated, that you apply to a wound or sore either directly or wrapped in cloth or paper.
- *Infusion or tisane or tea.* The preparation of medicinal herbs for internal or external application. You place a small quantity of a herb in a container, pour hot water over it, and let it steep (covered or uncovered) before use.
- *Decoction.* The extract of a boiled down or simmered herb leaf or root. You add herb leaf or root to water. You bring them to a sustained boil or simmer to draw their chemicals into the water. The average ratio is about 28 to 56 grams (1 to 2 ounces) of herb to 0.5 liter of water.
- *Expressed juice.* Liquids or saps squeezed from plant material and either applied to the wound or made into another medicine.

Many natural remedies work slower than the medicines you know. Therefore, start with smaller doses and allow more time for them to take effect. Naturally, some will act more rapidly than others.

Specific Remedies

The following remedies are for use only in a survival situation, not for routine use:

- *Diarrhea.* Drink tea made from the roots of blackberries and their relatives to stop diarrhea. White oak bark and other barks containing tannin are also effective. However, use them with caution when nothing else is available because of possible negative effects on the kidneys. You can also stop diarrhea by eating white clay or campfire ashes. Tea made from cowberry or cranberry or hazel leaves works too.
- *Antihemorrhagics.* Make medications to stop bleeding from a poultice of the puffball mushroom, from plantain leaves, or most effectively from the leaves of the common yarrow or woundwort *(Achillea millefolium).*
- *Antiseptics.* Use to cleanse wounds, sores, or rashes. You can make them from the expressed juice from wild onion or garlic, or expressed juice from chickweed leaves or the crushed leaves of dock. You can also make antiseptics from a decoction of burdock root, mallow leaves or roots, or white oak bark. All these medications are for external use only.
- *Fevers.* Treat a fever with a tea made from willow bark, an infusion of elder flowers or fruit, linden flower tea, or elm bark decoction.

- *Colds and sore throats.* Treat these illnesses with a decoction made from either plantain leaves or willow bark. You can also use a tea made from burdock roots, mallow or mullein flowers or roots, or mint leaves.
- *Aches, pains, and sprains.* Treat with externally applied poultices of dock, plantain, chickweed, willow bark, garlic, or sorrel. You can also use salves made by mixing the expressed juices of these plants in animal fat or vegetable oils.
- *Itching.* Relieve the itch from insect bites, sunburn, or plant poisoning rashes by applying a poultice of jewelweed *(Impatiens biflora)* or witch hazel leaves *(Hamamelis virginiana).* The jewelweed juice will help when applied to poison ivy rashes or insect stings. It works on sunburn as well as aloe vera.
- *Sedatives.* Get help in falling asleep by brewing a tea made from mint leaves or passionflower leaves.
- *Hemorrhoids.* Treat them with external washes from elm bark or oak bark tea, from the expressed juice of plantain leaves, or from a Solomon's seal root decoction.
- *Constipation.* Relieve constipation by drinking decoctions from dandelion leaves, rose hips, or walnut bark. Eating raw daylily flowers will also help.
- *Worms or intestinal parasites.* Using moderation, treat with tea made from tansy *(Tanacetum vulgare)* or from wild carrot leaves.
- *Gas and cramps.* Use a tea made from carrot seeds as an antiflatulent; use tea made from mint leaves to settle the stomach.
- *Antifungal washes.* Make a decoction of walnut leaves or oak bark or acorns to treat ringworm and athlete's foot. Apply frequently to the site, alternating with exposure to direct sunlight.

MISCELLANEOUS USES OF PLANTS

Make dyes from various plants to color clothing or to camouflage your skin. Usually, you will have to boil the plants to get the best results. Onion skins produce yellow, walnut hulls produce brown, and pokeberries provide a purple dye.

Make fibers and cordage from plant fibers. Most commonly used are the stems from nettles and milkweeds, yucca plants, and the inner bark of trees like the linden.

Make fish poison by immersing walnut hulls in a small area of quiet water. This poison makes it impossible for the fish to breathe but doesn't adversely affect their edibility.

Make tinder for starting fires from cattail fluff, cedar bark, lighter knot wood from pine trees, or hardened sap from resinous wood trees.

Make insulation by fluffing up female cattail heads or milkweed down.

Reprinted as permitted by U.S. Department of the Army

Make insect repellents by applying the expressed juice of wild garlic or onion to the skin, by placing sassafras leaves in your shelter, or by burning or smudging cattail seed hair fibers.

Plants can be your ally as long as you use them cautiously. *The key to the safe use of plants is positive identification* whether you use them as food or medicine or in constructing shelters or equipment.

CHAPTER 10 - POISONOUS PLANTS

Successful use of plants in a survival situation depends on positive identification. Knowing poisonous plants is as important to a survivor as knowing edible plants. Knowing the poisonous plants will help you avoid sustaining injuries from them.

HOW PLANTS POISON

Plants generally poison by--

- *Ingestion.* When a person eats a part of a poisonous plant.
- *Contact.* When a person makes contact with a poisonous plant that causes any type of skin irritation or dermatitis.
- *Absorption or inhalation.* When a person either absorbs the poison through the skin or inhales it into the respiratory system.

Plant poisoning ranges from minor irritation to death. A common question asked is, "How poisonous is this plant?" It is difficult to say how poisonous plants are because--

- Some plants require contact with a large amount of the plant before noticing any adverse reaction while others will cause death with only a small amount.
- Every plant will vary in the amount of toxins it contains due to different growing conditions and slight variations in subspecies.
- Every person has a different level of resistance to toxic substances.
- Some persons may be more sensitive to a particular plant.

Some common misconceptions about poisonous plants are--

- *Watch the animals and eat what they eat.* Most of the time this statement is true, but some animals can eat plants that are poisonous to humans.
- *Boil the plant in water and any poisons will be removed.* Boiling removes many poisons, but not all.
- *Plants with a red color are poisonous.* Some plants that are red are poisonous, but not all.

The point is there is no one rule to aid in identifying poisonous plants. You must make an effort to learn as much about them as possible.

ALL ABOUT PLANTS

It is to your benefit to learn as much about plants as possible. Many poisonous plants look like their edible relatives or like other edible plants. For example, poison hemlock appears very similar to wild carrot. Certain plants are safe to eat in certain seasons or stages of growth and poisonous in other stages. For example, the leaves of the pokeweed are edible when it first starts to grow, but it soon becomes poisonous. You can eat some plants and their fruits only when they are ripe. For example, the ripe fruit of mayapple is edible, but all other parts and the green fruit are poisonous. Some plants contain both edible and poisonous parts; potatoes and tomatoes are common plant foods, but their green parts are poisonous.

Some plants become toxic after wilting. For example, when the black cherry starts to wilt, hydrocyanic acid develops. Specific preparation methods make some plants edible that are poisonous raw. You can eat the thinly sliced and thoroughly dried corms (drying may take a year) of the jack-in-the-pulpit, but they are poisonous if not thoroughly dried.

Learn to identify and use plants before a survival situation. Some sources of information about plants are pamphlets, books, films, nature trails, botanical gardens, local markets, and local natives. Gather and cross-reference information from as many sources as possible, because many sources will not contain all the information needed.

RULES FOR AVOIDING POISONOUS PLANTS

Your best policy is to be able to look at a plant and identify it with absolute certainty and to know its uses or dangers. Many times this is not possible. If you have little or no knowledge of the local vegetation, use the rules to select plants for the "Universal Edibility Test." Remember, avoid --

- *All mushrooms.* Mushroom identification is very difficult and must be precise, even more so than with other plants. Some mushrooms cause death very quickly. Some mushrooms have no known antidote. Two general types of mushroom poisoning are gastrointestinal and central nervous system.
- *Contact with or touching plants unnecessarily.*

CONTACT DERMATITIS

Contact dermatitis from plants will usually cause the most trouble in the field. The effects may be persistent, spread by scratching, and are particularly dangerous if there is contact in or around the eyes. The principal toxin of these plants is usually an oil that gets on the skin upon contact with the plant. The oil can also get on equipment and then infect whoever touches the equipment. Never bum a contact poisonous plant because the smoke may be as harmful as the plant. There is a greater danger of being affected when overheated and sweating. The infection may be local or it may spread over the body. Symptoms may take from a few hours to several days to appear. Signs and symptoms can include burning, reddening, itching, swelling, and blisters.

When you first contact the poisonous plants or the first symptoms appear, try to remove the oil by washing with soap and cold water. If water is not available, wipe your skin repeatedly with dirt or sand. Do not use dirt if blisters have developed. The dirt may break open the blisters and leave the body open to infection. After you have removed the oil, dry the area. You can wash with a tannic acid solution and crush and rub jewelweed on the affected area to treat plant-caused rashes. You can make tannic acid from oak bark.

Poisonous plants that cause contact dermatitis are--

- Cowhage.
- Poison ivy.
- Poison oak.
- Poison sumac.
- Rengas tree.
- Trumpet vine.

INGESTION POISONING

Ingestion poisoning can be very serious and could lead to death very quickly. Do not eat any plant unless you have positively identified it first. Keep a log of all plants eaten.

Signs and symptoms of ingestion poisoning can include nausea, vomiting, diarrhea, abdominal cramps, depressed heartbeat and respiration, headaches, hallucinations, dry mouth, unconsciousness, coma, and death.

If you suspect plant poisoning, try to remove the poisonous material from the victim's mouth and stomach as soon as possible. Induce vomiting by tickling the back of his throat or by giving him warm saltwater, if he is conscious. Dilute the poison by administering large quantities of water or milk, if he is conscious. The following plants can cause ingestion poisoning if eaten:

- Castor bean.
- Chinaberry.
- Death camas.
- Lantana.
- Manchineel.
- Oleander.
- Pangi.
- Physic nut.
- Poison and water hemlocks.
- Rosary pea.
- Strychnine tree.

See Appendix C for photographs and descriptions of these plants.

CHAPTER 11 - DANGEROUS ANIMALS

Animals rarely are as threatening to the survivor as the rest of the environment. Common sense tells the survivor to avoid encounters with lions, bears, and other large or dangerous animals. You should also avoid large grazing animals with horns, hooves, and great weight. Your actions may prevent unexpected meetings. Move carefully through their environment. Do not attract large predators by leaving food lying around your camp. Carefully survey the scene before entering water or forests.

Smaller animals actually present more of a threat to the survivor than large animals. To compensate for their size, nature has given many small animals weapons such as fangs and stingers to defend themselves. Each year, a few people are bitten by sharks, mauled by alligators, and attacked by bears. Most of these incidents were in some way the victim's fault. However, each year more victims die from bites by relatively small venomous snakes than by large dangerous animals. Even more victims die from allergic reactions to bee stings. For this reason, we will pay more attention to smaller and potentially more dangerous creatures. These are the animals you are more likely to meet as you unwittingly move into their habitat, or they slip into your environment unnoticed.

Keeping a level head and an awareness of your surroundings will keep you alive if you use a few simple safety procedures. Do not let curiosity and carelessness kill or injure you.

INSECTS AND ARACHNIDS

You recognize and identify insects, except centipedes and millipedes, by their six legs while arachnids have eight. All these small creatures become pests when they bite, sting, or irritate you.
Although their venom can be quite painful, bee, wasp, and hornet stings rarely kill a survivor unless he is allergic to that particular toxin. Even the most dangerous spiders rarely kill, and the effects of tick-borne diseases are very slow-acting. However, in all cases, avoidance is the best defense. In environments known to have spiders and scorpions, check your footgear and clothing every morning. Also check your bedding and shelter for them. Use care when turning over rocks and logs. See Appendix D for examples of dangerous insects and arachnids.

Scorpions

You find scorpions (*Buthotus* species) in deserts, jungles, and forests of tropical, subtropical, and warm temperate areas of the world. They are mostly nocturnal in habit. You can find desert scorpions from below sea level in Death Valley to elevations as high as 3,600 meters in the Andes. Typically brown or black in moist areas, they may be yellow or light green in the desert. Their average size is about 2.5 centimeters. However, there are 20-centimeter giants in the jungles of Central America, New Guinea, and southern Africa. Fatalities from scorpion stings are rare, but they can occur in children, the elderly, and ill persons. Scorpions resemble small lobsters with raised, jointed tails bearing a stinger in the tip. Nature mimics the scorpions with whip scorpions or vinegar-roons. These are harmless and have a tail like a wire or whip, rather than the jointed tail and stinger of true scorpions.

Spiders

You recognize the brown recluse or fiddleback spider of North America *(Loxosceles reclusa)* by a prominent violin-shaped light spot on the back of its body. As its name suggests, this spider likes to hide in dark places. Though rarely fatal, its bite causes excessive tissue degeneration around the wound and can even lead to amputation of the digits if left untreated.

You find members of the widow family *(Latrodectus species)* worldwide, though the black widow of North America is perhaps the most well-known. Found in warmer areas of the world, the widows are small, dark spiders with often hourglass-shaped white, red, or orange spots on their abdomens.

Funnelwebs (*Atrax* species) are large, gray or brown Australian spiders. Chunky, with short legs, they are able to move easily up and down the cone-shaped webs from which they get their name. The local populace considers them deadly. Avoid them as they move about, usually at night, in search of prey. Symptoms of their bite are similar to those of the widow's--severe pain accompanied by sweating and shivering, weakness, and disabling episodes that can last a week.

Tarantulas are large, hairy spiders (*Theraphosidae* and *Lycosa* species) best known because they are often sold in pet stores. There is one species in Europe, but most come from tropical America. Some South American species do inject a dangerous toxin, but most simply produce a painful bite. Some tarantulas can be as large as a dinner plate. They all have large fangs for capturing food such as birds, mice, and lizards. If bitten by a tarantula, pain and bleeding are certain, and infection is likely.

Centipedes and Millipedes

Centipedes and millipedes are mostly small and harmless, although some tropical and desert species may reach 25 centimeters. A few varieties of centipedes have a poisonous bite, but infection is the greatest danger, as their sharp claws dig in and puncture the skin. To prevent skin punctures, brush them off in the direction they are traveling, if you find them crawling on your skin.

Bees, Wasps, and Hornets

We are all familiar with bees, wasps, and hornets. They come in many varieties and have a wide diversity of habits and habitats. You recognize bees by their hairy and usually thick body, while the wasps, hornets, and yellow jackets have more slender, nearly hairless, bodies. Some bees, such as honeybees, live in colonies. They may be either domesticated or living wild in caves or hollow trees. You may find other bees, such as carpenter bees, in individual nest holes in wood, or in the ground, like bumblebees. The main danger from bees is their barbed stinger located on their abdomens. When the bee stings you, it rips its stinger out of its abdomen along with the venom sac, and the bee dies. Except for killer bees, most bees tend to be more docile than wasps, hornets, and yellow jackets that have smooth stingers and are capable of repeated attacks.

Avoidance is the best tactic for self-protection. Watch out for flowers or fruit where bees may be feeding. Be careful of meat-eating yellow jackets when cleaning fish or game. The average person has a relatively minor and temporary reaction to bee stings and recovers in a couple of hours when the pain and headache go away. Those who are allergic to bee venom have severe reactions including anaphylactic shock, coma, and death. If antihistamine medicine is not available and you cannot find a substitute, an allergy sufferer in a survival situation is in grave danger.

Ticks

Ticks are common in the tropics and temperate regions. They are familiar to most of us. Ticks are small round arachnids with eight legs and can have either a soft or hard body. Ticks require a blood host to survive and reproduce. This makes them dangerous because they spread diseases like Lyme disease, Rocky Mountain spotted fever, encephalitis, and others that can ultimately be disabling or fatal. There is little you can do to treat these diseases once contracted, but time is your ally since they are slow-acting ailments. According to most authorities, it takes at least 6 hours of attachment to the host for the tick to transmit the disease organisms. Thus, you have time to thoroughly inspect your body for their presence. Beware of ticks when passing through the thick vegetation they cling to, when cleaning host animals for food, and when gathering natural materials to construct a shelter. Always use insect repellents, if possible.

LEECHES

Leeches are blood-sucking creatures with a wormlike appearance. You find them in the tropics and in temperate zones. You will certainly encounter them when swimming in infested waters or making expedient water crossings. You can find them when passing through swampy, tropical vegetation and bogs. You can also find them while cleaning food animals, such as turtles, found in fresh water. Leeches can crawl into small openings; therefore, avoid camping in their habitats when possible. Keep your trousers tucked in your boots. Check yourself frequently for leeches. Swallowed or eaten, leeches can be a great hazard. It is therefore essential to treat water from questionable sources by boiling or using chemical water treatments. Survivors have developed severe infections from wounds inside the throat or nose when sores from swallowed leeches became infected.

BATS

Despite the legends, bats (*Desmodus* species) are a relatively small hazard to the survivor. There are many bat varieties worldwide, but you find the true vampire bats only in Central and South America. They are small, agile fliers that land on their sleeping victims, mostly cows and horses, to lap a blood meal after biting their victim. Their saliva contains an anticoagulant that keeps the blood slowly flowing while they feed. Only a small percentage of these bats actually carry rabies; however, avoid any sick or injured bat. They can carry other diseases and infections and will bite readily when handled. Taking shelter in a cave occupied by bats, however, presents the much greater hazard of inhaling powdered bat dung, or guano. Bat dung carries many organisms that can cause diseases. Eating thoroughly cooked flying foxes or other bats presents no danger from rabies and other diseases, but again, the emphasis is on thorough cooking.

POISONOUS SNAKES

There are no infallible rules for expedient identification of poisonous snakes in the field, because the guidelines all require close observation or manipulation of the snake's body. The best strategy is to leave all snakes alone. Where snakes are plentiful and poisonous species are present, the risk of their bites negates their food value. Apply the following safety rules when traveling in areas where there are poisonous snakes:

- Walk carefully and watch where you step. Step onto logs rather than over them before looking and moving on.
- Look closely when picking fruit or moving around water.
- Do not tease, molest, or harass snakes. Snakes cannot close their eyes. Therefore, you cannot tell if they are asleep. Some snakes, such as mambas, cobras, and bushmasters, will attack aggressively when cornered or guarding a nest.
- Use sticks to turn logs and rocks.
- Wear proper footgear, particularly at night.
- Carefully check bedding, shelter, and clothing.
- Be calm when you encounter serpents. Snakes cannot hear and you can occasionally surprise them when they are sleeping or sunning. Normally, they will flee if given the opportunity.
- Use extreme care if you must kill snakes for food or safety. Although it is not common, warm, sleeping human bodies occasionally attract snakes.

See Appendix E for detailed descriptions of the <u>snakes</u> listed below.

Snake-Free Areas

The polar regions are free of snakes due to their inhospitable environments. Other areas considered to be free of poisonous snakes are New Zealand, Cuba, Haiti, Jamaica, Puerto Rico, Ireland, Polynesia, and Hawaii.

POISONOUS SNAKES OF THE AMERICAS

- American Copperhead *(Agkistrodon contortrix)*
- Bushmaster *(Lachesis mutus)*
- Coral snake *(Micrurus fulvius)*
- Cottonmouth *(Agkistrodon piscivorus)*
- Fer-de-lance *(Bothrops atrox)*
- Rattlesnake *(Crotalus species)*

POISONOUS SNAKES OF EUROPE

- Common adder *(Vipers berus)*
- Pallas' viper *(Agkistrodon halys)*

POISONOUS SNAKES OF AFRICA AND ASIA

- Boomslang *(Dispholidus typus)*
- Cobra *(Naja species)*
- Gaboon viper *(Bitis gabonica)*
- Green tree pit viper *(Trimeresurus gramineus)*
- Habu pit viper *(Trimeresurus flavoviridis)*
- Krait *(Bungarus caeruleus)*
- Malayan pit viper *(Callaselasma rhodostoma)*
- Mamba *(Dendraspis species)*
- Puff adder *(Bitis arietans)*
- Rhinoceros viper *(Bitis nasicornis)*
- Russell' s viper *(Vipera russellii)*
- Sand viper *(Cerastes vipera)*
- Saw-scaled viper *(Echis carinatus)*
- Wagler's pit viper *(Trimeresurus wagleri)*

POISONOUS SNAKES OF AUSTRALASIA

- Death adder *(Acanthophis antarcticus)*
- Taipan *(Oxyuranus scutellatus)*
- Tiger snake *(Notechis scutatus)*
- Yellow-bellied sea snake *(Pelamis platurus)*

DANGEROUS LIZARDS

The Gila monster and the Mexican beaded lizard are dangerous and poisonous lizards.

Gila Monster

The Gila monster *(Heloderma suspectrum)* of the American southwest, including Mexico, is a large lizard with dark, highly textured skin marked by pinkish mottling. It averages 35 to 45 centimeters in length and has a thick, stumpy tail. Unlikely to bite unless molested, it has a poisonous bite.

Mexican Beaded Lizard

The Mexican beaded lizard *(Heloderma horridum)* resembles its relative, the Gila monster. It has more uniform spots rather than bands of color (the Gila monster). It also is poisonous and has a docile nature. You find it from Mexico to Central America.

Komodo Dragon

This giant lizard *(Varanus komodoensis)* grows to more than 3 meters in length and can be dangerous if you try to capture it. This Indonesian lizard can weigh more than 135 kilograms.

DANGERS IN RIVERS

Common sense will tell you to avoid confrontations with hippopotami, alligators, crocodiles, and other large river creatures. There are, however, a few smaller river creatures with which you should be cautious.

Electric Eel

Electric eels *(Electrophorus electricus)* may reach 2 meters in length and 20 centimeters in diameter. Avoid them. They are capable of generating up to 500 volts of electricity in certain organs in their body. They use this shock to stun prey and enemies. Normally, you find these eels in the Orinoco and Amazon River systems in South America. They seem to prefer shallow waters that are more highly oxygenated and provide more food. They are bulkier than our native eels. Their upper body is dark gray or black, with a lighter-colored underbelly.

Piranha

Piranhas (*Serrasalmo* species) are another hazard of the Orinoco and Amazon River systems, as well as the Paraguay River Basin, where they are native. These fish vary greatly in size and coloration, but usually have a combination of orange undersides and dark tops. They have white, razor-sharp teeth that are clearly visible. They may be as long as 50 centimeters. Use great care when crossing waters where they live. Blood attracts them. They are most dangerous in shallow waters during the dry season.

Turtle

Be careful when handling and capturing large freshwater turtles, such as the snapping turtles and soft-shelled turtles of North America and the matamata and other turtles of South America. All of these turtles will bite in self-defense and can amputate fingers and toes.

Platypus

The platypus or duckbill *(Ornithorhyncus anatinus)* is the only member of its family and is easily recognized. It has a long body covered with grayish, short hair, a tail like a beaver, and a bill like a duck. Growing up to 60 centimeters in length, it may appear to be a good food source, but this egg-laying mammal, the only one in the world, is very dangerous. The male has a poisonous spur on each hind foot that can inflict intensely painful wounds. You find the platypus only in Australia, mainly along mud banks on waterways.

DANGERS IN BAYS AND ESTUARIES

In areas where seas and rivers come together, there are dangers associated with both fresh and salt water. In shallow salt waters, there are many creatures that can inflict pain and cause infection to develop. Stepping on sea urchins, for example, can produce pain and infection. When moving about in shallow water, wear some form of footgear and shuffle your feet along the bottom, rather than picking up your feet and stepping.

Stingrays *(Dasyatidae* species) are a real hazard in shallow waters, especially tropical waters. The type of bottom appears to be irrelevant. There is a great variance between species, but all have a sharp spike in their tail that may be venomous and can cause extremely painful wounds if stepped on. All rays have a typical shape that resembles a kite. You find them along the coasts of the Americas, Africa, and Australasia.

SALTWATER DANGERS

There are several fish that you should not handle, touch, or contact. There are others that you should not eat.

Fish Dangerous to Handle, Touch, or Contact

There are several fish you should not handle, touch, or contact that are identified below.

Shark
Sharks are the most feared animal in the sea. Usually, shark attacks cannot be avoided and are considered accidents. You, as a survivor, should take every precaution to avoid any contact with sharks. There are many shark species, but in general, dangerous sharks have wide mouths and visible teeth, while relatively harmless ones have small mouths on the underside of their heads. However, any shark can inflict painful and often fatal injuries, either through bites or through abrasions from their rough skin.

Rabbitfish
Rabbitfish or spinefoot *(Siganidae* species) occur mainly on coral reefs in the Indian and Pacific oceans. They have very sharp, possibly venomous spines in their fins. Handle them with care, if at all. This fish, like many others of the dangerous fish in this section, is considered edible by native peoples where the fish are found, but deaths occur from careless handling. Seek other nonpoisonous fish to eat if at all possible.

Tang
Tang or surgeonfish *(Acanthuridae* species) average 20 to 25 centimeters in length and often are beautifully colored. They are called surgeonfish because of the scalpellike spines located in the tail. The wounds inflicted by these spines can bring about death through infection, envenomation, and loss of blood, which may incidentally attract sharks.

Toadfish
Toadfish (Batrachoididae species) occur in tropical waters off the Gulf Coast of the United States and along both coasts of Central and South America. These dully colored fish average 18 to 25 centimeters in length. They typically bury themselves in the sand to await fish and other prey. They have sharp, very toxic spines along their backs.

Scorpion Fish
Poisonous scorpion fish or zebra fish *(Scorpaenidae* species) are mostly around reefs in the tropical Indian and Pacific oceans and occasionally in the Mediterranean and Aegean seas. They average 30 to 75 centimeters in length. Their coloration is highly variable, from reddish brown to almost purple or

brownish yellow. They have long, wavy fins and spines and their sting is intensively painful. Less poisonous relatives live in the Atlantic Ocean.

Stonefish

Stonefish (*Synanceja* species) are in the Pacific and Indian oceans. They can inject a painful venom from their dorsal spines when stepped on or handled carelessly. They are almost impossible to see because of their lumpy shape and drab colors. They range in size up to 40 centimeters.

Weever Fish

Weever fish (*Trachinidae* species) average 30 centimeters long. They are hard to see as they lie buried in the sand off the coasts of Europe, Africa, and the Mediterranean. Their color is usually a dull brown. They have venomous spines on the back and gills.

See Appendix F for more details on these venomous fish.

Animals and Fish Poisonous to Eat

Survival manuals often mention that the livers of polar bears are toxic due to their high concentrations of vitamin A. For this reason, we mention the chance of death after eating this organ. Another toxic meat is the flesh of the hawksbill turtle. You recognize them by their down-turned bill and yellow polka dots on their neck and front flippers. They weigh more than 275 kilograms and are unlikely to be captured.

Many fish living in reefs near shore, or in lagoons and estuaries, are poisonous to eat, though some are only seasonally dangerous. The majority are tropical fish; however, be wary of eating any unidentifiable fish wherever you are. Some predatory fish, such as barracuda and snapper, may become toxic if the fish they feed on in shallow waters are poisonous. The most poisonous types appear to have parrotlike beaks and hard shell-like skins with spines and often can inflate their bodies like balloons. However, at certain times of the year, indigenous populations consider the puffer a delicacy.

Blowfish

Blowfish or puffer (*Tetraodontidae* species) are more tolerant of cold water. You find them along tropical and temperate coasts worldwide, even in some of the rivers of Southeast Asia and Africa. Stout-bodied and round, many of these fish have short spines and can inflate themselves into a ball when alarmed or agitated. Their blood, liver, and gonads are so toxic that as little as 28 milligrams (1 ounce) can be fatal. These fish vary in color and size, growing up to 75 centimeters in length.

Triggerfish

The triggerfish (*Balistidae* species) occur in great variety, mostly in tropical seas. They are deep-bodied and compressed, resembling a seagoing pancake up to 60 centimeters in length, with large and sharp dorsal spines. Avoid them all, as many have poisonous flesh.

Barracuda

Although most people avoid them because of their ferocity, they occasionally eat barracuda *(Sphyraena barracuda)*. These predators of mostly tropical seas can reach almost 1.5 meters in length and have attacked humans without provocation. They occasionally carry the poison ciguatera in their flesh, making them deadly if consumed.

See Appendix F for more details on toxic fish and toxic mollusks.

Other Dangerous Sea Creatures

The blue-ringed octopus, jellyfish, and the cone and auger shells are other dangerous sea creatures.

Blue-Ringed Octopus

Most octopi are excellent when properly prepared. However, the blueringed octopus *(Hapalochlaena lunulata)* can inflict a deadly bite from its parrotlike beak. Fortunately, it is restricted to the Great Barrier Reef of Australia and is very small. It is easily recognized by its grayish white overall color and iridescent blue rings. Authorities warn that all tropical octopus species should be treated with caution, since many have poisonous bites, although the flesh is edible.

Jellyfish

Jellyfish-related deaths are rare, but the sting they inflict is extremely painful. The Portuguese man-of-war resembles a large pink or purple balloon floating on the sea. It has poisonous tentacles hanging up to 12 meters below its body. The huge tentacles are actually colonies of stinging cells. Most known deaths from jellyfish are attributed to the man-of-war. Other jellyfish can inflict very painful stings as well. Avoid the long tentacles of any jellyfish, even those washed up on the beach and apparently dead.

Cone Shell

The subtropical and tropical cone shells (*Conidae* species) have a venomous harpoonlike barb. All are cone-shaped and have a fine netlike pattern on the shell. A membrane may possibly obscure this coloration. There are some very poisonous cone shells, even some lethal ones in the Indian and Pacific oceans. Avoid any shell shaped like an ice cream cone.

Auger Shell

The auger shell or terebra (*Terebridae* species) are much longer and thinner than the cone shells, but can be nearly as deadly as the cone shells. They are found in temperate and tropical seas. Those in the Indian and Pacific oceans have a more toxic venom in their stinging barb. Do not eat these snails, as their flesh may be poisonous.

CHAPTER 12 - FIELD-EXPEDIENT WEAPONS, TOOLS, AND EQUIPMENT

As a soldier you know the importance of proper care and use of your weapons, tools, and equipment. This is especially true of your knife. You must always keep it sharp and ready to use. A knife is your most valuable tool in a survival situation. Imagine being in a survival situation without any weapons, tools, or equipment except your knife. It could happen! You might even be without a knife. You would probably feel helpless, but with the proper knowledge and skills, you can easily improvise needed items.

In survival situations, you may have to fashion any number and type of field-expedient tools and equipment to survive. Examples of tools and equipment that could make your life much easier are ropes, rucksacks, clothes, nets, and so on.

Weapons serve a dual purpose. You use them to obtain and prepare food and to provide self-defense. A weapon can also give you a feeling of security and provide you with the ability to hunt on the move.

CLUBS

You hold clubs, you do not throw them. As a field-expedient weapon, the club does not protect you from enemy soldiers. It can, however, extend your area of defense beyond your fingertips. It also serves to increase the force of a blow without injuring yourself. There are three basic types of clubs. They are the simple, weighted, and sling club.

Simple Club

A simple club is a staff or branch. It must be short enough for you to swing easily, but long enough and strong enough for you to damage whatever you hit. Its diameter should fit comfortably in your palm, but it should not be so thin as to allow the club to break easily upon impact. A straight-grained hardwood is best if you can find it.

Weighted Club

A weighted club is any simple club with a weight on one end. The weight may be a natural weight, such as a knot on the wood, or something added, such as a stone lashed to the club.
To make a weighted club, first find a stone that has a shape that will allow you to lash it securely to the club. A stone with a slight hourglass shape works well. If you cannot find a suitably shaped stone, you must fashion a groove or channel into the stone by a technique known as pecking. By repeatedly rapping the club stone with a smaller hard stone, you can get the desired shape.

Next, find a piece of wood that is the right length for you. A straight-grained hardwood is best. The length of the wood should feel comfortable in relation to the weight of the stone. Finally, lash the stone to the handle.

There are three techniques for lashing the stone to the handle: split handle, forked branch, and wrapped handle. The technique you use will depend on the type of handle you choose. See Figure 12-1.

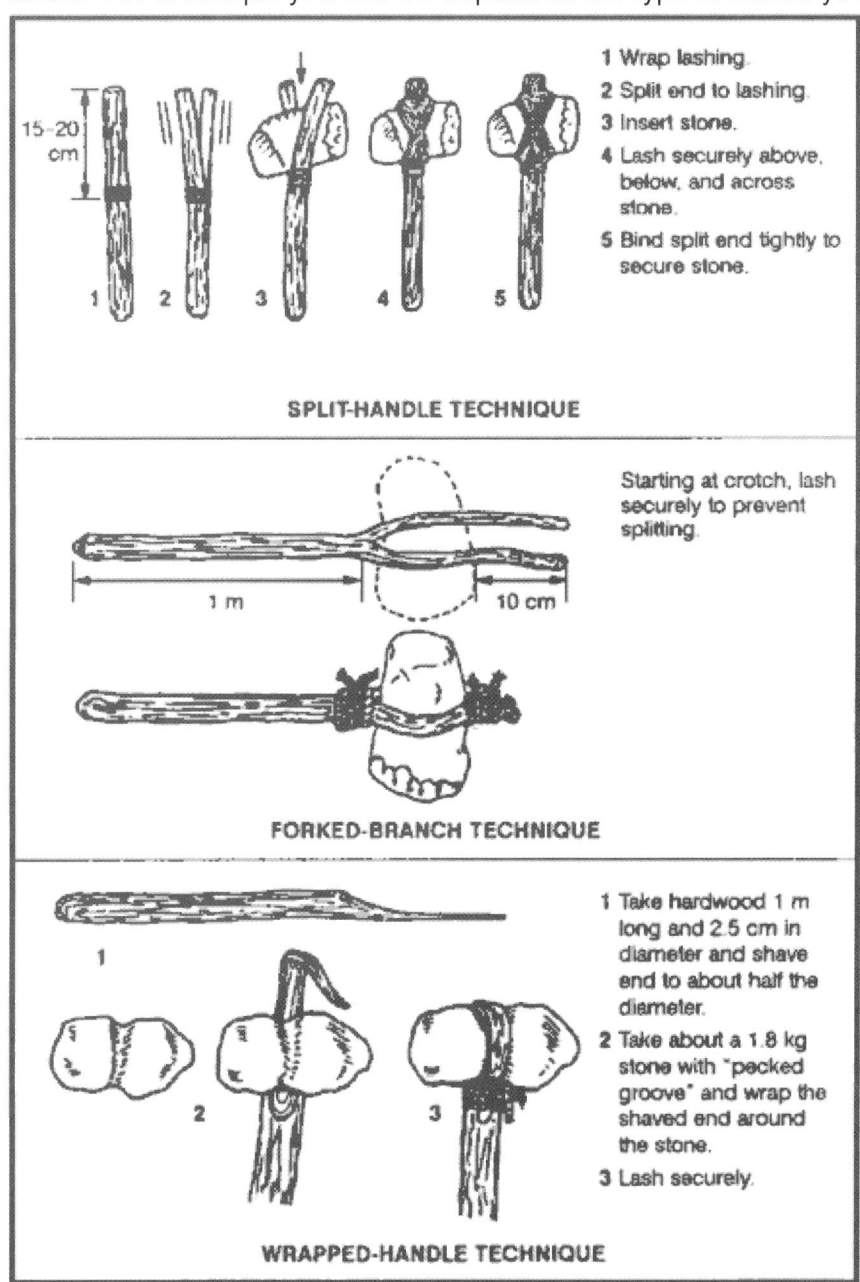

1 Wrap lashing.
2 Split end to lashing.
3 Insert stone.
4 Lash securely above, below, and across stone.
5 Bind split end tightly to secure stone.

15-20 cm

SPLIT-HANDLE TECHNIQUE

Starting at crotch, lash securely to prevent splitting.

1 m   10 cm

FORKED-BRANCH TECHNIQUE

1 Take hardwood 1 m long and 2.5 cm in diameter and shave end to about half the diameter.
2 Take about a 1.8 kg stone with "pecked groove" and wrap the shaved end around the stone.
3 Lash securely.

WRAPPED-HANDLE TECHNIQUE

Figure 12-1. Lashing clubs.

Sling Club

A sling club is another type of weighted club. A weight hangs 8 to 10 centimeters from the handle by a strong, flexible lashing (Figure 12-2). This type of club both extends the user's reach and multiplies the force of the blow.

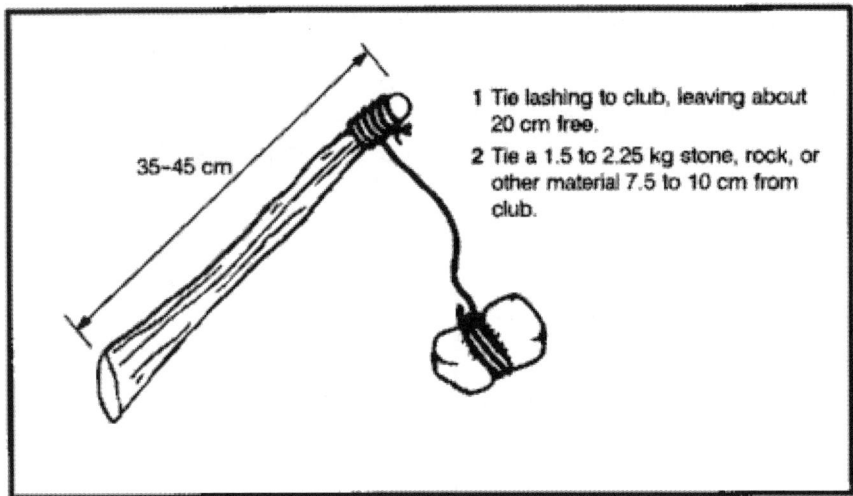

1 Tie lashing to club, leaving about 20 cm free.

2 Tie a 1.5 to 2.25 kg stone, rock, or other material 7.5 to 10 cm from club.

35–45 cm

Figure 12-2. Sling club.

EDGED WEAPONS

Knives, spear blades, and arrow points fall under the category of edged weapons. The following paragraphs will discuss the making of such weapons.

Knives

A knife has three basic functions. It can puncture, slash or chop, and cut. A knife is also an invaluable tool used to construct other survival items. You may find yourself without a knife or you may need another type knife or a spear. To improvise you can use stone, bone, wood, or metal to make a knife or spear blade.

Stone

To make a stone knife, you will need a sharp-edged piece of stone, a chipping tool, and a flaking tool. A chipping tool is a light, blunt-edged tool used to break off small pieces of stone. A flaking tool is a pointed tool used to break off thin, flattened pieces of stone. You can make a chipping tool from wood, bone, or metal, and a flaking tool from bone, antler tines, or soft iron (Figure 12-3).

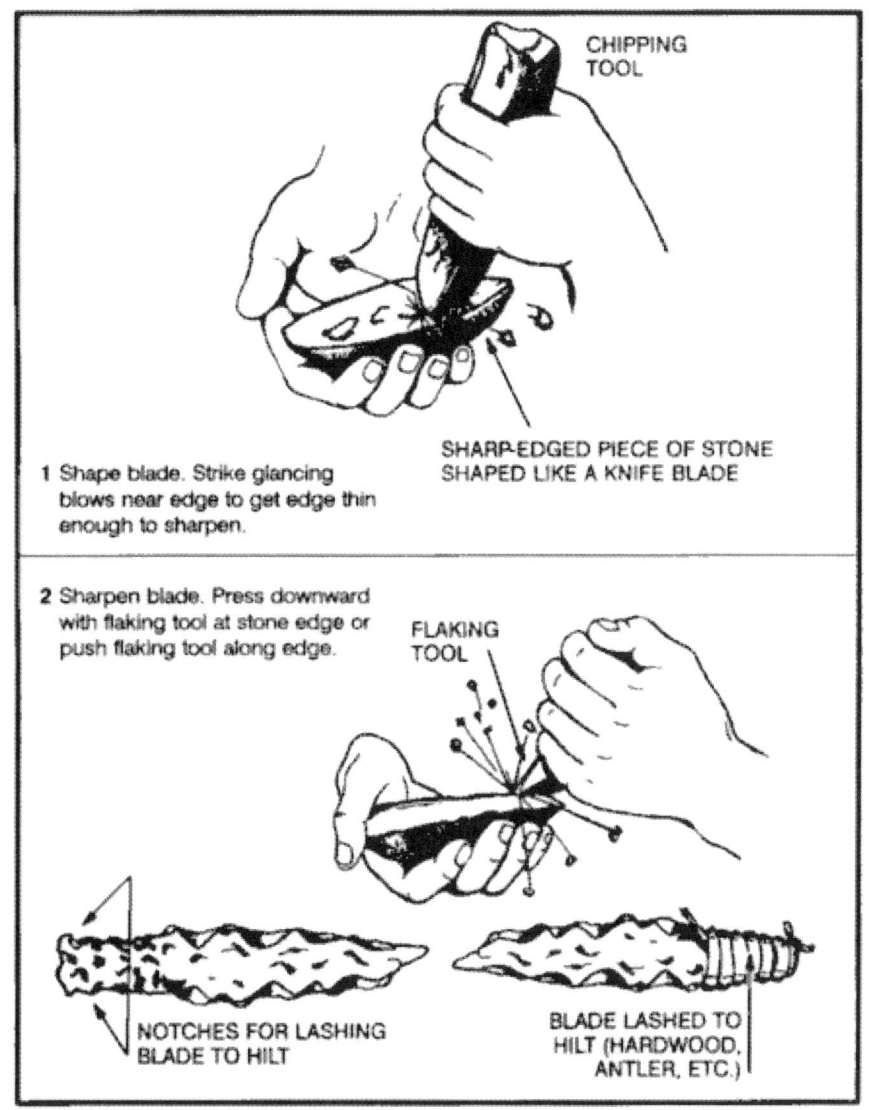

Figure 12-3. Making a stone knife.

Start making the knife by roughing out the desired shape on your sharp piece of stone, using the chipping tool. Try to make the knife fairly thin. Then, using the flaking tool, press it against the edges. This action will cause flakes to come off the opposite side of the edge, leaving a razor sharp edge. Use the flaking tool along the entire length of the edge you need to sharpen. Eventually, you will have a very sharp cutting edge that you can use as a knife.
Lash the blade to some type of hilt (Figure 12-3).

Note: Stone will make an excellent puncturing tool and a good chopping tool but will not hold a fine edge. Some stones such as chert or flint can have very fine edges.

Bone

You can also use bone as an effective field-expedient edged weapon. First, you will need to select a suitable bone. The larger bones, such as the leg bone of a deer or another medium-sized animal, are best. Lay the bone upon another hard object. Shatter the bone by hitting it with a heavy object, such as a rock. From the pieces, select a suitable pointed splinter. You can further shape and sharpen this splinter by rubbing it on a rough-surfaced rock. If the piece is too small to handle, you can still use it by adding a handle to it. Select a suitable piece of hardwood for a handle and lash the bone splinter securely to it.

Note: Use the bone knife only to puncture. It will not hold an edge and it may flake or break if used differently.

Wood

You can make field-expedient edged weapons from wood. Use these only to puncture. Bamboo is the only wood that will hold a suitable edge. To make a knife using wood, first select a straight-grained piece of hardwood that is about 30 centimeters long and 2.5 centimeters in diameter. Fashion the blade about 15 centimeters long. Shave it down to a point. Use only the straight-grained portions of the wood. Do not use the core or pith, as it would make a weak point.

Harden the point by a process known as fire hardening. If a fire is possible, dry the blade portion over the fire slowly until lightly charred. The drier the wood, the harder the point. After lightly charring the blade portion, sharpen it on a coarse stone. If using bamboo and after fashioning the blade, remove any other wood to make the blade thinner from the inside portion of the bamboo. Removal is done this way because bamboo's hardest part is its outer layer. Keep as much of this layer as possible to ensure the hardest blade possible. When charring bamboo over a fire, char only the inside wood; do not char the outside.

Metal

Metal is the best material to make field-expedient edged weapons. Metal, when properly designed, can fulfill a knife's three uses--puncture, slice or chop, and cut. First, select a suitable piece of metal, one that most resembles the desired end product. Depending on the size and original shape, you can obtain a point and cutting edge by rubbing the metal on a rough-surfaced stone. If the metal is soft enough, you can hammer out one edge while the metal is cold. Use a suitable flat, hard surface as an anvil and a smaller, harder object of stone or metal as a hammer to hammer out the edge. Make a knife handle from wood, bone, or other material that will protect your hand.

Other Materials

You can use other materials to produce edged weapons. Glass is a good alternative to an edged weapon or tool, if no other material is available. Obtain a suitable piece in the same manner as described for bone. Glass has a natural edge but is less durable for heavy work. You can also sharpen plastic--if it is thick enough or hard enough--into a durable point for puncturing.

Spear Blades

To make spears, use the same procedures to make the blade that you used to make a knife blade. Then select a shaft (a straight sapling) 1.2 to 1.5 meters long. The length should allow you to handle the spear easily and effectively. Attach the spear blade to the shaft using lashing. The preferred method is to split the handle, insert the blade, then wrap or lash it tightly. You can use other materials without adding a blade. Select a 1.2-to 1.5-meter long straight hardwood shaft and shave one end to a point. If possible, fire harden the point. Bamboo also makes an excellent spear. Select a piece 1.2 to 1.5 meters long. Starting 8 to 10 centimeters back from the end used as the point, shave down the end at a 45-degree angle (Figure 12-4). Remember, to sharpen the edges, shave only the inner portion.

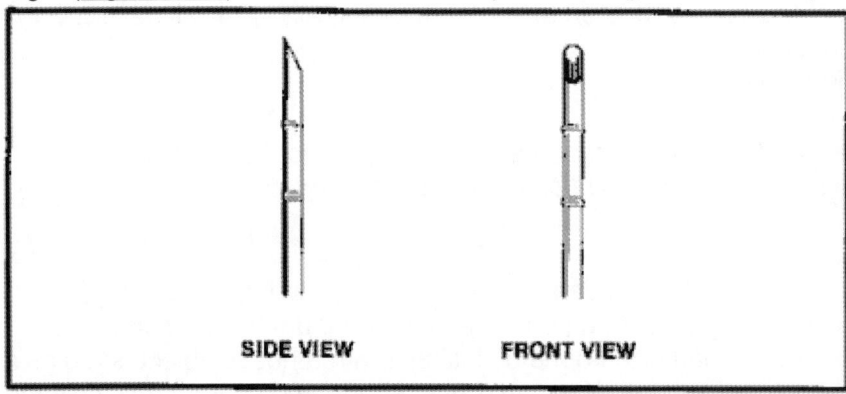

Figure 12-4. Bamboo spear.

Arrow Points

To make an arrow point, use the same procedures for making a stone knife blade. Chert, flint, and shell-type stones are best for arrow points. You can fashion bone like stone--by flaking. You can make an efficient arrow point using broken glass.

OTHER EXPEDIENT WEAPONS

You can make other field-expedient weapons such as the throwing stick, archery equipment, and the bola.

Throwing Stick

The throwing stick, commonly known as the rabbit stick, is very effective against small game (squirrels, chipmunks, and rabbits). The rabbit stick itself is a blunt stick, naturally curved at about a 45-degree angle. Select a stick with the desired angle from heavy hardwood such as oak. Shave off two opposite sides so that the stick is flat like a boomerang (Figure 12-5). You must practice the throwing technique for accuracy and speed. First, align the target by extending the nonthrowing arm in line with the mid to lower section of the target. Slowly and repeatedly raise the throwing arm up and back until the throwing stick crosses the back at about a 45-degree angle or is in line with the nonthrowing hip. Bring the throwing arm forward until it is just slightly above and parallel to the nonthrowing arm. This will be the throwing stick's release point. Practice slowly and repeatedly to attain accuracy.

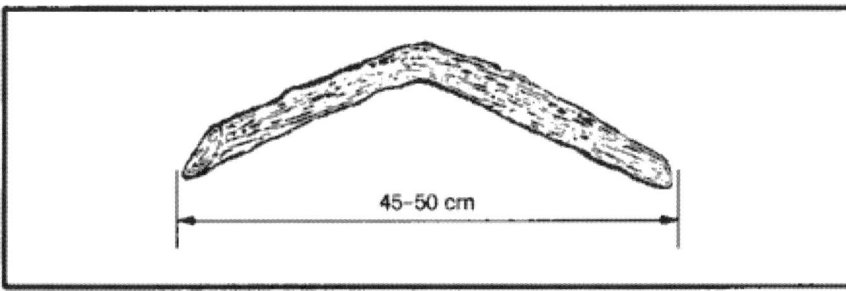

45-50 cm

Figure 12-5. Rabbit stick.

Archery Equipment

You can make a bow and arrow (Figure 12-6) from materials available in your survival area. To make a bow, use the procedure described under Killing Devices in Chapter 8.

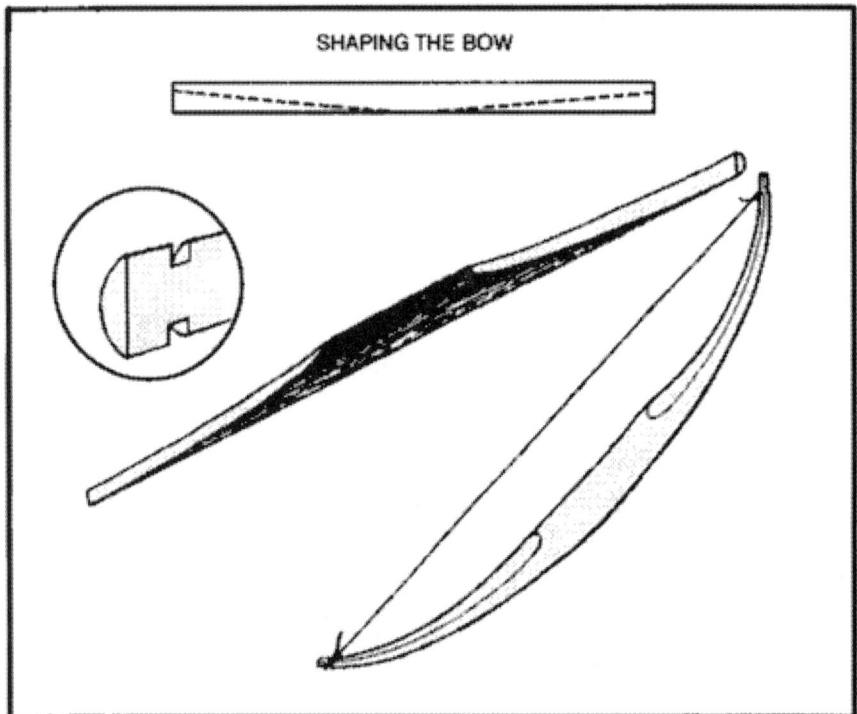

Figure 12-6. Archery equipment.

While it may be relatively simple to make a bow and arrow, it is not easy to use one. You must practice using it a long time to be reasonably sure that you will hit your target. Also, a field-expedient bow will not last very long before you have to make a new one. For the time and effort involved, you may well decide to use another type of field-expedient weapon.

Bola

The bola is another field-expedient weapon that is easy to make (Figure 12-7). It is especially effective for capturing running game or low-flying fowl in a flock. To use the bola, hold it by the center knot and twirl it above your head. Release the knot so that the bola flies toward your target. When you release the bola, the weighted cords will separate. These cords will wrap around and immobilize the fowl or animal that you hit.

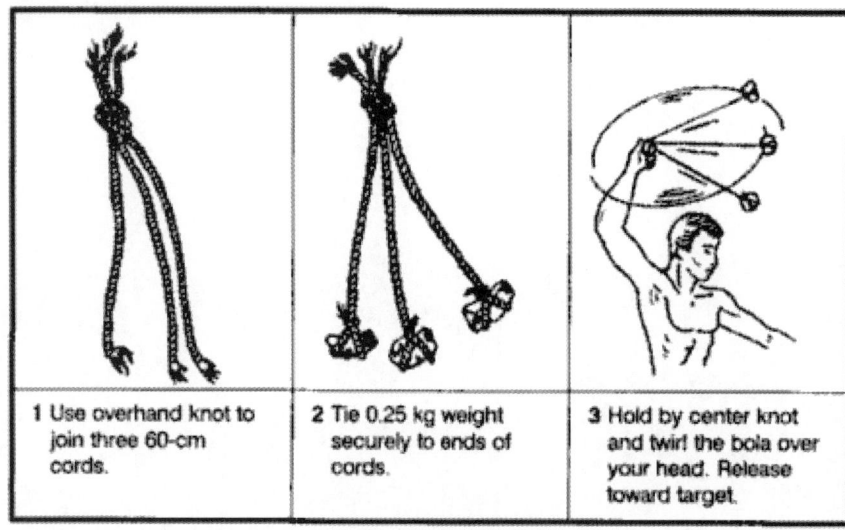

1 Use overhand knot to join three 60-cm cords.

2 Tie 0.25 kg weight securely to ends of cords.

3 Hold by center knot and twirl the bola over your head. Release toward target.

Figure 12-7. Bola.

ESSENTIAL ARMY MANUAL
(129) ARMY SURVIVAL MANUAL

LASHING AND CORDAGE

Many materials are strong enough for use as lashing and cordage. A number of natural and man-made materials are available in a survival situation. For example, you can make a cotton web belt much more useful by unraveling it. You can then use the string for other purposes (fishing line, thread for sewing, and lashing).

Natural Cordage Selection

Before making cordage, there are a few simple tests you can do to determine you material's suitability. First, pull on a length of the material to test for strength. Next, twist it between your fingers and roll the fibers together. If it withstands this handling and does not snap apart, tie an overhand knot with the fibers and gently tighten. If the knot does not break, the material is usable. Figure 12-8 shows various methods of making cordage.

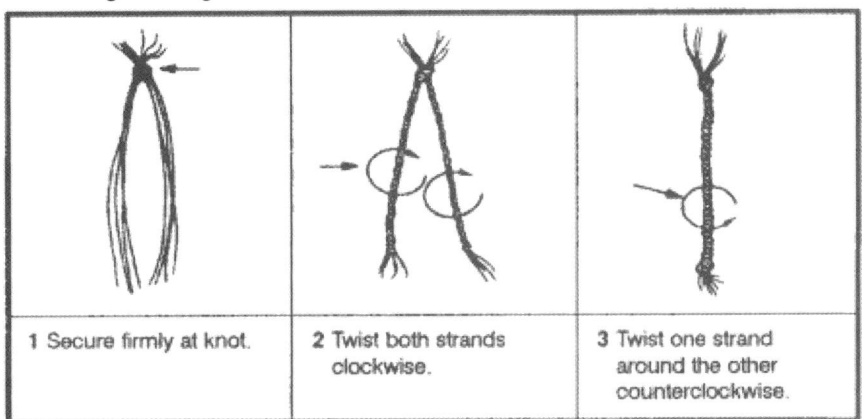

Figure 12-8. Making lines from plant fibers.

Lashing Material

The best natural material for lashing small objects is sinew. You can make sinew from the tendons of large game, such as deer. Remove the tendons from the game and dry them completely. Smash the dried tendons so that they separate into fibers. Moisten the fibers and twist them into a continuous strand. If you need stronger lashing material, you can braid the strands. When you use sinew for small lashings, you do not need knots as the moistened sinew is sticky and it hardens when dry.

You can shred and braid plant fibers from the inner bark of some trees to make cord. You can use the linden, elm, hickory, white oak, mulberry, chestnut, and red and white cedar trees. After you make the cord, test it to be sure it is strong enough for your purpose. You can make these materials stronger by braiding several strands together.

You can use rawhide for larger lashing jobs. Make rawhide from the skins of medium or large game. After skinning the animal, remove any excess fat and any pieces of meat from the skin. Dry the skin completely. You do not need to stretch it as long as there are no folds to trap moisture. You do not have to remove the hair from the skin. Cut the skin while it is dry. Make cuts about 6 millimeters wide. Start from the center of the hide and make one continuous circular cut, working clockwise to the hide's outer edge. Soak the rawhide for 2 to 4 hours or until it is soft. Use it wet, stretching it as much as possible while applying it. It will be strong and durable when it dries.

RUCKSACK CONSTRUCTION

The materials for constructing a rucksack or pack are almost limitless. You can use wood, bamboo, rope, plant fiber, clothing, animal skins, canvas, and many other materials to make a pack.

There are several construction techniques for rucksacks. Many are very elaborate, but those that are simple and easy are often the most readily made in a survival situation.

Horseshoe Pack

This pack is simple to make and use and relatively comfortable to carry over one shoulder. Lay available square-shaped material, such as poncho, blanket, or canvas, flat on the ground. Lay items on one edge of the material. Pad the hard items. Roll the material (with the items) toward the opposite edge and tie both ends securely. Add extra ties along the length of the bundle. You can drape the pack over one shoulder with a line connecting the two ends (Figure 12-9).

Figure 12-9. Horseshoe pack.

Square Pack

This pack is easy to construct if rope or cordage is available. Otherwise, you must first make cordage. To make this pack, construct a square frame from bamboo, limbs, or sticks. Size will vary for each person and the amount of equipment carried (Figure 12-10).

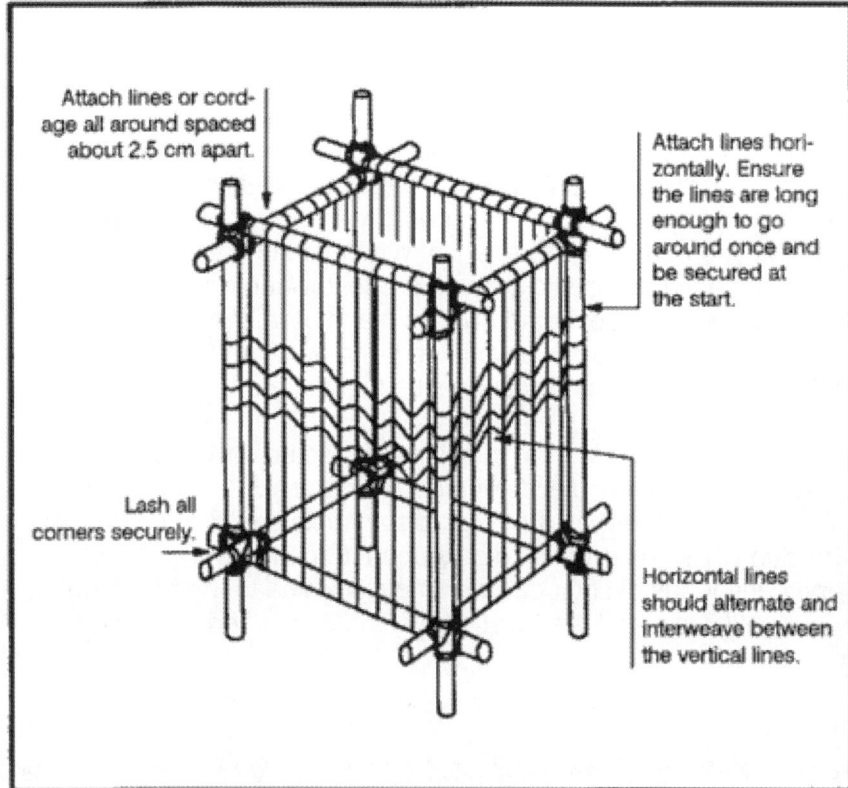

Figure 12-10. Square pack.

CLOTHING AND INSULATION

You can use many materials for clothing and insulation. Both man-made materials, such as parachutes, and natural materials, such as skins and plant materials, are available and offer significant protection.

Parachute Assembly

Consider the entire parachute assembly as a resource. Use every piece of material and hardware, to include the canopy, suspension lines, connector snaps, and parachute harness. Before disassembling the parachute, consider all of your survival requirements and plan to use different portions of the parachute accordingly. For example, consider shelter requirements, need for a rucksack, and so on, in addition to clothing or insulation needs.

Animal Skins

The selection of animal skins in a survival situation will most often be limited to what you manage to trap or hunt. However, if there is an abundance of wildlife, select the hides of larger animals with heavier coats and large fat content. Do not use the skins of infected or diseased animals if at all possible. Since they live in the wild, animals are carriers of pests such as ticks, lice, and fleas. Because of these pests, use water to thoroughly clean any skin obtained from any animal. If water is not available, at least shake out the skin thoroughly. As with rawhide, lay out the skin, and remove all fat and meat. Dry the skin completely. Use the hind quarter joint areas to make shoes and mittens or socks. Wear the hide with the fur to the inside for its insulating factor.

Plant Fibers

Several plants are sources of insulation from cold. Cattail is a marshland plant found along lakes, ponds, and the backwaters of rivers. The fuzz on the tops of the stalks forms dead air spaces and makes a good down-like insulation when placed between two pieces of material. Milkweed has pollenlike seeds that act as good insulation. The husk fibers from coconuts are very good for weaving ropes and, when dried, make excellent tinder and insulation.

COOKING AND EATING UTENSILS

Many materials may be used to make equipment for the cooking, eating, and storing of food.

Bowls

Use wood, bone, horn, bark, or other similar material to make bowls. To make wooden bowls, use a hollowed out piece of wood that will hold your food and enough water to cook it in. Hang the wooden container over the fire and add hot rocks to the water and food. Remove the rocks as they cool and add more hot rocks until your food is cooked.

> **CAUTION**
> Do not use rocks with air pockets, such as limestone and sandstone. They may explode while heating in the fire.

You can also use this method with containers made of bark or leaves. However, these containers will burn above the waterline unless you keep them moist or keep the fire low.
A section of bamboo works very well, if you cut out a section between two sealed joints (Figure 12-11).

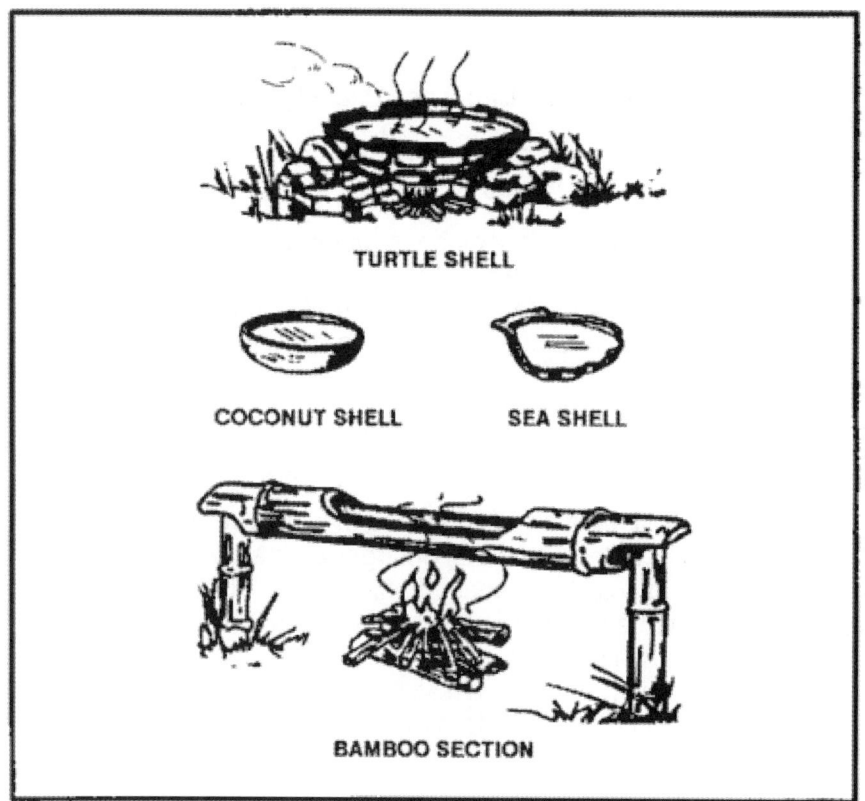

Figure 12-11. Containers for boiling food.

CAUTION
A sealed section of bamboo will explode if heated because of trapped air and water in the section.

Forks, Knives, and Spoons

Carve forks, knives, and spoons from nonresinous woods so that you do not get a wood resin aftertaste or do not taint the food. Nonresinous woods include oak, birch, and other hardwood trees.

Note: Do not use those trees that secrete a syrup or resinlike liquid on the bark or when cut.

Pots

You can make pots from turtle shells or wood. As described with bowls, using hot rocks in a hollowed out piece of wood is very effective. Bamboo is the best wood for making cooking containers.
To use turtle shells, first thoroughly boil the upper portion of the shell. Then use it to heat food and water over a flame (Figure 12-11).

Water Bottles

Make water bottles from the stomachs of larger animals. Thoroughly flush the stomach out with water, then tie off the bottom. Leave the top open, with some means of fastening it closed.

CHAPTER 13 - DESERT SURVIVAL

To survive and evade in arid or desert areas, you must understand and prepare for the environment you will face. You must determine your equipment needs, the tactics you will use, and how the environment will affect you and your tactics. Your survival will depend upon your knowledge of the terrain, basic climatic elements, your ability to cope with these elements, and your will to survive.

TERRAIN

Most arid areas have several types of terrain. The five basic desert terrain types are--

- Mountainous (High Altitude).
- Rocky plateau.
- Sand dunes.
- Salt marshes.
- Broken, dissected terrain ("gebel" or "wadi").

Desert terrain makes movement difficult and demanding. Land navigation will be extremely difficult as there may be very few landmarks. Cover and concealment may be very limited; therefore, the threat of exposure to the enemy remains constant.

Mountain Deserts

Scattered ranges or areas of barren hills or mountains separated by dry, flat basins characterize mountain deserts. High ground may rise gradually or abruptly from flat areas to several thousand meters above sea level. Most of the infrequent rainfall occurs on high ground and runs off rapidly in the form of flash floods. These floodwaters erode deep gullies and ravines and deposit sand and gravel around the edges of the basins. Water rapidly evaporates, leaving the land as barren as before, although there may be short-lived vegetation. If enough water enters the basin to compensate for the rate of evaporation, shallow lakes may develop, such as the Great Salt Lake in Utah, or the Dead Sea. Most of these lakes have a high salt content.

Rocky Plateau Deserts

Rocky plateau deserts have relatively slight relief interspersed with extensive flat areas with quantities of solid or broken rock at or near the surface. There may be steep-walled, eroded valleys, known as wadis in the Middle East and arroyos or canyons in the United States and Mexico. Although their flat bottoms may be superficially attractive as assembly areas, the narrower valleys can be extremely dangerous to men and material due to flash flooding after rains. The Golan Heights is an example of a rocky plateau desert.

Sandy or Dune Deserts

Sandy or dune deserts are extensive flat areas covered with sand or gravel. "Flat" is a relative term, as some areas may contain sand dunes that are over 300 meters high and 16 to 24 kilometers long. Trafficability in such terrain will depend on the windward or leeward slope of the dunes and the texture of the sand. Other areas, however, may be flat for 3,000 meters and more. Plant life may vary from none to scrub over 2 meters high. Examples of this type of desert include the edges of the Sahara, the empty quarter of the Arabian Desert, areas of California and New Mexico, and the Kalahari in South Africa.

Salt Marshes

Salt marshes are flat, desolate areas, sometimes studded with clumps of grass but devoid of other vegetation. They occur in arid areas where rainwater has collected, evaporated, and left large deposits of alkali salts and water with a high salt concentration. The water is so salty it is undrinkable. A crust that may be 2.5 to 30 centimeters thick forms over the saltwater.
In arid areas there are salt marshes hundreds of kilometers square. These areas usually support many insects, most of which bite. Avoid salt marshes. This type of terrain is highly corrosive to boots, clothing, and skin. A good example is the Shat-el-Arab waterway along the Iran-Iraq border.

Broken Terrain

All arid areas contain broken or highly dissected terrain. Rainstorms that erode soft sand and carve out canyons form this terrain. A wadi may range from 3 meters wide and 2 meters deep to several hundred meters wide and deep. The direction it takes varies as much as its width and depth. It twists and turns and forms a mazelike pattern. A wadi will give you good cover and concealment, but do not try to move through it because it is very difficult terrain to negotiate.

ENVIRONMENTAL FACTORS

Surviving and evading the enemy in an arid area depends on what you know and how prepared you are for the environmental conditions you will face. Determine what equipment you will need, the tactics you will use, and the environment's impact on them and you.
In a desert area there are seven environmental factors that you must consider--

- Low rainfall.
- Intense sunlight and heat.
- Wide temperature range.
- Sparse vegetation.
- High mineral content near ground surface.
- Sandstorms.
- Mirages.

Low Rainfall

Low rainfall is the most obvious environmental factor in an arid area. Some desert areas receive less than 10 centimeters of rain annually, and this rain comes in brief torrents that quickly run off the ground surface. You cannot survive long without water in high desert temperatures. In a desert survival situation, you must first consider "How much water do I have?" and "Where are other water sources?"

Intense Sunlight and Heat

Intense sunlight and heat are present in all arid areas. Air temperature can rise as high as 60 degrees C (140 degrees F) during the day. Heat gain results from direct sunlight, hot blowing winds, reflective heat (the sun's rays bouncing off the sand), and conductive heat from direct contact with the desert sand and rock (Figure 13-1).

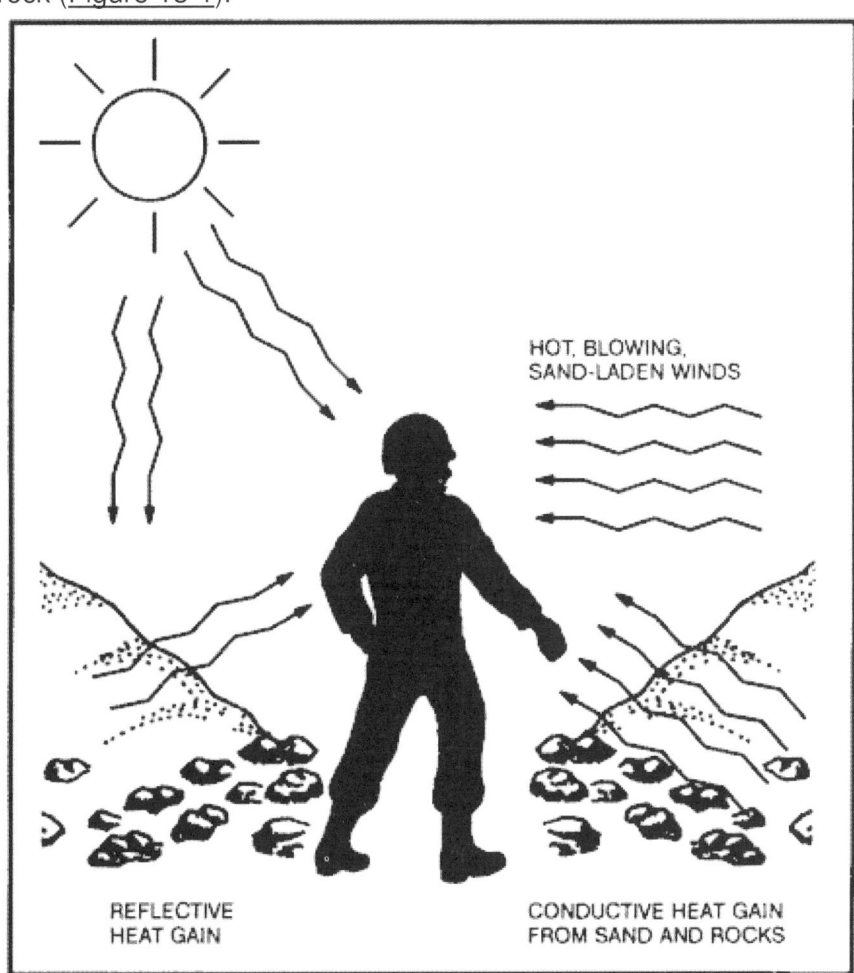

HOT, BLOWING,
SAND-LADEN WINDS

REFLECTIVE
HEAT GAIN

CONDUCTIVE HEAT GAIN
FROM SAND AND ROCKS

Figure 13-1. Types of heat gain.

The temperature of desert sand and rock averages 16 to 22 degrees C (30 to 40 degrees F) more than that of the air. For instance, when the air temperature is 43 degrees C (110 degrees F), the sand temperature may be 60 degrees C (140 degrees F).

Intense sunlight and heat increase the body's need for water. To conserve your body fluids and energy, you will need a shelter to reduce your exposure to the heat of the day. Travel at night to lessen your use of water.

Radios and sensitive items of equipment exposed to direct intense sunlight will malfunction.

Wide Temperature Range

Temperatures in arid areas may get as high as 55 degrees C during the day and as low as 10 degrees C during the night. The drop in temperature at night occurs rapidly and will chill a person who lacks warm clothing and is unable to move about. The cool evenings and nights are the best times to work or travel. If your plan is to rest at night, you will find a wool sweater, long underwear, and a wool stocking cap extremely helpful.

Sparse Vegetation

Vegetation is sparse in arid areas. You will therefore have trouble finding shelter and camouflaging your movements. During daylight hours large areas of terrain are visible and easily controlled by a small opposing force.

If traveling in hostile territory, follow the principles of desert camouflage--

- Hide or seek shelter in dry washes (wadis) with thicker growths of vegetation and cover from oblique observation.
- Use the shadows cast from brush, rocks, or outcropping. The temperature in shaded areas will be 11 to 17 degrees C cooler than the air temperature.
- Cover objects that will reflect the light from the sun.

Before moving, survey the area for sites that provide cover and concealment. You will have trouble estimating distance. The emptiness of desert terrain causes most people to underestimate distance by a factor of three: What appears to be 1 kilometer away is really 3 kilometers away.

High Mineral Content

All arid regions have areas where the surface soil has a high mineral content (borax, salt, alkali, and lime). Material in contact with this soil wears out quickly, and water in these areas is extremely hard and undrinkable. Wetting your uniform in such water to cool off may cause a skin rash. The Great Salt Lake area in Utah is an example of this type of mineral-laden water and soil. There is little or no plant life; there-fore, shelter is hard to find. Avoid these areas if possible.

Sandstorms

Sandstorms (sand-laden winds) occur frequently in most deserts. The "Seistan" desert wind in Iran and Afghanistan blows constantly for up to 120 days. Within Saudi Arabia, winds average 3.2 to 4.8 kilometers per hour (kph) and can reach 112 to 128 kph in early afternoon. Expect major sandstorms and dust storms at least once a week.
The greatest danger is getting lost in a swirling wall of sand. Wear goggles and cover your mouth and nose with cloth. If natural shelter is unavailable, mark your direction of travel, lie down, and sit out the storm.
Dust and wind-blown sand interfere with radio transmissions. Therefore, be ready to use other means for signaling, such as pyrotechnics, signal mirrors, or marker panels, if available.

Mirages

Mirages are optical phenomena caused by the refraction of light through heated air rising from a sandy or stony surface. They occur in the interior of the desert about 10 kilometers from the coast. They make objects that are 1.5 kilometers or more away appear to move.
This mirage effect makes it difficult for you to identify an object from a distance. It also blurs distant range contours so much that you feel surrounded by a sheet of water from which elevations stand out as "islands."
The mirage effect makes it hard for a person to identify targets, estimate range, and see objects clearly. However, if you can get to high ground (3 meters or more above the desert floor), you can get above the superheated air close to the ground and overcome the mirage effect. Mirages make land navigation difficult because they obscure natural features. You can survey the area at dawn, dusk, or by moonlight when there is little likelihood of mirage.
Light levels in desert areas are more intense than in other geographic areas. Moonlit nights are usually crystal clear, winds die down, haze and glare disappear, and visibility is excellent. You can see lights, red flash-lights, and blackout lights at great distances. Sound carries very far.
Conversely, during nights with little moonlight, visibility is extremely poor. Traveling is extremely hazardous. You must avoid getting lost, falling into ravines, or stumbling into enemy positions. Movement during such a night is practical only if you have a compass and have spent the day in a shelter, resting, observing and memorizing the terrain, and selecting your route.

NEED FOR WATER

The subject of man and water in the desert has generated considerable interest and confusion since the early days of World War II when the U. S. Army was preparing to fight in North Africa. At one time the U. S. Army thought it could condition men to do with less water by progressively reducing their water supplies during training. They called it water discipline. It caused hundreds of heat casualties.

A key factor in desert survival is understanding the relationship between physical activity, air temperature, and water consumption. The body requires a certain amount of water for a certain level of activity at a certain temperature. For example, a person performing hard work in the sun at 43 degrees C requires 19 liters of water daily. Lack of the required amount of water causes a rapid decline in an individual's ability to make decisions and to perform tasks efficiently.

Your body's normal temperature is 36.9 degrees C (98.6 degrees F). Your body gets rid of excess heat (cools off) by sweating. The warmer your body becomes--whether caused by work, exercise, or air temperature--the more you sweat. The more you sweat, the more moisture you lose. Sweating is the principal cause of water loss. If a person stops sweating during periods of high air temperature and heavy work or exercise, he will quickly develop heat stroke. This is an emergency that requires immediate medical attention.

Figure 13-2 shows daily water requirements for various levels of work. Understanding how the air temperature and your physical activity affect your water requirements allows you to take measures to get the most from your water supply. These measures are--

- Find shade! Get out of the sun!
- Place something between you and the hot ground.
- Limit your movements!
- Conserve your sweat. Wear your complete uniform to include T-shirt. Roll the sleeves down, cover your head, and protect your neck with a scarf or similar item. These steps will protect your body from hot-blowing winds and the direct rays of the sun. Your clothing will absorb your sweat, keeping it against your skin so that you gain its full cooling effect. By staying in the shade quietly, fully clothed, not talking, keeping your mouth closed, and breathing through your nose, your water requirement for survival drops dramatically.
- If water is scarce, do not eat. Food requires water for digestion; therefore, eating food will use water that you need for cooling.

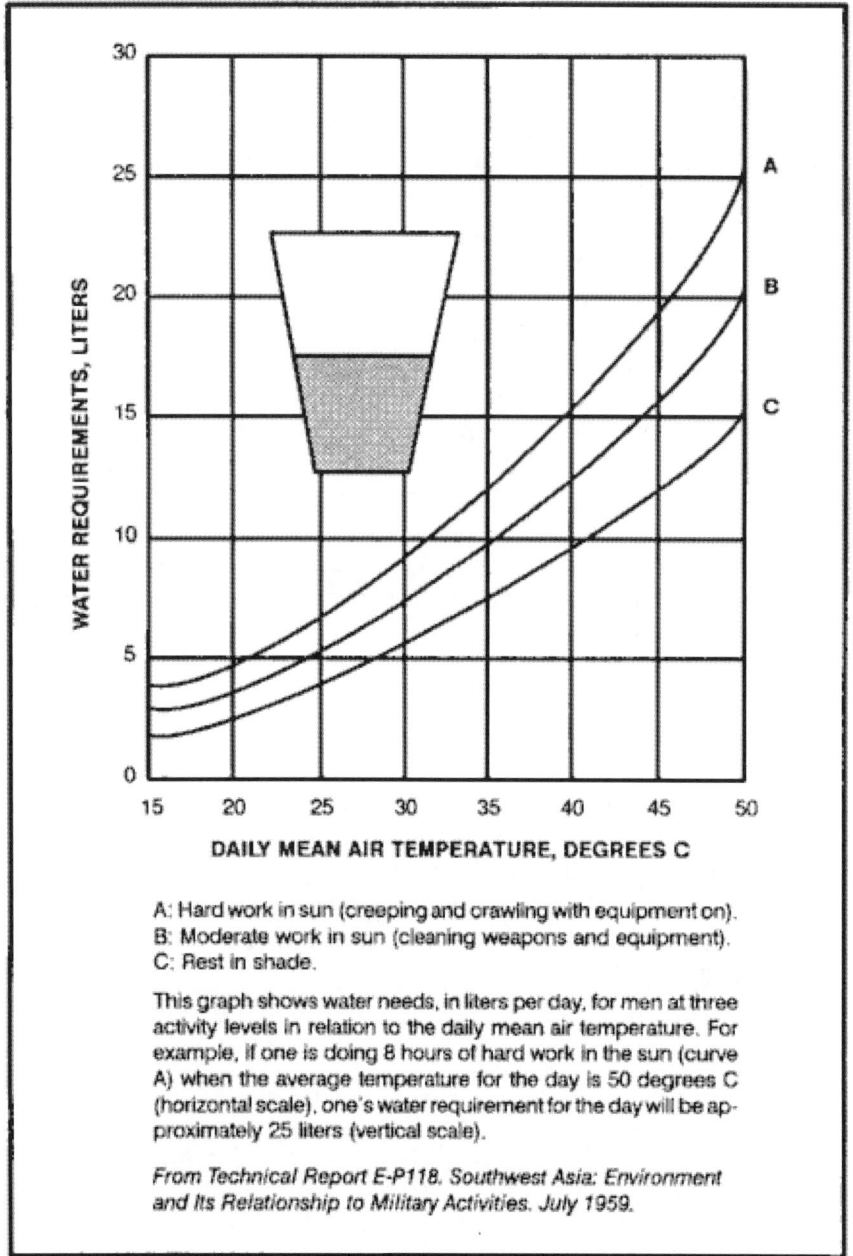

A: Hard work in sun (creeping and crawling with equipment on).
B: Moderate work in sun (cleaning weapons and equipment).
C: Rest in shade.

This graph shows water needs, in liters per day, for men at three activity levels in relation to the daily mean air temperature. For example, if one is doing 8 hours of hard work in the sun (curve A) when the average temperature for the day is 50 degrees C (horizontal scale), one's water requirement for the day will be approximately 25 liters (vertical scale).

From Technical Report E-P118. Southwest Asia: Environment and Its Relationship to Military Activities. July 1959.

Figure 13-2. Daily water requirements for three levels of activity.

Thirst is not a reliable guide for your need for water. A person who uses thirst as a guide will drink only two-thirds of his daily water requirement. To prevent this "voluntary" dehydration, use the following guide:

- At temperatures below 38 degrees C, drink 0.5 liter of water every hour.
- At temperatures above 38 degrees C, drink 1 liter of water every hour.

Drinking water at regular intervals helps your body remain cool and decreases sweating. Even when your water supply is low, sipping water constantly will keep your body cooler and reduce water loss through sweating. Conserve your fluids by reducing activity during the heat of day. **Do not** ration your water! If you try to ration water, you stand a good chance of becoming a heat casualty.

HEAT CASUALTIES

Your chances of becoming a heat casualty as a survivor are great, due to injury, stress, and lack of critical items of equipment. Following are the major <u>types</u> of heat casualties and their treatment *when little water and no medical help are available.*

Heat Cramps

The loss of salt due to excessive sweating causes heat cramps. Symptoms are moderate to severe muscle cramps in legs, arms, or abdomen. These symptoms may start as a mild muscular discomfort. You should now stop all activity, get in the shade, and drink water. If you fail to recognize the early symptoms and continue your physical activity, you will have severe muscle cramps and pain. Treat as for heat exhaustion, below.

Heat Exhaustion

A large loss of body water and salt causes heat exhaustion. Symptoms are headache, mental confusion, irritability, excessive sweating, weakness, dizziness, cramps, and pale, moist, cold (clammy) skin. Immediately get the patient under shade. Make him lie on a stretcher or similar item about 45 centimeters off the ground. Loosen his clothing. Sprinkle him with water and fan him. Have him drink small amounts of water every 3 minutes. Ensure he stays quiet and rests.

Heat Stroke

A severe heat injury caused by extreme loss of water and salt and the body's inability to cool itself. The patient may die if not cooled immediately. Symptoms are the lack of sweat, hot and dry skin, headache, dizziness, fast pulse, nausea and vomiting, and mental confusion leading to unconsciousness. Immediately get the person to shade. Lay him on a stretcher or similar item about 45 centimeters off the ground. Loosen his clothing. Pour water on him (it does not matter if the water is polluted or brackish) and fan him. Massage his arms, legs, and body. If he regains consciousness, let him drink small amounts of water every 3 minutes.

PRECAUTIONS

In a desert survival and evasion situation, it is unlikely that you will have a medic or medical supplies with you to treat heat injuries. Therefore, take extra care to avoid heat injuries. Rest during the day. Work during the cool evenings and nights. Use a buddy system to watch for heat injury, and observe the following guidelines:

- Make sure you tell someone where you are going and when you will return.
- Watch for signs of heat injury. If someone complains of tiredness or wanders away from the group, he may be a heat casualty.
- Drink water at least once an hour.
- Get in the shade when resting; do not lie directly on the ground.
- Do not take off your shirt and work during the day.
- Check the color of your urine. A light color means you are drinking enough water, a dark color means you need to drink more.

DESERT HAZARDS

There are several hazards unique to desert survival. These include insects, snakes, thorned plants and cacti, contaminated water, sunburn, eye irritation, and climatic stress.
Insects of almost every type abound in the desert. Man, as a source of water and food, attracts lice, mites, wasps, and flies. They are extremely unpleasant and may carry diseases. Old buildings, ruins, and caves are favorite habitats of spiders, scorpions, centipedes, lice, and mites. These areas provide protection from the elements and also attract other wild-life. Therefore, take extra care when staying in these areas. Wear gloves at all times in the desert. Do not place your hands anywhere without first looking to see what is there. Visually inspect an area before sitting or lying down. When you get up, shake out and inspect your boots and clothing. All desert areas have snakes. They inhabit ruins, native villages, garbage dumps, caves, and natural rock outcropping that offer shade. Never go barefoot or walk through these areas without carefully inspecting them for snakes. Pay attention to where you place your

feet and hands. Most snakebites result from stepping on or handling snakes. Avoid them. Once you see a snake, give it a wide berth.

CHAPTER 14 - TROPICAL SURVIVAL

Most people think of the tropics as a huge and forbidding tropical rain forest through which every step taken must be hacked out, and where every inch of the way is crawling with danger. Actually, over half of the land in the tropics is cultivated in some way.

A knowledge of field skills, the ability to improvise, and the application of the principles of survival will increase the prospects of survival. Do not be afraid of being alone in the jungle; fear will lead to panic. Panic will lead to exhaustion and decrease your chance of survival.

Everything in the jungle thrives, including disease germs and parasites that breed at an alarming rate. Nature will provide water, food, and plenty of materials to build shelters.

Indigenous peoples have lived for millennia by hunting and gathering. However, it will take an outsider some time to get used to the conditions and the nonstop activity of tropical survival.

TROPICAL WEATHER

High temperatures, heavy rainfall, and oppressive humidity characterize equatorial and subtropical regions, except at high altitudes. At low altitudes, temperature variation is seldom less than 10 degrees C and is often more than 35 degrees C. At altitudes over 1,500 meters, ice often forms at night. The rain has a cooling effect, but when it stops, the temperature soars.

Rainfall is heavy, often with thunder and lightning. Sudden rain beats on the tree canopy, turning trickles into raging torrents and causing rivers to rise. Just as suddenly, the rain stops. Violent storms may occur, usually toward the end of the summer months.

Hurricanes, cyclones, and typhoons develop over the sea and rush inland, causing tidal waves and devastation ashore. In choosing campsites, make sure you are above any potential flooding. Prevailing winds vary between winter and summer. The dry season has rain once a day and the monsoon has continuous rain. In Southeast Asia, winds from the Indian Ocean bring the monsoon, but it is dry when the wind blows from the landmass of China.

Tropical day and night are of equal length. Darkness falls quickly and daybreak is just as sudden.

JUNGLE TYPES

There is no standard jungle. The tropical area may be any of the following:

- Rain forests.
- Secondary jungles.
- Semievergreen seasonal and monsoon forests.

- Scrub and thorn forests.
- Savannas.
- Saltwater swamps.
- Freshwater swamps.

Tropical Rain Forests

The climate varies little in rain forests. You find these forests across the equator in the Amazon and Congo basins, parts of Indonesia, and several Pacific islands. Up to 3.5 meters of rain fall evenly throughout the year. Temperatures range from about 32 degrees C in the day to 21 degrees C at night. There are five layers of vegetation in this jungle (Figure 14-1). Where untouched by man, jungle trees rise from buttress roots to heights of 60 meters. Below them, smaller trees produce a canopy so thick that little light reaches the jungle floor. Seedlings struggle beneath them to reach light, and masses of vines and lianas twine up to the sun. Ferns, mosses, and herbaceous plants push through a thick carpet of leaves, and a great variety of fungi grow on leaves and fallen tree trunks.

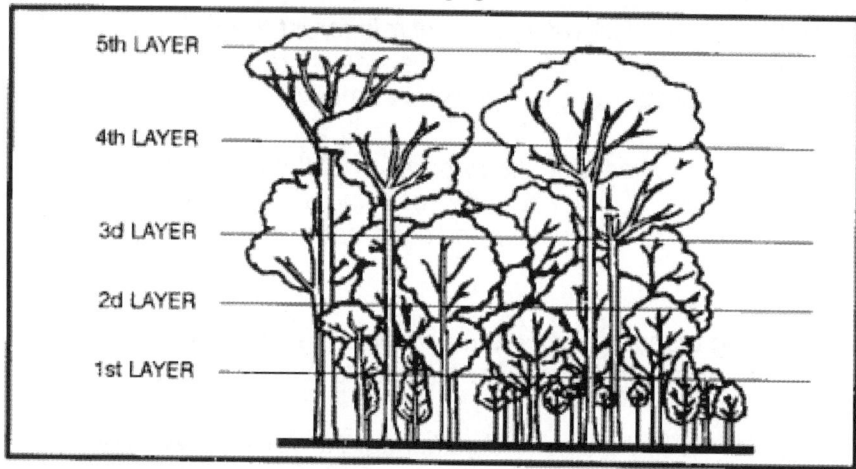

Figure 14-1. Five layers of tropical rain forest vegetation.

Because of the lack of light on the jungle floor, there is little undergrowth to hamper movement, but dense growth limits visibility to about 50 meters. You can easily lose your sense of direction in this jungle, and it is extremely hard for aircraft to see you.

Secondary Jungles

Secondary jungle is very similar to rain forest. Prolific growth, where sunlight penetrates to the jungle floor, typifies this type of forest. Such growth happens mainly along river banks, on jungle fringes, and where man has cleared rain forest. When abandoned, tangled masses of vegetation quickly reclaim these cultivated areas. You can often find cultivated food plants among this vegetation.

Semievergreen Seasonal and Monsoon Forests

The characteristics of the American and African semievergreen seasonal forests correspond with those of the Asian monsoon forests. These characteristics are--

- Their trees fall into two stories of tree strata. Those in the upper story average 18 to 24 meters; those in the lower story average 7 to 13 meters.
- The diameter of the trees averages 0.5 meter.
- Their leaves fall during a seasonal drought.

Except for the sago, nipa, and coconut palms, the same edible plants grow in these areas as in the tropical rain forests.

You find these forests in portions of Columbia and Venezuela and the Amazon basin in South America; in portions of southeast coastal Kenya, Tanzania, and Mozambique in Africa; in Northeastern India, much of Burma, Thailand, Indochina, Java, and parts of other Indonesian islands in Asia.

Tropical Scrub and Thorn Forests

The chief characteristics of tropical scrub and thorn forests are--

- There is a definite dry season.
- Trees are leafless during the dry season.
- The ground is bare except for a few tufted plants in bunches; grasses are uncommon.
- Plants with thorns predominate.
- Fires occur frequently.

You find tropical scrub and thorn forests on the west coast of Mexico, Yucatan peninsula, Venezuela, Brazil; on the northwest coast and central parts of Africa; and in Asia, in Turkestan and India.
Within the tropical scrub and thorn forest areas, you will find it hard to obtain food plants during the dry season. During the rainy season, plants are considerably more abundant.

Tropical Savannas

General characteristics of the savanna are--

- It is found within the tropical zones in South America and Africa.
- It looks like a broad, grassy meadow, with trees spaced at wide intervals.
- It frequently has red soil.
- It grows scattered trees that usually appear stunted and gnarled like apple trees. Palms also occur on savannas.

You find savannas in parts of Venezuela, Brazil, and the Guianas in South America. In Africa, you find them in the southern Sahara (north-central Cameroon and Gabon and southern Sudan), Benin, Togo, most of Nigeria, northeastern Zaire, northern Uganda, western Kenya, part of Malawi, part of Tanzania, southern Zimbabwe, Mozambique, and western Madagascar.

Saltwater Swamps

Saltwater swamps are common in coastal areas subject to tidal flooding. Mangrove trees thrive in these swamps. Mangrove trees can reach heights of 12 meters, and their tangled roots are an obstacle to movement. Visibility in this type of swamp is poor, and movement is extremely difficult. Sometimes, streams that you can raft form channels, but you usually must travel on foot through this swamp.
You find saltwater swamps in West Africa, Madagascar, Malaysia, the Pacific islands, Central and South America, and at the mouth of the Ganges River in India. The swamps at the mouths of the Orinoco and Amazon rivers and rivers of Guyana consist of mud and trees that offer little shade. Tides in saltwater swamps can vary as much as 12 meters.
Everything in a saltwater swamp may appear hostile to you, from leeches and insects to crocodiles and caimans. Avoid the dangerous animals in this swamp.
Avoid this swamp altogether if you can. If there are water channels through it, you may be able to use a raft to escape.

Freshwater Swamps

You find freshwater swamps in low-lying inland areas. Their characteristics are masses of thorny undergrowth, reeds, grasses, and occasional short palms that reduce visibility and make travel difficult. There are often islands that dot these swamps, allowing you to get out of the water. Wildlife is abundant in these swamps.

TRAVEL THROUGH JUNGLE AREAS

With practice, movement through thick undergrowth and jungle can be done efficiently. Always wear long sleeves to avoid cuts and scratches.

To move easily, you must develop "jungle eye," that is, you should not concentrate on the pattern of bushes and trees to your immediate front. You must focus on the jungle further out and find natural breaks in the foliage. Look *through* the jungle, not at it. Stop and stoop down occasionally to look along the jungle floor. This action may reveal game trails that you can follow.

Stay alert and move slowly and steadily through dense forest or jungle. Stop periodically to listen and take your bearings. Use a machete to cut through dense vegetation, but do not cut unnecessarily or you will quickly wear yourself out. If using a machete, stroke upward when cutting vines to reduce noise because sound carries long distances in the jungle. Use a stick to part the vegetation. Using a stick will also help dislodge biting ants, spiders, or snakes. **Do not** grasp at brush or vines when climbing slopes; they may have irritating spines or sharp thorns.

Many jungle and forest animals follow game trails. These trails wind and cross, but frequently lead to water or clearings. Use these trails if they lead in your desired direction of travel.

In many countries, electric and telephone lines run for miles through sparsely inhabited areas. Usually, the right-of-way is clear enough to allow easy travel. When traveling along these lines, be careful as you approach transformer and relay stations. In enemy territory, they may be guarded.

TRAVEL TIPS

Pinpoint your initial location as accurately as possible to determine a general line of travel to safety. If you do not have a compass, use a field-expedient direction finding method.

Take stock of water supplies and equipment.

Move in one direction, but not necessarily in a straight line. Avoid obstacles. In enemy territory, take advantage of natural cover and concealment.

Move smoothly through the jungle. Do not blunder through it since you will get many cuts and scratches. Turn your shoulders, shift your hips, bend your body, and shorten or lengthen your stride as necessary to slide between the undergrowth.

IMMEDIATE CONSIDERATIONS

There is less likelihood of your rescue from beneath a dense jungle canopy than in other survival situations. You will probably have to travel to reach safety.

If you are the victim of an aircraft crash, the most important items to take with you from the crash site are a machete, a compass, a first aid kit, and a parachute or other material for use as mosquito netting and shelter.

Take shelter from tropical rain, sun, and insects. Malaria-carrying mosquitoes and other insects are immediate dangers, so protect yourself against bites.
Do not leave the crash area without carefully blazing or marking your route. Use your compass. Know what direction you are taking.
In the tropics, even the smallest scratch can quickly become dangerously infected. Promptly treat any wound, no matter how minor.

WATER PROCUREMENT

Even though water is abundant in most tropical environments, you may, as a survivor, have trouble finding it. If you do find water, it may not be safe to drink. Some of the many sources are vines, roots, palm trees, and condensation. You can sometimes follow animals to water. Often you can get nearly clear water from muddy streams or lakes by digging a hole in sandy soil about 1 meter from the bank. Water will seep into the hole. You must purify any water obtained in this manner.

Animals as Signs of Water

Animals can often lead you to water. Most animals require water regularly. Grazing animals such as deer, are usually never far from water and usually drink at dawn and dusk. Converging game trails often lead to water. Carnivores (meat eaters) are not reliable indicators of water. They get moisture from the animals they eat and can go without water for long periods.
Birds can sometimes also lead you to water. Grain eaters, such as finches and pigeons, are never far from water. They drink at dawn and dusk. When they fly straight and low, they are heading for water. When returning from water, they are full and will fly from tree to tree, resting frequently. Do not rely on water birds to lead you to water. They fly long distances without stopping. Hawks, eagles, and other birds of prey get liquids from their victims; you cannot use them as a water indicator.
Insects can be good indicators of water, especially bees. Bees seldom range more than 6 kilometers from their nests or hives. They usually will have a water source in this range. Ants need water. A column of ants marching up a tree is going to a small reservoir of trapped water. You find such reservoirs even in arid areas. Most flies stay within 100 meters of water, especially the European mason fly, easily recognized by its iridescent green body.
Human tracks will usually lead to a well, bore hole, or soak. Scrub or rocks may cover it to reduce evaporation. Replace the cover after use.

Water From Plants

Plants such as vines, roots, and palm trees are good sources of water.

Vines

Vines with rough bark and shoots about 5 centimeters thick can be a useful source of water. You must learn by experience which are the water-bearing vines, because not all have drinkable water. Some may even have a poisonous sap. The poisonous ones yield a sticky, milky sap when cut. Nonpoisonous vines will give a clear fluid. Some vines cause a skin irritation on contact; therefore let the liquid drip into your mouth, rather than put your mouth to the vine. Preferably, use some type of container. Use the procedure described in Chapter 6 to obtain water from a vine.

Roots

In Australia, the water tree, desert oak, and bloodwood have roots near the surface. Pry these roots out of the ground and cut them into 30-centimeter lengths. Remove the bark and suck out the moisture, or shave the root to a pulp and squeeze it over your mouth.

Palm Trees

The buri, coconut, and nipa palms all contain a sugary fluid that is very good to drink. To obtain the liquid, bend a flowering stalk of one of these palms downward, and cut off its tip. If you cut a thin slice off the stalk every 12 hours, the flow will renew, making it possible to collect up to a liter per day. Nipa palm shoots grow from the base, so that you can work at ground level. On grown trees of other species, you may have to climb them to reach a flowering stalk. Milk from coconuts has a large water content, but may contain a strong laxative in ripe nuts. Drinking too much of this milk may cause you to lose more fluid than you drink.

Water From Condensation

Often it requires too much effort to dig for roots containing water. It may be easier to let a plant produce water for you in the form of condensation. Tying a clear plastic bag around a green leafy branch will cause water in the leaves to evaporate and condense in the bag. Placing cut vegetation in a plastic bag will also produce condensation. This is a solar still (see Chapter 6).

FOOD

Food is usually abundant in a tropical survival situation. To obtain animal food, use the procedures outlined in Chapter 8.

In addition to animal food, you will have to supplement your diet with edible plants. The best places to forage are the banks of streams and rivers. Wherever the sun penetrates the jungle, there will be a mass of vegetation, but river banks may be the most accessible areas.

If you are weak, do not expend energy climbing or felling a tree for food. There are more easily obtained sources of food nearer the ground. Do not pick more food than you need. Food spoils rapidly in tropical conditions. Leave food on the growing plant until you need it, and eat it fresh.

There are an almost unlimited number of edible plants from which to choose. Unless you can positively identify these plants, it may be safer at first to begin with palms, bamboos, and common fruits. The list below identifies some of the most common foods. Detailed descriptions and photographs are at Appendix B.

TROPICAL ZONE FOOD PLANTS

- Bael fruit *(Aegle marmelos)*
- Bamboo (various species)
- Banana or plantain (*Musa* species)
- Bignay *(Antidesma bunius)*
- Breadfruit *(Artrocarpus incisa)*
- Coconut palm *(Cocos nucifera)*
- Fishtail palm *(Caryota urens)*
- Horseradish tree *(Moringa pterygosperma)*
- Lotus (*Nelumbo* species)
- Mango *(Mangifera indica)*
- Manioc *(Manihot utillissima)*
- Nipa palm *(Nipa fruticans)*
- Papaya *(Carica papaya)*
- Persimmon *(Diospyros virginiana)*
- Rattan palm (*Calamus* species)
- Sago palm *(Metroxylon sagu)*
- Sterculia *(Sterculia foetida)*
- Sugarcane *(Saccharum officinarum)*
- Sugar palm *(Arenga pinnata)*
- Sweetsop *(Annona squamosa)*
- Taro (*Colocasia* and *Alocasia* species)

- Water lily *(Nymphaea odorata)*
- Wild fig (*Ficus* species)
- Wild rice *(Zizania aquatica)*
- Yam (*Dioscorea* species)

POISONOUS PLANTS

The proportion of poisonous plants in tropical regions is no greater than in any other area of the world. However, it may appear that most plants in the tropics are poisonous because of the great density of plant growth in some tropical areas. See Appendix C.

CHAPTER 15 - COLD WEATHER SURVIVAL

One of the most difficult survival situations is a cold weather scenario. Remember, cold weather is an adversary that can be as dangerous as an enemy soldier. Every time you venture into the cold, you are pitting yourself against the elements. With a little knowledge of the environment, proper plans, and appropriate equipment, you can overcome the elements. As you remove one or more of these factors, survival becomes increasingly difficult. Remember, winter weather is highly variable. Prepare yourself to adapt to blizzard conditions even during sunny and clear weather.

Cold is a far greater threat to survival than it appears. It decreases your ability to think and weakens your will to do anything except to get warm. Cold is an insidious enemy; as it numbs the mind and body, it subdues the will to survive.

Cold makes it very easy to forget your ultimate goal--to survive.

COLD REGIONS AND LOCATIONS

Cold regions include arctic and subarctic areas and areas immediately adjoining them. You can classify about 48 percent of the northern hemisphere's total landmass as a cold region due to the influence and extent of air temperatures. Ocean currents affect cold weather and cause large areas normally included in the temperate zone to fall within the cold regions during winter periods. Elevation also has a marked effect on defining cold regions.

Within the cold weather regions, you may face two types of cold weather environments--wet or dry. Knowing in which environment your area of operations falls will affect planning and execution of a cold weather operation.

Wet Cold Weather Environments

Wet cold weather conditions exist when the average temperature in a 24-hour period is -10 degrees C or above. Characteristics of this condition are freezing during the colder night hours and thawing during the day. Even though the temperatures are warmer during this condition, the terrain is usually very sloppy due to slush and mud. You must concentrate on protecting yourself from the wet ground and from freezing rain or wet snow.

Dry Cold Weather Environments

Dry cold weather conditions exist when the average temperature in a 24-hour period remains below -10 degrees C. Even though the temperatures in this condition are much lower than normal, you do not have to contend with the freezing and thawing. In these conditions, you need more layers of inner clothing to

ESSENTIAL ARMY SURVIVAL MANUAL [149] ARMY SURVIVAL MANUAL

protect you from temperatures as low as -60 degrees C. Extremely hazardous conditions exist when wind and low temperature combine.

WINDCHILL

Windchill increases the hazards in cold regions. Windchill is the effect of moving air on exposed flesh. For instance, with a 27.8-kph (15-knot) wind and a temperature of -10 degrees C, the equivalent windchill temperature is -23 degrees C. Figure 15-1 gives the windchill factors for various temperatures and wind speeds.

COOLING POWER OF WIND EXPRESSED AS "EQUIVALENT CHILL TEMPERATURE"

WIND SPEED KNOTS	WIND SPEED KPH	TEMPERATURE (DEGREES C) — EQUIVALENT CHILL TEMPERATURE																			
		4	2	-1	-4	-7	-9	-12	-15	-18	-21	-23	-26	-29	-32	-34	-37	-40	-43	-46	-48	-51
CALM	CALM	4	2	-1	-4	-7	-9	-12	-15	-18	-21	-23	-26	-29	-32	-34	-37	-40	-43	-46	-48	-51
4	8	2	-1	-4	-7	-9	-12	-15	-18	-21	-23	-26	-29	-32	-34	-37	-40	-43	-46	-48	-54	-57
9	16	-1	-4	-9	-12	-15	-18	-23	-26	-29	-32	-37	-40	-43	-46	-51	-54	-57	-59	-62	-68	-71
13	24	-4	-7	-12	-18	-21	-23	-29	-32	-34	-40	-43	-46	-51	-54	-57	-62	-65	-68	-73	-76	-79
17	32	-7	-9	-15	-18	-23	-26	-32	-34	-37	-43	-46	-51	-54	-59	-62	-65	-71	-73	-79	-82	-84
22	40	-9	-12	-18	-21	-26	-29	-34	-37	-43	-46	-51	-54	-59	-62	-68	-71	-76	-79	-84	-87	-93
26	48	-12	-15	-18	-23	-29	-32	-34	-40	-46	-48	-54	-57	-62	-65	-71	-73	-79	-82	-87	-90	-96
30	56	-12	-15	-21	-23	-29	-34	-37	-40	-46	-51	-54	-59	-62	-68	-73	-76	-82	-84	-90	-93	-98
35	64	-12	-18	-21	-26	-29	-34	-37	-43	-48	-51	-57	-59	-65	-71	-73	-79	-82	-87	-90	-96	-101
(Higher winds have little additional effects)		LITTLE DANGER					INCREASING DANGER (Flesh may freeze within 1 minute)							GREAT DANGER (Flesh may freeze within 30 seconds)							

DANGER OF FREEZING EXPOSED FLESH FOR PROPERLY CLOTHED PERSONS

Figure 15-1. Windchill table.

Remember, even when there is no wind, you will create the equivalent wind by skiing, running, being towed on skis behind a vehicle, working around aircraft that produce wind blasts.

BASIC PRINCIPLES OF COLD WEATHER SURVIVAL

It is more difficult for you to satisfy your basic water, food, and shelter needs in a cold environment than in a warm environment. Even if you have the basic requirements, you must also have adequate protective clothing and the will to survive. The will to survive is as important as the basic needs. There have been incidents when trained and well-equipped individuals have not survived cold weather situations because they lacked the will to live. Conversely, this will has sustained individuals less well-trained and equipped.

There are many different items of cold weather equipment and clothing issued by the U.S. Army today. Specialized units may have access to newer, lightweight gear such as polypropylene underwear, GORE-TEX outerwear and boots, and other special equipment. Remember, however, the older gear will keep you warm as long as you apply a few cold weather principles. If the newer types of clothing are available, use them. If not, then your clothing should be entirely wool, with the possible exception of a windbreaker. You must not only have enough clothing to protect you from the cold, you must also know how to maximize the warmth you get from it. For example, always keep your head covered. You can lose 40 to 45 percent of body heat from an unprotected head and even more from the unprotected neck, wrist, and ankles. These areas of the body are good radiators of heat and have very little insulating fat. The brain is very susceptible to cold and can stand the least amount of cooling. Because there is much blood circulation in the head, most of which is on the surface, you can lose heat quickly if you do not cover your head.

There are four basic principles to follow to keep warm. An easy way to remember these basic principles is to use the word COLD--

C - Keep clothing *clean.*

O - Avoid *overheating.*

L - Wear clothes *loose* and in *layers.*

D - Keep clothing *dry.*

C -   *Keep clothing clean.* This principle is always important for sanitation and comfort. In winter, it is also important from the standpoint of warmth. Clothes matted with dirt and grease lose much of their insulation value. Heat can escape more easily from the body through the clothing's crushed or filled up air pockets.

O -   *Avoid overheating.* When you get too hot, you sweat and your clothing absorbs the moisture. This affects your warmth in two ways: dampness decreases the insulation quality of clothing, and as sweat evaporates, your body cools. Adjust your clothing so that you do not sweat. Do this by partially opening your parka or jacket, by removing an inner layer of clothing, by removing heavy outer mittens, or by throwing back your parka hood or changing to lighter headgear. The head and hands act as efficient heat dissipaters when overheated.

L -   *Wear your clothing loose and in layers.* Wearing tight clothing and footgear restricts blood circulation and invites cold injury. It also decreases the volume of air trapped between the layers, reducing its insulating value. Several layers of lightweight clothing are better than one equally thick layer of clothing, because the layers have dead-air space between them. The dead-air space provides extra insulation. Also, layers of clothing allow you to take off or add clothing layers to prevent excessive sweating or to increase warmth.

D -   *Keep clothing dry.* In cold temperatures, your inner layers of clothing can become wet from sweat and your outer layer, if not water repellent, can become wet from snow and frost melted by body heat. Wear water repellent outer clothing, if available. It will shed most of the water collected from melting snow and frost. Before entering a heated shelter, brush off the snow and frost. Despite the precautions you take, there will be times when you cannot keep from getting wet. At such times, drying your clothing may become a major problem. On the march, hang your damp mittens and socks on your rucksack. Sometimes in freezing temperatures, the wind and sun will dry this clothing. You can also place damp socks or mittens, unfolded, near your body so that your body heat can dry them. In a campsite, hang damp clothing inside the shelter near the top, using drying lines or improvised racks. You may even be able to dry each item by holding it before an open fire. Dry leather items slowly. If no other means are available for drying your boots, put them between your sleeping bag shell and liner. Your body heat will help to dry the leather.

A heavy, down-lined sleeping bag is a valuable piece of survival gear in cold weather. Ensure the down remains dry. If wet, it loses a lot of its insulation value. If you do not have a sleeping bag, you can make one out of parachute cloth or similar material and natural dry material, such as leaves, pine needles, or moss. Place the dry material between two layers of the material.

Other important survival items are a knife; waterproof matches in a waterproof container, preferably one with a flint attached; a durable compass; map; watch; waterproof ground cloth and cover; flashlight; binoculars; dark glasses; fatty emergency foods; food gathering gear; and signaling items.

Remember, a cold weather environment can be very harsh. Give a good deal of thought to selecting the right equipment for survival in the cold. If unsure of an item you have never used, test it in an "overnight backyard" environment before venturing further. Once you have selected items that are essential for your survival, do not lose them after you enter a cold weather environment.

HYGIENE

Although washing yourself may be impractical and uncomfortable in a cold environment, you must do so. Washing helps prevent skin rashes that can develop into more serious problems.

In some situations, you may be able to take a snow bath. Take a handful of snow and wash your body where sweat and moisture accumulate, such as under the arms and between the legs, and then wipe yourself dry. If possible, wash your feet daily and put on clean, dry socks. Change your underwear at least twice a week. If you are unable to wash your underwear, take it off, shake it, and let it air out for an hour or two.

If you are using a previously used shelter, check your body and clothing for lice each night. If your clothing has become infested, use insecticide powder if you have any. Otherwise, hang your clothes in the cold, then beat and brush them. This will help get rid of the lice, but not the eggs.

If you shave, try to do so before going to bed. This will give your skin a chance to recover before exposing it to the elements.

MEDICAL ASPECTS

When you are healthy, your inner core temperature (torso temperature) remains almost constant at 37 degrees C (98.6 degrees F). Since your limbs and head have less protective body tissue than your torso, their temperatures vary and may not reach core temperature.

Your body has a control system that lets it react to temperature extremes to maintain a temperature balance. There are three main factors that affect this temperature balance--heat production, heat loss, and evaporation. The difference between the body's core temperature and the environment's temperature governs the heat production rate. Your body can get rid of heat better than it can produce it. Sweating helps to control the heat balance. Maximum sweating will get rid of heat about as fast as maximum exertion produces it.

Shivering causes the body to produce heat. It also causes fatigue that, in turn, leads to a drop in body temperature. Air movement around your body affects heat loss. It has been calculated that a naked man exposed to still air at or about 0 degrees C can maintain a heat balance if he shivers as hard as he can. However, he can't shiver forever.

It has also been calculated that a man at rest wearing the maximum arctic clothing in a cold environment can keep his internal heat balance during temperatures well below freezing. To withstand really cold conditions for any length of time, however, he will have to become active or shiver.

COLD INJURIES

The best way to deal with injuries and sicknesses is to take measures to prevent them from happening in the first place. Treat any injury or sickness that occurs as soon as possible to prevent it from worsening. The knowledge of signs and symptoms and the use of the buddy system are critical in maintaining health. Following are cold injuries that can occur.

Hypothermia

Hypothermia is the lowering of the body temperature at a rate faster than the body can produce heat. Causes of hypothermia may be general exposure or the sudden wetting of the body by falling into a lake or spraying with fuel or other liquids.

The initial symptom is shivering. This shivering may progress to the point that it is uncontrollable and interferes with an individual's ability to care for himself. This begins when the body's core (rectal) temperature falls to about 35.5 degrees C (96 degrees F). When the core temperature reaches 35 to 32 degrees C (95 to 90 degrees F), sluggish thinking, irrational reasoning, and a false feeling of warmth may occur. Core temperatures of 32 to 30 degrees C (90 to 86 degrees F) and below result in muscle rigidity, unconsciousness, and barely detectable signs of life. If the victim's core temperature falls below 25 degrees C (77 degrees F), death is almost certain.

To treat hypothermia, rewarm the entire body. If there are means available, rewarm the person by first immersing the trunk area only in warm water of 37.7 to 43.3 degrees C (100 to 110 degrees F).

CAUTION

Rewarming the total body in a warm water bath should be done only in a hospital environment because of the increased risk of cardiac arrest and rewarming shock.

One of the quickest ways to get heat to the inner core is to give warm water enemas. Such an action, however, may not be possible in a survival situation. Another method is to wrap the victim in a warmed sleeping bag with another person who is already warm; both should be naked.

CAUTION

The individual placed in the sleeping bag with victim could also become a hypothermia victim if left in the bag too long.

If the person is conscious, give him hot, sweetened fluids. One of the best sources of calories is honey or dextrose; if unavailable, use sugar, cocoa, or a similar soluble sweetener.

CAUTION

Do not force an unconscious person to drink.

There are two dangers in treating hypothermia--rewarming too rapidly and "after drop." Rewarming too rapidly can cause the victim to have circulatory problems, resulting in heart failure. After drop is the sharp body core temperature drop that occurs when taking the victim from the warm water. Its probable muse is the return of previously stagnant limb blood to the core (inner torso) area as recirculation occurs. Concentrating on warming the core area and stimulating peripheral circulation will lessen the effects of after drop. Immersing the torso in a warm bath, if possible, is the best treatment.

Frostbite

This injury is the result of frozen tissues. Light frostbite involves only the skin that takes on a dull whitish pallor. Deep frostbite extends to a depth below the skin. The tissues become solid and immovable. Your feet, hands, and exposed facial areas are particularly vulnerable to frostbite.

The best frostbite prevention, when you are with others, is to use the buddy system. Check your buddy's face often and make sure that he checks yours. If you are alone, periodically cover your nose and lower part of your face with your mittened hand.

The following pointers will aid you in keeping warm and preventing frostbite when it is extremely cold or when you have less than adequate clothing:

- *Face.* Maintain circulation by twitching and wrinkling the skin on your face making faces. Warm with your hands.
- *Ears.* Wiggle and move your ears. Warm with your hands.
- *Hands.* Move your hands inside your gloves. Warm by placing your hands close to your body.
- *Feet.* Move your feet and wiggle your toes inside your boots.

A loss of feeling in your hands and feet is a sign of frostbite. If you have lost feeling for only a short time, the frostbite is probably light. Otherwise, assume the frostbite is deep. To rewarm a light frostbite, use your hands or mittens to warm your face and ears. Place your hands under your armpits. Place your feet next to your buddy's stomach. A deep frostbite injury, if thawed and refrozen, will cause more damage than a nonmedically trained person can handle. Figure 15-2 lists some do's and don'ts regarding frostbite.

Do	Don't
• Periodically check for frostbite.	• Rub injury with snow.
• Rewarm light frostbite.	• Drink alcoholic beverages.
• Keep injured areas from refreezing.	• Smoke.
	• Try to thaw out a deep frostbite injury if you are away from definitive medical care.

Figure 15-2. Frostbite do's and don'ts.

Trench Foot and Immersion Foot

These conditions result from many hours or days of exposure to wet or damp conditions at a temperature just above freezing. The symptoms are a sensation of pins and needles, tingling, numbness, and then pain. The skin will initially appear wet, soggy, white, and shriveled. As it progresses and damage appears, the skin will take on a red and then a bluish or black discoloration. The feet become cold, swollen, and have a waxy appearance. Walking becomes difficult and the feet feel heavy and numb. The nerves and muscles sustain the main damage, but gangrene can occur. In extreme cases, the flesh dies and it may become necessary to have the foot or leg amputated. The best prevention is to keep your feet dry. Carry extra socks with you in a waterproof packet. You can dry wet socks against your torso (back or chest). Wash your feet and put on dry socks daily.

Dehydration

When bundled up in many layers of clothing during cold weather, you may be unaware that you are losing body moisture. Your heavy clothing absorbs the moisture that evaporates in the air. You must drink water to replace this loss of fluid. Your need for water is as great in a cold environment as it is in a warm environment (Chapter 13). One way to tell if you are becoming dehydrated is to check the color of your urine on snow. If your urine makes the snow dark yellow, you are becoming dehydrated and need to replace body fluids. If it makes the snow light yellow to no color, your body fluids have a more normal balance.

Cold Diuresis

Exposure to cold increases urine output. It also decreases body fluids that you must replace.

Sunburn

Exposed skin can become sunburned even when the air temperature is below freezing. The sun's rays reflect at all angles from snow, ice, and water, hitting sensitive areas of skin--lips, nostrils, and eyelids. Exposure to the sun results in sunburn more quickly at high altitudes than at low altitudes. Apply sunburn cream or lip salve to your face when in the sun.

Snow Blindness

The reflection of the sun's ultraviolet rays off a snow-covered area causes this condition. The symptoms of snow blindness are a sensation of grit in the eyes, pain in and over the eyes that increases with eyeball movement, red and teary eyes, and a headache that intensifies with continued exposure to light. Prolonged exposure to these rays can result in permanent eye damage. To treat snow blindness, bandage your eyes until the symptoms disappear.

You can prevent snow blindness by wearing sunglasses. If you don't have sunglasses, improvise. Cut slits in a piece of cardboard, thin wood, tree bark, or other available material (Figure 15-3). Putting soot under your eyes will help reduce shine and glare.

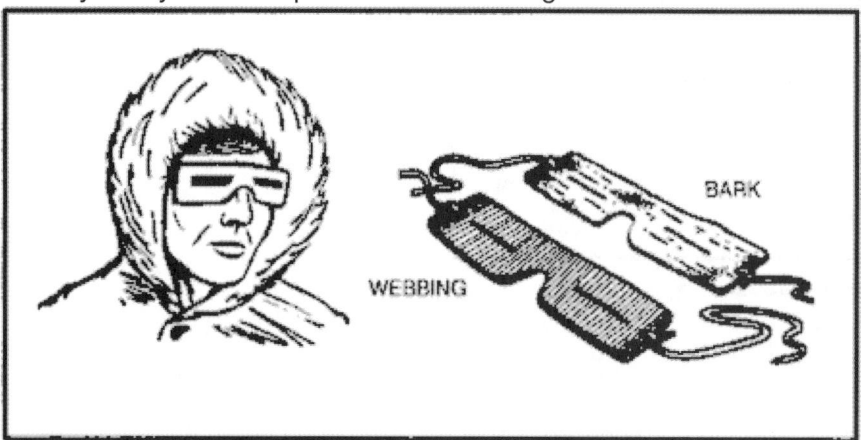

Figure 15-3. Improvised sunglasses.

Constipation

It is very important to relieve yourself when needed. Do not delay because of the cold condition. Delaying relieving yourself because of the cold, eating dehydrated foods, drinking too little liquid, and irregular eating habits can cause you to become constipated. Although not disabling, constipation can cause some discomfort. Increase your fluid intake to at least 2 liters above your normal 2 to 3 liters daily intake and, if available, eat fruit and other foods that will loosen the stool.

Insect Bites

Insect bites can become infected through constant scratching. Flies can carry various disease-producing germs. To prevent insect bites, use insect repellent, netting, and wear proper clothing. See Chapter 11 for information on insect bites and Chapter 4 for treatment.

SHELTERS

Your environment and the equipment you carry with you will determine the type of shelter you can build. You can build shelters in wooded areas, open country, and barren areas. Wooded areas usually provide the best location, while barren areas have only snow as building material. Wooded areas provide timber for shelter construction, wood for fire, concealment from observation, and protection from the wind.

> *Note: In extreme cold, do not use metal, such as an aircraft fuselage, for shelter. The metal will conduct away from the shelter what little heat you can generate.*

Shelters made from ice or snow usually require tools such as ice axes or saws. You must also expend much time and energy to build such a shelter. Be sure to ventilate an enclosed shelter, especially if you intend to build a fire in it. Always block a shelter's entrance, if possible, to keep the heat in and the wind out. Use a rucksack or snow block. Construct a shelter no larger than needed. This will reduce the amount of space to heat. A fatal error in cold weather shelter construction is making the shelter so large that it steals body heat rather than saving it. Keep shelter space small.

Never sleep directly on the ground. Lay down some pine boughs, grass, or other insulating material to keep the ground from absorbing your body heat.

Never fall asleep without turning out your stove or lamp. Carbon monoxide poisoning can result from a fire burning in an unventilated shelter. Carbon monoxide is a great danger. It is colorless and odorless. Any time you have an open flame, it may generate carbon monoxide. Always check your ventilation. Even in a ventilated shelter, incomplete combustion can cause carbon monoxide poisoning. Usually, there are no symptoms. Unconsciousness and death can occur without warning. Sometimes, however, pressure at the temples, burning of the eyes, headache, pounding pulse, drowsiness, or nausea may occur. The one characteristic, visible sign of carbon monoxide poisoning is a cherry red coloring in the tissues of the lips, mouth, and inside of the eyelids. Get into fresh air at once if you have any of these symptoms.

There are several types of field-expedient shelters you can quickly build or employ. Many use snow for insulation.

Snow Cave Shelter

The snow cave shelter (Figure 15-4) is a most effective shelter because of the insulating qualities of snow. Remember that it takes time and energy to build and that you will get wet while building it. First, you need to find a drift about 3 meters deep into which you can dig. While building this shelter, keep the roof arched for strength and to allow melted snow to drain down the sides. Build the sleeping platform higher than the entrance. Separate the sleeping platform from the snow cave's walls or dig a small trench between the platform and the wall. This platform will prevent the melting snow from wetting you and your equipment. This construction is especially important if you have a good source of heat in the snow cave. Ensure the roof is high enough so that you can sit up on the sleeping platform. Block the entrance with a snow block or other material and use the lower entrance area for cooking. The walls and ceiling should be at least 30 centimeters thick. Install a ventilation shaft. If you do not have a drift large enough to build a snow cave, you can make a variation of it by piling snow into a mound large enough to dig out.

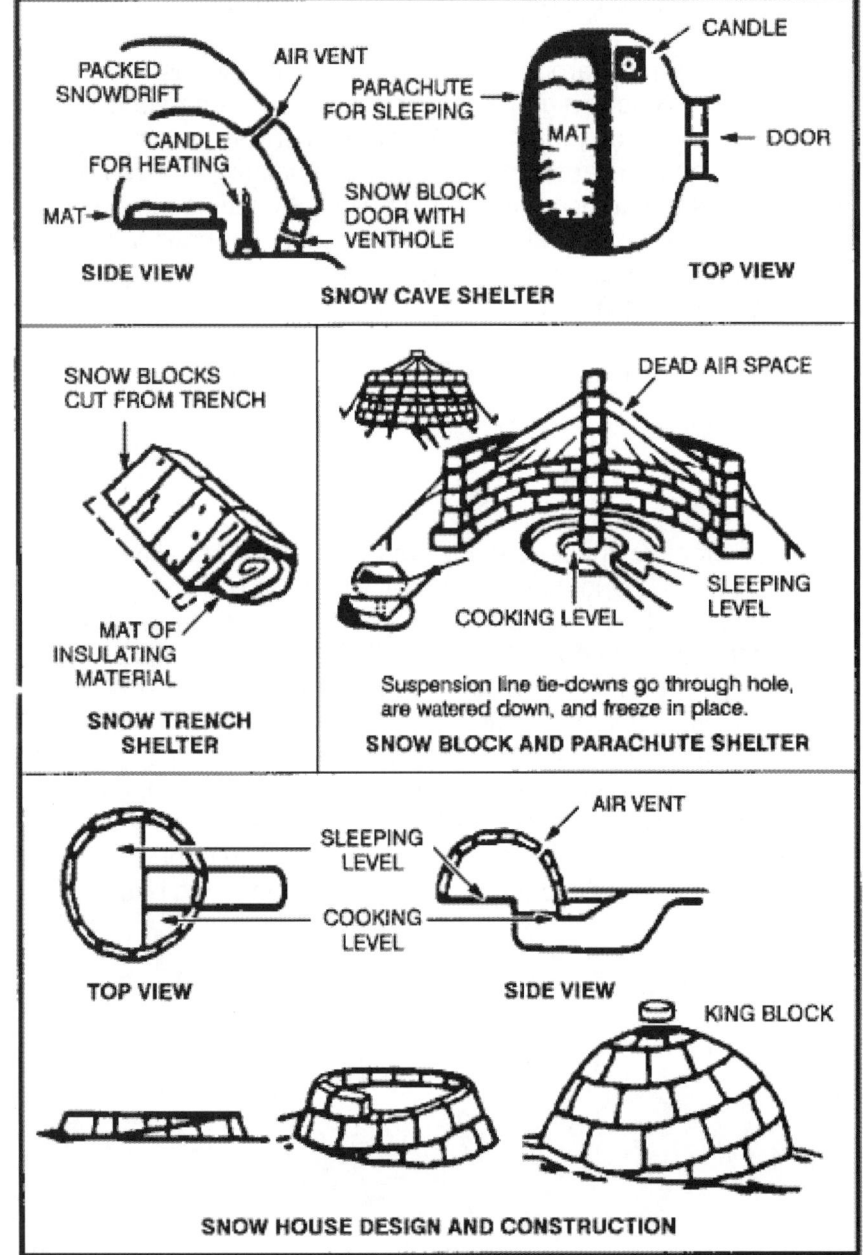

Figure 15-4. Snow houses.

Snow Trench Shelter

The idea behind this shelter (Figure 15-4) is to get you below the snow and wind level and use the snow's insulating qualities. If you are in an area of compacted snow, cut snow blocks and use them as overhead cover. If not, you can use a poncho or other material. Build only one entrance and use a snow block or rucksack as a door.

Snow Block and Parachute Shelter

Use snow blocks for the sides and parachute material for overhead cover (Figure 15-4). If snowfall is heavy, you will have to clear snow from the top at regular intervals to prevent the collapse of the parachute material.

Snow House or Igloo

In certain areas, the natives frequently use this type of shelter (Figure 15-4) as hunting and fishing shelters. They are efficient shelters but require some practice to make them properly. Also, you must be in an area that is suitable for cutting snow blocks and have the equipment to cut them (snow saw or knife).

Lean-To Shelter

Construct this shelter in the same manner as for other environments; however, pile snow around the sides for insulation (Figure 15-5).

Figure 15-5. Lean-to made from natural shelter.

Fallen Tree Shelter

To build this shelter, find a fallen tree and dig out the snow underneath it (Figure 15-6). The snow will not be deep under the tree. If you must remove branches from the inside, use them to line the floor.

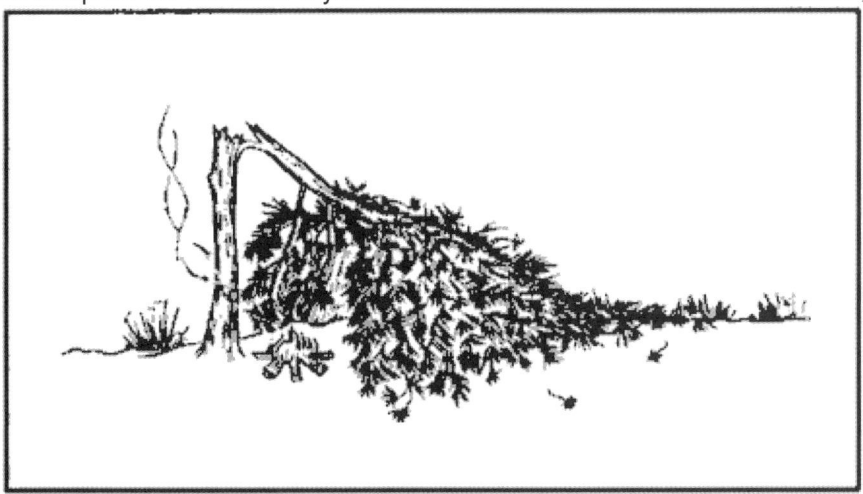

Figure 15-6. Fallen tree as shelter.

Tree-Pit Shelter

Dig snow out from under a suitable large tree. It will not be as deep near the base of the tree. Use the cut branches to line the shelter. Use a ground sheet as overhead cover to prevent snow from falling off the tree into the shelter. If built properly, you can have 360-degree visibility (Figure 5-12, Chapter 5).

20-Man Life Raft

This raft is the standard overwater raft on U.S. Air Force aircraft. You can use it as a shelter. Do not let large amounts of snow build up on the overhead protection. If placed in an open area, it also serves as a good signal to overhead aircraft.

FIRE

Fire is especially important in cold weather. It not only provides a means to prepare food, but also to get warm and to melt snow or ice for water. It also provides you with a significant psychological boost by making you feel a little more secure in your situation.

Use the techniques described in Chapter 7 to build and light your fire. If you are in enemy territory, remember that the smoke, smell, and light from your fire may reveal your location. Light reflects from surrounding trees or rocks, making even indirect light a source of danger. Smoke tends to go straight up in cold, calm weather, making it a beacon during the day, but helping to conceal the smell at night. In warmer weather, especially in a wooded area, smoke tends to hug the ground, making it less visible in the day, but making its odor spread.

If you are in enemy territory, cut low tree boughs rather than the entire tree for firewood. Fallen trees are easily seen from the air.

All wood will burn, but some types of wood create more smoke than others. For instance, coniferous trees that contain resin and tar create more and darker smoke than deciduous trees.

There are few materials to use for fuel in the high mountainous regions of the arctic. You may find some grasses and moss, but very little. The lower the elevation, the more fuel available. You may find some scrub willow and small, stunted spruce trees above the tree line. On sea ice, fuels are seemingly nonexistent. Driftwood or fats may be the only fuels available to a survivor on the barren coastlines in the arctic and subarctic regions.

Abundant fuels within the tree line are--

- Spruce trees are common in the interior regions. As a conifer, spruce makes a lot of smoke when burned in the spring and summer months. However, it burns almost smoke-free in late fall and winter.
- The tamarack tree is also a conifer. It is the only tree of the pine family that loses its needles in the fall. Without its needles, it looks like a dead spruce, but it has many knobby buds and cones on its bare branches. When burning, tamarack wood makes a lot of smoke and is excellent for signaling purposes.
- Birch trees are deciduous and the wood burns hot and fast, as if soaked with oil or kerosene. Most birches grow near streams and lakes, but occasionally you will find a few on higher ground and away from water.
- Willow and alder grow in arctic regions, normally in marsh areas or near lakes and streams. These woods burn hot and fast without much smoke.

Dried moss, grass, and scrub willow are other materials you can use for fuel. These are usually plentiful near streams in tundras (open, treeless plains). By bundling or twisting grasses or other scrub vegetation to form a large, solid mass, you will have a slower burning, more productive fuel.

If fuel or oil is available from a wrecked vehicle or downed aircraft, use it for fuel. Leave the fuel in the tank for storage, drawing on the supply only as you need it. Oil congeals in extremely cold temperatures, therefore, drain it from the vehicle or aircraft while still warm if there is no danger of explosion or fire. If you have no container, let the oil drain onto the snow or ice. Scoop up the fuel as you need it.

CAUTION
Do not expose flesh to petroleum, oil, and lubricants in extremely cold temperatures. The liquid state of

these products is deceptive in that it can cause frostbite.

Some plastic products, such as MRE spoons, helmet visors, visor housings, aid foam rubber will ignite quickly from a burning match. They will also burn long enough to help start a fire. For example, a plastic spoon will burn for about 10 minutes.

In cold weather regions, there are some hazards in using fires, whether to keep warm or to cook. For example--

- Fires have been known to burn underground, resurfacing nearby. Therefore, do not build a fire too close to a shelter.
- In snow shelters, excessive heat will melt the insulating layer of snow that may also be your camouflage.
- A fire inside a shelter lacking adequate ventilation can result in carbon monoxide poisoning.
- A person trying to get warm or to dry clothes may become careless and burn or scorch his clothing and equipment.
- Melting overhead snow may get you wet, bury you and your equipment, and possibly extinguish your fire.

In general, a small fire and some type of stove is the best combination for cooking purposes. A hobo stove (Figure 15-7) is particularly suitable to the arctic. It is easy to make out of a tin can, and it conserves fuel. A bed of hot coals provides the best cooking heat. Coals from a crisscross fire will settle uniformly. Make this type of fire by crisscrossing the firewood. A simple crane propped on a forked stick will hold a cooking container over a fire.

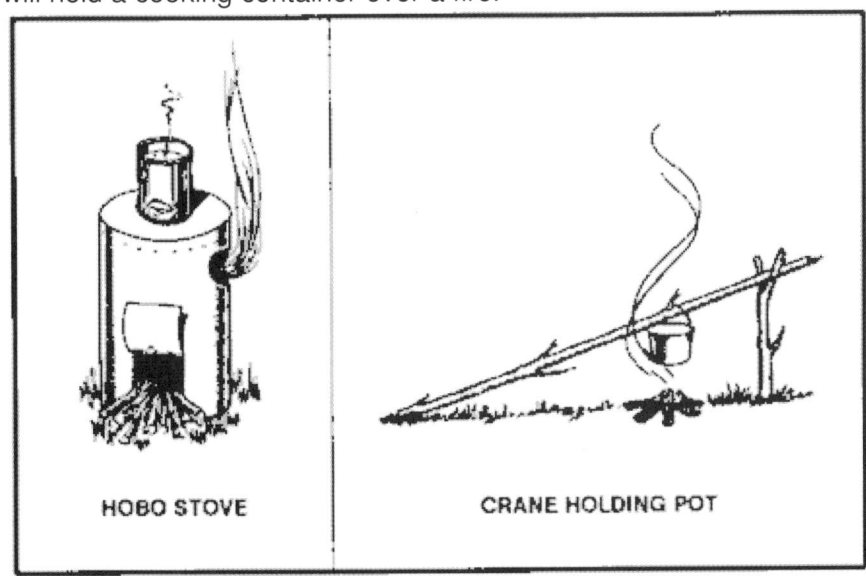

HOBO STOVE   CRANE HOLDING POT

Figure 15-7. Cooking fire/stove.

For heating purposes, a single candle provides enough heat to warm an enclosed shelter. A small fire about the size of a man's hand is ideal for use in enemy territory. It requires very little fuel, yet it generates considerable warmth and is hot enough to warm liquids.

WATER

There are many sources of water in the arctic and subarctic. Your location and the season of the year will determine where and how you obtain water.

Water sources in arctic and subarctic regions are more sanitary than in other regions due to the climatic and environmental conditions. However, *always purify* the water before drinking it. During the summer months, the best natural sources of water are freshwater lakes, streams, ponds, rivers, and springs. Water from ponds or lakes may be slightly stagnant, but still usable. Running water in streams, rivers, and bubbling springs is usually fresh and suitable for drinking.

The brownish surface water found in a tundra during the summer is a good source of water. However, you may have to filter the water before purifying it.

You can melt freshwater ice and snow for water. Completely melt both before putting them in your mouth. Trying to melt ice or snow in your mouth takes away body heat and may cause internal cold injuries. If on or near pack ice in the sea, you can use old sea ice to melt for water. In time, sea ice loses its salinity. You can identify this ice by its rounded corners and bluish color.

You can use body heat to melt snow. Place the snow in a water bag and place the bag between your layers of clothing. This is a slow process, but you can use it on the move or when you have no fire.

> Note: Do not waste fuel to melt ice or snow when drinkable water is available from other sources.

When ice is available, melt it, rather than snow. One cup of ice yields more water than one cup of snow. Ice also takes less time to melt. You can melt ice or snow in a water bag, MRE ration bag, tin can, or improvised container by placing the container near a fire. Begin with a small amount of ice or snow in the container and, as it turns to water, add more ice or snow.

Another way to melt ice or snow is by putting it in a bag made from porous material and suspending the bag near the fire. Place a container under the bag to catch the water.

During cold weather, avoid drinking a lot of liquid before going to bed. Crawling out of a warm sleeping bag at night to relieve yourself means less rest and more exposure to the cold.

Once you have water, keep it next to you to prevent refreezing. Also, do not fill your canteen completely. Allowing the water to slosh around will help keep it from freezing.

FOOD

There are several sources of food in the arctic and subarctic regions. The type of food--fish, animal, fowl, or plant--and the ease in obtaining it depend on the time of the year and your location.

Fish

During the summer months, you can easily get fish and other water life from coastal waters, streams, rivers, and lakes. Use the techniques described in Chapter 8 to catch fish.

The North Atlantic and North Pacific coastal waters are rich in seafood. You can easily find crawfish, snails, clams, oysters, and king crab. In areas where there is a great difference between the high and low tide water levels, you can easily find shellfish at low tide. Dig in the sand on the tidal flats. Look in tidal pools and on offshore reefs. In areas where there is a small difference between the high- and low-tide water levels, storm waves often wash shellfish onto the beaches.

The eggs of the spiny sea urchin that lives in the waters around the Aleutian Islands and southern Alaska are excellent food. Look for the sea urchins in tidal pools. Break the shell by placing it between two stones. The eggs are bright yellow in color.

Most northern fish and fish eggs are edible. Exceptions are the meat of the arctic shark and the eggs of the sculpins.

The bivalves, such as clams and mussels, are usually more palatable than spiral-shelled seafood, such as snails.

WARNING
The black mussel, a common mollusk of the far north, may be poisonous in any season. Toxins sometimes found in the mussel's tissue are as dangerous as strychnine.

The sea cucumber is another edible sea animal. Inside its body are five long white muscles that taste much like clam meat.

In early summer, smelt spawn in the beach surf. Sometimes you can scoop them up with your hands. You can often find herring eggs on the seaweed in midsummer. Kelp, the long ribbonlike seaweed, and other smaller seaweed that grow among offshore rocks are also edible.

Sea Ice Animals

You find polar bears in practically all arctic coastal regions, but rarely inland. Avoid them if possible. They are the most dangerous of all bears. They are tireless, clever hunters with good sight and an extraordinary sense of smell. If you must kill one for food, approach it cautiously. Aim for the brain; a bullet elsewhere will rarely kill one. Always cook polar bear meat before eating it.

CAUTION

Do not eat polar bear liver as it contains a toxic concentration of vitamin A.

Earless seal meat is some of the best meat available. You need considerable skill, however, to get close enough to an earless seal to kill it. In spring, seals often bask on the ice beside their breathing holes. They raise their heads about every 30 seconds, however, to look for their enemy, the polar bear.

To approach a seal, do as the Eskimos do--stay downwind from it, cautiously moving closer while it sleeps. If it moves, stop and imitate its movements by lying flat on the ice, raising your head up and down, and wriggling your body slightly. Approach the seal with your body side-ways to it and your arms close to your body so that you look as much like another seal as possible. The ice at the edge of the breathing hole is usually smooth and at an incline, so the least movement of the seal may cause it to slide into the water. Therefore, try to get within 22 to 45 meters of the seal and kill it instantly (aim for the brain). Try to reach the seal before it slips into the water. In winter, a dead seal will usually float, but it is difficult to retrieve from the water.

Keep the seal blubber and skin from coming into contact with any scratch or broken skin you may have. You could get "spekk-finger," that is, a reaction that causes the hands to become badly swollen.

Keep in mind that where there are seals, there are usually polar bears, and polar bears have stalked and killed seal hunters.

You can find porcupines in southern subarctic regions where there are trees. Porcupines feed on bark; if you find tree limbs stripped bare, you are likely to find porcupines in the area.

Ptarmigans, owls, Canadian jays, grouse, and ravens are the only birds that remain in the arctic during the winter. They are scarce north of the tree line. Ptarmigans and owls are as good for food as any game bird. Ravens are too thin to be worth the effort it takes to catch them. Ptarmigans, which change color to blend with their surroundings, are hard to spot. Rock ptarmigans travel in pairs and you can easily approach them. Willow ptarmigans live among willow clumps in bottom-lands. They gather in large flocks and you can easily snare them. During the summer months all arctic birds have a 2- to 3-week molting period during which they cannot fly and are easy to catch. Use one of the techniques described in Chapter 8 to catch them.

Skin and butcher game (see Chapter 8) while it is still warm. If you do not have time to skin the game, at least remove its entrails, musk glands, and genitals before storing. If time allows, cut the meat into usable pieces and freeze each separately so that you can use the pieces as needed. Leave the fat on all animals except seals. During the winter, game freezes quickly if left in the open. During the summer, you can store it in underground ice holes.

Plants

Although tundras support a variety of plants during the warm months, all are small, however, when compared to plants in warmer climates. For instance, the arctic willow and birch are shrubs rather than trees. The following is a list of some plant foods found in arctic and subarctic regions (see Appendix B for descriptions).

ARCTIC FOOD PLANTS

- Arctic raspberry and blueberry
- Arctic willow
- Bearberry
- Cranberry
- Crowberry
- Dandelion
- Eskimo potato

- Fireweed
- Iceland moss
- Marsh marigold
- Reindeer moss
- Rock tripe
- Spatterdock

There are some plants growing in arctic and subarctic regions that are poisonous if eaten (see Appendix C). Use the plants that you know are edible. When in doubt, follow the Universal Edibility Test in Chapter 9, Figure 9-5.

TRAVEL

As a survivor or an evader in an arctic or subarctic region, you will face many obstacles. Your location and the time of the year will determine the types of obstacles and the inherent dangers. You should--

- Avoid traveling during a blizzard.
- Take care when crossing thin ice. Distribute your weight by lying flat and crawling.
- Cross streams when the water level is lowest. Normal freezing and thawing action may cause a stream level to vary as much as 2 to 2.5 meters per day. This variance may occur any time during the day, depending on the distance from a glacier, the temperature, and the terrain. Consider this variation in water level when selecting a campsite near a stream.
- Consider the clear arctic air. It makes estimating distance difficult. You more frequently underestimate than overestimate distances.
- Do not travel in "whiteout" conditions. The lack of contrasting colors makes it impossible to judge the nature of the terrain.
- Always cross a snow bridge at right angles to the obstacle it crosses. Find the strongest part of the bridge by poking ahead of you with a pole or ice axe. Distribute your weight by crawling or by wearing snowshoes or skis.
- Make camp early so that you have plenty of time to build a shelter.
- Consider frozen or unfrozen rivers as avenues of travel. However, some rivers that appear frozen may have soft, open areas that make travel very difficult or may not allow walking, skiing, or sledding.
- Use snowshoes if you are traveling over snow-covered terrain. Snow 30 or more centimeters deep makes traveling difficult. If you do not have snowshoes, make a pair using willow, strips of cloth, leather, or other suitable material.

It is almost impossible to travel in deep snow without snowshoes or skis. Traveling by foot leaves a well-marked trail for any pursuers to follow. If you must travel in deep snow, avoid snow-covered streams. The snow, which acts as an insulator, may have prevented ice from forming over the water. In hilly terrain, avoid areas where avalanches appear possible. Travel in the early morning in areas where there is danger of avalanches. On ridges, snow gathers on the lee side in overhanging piles called cornices. These often extend far out from the ridge and may break loose if stepped on.

WEATHER SIGNS

There are several good indicators of climatic changes.

Wind

You can determine wind direction by dropping a few leaves or grass or by watching the treetops. Once you determine the wind direction, you can predict the type of weather that is imminent. Rapidly shifting winds indicate an unsettled atmosphere and a likely change in the weather.

Clouds

Clouds come in a variety of shapes and patterns. A general knowledge of clouds and the atmospheric conditions they indicate can help you predict the weather. See Appendix G for details.

Smoke

Smoke rising in a thin vertical column indicates fair weather. Low rising or "flattened out" smoke indicates stormy weather.

Birds and Insects

Birds and insects fly lower to the ground than normal in heavy, moisture-laden air. Such flight indicates that rain is likely. Most insect activity increases before a storm, but bee activity increases before fair weather.

Low-Pressure Front

Slow-moving or imperceptible winds and heavy, humid air often indicate a low-pressure front. Such a front promises bad weather that will probably linger for several days. You can "smell" and "hear" this front. The sluggish, humid air makes wilderness odors more pronounced than during high-pressure conditions. In addition, sounds are sharper and carry farther in low-pressure than high-pressure conditions.

CHAPTER 16 - SEA SURVIVAL

Perhaps the most difficult survival situation to be in is sea survival. Short-or long-term survival depends upon rations and equipment available and your ingenuity. You must be resourceful to survive.

Water covers about 75 percent of the earth's surface, with about 70 percent being oceans and seas. You can assume that you will sometime cross vast expanses of water. There is always the chance that the plane or ship you are on will become crippled by such hazards as storms, collision, fire, or war.

THE OPEN SEA

As a survivor on the open sea, you will face waves and wind. You may also face extreme heat or cold. To keep these environmental hazards from becoming serious problems, take precautionary measures as soon as possible. Use the available resources to protect yourself from the elements and from heat or extreme cold and humidity.

Protecting yourself from the elements meets only one of your basic needs. You must also be able to obtain water and food. Satisfying these three basic needs will help prevent serious physical and psychological problems. However, you must know how to treat health problems that may result from your situation.

Precautionary Measures

Your survival at sea depends upon--

- Your knowledge of and ability to use the available survival equipment.
- Your special skills and ability to apply them to cope with the hazards you face.
- Your will to live.

When you board a ship or aircraft, find out what survival equipment is on board, where it is stowed, and what it contains. For instance, how many life preservers and lifeboats or rafts are on board? Where are they located? What type of survival equipment do they have? How much food, water, and medicine do they contain? How many people are they designed to support?

If you are responsible for other personnel on board, make sure you know where they are and they know where you are.

Down at Sea

If you are in an aircraft that goes down at sea, take the following actions once you clear the aircraft. Whether you are in the water or in a raft --

- Get clear and upwind of the aircraft as soon as possible, but stay in the vicinity until the aircraft sinks.
- Get clear of fuel-covered water in case the fuel ignites.
- Try to find other survivors.

A search for survivors usually takes place around the entire area of and near the crash site. Missing personnel may be unconscious and floating low in the water. Figure 16-1 illustrates rescue procedures.

Figure 16-1. Rescue from water.

The best technique for rescuing personnel from the water is to throw them a life preserver attached to a line. Another is to send a swimmer (rescuer) from the raft with a line attached to a flotation device that will support the rescuer's weight. This device will help conserve a rescuer's energy while recovering the survivor. The least acceptable technique is to send an attached swimmer without flotation devices to retrieve a survivor. In all cases, the rescuer wears a life preserver. A rescuer should not underestimate the strength of a panic-stricken person in the water. A careful approach can prevent injury to the rescuer.

When the rescuer approaches a survivor in trouble from behind, there is little danger the survivor will kick, scratch, or grab him. The rescuer swims to a point directly behind the survivor and grasps the life preserver's backstrap. The rescuer uses the sidestroke to drag the survivor to the raft.

If you are in the water, make your way to a raft. If no rafts are available, try to find a large piece of floating debris to cling to. Relax; a person who knows how to relax in ocean water is in very little danger of drowning. The body's natural buoyancy will keep at least the top of the head above water, but some movement is needed to keep the face above water.

Floating on your back takes the least energy. Lie on your back in the water, spread your arms and legs, and arch your back. By controlling your breathing in and out, your face will always be out of the water and you may even sleep in this position for short periods. Your head will be partially submerged, but your face will be above water. If you cannot float on your back or if the sea is too rough, float facedown in the water as shown in Figure 16-2.

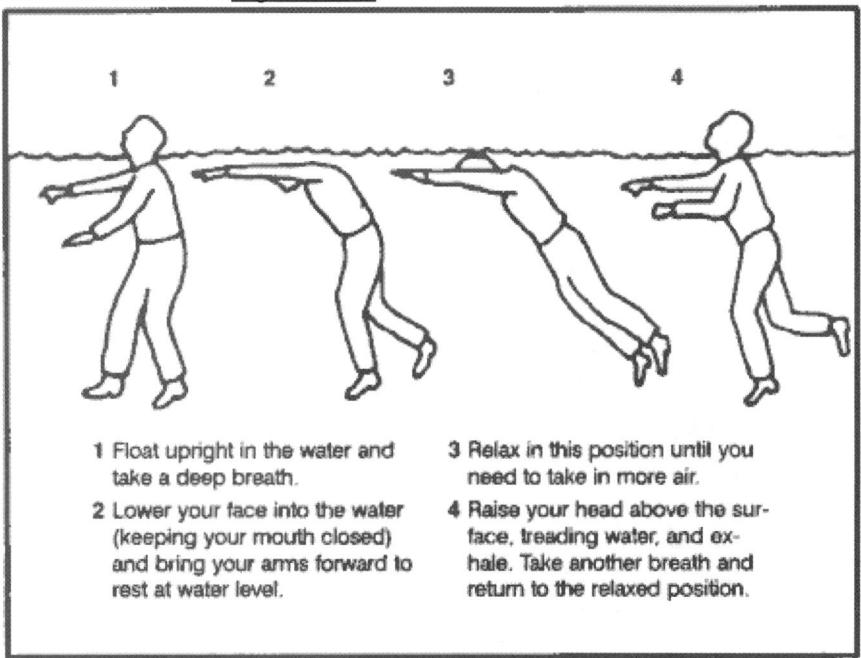

1 Float upright in the water and take a deep breath.

2 Lower your face into the water (keeping your mouth closed) and bring your arms forward to rest at water level.

3 Relax in this position until you need to take in more air.

4 Raise your head above the surface, treading water, and exhale. Take another breath and return to the relaxed position.

Figure 16-2. Floating position.

The following are the best swimming strokes during a survival situation:

- *Dog paddle.* This stroke is excellent when clothed or wearing a life jacket. Although slow in speed, it requires very little energy.
- *Breaststroke.* Use this stroke to swim underwater, through oil or debris, or in rough seas. It is probably the best stroke for long-range swimming: it allows you to conserve your energy and maintain a reasonable speed.
- *Sidestroke.* It is a good relief stroke because you use only one arm to maintain momentum and buoyancy.
- *Backstroke.* This stroke is also an excellent relief stroke. It relieves the muscles that you use for other strokes. Use it if an underwater explosion is likely.

If you are in an area where surface oil is burning--

- Discard your shoes and buoyant life preserver.

    Note: If you have an uninflated life preserver, keep it.

- Cover your nose, mouth, and eyes and quickly go underwater.
- Swim underwater as far as possible before surfacing to breathe.

- Before surfacing to breathe and while still underwater, use your hands to push burning fluid away from the area where you wish to surface. Once an area is clear of burning liquid, you can surface and take a few breaths. Try to face downwind before inhaling.
- Submerge feet first and continue as above until clear of the flames.

If you are in oil-covered water that is free of fire, hold your head high to keep the oil out of your eyes. Attach your life preserver to your wrist and then use it as a raft.

If you have a life preserver, you can stay afloat for an indefinite period. In this case, use the "HELP" body position: Heat Escaping Lessening Posture (HELP). Remain still and assume the fetal position to help you retain body heat. You lose about 50 percent of your body heat through your head. Therefore, keep your head out of the water. Other areas of high heat loss are the neck, the sides, and the groin. Figure 16-3 illustrates the HELP position.

Figure 16-3. HELP position.

If you are in a raft--

- Check the physical condition of all on board. Give first aid if necessary. Take seasickness pills if available. The best way to take these pills is to place them under the tongue and let them dissolve. There are also suppositories or injections against seasickness. Vomiting, whether from seasickness or other causes, increases the danger of dehydration.
- Try to salvage all floating equipment--rations; canteens, thermos jugs, and other containers; clothing; seat cushions; parachutes; and anything else that will be useful to you. Secure the salvaged items in or to your raft. Make sure the items have no sharp edges that can puncture the raft.
- If there are other rafts, lash the rafts together so they are about 7.5 meters apart. Be ready to draw them closer together if you see or hear an aircraft. It is easier for an aircrew to spot rafts that are close together rather than scattered.
- Remember, rescue at sea is a cooperative effort. Use all available visual or electronic signaling devices to signal and make contact with rescuers. For example, raise a flag or reflecting material on an oar as high as possible to attract attention.
- Locate the emergency radio and get it into operation. Operating instructions are on it. Use the emergency transceiver only when friendly aircraft are likely to be in the area.

- Have other signaling devices ready for instant use. If you are in enemy territory, avoid using a signaling device that will alert the enemy. However, if your situation is desperate, you may have to signal the enemy for rescue if you are to survive.

Check the raft for inflation, leaks, and points of possible chafing. Make sure the main buoyancy chambers are firm (well rounded) but not overly tight (Figure 16-4). Check inflation regularly. Air expands with heat; therefore, on hot days, release some air and add air when the weather cools.

- Decontaminate the raft of all fuel. Petroleum will weaken its surfaces and break down its glued joints.

Throw out the sea anchor, or improvise a drag from the raft's case, bailing bucket, or a roll of clothing. A sea anchor helps you stay close to your ditching site, making it easier for searchers to find you if you have relayed your location. Without a sea anchor, your raft may drift over 160 kilometers in a day, making it much harder to find you. You can adjust the sea anchor to act as a drag to slow down the rate of travel with the current, or as a means to travel with the current. You make this adjustment by opening or closing the sea anchor's apex. When open, the sea anchor (Figure 16-5) acts as a drag that keeps you in the general area. When closed, it forms a pocket for the current to strike and propels the raft in the current's direction.

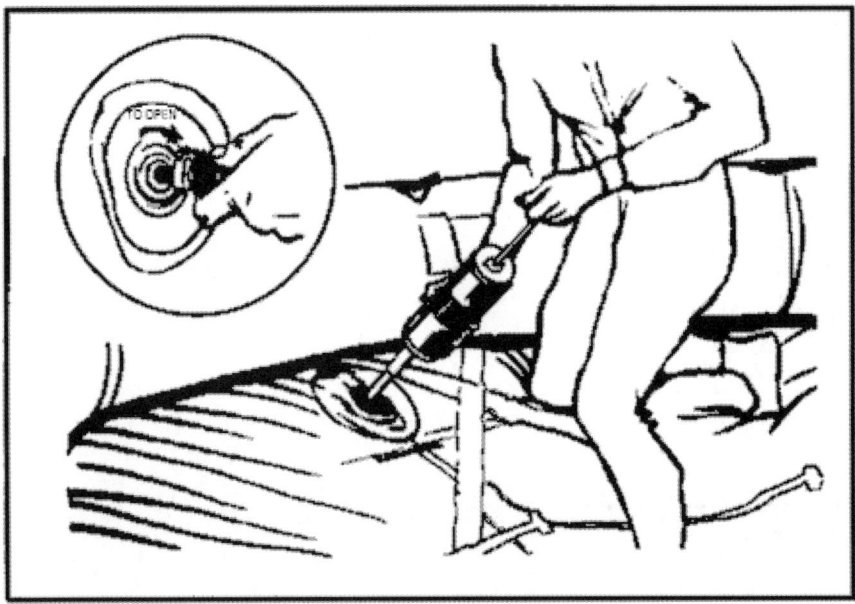

Figure 16-4. Inflating the 20-man raft.

ESSENTIAL ARMY MANUAL
[169] ARMY SURVIVAL MANUAL

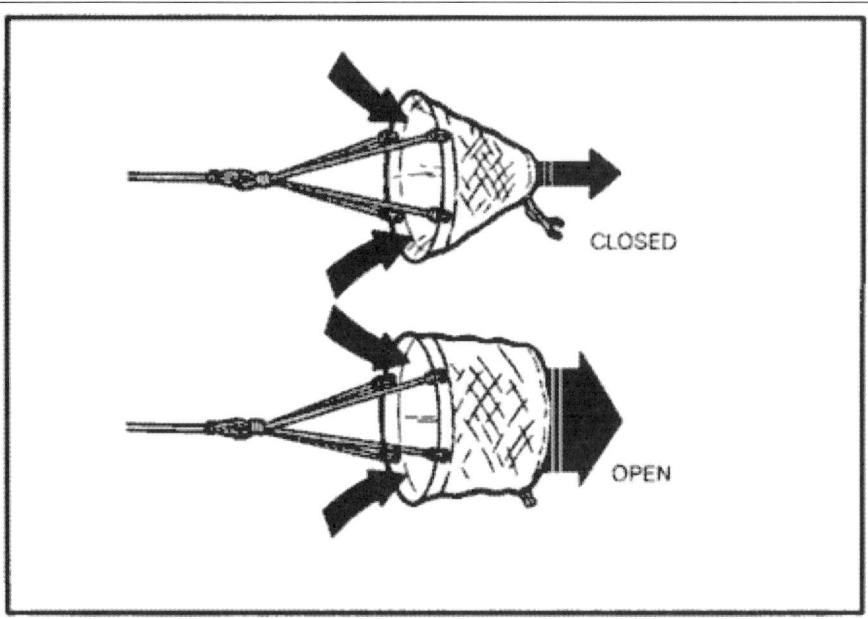

Figure 16-5. Sea anchor.

Additionally, adjust the sea anchor so that when the raft is on the wave's crest, the sea anchor is in the wave's trough (Figure 16-6).

- Wrap the sea anchor rope with cloth to prevent its chafing the raft. The anchor also helps to keep the raft headed into the wind and waves.
- In stormy water, rig the spray and windshield at once. In a 20-man raft, keep the canopy erected at all times. Keep your raft as dry as possible. Keep it properly balanced. All personnel should stay seated, the heaviest one in the center.
- Calmly consider all aspects of your situation and determine what you and your companions must do to survive. Inventory all equipment, food, and water. Waterproof items that salt water may affect. These include compasses, watches, sextant, matches, and lighters. Ration food and water.
- Assign a duty position to each person: for example, water collector, food collector, lookout, radio operator, signaler, and water bailers.

  Note: Lookout duty should not exceed 2 hours. Keep in mind and remind others that cooperation is one of the keys to survival.

- Keep a log. Record the navigator's last fix, the time of ditching, the names and physical condition of personnel, and the ration schedule. Also record the winds, weather, direction of swells, times of sunrise and sunset, and other navigational data.
- If you are down in unfriendly waters, take special security measures to avoid detection. Do not travel in the daytime. Throw out the sea anchor and wait for nightfall before paddling or hoisting sail. Keep low in the raft; stay covered with the blue side of the camouflage cloth up. Be sure a passing ship or aircraft is friendly or neutral be-fore trying to attract its attention. If the enemy detects you and you are close to capture, destroy the log book, radio, navigation equipment, maps, signaling equipment, and firearms. Jump overboard and submerge if the enemy starts strafing.
- Decide whether to stay in position or to travel. Ask yourself, "How much information was signaled before the accident? Is your position known to rescuers? Do you know it yourself? Is the weather favorable for a search? Are other ships or aircraft likely to pass your present position? How many days supply of food and water do you have?"

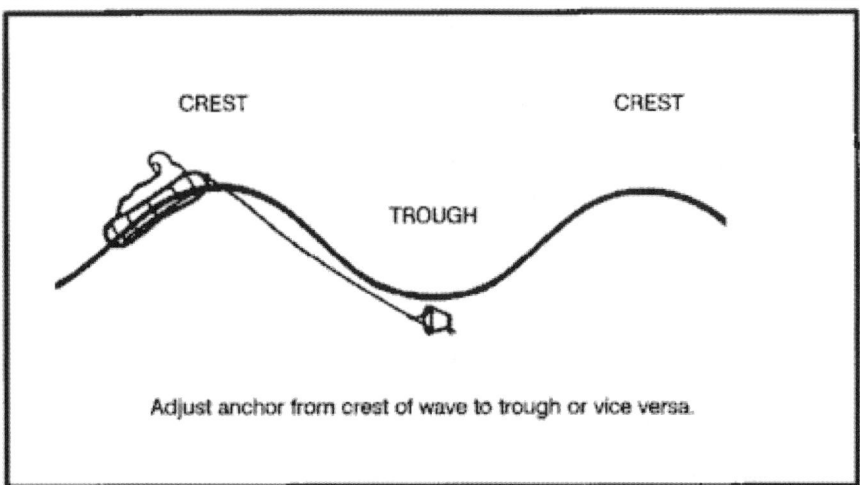

CREST   CREST

TROUGH

Adjust anchor from crest of wave to trough or vice versa.

Figure 16-6. Deployment of the sea anchor.

Cold Weather Considerations

If you are in a cold climate--

- Put on an antiexposure suit. If unavailable, put on any extra clothing available. Keep clothes loose and comfortable.
- Take care not to snag the raft with shoes or sharp objects. Keep the repair kit where you can readily reach it.
- Rig a windbreak, spray shield, and canopy.
- Try to keep the floor of the raft dry. Cover it with canvas or cloth for insulation.
- Huddle with others to keep warm, moving enough to keep the blood circulating. Spread an extra tarpaulin, sail, or parachute over the group.
- Give extra rations, if available, to men suffering from exposure to cold.

The greatest problem you face when submerged in cold water is death due to hypothermia. When you are immersed in cold water, hypothermia occurs rapidly due to the decreased insulating quality of wet clothing and the result of water displacing the layer of still air that normally surrounds the body. The rate of heat exchange in water is about 25 times greater than it is in air of the same temperature. Figure 16-7 lists life expectancy times for immersion in water.

Water Temperature	Time
21.0–15.5 degrees C (70–60 degrees F)	12 hours
15.5–10.0 degrees C (60–50 degrees F)	6 hours
10.0–4.5 degrees C (50–40 degrees F)	1 hour
4.5 degrees C (40 degrees F) and below	less than 1 hour
Note: Wearing an antiexposure suit may increase these times up to a maximum of 24 hours.	

Figure 16-7. Life expectancy times for immersion in water.

Your best protection against the effects of cold water is to get into the life raft, stay dry, and insulate your body from the cold surface of the bottom of the raft. If these actions are not possible, wearing an antiexposure suit will extend your life expectancy considerably. Remember, keep your head and neck out of the water and well insulated from the cold water's effects when the temperature is below 19 degrees C. Wearing life preservers increases the predicted survival time as body position in the water increases the chance of survival.

Hot Weather Considerations

If you are in a hot climate--

- Rig a sunshade or canopy. Leave enough space for ventilation.
- Cover your skin, where possible, to protect it from sunburn. Use sunburn cream, if available, on all exposed skin. Your eyelids, the back of your ears, and the skin under your chin sunburn easily.

Raft Procedures

Most of the rafts in the U. S. Army and Air Force inventories can satisfy the needs for personal protection, mode of travel, and evasion and camouflage.

Note: Before boarding any raft, remove and tether (attach) your life preserver to yourself or the raft. Ensure there are no other metallic or sharp objects on your clothing or equipment that could damage the raft. After boarding the raft, don your life preserver again.

One-Man Raft

The one-man raft has a main cell inflation. If the CO_2 bottle should malfunction or if the raft develops a leak, you can inflate it by mouth.

The spray shield acts as a shelter from the cold, wind, and water. In some cases, this shield serves as insulation. The raft's insulated bottom limits the conduction of cold thereby protecting you from hypothermia (Figure 16-8).

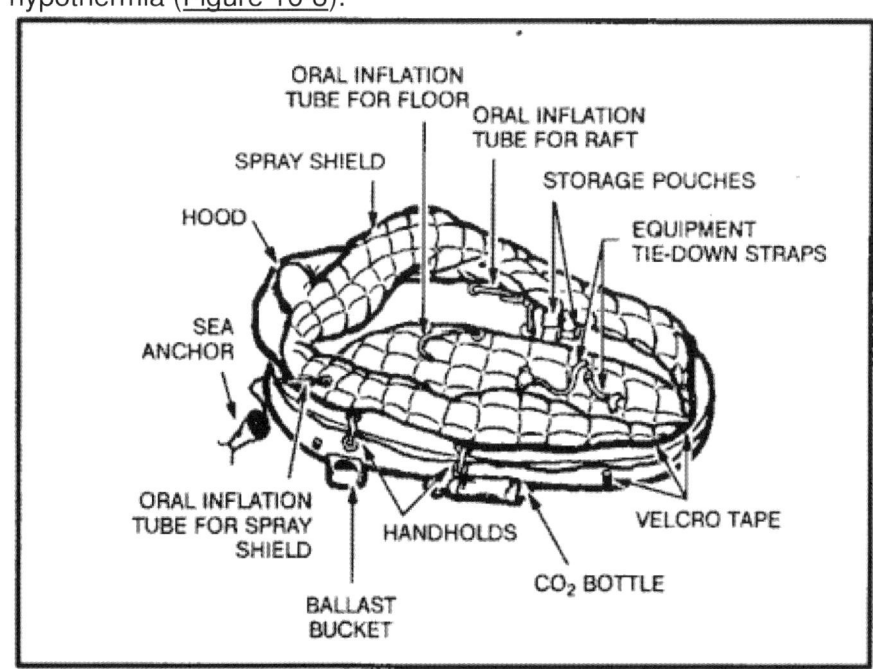

Figure 16-8. One-man raft with spray shield.

You can travel more effectively by inflating or deflating the raft to take advantage of the wind or current. You can use the spray shield as a sail white the ballast buckets serve to increase drag in the water. You may use the sea anchor to control the raft's speed and direction.

There are rafts developed for use in tactical areas that are black. These rafts blend with the sea's background. You can further modify these rafts for evasion by partially deflating them to obtain a lower profile.

A lanyard connects the one-man raft to a parachutist (survivor) landing in the water. You (the survivor) inflate it upon landing. You do not swim to the raft, but pull it to you via the lanyard. The raft may hit the water upside down, but you can right it by approaching the side to which the bottle is attached and flipping the raft over. The spray shield must be in the raft to expose the boarding handles. Follow the steps outlined in the <u>note</u> under raft procedures above when boarding the raft (<u>Figure 16-9</u>).

Figure 16-9. Boarding the one-man raft.

If you have an arm injury, the best way to board is by turning your back to the small end of the raft, pushing the raft under your buttocks, and lying back. Another way to board the raft is to push down on its small end until one knee is inside and lie forward (<u>Figure 16-10</u>).

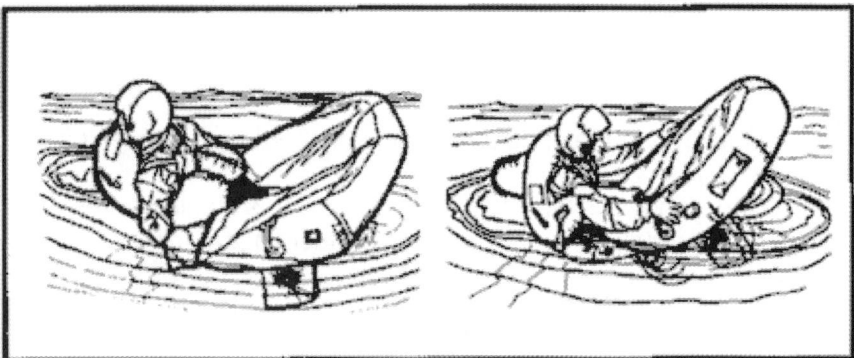

Figure 16-10. Boarding the one-man raft (other methods).

In rough seas, it may be easier for you to grasp the small end of the raft and, in a prone position, to kick and pull yourself into the raft. When you are lying face down in the raft, deploy and adjust the sea anchor. To sit upright, you may have to disconnect one side of the seat kit and roll to that side. Then you adjust the spray shield. There are two variations of the one-man raft; the improved model incorporates an inflatable spray shield and floor that provide additional insulation. The spray shield helps keep you dry and warm in cold oceans and protects you from the sun in the hot climates (<u>Figure 16-11</u>).

Figure 16-11. One-man raft with spray shield inflated.

Seven-Man Raft

Some multiplace aircraft carry the seven-man raft. It is a component of the survival drop kit (Figure 16-12). This raft may inflate upside down and require you to right the raft before boarding. Always work from the bottle side to prevent injury if the raft turns over. Facing into the wind, the wind provides additional help in righting the raft. Use the handles on the inside bottom of the raft for boarding (Figure 16-13).

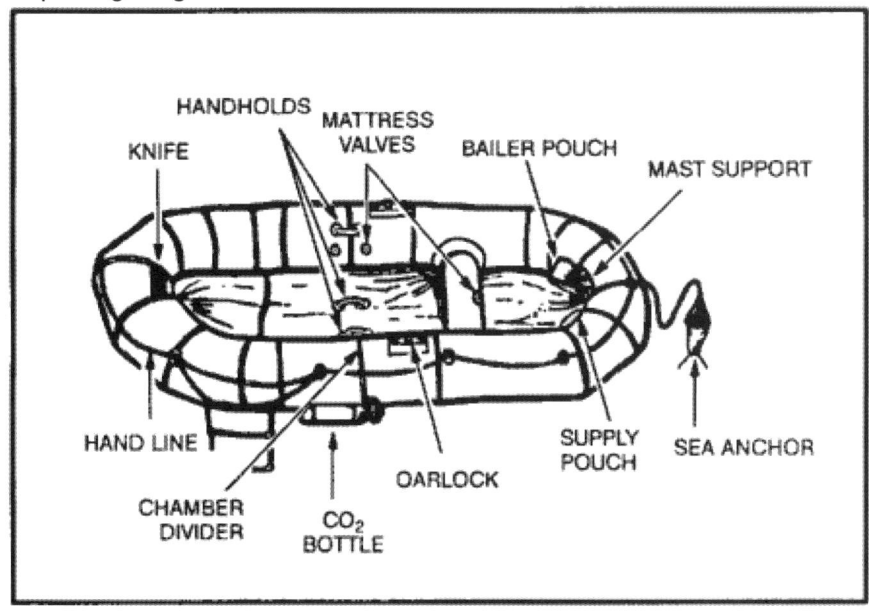

Figure 16-12. Seven-man raft.

Figure 16-13. Method of righting raft.

Use the boarding ramp if someone holds down the raft's opposite side. If you don't have help, again work from the bottle side with the wind at your back to help hold down the raft. Follow the steps outlined in the note under raft procedures above. Then grasp an oarlock and boarding handle, kick your legs to get your body prone on the water, and then kick and pull yourself into the raft. If you are weak or injured, you may partially deflate the raft to make boarding easier (Figure 16-14).

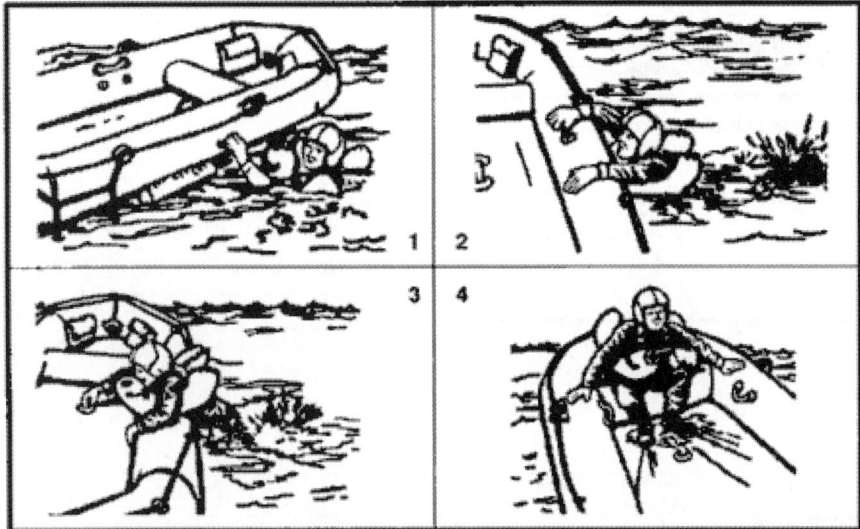

Figure 16-14. Method of boarding seven-man raft.

Use the hand pump to keep the buoyancy chambers and cross seat firm. Never overinflate the raft.
Twenty- or Twenty-Five-Man Rafts
You may find 20- or 25-man rafts in multiplace aircraft (Figures 16-15 and 16-16). You will find them in accessible areas of the fuselage or in raft compartments. Some may be automatically deployed from the cock-pit, while others may need manual deployment. No matter how the raft lands in the water, it is ready for boarding. A lanyard connects the accessory kit to the raft and you retrieve the kit by hand. You must manually inflate the center chamber with the hand pump. Board the 20- or 25-man raft from the aircraft, if possible. If not, board in the following manner:

- Approach the lower boarding ramp.
- Remove your life preserver and tether it to yourself so that it trails behind you.
- Grasp the boarding handles and kick your legs to get your body into a prone position on the water's surface; then kick and pull until you are inside the raft.

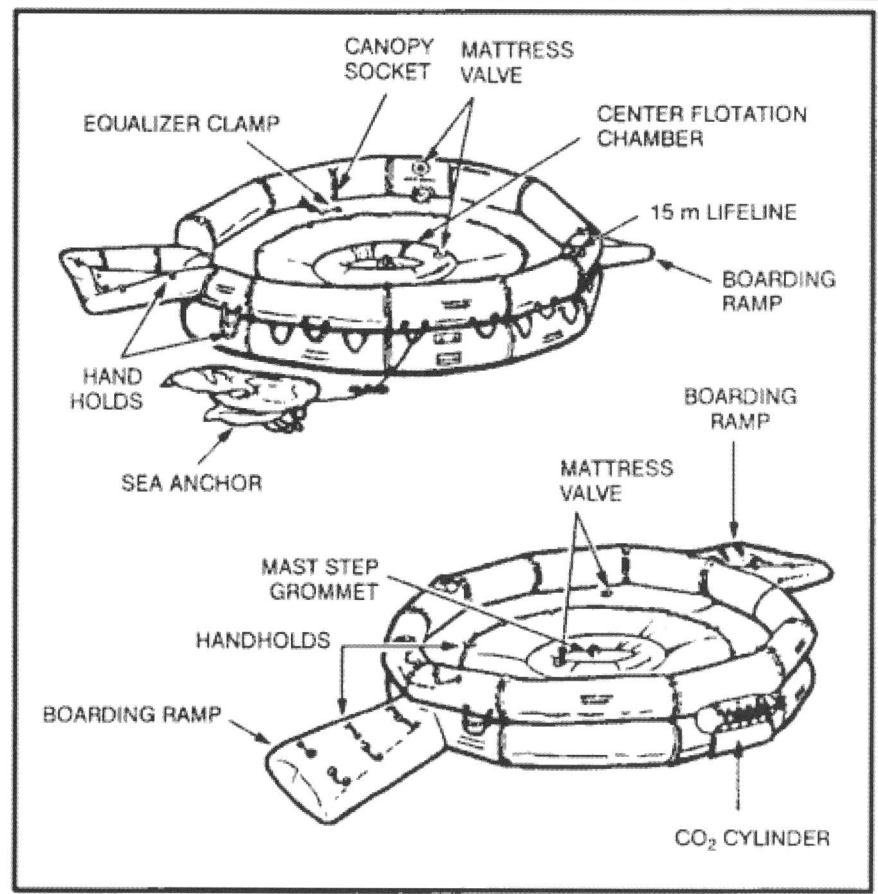

Figure 16-15. Twenty-man raft.

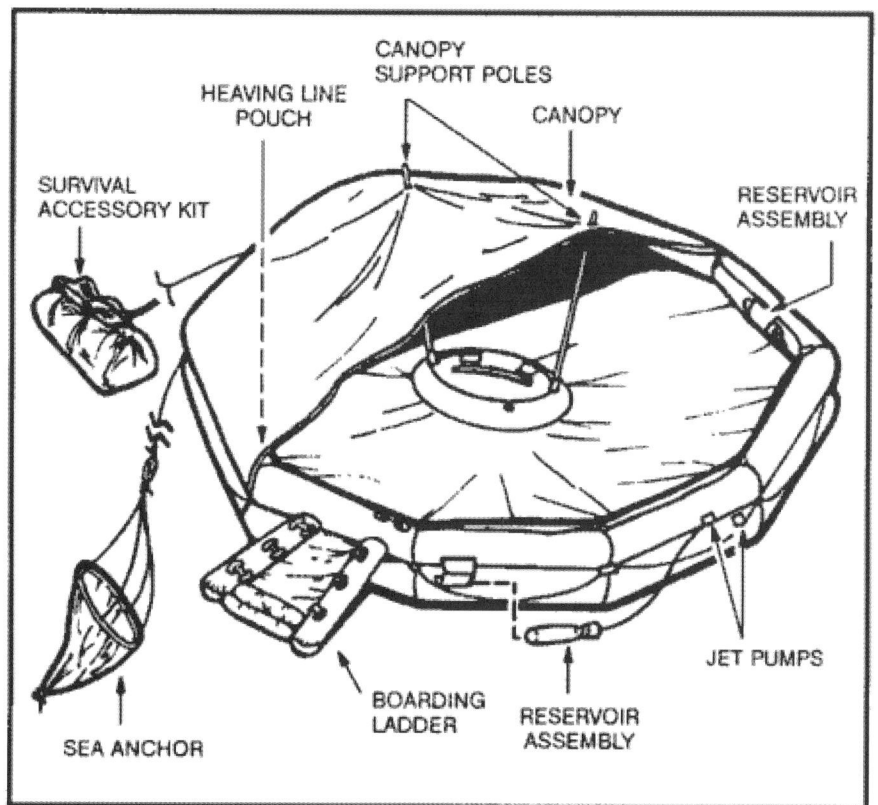

Figure 16-16. Twenty-five-man raft.

An incompletely inflated raft will make boarding easier. Approach the intersection of the raft and ramp, grasp the upper boarding handle, and swing one leg onto the center of the ramp, as in mounting a horse (Figure 16-17).

Figure 16-17. Boarding the 20-man raft.

Immediately tighten the equalizer clamp upon entering the raft to prevent deflating the entire raft in case of a puncture (Figure 16-18).

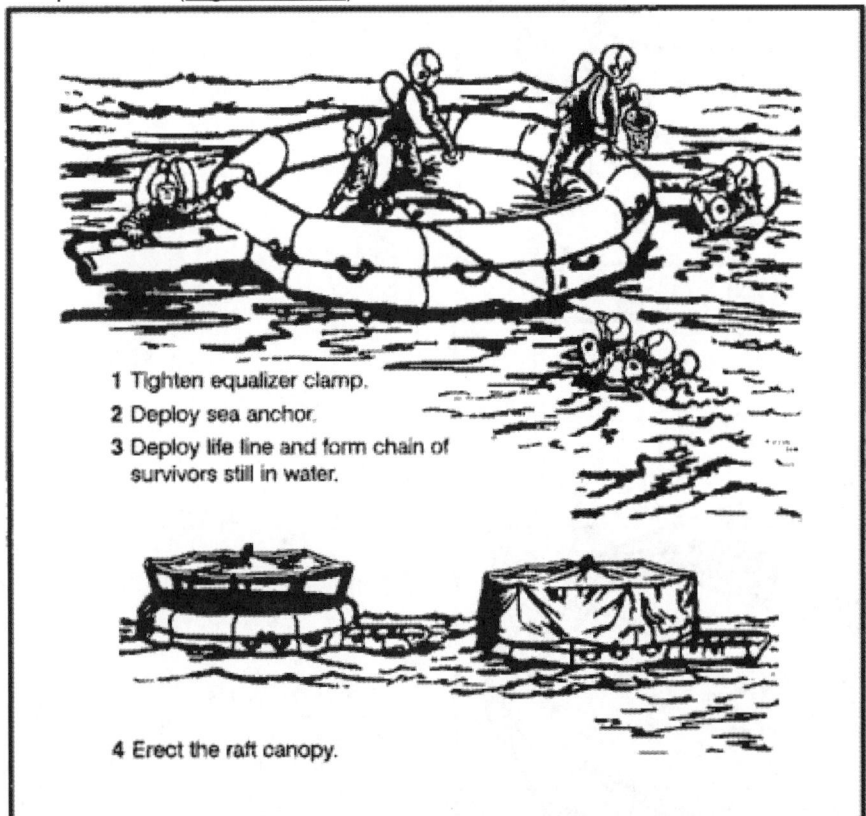

1 Tighten equalizer clamp.
2 Deploy sea anchor.
3 Deploy life line and form chain of survivors still in water.
4 Erect the raft canopy.

Figure 16-18. Immediate action—multiplace raft.

Use the pump to keep these rafts' chambers and center ring firm. They should be well rounded but not overly tight.

Sailing Rafts

Rafts do not have keels, therefore, you can't sail them into the wind. However, anyone can sail a raft downwind. You can successfully sail multiplace (except 20- to 25-man) rafts 10 degrees off from the direction of the wind. Do not try to sail the raft unless land is near. If you decide to sail and the wind is blowing toward a desired destination, fully inflate the raft, sit high, take in the sea anchor, rig a sail, and use an oar as a rudder.

In a multiplace (except 20- to 25-man) raft, erect a square sail in the bow using the oars and their extensions as the mast and crossbar (Figure 16-19). You may use a waterproof tarpaulin or parachute material for the sail. If the raft has no regular mast socket and step, erect the mast by tying it securely to the front cross seat using braces. Pad the bottom of the mast to prevent it from chafing or punching a hole through the floor, whether or not there is a socket. The heel of a shoe, with the toe wedged under the seat, makes a good improvised mast step. Do not secure the comers of the lower edge of the sail. Hold the lines attached to the comers with your hands so that a gust of wind will not rip the sail, break the mast, or capsize the raft.

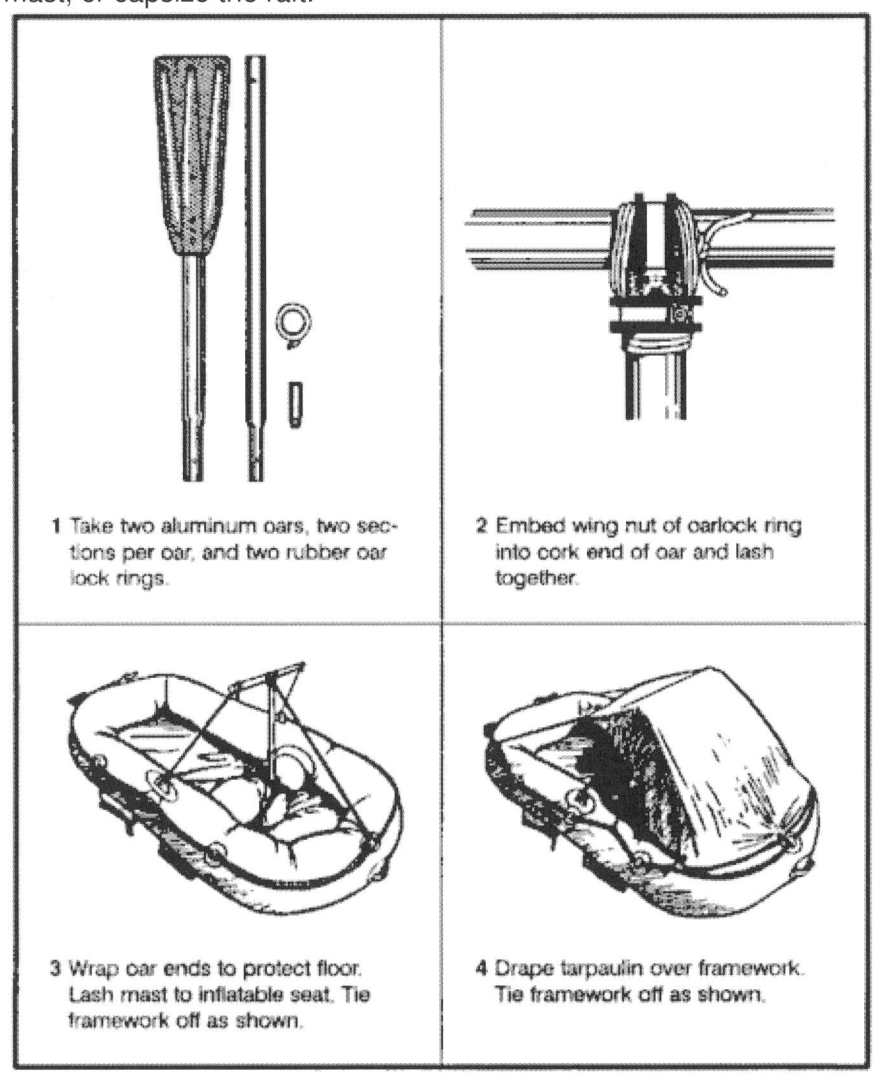

1 Take two aluminum oars, two sections per oar, and two rubber oar lock rings.

2 Embed wing nut of oarlock ring into cork end of oar and lash together.

3 Wrap oar ends to protect floor. Lash mast to inflatable seat. Tie framework off as shown.

4 Drape tarpaulin over framework. Tie framework off as shown.

Figure 16-19. Sail construction.

Take every precaution to prevent the raft from turning over. In rough weather, keep the sea anchor away from the bow. Have the passengers sit low in the raft, with their weight distributed to hold the upwind side down. To prevent falling out, they should also avoid sitting on the sides of the raft or standing up. Avoid sudden movements without warning the other passengers. When the sea anchor is not in use, tie it to the raft and stow it in such a manner that it will hold immediately if the raft capsizes.

Water

Water is your most important need. With it alone, you can live for ten days or longer, depending on your will to live. When drinking water, moisten your lips, tongue, and throat before swallowing.
Short Water Rations
When you have a limited water supply and you can't replace it by chemical or mechanical means, use the water efficiently. Protect freshwater supplies from seawater contamination. Keep your body well

shaded, both from overhead sun and from reflection off the sea surface. Allow ventilation of air; dampen your clothes during the hottest part of the day. Do not exert yourself. Relax and sleep when possible. Fix your daily water ration after considering the amount of water you have, the output of solar stills and desalting kit, and the number and physical condition of your party.

If you don't have water, don't eat. If your water ration is two liters or more per day, eat any part of your ration or any additional food that you may catch, such as birds, fish, shrimp. The life raft's motion and anxiety may cause nausea. If you eat when nauseated, you may lose your food immediately. If nauseated, rest and relax as much as you can, and take only water.

To reduce your loss of water through perspiration, soak your clothes in the sea and wring them out before putting them on again. Don't overdo this during hot days when no canopy or sun shield is available. This is a trade-off between cooling and saltwater boils and rashes that will result. Be careful not to get the bottom of the raft wet.

Watch the clouds and be ready for any chance of showers. Keep the tarpaulin handy for catching water. If it is encrusted with dried salt, wash it in seawater. Normally, a small amount of seawater mixed with rain will hardly be noticeable and will not cause any physical reaction. In rough seas you cannot get uncontaminated fresh water.

At night, secure the tarpaulin like a sunshade, and turn up its edges to collect dew. It is also possible to collect dew along the sides of the raft using a sponge or cloth. When it rains, drink as much as you can hold.

Solar Still

When solar stills are available, read the instructions and set them up immediately. Use as many stills as possible, depending on the number of men in the raft and the amount of sunlight available. Secure solar stills to the raft with care. This type of solar still only works on flat, calm seas.

Desalting Kits

When desalting kits are available in addition to solar stills, use them only for immediate water needs or during long overcast periods when you cannot use solar stills. In any event, keep desalting kits and emergency water stores for periods when you cannot use solar stills or catch rainwater.

Water From Fish

Drink the aqueous fluid found along the spine and in the eyes of large fish. Carefully cut the fish in half to get the fluid along the spine and suck the eye. If you are so short of water that you need to do this, then **do not** drink any of the other body fluids. These other fluids are rich in protein and fat and will use up more of your reserve water in digestion than they supply.

Sea Ice

In arctic waters, use old sea ice for water. This ice is bluish, has rounded comers, and splinters easily. It is nearly free of salt. New ice is gray, milky, hard, and salty. Water from icebergs is fresh, but icebergs are dangerous to approach. Use them as a source of water only in emergencies.

REMEMBER!

Do not drink seawater.

Do not drink urine.

Do not drink alcohol.

Do not smoke.

Do not eat, unless water is available.

Sleep and rest are the best ways of enduring periods of reduced water and food intake. However, make sure that you have enough shade when napping during the day. If the sea is rough, tie yourself to the raft, close any cover, and ride out the storm as best you can. *Relax* is the key word--at least try to relax.

Food Procurement

In the open sea, fish will be the main food source. There are some poisonous and dangerous ocean fish, but, in general, when out of sight of land, fish are safe to eat. Nearer the shore there are fish that are both dangerous and poisonous to eat. There are some fish, such as the red snapper and barracuda, that are normally edible but poisonous when taken from the waters of atolls and reefs. Flying fish will even jump into your raft!

Fish

When fishing, do not handle the fishing line with bare hands and never wrap it around your hands or tie it to a life raft. The salt that adheres to it can make it a sharp cutting edge, an edge dangerous both to the raft and your hands. Wear gloves, if they are available, or use a cloth to handle fish and to avoid injury from sharp fins and gill covers.

In warm regions, gut and bleed fish immediately after catching them. Cut fish that you do not eat immediately into thin, narrow strips and hang them to dry. A well-dried fish stays edible for several days. Fish not cleaned and dried may spoil in half a day. Fish with dark meat are very prone to decomposition. If you do not eat them all immediately, do not eat any of the leftovers. Use the leftovers for bait.

Never eat fish that have pale, shiny gills, sunken eyes, flabby skin and flesh, or an unpleasant odor. Good fish show the opposite characteristics. Sea fish have a saltwater or clean fishy odor. Do not confuse eels with sea snakes that have an obviously scaly body and strongly compressed, paddle-shaped tail. Both eels and sea snakes are edible, but you must handle the latter with care because of their poisonous bites. The heart, blood, intestinal wall, and liver of most fish are edible. Cook the intestines. Also edible are the partly digested smaller fish that you may find in the stomachs of large fish. In addition, sea turtles are edible.

Shark meat is a good source of food whether raw, dried, or cooked. Shark meat spoils very rapidly due to the high concentration of urea in the blood, therefore, bleed it immediately and soak it in several changes of water. People prefer some shark species over others. Consider them all edible except the Greenland shark whose flesh contains high quantities of vitamin A. Do not eat the livers, due to high vitamin A content.

Fishing Aids

You can use different materials to make fishing aids as described in the following paragraphs:

- *Fishing line.* Use pieces of tarpaulin or canvas. Unravel the threads and tie them together in short lengths in groups of three or more threads. Shoelaces and parachute suspension line also work well.
- *Fish hooks.* No survivor at sea should be without fishing equipment but if you are, improvise hooks as shown in Chapter 8.
- *Fish lures.* You can fashion lures by attaching a double hook to any shiny piece of metal.
- *Grapple.* Use grapples to hook seaweed. You may shake crabs, shrimp, or small fish out of the seaweed. These you may eat or use for bait. You may eat seaweed itself, but only when you have plenty of drinking water. Improvise grapples from wood. Use a heavy piece of wood as the main shaft, and lash three smaller pieces to the shaft as grapples.
- *Bait.* You can use small fish as bait for larger ones. Scoop the small fish up with a net. If you don't have a net, make one from cloth of some type. Hold the net under the water and scoop upward. Use all the guts from birds and fish for bait. When using bait, try to keep it moving in the water to give it the appearance of being alive.

Helpful Fishing Hints

Your fishing should be successful if you remember the following important hints:

- Be extremely careful with fish that have teeth and spines.
- Cut a large fish loose rather than risk capsizing the raft. Try to catch small rather than large fish.
- Do not puncture your raft with hooks or other sharp instruments.
- Do not fish when large sharks are in the area.
- Watch for schools of fish; try to move close to these schools.
- Fish at night using a light. The light attracts fish.
- In the daytime, shade attracts some fish. You may find them under your raft.

- Improvise a spear by tying a knife to an oar blade. This spear can help you catch larger fish, but you must get them into the raft quickly or they will slip off the blade. Also, tie the knife very securely or you may lose it.
- Always take care of your fishing equipment. Dry your fishing lines, clean and sharpen the hooks, and do not allow the hooks to stick into the fishing lines.

Birds

As stated in Chapter 8, all birds are edible. Eat any birds you can catch. Sometimes birds may land on your raft, but usually they are cautious. You may be able to attract some birds by towing a bright piece of metal behind the raft. This will bring the bird within shooting range, provided you have a firearm.
If a bird lands within your reach, you may be able to catch it. If the birds do not land close enough or land on the other end of the raft, you may be able to catch them with a bird noose. Bait the center of the noose and wait for the bird to land. When the bird's feet are in the center of the noose, pull it tight.
Use all parts of the bird. Use the feathers for insulation, the entrails and feet for bait, and so on. Use your imagination.

Medical Problems Associated With Sea Survival

At sea, you may become seasick, get saltwater sores, or face some of the same medical problems that occur on land, such as dehydration or sunburn. These problems can become critical if left untreated.
Seasickness
Seasickness is the nausea and vomiting caused by the motion of the raft. It can result in--

- Extreme fluid loss and exhaustion.
- Loss of the will to survive.
- Others becoming seasick.
- Attraction of sharks to the raft.
- Unclean conditions.

To treat seasickness--

- Wash both the patient and the raft to remove the sight and odor of vomit.
- Keep the patient from eating food until his nausea is gone.
- Have the patient lie down and rest.
- Give the patient seasickness pills if available. If the patient is unable to take the pills orally, insert them rectally for absorption by the body.

  Note: Some survivors have said that erecting a canopy or using the horizon as a focal point helped overcome seasickness. Others have said that swimming alongside the raft for short periods helped, but extreme care must be taken if swimming.

Saltwater Sores
These sores result from a break in skin exposed to saltwater for an extended period. The sores may form scabs and pus. Do not open or drain. Flush the sores with fresh water, if available, and allow to dry. Apply an antiseptic, if available.
Immersion Rot, Frostbite, and Hypothermia
These problems are similar to those encountered in cold weather environments. Symptoms and treatment are the same as covered in Chapter 15.
Blindness/Headache
If flame, smoke, or other contaminants get in the eyes, flush them immediately with salt water, then with fresh water, if available. Apply ointment, if available. Bandage both eyes 18 to 24 hours, or longer if damage is severe. If the glare from the sky and water causes your eyes to become bloodshot and

inflamed, bandage them lightly. Try to prevent this problem by wearing sunglasses. Improvise sunglasses if necessary.

Constipation

This condition is a common problem on a raft. Do not take a laxative, as this will cause further dehydration. Exercise as much as possible and drink an adequate amount of water, if available.

Difficult Urination

This problem is not unusual and is due mainly to dehydration. It is best not to treat it, as it could cause further dehydration.

Sunburn

Sunburn is a serious problem in sea survival. Try to prevent sunburn by staying in shade and keeping your head and skin covered. Use cream or Chap Stick from your first aid kit. Remember, reflection from the water also causes sunburn.

Sharks

Whether you are in the water or in a boat or raft, you may see many types of sea life around you. Some may be more dangerous than others. Generally, sharks are the greatest danger to you. Other animals such as whales, porpoises, and stingrays may look dangerous, but really pose little threat in the open sea.

Of the many hundreds of shark species, only about 20 species are known to attack man. The most dangerous are the great white shark, the hammerhead, the mako, and the tiger shark. Other sharks known to attack man include the gray, blue, lemon, sand, nurse, bull, and oceanic white tip sharks. Consider any shark longer than 1 meter dangerous.

There are sharks in all oceans and seas of the world. While many live and feed in the depths of the sea, others hunt near the surface. The sharks living near the surface are the ones you will most likely see. Their dorsal fins frequently project above the water. Sharks in the tropical and subtropical seas are far more aggressive than those in temperate waters.

All sharks are basically eating machines. Their normal diet is live animals of any type, and they will strike at injured or helpless animals. Sight, smell, or sound may guide them to their prey. Sharks have an acute sense of smell and the smell of blood in the water excites them. They are also very sensitive to any abnormal vibrations in the water. The struggles of a wounded animal or swimmer, underwater explosions, or even a fish struggling on a fishline will attract a shark.

Sharks can bite from almost any position; they do not have to turn on their side to bite. The jaws of some of the larger sharks are so far forward that they can bite floating objects easily without twisting to the side.

Sharks may hunt alone, but most reports of attacks cite more than one shark present. The smaller sharks tend to travel in schools and attack in mass. Whenever one of the sharks finds a victim, the other sharks will quickly join it. Sharks will eat a wounded shark as quickly as their prey.

Sharks feed at all hours of the day and night. Most reported shark contacts and attacks were during daylight, and many of these have been in the late afternoon. Some of the measures that you can take to protect yourself against sharks when you are in the water are--

- *Stay with other swimmers.* A group can maintain a 360-degree watch. A group can either frighten or fight off sharks better than one man.
- *Always watch for sharks.* Keep all your clothing on, to include your shoes. Historically, sharks have attacked the unclothed men in groups first, mainly in the feet. Clothing also protects against abrasions should the shark brush against you.
- *Avoid urinating.* If you must, only do so in small amounts. Let it dissipate between discharges. If you must defecate, do so in small amounts and throw it as far away from you as possible. Do the same if you must vomit.

If a shark attack is imminent while you are in the water, splash and yell just enough to keep the shark at bay. Sometimes yelling underwater or slapping the water repeatedly will scare the shark away. Conserve your strength for fighting in case the shark attacks.

If attacked, kick and strike the shark. Hit the shark on the gills or eyes if possible. If you hit the shark on the nose, you may injure your hand if it glances off and hits its teeth.
When you are in a raft and see sharks--

- Do not fish. If you have hooked a fish, let it go. Do not clean fish in the water.
- Do not throw garbage overboard.
- Do not let your arms, legs, or equipment hang in the water.
- Keep quiet and do not move around.
- Bury all dead as soon as possible. If there are many sharks in the area, conduct the burial at night.

When you are in a raft and a shark attack is imminent, hit the shark with anything you have, except your hands. You will do more damage to your hands than the shark. If you strike with an oar, be careful not to lose or break it.

Detecting Land

You should watch carefully for any signs of land. There are many indicators that land is near.
A fixed cumulus cloud in a clear sky or in a sky where all other clouds are moving often hovers over or slightly downwind from an island.
In the tropics, the reflection of sunlight from shallow lagoons or shelves of coral reefs often causes a greenish tint in the sky.
In the arctic, light-colored reflections on clouds often indicate ice fields or snow-covered land. These reflections are quite different from the dark gray ones caused by open water.
Deep water is dark green or dark blue. Lighter color indicates shallow water, which may mean land is near.
At night, or in fog, mist, or rain, you may detect land by odors and sounds. The musty odor of mangrove swamps and mud flats carry a long way. You hear the roar of surf long before you see the surf. The continued cries of seabirds coming from one direction indicate their roosting place on nearby land.
There usually are more birds near land than over the open sea. The direction from which flocks fly at dawn and to which they fly at dusk may indicate the direction of land. During the day, birds are searching for food and the direction of flight has no significance.
Mirages occur at any latitude, but they are more likely in the tropics, especially during the middle of the day. Be careful not to mistake a mirage for nearby land. A mirage disappears or its appearance and elevation change when viewed from slightly different heights.
You may be able to detect land by the pattern of the waves (refracted) as they approach land (Figure 16-20). By traveling with the waves and parallel to the slightly turbulent area marked "X" on the illustration, you should reach land.

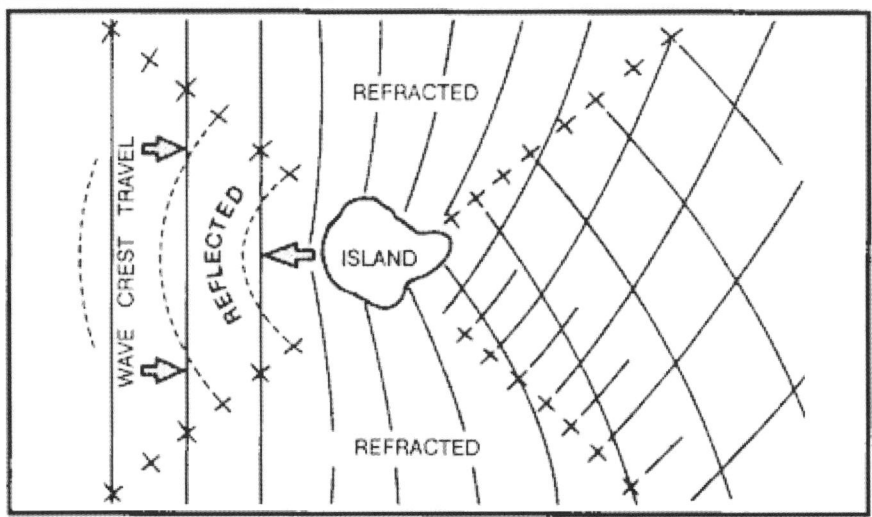

Figure 16-20. Wave patterns about an island.

Rafting or Beaching Techniques

Once you have found land, you must get ashore safely. To raft ashore, you can usually use the one-man raft without danger. However, going ashore in a strong surf is dangerous. Take your time. Select your landing point carefully. Try not to land when the sun is low and straight in front of you. Try to land on the lee side of an island or on a point of land jutting out into the water. Keep your eyes open for gaps in the surf line, and head for them. Avoid coral reefs and rocky cliffs. There are no coral reefs near the mouths of freshwater streams. Avoid rip currents or strong tidal currents that may carry you far out to sea. Either signal ashore for help or sail around and look for a sloping beach where the surf is gentle.

If you have to go through the surf to reach shore, take down the mast. Keep your clothes and shoes on to avoid severe cuts. Adjust and inflate your life vest. Trail the sea anchor over the stem using as much line as you have. Use the oars or paddles and constantly adjust the sea anchor to keep a strain on the anchor line. These actions will keep the raft pointed toward shore and prevent the sea from throwing the stern around and capsizing you. Use the oars or paddles to help ride in on the seaward side of a large wave.

The surf may be irregular and velocity may vary, so modify your procedure as conditions demand. A good method of getting through the surf is to have half the men sit on one side of the raft, half on the other, facing away from each other. When a heavy sea bears down, half should row (pull) toward the sea until the crest passes; then the other half should row (pull) toward the shore until the next heavy sea comes along.

Against a strong wind and heavy surf, the raft must have all possible speed to pass rapidly through the oncoming crest to avoid being turned broadside or thrown end over end. If possible, avoid meeting a large wave at the moment it breaks.

If in a medium surf with no wind or offshore wind, keep the raft from passing over a wave so rapidly that it drops suddenly after topping the crest. If the raft turns over in the surf, try to grab hold of it and ride it in. As the raft nears the beach, ride in on the crest of a large wave. Paddle or row hard and ride in to the beach as far as you can. Do not jump out of the raft until it has grounded, then quickly get out and beach it.

If you have a choice, do not land at night. If you have reason to believe that people live on the shore, lay away from the beach, signal, and wait for the inhabitants to come out and bring you in.

If you encounter sea ice, land only on large, stable floes. Avoid icebergs that may capsize and small floes or those obviously disintegrating. Use oars and hands to keep the raft from rubbing on the edge of the ice. Take the raft out of the water and store it well back from the floe's edge. You may be able to use it for shelter. Keep the raft inflated and ready for use. Any floe may break up without warning.

Swimming Ashore

If rafting ashore is not possible and you have to swim, wear your shoes and at least one thickness of clothing. Use the sidestroke or breaststroke to conserve strength.

If the surf is moderate, ride in on the back of a small wave by swimming forward with it. Dive to a shallow depth to end the ride just before the wave breaks.

In high surf, swim toward shore in the trough between waves. When the seaward wave approaches, face it and submerge. After it passes, work toward shore in the next trough. If caught in the undertow of a large wave, push off the bottom or swim to the surface and proceed toward shore as above.

If you must land on a rocky shore, look for a place where the waves rush up onto the rocks. Avoid places where the waves explode with a high, white spray. Swim slowly when making your approach. You will need your strength to hold on to the rocks. You should be fully clothed and wear shoes to reduce injury. After selecting your landing point, advance behind a large wave into the breakers. Face toward shore and take a sitting position with your feet in front, 60 to 90 centimeters (2 or 3 feet) lower than your head. This position will let your feet absorb the shock when you land or strike sub-merged boulders or reefs. If you do not reach shore behind the wave you picked, swim with your hands only. As the next wave approaches, take a sitting position with your feet forward. Repeat the procedure until you land.

Water is quieter in the lee of a heavy growth of seaweed. Take advantage of such growth. Do not swim through the seaweed; crawl over the top by grasping the vegetation with overhand movements.

Cross a rocky or coral reef as you would land on a rocky shore. Keep your feet close together and your knees slightly bent in a relaxed sitting posture to cushion the blows against the coral.

Pickup or Rescue

On sighting rescue craft approaching for pickup (boat, ship, conventional aircraft, or helicopter), quickly clear any lines (fishing lines, desalting kit lines) or other gear that could cause entanglement during rescue. Secure all loose items in the raft. Take down canopies and sails to ensure a safer pickup. After securing all items, put on your helmet, if available. Fully inflate your life preserver. Remain in the raft, unless otherwise instructed, and remove all equipment except the preservers. If possible, you will receive help from rescue personnel lowered into the water. Remember, follow all instructions given by the rescue personnel.

If the helicopter recovery is unassisted, do the following before pickup:

- Secure all the loose equipment in the raft, accessory bag, or in pockets.
- Deploy the sea anchor, stability bags, and accessory bag.
- Partially deflate the raft and fill it with water.
- Unsnap the survival kit container from the parachute harness.
- Grasp the raft handhold and roll out of the raft.
- Allow the recovery device or the cable to ground out on the water's surface.
- Maintain the handhold until the recovery device is in your other hand.
- Mount the recovery device, avoiding entanglement with the raft.
- Signal the hoist operator for pickup.

SEASHORES

Search planes or ships do not always spot a drifting raft or swimmer. You may have to land along the coast before being rescued. Surviving along the seashore is different from open sea survival. Food and water are more abundant and shelter is obviously easier to locate and construct.

If you are in friendly territory and decide to travel, it is better to move along the coast than to go inland. Do not leave the coast except to avoid obstacles (swamps and cliffs) or unless you find a trail that you know leads to human habitation.

In time of war, remember that the enemy patrols most coastlines. These patrols may cause problems for you if you land on a hostile shore. You will have extremely limited travel options in this situation. Avoid all contact with other humans, and make every effort to cover all tracks you leave on the shore.

Special Health Hazards

Coral, poisonous and aggressive fish, crocodiles, sea urchins, sea biscuits, sponges, anemones, and tides and undertow pose special health hazards.

Coral

Coral, dead or alive, can inflict painful cuts. There are hundreds of water hazards that can cause deep puncture wounds, severe bleeding, and the danger of infection. Clean all coral cuts thoroughly. Do not use iodine to disinfect any coral cuts. Some coral polyps feed on iodine and may grow inside your flesh if you use iodine.

Poisonous Fish

Many reef fish have toxic flesh. For some species, the flesh is always poisonous, for other species, only at certain times of the year. The poisons are present in all parts of the fish, but especially in the liver, intestines, and eggs.

Fish toxins are water soluble--no amount of cooking will neutralize them. They are tasteless, therefore the standard edibility tests are use-less. Birds are least susceptible to the poisons. Therefore, do not think that because a bird can eat a fish, it is a safe species for you to eat.

The toxins will produce a numbness of the lips, tongue, toes, and tips of the fingers, severe itching, and a clear reversal of temperature sensations. Cold items appear hot and hot items cold. There will probably also be nausea, vomiting, loss of speech, dizziness, and a paralysis that eventually brings death.

In addition to fish with poisonous flesh, there are those that are dangerous to touch. Many stingrays have a poisonous barb in their tail. There are also species that can deliver an electric shock. Some reef fish, such as stonefish and toadfish, have venomous spines that can cause very painful although seldom fatal injuries. The venom from these spines causes a burning sensation or even an agonizing pain that is out of proportion to the apparent severity of the wound. Jellyfish, while not usually fatal, can inflict a very painful sting if it touches you with its tentacles. See Chapter 11 and Appendix F for details on particularly dangerous fish of the sea and seashore.

Aggressive Fish

You should also avoid some ferocious fish. The bold and inquisitive barracuda has attacked men wearing shiny objects. It may charge lights or shiny objects at night. The sea bass, which can grow to 1.7 meters, is another fish to avoid. The moray eel, which has many sharp teeth and grows to 1.5 meters, can also be aggressive if disturbed.

Sea Snakes

Sea snakes are venomous and sometimes found in mid ocean. They are unlikely to bite unless provoked. **Avoid** them.

Crocodiles

Crocodiles inhabit tropical saltwater bays and mangrove-bordered estuaries and range up to 65 kilometers into the open sea. Few remain near inhabited areas. You commonly find crocodiles in the remote areas of the East Indies and Southeast Asia. Consider specimens over 1 meter long dangerous, especially females guarding their nests. Crocodile meat is an excellent source of food when available.

Sea Urchins, Sea Biscuits, Sponges, and Anemones

These animals can cause extreme, though seldom fatal, pain. Usually found in tropical shallow water near coral formations, sea urchins resemble small, round porcupines. If stepped on, they slip fine needles of lime or silica into the skin, where they break off and fester. If possible, remove the spines and treat the injury for infection. The other animals mentioned inflict injury similarly.

Tides and Undertow

These are another hazard to contend with. If caught in a large wave's undertow, push off the bottom or swim to the surface and proceed shoreward in a trough between waves. Do not fight against the pull of the undertow. Swim with it or perpendicular to it until it loses strength, then swim for shore.

Food

Obtaining food along a seashore should not present a problem. There are many types of seaweed and other plants you can easily find and eat. See Chapter 9 and Appendix B for a discussion of these plants. There is a great variety of animal life that can supply your need for food in this type of survival situation.

Mollusks

Mussels, limpets, clams, sea snails, octopuses, squids, and sea slugs are all edible. Shellfish will usually supply most of the protein eaten by coastal survivors. Avoid the blue-ringed octopus and cone shells (described in Chapter 11 and Appendix F). Also beware of "red tides" that make mollusks poisonous. Apply the edibility test on each species before eating.

Worms

Coastal worms are generally edible, but it is better to use them for fish bait. Avoid bristle worms that look like fuzzy caterpillars. Also avoid tubeworms that have sharp-edged tubes. Arrowworms, alias amphioxus, are not true worms. You find them in the sand and are excellent either fresh or dried.

Crabs, Lobsters, and Barnacles

These animals are seldom dangerous to man and are an excellent food source. The pincers of larger crabs or lobsters can crush a man's finger. Many species have spines on their shells, making it preferable to wear gloves when catching them. Barnacles can cause scrapes or cuts and are difficult to detach from their anchor, but the larger species are an excellent food source.

Sea Urchins

These are common and can cause painful injuries when stepped on or touched. They are also a good source of food. Handle them with gloves, and remove all spines.

Sea Cucumbers

This animal is an important food source in the Indo-Pacific regions. Use them whole after evisceration or remove the five muscular strips that run the length of its body. Eat them smoked, pickled, or cooked.

CHAPTER 17 - EXPEDIENT WATER CROSSINGS

In a survival situation, you may have to cross a water obstacle. It may be in the form of a river, a stream, a lake, a bog, quicksand, quagmire, or muskeg. Even in the desert, flash floods occur, making streams an obstacle. Whatever it is, you need to know how to cross it safely.

RIVERS AND STREAMS

You can apply almost every description to rivers and streams. They may be shallow or deep, slow or fast moving, narrow or wide. Before you try to cross a river or stream, develop a good plan.
Your first step is to look for a high place from which you can get a good view of the river or stream. From this place, you can look for a place to cross. If there is no high place, climb a tree. Good crossing locations include--

- A level stretch where it breaks into several channels. Two or three narrow channels are usually easier to cross than a wide river.
- A shallow bank or sandbar. If possible, select a point upstream from the bank or sandbar so that the current will carry you to it if you lose your footing.
- A course across the river that leads downstream so that you will cross the current at about a 45-degree angle.

The following areas possess potential hazards; avoid them, if possible:

- *Obstacles on the opposite side of the river that might hinder your travel.* Try to select the spot from which travel will be the safest and easiest.
- *A ledge of rocks that crosses the river.* This often indicates dangerous rapids or canyons.
- *A deep or rapid waterfall or a deep channel.* Never try to ford a stream directly above or even close to such hazards.
- *Rocky places.* You may sustain serious injuries from slipping or falling on rocks. Usually, submerged rocks are very slick, making balance extremely difficult. An occasional rock that breaks the current, however, may help you.
- *An estuary of a river.* An estuary is normally wide, has strong currents, and is subject to tides. These tides can influence some rivers many kilometers from their mouths. Go back upstream to an easier crossing site.
- *Eddies.* An eddy can produce a powerful backward pull downstream of the obstruction causing the eddy and pull you under the surface.

The depth of a fordable river or stream is no deterrent if you can keep your footing. In fact, deep water sometimes runs more slowly and is therefore safer than fast-moving shallow water. You can always dry your clothes later, or if necessary, you can make a raft to carry your clothing and equipment across the river.

You must not try to swim or wade across a stream or river when the water is at very low temperatures. This swim could be fatal. Try to make a raft of some type. Wade across if you can get only your feet wet. Dry them vigorously as soon as you reach the other bank.

RAPIDS

If necessary, you can safely cross a deep, swift river or rapids. To swim across a deep, swift river, swim with the current, never fight it. Try to keep your body horizontal to the water. This will reduce the danger of being pulled under.

In fast, shallow rapids, lie on your back, feet pointing downstream, finning your hands alongside your hips. This action will increase buoyancy and help you steer away from obstacles. Keep your feet up to avoid getting them bruised or caught by rocks.

In deep rapids, lie on your stomach, head downstream, angling toward the shore whenever you can. Watch for obstacles and be careful of backwater eddies and converging currents, as they often contain dangerous swirls. Converging currents occur where new watercourses enter the river or where water has been diverted around large obstacles such as small islands.

To ford a swift, treacherous stream, apply the following steps:

- Remove your pants and shirt to lessen the water's pull on you. Keep your footgear on to protect your feet and ankles from rocks. It will also provide you with firmer footing.
- Tie your pants and other articles to the top of your rucksack or in a bundle, if you have no pack. This way, if you have to release your equipment, all your articles will be together. It is easier to find one large pack than to find several small items.
- Carry your pack well up on your shoulders and be sure you can easily remove it, if necessary. Not being able to get a pack off quickly enough can drag even the strongest swimmers under.

Find a strong pole about 7.5 centimeters in diameter and 2.1 to 2.4 meters long to help you ford the stream. Grasp the pole and plant it firmly on your upstream side to break the current. Plant your feet firmly with each step, and move the pole forward a little downstream from its previous position, but still upstream from you. With your next step, place your foot below the pole. Keep the pole well slanted so that the force of the current keeps the pole against your shoulder (Figure 17-1).

- Cross the stream so that you will cross the downstream current at a 45-degree angle.

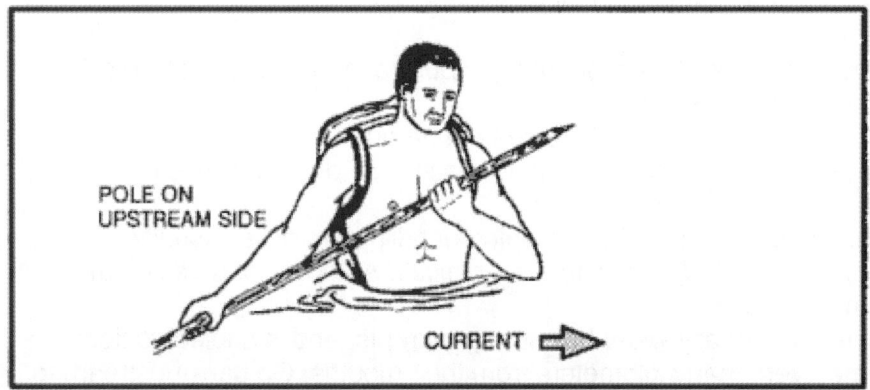

Figure 17-1. One man crossing swift stream.

Using this method, you can safely cross currents usually too strong for one person to stand against. Do not concern yourself about your pack's weight, as the weight will help rather than hinder you in fording the stream.

If there are other people with you, cross the stream together. Ensure that everyone has prepared their pack and clothing as outlined above. Position the heaviest person on the downstream end of the pole and the lightest on the upstream end. In using this method, the upstream person breaks the current, and those below can move with relative ease in the eddy formed by the upstream person. If the upstream person gets temporarily swept off his feet, the others can hold steady while he regains his footing (Figure 17-2).

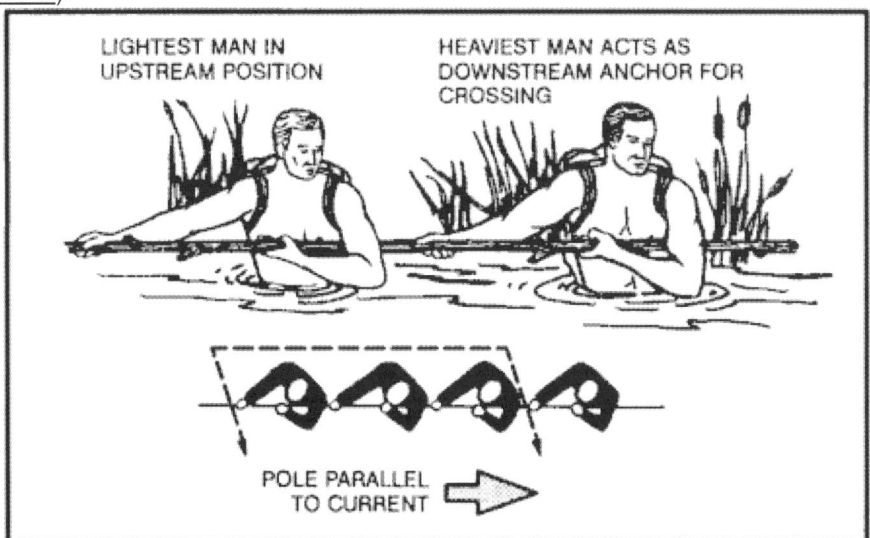

LIGHTEST MAN IN
UPSTREAM POSITION

HEAVIEST MAN ACTS AS
DOWNSTREAM ANCHOR FOR
CROSSING

POLE PARALLEL
TO CURRENT

Figure 17-2. Several men crossing swift stream.

If you have three or more people and a rope available, you can use the technique shown in Figure 17-3 to cross the stream. The length of the rope must be three times the width of the stream.

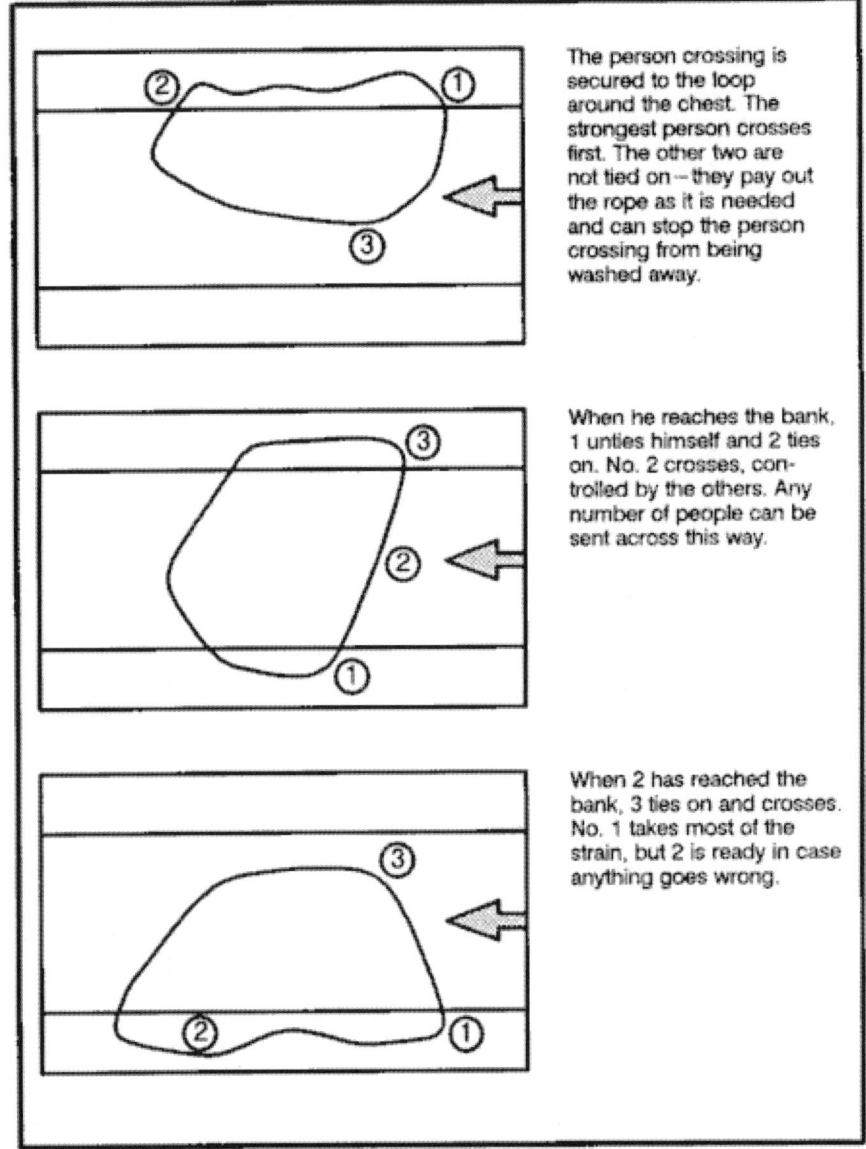

The person crossing is secured to the loop around the chest. The strongest person crosses first. The other two are not tied on—they pay out the rope as it is needed and can stop the person crossing from being washed away.

When he reaches the bank, 1 unties himself and 2 ties on. No. 2 crosses, controlled by the others. Any number of people can be sent across this way.

When 2 has reached the bank, 3 ties on and crosses. No. 1 takes most of the strain, but 2 is ready in case anything goes wrong.

Figure 17-3. Individuals tied together to cross stream.

RAFTS

If you have two ponchos, you can construct a brush raft or an Australian poncho raft. With either of these rafts, you can safely float your equipment across a slow-moving stream or river.

Brush Raft

The brush raft, if properly constructed, will support about 115 kilograms. To construct it, use ponchos, fresh green brush, two small saplings, and rope or vine as follows (Figure 17-4):

- Push the hood of each poncho to the inner side and tightly tie off the necks using the drawstrings.
- Attach the ropes or vines at the corner and side grommets of each poncho. Make sure they are long enough to cross to and tie with the others attached at the opposite corner or side.
- Spread one poncho on the ground with the inner side up. Pile fresh, green brush (no thick branches) on the poncho until the brush stack is about 45 centimeters high. Pull the drawstring up through the center of the brush stack.
- Make an X-frame from two small saplings and place it on top of the brush stack. Tie the X-frame securely in place with the poncho drawstring.

- Pile another 45 centimeters of brush on top of the X-frame, then compress the brush slightly.
- Pull the poncho sides up around the brush and, using the ropes or vines attached to the comer or side grommets, tie them diagonally from comer to corner and from side to side.
- Spread the second poncho, inner side up, next to the brush bundle.
- Roll the brush bundle onto the second poncho so that the tied side is down. Tie the second poncho around the brush bundle in the same manner as you tied the first poncho around the brush.
- Place it in the water with the tied side of the second poncho facing up.

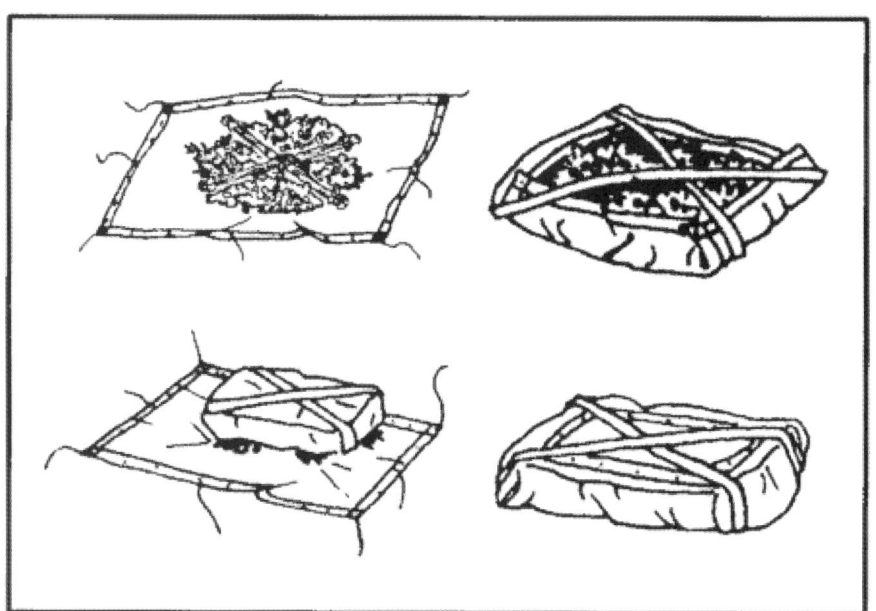

Figure 17-4. Brush raft.

Australian Poncho Raft

If you do not have time to gather brush for a brush raft, you can make an Australian poncho raft. This raft, although more waterproof than the poncho brush raft, will only float about 35 kilograms of equipment. To construct this raft, use two ponchos, two rucksacks, two 1.2-meter poles or branches, and ropes, vines, bootlaces, or comparable material as follows (Figure 17-5):

- Push the hood of each poncho to the inner side and tightly tie off the necks using the drawstrings.
- Spread one poncho on the ground with the inner side up. Place and center the two 1.2-meter poles on the poncho about 45 centimeters apart.
- Place your rucksacks or packs or other equipment between the poles. Also place other items that you want to keep dry between the poles. Snap the poncho sides together.
- Use your buddy's help to complete the raft. Hold the snapped portion of the poncho in the air and roll it tightly down to the equipment. Make sure you roll the full width of the poncho.
- Twist the ends of the roll to form pigtails in opposite directions. Fold the pigtails over the bundle and tie them securely in place using ropes, bootlaces, or vines.
- Spread the second poncho on the ground, inner side up. If you need more buoyancy, place some fresh green brush on this poncho.
- Place the equipment bundle, tied side down, on the center of the second poncho. Wrap the second poncho around the equipment bundle following the same procedure you used for wrapping the equipment in the first poncho.
- Tie ropes, bootlaces, vines, or other binding material around the raft about 30 centimeters from the end of each pigtail. Place and secure weapons on top of the raft.

- Tie one end of a rope to an empty canteen and the other end to the raft. This will help you to tow the raft.

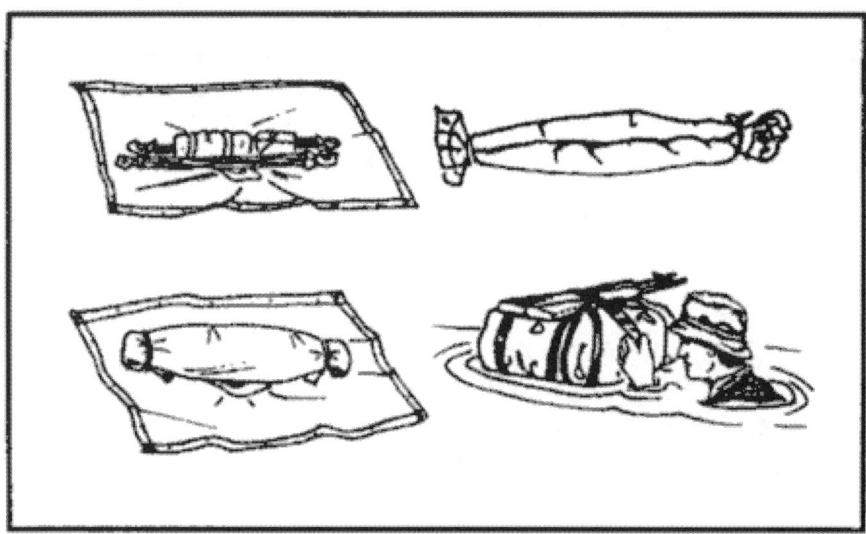

Figure 17-5. Australian poncho raft.

Poncho Donut Raft

Another type of raft is the poncho donut raft. It takes more time to construct than the brush raft or Australian poncho raft, but it is effective. To construct it, use one poncho, small saplings, willow or vines, and rope, bootlaces, or other binding material (Figure 17-6) as follows:

- Make a framework circle by placing several stakes in the ground that roughly outline an inner and outer circle.
- Using young saplings, willow, or vines, construct a donut ring within the circles of stakes.
- Wrap several pieces of cordage around the donut ring about 30 to 60 centimeters apart and tie them securely.
- Push the poncho's hood to the inner side and tightly tie off the neck using the drawstring.
- Place the poncho on the ground, inner side up. Place the donut ring on the center of the poncho. Wrap the poncho up and over the donut ring and tie off each grommet on the poncho to the ring.
- Tie one end of a rope to an empty canteen and the other end to the raft. This rope will help you to tow the raft.

Figure 17-6. Poncho donut raft.

When launching any of the above rafts, take care not to puncture or tear it by dragging it on the ground. Before you start to cross the river or stream, let the raft lay on the water a few minutes to ensure that it floats.

If the river is too deep to ford, push the raft in front of you while you are swimming. The design of the above <u>rafts</u> does not allow them to carry a person's full body weight. Use them as a float to get you and your equipment safely across the river or stream.

Be sure to check the water temperature before trying to cross a river or water obstacle. If the water is extremely cold and you are unable to find a shallow fording place in the river, do not try to ford it. Devise other means for crossing. For instance, you might improvise a bridge by felling a tree over the river. Or you might build a raft large enough to carry you and your equipment. For this, however, you will need an axe, a knife, a rope or vines, and time.

Log Raft

You can make a raft using any dry, dead, standing trees for logs. However, spruce trees found in polar and subpolar regions make the best rafts. A simple method for making a raft is to use pressure bars lashed securely at each end of the raft to hold the logs together (<u>Figure 17-7</u>).

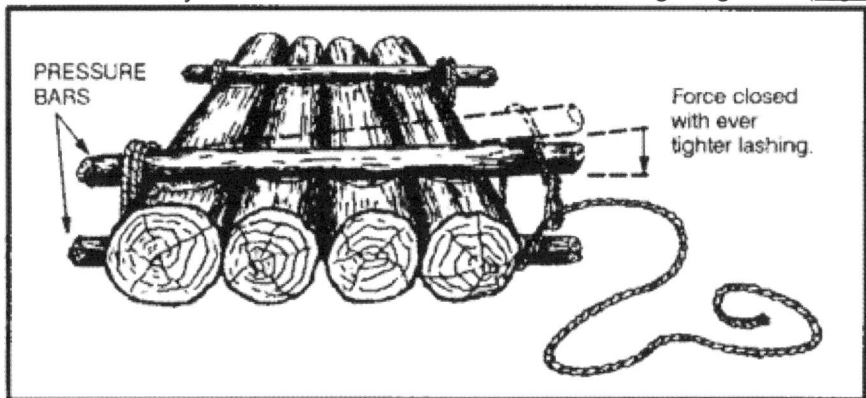

Figure 17-7. Use of pressure bars.

FLOTATION DEVICES

If the water is warm enough for swimming and you do not have the time or materials to construct one of the poncho-type rafts, you can use various flotation devices to negotiate the water obstacle. Some items you can use for flotation devices are--

* *Trousers.* Knot each trouser leg at the bottom and close the fly. With both hands, grasp the waistband at the sides and swing the trousers in the air to trap air in each leg. Quickly press the sides of the waistband together and hold it underwater so that the air will not escape. You now have water wings to keep you afloat as you cross the body of water.

  Note: Wet the trousers before inflating to trap the air better You may have to reinflate the trousers several times when crossing a large body of water.

* *Empty containers.* Lash together her empty gas cans, water jugs, ammo cans, boxes, or other items that will trap or hold air. Use them as water wings. Use this type of flotation device only in a slow-moving river or stream.
* *Plastic bags and ponchos.* Fill two or more plastic bags with air and secure them together at the opening. Use your poncho and roll green vegetation tightly inside it so that you have a roll at least 20 centimeters in diameter. Tie the ends of the roll securely. You can wear it around your waist or across one shoulder and under the opposite arm.

Logs. Use a stranded drift log if one is available, or find a log near the water to use as a float. Be sure to test the log before starting to cross. Some tree logs, palm for example, will sink even when the wood is dead. Another method is to tie two logs about 60 centimeters apart. Sit between the logs with your back against one and your legs over the other (<u>Figure 17-8</u>).

- *Cattails.* Gather stalks of cattails and tie them in a bundle 25 centimeters or more in diameter. The many air cells in each stalk cause a stalk to float until it rots. Test the cattail bundle to be sure it will support your weight before trying to cross a body of water.

Figure 17-8. Log flotation.

There are many other flotation devices that you can devise by using some imagination. Just make sure to test the device before trying to use it.

OTHER WATER OBSTACLES

Other water obstacles that you may face are bogs, quagmire, muskeg, or quicksand. Do not try to walk across these. Trying to lift your feet while standing upright will make you sink deeper. Try to bypass these obstacles. If you are unable to bypass them, you may be able to bridge them using logs, branches, or foliage.

A way to cross a bog is to lie face down, with your arms and legs spread. Use a flotation device or form pockets of air in your clothing. Swim or pull your way across moving slowly and trying to keep your body horizontal.

In swamps, the areas that have vegetation are usually firm enough to support your weight. However, vegetation will usually not be present in open mud or water areas. If you are an average swimmer, however, you should have no problem swimming, crawling, or pulling your way through miles of bog or swamp.

Quicksand is a mixture of sand and water that forms a shifting mass. It yields easily to pressure and sucks down and engulfs objects resting on its surface. It varies in depth and is usually localized. Quicksand commonly occurs on flat shores, in silt-choked rivers with shifting watercourses, and near the mouths of large rivers. If you are uncertain whether a sandy area is quicksand, toss a small stone on it. The stone will sink in quicksand. Although quicksand has more suction than mud or muck, you can cross it just as you would cross a bog. Lie face down, spread your arms and legs, and move slowly across.

VEGETATION OBSTACLES

Some water areas you must cross may have underwater and floating plants that will make swimming difficult. However, you can swim through relatively dense vegetation if you remain calm and do not thrash about. Stay as near the surface as possible and use the breaststroke with shallow leg and arm motion. Remove the plants around you as you would clothing. When you get tired, float or swim on your back until you have rested enough to continue with the breaststroke.

The mangrove swamp is another type of obstacle that occurs along tropical coastlines. Mangrove trees or shrubs throw out many prop roots that form dense masses. To get through a mangrove swamp, wait for low tide. If you are on the inland side, look for a narrow grove of trees and work your way seaward through these. You can also try to find the bed of a waterway or creek through the trees and follow it to the sea. If you are on the seaward side, work inland along streams or channels. Be on the lookout for crocodiles that you find along channels and in shallow water. If there are any near you, leave the water and scramble over the mangrove roots. While crossing a mangrove swamp, it is possible to gather food from tidal pools or tree roots.

To cross a large swamp area, construct some type of raft.

CHAPTER 18 - FIELD-EXPEDIENT DIRECTION FINDING

In a survival situation, you will be extremely fortunate if you happen to have a map and compass. If you do have these two pieces of equipment, you will most likely be able to move toward help. If you are not proficient in using a map and compass, you must take the steps to gain this skill.

There are several methods by which you can determine direction by using the sun and the stars. These methods, however, will give you only a general direction. You can come up with a more nearly true direction if you know the terrain of the territory or country.

You must learn all you can about the terrain of the country or territory to which you or your unit may be sent, especially any prominent features or landmarks. This knowledge of the terrain together with using the <u>methods</u> explained below will let you come up with fairly true directions to help you navigate.

USING THE SUN AND SHADOWS

The earth's relationship to the sun can help you to determine direction on earth. The sun always rises in the east and sets in the west, but not exactly due east or due west. There is also some seasonal variation. In the northern hemisphere, the sun will be due south when at its highest point in the sky, or when an object casts no appreciable shadow. In the southern hemisphere, this same noonday sun will mark due north. In the northern hemisphere, shadows will move clockwise. Shadows will move counterclockwise in the southern hemisphere. With practice, you can use shadows to determine both direction and time of day. The shadow methods used for direction finding are the shadow-tip and watch methods.

Shadow-Tip Methods

In the first shadow-tip method, find a straight stick 1 meter long, and a level spot free of brush on which the stick will cast a definite shadow. This method is simple and accurate and consists of four steps:

- *Step 1.* Place the stick or branch into the ground at a level spot where it will cast a distinctive shadow. Mark the shadow's tip with a stone, twig, or other means. This first shadow mark is always west--**everywhere** on earth.
- *Step 2.* Wait 10 to 15 minutes until the shadow tip moves a few centimeters. Mark the shadow tip's new position in the same way as the first.
- *Step 3.* Draw a straight line through the two marks to obtain an approximate east-west line.
- *Step 4.* Stand with the first mark (west) to your left and the second mark to your right--you are now facing north. This fact is true **everywhere** on earth.

An alternate method is more accurate but requires more time. Set up your shadow stick and mark the first shadow in the morning. Use a piece of string to draw a clean arc through this mark and around the stick. At midday, the shadow will shrink and disappear. In the afternoon, it will lengthen again and at the point where it touches the arc, make a second mark. Draw a line through the two marks to get an accurate east-west line (see Figure 18-1).

Figure 18-1. Shadow-tip method.

The Watch Method

You can also determine direction using a common or analog watch--one that has hands. The direction will be accurate if you are using true local time, without any changes for daylight savings time. Remember, the further you are from the equator, the more accurate this method will be. If you only have a digital watch, you can overcome this obstacle. Quickly draw a watch on a circle of paper with the correct time on it and use it to determine your direction at that time.

In the northern hemisphere, hold the watch horizontal and point the hour hand at the sun. Bisect the angle between the hour hand and the 12 o'clock mark to get the north-south line (Figure 18-2). If there is any doubt as to which end of the line is north, remember that the sun rises in the east, sets in the west, and is due south at noon. The sun is in the east before noon and in the west after noon.

Note: If your watch is set on daylight savings time, use the midway point between the hour hand and 1 o'clock to determine the north-south line.

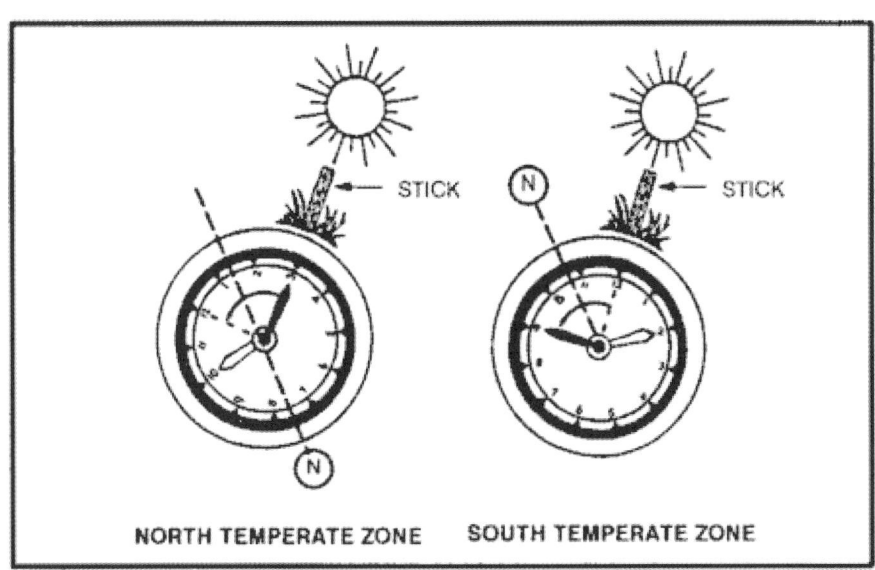

NORTH TEMPERATE ZONE   SOUTH TEMPERATE ZONE

Figure 18-2. Watch method.

In the southern hemisphere, point the watch's 12 o'clock mark toward the sun and a midpoint halfway between 12 and the hour hand will give you the north-south line (Figure 18-2).

USING THE MOON

Because the moon has no light of its own, we can only see it when it reflects the sun's light. As it orbits the earth on its 28-day circuit, the shape of the reflected light varies according to its position. We say there is a new moon or no moon when it is on the opposite side of the earth from the sun. Then, as it moves away from the earth's shadow, it begins to reflect light from its right side and waxes to become a full moon before waning, or losing shape, to appear as a sliver on the left side. You can use this information to identify direction.

If the moon rises before the sun has set, the illuminated side will be the west. If the moon rises after midnight, the illuminated side will be the east. This obvious discovery provides us with a rough east-west reference during the night.

USING THE STARS

Your location in the Northern or Southern Hemisphere determines which constellation you use to determine your north or south direction.

The Northern Sky

The main constellations to learn are the Ursa Major, also known as the Big Dipper or the Plow, and Cassiopeia (Figure 18-3). Neither of these constellations ever sets. They are always visible on a clear night. Use them to locate Polaris, also known as the polestar or the North Star. The North Star forms part of the Little Dipper handle and can be confused with the Big Dipper. Prevent confusion by using both the Big Dipper and Cassiopeia together. The Big Dipper and Cassiopeia are always directly opposite each. other and rotate counterclockwise around Polaris, with Polaris in the center. The Big Dipper is a seven star constellation in the shape of a dipper. The two stars forming the outer lip of this dipper are the "pointer stars" because they point to the North Star. Mentally draw a line from the outer bottom star to the outer top star of the Big Dipper's bucket. Extend this line about five times the distance between the pointer stars. You will find the North Star along this line.

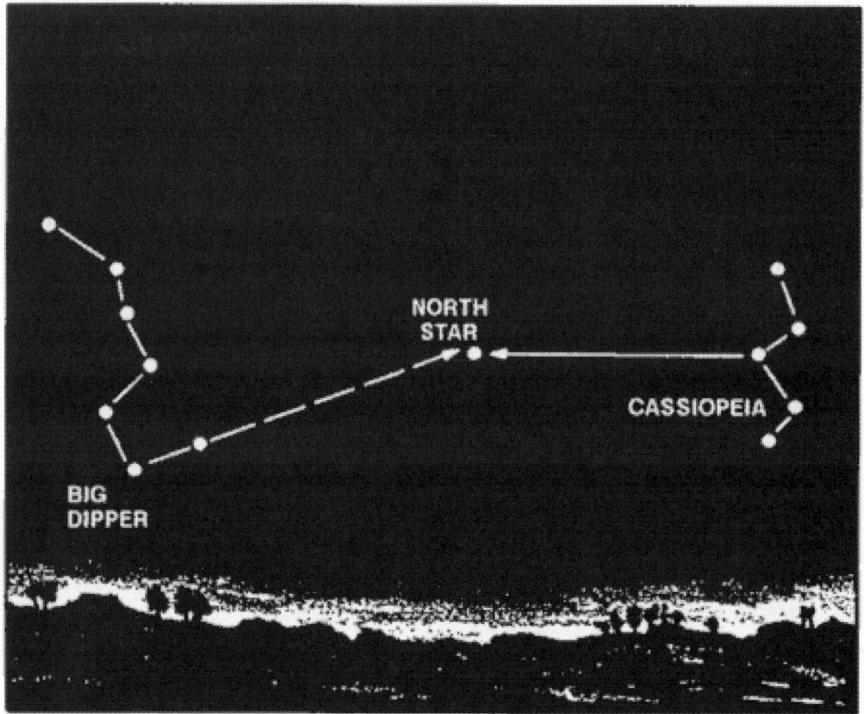

Figure 18-3. The Big Dipper and Cassiopeia.

Cassiopeia has five stars that form a shape like a "W" on its side. The North Star is straight out from Cassiopeia's center star.

After locating the North Star, locate the North Pole or true north by drawing an imaginary line directly to the earth.

The Southern Sky

Because there is no star bright enough to be easily recognized near the south celestial pole, a constellation known as the Southern Cross is used as a signpost to the South (Figure 18-4). The Southern Cross or Crux has five stars. Its four brightest stars form a cross that tilts to one side. The two stars that make up the cross's long axis are the pointer stars. To determine south, imagine a distance five times the distance between These stars and the point where this imaginary line ends is in the general direction of south. Look down to the horizon from this imaginary point and select a landmark to steer by. In a static survival situation, you can fix this location in daylight if you drive stakes in the ground at night to point the way.

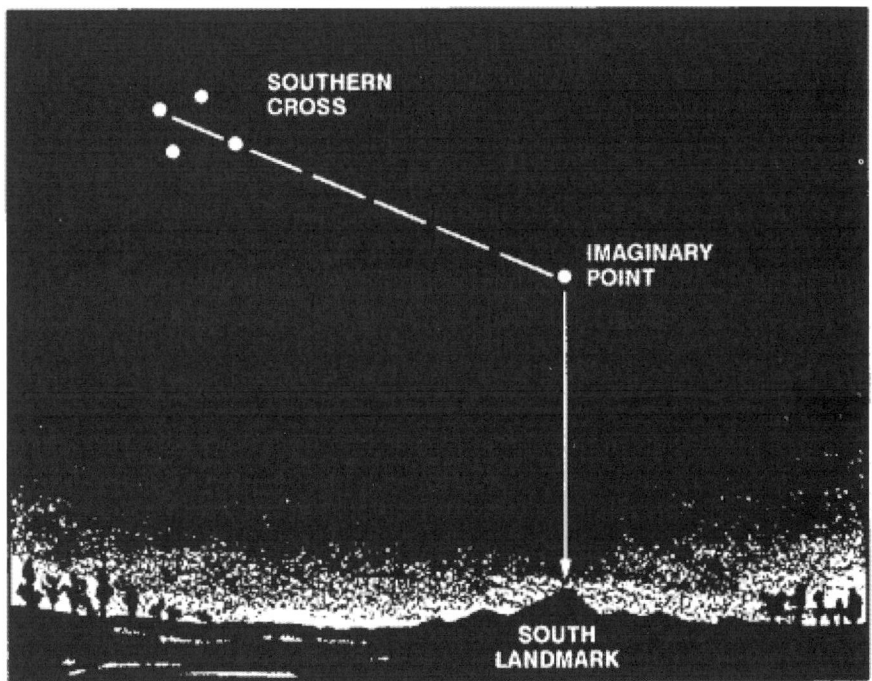

Figure 18-4. Southern Cross.

MAKING IMPROVISED COMPASSES

You can construct improvised compasses using a piece of ferrous metal that can be needle shaped or a flat double-edged razor blade and a piece of nonmetallic string or long hair from which to suspend it. You can magnetize or polarize the metal by slowly stroking it in one direction on a piece of silk or carefully through your hair using deliberate strokes. You can also polarize metal by stroking it repeatedly at one end with a magnet. Always rub in one direction only. If you have a battery and some electric wire, you can polarize the metal electrically. The wire should be insulated. If not insulated, wrap the metal object in a single, thin strip of paper to prevent contact. The battery must be a minimum of 2 volts. Form a coil with the electric wire and touch its ends to the battery's terminals. Repeatedly insert one end of the metal object in and out of the coil. The needle will become an electromagnet. When suspended from a piece of nonmetallic string, or floated on a small piece of wood in water, it will align itself with a north-south line. You can construct a more elaborate improvised compass using a sewing needle or thin metallic object, a nonmetallic container (for example, a plastic dip container), its lid with the center cut out and waterproofed, and the silver tip from a pen. To construct this compass, take an ordinary sewing needle and break in half. One half will form your direction pointer and the other will act as the pivot point. Push the portion used as the pivot point through the bottom center of your container; this portion should be flush on the bottom and not interfere with the lid. Attach the center of the other portion (the pointer) of the needle on the pen's silver tip using glue, tree sap, or melted plastic. Magnetize one end of the pointer and rest it on the pivot point.

OTHER MEANS OF DETERMINING DIRECTION

The old saying about using moss on a tree to indicate north is not accurate because moss grows completely around some trees. Actually, growth is more lush on the side of the tree facing the south in the Northern Hemisphere and vice versa in the Southern Hemisphere. If there are several felled trees around for comparison, look at the stumps. Growth is more vigorous on the side toward the equator and the tree growth rings will be more widely spaced. On the other hand, the tree growth rings will be closer together on the side toward the poles.

Wind direction may be helpful in some instances where there are prevailing directions and you know what they are.

Recognizing the differences between vegetation and moisture patterns on north- and south-facing slopes can aid in determining direction. In the northern hemisphere, north-facing slopes receive less sun than south-facing slopes and are therefore cooler and damper. In the summer, north-facing slopes retain patches of snow. In the winter, the trees and open areas on south-facing slopes are the first to lose their snow, and ground snowpack is shallower.

CHAPTER 19 - SIGNALING TECHNIQUES

One of your first concerns when you find yourself in a survival situation is to communicate with your friends or allies. Generally, communication is the giving and receiving of information. As a survivor, you must get your rescuer's attention first, and second, send a message your rescuer understands. Some attention-getters are man-made geometric patterns such as straight lines, circles, triangles, or X's displayed in uninhabited areas; a large fire or flash of light; a large, bright object moving slowly; or contrast, whether from color or shadows. The type of signal used will depend on your environment and the enemy situation.

APPLICATION

If in a noncombat situation, you need to find the largest available clear and flat area *on the highest possible terrain.* Use as obvious a signal as you can create. On the other hand, you will have to be more discreet in combat situations. You do not want to signal and attract the enemy. Pick an area that is visible from the air, but ensure there are hiding places nearby. Try to have a hill or other object between the signal site and the enemy to mask your signal from the enemy. Perform a thorough reconnaissance of the area to ensure there are no enemy forces nearby.

Whatever signaling technique or device you plan to use, know how to use it and be ready to put it into operation on short notice. If possible, avoid using signals or signaling techniques that can physically endanger you. Keep in mind that signals to your **friends** may alert the enemy of your presence and location. Before signaling, carefully weigh your rescue chances by **friends** against the danger of capture by the enemy.

A radio is probably the surest and quickest way to let others know where you are and to let you receive their messages. Become familiar with the radios in your unit. Learn how to operate them and how to send and receive messages.

You will find descriptions of other signaling techniques, devices, and articles you can use. Learn how to use them. Think of ways in which you can adapt or change them for different environments. Practice using these signaling techniques, devices, and articles before you need them. Planned, prearranged signaling techniques may improve your chance of rescue.

MEANS FOR SIGNALING

There are two main ways to get attention or to communicate--visual and audio. The means you use will depend on your situation and the material you have available. Whatever the means, always have visual and audio signals ready for use.

Visual Signals

These signals are materials or equipment you use to make your presence known to rescuers.
Fire

During darkness, fire is the most effective visual means for signaling. Build three fires in a triangle (the international distress signal) or in a straight line with about 25 meters between the fires. Build them as soon as time and the situation permit and protect them until you need them. If you are alone, maintaining three fires may be difficult. If so, maintain one signal fire.

When constructing signal fires, consider your geographic location. If in a jungle, find a natural clearing or the edge of a stream where you can build fires that the jungle foliage will not hide. You may even have to clear an area. If in a snow-covered area, you may have to clear the ground of snow or make a platform on which to build the fire so that melting snow will not extinguish it.

A burning tree (tree torch) is another way to attract attention (Figure 19-1). You can set pitch-bearing trees afire, even when green. You can get other types of trees to burn by placing dry wood in the lower branches and igniting it so that the flames flare up and ignite the foliage. Before the primary tree is consumed, cut and add more small green trees to the fire to produce more smoke. Always select an isolated tree so that you do not start a forest fire and endanger yourself.

Figure 19-1. Tree torch.

Smoke

During daylight, build a smoke generator and use smoke to gain attention (Figure 19-2). The international distress signal is three columns of smoke. Try to create a color of smoke that contrasts with the background; dark smoke against a light background and vice versa. If you practically smother a large fire with green leaves, moss, or a little water, the fire will produce white smoke. If you add rubber or oil-soaked rags to a fire, you will get black smoke.

ESSENTIAL ARMY MANUAL
(203) ARMY SURVIVAL MANUAL

LOTS OF DEAD, DRY
TWIGS OR KINDLING
FOR QUICK-STARTING,
FAST-BURNING FIRE

EVERGREEN
BOUGHS

SMALL OPENING FOR LIGHTING FIRE

Figure 19-2. Smoke generator—ground.

In a desert environment, smoke hangs close to the ground, but a pilot can spot it in open desert terrain. Smoke signals are effective only on comparatively calm, clear days. High winds, rain, or snow disperse smoke, lessening its chances of being seen.

Smoke Grenades

If you have smoke grenades with you, use them in the same pattern as described for fires. Keep them dry so that they will work when you need them. Take care not to ignite the vegetation in the area when you use them.

Pen Flares

These flares are part of an aviator's survival vest. The device consists of a pen-shaped gun with a flare attached by a nylon cord. When fired, the pen flare sounds like a pistol shot and fires the flare about 150 meters high. It is about 3 centimeters in diameter.

To have the pen flare ready for immediate use, take it out of its wrapper, attach the flare, leave the gun uncocked, and wear it on a cord or chain around your neck. Be ready to fire it in front of search aircraft and be ready with a secondary signal. Also, be ready to take cover in case the pilot mistakes the flare for enemy fire.

Tracer Ammunition

You may use rifle or pistol tracer ammunition to signal search aircraft. **Do not** fire the ammunition in front of the aircraft. As with pen flares, be ready to take cover if the pilot mistakes your tracers for enemy fire.

Star Clusters

Red is the international distress color; therefore, use a red star cluster whenever possible. Any color, however, will let your rescuers know where you are. Star clusters reach a height of 200 to 215 meters, burn an average of 6 to 10 seconds, and descend at a rate of 14 meters per second.

Star Parachute Flares

These flares reach a height of 200 to 215 meters and descend at a rate of 2.1 meters per second. The M126 (red) burns about 50 seconds and the M127 (white) about 25 seconds. At night you can see these flares at 48 to 56 kilometers.

Mirrors or Shiny Objects

On a sunny day, a mirror is your best signaling device. If you don't have a mirror, polish your canteen cup, your belt buckle, or a similar object that will reflect the sun's rays. Direct the flashes in one area so that they are secure from enemy observation. Practice using a mirror or shiny object for signaling *now;* do not wait until you need it. If you have an MK-3 signal mirror, follow the instructions on its back (Figure 19-3).

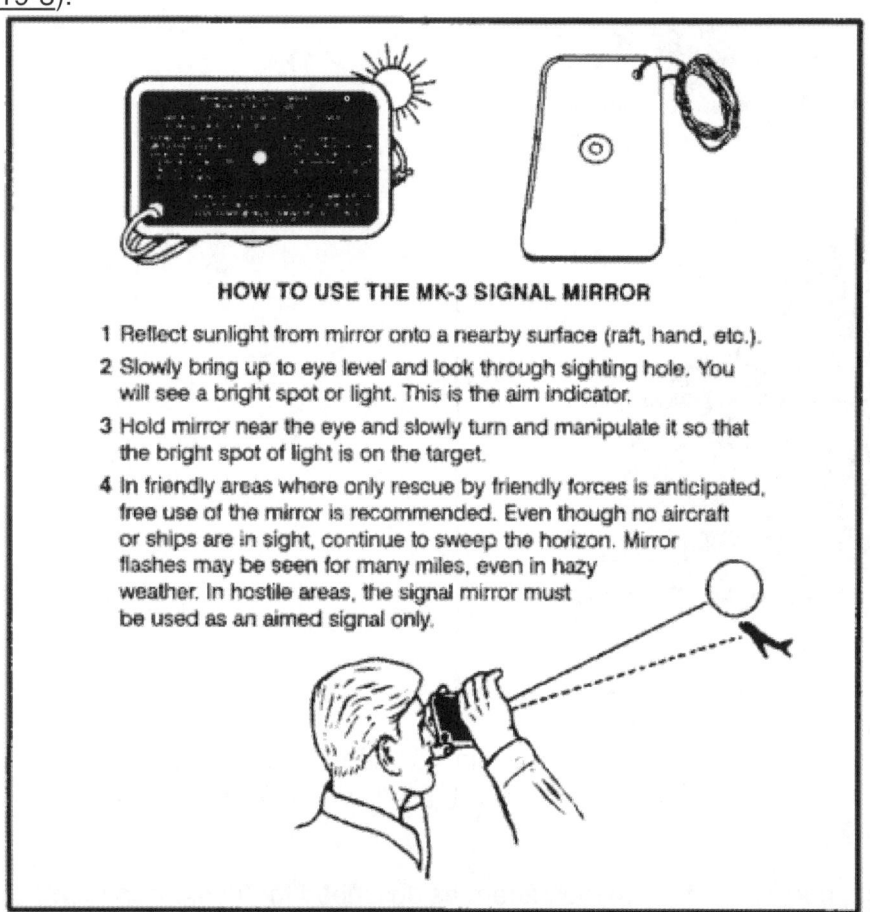

Figure 19-3. Signal mirror.

Wear the signal mirror on a cord or chain around your neck so that it is ready for immediate use. However, be sure the glass side is against your body so that it will not flash; the enemy can see the flash.

CAUTION
Do not flash a signal mirror rapidly because a pilot may mistake the flashes for enemy fire. Do not direct the beam in the aircraft's cockpit for more than a few seconds as it may blind the pilot.

Haze, ground fog, and mirages may make it hard for a pilot to spot signals from a flashing object. So, if possible, get to the highest point in your area when signaling. If you can't determine the aircraft's location, flash your signal in the direction of the aircraft noise.

> Note: Pilots have reported seeing mirror flashes up to 160 kilometers away under ideal conditions.

Figures 19-4 and 19-5 show methods of aiming a signal mirror for signaling.

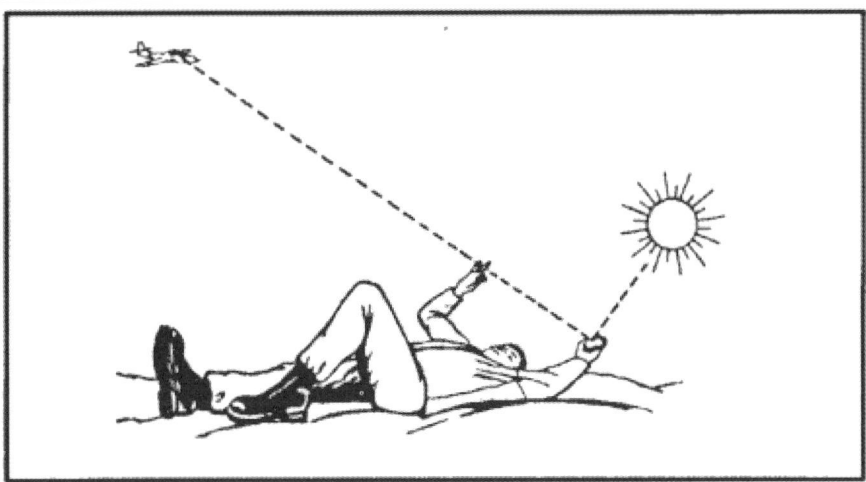

Figure 19-4. Aiming an improvised signal mirror.

Figure 19-5. Aiming an improvised signal mirror—stationary object.

Flashlight or Strobe Light

At night you can use a flashlight or a strobe light to send an SOS to an aircraft. When using a strobe light, take care to prevent the pilot from mistaking it for incoming ground fire. The strobe light flashes 60 times per minute. Some strobe lights have infrared covers and lenses. Blue flash collimators are also available for strobe lights.

VS-17 Panel

During daylight you can use a VS-17 panel to signal. Place the orange side up as it is easier to see from the air than the violet side. Flashing the panel will make it easier for the aircrew to spot. You can use any bright orange or violet cloth as a substitute for the VS-17.

Clothing

Spreading clothing on the ground or in the top of a tree is another way to signal. Select articles whose color will contrast with the natural surroundings. Arrange them in a large geometric pattern to make them more likely to attract attention.

Natural Material

If you lack other means, you can use natural materials to form a symbol or message that can be seen from the air. Build mounds that cast shadows; you can use brush, foliage of any type, rocks, or snow blocks.

In snow-covered areas, tramp the snow to form letters or symbols and fill the depression with contrasting material (twigs or branches). In sand, use boulders, vegetation, or seaweed to form a symbol or message. In brush-covered areas, cut out patterns in the vegetation or sear the ground. In tundra, dig trenches or turn the sod upside down.

In any terrain, use contrasting materials that will make the symbols visible to the aircrews.

Sea Dye Markers

All Army aircraft involved in operations near or over water will normally carry a water survival kit that contains sea dye markers. If you are in a water survival situation, use sea dye markers during daylight to indicate your location. These spots of dye stay conspicuous for about 3 hours, except in very rough seas. Use them only if you are in a friendly area. Keep the markers wrapped until you are ready to use them. Use them only when you hear or sight an aircraft. Sea dye markers are also very effective on snow-covered ground; use them to write distress code letters.

Audio Signals

Radios, whistles, and gunshots are some of the methods you can use to signal your presence to rescuers.

Radio Equipment
The AN/PRC-90 survival radio is a part of the Army aviator's survival vest. The AN/PRC-112 will eventually replace the AN/PRC-90. Both radios can transmit either tone or voice. Any other type of Army radio can do the same. The ranges of the different radios vary depending on the altitude of the receiving aircraft, terrain, vegetation density, weather, battery strength, type of radio, and interference. To obtain maximum performance from radios, use the following procedures:

- Try to transmit only in clear, unobstructed terrain. Since radios are line-of-sight communications devices, any terrain between the radio and the receiver will block the signal.
- Keep the antenna at right angles to the rescuing aircraft. There is no signal from the tip of the antenna.
- If the radio has tone capability, place it upright on a flat, elevated surface so that you can perform other survival tasks.
- Never let the antenna touch your clothing, body, foliage, or the ground. Such contact greatly reduces the range of the signal.
- Conserve battery power. Turn the radio off when you are not using it. Do not transmit or receive constantly. In hostile territory, keep transmissions short to avoid enemy radio direction finding.
- In cold weather, keep the battery inside your clothing when not using the radio. Cold quickly drains the battery's power. Do not expose the battery to extreme heat such as desert sun. High heat may cause the battery to explode. Try to keep the radio and battery as dry as possible, as water may destroy the circuitry.

Whistles
Whistles provide an excellent way for close up signaling. In some documented cases, they have been heard up to 1.6 kilometers away. Manufactured whistles have more range than a human whistle.
Gunshots
In some situations you can use firearms for signaling. Three shots fired at distinct intervals usually indicate a distress signal. Do not use this technique in enemy territory. The enemy will surely come to investigate shots.

CODES AND SIGNALS

Now that you know how to let people know where you are, you need to know how to give them more information. It is easier to form one symbol than to spell out an entire message. Therefore, learn the codes and symbols that all aircraft pilots understand.

SOS

You can use lights or flags to send an SOS--three dots, three dashes, three dots. The SOS is the internationally recognized distress signal in radio Morse code. A dot is a short, sharp pulse; a dash is a longer pulse. Keep repeating the signal. When using flags, hold flags on the left side for dashes and on the right side for dots.

Ground-to-Air Emergency Code

This code (Figure 19-6) is actually five definite, meaningful symbols. Make these symbols a minimum of 1 meter wide and 6 meters long. If you make them larger, keep the same 1: 6 ratio. Ensure the signal contrasts greatly with the ground it is on. Place it in an open area easily spotted from the air.

Number	Message	Code symbol
1	Require assistance.	V
2	Require medical assistance.	X
3	No or negative.	N
4	Yes or affirmative.	Y
5	Proceed in this direction.	↑

Figure 19-6. Ground-to-air emergency code (pattern signals).

Body Signals

When an aircraft is close enough for the pilot to see you clearly, use body movements or positions (Figure 19-7) to convey a message.

Figure 19-7. Body signals.

Panel Signals

If you have a life raft cover or sail, or a suitable substitute, use the symbols shown in Figure 19-8 to convey a message.

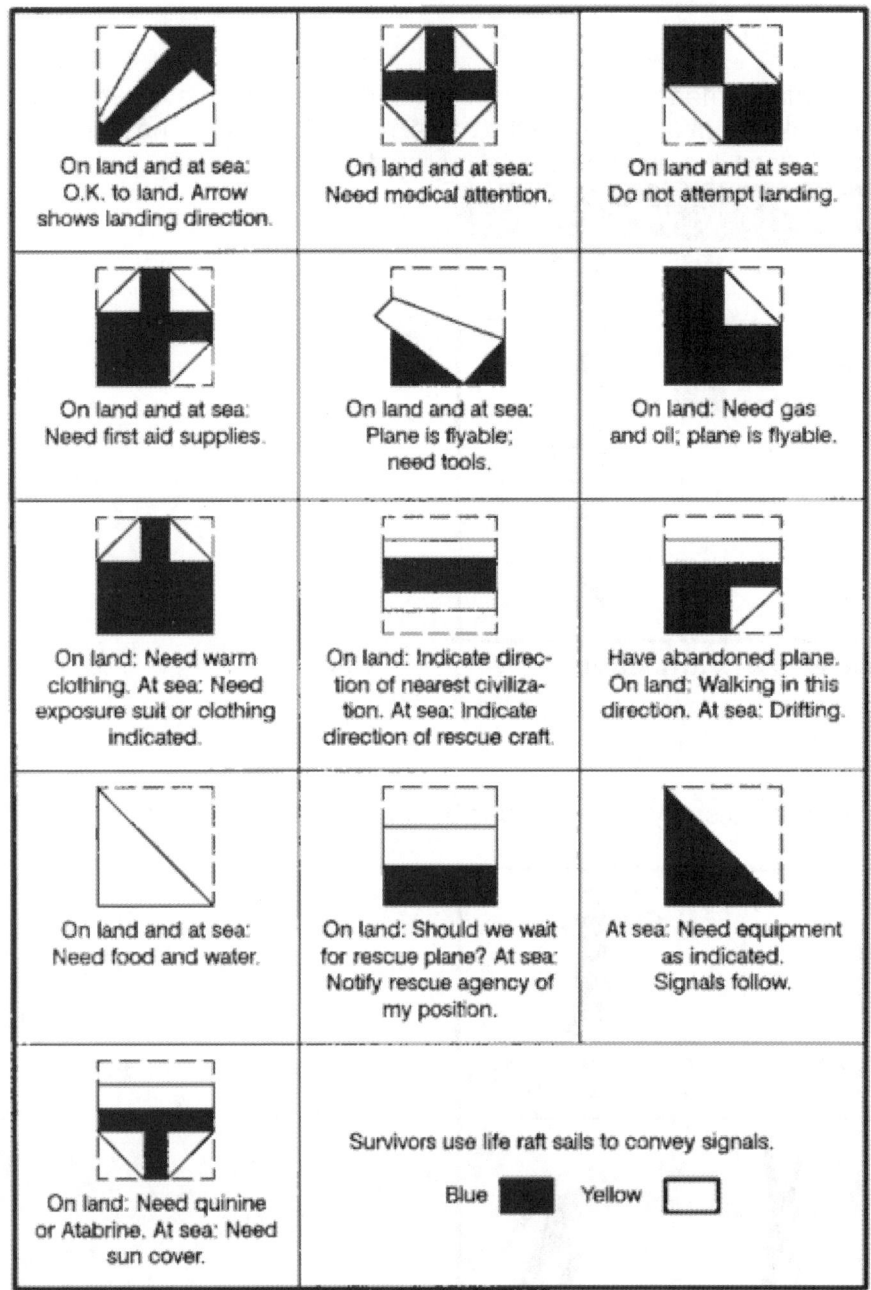

Figure 19-8. Panel signals.

Aircraft Acknowledgments

Once the pilot of a fixed-wing aircraft has sighted you, he will normally indicate he has seen you by flying low, moving the plane, and flashing lights as shown in Figure 19-9. Be ready to relay other messages to the pilot once he acknowledges that he received and understood your first message. Use a radio, if possible, to relay further messages. If no radio is available, use the codes covered in the previous paragraphs.

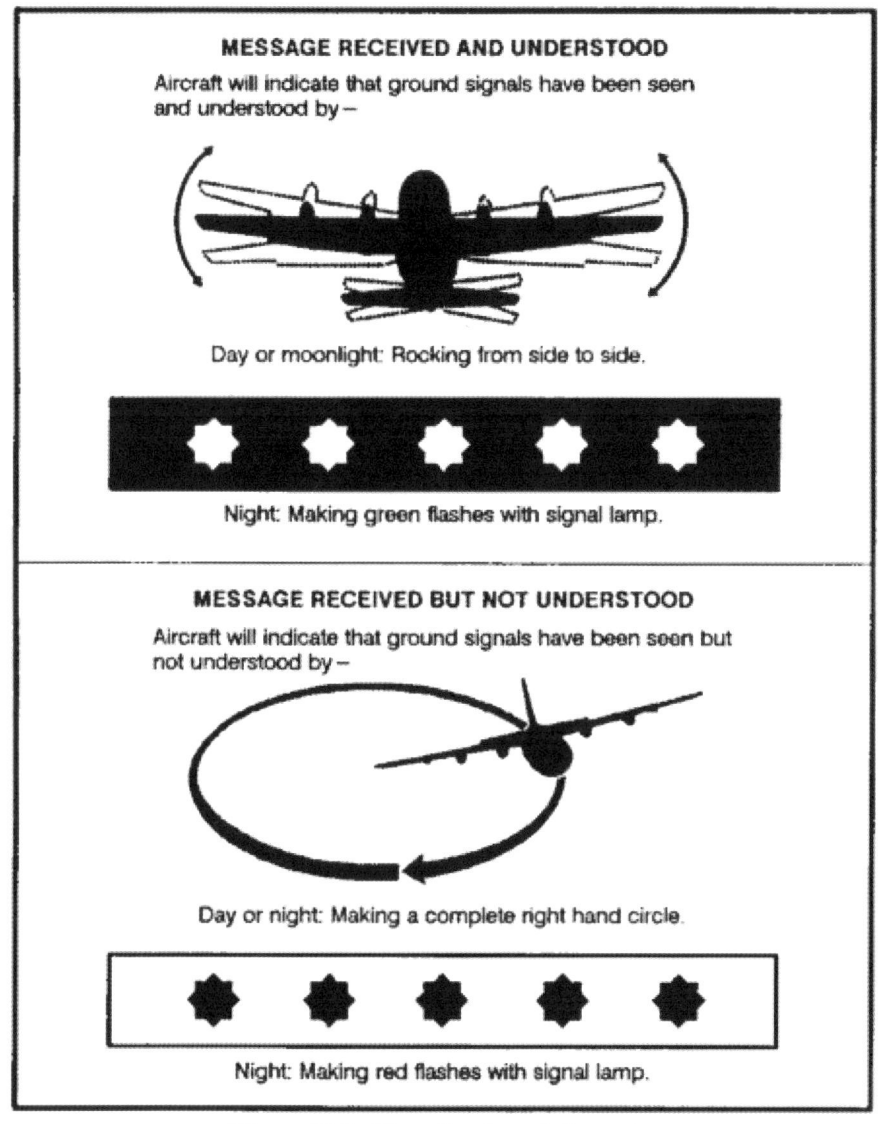

MESSAGE RECEIVED AND UNDERSTOOD

Aircraft will indicate that ground signals have been seen and understood by —

Day or moonlight: Rocking from side to side.

Night: Making green flashes with signal lamp.

MESSAGE RECEIVED BUT NOT UNDERSTOOD

Aircraft will indicate that ground signals have been seen but not understood by —

Day or night: Making a complete right hand circle.

Night: Making red flashes with signal lamp.

Figure 19-9. Aircraft acknowledgments.

AIRCRAFT VECTORING PROCEDURES

If you can contact a friendly aircraft with a radio, guide the pilot to your location. Use the following general format to guide the pilot:

- Mayday, Mayday.
- Call sign (if any).
- Name.
- Location.
- Number of survivors.
- Available landing sites.
- Any remarks such as medical aid or other specific types of help needed immediately.

Simply because you have made contact with rescuers does not mean you are safe. Follow instructions and continue to use sound survival and evasion techniques until you are actually rescued.

CHAPTER 20 - SURVIVAL MOVEMENT IN HOSTILE AREAS

The "rescue at any cost" philosophy of previous conflicts is not likely to be possible in future conflicts. Our potential adversaries have made great progress in air defense measures and radio direction finding (RDF) techniques. We must assume that U.S. military forces trapped behind enemy lines in future conflicts may not experience quick recovery by friendly elements. Soldiers may have to move for extended times and distances to places less threatening to the recovery forces. The soldier will not likely know the type of recovery to expect. Each situation and the available resources determine the type of recovery possible. Since no one can be absolutely sure until the recovery effort begins, soldiers facing a potential cutoff from friendly forces should be familiar with all the possible types of recovery, their related problems, and their responsibilities to the recovery effort. Preparation and training can improve the chances of success.

PHASES OF PLANNING

Preparation is a requirement for all missions. When planning, you must consider how to avoid capture and return to your unit. Contingency plans must be prepared in conjunction with unit standing operating procedures (SOPs). Courses of action you or your unit will take must also be considered.

Contingency Plan of Action (CPA)

Intelligence sections can help prepare personnel for contingency actions through information supplied in area studies, SERE (survival, evasion, resistance, and escape) contingency guides, threat briefings, current intelligence reports, and current contact and authentication procedures. Pre-mission preparation includes the completion of a CPA. The study and research needed to develop the CPA will make you aware of the current situation in your mission area. Your CPA will let recovery forces know your probable actions should you have to move to avoid capture.
Start preparing even before pre-mission planning. Many parts of the CPA are SOP for your unit. Include the CPA in your training. Planning starts in your daily training.
The CPA is your entire plan for your return to friendly control. It consists of five paragraphs written in the operation order format. You can take most of paragraph 1, Situation, with you on the mission. Appendix H contains the CPA format. It also indicates what portion of the CPA you can take with you.
A comprehensive CPA is a valuable asset to the soldier trapped behind enemy lines who must try to avoid capture. To complete paragraph 1, know your unit's assigned area or concentrate on potential mission areas of the world. Many open or closed sources contain the information you need to complete a CPA. Open sources may include newspapers, magazines, country or area handbooks, area studies, television, radio, persons familiar with the area, and libraries. Closed sources may include area studies, area assessments, SERE contingency guides, various classified field manuals, and intelligence reports.
Prepare your CPA in three phases. During your normal training, prepare paragraph 1, Situation. Prepare paragraphs 2, 3, 4, and 5 during your pre-mission planning. After deployment into an area, continually update your CPA based on mission changes and intelligence updates.

The CPA is a guide. You may add or delete certain portions based on the mission. The CPA may be a recovery force's only means of determining your location and intentions after you start to move. It is an essential tool for your survival and return to friendly control.

Standing Operating Procedures

Unit SOPs are valuable tools your unit has that will help your planning. When faced with a dangerous situation requiring immediate action, it is not the time to discuss options; it is the time to act. Many of the techniques used during small unit movement can be carried over to fit requirements for moving and returning to friendly control. Items from the SOP should include, but are not limited to--

- Movement team size (three to four persons per team).
- Team communications (technical and nontechnical).
- Essential equipment.
- Actions at danger areas.
- Signaling techniques.
- Immediate action drills.
- Linkup procedures.
- Helicopter recovery devices and procedures.
- Security procedures during movement and at hide sites.
- Rally points.

Rehearsals work effectively for reinforcing these SOP skills and also provide opportunities for evaluation and improvement.

Notification to Move and Avoid Capture

An isolated unit has several general courses of action it can take to avoid the capture of the group or individuals. These courses of action are not courses the commander can choose instead of his original mission. He cannot arbitrarily abandon the assigned mission. Rather, he may adopt these courses of action after completing his mission when his unit cannot complete its assigned mission (because of combat power losses) or when he receives orders to extract his unit from its current position. If such actions are not possible, the commander may decide to have the unit try to move to avoid capture and return to friendly control. In either case, as long as there is communication with higher headquarters, that headquarters will make the decision.
If the unit commander loses contact with higher headquarters, he must make the decision to move or wait. He bases his decision on many factors, including the mission, rations and ammunition on hand, casualties, the chance of relief by friendly forces, and the tactical situation. The commander of an isolated unit faces other questions. What course of action will inflict maximum damage on the enemy? What course of action will assist in completing the higher headquarters' overall mission?
Movement teams conduct the execution portion of the plan when notified by higher headquarters or, if there is no contact with higher headquarters, when the highest ranking survivor decides that the situation requires the unit to try to escape capture or destruction. Movement team leaders receive their notification through prebriefed signals. Once the signal to try to avoid capture is given, it must be passed rapidly to all personnel. Notify higher headquarters, if possible. If unable to communicate with higher headquarters, leaders must recognize that organized resistance has ended, and that organizational control has ceased. Command and control is now at the movement team or individual level and is returned to higher organizational control only after reaching friendly lines.

EXECUTION

Upon notification to avoid capture, all movement team members will try to link up at the initial movement point. This point is where team members rally and actually begin their movement. Tentatively select the initial movement point during your planning phase through a map recon. Once on the ground, the team

verifies this location or selects a better one. All team members must know its location. The initial movement point should be easy to locate and occupy for a minimum amount of time.
Once the team has rallied at the initial movement point, it must--

- Give first aid.
- Inventory its equipment (decide what to abandon, destroy, or take along).
- Apply camouflage.
- Make sure everyone knows the tentative hide locations.
- Ensure everyone knows the primary and alternate routes and rally points en route to the hide locations.
- Always maintain security.
- Split the team into smaller elements. The ideal element should have two to three members; however, it could include more depending on team equipment and experience.

The movement portion of returning to friendly control is the most dangerous as you are now most vulnerable. It is usually better to move at night because of the concealment darkness offers. Exceptions to such movement would be when moving through hazardous terrain or dense vegetation (for example, jungle or mountainous terrain). When moving, avoid the following even if it takes more time and energy to bypass:

- Obstacles and barriers.
- Roads and trails.
- Inhabited areas.
- Waterways and bridges.
- Natural lines of drift.
- Man-made structures.
- All civilian and military personnel.

Movement in enemy-held territory is a very slow and deliberate process. The slower you move and the more careful you are, the better. Your best security will be using your senses. Use your eyes and ears to detect people before they detect you. Make frequent listening halts. In daylight, observe a section of your route before you move along it. The distance you travel before you hide will depend on the enemy situation, your health, the terrain, the availability of cover and concealment for hiding, and the amount of darkness left.
Once you have moved into the area in which you want to hide (hide area), select a hide site. Keep the following formula in mind when selecting a hide site: BLISS.
B - Blends in with the surroundings.
L - Low in silhouette.
I - Irregular in shape.
S - Small in size.
S - Secluded.
Avoid the use of existing buildings or shelters. Usually, your best option will be to crawl into the thickest vegetation you can find. Construct any type of shelter within the hide area only in cold weather and desert environments. If you build a shelter, follow the BLISS formula.

Hide Site Activities

After you have located your hide site, do not move straight into it. Use a button hook or other deceptive technique to move to a position outside of the hide site. Conduct a listening halt before moving individually into the hide site. Be careful not to disturb or cut any vegetation. Once you have occupied the hide site, limit your activities to maintaining security, resting, camouflaging, and planning your next moves.

Maintain your security through visual scanning and listening. Upon detection of the enemy, the security personnel alert all personnel, even if the team's plan is to stay hidden and not move upon sighting the enemy. Take this action so that everyone is aware of the danger and ready to react.

If any team member leaves the team, give him a five-point contingency plan. Take such steps especially when a recon team or a work party is out of the hole-up or hide site.

It is extremely important to stay healthy and alert when trying to avoid capture. Take every opportunity to rest, but do not sacrifice security. Rotate security so that all members of your movement team can rest. Treat all injuries, no matter how minor. Loss of your health will mean loss of your ability to continue to avoid capture.

Camouflage is an important aspect of both moving and securing a hide site. Always use a buddy system to ensure that camouflage is complete. Ensure that team members blend with the hide site. Use natural or man-made materials. If you add any additional camouflage material to the hide site, do not cut vegetation in the immediate area.

Plan your next actions while at the hide site. Start your planning process immediately upon occupying the hide site. Inform all team members of their current location and designate an alternate hide site location. Once this is done, start planning for the team's next movement.

Planning the team's movement begins with a map recon. Choose the next hide area first. Then choose a primary and an alternate route to the hide area. In choosing the routes, do not use straight lines. Use one or two radical changes in direction. Pick the routes that offer the best cover and concealment, the fewest obstacles, and the least likelihood of contact with humans. There should be locations along the route where the team can get water. To aid team navigation, use azimuths, distances, checkpoints or steering marks, and corridors. Plan rally points and rendezvous points at intervals along the route.

Other planning considerations may fall under what the team already has in the team SOP. Examples are immediate action drills, actions on sighting the enemy, and hand-and-arm signals.

Once planning is complete, ensure everyone knows and memorizes the entire plan. The team members should know the distances and azimuths for the entire route to the next hide area. They should study the map and know the various terrain they will be moving across so that they can move without using the map.

Do not occupy a hide site for more than 24 hours. In most situations, hide during the day and move at night. Limit your actions in the hide site to those discussed above. Once in the hide site, restrict all movement to less than 45 centimeters above the ground. Do not build fires or prepare food. Smoke and food odors will reveal your location. Before leaving the hide site, sterilize it to prevent tracking.

Hole-Up Areas

After moving and hiding for several days, usually three or four, you or the movement team will have to move into a hole-up area. This is an area where you can rest, recuperate, and get and prepare food. Choose an area near a water source. You then have a place to get water, to place fishing devices, and to trap game. Since waterways are a line of communication, locate your hide site well away from the water.

The hole-up area should offer plenty of cover and concealment for movement in and around the area. Always maintain security while in the hole-up area. Always man the hole-up area. Actions in the hole-up area are the same as in hide site, except that you can move away from the hole-up area to get and prepare food. Actions in the hole-up area include--

- Selecting and occupying the next hide site (remember you are still in a dangerous situation; this is not a friendly area).
- Reconnoitering the area for resources and potential concealed movement routes to the alternate hide site.
- Gathering food (nuts, berries, vegetables). When moving around the area for food, maintain security and avoid leaving tracks or other signs. When setting traps and snares, keep them well-camouflaged and in areas where people are not likely to discover them. Remember, the local population sometimes heavily travels trails near water sources.

- Getting water from sources within the hide area. Be careful not to leave tracks of signs along the banks of water sources when getting water. Moving on hard rocks or logs along the banks to get water will reduce the signs you leave.
- Setting clandestine fishing devices, such as stakeouts, below the surface of the water to avoid detection.
- Locating a fire site well away from the hide site. Use this site to prepare food or boil water. Camouflage and sterilize the fire site after each use. Be careful that smoke and light from the fire does not compromise the hole-up area.

While in the hole-up area, security is still your primary concern. Designate team members to perform specific tasks. To limit movement around the area, you may have a two-man team perform more than one task. For example, the team getting water could also set the fishing devices. Do not occupy the hole-up area longer than 72 hours.

RETURN TO FRIENDLY CONTROL

Establishing contact with friendly lines or patrols is the most crucial part of movement and return to friendly control. All your patience, planning, and hardships will be in vain if you do not exercise caution when contacting friendly frontline forces. Friendly patrols have killed personnel operating behind enemy lines because they did not make contact properly. Most of the casualties could have been avoided if caution had been exercised and a few simple procedures followed. The normal tendency is to throw caution to the winds when in sight of friendly forces. You must overcome this tendency and understand that linkup is a very sensitive situation.

Border Crossings

If you have made your way to a friendly or neutral country, use the following procedures to cross the border and link up with friendly forces on the other side:

- Occupy a hide site on the near side of the border and send a team out to reconnoiter the potential crossing site.
- Surveil the crossing site for at least 24 hours, depending on the enemy situation.
- Make a sketch of the site, taking note of terrain, obstacles, guard routines and rotations, and any sensor devices or trip wires. Once the recon is complete, the team moves to the hide site, briefs the rest of the team, and plans to cross the border at night.
- After crossing the border, set up a hide site on the far side of the border and try to locate friendly positions. Do not reveal your presence.
- Depending on the size of your movement team, have two men surveil the potential linkup site with friendly forces until satisfied that the personnel are indeed friendly.
- Make contact with the friendly forces during daylight. Personnel chosen to make contact should be unarmed, have no equipment, and have positive identification readily available. The person who actually makes the linkup should be someone who looks least like the enemy.
- During the actual contact, have only one person make the contact. The other person provides the security and observes the linkup area from a safe distance. The observer should be far enough away so that he can warn the rest of the movement team if something goes wrong.
- Wait until the party he is contacting looks in his direction so that he does not surprise the contact. He stands up from behind cover, with hands overhead and states that he is an American. After this, he follows any instructions given him. He avoids answering any tactical questions and does not give any indication that there are other team members.
- Reveal that there are other personnel with him only after verifying his identity and satisfying himself he has made contact with friendly forces.

Language problems or difficulties confirming identities may arise. The movement team should maintain security, be patient, and have a contingency plan.

ESSENTIAL ARMY MANUAL
(215) ARMY SURVIVAL MANUAL

Note: If you are moving to a neutral country, you are surrendering to that power and become a detained person.

Linkup at the FEBA/FLOT

If caught between friendly and enemy forces and there is heavy fighting in the area, you may choose to hide and let the friendly lines pass over you. If overrun by friendly forces, you may try to link up from their rear during daylight hours. If overrun by enemy forces, you may move further to the enemy rear, try to move to the forward edge of the battle area (FEBA)/forward line of own troops (FLOT) during a lull in the fighting, or move to another area along the front.

The actual linkup will be done as for linkup during a border crossing. The only difference is that you must be more careful on the initial contact. Frontline personnel are more likely to shoot first and ask questions later, especially in areas of heavy fighting. You should be near or behind cover before trying to make contact.

Linkup With Friendly Patrols

If friendly lines are a circular perimeter or an isolated camp, for example, any direction you approach from will be considered enemy territory. You do not have the option of moving behind the lines and trying to link up. This move makes the linkup extremely dangerous. One option you have is to place the perimeter under observation and wait for a friendly patrol to move out in your direction, providing a chance for a linkup. You may also occupy a position outside of the perimeter and call out to get the attention of the friendly forces. Ideally, display anything that is white while making contact. If nothing else is available, use any article of clothing. The idea is to draw attention while staying behind cover. Once you have drawn attention to your signal and called out, follow instructions given to you.

Be constantly on the alert for friendly patrols because these provide a means for return to friendly control. Find a concealed position that allows you maximum visual coverage of the area. Try to memorize every terrain feature so that, if necessary, you can infiltrate to friendly positions under the cover of darkness. Remember, trying to infiltrate in darkness is extremely dangerous.

Because of the missions of combat and recon patrols and where they are operating, making contact can be dangerous. If you decide not to make contact, you can observe their route and approach friendly lines at about the same location. Such observation will enable you to avoid mines and booby traps.

Once you have spotted a patrol, remain in position and, if possible, allow the patrol to move toward you. When the patrol is 25 to 50 meters from your position, signal them and call out a greeting that is clearly and unmistakably of American origin.

If you have nothing white, an article of clothing will suffice to draw attention. If the distance is greater than 50 meters, a recon patrol may avoid contact and bypass your position. If the distance is less than 25 meters, a patrol member may react instantly by firing a fatal shot.

It is crucial, at the time of contact, that there is enough light for the patrol to identify you as an American. Whatever linkup technique you decide to use, use extreme caution. From the perspective of the friendly patrol or friendly personnel occupying a perimeter, you are hostile until they make positive identification.

CHAPTER 21 - CAMOUFLAGE

In a survival situation, especially in a hostile environment, you may find it necessary to camouflage yourself, your equipment, and your movement. It may mean the difference between survival and capture by the enemy. Camouflage and movement techniques, such as stalking, will also help you get animals or game for food using primitive weapons and skills.

PERSONAL CAMOUFLAGE

When camouflaging yourself, consider that certain shapes are particular to humans. The enemy will look for these shapes. The shape of a hat, helmet, or black boots can give you away. Even animals know and run from the shape of a human silhouette. Break up your outline by placing small amounts of vegetation from the surrounding area in your uniform, equipment, and headgear. Try to reduce any shine from skin or equipment. Blend in with the surrounding colors and simulate the texture of your surroundings.

Shape and Outline

Change the outline of weapons and equipment by tying vegetation or strips of cloth onto them. Make sure the added camouflage does not hinder the equipment's operation. When hiding, cover yourself and your equipment with leaves, grass, or other local debris. Conceal any signaling devices you have prepared, but keep them ready for use.

Color and Texture

Each area of the world and each climatic condition (arctic/winter, temperate/jungle, or swamp/desert) has color patterns and textures that are natural for that area. While color is self-explanatory, texture defines the surface characteristics of something when looking at it. For example, surface textures may be smooth, rough, rocky, leafy, or many other possible combinations. Use color and texture together to camouflage yourself effectively. It makes little sense to cover yourself with dead, brown vegetation in the middle of a large grassy field. Similarly, it would be useless to camouflage yourself with green grass in the middle of a desert or rocky area.

To hide and camouflage movement in any specific area of the world, you must take on the color and texture of the immediate surroundings. Use natural or man-made materials to camouflage yourself. Camouflage paint, charcoal from burned paper or wood, mud, grass, leaves, strips of cloth or burlap, pine boughs, and camouflaged uniforms are a few examples.

Cover all areas of exposed skin, including face, hands, neck, and ears. Use camouflage paint, charcoal, or mud to camouflage yourself. Cover with a darker color areas that stick out more and catch more light (forehead, nose, cheekbones, chin, and ears). Cover other areas, particularly recessed or shaded areas (around the eyes and under the chin), with lighter colors. Be sure to use an irregular pattern. Attach vegetation from the area or strips of cloth of the proper color to clothing and equipment. If you use vegetation, replace it as it wilts. As you move through an area, be alert to the color changes and modify your camouflage colors as necessary.

Figure 21-1 gives a general idea of how to apply camouflage for various areas and climates. Use appropriate colors for your surroundings. The blotches or slashes will help to simulate texture.

Area	Method
Temperate deciduous forest	Blotches
Coniferous forest	Broad slash
Jungle	Broad slash
Desert	Slash
Arctic	Blotches
Grass or open area	Slash

Figure 21-1. Camouflage methods for specific areas.

Shine

As skin gets oily, it becomes shiny. Equipment with worn off paint is also shiny. Even painted objects, if smooth, may shine. Glass objects such as mirrors, glasses, binoculars, and telescopes shine. You must cover these glass objects when not in use. Anything that shines automatically attracts attention and will give away your location.

Whenever possible, wash oily skin and reapply camouflage. Skin oil will wash off camouflage, so reapply it frequently. If you must wear glasses, camouflage them by applying a thin layer of dust to the outside of the lenses. This layer of dust will reduce the reflection of light. Cover shiny spots on equipment by painting, covering with mud, or wrapping with cloth or tape. Pay particular attention to covering boot eyelets, buckles on equipment, watches and jewelry, zippers, and uniform insignia. Carry a signal mirror in its designed pouch or in a pocket with the mirror portion facing your body.

Shadow

When hiding or traveling, stay in the deepest part of the shadows. The outer edges of the shadows are lighter and the deeper parts are darker. Remember, if you are in an area where there is plenty of vegetation, keep as much vegetation between you and a potential enemy as possible. This action will make it very hard for the enemy to see you as the vegetation will partially mask you from his view. Forcing an enemy to look through many layers of masking vegetation will fatigue his eyes very quickly. When traveling, especially in built-up areas at night, be aware of where you cast your shadow. It may extend out around the comer of a building and give away your position. Also, if you are in a dark shadow and there is a light source to one side, an enemy on the other side can see your silhouette against the light.

Movement

Movement, especially fast movement, attracts attention. If at all possible, avoid movement in the presence of an enemy. If capture appears imminent in your present location and you must move, move away slowly, making as little noise as possible. By moving slowly in a survival situation, you decrease the chance of detection and conserve energy that you may need for long-term survival or long-distance evasion.

When moving past obstacles, avoid going over them. If you must climb over an obstacle, keep your body level with its top to avoid silhouetting yourself. Do not silhouette yourself against the skyline when crossing hills or ridges. When you are moving, you will have difficulty detecting the movement of others. Stop frequently, listen, and look around slowly to detect signs of hostile movement.

Noise

Noise attracts attention, especially if there is a sequence of loud noises such as several snapping twigs. If possible, avoid making any noise at all. Slow down your pace as much as necessary to avoid making noise when moving around or away from possible threats.
Use background noises to cover the noise of your movement. Sounds of aircraft, trucks, generators, strong winds, and people talking will cover some or all the sounds produced by your movement. Rain will mask a lot of movement noise, but it also reduces your ability to detect potential enemy noise.

Scent

Whether hunting animals or avoiding the enemy, it is always wise to camouflage the scent associated with humans. Start by washing yourself and your clothes without using soap. This washing method removes soap and body odors. Avoiding strong smelling foods, such as garlic, helps reduce body odors. Do not use tobacco products, candy, gum, or cosmetics.
You can use aromatic herbs or plants to wash yourself and your clothing, to rub on your body and clothing, or to chew on to camouflage your breath. Pine needles, mint, or any similar aromatic plant will help camouflage your scent from both animals and humans. Standing in smoke from a fire can help mask your scent from animals. While animals are afraid of fresh smoke from a fire, older smoke scents are normal smells after forest fires and do not scare them.
While traveling, use your sense of smell to help you find or avoid humans. Pay attention to smells associated with humans, such as fire, cigarettes, gasoline, oil, soap, and food. Such smells may alert you to their presence long before you can see or hear them, depending on wind speed and direction. Note the wind's direction and, when possible, approach from or skirt around on the downwind side when nearing humans or animals.

METHODS OF STALKING

Sometimes you need to move, undetected, to or from a location. You need more than just camouflage to make these moves successfully. The ability to stalk or move without making any sudden quick movement or loud noise is essential to avoiding detection.
You must practice stalking if it is to be effective. Use the following techniques when practicing.

Upright Stalking

Take steps about half your normal stride when stalking in the upright position. Such strides help you to maintain your balance. You should be able to stop at any point in that movement and hold that position as long as necessary. Curl the toes up out of the way when stepping down so the outside edge of the ball of the foot touches the ground. Feel for sticks and twigs that may snap when you place your weight on them. If you start to step on one, lift your foot and move it. After making contact with the outside edge of the ball of your foot, roll to the inside ball of your foot, place your heel down, followed by your toes. Then gradually shift your weight forward to the front foot. Lift the back foot to about knee height and start the process over again.
Keep your hands and arms close to your body and avoid waving them about or hitting vegetation. When moving in a crouch, you gain extra support by placing your hands on your knees. One step usually takes 1 minute to complete, but the time it takes will depend on the situation.

Crawling

Crawl on your hands and knees when the vegetation is too low to allow you to walk upright without being seen. Move one limb at a time and be sure to set it down softly, feeling for anything that may snap and make noise. Be careful that your toes and heels do not catch on vegetation.

Prone Stalking

To stalk in the prone position, you do a low, modified push-up on your hands and toes, moving yourself forward slightly, and then lowering yourself again slowly. Avoid dragging and scraping along the ground as this makes excessive noise and leaves large trails for trackers to follow.

Animal Stalking

Before stalking an animal, select the best route. If the animal is moving, you will need an intercepting route. Pick a route that puts objects between you and the animal to conceal your movement from it. By positioning yourself in this way, you will be able to move faster, until you pass that object. Some objects, such as large rocks and trees, may totally conceal you, and others, such as small bushes and grass, may only partially conceal you. Pick the route that offers the best concealment and requires the least amount of effort.
Keep your eyes on the animal and stop when it looks your way or turns its ears your way, especially if it suspects your presence. As you get close, squint your eyes slightly to conceal both the light-dark contrast of the whites of the eyes and any shine from your eyes. Keep your mouth closed so that the animal does not see the whiteness or shine of your teeth.

CHAPTER 22 - CONTACT WITH PEOPLE

Some of the best and most frequently given advice, when dealing with local peoples, is for the survivor to accept, respect, and adapt to their ways. Thus, "when in Rome, do as the Romans do." This is excellent advice, but there are several considerations involved in putting this advice into practice.

CONTACT WITH LOCAL PEOPLE

You must give serious consideration to dealing with the local people. Do they have a primitive culture? Are they farmers, fishermen, friendly people, or enemy? As a survivor, "cross-cultural communication" can vary radically from area to area and from people to people. It may mean interaction with people of an extremely primitive culture or contact with people who have a relatively modem culture. A culture is identified by standards of behavior that its members consider proper and acceptable but may or may not conform to your idea of what is proper. No matter who these people are, you can expect they will have laws, social and economic values, and political and religious beliefs that may be radically different from yours. Before deploying into your area of operations, study these different cultural aspects. Prior study and preparation will help you make or avoid contact if you have to deal with the local population.

People will be friendly, unfriendly, or they will choose to ignore you. Their attitude may be unknown. If the people are known to be friendly, try to keep them friendly through your courtesy and respect for their religion, politics, social customs, habits, and all other aspects of their culture. If the people are known to be enemies or are unknowns, make every effort to avoid any contact and leave no sign of your presence. A basic knowledge of the daily habits of the local people will be essential in this attempt. If after careful observation you determine that an unknown people are friendly, you may contact them if you absolutely need their help.

Usually, you have little to fear and much to gain from cautious and respectful contact with local people of friendly or neutral countries. If you become familiar with the local customs, display common decency, and most important, show respect for their customs, you should be able to avoid trouble and possibly gain needed help. To make contact, wait until only one person is near and, if possible, let that person make the initial approach. Most people will be willing to help a survivor who appears to be in need. However, local political attitudes, instruction, or propaganda efforts may change the attitudes of otherwise friendly people. Conversely, in unfriendly countries, many people, especially in remote areas, may feel animosity toward their politicians and may be more friendly toward a survivor.

The key to successful contact with local peoples is to be friendly, courteous, and patient. Displaying fear, showing weapons, and making sudden or threatening movements can cause a local person to fear you. Such actions can prompt a hostile response. When attempting a contact, smile as often as you can. Many local peoples are shy and seem unapproachable, or they may ignore you. Approach them slowly and do not rush your contact.

THE SURVIVOR'S BEHAVIOR

Use salt, tobacco, silver money, and similar items discreetly when trading with local people. Paper money is well-known worldwide. Do not overpay; it may lead to embarrassment and even danger. Always treat people with respect. Do not bully them or laugh at them.

Using sign language or acting out needs or questions can be very effective. Many people are used to such language and communicate using nonverbal sign language. Try to learn a few words and phrases of the local language in and around your potential area of operations. Trying to speak someone's language is one of the best ways to show respect for his culture. Since English is widely used, some of the local people may understand a few words of English.

Some areas may be taboo. They range from religious or sacred places to diseased or danger areas. In some areas, certain animals must not be killed. Learn the rules and follow them. Watch and learn as much as possible. Such actions will help to strengthen relations and provide new knowledge and skills that may be very important later. Seek advice on local hazards and find out from friendly people where the hostile people are. Always remember that people frequently insist that other peoples are hostile, simply because they do not understand different cultures and distant peoples. The people they can usually trust are their immediate neighbors--much the same as in our own neighborhood.

Frequently, local people, like ourselves, will suffer from contagious diseases. Build a separate shelter, if possible, and avoid physical contact without giving the impression of doing so. Personally prepare your food and drink, if you can do so without giving offense. Frequently, the local people will accept the use of "personal or religious custom" as an explanation for isolationist behavior.

Barter, or trading, is common in more primitive societies. Hard coin is usually good, whether for its exchange value or as jewelry or trinkets. In isolated areas, matches, tobacco, salt, razor blades, empty containers, or cloth may be worth more than any form of money.

Be very cautious when touching people. Many people consider "touching" taboo and such actions may be dangerous. Avoid sexual contact.

Hospitality among some people is such a strong cultural trait that they may seriously reduce their own supplies to feed a stranger. Accept what they offer and share it equally with all present. Eat in the same way they eat and, most important, try to eat all they offer.

If you make any promises, keep them. Respect personal property and local customs and manners, even if they seem odd. Make some kind of payment for food, supplies, and so forth. Respect privacy. Do not enter a house unless invited.

CHANGES TO POLITICAL ALLEGIANCE

In today's world of fast-paced international politics, political attitudes and commitments within nations are subject to rapid change. The population of many countries, especially politically hostile countries, must not be considered friendly just because they do not demonstrate open hostility. Unless briefed to the contrary; avoid all contact with such people.

CHAPTER 23 - SURVIVAL IN MAN-MADE HAZARDS

Nuclear, chemical, and biological weapons have become potential realities on any modern battlefield. Recent experience in Afghanistan, Cambodia, and other areas of conflict has proved the use of chemical and biological weapons (such as mycotoxins). The warfighting doctrine of the NATO and Warsaw Pact nations addresses the use of both nuclear and chemical weapons. The potential use of these weapons intensifies the problems of survival because of the serious dangers posed by either radioactive fallout or contamination produced by persistent biological or chemical agents.

You must use special precautions if you expect to survive in these man-made hazards. If you are subjected to any of the effects of nuclear, chemical, or biological warfare, the survival procedures recommended in this chapter may save your life. This chapter presents some background information on each type of hazard so that you may better understand the true nature of the hazard. Awareness of the hazards, knowledge of this chapter, and application of common sense should keep you alive.

THE NUCLEAR ENVIRONMENT
Prepare yourself to survive in a nuclear environment. Know how to react to a nuclear hazard.

Effects of Nuclear Weapons

The effects of nuclear weapons are classified as either initial or residual. Initial effects occur in the immediate area of the explosion and are hazardous in the first minute after the explosion. Residual effects can last for days or years and cause death. The principal initial effects are blast and radiation.
Blast
Defined as the brief and rapid movement of air away from the explosion's center and the pressure accompanying this movement. Strong winds accompany the blast. Blast hurls debris and personnel, collapses lungs, ruptures eardrums, collapses structures and positions, and causes immediate death or injury with its crushing effect.
Thermal Radiation
The heat and light radiation a nuclear explosion's fireball emits. Light radiation consists of both visible light and ultraviolet and infrared light. Thermal radiation produces extensive fires, skin burns, and flash blindness.
Nuclear Radiation
Nuclear radiation breaks down into two categories-initial radiation and residual radiation.
Initial nuclear radiation consists of intense gamma rays and neutrons produced during the first minute after the explosion. This radiation causes extensive damage to cells throughout the body. Radiation damage may cause headaches, nausea, vomiting, diarrhea, and even death, depending on the radiation dose received. The major problem in protecting yourself against the initial radiation's effects is that you

may have received a lethal or incapacitating dose before taking any protective action. Personnel exposed to lethal amounts of initial radiation may well have been killed or fatally injured by blast or thermal radiation.

Residual radiation consists of all radiation produced after one minute from the explosion. It has more effect on you than initial radiation. A discussion of <u>residual radiation</u> takes place in a subsequent paragraph.

Types of Nuclear Bursts

There are three types of nuclear bursts--airburst, surface burst, and subsurface burst. The type of burst directly affects your chances of survival. A subsurface burst occurs completely underground or underwater. Its effects remain beneath the surface or in the immediate area where the surface collapses into a crater over the burst's location. Subsurface bursts cause you little or no radioactive hazard unless you enter the immediate area of the crater. No further discussion of this type of burst will take place.

An airburst occurs in the air above its intended target. The airburst provides the maximum radiation effect on the target and is, therefore, most dangerous to you in terms of *immediate* nuclear effects.

A surface burst occurs on the ground or water surface. Large amounts of fallout result, with serious long-term effects for you. This type of burst is your *greatest* nuclear hazard.

Nuclear Injuries

Most injuries in the nuclear environment result from the initial nuclear effects of the detonation. These injuries are classed as blast, thermal, or radiation injuries. Further radiation injuries may occur if you do not take proper precautions against fallout. Individuals in the area near a nuclear explosion will probably suffer a combination of all three types of injuries.

Blast Injuries

Blast injuries produced by nuclear weapons are similar to those caused by conventional high-explosive weapons. Blast overpressure can produce collapsed lungs and ruptured internal organs. Projectile wounds occur as the explosion's force hurls debris at you. Large pieces of debris striking you will cause fractured limbs or massive internal injuries. Blast over-pressure may throw you long distances, and you will suffer severe injury upon impact with the ground or other objects. Substantial cover and distance from the explosion are the best protection against blast injury. Cover blast injury wounds as soon as possible to prevent the entry of radioactive dust particles.

Thermal Injuries

The heat and light the nuclear fireball emits causes thermal injuries. First-, second-, or third-degree burns may result. Flash blindness also occurs. This blindness may be permanent or temporary depending on the degree of exposure of the eyes. Substantial cover and distance from the explosion can prevent thermal injuries. Clothing will provide significant protection against thermal injuries. Cover as much exposed skin as possible before a nuclear explosion. First aid for thermal injuries is the same as first aid for burns. Cover open burns (second-or third-degree) to prevent the entry of radioactive particles. Wash all burns before covering.

Radiation Injuries

Neutrons, gamma radiation, alpha radiation, and beta radiation cause radiation injuries. Neutrons are high-speed, extremely penetrating particles that actually smash cells within your body. Gamma radiation is similar to X rays and is also a highly penetrating radiation. During the initial fireball stage of a nuclear detonation, initial gamma radiation and neutrons are the most serious threat. Beta and alpha radiation are radioactive particles normally associated with radioactive dust from fallout. They are short-range particles and you can easily protect yourself against them if you take precautions. See *Bodily Reactions to Radiation,* below, for the symptoms of radiation injuries.

Residual Radiation

Residual radiation is all radiation emitted after 1 minute from the instant of the nuclear explosion. Residual radiation consists of induced radiation and fallout.

Induced Radiation

It describes a relatively small, intensely radioactive area directly underneath the nuclear weapon's fireball. The irradiated earth in this area will remain highly radioactive for an extremely long time. You should not travel into an area of induced radiation.

Fallout

Fallout consists of radioactive soil and water particles, as well as weapon fragments. During a surface detonation, or if an airburst's nuclear fireball touches the ground, large amounts of soil and water are vaporized along with the bomb's fragments, and forced upward to altitudes of 25,000 meters or more. When these vaporized contents cool, they can form more than 200 different radioactive products. The vaporized bomb contents condense into tiny radioactive particles that the wind carries and they fall back to earth as radioactive dust. Fallout particles emit alpha, beta, and gamma radiation. Alpha and beta radiation are relatively easy to counteract, and residual gamma radiation is much less intense than the gamma radiation emitted during the first minute after the explosion. Fallout is your most significant radiation hazard, provided you have not received a lethal radiation dose from the initial radiation.

Bodily Reactions to Radiation

The effects of radiation on the human body can be broadly classed as either chronic or acute. Chronic effects are those that occur some years after exposure to radiation. Examples are cancer and genetic defects. Chronic effects are of minor concern insofar as they affect your immediate survival in a radioactive environment. On the other hand, acute effects are of primary importance to your survival. Some acute effects occur within hours after exposure to radiation. These effects result from the radiation's direct physical damage to tissue. Radiation sickness and beta burns are examples of acute effects. Radiation sickness symptoms include nausea, diarrhea, vomiting, fatigue, weakness, and loss of hair. Penetrating beta rays cause radiation burns; the wounds are similar to fire burns.

Recovery Capability

The extent of body damage depends mainly on the part of the body exposed to radiation and how long it was exposed, as well as its ability to recover. The brain and kidneys have little recovery capability. Other parts (skin and bone marrow) have a great ability to recover from damage. Usually, a dose of 600 centigrams (cgys) to the entire body will result in almost certain death. If only your hands received this same dose, your overall health would not suffer much, although your hands would suffer severe damage.

External and Internal Hazards

An external or an internal hazard can cause body damage. Highly penetrating gamma radiation or the less penetrating beta radiation that causes burns can cause external damage. The entry of alpha or beta radiation-emitting particles into the body can cause internal damage. The external hazard produces overall irradiation and beta burns. The internal hazard results in irradiation of critical organs such as the gastrointestinal tract, thyroid gland, and bone. A very small amount of radioactive material can cause extreme damage to these and other internal organs. The internal hazard can enter the body either through consumption of contaminated water or food or by absorption through cuts or abrasions. Material that enters the body through breathing presents only a minor hazard. You can greatly reduce the internal radiation hazard by using good personal hygiene and carefully decontaminating your food and water.

Symptoms

The symptoms of radiation injuries include nausea, diarrhea, and vomiting. The severity of these symptoms is due to the extreme sensitivity of the gastrointestinal tract to radiation. The severity of the symptoms and the speed of onset after exposure are good indicators of the degree of radiation damage. The gastrointestinal damage can come from either the external or the internal radiation hazard.

Countermeasures Against Penetrating External Radiation

Knowledge of the radiation hazards discussed earlier is extremely important in surviving in a fallout area. It is also critical to know how to protect yourself from the most dangerous form of residual radiation-- penetrating external radiation.

The means you can use to protect yourself from penetrating external radiation are time, distance, and shielding. You can reduce the level of radiation and help increase your chance of survival by controlling the duration of exposure. You can also get as far away from the radiation source as possible. Finally you can place some radiation-absorbing or shielding material between you and the radiation.

Time

Time is important to you, as the survivor, in two ways. First, radiation dosages are cumulative. The longer you are exposed to a radioactive source, the greater the dose you will receive. Obviously, spend as little time in a radioactive area as possible. Second, radioactivity decreases or decays over time. This concept is known as radioactive *half-life*. Thus, a radioactive element decays or loses half of its radioactivity within a certain time. The rule of thumb for radioactivity decay is that it decreases in intensity by a factor of ten for every sevenfold increase in time following the peak radiation level. For example, if a nuclear fallout area had a maximum radiation rate of 200 cgys per hour when fallout is complete, this rate would fall to 20 cgys per hour after 7 hours; it would fall still further to 2 cgys per hour after 49 hours. Even an untrained observer can see that the greatest hazard from fallout occurs immediately after detonation, and that the hazard decreases quickly over a relatively short time. As a survivor, try to avoid fallout areas until the radioactivity decays to safe levels. If you can avoid fallout areas long enough for most of the radioactivity to decay, you enhance your chance of survival.

Distance

Distance provides very effective protection against penetrating gamma radiation because radiation intensity decreases by the square of the distance from the source. For example, if exposed to 1,000 cgys of radiation standing 30 centimeters from the source, at 60 centimeters, you would only receive 250 cgys. Thus, when you double the distance, radiation decreases to $(0.5)^2$ or 0.25 the amount. While this formula is valid for concentrated sources of radiation in small areas, it becomes more complicated for large areas of radiation such as fallout areas.

Shielding

Shielding is the most important method of protection from penetrating radiation. Of the three countermeasures against penetrating radiation, shielding provides the greatest protection and is the easiest to use under survival conditions. Therefore, it is the most desirable method.

If shielding is not possible, use the other two methods to the maximum extent practical.

Shielding actually works by absorbing or weakening the penetrating radiation, thereby reducing the amount of radiation reaching your body. The denser the material, the better the shielding effect. Lead, iron, concrete, and water are good examples of shielding materials.

Special Medical Aspects

The presence of fallout material in your area requires slight changes in first aid procedures. You must cover all wounds to prevent contamination and the entry of radioactive particles. You must first wash burns of beta radiation, then treat them as ordinary burns. Take extra measures to prevent infection. Your body will be extremely sensitive to infections due to changes in your blood chemistry. Pay close attention to the prevention of colds or respiratory infections. Rigorously practice personal hygiene to prevent infections. Cover your eyes with improvised goggles to prevent the entry of particles.

Shelter

As stated earlier, the shielding material's effectiveness depends on its thickness and density. An ample thickness of shielding material will reduce the level of radiation to negligible amounts.

The primary reason for finding and building a shelter is to get protection against the high-intensity radiation levels of early gamma fallout as fast as possible. Five minutes to locate the shelter is a good guide. Speed in finding shelter is absolutely essential. Without shelter, the dosage received in the first few hours will exceed that received during the rest of a week in a contaminated area. The dosage received in this first week will exceed the dosage accumulated during the rest of a lifetime spent in the same contaminated area.

Shielding Materials

The thickness required to weaken gamma radiation from fallout is far less than that needed to shield against initial gamma radiation. Fallout radiation has less energy than a nuclear detonation's initial radiation. For fallout radiation, a relatively small amount of shielding material can provide adequate protection. Figure 23-1 gives an idea of the thickness of various materials needed to reduce residual gamma radiation transmission by 50 percent.

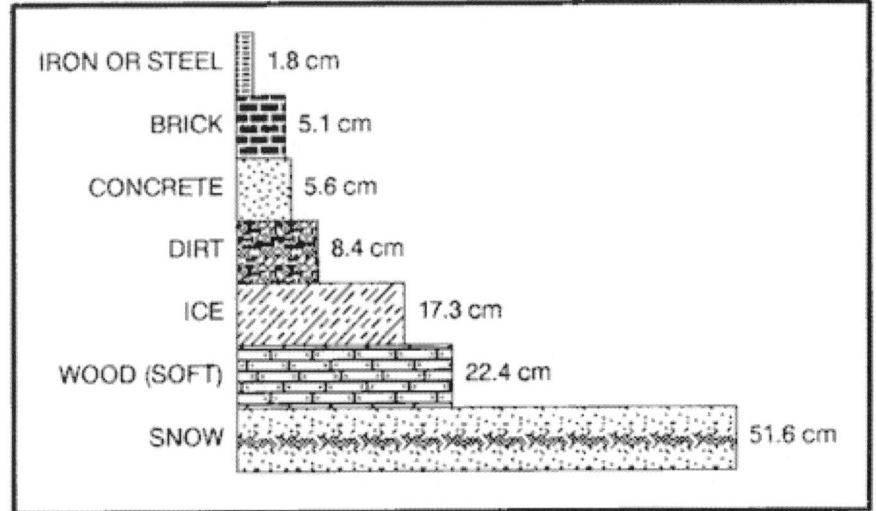

Figure 23-1. Thickness of materials to reduce gamma radiation.

The principle of *half-value layer thickness* is useful in understanding the absorption of gamma radiation by various materials. According to this principle, if 5 centimeters of brick reduce the gamma radiation level by one-half, adding another 5 centimeters of brick (another half-value layer) will reduce the intensity by another half, namely, to one-fourth the original amount. Fifteen centimeters will reduce gamma radiation fallout levels to one-eighth its original amount, 20 centimeters to one-sixteenth, and so on. Thus, a shelter protected by 1 meter of dirt would reduce a radiation intensity of 1,000 cgys per hour on the outside to about 0.5 cgy per hour inside the shelter.

Natural Shelters

Terrain that provides natural shielding and easy shelter construction is the ideal location for an emergency shelter. Good examples are ditches, ravines, rocky outcropping, hills, and river banks. In level areas without natural protection, dig a fighting position or slit trench.

Trenches

When digging a trench, work from inside the trench as soon as it is large enough to cover part of your body thereby not exposing all your body to radiation. In open country, try to dig the trench from a prone position, stacking the dirt carefully and evenly around the trench. On level ground, pile the dirt around your body for additional shielding. Depending upon soil conditions, shelter construction time will vary from a few minutes to a few hours. If you dig as quickly as possible, you will reduce the dosage you receive.

Other Shelters

While an underground shelter covered by 1 meter or more of earth provides the best protection against fallout radiation, the following unoccupied structures (in order listed) offer the next best protection:

- Caves and tunnels covered by more than 1 meter of earth.
- Storm or storage cellars.
- Culverts.
- Basements or cellars of abandoned buildings.
- Abandoned buildings made of stone or mud.

Roofs

It is not mandatory that you build a roof on your shelter. Build one only if the materials are readily available with only a brief exposure to outside contamination. If building a roof would require extended exposure to penetrating radiation, it would be wiser to leave the shelter roofless. A roof's sole function is to reduce radiation from the fallout source to your body. Unless you use a thick roof, a roof provides very little shielding.

You can construct a simple roof from a poncho anchored down with dirt, rocks, or other refuse from your shelter. You can remove large particles of dirt and debris from the top of the poncho by beating it off from the inside at frequent intervals. This cover will not offer shielding from the radioactive particles deposited on the surface, but it will increase the distance from the fallout source and keep the shelter area from further contamination.

Shelter Site Selection and Preparation

To reduce your exposure time and thereby reduce the dosage received, remember the following factors when selecting and setting up a shelter:

- Where possible, seek a crude, existing shelter that you can improve. If none is available, dig a trench.
- Dig the shelter deep enough to get good protection, then enlarge it as required for comfort.
- Cover the top of the fighting position or trench with any readily available material and a thick layer of earth, if you can do so without leaving the shelter. While a roof and camouflage are both desirable, it is probably safer to do without them than to expose yourself to radiation outside your fighting position.
- While building your shelter, keep all parts of your body covered with clothing to protect it against beta burns.
- Clean the shelter site of any surface deposit using a branch or other object that you can discard. Do this cleaning to remove contaminated materials from the area you will occupy. The cleaned area should extend at least 1.5 meters beyond your shelter's area.
- Decontaminate any materials you bring into the shelter. These materials include grass or foliage that you use as insulation or bedding, and your outer clothing (especially footgear). If the weather permits and you have heavily contaminated outer clothing, you may want to remove it and bury it under a foot of earth at the end of your shelter. You may retrieve it later (after the radioactivity decays) when leaving the shelter. If the clothing is dry, you may decontaminate it by beating or shaking it outside the shelter's entrance to remove the radioactive dust. You may use any body of water, even though contaminated, to rid materials of excess fallout particles. Simply dip the material into the water and shake it to get rid of the excess water. Do not wring it out, this action will trap the particles.
- If at all possible and without leaving the shelter, wash your body thoroughly with soap and water, even if the water on hand may be contaminated. This washing will remove most of the harmful radioactive particles that are likely to cause beta burns or other damage. If water is not available, wipe your face and any other exposed skin surface to remove contaminated dust and dirt. You may wipe your face with a clean piece of cloth or a handful of uncontaminated dirt. You get this uncontaminated dirt by scraping off the top few inches of soil and using the "clean" dirt.
- Upon completing the shelter, lie down, keep warm, and sleep and rest as much as possible while in the shelter.
- When not resting, keep busy by planning future actions, studying your maps, or making the shelter more comfortable and effective.
- Don't panic if you experience nausea and symptoms of radiation sickness. Your main danger from radiation sickness is infection. There is no first aid for this sickness. Resting, drinking fluids, taking any medicine that prevents vomiting, maintaining your food intake, and preventing additional exposure will help avoid infection and aid recovery. Even small doses of radiation can cause these symptoms which may disappear in a short time.

Exposure Timetable

The following timetable provides you with the information needed to avoid receiving serious dosage and still let you cope with survival problems:

- Complete isolation from 4 to 6 days following delivery of the last weapon.
- A very brief exposure to procure water on the third day is permissible, but exposure should not exceed 30 minutes.
- One exposure of not more than 30 minutes on the seventh day.

- One exposure of not more than 1 hour on the eighth day.
- Exposure of 2 to 4 hours from the ninth day through the twelfth day.
- Normal operation, followed by rest in a protected shelter, from the thirteenth day on.
- In all instances, make your exposures as brief as possible. Consider only mandatory requirements as valid reasons for exposure. Decontaminate at every stop.

The times given above are conservative. If forced to move after the first or second day, you may do so, Make sure that the exposure is no longer than absolutely necessary.

Water Procurement

In a fallout-contaminated area, available water sources may be contaminated. If you wait at least 48 hours before drinking any water to allow for radioactive decay to take place and select the safest possible water source, you will greatly reduce the danger of ingesting harmful amounts of radioactivity. Although many factors (wind direction, rainfall, sediment) will influence your choice in selecting water sources, consider the following guidelines.

Safest Water Sources
Water from springs, wells, or other underground sources that undergo natural filtration will be your safest source. Any water found in the pipes or containers of abandoned houses or stores will also be free from radioactive particles. This water will be safe to drink, although you will have to take precautions against bacteria in the water.
Snow taken from 15 or more centimeters below the surface during the fallout is also a safe source of water.

Streams and Rivers
Water from streams and rivers will be relatively free from fallout within several days after the last nuclear explosion because of dilution. If at all possible, filter such water before drinking to get rid of radioactive particles. The best filtration method is to dig sediment holes or seepage basins along the side of a water source. The water will seep laterally into the hole through the intervening soil that acts as a filtering agent and removes the contaminated fallout particles that settled on the original body of water. This method can remove up to 99 percent of the radioactivity in water. You must cover the hole in some way in order to prevent further contamination. See Figure 6-9 for an example of a water filter.

Standing Water
Water from lakes, pools, ponds, and other standing sources is likely to be heavily contaminated, though most of the heavier, long-lived radioactive isotopes will settle to the bottom. Use the settling technique to purify this water. First, fill a bucket or other deep container three-fourths full with contaminated water. Then take dirt from a depth of 10 or more centimeters below the ground surface and stir it into the water. Use about 2.5 centimeters of dirt for every 10 centimeters of water. Stir the water until you see most dirt particles suspended in the water. Let the mixture settle for at least 6 hours. The settling dirt particles will carry most of the suspended fallout particles to the bottom and cover them. You can then dip out the clear water. Purify this water using a filtration device.

Additional Precautions
As an additional precaution against disease, treat all water with water purification tablets from your survival kit or boil it.

Food Procurement

Although it is a serious problem to obtain edible food in a radiation-contaminated area, it is not impossible to solve. You need to follow a few special procedures in selecting and preparing rations and local foods for use. Since secure packaging protects your combat rations, they will be perfectly safe for use. Supplement your rations with any food you can find on trips outside your shelter. Most processed foods you may find in abandoned buildings are safe for use after decontaminating them. These include canned and packaged foods after removing the containers or wrappers or washing them free of fallout particles. These processed foods also include food stored in any closed container and food stored in protected areas (such as cellars), if you wash them before eating. Wash all food containers or wrappers before handling them to prevent further contamination.

If little or no processed food is available in your area, you may have to supplement your diet with local food sources. Local food sources are animals and plants.

Animals as a Food Source

Assume that all animals, regardless of their habitat or living conditions, were exposed to radiation. The effects of radiation on animals are similar to those on humans. Thus, most of the wild animals living in a fallout area are likely to become sick or die from radiation during the first month after the nuclear explosion. Even though animals may not be free from harmful radioactive materials, you can and must use them in survival conditions as a food source if other foods are not available. With careful preparation and by following several important principles, animals can be safe food sources.

First, do not eat an animal that appears to be sick. It may have developed a bacterial infection as a result of radiation poisoning. Contaminated meat, even if thoroughly cooked, could cause severe illness or death if eaten.

Carefully skin all animals to prevent any radioactive particles on the skin or fur from entering the body. Do not eat meat close to the bones and joints as an animal's skeleton contains over 90 percent of the radioactivity. The remaining animal muscle tissue, however, will be safe to eat. Before cooking it, cut the meat away from the bone, leaving at least a 3-millimeter thickness of meat on the bone. Discard all internal organs (heart, liver, and kidneys) since they tend to concentrate beta and gamma radioactivity. Cook all meat until it is very well done. To be sure the meat is well done, cut it into less than 13-millimeter-thick pieces before cooking. Such cuts will also reduce cooking time and save fuel.

The extent of contamination in fish and aquatic animals will be much greater than that of land animals. This is also true for water plants, especially in coastal areas. Use aquatic food sources only in conditions of extreme emergency.

All eggs, even if laid during the period of fallout, will be safe to eat. Completely avoid milk from any animals in a fallout area because animals absorb large amounts of radioactivity from the plants they eat.

Plants as a Food Source

Plant contamination occurs by the accumulation of fallout on their outer surfaces or by absorption of radioactive elements through their roots. Your first choice of plant food should be vegetables such as potatoes, turnips, carrots, and other plants whose edible portion grows underground. These are the safest to eat once you scrub them and remove their skins.

Second in order of preference are those plants with edible parts that you can decontaminate by washing and peeling their outer surfaces. Examples are bananas, apples, tomatoes, prickly pears, and other such fruits and vegetables.

Any smooth-skinned vegetable, fruit, or plant that you cannot easily peel or effectively decontaminate by washing will be your third choice of emergency food.

The effectiveness of decontamination by scrubbing is inversely proportional to the roughness of the fruit's surface. Smooth-surfaced fruits have lost 90 percent of their contamination after washing, while washing rough-surfaced plants removes only about 50 percent of the contamination.

You eat rough-surfaced plants (such as lettuce) only as a last resort because you cannot effectively decontaminate them by peeling or washing. Other difficult foods to decontaminate by washing with water include dried fruits (figs, prunes, peaches, apricots, pears) and soya beans.

In general, you can use any plant food that is ready for harvest if you can effectively decontaminate it. Growing plants, however, can absorb some radioactive materials through their leaves as well as from the soil, especially if rains have occurred during or after the fallout period. Avoid using these plants for food except in an emergency.

BIOLOGICAL ENVIRONMENTS

The use of biological agents is real. Prepare yourself for survival by being proficient in the tasks identified in your Soldier's Manuals of Common Tasks (SMCTs). Know what to do to protect yourself against these agents.

Biological Agents and Effects

Biological agents are microorganisms that can cause disease among personnel, animals, or plants. They can also cause the deterioration of material. These agents fall into two broad categories-pathogens (usually called germs) and toxins. Pathogens are living microorganisms that cause lethal or incapacitating diseases. Bacteria, rickettsiae, fungi, and viruses are included in the pathogens. Toxins are poisons that plants, animals, or microorganisms produce naturally. Possible biological war-fare toxins include a variety of neurotoxic (affecting the central nervous system) and cytotoxic (causing cell death) compounds.

Germs

Germs are living organisms. Some nations have used them in the past as weapons. Only a few germs can start an infection, especially if inhaled into the lungs. Because germs are so small and weigh so little, the wind can spread them over great distances; they can also enter unfiltered or nonairtight places. Buildings and bunkers can trap them thus causing a higher concentration. Germs do not affect the body immediately. They must multiply inside the body and overcome the body's defenses--a process called the incubation period. Incubation periods vary from several hours to several months, depending on the germ. Most germs must live within another living organism (host), such as your body, to survive and grow. Weather conditions such as wind, rain, cold, and sunlight rapidly kill germs.

Some germs can form protective shells, or spores, to allow survival outside the host. Spore-producing agents are a long-term hazard you must neutralize by decontaminating infected areas or personnel. Fortunately, most live agents are not spore-producing. These agents must find a host within roughly a day of their delivery or they die. Germs have three basic routes of entry into your body: through the respiratory tract, through a break in the skin, and through the digestive tract. Symptoms of infection vary according to the disease.

Toxins

Toxins are substances that plants, animals, or germs produce naturally. These toxins are what actually harm man, not bacteria. Botulin, which produces botulism, is an example. Modern science has allowed large-scale production of these toxins without the use of the germ that produces the toxin. Toxins may produce effects similar to those of chemical agents. Toxic victims may not, however, respond to first aid measures used against chemical agents. Toxins enter the body in the same manner as germs. However, some toxins, unlike germs, can penetrate unbroken skin. Symptoms appear almost immediately, since there is no incubation period. Many toxins are extremely lethal, even in very small doses. Symptoms may include any of the following:

- Dizziness.
- Mental confusion.
- Blurred or double vision.
- Numbness or tingling of skin.
- Paralysis.
- Convulsions.
- Rashes or blisters.
- Coughing.
- Fever.
- Aching muscles.
- Tiredness.
- Nausea, vomiting, and/or diarrhea.
- Bleeding from body openings.
- Blood in urine, stool, or saliva.
- Shock.
- Death.

Detection of Biological Agents

Biological agents are, by nature, difficult to detect. You cannot detect them by any of the five physical senses. Often, the first sign of a biological agent will be symptoms of the victims exposed to the agent.

Your best chance of detecting biological agents before they can affect you is to recognize their means of delivery. The three main means of delivery are--

- *Bursting-type munitions.* These may be bombs or projectiles whose burst causes very little damage. The burst will produce a small cloud of liquid or powder in the immediate impact area. This cloud will disperse eventually; the rate of dispersion depends on terrain and weather conditions.
- *Spray tanks or generators.* Aircraft or vehicle spray tanks or ground-level aerosol generators produce an aerosol cloud of biological agents.
- *Vectors.* Insects such as mosquitoes, fleas, lice, and ticks deliver pathogens. Large infestations of these insects may indicate the use of biological agents.

Another sign of a possible biological attack is the presence of unusual substances on the ground or on vegetation, or sick-looking plants, crops, or animals.

Influence of Weather and Terrain

Your knowledge of how weather and terrain affect the agents can help you avoid contamination by biological agents. Major weather factors that affect biological agents are sunlight, wind, and precipitation. Aerosol sprays will tend to concentrate in low areas of terrain, similar to early morning mist.
Sunlight contains visible and ultraviolet solar radiation that rapidly kills most germs used as biological agents. However, natural or man-made cover may protect some agents from sunlight. Other man-made mutant strains of germs may be resistant to sunlight.
High wind speeds increase the dispersion of biological agents, dilute their concentration, and dehydrate them. The further downwind the agent travels, the less effective it becomes due to dilution and death of the pathogens. However, the downwind hazard area of the biological agent is significant and you cannot ignore it.
Precipitation in the form of moderate to heavy rain tends to wash biological agents out of the air, reducing downwind hazard areas. However, the agents may still be very effective where they were deposited on the ground.

Protection Against Biological Agents

While you must maintain a healthy respect for biological agents, there is no reason for you to panic. You can reduce your susceptibility to biological agents by maintaining current immunizations, avoiding contaminated areas, and controlling rodents and pests. You must also use proper first aid measures in the treatment of wounds and only safe or properly decontaminated sources of food and water. You must ensure that you get enough sleep to prevent a run-down condition. You must always use proper field sanitation procedures.
Assuming you do not have a protective mask, always try to keep your face covered with some type of cloth to protect yourself against biological agent aerosols. Dust may contain biological agents; wear some type of mask when dust is in the air.
Your uniform and gloves will protect you against bites from vectors (mosquitoes and ticks) that carry diseases. Completely button your clothing and tuck your trousers tightly into your boots. Wear a chemical protective overgarment, if available, as it provides better protection than normal clothing. Covering your skin will also reduce the chance of the agent entering your body through cuts or scratches. Always practice high standards of personal hygiene and sanitation to help prevent the spread of vectors.
Bathe with soap and water whenever possible. Use germicidal soap, if available. Wash your hair and body thoroughly, and clean under your fingernails. Clean teeth, gums, tongue, and the roof of your mouth frequently. Wash your clothing in hot, soapy water if you can. If you cannot wash your clothing, lay it out in an area of bright sunlight and allow the light to kill the microorganisms. After a toxin attack, decontaminate yourself as if for a chemical attack using the M258A2 kit (if available) or by washing with soap and water.

ESSENTIAL ARMY MANUAL
(231) ARMY SURVIVAL MANUAL

Shelter

You can build expedient shelters under biological contamination conditions using the same techniques described in Chapter 5. However, you must make slight changes to reduce the chance of biological contamination. Do not build your shelter in depressions in the ground. Aerosol sprays tend to concentrate in these depressions. Avoid building your shelter in areas of vegetation, as vegetation provides shade and some degree of protection to biological agents. Avoid using vegetation in constructing your shelter. Place your shelter's entrance at a 90-degree angle to the prevailing winds. Such placement will limit the entry of airborne agents and prevent air stagnation in your shelter. Always keep your shelter clean.

Water Procurement

Water procurement under biological conditions is difficult but not impossible. Whenever possible, try to use water that has been in a sealed container. You can assume that the water inside the sealed container is not contaminated. Wash the water container thoroughly with soap and water or boil it for at least 10 minutes before breaking the seal.

If water in sealed containers is not available, your next choice, *only under emergency conditions,* is water from springs. Again, boil the water for at least 10 minutes before drinking. Keep the water covered while boiling to prevent contamination by airborne pathogens. Your *last choice, only in an extreme emergency,* is to use standing water. Vectors and germs can survive easily in stagnant water. Boil this water as long as practical to kill all organisms. Filter this water through a cloth to remove the dead vectors. Use water purification tablets in all cases.

Food Procurement

Food procurement, like water procurement, is not impossible, but you must take special precautions. Your combat rations are sealed, and you can assume they are not contaminated. You can also assume that sealed containers or packages of processed food are safe. To ensure safety, decontaminate all food containers by washing with soap and water or by boiling the container in water for 10 minutes.

You consider supplementing your rations with local plants or animals only in extreme emergencies. No matter what you do to prepare the food, there is no guarantee that cooking will kill all the biological agents. Use local food only in life or death situations. Remember, you can survive for a long time without food, especially if the food you eat may kill you!

If you must use local food, select only healthy-looking plants and animals. Do not select known carriers of vectors such as rats or other vermin. Select and prepare plants as you would in radioactive areas. Prepare animals as you do plants. Always use gloves and protective clothing when handling animals or plants. Cook all plant and animal food by boiling only. Boil all food for at least 10 minutes to kill all pathogens. Do not try to fry, bake, or roast local food. There is no guarantee that all infected portions have reached the required temperature to kill all pathogens. Do not eat raw food.

CHEMICAL ENVIRONMENTS

Chemical agent warfare is real. It can create extreme problems in a survival situation, but you can overcome the problems with the proper equipment, knowledge, and training. As a survivor, your first line of defense against chemical agents is your proficiency in individual nuclear, biological, and chemical (NBC) training, to include donning and wearing the protective mask and overgarment, personal decontamination, recognition of chemical agent symptoms, and individual first aid for chemical agent contamination. The SMCTs cover these subjects. If you are not proficient in these skills, you will have little chance of surviving a chemical environment.

The subject matter covered below is not a substitute for any of the individual tasks in which you must be proficient. The SMCTs address the various chemical agents, their effects, and first aid for these agents. The following information is provided under the assumption that you are proficient in the use of chemical protective equipment and know the symptoms of various chemical agents.

Detection of Chemical Agents

The best method for detecting chemical agents is the use of a chemical agent detector. If you have one, use it. However, in a survival situation, you will most likely have to rely solely on the use of all of your physical senses. You must be alert and able to detect any clues indicating the use of chemical warfare. General indicators of the presence of chemical agents are tears, difficult breathing, choking, itching, coughing, and dizziness. With agents that are very hard to detect, you must watch for symptoms in fellow survivors. Your surroundings will provide valuable clues to the presence of chemical agents; for example, dead animals, sick people, or people and animals displaying abnormal behavior.

Your sense of smell may alert you to some chemical agents, but most will be odorless. The odor of newly cut grass or hay may indicate the presence of choking agents. A smell of almonds may indicate blood agents.

Sight will help you detect chemical agents. Most chemical agents in the solid or liquid state have some color. In the vapor state, you can see some chemical agents as a mist or thin fog immediately after the bomb or shell bursts. By observing for symptoms in others and by observing delivery means, you may be able to have some warning of chemical agents. Mustard gas in the liquid state will appear as oily patches on leaves or on buildings.

The sound of enemy munitions will give some clue to the presence of chemical weapons. Muffled shell or bomb detonations are a good indicator.

Irritation in the nose or eyes or on the skin is an urgent warning to protect your body from chemical agents. Additionally, a strange taste in food, water, or cigarettes may serve as a warning that they have been contaminated.

Protection Against Chemical Agents

As a survivor, always use the following general steps, in the order listed, to protect yourself from a chemical attack:

* Use protective equipment.
* Give quick and correct self-aid when contaminated.
* Avoid areas where chemical agents exist.
* Decontaminate your equipment and body as soon as possible.

Your protective mask and overgarment are the key to your survival. Without these, you stand very little chance of survival. You must take care of these items and protect them from damage. You must practice and know correct self-aid procedures before exposure to chemical agents. The detection of chemical agents and the avoidance of contaminated areas is extremely important to your survival. Use whatever detection kits may be available to help in detection. Since you are in a survival situation, avoid contaminated areas at all costs. You can expect no help should you become contaminated. If you do become contaminated, decontaminate yourself as soon as possible using proper procedures.

Shelter

If you find yourself in a contaminated area, try to move out of the area as fast as possible. Travel crosswind or upwind to reduce the time spent in the downwind hazard area. If you cannot leave the area immediately and have to build a shelter, use normal shelter construction techniques, with a few changes. Build the shelter in a clearing, away from all vegetation. Remove all topsoil in the area of the shelter to decontaminate the area. Keep the shelter's entrance closed and oriented at a 90-degree angle to the prevailing wind. Do not build a fire using contaminated wood--the smoke will be toxic. Use extreme caution when entering your shelter so that you will not bring contamination inside.

Water Procurement

As with biological and nuclear environments, getting water in a chemical environment is difficult. Obviously, water in sealed containers is your best and safest source. You must protect this water as much as possible. Be sure to decontaminate the containers before opening.

If you cannot get water in sealed containers, try to get it from a closed source such as underground water pipes. You may use rainwater or snow if there is no evidence of contamination. Use water from slow-moving streams, if necessary, but always check first for signs of contamination, and always filter the water as described under nuclear conditions. Signs of water source contamination are foreign odors such as garlic, mustard, geranium, or bitter almonds; oily spots on the surface of the water or nearby; and the presence of dead fish or animals. If these signs are present, do not use the water. Always boil or purify the water to prevent bacteriological infection.

Food Procurement

It is extremely difficult to eat while in a contaminated area. You will have to break the seal on your protective mask to eat. If you eat, find an area in which you can safely unmask. The safest source of food is your sealed combat rations. Food in sealed cans or bottles will also be safe. Decontaminate all sealed food containers before opening, otherwise you will contaminate the food.

If you must supplement your combat rations with local plants or animals, *do not* use plants from contaminated areas or animals that appear to be sick. When handling plants or animals, always use protective gloves and clothing.

FM 22-100

ARMY LEADERSHIP

BE, KNOW, DO

August 1999

Headquarters, Department of the Army

Field Manual
No. 22-100

Headquarters
Department of the Army
Washington, DC, 31 August 1999

Army Leadership

Contents

Page

DISTRIBUTION RESTRICTION: Approved for public release; distribution is unlimited.

*This publication supersedes FM 22-100, 31 July 1990; FM 22-101, 3 June 1985; FM 22-102, 2 March 1987; FM 22-103, 21 June 1987; DA Pam 600-80, 9 June 1987; and DA Form 4856, June 1985.

Examples

Preface

The Army consists of the active component, Army National Guard, Army Reserve, and Department of the Army (DA) civilians. It's the world's premier land combat force—a full-spectrum force trained and ready to answer the nation's call. The Army's foundation is confident and competent leaders of character. This manual is addressed to them and to those who train and develop them.

PURPOSE

FM 22-100 is a single-source reference for all Army leaders. Its purpose is three-fold:

- To provide leadership doctrine for meeting mission requirements under all conditions.

- To establish a unified leadership theory for all Army leaders: military and civilian, active and reserve, officer and enlisted.

- To provide a comprehensive and adaptable leadership resource for the Army of the 21st century.

As the capstone leadership manual for the Army, FM 22-100 establishes the Army's leadership doctrine, the fundamental principles by which Army leaders act to accomplish the mission and take care of their people. The doctrine discusses how Army values form the basis of character. In addition, it links a suite of instruments, publications, and initiatives that the Army uses to develop leaders. Among these are—

- AR 600-100, which establishes the basis for leader development doctrine and training.

- DA Pam 350-58, which describes the Army's leader development model.

- DA Pam 600-3, which discusses qualification criteria and outlines development and career management programs for commissioned officers.

- DA Pam 600-11, which discusses qualification criteria and outlines development and career management programs for warrant officers.

- DA Pam 600-25, which discusses noncommissioned officer (NCO) career development.

- DA Pam 690-46, which discusses mentoring of DA civilians.

- The TRADOC Common Core, which lists tasks that military and DA civilian leaders must perform and establishes who is responsible for training leaders to perform them.

- Officer, NCO, and DA civilian evaluation reports.

FM 22-100 also serves as the basis for future leadership and leader development initiatives associated with the three pillars of the Army's leader development model. Specifically, FM 22-100 serves as—

- The basis for leadership assessment.

- The basis for developmental counseling and leader development.
- The basis for leadership evaluation.
- A reference for leadership development in operational assignments.
- A guide for institutional instruction at proponent schools.
- A resource for individual leaders' self-development goals and initiatives.

FM 22-100 directly supports the Army's keystone manuals, FM 100-1 and FM 100-5, which describe the Army and its missions. It contains principles all Army leaders use when they apply the doctrine, tactics, techniques, and procedures established in the following types of doctrinal publications:

- Combined arms publications, which describe the tactics and techniques of combined arms forces.
- Proponency publications, which describe doctrinal principles, tactics, techniques, and collective training tasks for branch-oriented or functional units.
- Employment procedure publications, which address the operation, employment, and maintenance of specific systems.
- Soldier publications, which address soldier duties.
- Reference publications, which focus on procedures (as opposed to doctrine, tactics, or techniques) for managing training, operating in special environments or against specific threats, providing leadership, and performing fundamental tasks.

This edition of FM 22-100 establishes a unified leadership theory for all Army leaders based on the Army leadership framework and three leadership levels. Specifically, it—

- Defines and discusses Army values and leader attributes.
- Discusses character-based leadership.
- Establishes leader attributes as part of character.
- Focuses on improving people and organizations for the long term.
- Outlines three levels of leadership—direct, organizational, and strategic.
- Identifies four skill domains that apply at all levels.
- Specifies leadership actions for each level.

The Army leadership framework brings together many existing leadership concepts by establishing leadership dimensions and showing how they relate to each other. Solidly based on BE, KNOW, DO—that is, character, competence, and action—the Army leadership framework provides a single instrument for leader development. Individuals can use it for self-development. Leaders can use it to develop subordinates. Commanders can use it to focus their programs. By establishing leadership dimensions grouped under the skill domains of values, attributes, skills, and actions, the Army leadership framework provides a simple way to think about and discuss leadership.

The Army is a values-based institution. FM 22-100 establishes and clarifies those values. Army leaders must set high standards, lead by example, do what is legally and morally right, and influence other people to do the same. They must establish and sustain a climate that ensures people are treated with dignity and respect and create an environment in which people are challenged and motivated to be all they can be. FM 22-100 discusses these aspects of leadership and how they contribute to developing leaders of character and competence. These are the leaders who make the Army a trained and ready force prepared to fight and win the nation's wars.

The three leadership levels—direct, organizational, and strategic—reflect the different challenges facing leaders as they move into positions of increasing responsibility. Direct leaders lead face to face: they are the Army's first-line leaders. Organizational leaders lead large organizations, usually brigade-sized and larger. Strategic leaders are the Army's most senior leaders. They lead at the major command and national levels.

Unlike previous editions of FM 22-100—which focused exclusively on leadership by uniformed leaders at battalion level and below—this edition addresses leadership at all levels and is addressed to all Army leaders, military and DA civilian. It supersedes four publications—FM 22-101, *Leadership Counseling*; FM 22-102, *Soldier Team Development*; FM 22-103, *Leadership and Command at Senior Levels*; and DA Pam 600-80 *Executive Leadership*—as well as the previous edition of FM 22-100. A comprehensive reference, this manual shows how leader skills, actions, and concerns at the different levels are linked and allows direct leaders to read about issues that affect organizational and strategic leaders. This information can assist leaders serving in positions supporting organizational and strategic leaders and to other leaders who must work with members of organizational- and strategic-level staffs.

FM 22-100 emphasizes self-development and development of subordinates. It includes performance indicators to help leaders assess the values, attributes, skills, and actions that the rest of the manual discusses. It discusses developmental counseling, a skill all Army leaders must perfect so they can mentor their subordinates and leave their organization and people better than they found them. FM 22-100 prescribes DA Form 4856-E (Developmental Counseling Form), which supersedes DA Form 4856 (General Counseling Form). DA Form 4856-E is designed to support leader development. Its format follows the counseling steps outlined in Appendix C.

FM 22-100 offers a framework for how to lead and provides points for Army leaders to consider when assessing and developing themselves, their people, and their organizations. It doesn't presume to tell Army leaders exactly how they should lead every step of the way. They must be themselves and apply this leadership doctrine as appropriate to the situations they face.

SCOPE

FM 22-100 is divided into three parts. Part I (Chapters 1, 2, and 3) discusses leadership aspects common to all Army leaders. Part II (Chapters 4 and 5) addresses the skills and actions required of direct leaders. Part III (Chapters 6 and 7) discusses the skills and actions required of organizational and strategic leaders. The manual also includes six appendixes.

Chapter 1 defines Army leadership, establishes the Army leadership framework, and describes the three Army leadership levels. It addresses the characteristics of an Army leader (BE, KNOW, DO), the importance of being a good subordinate, and how all Army leaders lead other leaders. Chapter 1 concludes with a discussion of moral and collective excellence.

Chapter 2 examines character, competence, and leadership—what an Army leader must BE, KNOW, and DO. The chapter addresses character in terms of Army values and leader attributes. In addition, it describes character development and how character is related to ethics, orders—to include illegal orders—and beliefs. Chapter 2 concludes by introducing the categories of leader skills—interpersonal, conceptual, technical, and tactical—and the categories of leader actions—influencing, operating, and improving.

Chapter 3 covers the human dimension of leadership. The chapter begins by discussing discipline, morale, and care of subordinates. It then addresses stress, both combat- and change-related. Discussions of organizational climate, institutional culture, and leadership styles follow. Chapter 3 concludes by examining intended and unintended consequences of decisions and leader actions.

Chapters 4 and 5 discuss the skills and actions required of direct leaders. The skills and actions are grouped under the categories introduced at the end of Chapter 2.

Chapters 6 and 7 provide an overview of the skills and actions required of organizational and strategic leaders. These chapters introduce direct leaders to the concerns faced by leaders and staffs operating at the organizational and strategic levels. Like Chapters 4 and 5, Chapters 6 and 7 group skills and actions under the categories introduced in Chapter 2.

Appendix A outlines the roles and relationships of commissioned, warrant, and noncommissioned officers. It includes discussions of authority, responsibility, the chain of command, the NCO support channel, and DA civilian support.

Appendix B lists performance indicators for Army values and leader attributes, skills, and actions. It provides general examples of what Army leaders must BE, KNOW, and DO.

Appendix C addresses developmental counseling in detail. It begins with a discussion of the characteristics of a good counselor, the skills a counselor requires, and the limitations leaders face when they counsel subordinates. The appendix then examines the types of developmental counseling, counseling approaches, and counseling techniques. Appendix C concludes by describing the counseling process and explaining how to use DA Form 4856-E, the Developmental Counseling Form.

Appendix D explains how to prepare a leader plan of action and provides an example of a direct leader preparing a leader plan of action based on information gathered using an ethical climate assessment survey (ECAS). The example explains how to conduct an ECAS.

Appendix E discusses how Army values contribute to character development and the importance of developing the character of subordinates.

Appendix F contains a copy of the Constitution of the United States. All members of the Army take an oath to "support and defend the Constitution of the United States." It is included so it will be immediately available for Army leaders.

This manual is a tool to help you answer these questions, to begin or continue becoming a leader of character and competence, an Army leader. Chapter 1 starts with an overview of what the Army requires of you as an Army leader. This is the Army leadership framework; it forms the structure of the Army's leadership doctrine. Chapter 1 also discusses the three levels of Army leadership: direct, organizational, and strategic. Chapter 2 discusses character, competence, and leadership—what you must BE, KNOW, and DO as an Army leader. Chapter 3 talks about the human dimension, the many factors that affect the people and teams that you lead and the institution of which you and they are a part.

CHAPTER 1

The Army Leadership Framework

Just as the diamond requires three properties for its formation—carbon, heat, and pressure—successful leaders require the interaction of three properties—character, knowledge, and application. Like carbon to the diamond, character is the basic quality of the leader....But as carbon alone does not create a diamond, neither can character alone create a leader. The diamond needs heat. Man needs knowledge, study, and preparation....The third property, pressure—acting in conjunction with carbon and heat—forms the diamond. Similarly, one's character, attended by knowledge, blooms through application to produce a leader.

General Edward C. Meyer
Former Army Chief of Staff

1-1. The Army's ultimate responsibility is to win the nation's wars. For you as an Army leader, leadership in combat is your primary mission and most important challenge. To meet this challenge, you must develop character and competence while achieving excellence. This manual is about leadership. It focuses on character, competence, and excellence. It's about accomplishing the mission and taking care of people. It's about living up to your ultimate responsibility, leading your soldiers in combat and winning our nation's wars.

1-2. Figure 1-1 shows the Army leadership framework. The top of the figure shows the four categories of things leaders must BE, KNOW, and DO. The bottom of the figure lists dimensions of Army leadership, grouped under these four categories. The dimensions consist of Army values and subcategories under attributes, skills, and actions.

1-3. Leadership starts at the top, with the character of the leader, with your character. In order to lead others, you must first make sure your own house is in order. For example, the first line of *The Creed of the Noncommissioned Officer* states, "No one is more professional than I." But it takes a remarkable person to move from memorizing a creed to actually living that creed; a true leader *is* that remarkable person.

1-4. Army leadership begins with what the leader must BE, the values and attributes that shape a leader's character. It may be helpful to think of these as internal qualities: you possess them all the time, alone and with others. They define who you are; they give you a solid footing. These values and attributes are the same for all leaders, regardless of position, although

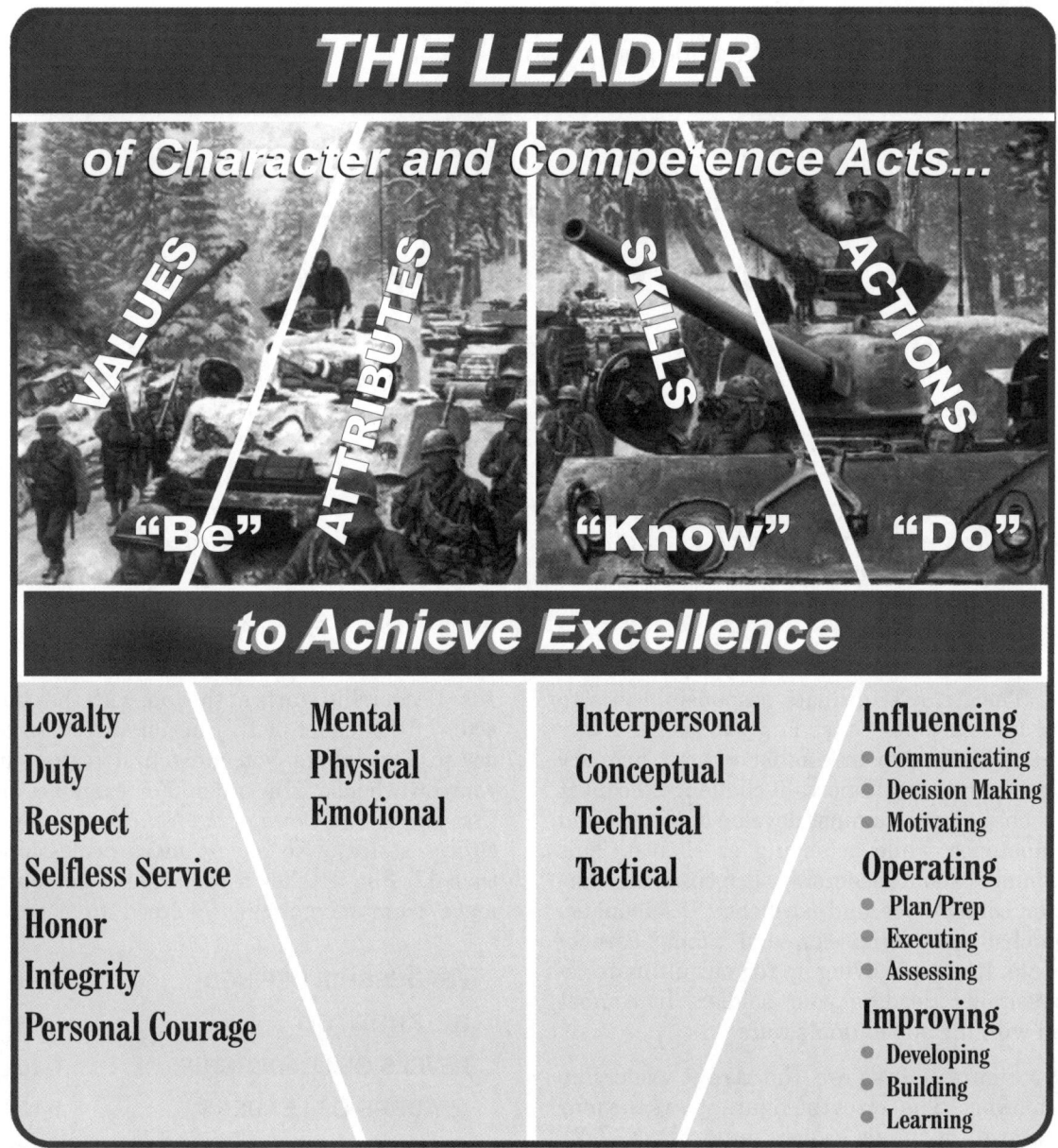

THE LEADER

of Character and Competence Acts...

VALUES   ATTRIBUTES   SKILLS   ACTIONS

"Be"   "Know"   "Do"

to Achieve Excellence

Loyalty	Mental	Interpersonal	**Influencing**
Duty	Physical	Conceptual	• Communicating
Respect	Emotional	Technical	• Decision Making
Selfless Service		Tactical	• Motivating
Honor			**Operating**
Integrity			• Plan/Prep
Personal Courage			• Executing
			• Assessing
			Improving
			• Developing
			• Building
			• Learning

Leaders of character and competence act to achieve excellence by developing a force that can fight and win the nation's wars and serve the common defense of the United States.

Figure 1-1. The Army Leadership Framework

you certainly refine your understanding of them as you become more experienced and assume positions of greater responsibility. For example, a sergeant major with combat experience has a deeper understanding of selfless service and personal courage than a new soldier does.

1-5. Your skills are those things you KNOW how to do, your competence in everything from the technical side of your job to the people skills a leader requires. The skill categories of the Army leadership framework apply to all leaders. However, as you assume positions of greater responsibility, you must master additional skills in each category. Army

leadership positions fall into one of three levels: direct, organizational, and strategic. These levels are described later in this chapter. Chapters 4, 6, and 7 describe the skills leaders at each level require.

1-6. But character and knowledge—while absolutely necessary—are not enough. You cannot be effective, you cannot be a leader, until you *apply* what you know, until you act and DO what you must. As with skills, you will learn more leadership actions as you serve in different positions. Because actions are the essence of leadership, the discussion begins with them.

LEADERSHIP DEFINED

Leadership is **influencing** people—by providing purpose, direction, and motivation—while **operating** to accomplish the mission and **improving** the organization.

INFLUENCING

1-7. Influencing means getting people to do what you want them to do. It is the means or method to achieve two ends: operating and improving. But there's more to influencing than simply passing along orders. The example you set is just as important as the words you speak. And you set an example—good or bad—with every action you take and word you utter, on or off duty. Through your words and example, you must communicate purpose, direction, and motivation.

Purpose

1-8. Purpose gives people a reason to do things. This does not mean that as a leader you must explain every decision to the satisfaction of your subordinates. It does mean you must earn their trust: they must know from experience that you care about them and would not ask them to do something—particularly something dangerous—unless there was a good reason, unless the task was essential to mission accomplishment.

1-9. Look, for example, at a battalion maintenance section. Its motor sergeant always takes the time—and has the patience—to explain to the mechanics what is required of them. Nothing fancy; the motor sergeant usually just calls them together for a few minutes to talk about the workload and the time crunch. The soldiers may get tired of hearing "And, of course, unless we get the work finished, this unit doesn't roll and the mission doesn't get done," but they know it's true. And every time he passes information this way, the motor sergeant sends this signal to the soldiers: that he cares about their time and work and what they think, that they are members of a team, not cogs in the "green machine."

1-10. Then one day the unit is alerted for an emergency deployment. Things are happening at breakneck speed; there is no time to pause, and everything and everyone is under stress. The motor sergeant cannot stop to explain things, pat people on the back, or talk them up. But the soldiers will work themselves to exhaustion, if need be, because the motor sergeant has earned their trust. They know and appreciate their leader's normal way of operating, and they will assume there is a good reason the leader is doing things differently this time. And should the deployment lead to a combat mission, the team will be better prepared to accomplish their mission under fire. Trust is a basic bond of leadership, and it must be developed over time.

Direction

1-11. When providing direction, you communicate the way you want the mission accomplished. You prioritize tasks, assign responsibility for completing them (delegating authority when necessary), and make sure your people understand the standard. In short, you figure out how to get the work done right with the available people, time, and other resources; then you communicate that information to your subordinates: "We'll do these things first. You people work here; you people work there." As you think the job through, you can better aim your effort and resources at the right targets.

1-12. People want direction. They want to be given challenging tasks, training in how to accomplish them, and the resources necessary to do them well. Then they want to be left alone to do the job.

Motivation

1-13. Motivation gives subordinates the will to do everything they can to accomplish a mission. It results in their acting on their own initiative when they see something needs to be done.

1-14. To motivate your people, give them missions that challenge them. After all, they did not join the Army to be bored. Get to know your people and their capabilities; that way you can tell just how far to push each one. Give them as much responsibility as they can handle; then let them do the work without looking over their shoulders and nagging them. When they succeed, praise them. When they fall short, give them credit for what they have done and coach or counsel them on how to do better next time.

1-15. People who are trained this way will accomplish the mission, even when no one is watching. They will work harder than they thought they could. And when their leader

notices and gives them credit (with something more than the offhand comment "good job"), they will be ready to take on even more next time.

1-16. But Army leaders motivate their people by more than words. The example you set is at least as important as what you say and how well you manage the work. As the unit prepares for the rollout, the motor sergeant you just read about is in the motor pool with the mechanics on Friday night and Saturday morning. If his people are working in the rain, the NCO's uniform will be wet too. If they have missed breakfast, the leader's stomach will be growling just as loudly. The best leaders lead from the front. Don't underestimate the importance of being where the action is.

OPERATING

1-17. Actions taken to influence others serve to accomplish operating actions, those actions you take to achieve the short-term goal of accomplishing the mission. The motor sergeant will make sure the vehicles roll out, on time and combat ready, through planning and preparing (laying out the work and making the necessary arrangements), executing (doing the job), and assessing (learning how to work smarter next time). The motor sergeant provides an example of how direct leaders perform operating actions. All leaders execute these operating actions, which become more complex as they assume positions of increasing responsibility.

IMPROVING

1-18. The motor sergeant's job is not complete when the last vehicle clears the gate. While getting the job done is key, the Army also expects him to do far more than just accomplish the day's work. Army leaders also strive to improve everything entrusted to them: their people, facilities, equipment, training, and resources. There will be a new mission, of course, but part of finishing the old one is improving the organization.

1-19. After checking to be sure the tools are repaired, cleaned, accounted for, and put away, the motor sergeant conducts an informal after-action review (AAR) with the section. (An AAR is a professional discussion of an event, focused on performance standards, that allows participants to discover for themselves what happened, why it happened, and how to sustain strengths and improve on weaknesses. Chapter 5 discusses AARs.) The motor sergeant is self-confident enough to ask subordinates for their ideas on how to make things work better (always a key goal). He then acts based on his own and team members' observations. The motor sergeant looks for strong areas to sustain and praises team members as appropriate; however if the motor sergeant saw the team members spend too much time on some tasks and not enough on others, he changes the section standing operating procedures (SOP) or counsels the people involved. (Developmental counseling is not an adverse action; it is a skill you use to help your subordinates become better team members, improve performance, and prepare for the future. Counseling should address strong areas as well as weak ones and successes as well as failures. Appendix C discusses developmental counseling.) If the motor sergeant discovers gaps in individual or collective skills, he plans and conducts the training necessary to fill them. If something the motor sergeant did or a decision he made didn't turn out quite right, he will not make the same error again. More than that, the motor sergeant lets his people know what went wrong, finds out their impressions of why it happened, and determines how they will make it work next time.

1-20. By doing these things, the motor sergeant is creating a better organization, one that will work smarter the next time. His example sends an important message. The soldiers see their leader look at his own and the organization's performance, evaluate it, identify strong areas to sustain as well as mistakes and shortcomings, and commit to a better way of doing things. These actions are more powerful than any lecture on leadership.

BE, KNOW, DO

1-21. BE, KNOW, DO clearly and concisely state the characteristics of an Army leader. You have just read about leader actions, the DO of BE, KNOW, DO. Leadership is about taking action, but there's more to being a leader than just what you do. Character and competence, the BE and the KNOW, underlie everything a leader does. So becoming a leader involves developing all aspects of yourself. This includes adopting and living Army values. It means developing the attributes and learning the skills of an Army leader. Only by this self-development will you become a confident and competent leader of character. Being an Army leader is not easy. There are no cookie-cutter solutions to leadership challenges, and there are no shortcuts to success. However, the tools are available to every leader. It is up to you to master and use them.

more than that, it links that knowledge to action. Character gives you the courage to do what is right regardless of the circumstances or the consequences. (Appendix E discusses character development.)

1-23. You demonstrate character through your behavior. One of your key responsibilities as a leader is to teach Army values to your subordinates. The old saying that actions speak louder than words has never been more true than here. Leaders who talk about honor, loyalty, and selfless service but do not live these values—both on and off duty—send the wrong message, that this "values stuff" is all just talk.

1-24. Understanding Army values and leader attributes (which Chapter 2 discusses) is only the first step. You also must embrace Army values and develop leader attributes, living them until they become habit. You must teach Army values to your subordinates through action and example and help them develop leader attributes in themselves.

BE

1-22. Character describes a person's inner strength, the BE of BE, KNOW, DO. Your character helps you know what is right;

KNOW

1-25. A leader must have a certain level of knowledge to be competent. That knowledge is spread across four skill domains. You must develop **interpersonal skills**, knowledge of your people and how to work with them. You must have **conceptual skills**, the ability to understand and apply the doctrine and other ideas required to do your job. You must learn **technical skills**, how to use your equipment. Finally, warrior leaders must master **tactical skills**, the ability to make the right decisions concerning employment of units in combat. Tactical skills include mastery of the art of tactics appropriate to the leader's level of responsibility and unit type. They're amplified by the other skills—interpersonal, conceptual, and technical—and are the most important skills for warfighters. (FM 100-40 discusses the art of tactics.)

1-26. Mastery of different skills in these domains is essential to the Army's success in peace and war. But a true leader is not satisfied with knowing only how to do what will get the organization through today; you must also be concerned about what it will need tomorrow. You must strive to master your job and prepare to take over your boss's job. In addition, as you move to jobs of increasing responsibility, you'll face new equipment, new ideas, and new ways of thinking and doing things. You must learn to apply all these to accomplish your mission.

1-27. Army schools teach you basic job skills, but they are only part of the learning picture. You'll learn even more on the job. Good leaders add to their knowledge and skills every day. True leaders seek out opportunities; they're always looking for ways to increase their professional knowledge and skills. Dedicated squad leaders jump at the chance to fill in as acting platoon sergeant, not because they've mastered the platoon sergeant's job but because they know the best place to learn about it is in the thick of the action. Those squad leaders challenge themselves and will learn through doing; what's more, with coaching, they'll learn as much from their mistakes as from their successes.

DO

1-28. You read about leader actions, the DO of Army leadership doctrine, at the beginning of this chapter. Leader actions include—

- **Influencing:** making decisions, communicating those decisions, and motivating people.

- **Operating:** the things you do to accomplish your organization's immediate mission.

- **Improving:** the things you do to increase the organization's capability to accomplish current or future missions.

1-29. Earlier in this chapter, you read about a motor sergeant who lives Army values, has developed leader attributes, and routinely performs leader actions. But that was an example, and a garrison example at that. What about reality? What about combat? Trained soldiers know what they are supposed to do, but under stress, their instincts might tell them to do something different. The exhausted, hungry, cold, wet, disoriented, and frightened soldier is more likely to do the wrong thing—stop moving, lie down, retreat—than one not under that kind of stress. This is when the leader must step in—when things are falling apart, when there seems to be no hope—and get the job done.

1-30. The fight between the 20th Regiment of Maine Volunteers and the 15th and 47th Regiments of Alabama Infantry during the Civil War illustrates what can happen when a leader acts decisively. It shows how the actions of one leader, in a situation that looked hopeless, not only saved his unit, but allowed the entire Union Army to maintain its position and defeat the Confederate invasion of Pennsylvania. The story's hero is a colonel—but it could have been a captain, or a sergeant, or a corporal. At other times and in other places it has been.

COL Chamberlain at Gettysburg

In late June 1863 GEN Robert E. Lee's Army of Northern Virginia passed through western Maryland and invaded Pennsylvania. For five days, the Army of the Potomac hurried to get between the Confederates and the national capital. On 1 July the 20th Maine received word to press on to Gettysburg. The Union Army had engaged the Confederates there, and Union commanders were hurrying all available forces to the hills south of the little town.

The 20th Maine arrived at Gettysburg near midday on 2 July, after marching more than one hundred miles in five days. They had had only two hours sleep and no hot food during the previous 24 hours. The regiment was preparing to go into a defensive position as part of the brigade commanded by COL Strong Vincent when a staff officer rode up to COL Vincent and began gesturing towards a little hill at the extreme southern end of the Union line. The hill, Little Round Top, dominated the Union position and, at that moment, was unoccupied. If the Confederates placed artillery on it, they could force the entire Union Army to withdraw. The hill had been left unprotected through a series of mistakes—wrong assumptions, the failure to communicate clearly, and the failure to check—and the situation was critical.

Realizing the danger, COL Vincent ordered his brigade to occupy Little Round Top. He positioned the 20th Maine, commanded by COL Joshua L. Chamberlain, on his brigade's left flank, the extreme left of the Union line. COL Vincent told COL Chamberlain to "hold at all hazards."

On Little Round Top, COL Chamberlain told his company commanders the purpose and importance of their mission. He ordered the right flank company to tie in with the 83d Pennsylvania and the left flank company to anchor on a large boulder. His thoughts turned to his left flank. There was nothing there except a small hollow and the rising slope of Big Round Top. The 20th Maine was literally at the end of the line.

COL Chamberlain then showed a skill common to good tactical leaders. He imagined threats to his unit, did what he could to guard against them, and considered what he would do to meet other possible threats. Since his left flank was open, COL Chamberlain sent B Company, commanded by CPT Walter G. Morrill, off to guard it and "act as the necessities of battle required." The captain positioned his men behind a stone wall that would face the flank of any Confederate advance. There, fourteen soldiers from the 2d US Sharpshooters, who had been separated from their unit, joined them.

The 20th Maine had been in position only a few minutes when the soldiers of the 15th and 47th Alabama attacked. The Confederates had also marched all night and were tired and thirsty. Even so, they attacked ferociously.

The Maine men held their ground, but then one of COL Chamberlain's officers reported seeing a large body of Confederate soldiers moving laterally behind the attacking force. COL Chamberlain climbed on a rock—exposing himself to enemy fire—and saw a Confederate unit moving around his exposed left flank. If they outflanked him, his unit would be pushed off its position and destroyed. He would have failed his mission.

COL Chamberlain had to think fast. The tactical manuals he had so diligently studied called for a maneuver that would not work on this terrain. The colonel had to create a new maneuver, one that his soldiers could execute, and execute now.

The 20th Maine was in a defensive line, two ranks deep. It was threatened by an attack around its left flank. So the colonel ordered his company commanders to stretch the line to the left and bend it back to form an angle, concealing the maneuver by keeping up a steady rate of fire. The corner of the angle would be the large boulder he had pointed out earlier. The sidestep maneuver was tricky, but it was a combination of other battle drills his soldiers knew. In spite of the terrible noise that made voice commands useless, in spite of the blinding smoke, the cries of the wounded, and the continuing Confederate attack, the Maine men were able to pull it off. Now COL Chamberlain's thin line was only

---

COL Chamberlain at Gettysburg (continued)

one rank deep. His units, covering twice their normal frontage, were bent back into an L shape. Minutes after COL Chamberlain repositioned his force, the Confederate infantry, moving up what they thought was an open flank, were thrown back by the redeployed left wing of the 20th Maine. Surprised and angry, they nonetheless attacked again.

The Maine men rallied and held; the Confederates regrouped and attacked. "The Alabamians drove the Maine men from their positions five times. Five times they fought their way back again. At some places, the muzzles of the opposing guns almost touched." After these assaults, the Maine men were down to one or two rounds per man, and the determined Confederates were regrouping for another try. COL Chamberlain saw that he could not stay where he was and could not withdraw. So he decided to counterattack. His men would have the advantage of attacking down the steep hill, he reasoned, and the Confederates would not be expecting it. Clearly he was risking his entire unit, but the fate of the Union Army depended on his men.

The decision left COL Chamberlain with another problem: there was nothing in the tactics book about how to get his unit from their L-shaped position into a line of advance. Under tremendous fire and in the midst of the battle, COL Chamberlain again called his commanders together. He explained that the regiment's left wing would swing around "like a barn door on a hinge" until it was even with the right wing. Then the entire regiment, bayonets fixed, would charge downhill, staying anchored to the 83d Pennsylvania on its right. The explanation was clear and the situation clearly desperate.

When COL Chamberlain gave the order, 1LT Holman Melcher of F Company leaped forward and led the left wing downhill toward the surprised Confederates. COL Chamberlain had positioned himself at the boulder at the center of the L. When the left wing was abreast of the right wing, he jumped off the rock and led the right wing down the hill. The entire regiment was now charging on line, swinging like a great barn door—just as its commander had intended.

The Alabama soldiers, stunned at the sight of the charging Union troops, fell back on the positions behind them. There the 20th Maine's charge might have failed if not for a surprise resulting from COL Chamberlain's foresight. Just then CPT Morrill's B Company and the sharpshooters opened fire on the Confederate flank and rear. The exhausted and shattered Alabama regiments thought they were surrounded. They broke and ran, not realizing that one more attack would have carried the hill.

The slopes of Little Round Top were littered with bodies. Saplings halfway up the hill had been sawed in half by weapons fire. A third of the 20th Maine had fallen, 130 men out of 386. Nonetheless, the farmers, woodsmen, and fishermen from Maine—under the command of a brave and creative leader who had anticipated enemy actions, improvised under fire, and applied disciplined initiative in the heat of battle—had fought through to victory.

1-31. COL Joshua Chamberlain was awarded the Medal of Honor for his actions on 2 July 1863. After surviving terrible wounds at Petersburg, Virginia, he and his command were chosen to receive the surrender of Confederate units at Appomattox in April 1865. His actions there contributed to national reconciliation and are described in Chapter 7.

**PUTTING
IT TOGETHER**

1-32. Study the Army leadership framework; it is the Army's common basis for thinking about leadership. With all the day-to-day tasks you

must do, it's easy to get lost in particulars. The Army leadership framework is a tool that allows you to step back and think about leadership as a whole. It is a canopy that covers the hundreds of things you do every day. The Army leadership framework gives you the big picture and can help you put your job, your people, and your organization in perspective.

1-33. The dimensions of the Army leadership framework shown in Figure 1-1—the values, attributes, skills, and actions that support BE, KNOW, and DO—each contain components. All are interrelated; none stands alone. For example, *will* is very important, as you saw in the case of COL Chamberlain. It's discussed in Chapter 2 under mental attributes. Yet will cannot stand by itself. Left unchecked and without moral boundaries, will can be dangerous. The case of Adolf Hitler shows this fact. Will misapplied can also produce disastrous results. Early in World War I, French forces attacked German machine gun positions across open fields, believing their élan (unit morale and will to win) would overcome a technologically advanced weapon. The cost in lives was catastrophic. Nevertheless, the will of leaders of character and competence—like the small unit leaders at Normandy that you'll read about later in this chapter—can make the difference between victory and defeat.

1-34. This is how you should think about the Army leadership framework: all its pieces work in combination to produce something bigger and better than the sum of the parts. BE the leader of character: embrace Army values and demonstrate leader attributes. Study and practice so that you have the skills to KNOW your job. Then act, DO what's right to achieve excellence.

1-35. The Army leadership framework applies to all Army leaders. However, as you assume positions of increasing responsibility, you'll need to develop additional attributes and master more skills and actions. Part of this knowledge includes understanding what your bosses are doing—the factors that affect their decisions and the environment in which they work. To help you do this, Army leadership positions are divided into three levels—direct, organizational, and strategic.

LEVELS OF LEADERSHIP

NCOs like to make a decision right away and move on to the next thing…so the higher up the flagpole you go, the more you have to learn a very different style of leadership.

Command Sergeant Major Douglas E. Murray
United States Army Reserve

Figure 1-2. Army Leadership Levels

1-36. Figure 1-2 shows the perspectives of the three levels of Army leadership: direct, organizational, and strategic. Factors that determine a position's leadership level can include the position's span of control, its headquarters level, and the extent of the influence the leader holding the position exerts. Other factors include the size of the unit or organization, the type of operations it conducts, the number of people assigned, and its planning horizon.

1-37. Sometimes the rank or grade of the leader holding a position does not indicate the position's leadership level. That's why Figure 1-2 does not show rank. A sergeant first class serving as a platoon sergeant works at the direct leadership level. If the same NCO holds a headquarters job dealing with issues and policy affecting a brigade-sized or larger organization, the NCO works at the organizational leadership level. However, if the NCO's primary duty is running a staff section that supports the leaders who run the organization, the NCO is a direct leader. In fact, most leadership positions are direct leadership positions, and every leader at every level acts as a direct leader when dealing with immediate subordinates.

1-38. The headquarters echelon alone doesn't determine a position's leadership level. Soldiers and DA civilians of all ranks and grades serve in strategic-level headquarters, but they are not all strategic-level leaders. The responsibilities of a duty position, together with the other factors paragraph 1-36 lists, determine its leadership level. For example, a DA civilian at a training area range control with a dozen subordinates works at the direct leadership level while a DA civilian deputy garrison commander with a span of influence over several thousand people works at the organizational leadership level. Most NCOs, company grade officers, field grade officers, and DA civilian leaders serve at the direct leadership level. Some senior NCOs, field grade officers, and higher-grade DA civilians serve at the organizational leadership level. Most general officers and equivalent Senior Executive Service DA civilians serve at the organizational or strategic leadership levels.

DIRECT LEADERSHIP

1-39. Direct leadership is face-to-face, first-line leadership. It takes place in those organizations where subordinates are used to seeing their leaders all the time: teams and squads, sections and platoons, companies, batteries, and troops—even squadrons and battalions. The direct leader's span of influence, those whose lives he can reach out and touch, may range from a handful to several hundred people.

1-40. Direct leaders develop their subordinates one-on-one; however, they also influence their organization through their subordinates. For instance, a cavalry squadron commander is close enough to his soldiers to have a direct influence on them. They're used to seeing him regularly, even if it is only once a week in garrison; they expect to see him from time to time in the field. Still, during daily operations, the commander guides the organization primarily through his subordinate officers and NCOs.

1-41. For direct leaders there is more certainty and less complexity than for organizational and strategic leaders. Direct leaders are close enough to see—very quickly—how things work, how things don't work, and how to address any problems. (Chapter 4 discusses direct leader skills. Chapter 5 discusses direct leader actions.)

ORGANIZATIONAL LEADERSHIP

1-42. Organizational leaders influence several hundred to several thousand people. They do this indirectly, generally through more levels of subordinates than do direct leaders. The additional levels of subordinates can make it more difficult for them to see results. Organizational leaders have staffs to help them lead their people and manage their organizations' resources. They establish policies and the organizational climate that support their subordinate leaders. (Chapter 3 introduces climate and culture and explains the role of direct leaders in setting the organizational climate. Chapters 6 and 7 discuss the roles of organizational and strategic leaders in establishing and maintaining the organizational climate and institutional culture.)

1-43. Organizational leadership skills differ from direct leadership skills in degree, but not in kind. That is, the skill domains are the same, but organizational leaders must deal with more complexity, more people, greater uncertainty, and a greater number of unintended consequences. They find themselves influencing people more through policymaking and systems integration than through face-to-face contact.

1-44. Organizational leaders include military leaders at the brigade through corps levels, military and DA civilian leaders at directorate through installation levels, and DA civilians at the assistant through undersecretary of the Army levels. They focus on planning and mission accomplishment over the next two to ten years.

1-45. Getting out of their offices and visiting the parts of their organizations where the work is done is especially important for organizational leaders. They must make time to get to the field to compare the reports their staff gives them with the actual conditions their people face and the perceptions of the organization and mission they hold. Because of their less-frequent presence among their soldiers and DA civilians, organizational leaders must use those visits they are able to make to assess how well the commander's intent is understood and to reinforce the organization's priorities.

STRATEGIC LEADERSHIP

1-46. Strategic leaders include military and DA civilian leaders at the major command through Department of Defense levels. Strategic leaders are responsible for large organizations and influence several thousand to hundreds of thousands of people. They establish force structure, allocate resources, communicate strategic vision, and prepare their commands and the Army as a whole for their future roles.

1-47. Strategic leaders work in an uncertain environment on highly complex problems that affect and are affected by events and organizations outside the Army. Actions of a theater commander in chief (CINC), for example, may even have an impact on global politics. (CINCs command combatant commands, very large, joint organizations assigned broad, continuing missions. Theater CINCs are assigned responsibilities for a geographic area (a theater); for example, the CINC of the US Central Command is responsible for most of southwestern Asia and part of eastern Africa. Functional CINCs are assigned responsibilities not bounded by geography; for example, the CINC of the US Transportation Command is responsible for providing integrated land, sea, and air transportation to all services. (JP 0-2, JP 3-0, and FM 100-7 discuss combatant commands.) Although civilian leaders make national policy, decisions a CINC makes while carrying out that policy may affect whether or not a national objective is achieved. Strategic leaders apply many of the same leadership skills and actions they mastered as direct and organizational leaders; however, strategic leadership requires others that are more complex and indirectly applied.

1-48. Strategic leaders concern themselves with the total environment in which the Army functions; their decisions take into account such things as congressional hearings, Army budgetary constraints, new systems acquisition, civilian programs, research, development, and interservice cooperation—just to name a few.

1-49. Strategic leaders, like direct and organizational leaders, process information quickly, assess alternatives based on incomplete data, make decisions, and generate support. However, strategic leaders' decisions affect more people, commit more resources, and have wider-ranging consequences in both space and time than do decisions of organizational and direct leaders.

1-50. Strategic leaders often do not see their ideas come to fruition during their "watch"; their initiatives may take years to plan, prepare, and execute. In-process reviews (IPRs) might not even begin until after the leader has left the job. This has important implications for long-range planning. On the other hand, some strategic decisions may become a front-page headline of the next morning's newspaper. Strategic leaders have very few opportunities to visit the lowest-level organizations of their commands; thus, their sense of when and where to visit is crucial. Because they exert influence primarily through subordinates, strategic leaders must develop strong skills in picking and developing good ones. This is an important improving skill, which Chapter 7 discusses.

LEADERS OF LEADERS

More than anything else, I had confidence in my soldiers, junior leaders, and staff. They were trained, and I knew they would carry the fight to the enemy. I trusted them, and they knew I trusted them. I think in Just Cause, which was a company commander's war, being a decentralized commander paid big dividends because I wasn't in the knickers of my company commanders all the time. I gave them the mission and let them do it. I couldn't do it for them.

A Battalion Commander, Operation Just Cause
Panama, 1989

1-51. At any level, anyone responsible for supervising people or accomplishing a mission that involves other people is a leader. Anyone who influences others, motivating them to action or influencing their thinking or decision making, is a leader. It's not a function only of position; it's also a function of role. In addition, everyone in the Army—including every leader—fits somewhere in a chain of command. Everyone in the Army is also a follower or subordinate. There are, obviously, many leaders in an organization, and it's important to understand that you don't just lead subordinates—you lead other leaders. Even at the lowest level, you are a leader of leaders.

1-52. For example, a rifle company has four leadership levels: the company commander leads through platoon leaders, the platoon leaders through squad leaders, and the squad leaders through team leaders. At each level, the leader must let subordinate leaders do their jobs. Practicing this kind of decentralized execution based on mission orders in peacetime trains subordinates who will, in battle, exercise disciplined initiative in the absence of orders. They'll continue to fight when the radios are jammed, when the plan falls apart, when the enemy does something unexpected. (Appendix A discusses leader roles and relationships. FM 100-34 discusses mission orders and initiative.)

1-53. This decentralization does not mean that a commander never steps in and takes direct control. There will be times when a leader has to stop leading through subordinates, step forward, and say, "Follow me!" A situation like this may occur in combat, when things are falling apart and, like BG Thomas J. Jackson, you'll need to "stand like a stone wall" and

save victory. (You'll read about BG Jackson in Chapter 2.) Or it may occur during training, when a subordinate is about to make a mistake that could result in serious injury or death and you must act to prevent disaster.

1-54. More often, however, you should empower your subordinate leaders: give them a task, delegate the necessary authority, and let them do the work. Of course you need to check periodically. How else will you be able to critique, coach, and evaluate them? But the point is to "power down without powering off." Give your subordinate leaders the authority they need to get the job done. Then check on them frequently enough to keep track of what is going on but not so often that you get in their way. You can develop this skill through experience.

1-55. It takes personal courage to operate this way. But a leader must let subordinate leaders learn by doing. Is there a risk that, for instance, a squad leader—especially an inexperienced one—will make mistakes? Of course there is. But if your subordinate leaders are to grow, you must let them take risks. This means you must let go of some control and let your subordinate leaders do things on their own—within bounds established by mission orders and your expressed intent.

1-56. A company commander who routinely steps in and gives orders directly to squad leaders weakens the whole chain of command, denies squad leaders valuable learning experiences, and sends a signal to the whole company that the chain of command and NCO support channel can be bypassed at any time. On the other hand, successful accomplishment of specified and implied missions results from subordinate leaders at all levels exercising disciplined initiative within the commander's intent. Effective leaders strive to create an environment of trust and understanding that encourages their subordinates to seize the initiative and act. (Appendix A discusses authority, the chain of command, and the NCO support channel. FM 100-34 contains information about building trust up and down the chain of command.)

1-57. Weak leaders who have not trained their subordinates sometimes say, "My organization can't do it without me." Many people, used to being at the center of the action, begin to feel as if they're indispensable. You have heard them: "I can't take a day off. I have to be here all the time. I must watch my subordinates' every move, or who knows what will happen?" But no one is irreplaceable. The Army is not going to stop functioning because one leader—no matter how senior, no matter how central—steps aside. In combat, the loss of a leader is a shock to a unit, but the unit must continue its mission. If leaders train their subordinates properly, one of them will take charge.

1-58. Strong commanders—those with personal courage—realize their subordinate leaders need room to work. This doesn't mean that you should let your subordinates make the same mistake over and over. Part of your responsibility as a leader is to help your subordinates succeed. You can achieve this through empowering and coaching. Train your subordinates to plan, prepare, execute, and assess well enough to operate independently. Provide sufficient purpose, direction, and motivation for them to operate in support of the overall plan.

1-59. Finally, check and make corrections. Take time to help your subordinates sort out what happened and why. Conduct AARs so your people don't just make mistakes, but learn from them. There is not a soldier out there, from private to general, who has not slipped up from time to time. Good soldiers, and especially good leaders, learn from those mistakes. Good leaders help their subordinates grow by teaching, coaching, and counseling.

LEADERSHIP AND COMMAND

When you are commanding, leading [soldiers] under conditions where physical exhaustion and privations must be ignored, where the lives of [soldiers] may be sacrificed, then, the efficiency of your leadership will depend only to a minor degree on your tactical ability. It will primarily be determined by your character, your reputation, not much for courage—which will be accepted as a matter of course—but by the previous reputation you have established for fairness, for that high-minded patriotic purpose, that quality of unswerving determination to carry through any military task assigned to you.

General of the Army George C. Marshall
Speaking to officer candidates in September, 1941

1-60. Command is a specific and legal position unique to the military. It's where the buck stops. Like all leaders, commanders are responsible for the success of their organizations, but commanders have special accountability to their superiors, the institution, and the nation. Commanders must think deeply and creatively, for their concerns encompass

yesterday's heritage, today's mission, and tomorrow's force. To maintain their balance among all the demands on them, they must exemplify Army values. The nation, as well as the members of the Army, hold commanders accountable for accomplishing the mission, keeping the institution sound, and caring for its people.

1-61. Command is a sacred trust. The legal and moral responsibilities of commanders exceed those of any other leader of similar position or authority. Nowhere else does a boss have to answer for how subordinates live and what they do after work. Our society and the institution look to commanders to make sure that missions succeed, that people receive the proper training and care, that values survive. On the one hand, the nation grants commanders special authority to be good stewards of its most precious resources: freedom and people. On the other hand, those citizens serving in the Army also trust their commanders to lead them well. NCOs probably have a more immediate impact on their people, but commanders set the policies that reward superior performance and personally punish misconduct. It's no wonder that organizations take on the personal stamp of their commanders. Those selected to command offer something beyond their formal authority: their personal example and public actions have tremendous moral force. Because of that powerful aspect of their position, people inside and outside the Army see a commander as the human face of "the system"—the person who embodies the commitment of the Army to operational readiness and care of its people.

SUBORDINATES

To our subordinates we owe everything we are or hope to be. For it is our subordinates, not our superiors, who raise us to the dizziest of professional heights, and it is our subordinates who can and will, if we deserve it, bury us in the deepest mire of disgrace. When the chips are down and our subordinates have accepted us as their leader, we don't need any superior to tell us; we see it in their eyes and in their faces, in the barracks, on the field, and on the battle line. And on that final day when we must be ruthlessly demanding, cruel and heartless, they will rise as one to do our bidding, knowing full well that it may be their last act in this life.

Colonel Albert G. Jenkins, CSA
8th Virginia Cavalry

1-62. No one is only a leader; each of you is also a subordinate, and all members of the Army are part of a team. A technical supervisor leading a team of DA civilian specialists, for instance, isn't just the leader of that group. The team chief also works for someone else, and the team has a place in a larger organization.

1-63. Part of being a good subordinate is supporting your chain of command. And it's your responsibility to make sure your team supports the larger organization. Consider a leader whose team is responsible for handling the pay administration of a large organization. The chief knows that when the team makes a mistake or falls behind in its work, its customers—soldiers and DA civilians—pay the price in terms of late pay actions. One day a message from the boss introducing a new computer system for handling payroll changes arrives. The team chief looks hard at the new system and decides it will not work as well as the old one. The team will spend a lot of time installing the new system, all the while keeping up with their regular workload. Then they'll have to spend more time undoing the work once the new system fails. And the team chief believes it will fail—all his experience points to that.

1-64. But the team chief cannot simply say, "We'll let these actions pile up; that'll send a signal to the commander about just how bad the new system is and how important we are down here." The team does not exist in a vacuum; it's part of a larger organization that serves soldiers and DA civilians. For the good of the organization and the people in it, the team chief must make sure the job gets done.

1-65. Since the team chief disagrees with the boss's order and it affects both the team's mission and the welfare of its members, the team chief must tell the boss; he must have the moral courage to make his opinions known. Of course, the team chief must also have the right attitude; disagreement doesn't mean it's okay to be disrespectful. He must choose the time and place—usually in private—to explain his concerns to the boss fully and clearly. In addition, the team chief must go into the meeting knowing that, at some point, the discussion will be over and he must execute the boss's decision, whatever it is.

1-66. Once the boss has listened to all the arguments and made a decision, the team chief must support that decision as if it were his own. If he goes to the team and says, "I still don't think this is a good idea, but we're going to do it anyway," the team chief undermines the chain of command and teaches his people a bad lesson. Imagine what it would do to an organization's effectiveness if subordinates chose which orders to pursue vigorously and which ones to half step.

1-67. Such an action would also damage the team chief himself: in the future the team may treat his orders as he treated the boss's. And there is no great leap between people thinking their leader is disloyal to the boss to the same people thinking their leader will be disloyal to them as well. The good leader executes the boss's decision with energy and enthusiasm; looking at their leader, subordinates will believe the leader thinks it's absolutely the best possible solution. The only exception to this involves your duty to disobey obviously illegal orders. This is not a privilege you can claim, but a duty you must perform. (Chapter 2 discusses character and illegal orders. Chapter 4 discusses ethical reasoning.)

1-68. Loyalty to superiors and subordinates does more than ensure smooth-running peacetime organizations. It prepares units for combat by building soldiers' trust in leaders and leaders' faith in soldiers. The success of the airborne assault prior to the 1944 Normandy invasion is one example of how well-trained subordinate leaders can make the difference between victory and defeat.

Small Unit Leaders' Initiative in Normandy

The amphibious landings in Normandy on D-Day, 1944, were preceded by a corps-sized, night parachute assault by American and British airborne units. Many of the thousands of aircraft that delivered the 82d and 101st (US) Airborne Divisions to Normandy on the night of 5-6 June 1944 were blown off course. Some wound up in the wrong place because of enemy fire; others were simply lost. Thousands of paratroopers, the spearhead of the Allied invasion of Western Europe, found themselves scattered across unfamiliar countryside, many of them miles from their drop zones. They wandered about in the night, searching for their units, their buddies, their leaders, and their objectives. In those first few hours, the fate of the invasion hung in the balance; if the airborne forces did not cut the roads leading to the beaches, the Germans could counterattack the landing forces at the water's edge, crushing the invasion before it even began.

Fortunately for the Allies and the soldiers in the landing craft, the leaders in these airborne forces had trained their subordinate leaders well, encouraging their initiative, allowing them to do their jobs. Small unit leaders scattered around the darkened, unfamiliar countryside knew they were part of a larger effort, and they knew its success was up to them. They had been trained to act instead of waiting to be told what to do; they knew that if the invasion was to succeed, their small units had to accomplish their individual missions.

Among these leaders were men like CPT Sam Gibbons of the 505th Parachute Infantry Regiment. He gathered a group of 12 soldiers—from different commands—and liberated a tiny village—which turned out to be outside the division area of operations—before heading south toward his original objective, the Douve River bridges. CPT Gibbons set off with a dozen people he had never seen before and no demolition equipment to destroy a bridge nearly 15 kilometers away. Later, he remarked,

---

Small Unit Leaders' Initiative in Normandy (continued)

"This certainly wasn't the way I had thought the invasion would go, nor had we ever rehearsed it in this manner." But he was moving out to accomplish the mission. Throughout the Cotentin Peninsula, small unit leaders from both divisions were doing the same.

This was the payoff for hard training and leaders who valued soldiers, communicated the importance of the mission, and trusted their subordinate leaders to accomplish it. As they trained their commands for the invasion, organizational leaders focused downward as well as upward. They took care of their soldiers' needs while providing the most realistic training possible. This freed their subordinate leaders to focus upward as well as downward. Because they knew their units were well-trained and their leaders would do everything in their power to support them, small unit leaders were able to focus on the force's overall mission. They knew and understood the commander's intent. They believed that if they exercised disciplined initiative within that intent, things would turn out right. Events proved them correct.

1-69. You read earlier about how COL Joshua Chamberlain accomplished his mission and took care of his soldiers at Little Round Top. Empower subordinates to take initiative and be the subordinate leader who stands up and makes a difference. That lesson applies in peace and in combat, from the smallest organization to the largest. Consider the words of GEN Edward C. Meyer, former Army Chief of Staff:

When I became chief of staff, I set two personal goals for myself. The first was to ensure that the Army was continually prepared to go to war, and the second was to create a climate in which each member could find personal meaning and fulfillment. It is my belief that only by attainment of the second goal will we ensure the first.

1-70. GEN Meyer's words and COL Chamberlain's actions both say the same thing: leaders must accomplish the mission and take care of their soldiers. For COL Chamberlain, this meant he had to personally lead his men in a bayonet charge and show he believed they could do what he asked of them. For GEN Meyer the challenge was on a larger scale: his task was to make sure the entire Army was ready to fight and win. He knew—and he tells us—that the only way to accomplish such a huge goal is to pay attention to the smallest parts of the machine, the individual soldiers and DA civilians. Through his subordinate leaders, GEN Meyer offered challenges and guidance and set the example so that every member of the Army felt a part of the team and knew that the team was doing important work.

1-71. Both leaders understood the path to excellence: disciplined leaders with strong values produce disciplined soldiers with strong values. Together they become disciplined, cohesive units that train hard, fight honorably, and win decisively.

THE PAYOFF: EXCELLENCE

Leaders of character and competence act to achieve excellence by developing a force that can fight and win the nation's wars and serve the common defense of the United States.

1-72. You achieve excellence when your people are disciplined and committed to Army values. Individuals and organizations pursue excellence to improve, to get better and better. The Army is led by leaders of character who are good role models, consistently set the example, and accomplish the mission while improving their units. It is a cohesive organization of high-performing units characterized by the warrior ethos.

1-73. Army leaders get the job done. Sometimes it's on a large scale, such as GEN Meyer's role in making sure the Army was ready to fight. Other times it may be amid the terror of combat, as with COL Chamberlain at Gettysburg. However, most of you will not become Army Chief of Staff. Not all of you will face the challenge of combat. So it would be a mistake to think that the only time mission accomplishment and leadership are important is with the obvious examples—the general officer, the combat leader. The Army cannot accomplish its mission unless all Army leaders, soldiers, and DA civilians accomplish theirs—whether that means filling out a status report, repairing a vehicle, planning a budget, packing a parachute, maintaining pay records, or walking guard duty. The Army isn't a single general or a handful of combat heroes; it's hundreds of thousands of soldiers and DA civilians, tens of thousands of leaders, all striving to do the right things. Every soldier, every DA civilian, is important to the success of the Army.

MORAL EXCELLENCE: ACCOMPLISHING THE MISSION WITH CHARACTER

To the brave men and women who wear the uniform of the United States of America—thank you. Your calling is a high one—to be the defenders of freedom and the guarantors of liberty.

George Bush
41st President of the United States

1-74 The ultimate end of war, at least as America fights it, is to restore peace. For this reason the Army must accomplish its mission honorably. The Army fights to win, but with one eye on the kind of peace that will follow the war. The actions of Ulysses S. Grant, general in chief of the Union Army at the end of the Civil War, provide an example of balancing fighting to win with restoring the peace.

1-75. In combat GEN Grant had been a relentless and determined commander. During the final days of campaigning in Virginia, he hounded his exhausted foes and pushed his own troops on forced marches of 30 and 40 miles to end the war quickly. GEN Grant's approach to war was best summed up by President Lincoln, who said simply, "He fights."

1-76. Yet even before the surrender was signed, GEN Grant had shifted his focus to the peace. Although some of his subordinates wanted the Confederates to submit to the humiliation of an unconditional surrender, GEN Grant treated his former enemies with respect and considered the long-term effects of his decisions. Rather than demanding an unconditional surrender, GEN Grant negotiated terms with GEN Lee. One of those was allowing his former enemies to keep their horses because they needed them for spring plowing. GEN Grant reasoned that peace would best be served if the Southerners got back to a normal existence as quickly as possible. GEN Grant's decisions and actions sent a message to every man in the Union Army: that it was time to move on, to get back to peacetime concerns.

1-77. At the same time, the Union commander insisted on a formal surrender. He realized that for a true peace to prevail, the Confederates had to publicly acknowledge that organized hostility to the Union had ended. GEN Grant knew that true peace would come about only if both victors and vanquished put the war behind them—a timeless lesson.

1-78. The Army must accomplish its mission honorably. FM 27-10 discusses the law of war and reminds you of the importance of restoring peace. The Army minimizes collateral damage and avoids harming noncombatants for practical as well as honorable reasons. No matter what, though, soldiers fight to win, to live or die with honor for the benefit of our country and its citizens.

1-79. Army leaders often make decisions amid uncertainty, without guidance or precedent, in situations dominated by fear and risk, and sometimes under the threat of sudden, violent death. At those times leaders fall back on their values and Army values and ask, What is right? The question is simple; the answer, often, is not. Having made the decision, the leader depends on self-discipline to see it through.

ACHIEVING COLLECTIVE EXCELLENCE

1-80. Some examples of excellence are obvious: COL Chamberlain's imaginative defense of Little Round Top, GA Dwight Eisenhower drafting his D-Day message (you'll read about it in Chapter 2), MSG Gary Gordon and SFC Randall Shughart putting their lives on the line to save other soldiers in Somalia (their story is in Chapter 3). Those examples of excellence shine, and good leaders teach these stories; soldiers must know they are part of a long tradition of excellence and valor.

1-81. But good leaders see excellence wherever and whenever it happens. Excellent leaders make certain all subordinates know the important roles they play. Look for everyday examples that occur under ordinary circumstances: the way a soldier digs a fighting position, prepares for guard duty, fixes a radio, lays an artillery battery; the way a DA civilian handles an action, takes care of customers, meets a deadline on short notice. Good leaders know that each of these people is contributing in a small but important way to the business of the Army. An excellent Army is the collection of small tasks done to standard, day in and day out. At the end of the day, at the end of a career, those leaders, soldiers and DA civilians—the ones whose excellent work created an excellent Army—can look back confidently. Whether they commanded an invasion armada of thousands of soldiers or supervised a technical section of three people, they know they did the job well and made a difference.

1-82. Excellence in leadership does not mean perfection; on the contrary, an excellent leader allows subordinates room to learn from their mistakes as well as their successes. In such a climate, people work to improve and take the risks necessary to learn. They know that when they fall short—as they will—their leader will pick them up, give them new or more detailed instructions, and send them on their way again. This is the only way to improve the force, the only way to train leaders.

1-83. A leader who sets a standard of "zero defects, no mistakes" is also saying "Don't take any chances. Don't try anything you can't already do perfectly, and for heaven's sake, don't try anything new." That organization will not improve; in fact, its ability to perform the mission will deteriorate rapidly. Accomplishing the Army's mission requires leaders who are imaginative, flexible, and daring. Improving the Army for future missions requires leaders who are thoughtful and reflective. These qualities are incompatible with a "zero-defects" attitude.

1-84. Competent, confident leaders tolerate honest mistakes that do not result from negligence. The pursuit of excellence is not a game to achieve perfection; it involves trying, learning, trying again, and getting better each time. This in no way justifies or excuses failure. Even the best efforts and good intentions cannot take away an individual's responsibility for his actions.

SUMMARY

1-85. Leadership in combat is your primary and most important challenge. It requires you to accept a set of values that contributes to a core of motivation and will. If you fail to accept and live these Army values, your soldiers may die unnecessarily. Army leaders of character and competence act to achieve excellence by developing a force that can fight and win the nation's wars and serve the common defense of the United States. The Army leadership framework identifies the dimensions of Army leadership: what the Army expects you, as one of its leaders, to BE, KNOW, and DO.

1-86. Leadership positions fall into one of three leadership levels: direct, organizational, and strategic. The perspective and focus of leaders change and the actions they must DO become more complex with greater consequences as they assume positions of greater responsibility. Nonetheless, they must still live Army values and possess leader attributes.

1-87. Being a good subordinate is part of being a good leader. Everyone is part of a team, and all members have responsibilities that go with belonging to that team. But every soldier and DA civilian who is responsible for supervising people or accomplishing a mission that involves other people is a leader. All soldiers and DA civilians at one time or another must act as leaders.

1-88. Values and attributes make up a leader's character, the BE of Army leadership. Character embodies self-discipline and the will to win, among other things. It contributes to the motivation to persevere. From this motivation comes the lifelong work of mastering the skills that define competence, the KNOW of Army leadership. As you reflect on Army values and leadership attributes and develop the skills your position and experience require, you become a leader of character and competence, one who can act to achieve excellence, who can DO what is necessary to accomplish the mission and take care of your people. That is leadership—influencing people by providing purpose, direction, and motivation while operating to accomplish the mission and improving the organization. That is what makes a successful leader, one who lives the principles of BE, KNOW, DO.

CHAPTER 2

The Leader and Leadership: What the Leader Must Be, Know, and Do

I do solemnly swear (or affirm) that I will support and defend the Constitution of the United States against all enemies, foreign and domestic; that I will bear true faith and allegiance to the same; and that I will obey the orders of the President of the United States and the orders of the officers appointed over me, according to regulations and the Uniform Code of Military Justice. So help me God.

Oath of Enlistment

I [full name], having been appointed a [rank] in the United States Army, do solemnly swear (or affirm) that I will support and defend the Constitution of the United States against all enemies, foreign and domestic; that I will bear true faith and allegiance to the same; that I take this obligation freely, without any mental reservation or purpose of evasion, and that I will well and faithfully discharge the duties of the office upon which I am about to enter. So help me God.

Oath of office taken by commissioned officers and DA civilians

2-1. Beneath the Army leadership framework shown in Figure 1-1, 30 words spell out your job as a leader: **Leaders of character and competence act to achieve excellence by developing a force that can fight and win the nation's wars and serve the common defense of the United States.** There's a lot in that sentence. This chapter looks at it in detail.

2-2. Army leadership doctrine addresses what makes leaders of character and competence and what makes leadership. Figure 2-1 highlights these values and attributes. Remember from Chapter 1 that character describes what leaders must BE; competence refers to what leaders must KNOW; and action is what leaders must DO. Although this chapter discusses these concepts one at a time, they don't stand alone; they are closely connected and together make up who you seek to be (a leader of character and competence) and what you need to do (leadership).

SECTION I

CHARACTER: WHAT A LEADER MUST BE

Everywhere you look—on the fields of athletic competition, in combat training, operations, and in civilian communities—soldiers are doing what is right.

Former Sergeant Major of the Army
Julius W. Gates

2-3. Character—who you are—contributes significantly to how you act. Character helps you know what's right and do what's right, all the time and at whatever the cost. Character is made up of two interacting parts: values and attributes. Stephen Ambrose, speaking about the Civil War, says that "at the pivotal point in the war it was always the character of individuals that made the difference." Army leaders must be those critical individuals of character themselves and in turn develop character in those they lead. (Appendix E discusses character development.)

ARMY VALUES

Figure 2-1. Army Values

2-4. Your attitudes about the worth of people, concepts, and other things describe your values. Everything begins there. Your subordinates enter the Army with their own values, developed in childhood and nurtured through experience. All people are all shaped by what they've seen, what they've learned, and whom they've met.

But when soldiers and DA civilians take the oath, they enter an institution guided by Army values. These are more than a system of rules. They're not just a code tucked away in a drawer or a list in a dusty book. These values tell you what you need to be, every day, in every action you take. Army values form the very identity of the Army, the solid rock upon which everything else stands, especially in combat. They are the glue that binds together the members of a noble profession. As a result, the whole is much greater than the sum of its parts. Army values are nonnegotiable: they apply to everyone and in every situation throughout the Army.

2-5. Army values remind us and tell the rest of the world—the civilian government we serve, the nation we protect, even our enemies—who we are and what we stand for. The trust soldiers and DA civilians have for each other and the trust the American people have in us depends on how well we live up to Army values. They are the fundamental building blocks that enable us to discern right from wrong in any situation. Army values are consistent; they support one another. You can't follow one value and ignore another.

2-6. Here are the Army values that guide you, the leader, and the rest of the Army. They form the acronym LDRSHIP:

Loyalty
 Duty
  Respect
   Selfless Service
    Honor
     Integrity
      Personal Courage

2-7. The following discussions can help you understand Army values, but understanding is only the first step. As a leader, you must not only understand them; you must believe in them, model them in your own actions, and teach others to accept and live by them.

LOYALTY

Bear true faith and allegiance to the US Constitution, the Army, your unit, and other soldiers.

Loyalty is the big thing, the greatest battle asset of all. But no man ever wins the loyalty of troops by preaching loyalty. It is given to him as he proves his possession of the other virtues.

Brigadier General S. L. A. Marshall
Men Against Fire

2-8. Since before the founding of the republic, the Army has respected its subordination to its civilian political leaders. This subordination is fundamental to preserving the liberty of all Americans. You began your Army career by swearing allegiance to the Constitution, the basis of our government and laws. If you've never read it or if it has been a while, the Constitution is in Appendix F. Pay particular attention to Article I, Section 8, which outlines congressional responsibilities regarding the armed forces, and Article II, Section 2, which designates the president as commander in chief. Beyond your allegiance to the Constitution, you have an obligation to be faithful to the Army—the institution and its people—and to your unit or organization. Few examples illustrate loyalty to country and institution as well as the example of GEN George Washington in 1782.

2-9. GEN Washington's example shows how the obligation to subordinates and peers fits in the context of loyalty to the chain of command and the institution at large. As commander of the Continental Army, GEN Washington was obligated to see that his soldiers were taken care of. However, he also was obligated to ensure that the new nation remained secure and that the Continental Army remained able to fight if necessary. If the Continental Army had marched on the seat of government, it may well have destroyed the nation by undermining the law that held it together. It also would have destroyed the Army as an institution by destroying the basis for the authority under which it served. GEN Washington realized these things and acted based on his knowledge. Had he done nothing else, this single act would have been enough to establish GEN George Washington as the father of his country.

GEN Washington at Newburgh

Following its victory at Yorktown in 1781, the Continental Army set up camp at Newburgh, New York, to wait for peace with Great Britain. The central government formed under the Articles of Confederation proved weak and unwilling to supply the Army properly or even pay the soldiers who had won the war for independence. After months of waiting many officers, angry and impatient, suggested that the Army march on the seat of government in Philadelphia, Pennsylvania, and force Congress to meet the Army's demands. One colonel even suggested that GEN Washington become King George I.

Upon hearing this, GEN Washington assembled his officers and publicly and emphatically rejected the suggestion. He believed that seizing power by force would have destroyed everything for which the Revolutionary War had been fought. By this action, GEN Washington firmly established an enduring precedent: America's armed forces are subordinate to civilian authority and serve the democratic principles that are now enshrined in the Constitution. GEN Washington's action demonstrated the loyalty to country that the Army must maintain in order to protect the freedom enjoyed by all Americans.

2-10. Loyalty is a two-way street: you should not expect loyalty without being prepared to give it as well. Leaders can neither demand loyalty nor win it from their people by talking about it. The loyalty of your people is a gift they give you when, and only when, you deserve it—when you train them well, treat them fairly, and live by the concepts you talk about. Leaders who are loyal to their subordinates never let them be misused.

2-11. Soldiers fight for each other—loyalty is commitment. Some of you will encounter the most important way of earning this loyalty: leading your soldiers well in combat. There's no loyalty fiercer than that of soldiers who trust their leader to take them through the dangers of combat. However, loyalty extends to all members of an organization—to your superiors and subordinates, as well as your peers.

2-12. Loyalty extends to all members of all components of the Army. The reserve components—Army National Guard and Army Reserve—play an increasingly active role in the Army's mission. Most DA civilians will not be called upon to serve in combat theaters, but their contributions to mission accomplishment are nonetheless vital. As an Army leader, you'll serve throughout your career with soldiers of the active and reserve components as well as

DA civilians. All are members of the same team, loyal to one another.

DUTY

Fulfill your obligations.

The essence of duty is acting in the absence of orders or direction from others, based on an inner sense of what is morally and professionally right....

General John A. Wickham Jr.
Former Army Chief of Staff

2-13. Duty begins with everything required of you by law, regulation, and orders; but it includes much more than that. Professionals do their work not just to the minimum standard, but to the very best of their ability. Soldiers and DA civilians commit to excellence in all aspects of their professional responsibility so that when the job is done they can look back and say, "I couldn't have given any more."

2-14. Army leaders take the initiative, figuring out what needs to be done before being told what to do. What's more, they take full responsibility for their actions and those of their subordinates. Army leaders never shade the truth to make the unit look good—or even to make their subordinates feel good. Instead, they follow their higher duty to the Army and the nation.

Duty in Korea

CPT Viola B. McConnell was the only Army nurse on duty in Korea in July of 1950. When hostilities broke out, she escorted nearly 700 American evacuees from Seoul to Japan aboard a freighter designed to accommodate only 12 passengers. CPT McConnell assessed priorities for care of the evacuees and worked exhaustively with a medical team to care for them. Once in Japan, she requested reassignment back to Korea. After all she had already done, CPT McConnell returned to Taejon to care for and evacuate wounded soldiers of the 24th Infantry Division.

2-15. CPT McConnell understood and fulfilled her duty to the Army and to the soldiers she supported in ways that went beyond her medical training. A leader's duty is to take charge, even in unfamiliar circumstances. But duty isn't reserved for special occasions. When a platoon sergeant tells a squad leader to inspect weapons, the squad leader has fulfilled his

minimum obligation when he has checked the weapons. He's done what he was told to do. But if the squad leader finds weapons that are not clean or serviced, his sense of duty tells him to go beyond the platoon sergeant's instructions. The squad leader does his duty when he corrects the problem and ensures the weapons are up to standard.

2-16. In extremely rare cases, you may receive an illegal order. Duty requires that you refuse to obey it. You have no choice but to do what's ethically and legally correct. Paragraphs 2-97 through 2-99 discuss illegal orders.

RESPECT

Treat people as they should be treated.

The discipline which makes the soldiers of a free country reliable in battle is not to be gained by harsh or tyrannical treatment. On the contrary, such treatment is far more likely to destroy than to make an army. It is possible to impart instruction and to give commands in such manner and such a tone of voice to inspire in the soldier no feeling but an intense desire to obey, while the opposite manner and tone of voice cannot fail to excite strong resentment and a desire to disobey. The one mode or the other of dealing with subordinates springs from a corresponding spirit in the breast of the commander. He who feels the respect which is due to others cannot fail to inspire in them regard for himself, while he who feels, and hence manifests, disrespect toward others, especially his inferiors, cannot fail to inspire hatred against himself.

Major General John M. Schofield
Address to the United States Corps of Cadets
11 August 1879

2-17. Respect for the individual forms the basis for the rule of law, the very essence of what makes America. In the Army, respect means recognizing and appreciating the inherent dignity and worth of all people. This value reminds you that your people are your greatest resource. Army leaders honor everyone's individual worth by treating all people with dignity and respect.

2-18. As America becomes more culturally diverse, Army leaders must be aware that they will deal with people from a wider range of ethnic, racial, and religious backgrounds. Effective leaders are tolerant of beliefs different from their own as long as those beliefs don't conflict with Army values, are not illegal, and are not unethical. As an Army leader, you need to avoid misunderstandings arising from cultural differences. Actively seeking to learn about people and cultures different from your own can help you do this. Being sensitive to other cultures can also aid you in counseling your people more effectively. You show respect when you seek to understand your people's background, see things from their perspective, and appreciate what's important to them.

2-19. As an Army leader, you must also foster a climate in which everyone is treated with dignity and respect regardless of race, gender, creed, or religious belief. Fostering this climate begins with your example: how you live Army values shows your people how they should live them. However, values training is another major contributor. Effective training helps create a common understanding of Army values and the standards you expect. When you conduct it as part of your regular routine—such as during developmental counseling sessions—you reinforce the message that respect for others is part of the character of every soldier and DA civilian. Combined with your example, such training creates an organizational climate that promotes consideration for others, fairness in all dealings, and equal opportunity. In essence, Army leaders treat others as they wish to be treated.

2-20. As part of this consideration, leaders create an environment in which subordinates are challenged, where they can reach their full potential and be all they can be. Providing tough training doesn't demean subordinates; in fact, building their capabilities and showing faith in their potential is the essence of respect. Effective leaders take the time to learn what their subordinates want to accomplish. They advise their people on how they can grow, personally and professionally. Not all of your subordinates will succeed equally, but they all deserve respect.

2-21. Respect is also an essential component for the development of disciplined, cohesive, and effective warfighting teams. In the deadly confusion of combat, soldiers often overcome incredible odds to accomplish the mission and protect the lives of their comrades. This spirit of selfless service and duty is built on a soldier's personal trust and regard for fellow soldiers. A leader's willingness to tolerate discrimination or harassment on any basis, or a failure to cultivate a climate of respect, eats away at this trust and erodes unit cohesion. But respect goes beyond issues of discrimination and harassment; it includes the broader issue of civility, the way people treat each other and those they come in contact with. It involves being sensitive to diversity and one's own behaviors that others may find insensitive, offensive, or abusive. Soldiers and DA civilians, like their leaders, treat everyone with dignity and respect.

SELFLESS SERVICE

Put the welfare of the nation, the Army, and subordinates before your own.

The nation today needs men who think in terms of service to their country and not in terms of their country's debt to them.

General of the Army Omar N. Bradley

2-22. You have often heard the military referred to as "the service." As a member of the Army, you serve the United States. Selfless service means doing what's right for the nation, the Army, your organization, and your people—and putting these responsibilities above your own interests. The needs of the Army and the nation come first. This doesn't mean that you neglect your family or yourself; in fact, such neglect weakens a leader and can cause the Army more harm than good. Selfless service doesn't mean that you can't have a strong ego, high self-esteem, or even healthy ambition. Rather, selfless service means that you don't make decisions or take actions that help your image or your career but hurt others or sabotage the mission. The selfish superior claims credit for work his subordinates do; the selfless leader gives credit to those who earned it. The Army can't function except as a team, and for a team to work, the individual has to give up self-interest for the good of the whole.

2-23. Soldiers are not the only members of the Army who display selfless service. DA civilians display this value as well. Then Army Chief of Staff, Gordon R. Sullivan assessed the DA civilian contribution to Operation Desert Storm this way:

Not surprisingly, most of the civilians deployed to Southwest Asia volunteered to serve there. But the civilian presence in the Gulf region meant more than moral support and filling in for soldiers. Gulf War veterans say that many of the combat soldiers could owe their lives to the DA civilians who helped maintain equipment by speeding up the process of getting parts and other support from 60 logistics agencies Army-wide.

2-24. As GEN Sullivan's comment indicates, selfless service is an essential component of teamwork. Team members give of themselves so that the team may succeed. In combat some soldiers give themselves completely so that their comrades can live and the mission can be accomplished. But the need for selflessness isn't limited to combat situations. Requirements for individuals to place their own needs below those of their organization can occur during peacetime as well. And the requirement for selflessness doesn't decrease with one's rank; it increases. Consider this example of a soldier of long service and high rank who demonstrated the value of selfless service.

GA Marshall Continues to Serve

GA George C. Marshall served as Army Chief of Staff from 1939 until 1945. He led the Army through the buildup, deployment, and worldwide operations of World War II. Chapter 7 outlines some of his contributions to the Allied victory. In November 1945 he retired to a well-deserved rest at his home in Leesburg, Virginia. Just six days later President Harry S Truman called on him to serve as Special Ambassador to China. From the White House President Truman telephoned GA Marshall at his home: "General, I want you to go to China for me," the president said. "Yes, Mr. President," GA Marshall replied. He then hung up the telephone, informed his wife of the president's request and his reply, and prepared to return to government service.

> ### GA Marshall Continues to Serve (continued)
>
> President Truman didn't appoint GA Marshall a special ambassador to reward his faithful service; he appointed GA Marshall because there was a tough job in China that needed to be done. The Chinese communists under Mao Tse-tung were battling the Nationalists under Chiang Kai-shek, who had been America's ally against the Japanese; GA Marshall's job was to mediate peace between them. In the end, he was unsuccessful in spite of a year of frustrating work; the scale of the problem was more than any one person could handle. However, in January 1947 President Truman appointed GA Marshall Secretary of State. The Cold War had begun and the president needed a leader Americans trusted. GA Marshall's reputation made him the one; his selflessness led him to continue to serve.

2-25. When faced with a request to solve a difficult problem in an overseas theater after six years of demanding work, GA Marshall didn't say, "I've been in uniform for over thirty years, we just won a world war, and I think I've done enough." Instead, he responded to his commander in chief the only way a professional could. He said yes, took care of his family, and prepared to accomplish the mission. After a year overseas, when faced with a similar question, he gave the same answer. GA Marshall always placed his country's interests first and his own second. Army leaders who follow his example do the same.

HONOR

Live up to all the Army values.

What is life without honor? Degradation is worse than death.

Lieutenant General Thomas J. "Stonewall" Jackson

2-26. Honor provides the "moral compass" for character and personal conduct in the Army. Though many people struggle to define the term, most recognize instinctively those with a keen sense of right and wrong, those who live such that their words and deeds are above reproach. The expression "honorable person," therefore, refers to both the character traits an individual actually possesses and the fact that the community recognizes and respects them.

2-27. Honor holds Army values together while at the same time being a value itself. Together, Army values describe the foundation essential to develop leaders of character. Honor means demonstrating an understanding of what's right and taking pride in the community's acknowledgment of that reputation. Military ceremonies recognizing individual and unit achievement demonstrate and reinforce the importance the Army places on honor.

2-28. For you as an Army leader, demonstrating an understanding of what's right and taking pride in that reputation means this: **Live up to all the Army values**. Implicitly, that's what you promised when you took your oath of office or enlistment. You made this promise publicly, and the standards—Army values—are also public. To be an honorable person, you must be true to your oath and live Army values in all you do. Living honorably strengthens Army values, not only for yourself but for others as well: all members of an organization contribute to the organization's climate (which you'll read about in Chapter 3). By what they do, people living out Army values contribute to a climate that encourages all members of the Army to do the same.

2-29. How you conduct yourself and meet your obligations defines who you are as a person; how the Army meets the nation's commitments defines the Army as an institution. For you as an Army leader, honor means putting Army values above self-interest, above career and comfort. For all soldiers, it means putting Army values above self-preservation as well. This honor is essential for creating a bond of trust among members of the Army and between the Army and the nation it serves. Army leaders have the strength of will to live according to Army values, even though the temptations to do otherwise are strong, especially in the face of personal danger. The military's highest award is the Medal of Honor. Its recipients didn't do just what was required of them; they went beyond the expected, above and beyond the call of duty. Some gave their own lives so that others could live. It's fitting that the word we use to describe their achievements is "honor."

MSG Gordon and SFC Shughart in Somalia

During a raid in Mogadishu in October 1993, MSG Gary Gordon and SFC Randall Shughart, leader and member of a sniper team with Task Force Ranger in Somalia, were providing precision and suppressive fires from helicopters above two helicopter crash sites. Learning that no ground forces were available to rescue one of the downed aircrews and aware that a growing number of enemy were closing in on the site, MSG Gordon and SFC Shughart volunteered to be inserted to protect their critically wounded comrades. Their initial request was turned down because of the danger of the situation. They asked a second time; permission was denied. Only after their third request were they inserted.

MSG Gordon and SFC Shughart were inserted one hundred meters south of the downed chopper. Armed only with their personal weapons, the two NCOs fought their way to the downed fliers through intense small arms fire, a maze of shanties and shacks, and the enemy converging on the site. After MSG Gordon and SFC Shughart pulled the wounded from the wreckage, they established a perimeter, put themselves in the most dangerous position, and fought off a series of attacks. The two NCOs continued to protect their comrades until they had depleted their ammunition and were themselves fatally wounded. Their actions saved the life of an Army pilot.

2-30. No one will ever know what was running through the minds of MSG Gordon and SFC Shughart as they left the comparative safety of their helicopter to go to the aid of the downed aircrew. The two NCOs knew there was no ground rescue force available, and they certainly knew there was no going back to their helicopter. They may have suspected that things would turn out as they did; nonetheless, they did what they believed to be the right thing. They acted based on Army values, which they had clearly made their own: *loyalty* to their fellow soldiers; the *duty* to stand by them, regardless of the circumstances; the *personal courage* to act, even in the face of great danger; *selfless service*, the willingness to give their all. MSG Gary I. Gordon and SFC Randall D. Shughart lived Army values to the end; they were posthumously awarded Medals of Honor.

INTEGRITY

Do what's right—legally and morally.

The American people rightly look to their military leaders not only to be skilled in the technical aspects of the profession of arms, but also to be men of integrity.

General J. Lawton Collins
Former Army Chief of Staff

2-31. People of integrity consistently act according to principles—not just what might work at the moment. Leaders of integrity make their principles known and consistently act in accordance with them. The Army requires leaders of integrity who possess high moral standards and are honest in word and deed. Being honest means being truthful and upright all the time, despite pressures to do otherwise. Having integrity means being both morally complete and true to yourself. As an Army leader, you're honest to yourself by committing to and consistently living Army values; you're honest to others by not presenting yourself or your actions as anything other than what they are. Army leaders say what they mean and do what they say. If you can't accomplish a mission, inform your chain of command. If you inadvertently pass on bad information, correct it as soon as you find out it's wrong. People of integrity do the right thing not because it's convenient or because they have no choice. They choose the right thing because their character permits no less. Conducting yourself with integrity has three parts:

- Separating what's right from what's wrong.

- Always acting according to what you know to be right, even at personal cost.

- Saying openly that you're acting on your understanding of right versus wrong.

2-32. Leaders can't hide what they do: that's why you must carefully decide how you act. As an Army leader, you're always on display. If you want to instill Army values in others, you must internalize and demonstrate them yourself. Your personal values may and probably do extend beyond the Army values, to include such things as political, cultural, or religious beliefs. However, if you're to be an Army leader *and* a person of integrity, these values must reinforce, not contradict, Army values.

2-33. Any conflict between your personal values and Army values must be resolved before you can become a morally complete Army leader. You may need to consult with someone whose values and judgment you respect. You would not be the first person to face this issue, and as a leader, you can expect others to come to you, too. Chapter 5 contains the story of how SGT Alvin York and his leaders confronted and resolved a conflict between SGT York's personal values and Army values. Read it and reflect on it. If one of your subordinates asks you to help resolve a similar conflict, you must be prepared by being sure your own values align with Army values. Resolving such conflicts is necessary to become a leader of integrity.

PERSONAL COURAGE

Face fear, danger, or adversity (physical or moral).

The concept of professional courage does not always mean being as tough as nails either. It also suggests a willingness to listen to the soldiers' problems, to go to bat for them in a tough situation, and it means knowing just how far they can go. It also means being willing to tell the boss when he's wrong.

Former Sergeant Major of the Army William Connelly

2-34. Personal courage isn't the absence of fear; rather, it's the ability to put fear aside and do what's necessary. It takes two forms, physical and moral. Good leaders demonstrate both.

2-35. Physical courage means overcoming fears of bodily harm and doing your duty. It's the bravery that allows a soldier to take risks in combat in spite of the fear of wounds or death. Physical courage is what gets the soldier at Airborne School out the aircraft door. It's what allows an infantryman to assault a bunker to save his buddies.

2-36. In contrast, moral courage is the willingness to stand firm on your values, principles, and convictions—even when threatened. It enables leaders to stand up for what they believe is right, regardless of the consequences. Leaders who take responsibility for their decisions and actions, even when things go wrong, display moral courage. Courageous leaders are willing to look critically inside themselves, consider new ideas, and change what needs changing.

2-37. Moral courage is sometimes overlooked, both in discussions of personal courage and in the everyday rush of business. A DA civilian at a meeting heard *courage* mentioned several times in the context of combat. The DA civilian pointed out that consistent moral courage is every bit as important as momentary physical courage. Situations requiring physical courage are rare; situations requiring moral courage can occur frequently. Moral courage is essential to living the Army values of integrity and honor every day.

2-38. Moral courage often expresses itself as candor. Candor means being frank, honest, and sincere with others while keeping your words free from bias, prejudice, or malice. Candor means calling things as you see them, even when it's uncomfortable or you think it might be better for you to just keep quiet. It means not allowing your feelings to affect what you say about a person or situation. A candid company commander calmly points out the first sergeant's mistake. Likewise, the candid first sergeant respectfully points out when the company commander's pet project isn't working and they need to do something different. For trust to exist between leaders and subordinates, candor is essential. Without it, subordinates won't know if they've met the standard and leaders won't know what's going on.

2-39. In combat physical and moral courage may blend together. The right thing to do may not only be unpopular, but dangerous as well. Situations of that sort reveal who's a leader of character and who's not. Consider this example.

WO1 Thompson at My Lai

Personal courage—whether physical, moral, or a combination of the two—may be manifested in a variety of ways, both on and off the battlefield. On March 16, 1968 Warrant Officer (WO1) Hugh C. Thompson Jr. and his two-man crew were on a reconnaissance mission over the village of My Lai, Republic of Vietnam. WO1 Thompson watched in horror as he saw an American soldier shoot an injured Vietnamese child. Minutes later, when he observed American soldiers advancing on a number of civilians in a ditch, WO1 Thompson landed his helicopter and questioned a young officer about what was happening on the ground. Told that the ground action was none of his business, WO1 Thompson took off and continued to circle the area.

When it became apparent that the American soldiers were now firing on civilians, WO1 Thompson landed his helicopter between the soldiers and a group of 10 villagers who were headed for a homemade bomb shelter. He ordered his gunner to train his weapon on the approaching American soldiers and to fire if necessary. Then he personally coaxed the civilians out of the shelter and airlifted them to safety. WO1 Thompson's radio reports of what was happening were instrumental in bringing about the cease-fire order that saved the lives of more civilians. His willingness to place himself in physical danger in order to do the morally right thing is a sterling example of personal courage.

LEADER ATTRIBUTES

Leadership is not a natural trait, something inherited like the color of eyes or hair...Leadership is a skill that can be studied, learned, and perfected by practice.

The Noncom's Guide, 1962

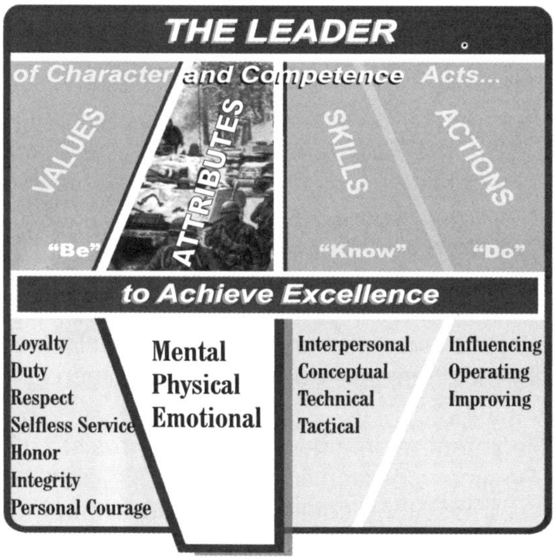

Figure 2-2. Leader Attributes

2-40. Values tell us part of what the leader must BE; the other side of what a leader must BE are the attributes listed in Figure 2-2. Leader attributes influence leader actions; leader actions, in turn, always influence the unit or organization. As an example, if you're physically fit, you're more likely to inspire your subordinates to be physically fit.

2-41. Attributes are a person's fundamental qualities and characteristics. People are born with some attributes; for instance, a person's genetic code determines eye, hair, and skin color. However, other attributes—including leader attributes—are learned and can be changed. Leader attributes can be characterized as mental, physical, and emotional. Successful leaders work to improve those attributes.

MENTAL ATTRIBUTES

2-42. The mental attributes of an Army leader include will, self-discipline, initiative, judgment, self-confidence, intelligence, and cultural awareness.

Will

The will of soldiers is three times more important than their weapons.

Colonel Dandridge M. "Mike" Malone
Small Unit Leadership: A Commonsense Approach

2-43. Will is the inner drive that compels soldiers and leaders to keep going when they are exhausted, hungry, afraid, cold, and wet—when it would be easier to quit. Will enables soldiers to press the fight to its conclusion. Yet will without competence is useless. It's not enough that soldiers are willing, or even eager, to fight; they must know how to fight. Likewise, soldiers who have competence but no will don't fight. The leader's task is to develop a winning spirit by building their subordinates' will as well as their skill. That begins with hard, realistic training.

2-44. Will is an attribute essential to all members of the Army. Work conditions vary among branches and components, between those deployed and those closer to home. In the Army, personal attitude must prevail over any adverse external conditions. All members of the Army—active, reserve, and DA civilian—will experience situations when it would be easier to quit rather than finish the task at hand. At those times, everyone needs that inner drive to press on to mission completion.

2-45. It's easy to talk about will when things go well. But the test of your will comes when things go badly— when events seem to be out of control, when you think your bosses have forgotten you, when the plan doesn't seem to work and it looks like you're going to lose. It's then that you must draw on your inner reserves to persevere—to do your job until there's nothing left to do it with and then to remain faithful to your people, your organization, and your country. The story of the American and Filipino stand on the Bataan Peninsula and their subsequent captivity is one of individuals, leaders, and units deciding to remain true to the end—and living and dying by that decision.

The Will to Persevere

On 8 December 1941, hours after the attack on Pearl Harbor, Japanese forces attacked the American and Filipino forces defending the Philippines. With insufficient combat power to launch a counterattack, GEN Douglas MacArthur, the American commander, ordered his force to consolidate on the Bataan Peninsula and hold as long as possible. Among his units was the 12th Quartermaster (QM) Regiment, which had the mission of supporting the force.

Completely cut off from outside support, the Allies held against an overwhelming Japanese army for the next three and a half months. Soldiers of the 12th QM Regiment worked in the debris of warehouses and repair shops under merciless shelling and bombing, fighting to make the meager supplies last. They slaughtered water buffaloes for meat, caught fish with traps they built themselves, and distilled salt from sea water. In coffeepots made from oil drums they boiled and reboiled the tiny coffee supply until the grounds were white. As long as an ounce of food existed, it was used. In the last desperate days, they resorted to killing horses and pack mules. More important, these supporters delivered rations to the foxholes on the front lines—fighting their way in when necessary. After Bataan and Corregidor fell, members of the 12th QM Regiment were prominent among the 7,000 Americans and Filipinos who died on the infamous Bataan Death March.

Though captured, the soldiers of the 12th QM Regiment maintained their will to resist. 1LT Beulah Greenwalt, a nurse assigned to the 12th QM Regiment, personified this will. Realizing the regimental colors represent the soul of a regiment and that they could serve as a symbol for resistance, 1LT Greenwalt assumed the mission of protecting the colors from the Japanese. She carried the colors to the prisoner of war (PW) camp in Manila by wrapping them around her

---

The Will to Persevere (continued)

shoulders and convincing her Japanese captors that they were "only a shawl." For the next 33 months 1LT Greenwalt and the remains of the regiment remained PWs, living on starvation diets and denied all comforts. But through it all, 1LT Greenwalt held onto the flag. The regimental colors were safeguarded: the soul of the regiment remained with the regiment, and its soldiers continued to resist.

When the war ended in 1945 and the surviving PWs were released, 1LT Greenwalt presented the colors to the regimental commander. She and her fellow PWs had persevered. They had resisted on Bataan until they had no more means to resist. They continued to resist through three long years of captivity. They decided on Bataan to carry on, and they renewed that decision daily until they were liberated. The 12th QM Regiment—and the other units that had fought and resisted with them—remained true to themselves, the Army, and their country. Their will allowed them to see events through to the end.

Self-Discipline

The core of a soldier is moral discipline. It is intertwined with the discipline of physical and mental achievement. Total discipline overcomes adversity, and physical stamina draws on an inner strength that says "drive on."

Former Sergeant Major of the Army
William G. Bainbridge

2-46. Self-disciplined people are masters of their impulses. This mastery comes from the habit of doing the right thing. Self-discipline allows Army leaders to do the right thing regardless of the consequences for them or their subordinates. Under the extreme stress of combat, you and your team might be cut off and alone, fearing for your lives, and having to act without guidance or knowledge of what's going on around you. Still, you—the leader—must think clearly and act reasonably. Self-discipline is the key to this kind of behavior.

2-47. In peacetime, self-discipline gets the unit out for the hard training. Self-discipline makes the tank commander demand another run-through of a battle drill if the performance doesn't meet the standard—even though everyone is long past ready to quit. Self-discipline doesn't mean that you never get tired or discouraged—after all, you're only human. It does mean that you do what needs to be done regardless of your feelings.

Initiative

The leader must be an aggressive thinker—always anticipating and analyzing.

He must be able to make good assessments and solid tactical judgments.

Brigadier General John. T. Nelson II

2-48. Initiative is the ability to be a self-starter—to act when there are no clear instructions, to act when the situation changes or when the plan falls apart. In the operational context, it means setting and dictating the terms of action throughout the battle or operation. An individual leader with initiative is willing to decide and initiate independent actions when the concept of operations no longer applies or when an unanticipated opportunity leading to accomplishment of the commander's intent presents itself. Initiative drives the Army leader to seek a better method, anticipate what must be done, and perform without waiting for instructions. Balanced with good judgment, it becomes *disciplined* initiative, an essential leader attribute. (FM 100-5 discusses initiative as it relates to military actions at the operational level. FM 100-34 discusses the relationship of initiative to command and control. FM 100-40 discusses the place of initiative in the art of tactics.)

2-49. As an Army leader, you can't just give orders: you must make clear the intent of those orders, the final goal of the mission. In combat, it's critically important for subordinates to understand their commander's intent. When they are cut off or enemy actions derail the original plan, well-trained soldiers who understand the commander's intent will apply disciplined initiative to accomplish the mission.

2-50. Disciplined initiative doesn't just appear; you must develop it within your subordinates. Your leadership style and the organizational climate you establish can either encourage or discourage initiative: you can instill initiative in your subordinates or you can drive it out. If you underwrite honest mistakes, your subordinates will be more likely to develop initiative. If you set a "zero defects" standard, you risk strangling initiative in its cradle, the hearts of your subordinates. (Chapter 5 discusses "zero defects" and learning.)

The Quick Reaction Platoon

On 26 December 1994 a group of armed and disgruntled members of the Haitian Army entered the Haitian Army Headquarters in Port-au-Prince demanding back pay. A gunfight ensued less than 150 meters from the grounds of the Haitian Palace, seat of the new government. American soldiers from C Company, 1-22 Infantry, who had deployed to Haiti as part of Operation Uphold Democracy, were guarding the palace grounds. The quick reaction platoon leader deployed and immediately maneuvered his platoon towards the gunfire. The platoon attacked, inflicting at least four casualties and causing the rest of the hostile soldiers to flee. The platoon quelled a potentially explosive situation by responding correctly and aggressively to the orders of their leader, who knew his mission and the commander's intent.

Judgment

I learned that good judgment comes from experience and that experience grows out of mistakes.

General of the Army Omar N. Bradley

2-51. Leaders must often juggle hard facts, questionable data, and gut-level intuition to arrive at a decision. Good judgment means making the best decision for the situation. It's a key attribute of the art of command and the transformation of knowledge into understanding. (FM 100-34 discusses how leaders convert data and information into knowledge and understanding.)

2-52. Good judgment is the ability to size up a situation quickly, determine what's important, and decide what needs to be done. Given a problem, you should consider a range of alternatives before you act. You need to think through the consequences of what you're about to do before you do it. In addition to considering the consequences, you should also think methodically. Some sources that aid judgment are the boss's intent, the desired goal, rules, laws, regulations, experience, and values. Good judgment also includes the ability to size up subordinates, peers, and the enemy for strengths, weaknesses, and potential actions. It's a critical part of problem solving and decision making. (Chapter 5 discusses problem solving and decision making).

2-53. Judgment and initiative go hand in hand. As an Army leader, you must weigh what you know and make decisions in situations where others do nothing. There will be times when you'll have to make decisions under severe time constraints. In all cases, however, you must take responsibility for your actions. In addition, you must encourage disciplined initiative in, and teach good judgment to, your subordinates. Help your subordinates learn from mistakes by coaching and mentoring them along the way. (Chapter 5 discusses mentoring.)

Self-Confidence

2-54. Self-confidence is the faith that you'll act correctly and properly in any situation, even one in which you're under stress and don't have all the information you want. Self-confidence comes from competence: it's based on mastering skills, which takes hard work and dedication. Leaders who know their own capabilities and believe in themselves are self-confident. Don't mistake bluster—loudmouthed bragging or self-promotion—for self-confidence. Truly self-confident leaders don't need to advertise; their actions say it all.

2-55. Self-confidence is important for leaders and teams. People want self-confident leaders, leaders who understand the situation, know what needs to be done, and demonstrate that understanding and knowledge. Self-confident leaders instill self-confidence in their people. In combat, self-confidence helps soldiers control doubt and reduce anxiety. Together with will and self-discipline, self-confidence helps leaders act—do what must be done in circumstances where it would be easier to do nothing—and to convince their people to act as well.

Intelligence

2-56. Intelligent leaders think, learn, and reflect; then they apply what they learn. Intelligence is more than knowledge, and the ability to think isn't the same as book learning. All people have some intellectual ability that, when developed, allows them to analyze and understand a situation. And although some people are smarter than others, all people can develop the capabilities they have. Napoleon himself observed how a leader's intellectual development applies directly to battlefield success:

It is not genius which reveals to me suddenly and secretly what I should do in circumstances unexpected by others; it is thought and meditation.

2-57. Knowledge is only part of the equation. Smart decisions result when you combine professional skills (which you learn through study) with experience (which you gain on the job) and your ability to reason through a problem based on the information available. Reflection is also important. From time to time, you find yourself carefully and thoughtfully considering how leadership, values, and other military principles apply to you and your job. When things don't go quite the way they intended, intelligent leaders are confident enough to step back and ask, "Why did things turn out that way?" Then they are smart enough to build on their strengths and avoid making the same mistake again.

2-58. Reflection also contributes to your originality (the ability to innovate, rather than only adopt others' methods) and intuition (direct, immediate insight or understanding of important

factors without apparent rational thought or inference). Remember COL Chamberlain at Little Round Top. To his soldiers, it sometimes appeared that he could "see through forests and hills and know what was coming." But this was no magical ability. Through study and reflection, the colonel had learned how to analyze terrain and imagine how the enemy might attempt to use it to his advantage. He had applied his intelligence and developed his intellectual capabilities. Good leaders follow COL Chamberlain's example.

Cultural Awareness

2-59. Culture is a group's shared set of beliefs, values, and assumptions about what's important. As an Army leader, you must be aware of cultural factors in three contexts:

- You must be sensitive to the different backgrounds of your people.

- You must be aware of the culture of the country in which your organization is operating.

- You must take into account your partners' customs and traditions when you're working with forces of another nation.

2-60. Within the Army, people come from widely different backgrounds: they are shaped by their schooling, race, gender, and religion as well as a host of other influences. Although they share Army values, an African-American man from rural Texas may look at many things differently from, say, a third-generation Irish-American man who grew up in Philadelphia or a Native American woman from the Pacific Northwest. But be aware that perspectives vary within groups as well. That's why you should try to understand individuals based on their own ideas, qualifications, and contributions and not jump to conclusions based on stereotypes.

2-61. Army values are part of the Army's institutional culture, a starting point for how you as a member of the Army should think and act. Beyond that, Army leaders not only recognize that people are different; they value them because of their differences, because they are people. Your job as a leader isn't to make everyone the same.

Instead, your job is to take advantage of the fact that everyone is different and build a cohesive team. (Chapter 7 discusses the role strategic leaders play in establishing and maintaining the Army's institutional culture.)

2-62. There's great diversity in the Army—religious, ethnic, and social—and people of different backgrounds bring different talents to the table. By joining the Army, these people have agreed to adopt the Army culture. Army leaders make this easier by embracing and making use of everyone's talents. What's more, they create a team where subordinates know they are valuable and their talents are important.

2-63. You never know how the talents of an individual or group will contribute to mission accomplishment. For example, during World War II US Marines from the Navajo nation formed a group of radio communications specialists dubbed the Navajo Code Talkers. The code talkers used their native language—a unique talent—to handle command radio traffic. Not even the best Japanese code breakers could decipher what was being said.

2-64. Understanding the culture of your adversaries and of the country in which your organization is operating is just as important as understanding the culture of your own country and organization. This aspect of cultural awareness has always been important, but today's operational environment of frequent deployments—often conducted by small units under constant media coverage—makes it even more so. As an Army leader, you need to remain aware of current events—particularly those in areas where America has national interests. You may have to deal with people who live in those areas, either as partners, neutrals, or adversaries. The more you know about them, the better prepared you'll be.

2-65. You may think that understanding other cultures applies mostly to stability operations and support operations. However, it's critical to planning offensive and defensive operations as well. For example, you may employ different tactics against an adversary who considers surrender a dishonor worse than death than against those for whom surrender is an honorable option. Likewise, if your organization is operating as part of a multinational team, how well you understand your partners will affect how well the team accomplishes its mission.

2-66. Cultural awareness is crucial to the success of multinational operations. In such situations Army leaders take the time to learn the customs and traditions of the partners' cultures. They learn how and why others think and act as they do. In multinational forces, effective leaders create a "third culture," which is the bridge or the compromise among partners. This is what GA Eisenhower did in the following example.

GA Eisenhower Forms SHAEF

During World War II, one of GA Eisenhower's duties as Supreme Allied Commander in the European Theater of Operations (ETO) was to form his theater headquarters, the Supreme Headquarters, Allied Expeditionary Force (SHAEF). GA Eisenhower had to create an environment in this multinational headquarters in which staff members from the different Allied armies could work together harmoniously. It was one of GA Eisenhower's toughest jobs.

The forces under his command—American, British, French, Canadian, and Polish—brought not only different languages, but different ways of thinking, different ideas about what was important, and different strategies. GA Eisenhower could have tried to bend everyone to his will and his way of thinking; he was the boss, after all. But it's doubtful the Allies would have fought as well for a bullying commander or that a bullying commander would have survived politically. Instead, he created a positive organizational climate that made best use of the various capabilities of his subordinates. This kind of work takes tact, patience, and trust. It doesn't destroy existing cultures but creates a new one. (Chapter 7 discusses how building this coalition contributed to the Allied victory in the ETO.)

PHYSICAL ATTRIBUTES

2-67. Physical attributes—health fitness, physical fitness, and military and professional bearing—can be developed. Army leaders maintain the appropriate level of physical fitness and military bearing.

Health Fitness

Disease was the chief killer in the [American Civil] war. Two soldiers died of it for every one killed in battle...In one year, 995 of every thousand men in the Union army contracted diarrhea and dysentery.

Geoffrey C. Ward
The Civil War

2-68. Health fitness is everything you do to maintain good health, things such as undergoing routine physical exams, practicing good dental hygiene, maintaining deployability standards, and even personal grooming and cleanliness. A soldier unable to fight because of dysentery is as much a loss as one who's wounded. Healthy soldiers can perform under extremes in temperature, humidity, and other conditions better than unhealthy ones. Health fitness also includes avoiding things that degrade your health, such as substance abuse, obesity, and smoking.

Physical Fitness

Fatigue makes cowards of us all.

General George S. Patton Jr.
Commanding General, Third Army, World War II

2-69. Unit readiness begins with physically fit soldiers and leaders. Combat drains soldiers physically, mentally, and emotionally. To minimize those effects, Army leaders are physically fit, and they make sure their subordinates are fit as well. Physically fit soldiers perform better in all areas, and physically fit leaders are better able to think, decide, and act appropriately under pressure. Physical readiness provides a foundation for combat readiness, and it's up to you, the leader, to get your soldiers ready.

2-70. Although physical fitness is a crucial element of success in battle, it's not just for frontline soldiers. Wherever they are, people who are physically fit feel more competent and confident. That attitude reassures and inspires those around them. Physically fit soldiers and DA civilians can handle stress better, work longer and harder, and recover faster than ones who are not fit. These payoffs are valuable in both peace and war.

2-71. The physical demands of leadership positions, prolonged deployments, and continuous operations can erode more than just physical attributes. Soldiers must show up ready for deprivations because it's difficult to maintain high levels of fitness during deployments and demanding operations. Trying to get fit under those conditions is even harder. If a person isn't physically fit, the effects of additional stress snowball until their mental and emotional fitness are compromised as well. Army leaders' physical fitness has significance beyond their personal performance and well-being. Since leaders' decisions affect their organizations' combat effectiveness, health, and safety and not just their own, maintaining physical fitness is an ethical as well as a practical imperative.

2-72. The Army Physical Fitness Test (APFT) measures a baseline level of physical fitness. As an Army leader, you're required to develop a physical fitness program that enhances your soldiers' ability to complete soldier and leader tasks that support the unit's mission essential task list (METL). (FM 25-101 discusses METL-based integration of soldier, leader, and collective training.) Fitness programs that emphasize training specifically for the APFT are boring and don't prepare soldiers for the varied stresses of combat. Make every effort to design a physical fitness program that prepares your people for what you expect them to do in combat. Readiness should be your program's primary focus; preparation for the APFT itself is secondary. (FM 21-20 is your primary physical fitness resource.)

You have to lead men in war by requiring more from the individual than he thinks he can do. You have to [bring] them along to endure and to display qualities of fortitude that are beyond the average man's thought of what he should be expected to do. You have to

> *inspire them when they are hungry and exhausted and desperately uncomfortable and in great danger; and only a man of positive characteristics of leadership, with the physical stamina [fitness] that goes with it, can function under those conditions.*
>
> General of the Army George C. Marshall
> Army Chief of Staff, World War II

Military and Professional Bearing

Our...soldiers should look as good as they are.

Sergeant Major of the Army Julius W. Gates

2-73. As an Army leader, you're expected to look like a soldier. Know how to wear the uniform and wear it with pride at all times. Meet height and weight standards. By the way you carry yourself and through your military courtesy and appearance, you send a signal: I am proud of my uniform, my unit, and myself. Skillful use of your professional bearing—fitness, courtesy, and military appearance—can often help you manage difficult situations. A professional—DA civilian or soldier—presents a professional appearance, but there's more to being an Army professional than looking good. Professionals are competent as well; the Army requires you to both *look* good and *be* good.

EMOTIONAL ATTRIBUTES

Anyone can become angry—that is easy. But to be angry with the right person, to the right degree, at the right time, for the right purpose, and in the right way—that is not easy.

Aristotle
Greek philosopher and tutor to Alexander the Great

2-74. As an Army leader, your emotional attributes—self-control, balance, and stability—contribute to how you feel and therefore to how you interact with others. Your people are human beings with hopes, fears, concerns, and dreams. When you understand that will and endurance come from emotional energy, you possess a powerful leadership tool. The feedback you give can help your subordinates use their emotional energy to accomplish amazing feats in tough times.

Self-Control in Combat

An American infantry company in Vietnam had been taking a lot of casualties from booby traps. The soldiers were frustrated because they could not fight back. One night, snipers ambushed the company near a village, killing two soldiers. The rest of the company—scared, anguished, and frustrated—wanted to enter the village, but the commander—who was just as angry—knew that the snipers were long gone. Further, he knew that there was a danger his soldiers would let their emotions get the upper hand, that they might injure or kill some villagers out of a desire to strike back at something. Besides being criminal, such killings would drive more villagers to the Viet Cong. The commander maintained control of his emotions, and the company avoided the village.

2-75. Self-control, balance, and stability also help you make the right ethical choices. Chapter 4 discusses the steps of ethical reasoning. However, in order to follow those steps, you must remain in control of yourself; you can't be at the mercy of your impulses. You must remain calm under pressure, "watch your lane," and expend energy on things you can fix. Inform your boss of things you can't fix and don't worry about things you can't affect.

2-76. Leaders who are emotionally mature also have a better awareness of their own strengths and weaknesses. Mature leaders spend their energy on self-improvement; immature leaders spend their energy denying there's anything wrong. Mature, less defensive leaders benefit from constructive criticism in ways that immature people cannot.

Self-Control

Sure I was scared, but under the circumstances, I'd have been crazy not to be scared....There's nothing wrong with fear. Without fear, you can't have acts of courage.

Sergeant Theresa Kristek
Operation Just Cause, Panama

2-77. Leaders control their emotions. No one wants to work for a hysterical leader who might lose control in a tough situation. This doesn't mean you never show emotion. Instead, you must display the proper amount of emotion and passion—somewhere between too much and too little—required to tap into your subordinates' emotions. Maintaining self-control inspires calm confidence in subordinates, the coolness under fire so essential to a successful unit. It also encourages feedback from your subordinates that can expand your sense of what's really going on.

Balance

An officer or noncommissioned officer who loses his temper and flies into a tantrum has failed to obtain his first triumph in discipline.

Noncommissioned Officer's Manual, 1917

2-78. Emotionally balanced leaders display the right emotion for the situation and can also read others' emotional state. They draw on their experience and provide their subordinates the proper perspective on events. They have a range of attitudes—from relaxed to intense—with which to approach situations and can choose the one appropriate to the circumstances. Such leaders know when it's time to send a message that things are urgent and how to do that without throwing the organization into chaos. They also know how to encourage people at the toughest moments and keep them driving on.

Stability

Never let yourself be driven by impatience or anger. One always regrets having followed the first dictates of his emotions.

Marshal de Belle-Isle
French Minister of War, 1757-1760

2-79. Effective leaders are steady, levelheaded under pressure and fatigue, and calm in the face of danger. These characteristics calm their subordinates, who are always looking to their leader's example. Display the emotions you want your people to display; don't give in to the temptation to do what feels good for you. If you're under great stress, it might feel better to vent—scream, throw things, kick furniture—but that will not help the organization. If you want your subordinates to be calm and rational under pressure, you must be also.

BG Jackson at First Bull Run

At a crucial juncture in the First Battle of Bull Run, the Confederate line was being beaten back from Matthews Hill by Union forces. Confederate BG Thomas J. Jackson and his 2,000-man brigade of Virginians, hearing the sounds of battle to the left of their position, pressed on to the action. Despite a painful shrapnel wound, BG Jackson calmly placed his men in a defensive position on Henry Hill and assured them that all was well.

As men of the broken regiments flowed past, one of their officers, BG Barnard E. Bee, exclaimed to BG Jackson, "General, they are driving us!" Looking toward the direction of the enemy, BG Jackson replied, "Sir, we will give them the bayonet." Impressed by BG Jackson's confidence and self-control, BG Bee rode off towards what was left of the officers and men of his brigade. As he rode into the throng he gestured with his sword toward Henry Hill and shouted, "Look, men! There is Jackson standing like a stone wall! Let us determine to die here, and we will conquer! Follow me!"

BG Bee would later be mortally wounded, but the Confederate line stiffened and the nickname he gave to BG Jackson would live on in American military history. This example shows how one leader's self-control under fire can turn the tide of battle by influencing not only the leader's own soldiers, but the leaders and soldiers of other units as well.

FOCUS ON CHARACTER

Just as fire tempers iron into fine steel, so does adversity temper one's character into firmness, tolerance, and determination.

Margaret Chase Smith
Lieutenant Colonel, US Air Force Reserve
and United States Senator

2-80. Earlier in this chapter, you read how character is made up of two interacting sets of characteristics: values and attributes. People enter the Army with values and attributes they've developed over the course of a lifetime, but those are just the starting points for further character development. Army leaders continuously develop in themselves and their subordinates the Army values and leader attributes that this chapter discusses and Figure 1-1 shows. This isn't just an academic exercise, another mandatory training topic to address once a year. Your character shows through in your actions—on and off duty.

2-81. Character helps you determine what's right and motivates you to do it, regardless of the circumstances or the consequences. What's more, an informed ethical conscience consistent with Army values steels you for making the right choices when faced with tough questions. Since Army leaders seek to do what's right and inspire others to do the same, you must be concerned with character development. Examine the actions in this example, taken from the report of a platoon sergeant during Operation Desert Storm. Consider the aspects of character that contributed to them.

Character and Prisoners

The morning of [28 February 1991], about a half-hour prior to the cease-fire, we had a T-55 tank in front of us and we were getting ready [to engage it with a TOW]. We had the TOW up and we were tracking him and my wingman saw him just stop and a head pop up out of it. And Neil started calling me saying, "Don't shoot, don't shoot, I think they're getting off the tank." And they did. Three of them jumped off the tank and ran around a sand dune. I told my wingman, "I'll cover the tank, you go on down and check around the back side and see what's down there." He went down there and found about 150 PWs....

[T]he only way we could handle that many was just to line them up and run them through...a little gauntlet...[W]e had to check them for weapons and stuff and we lined them up and called for the PW handlers to pick them up. It was just amazing.

We had to blow the tank up. My instructions were to destroy the tank, so I told them to go ahead and move it around the back side of the berm a little bit to safeguard us, so we wouldn't catch any shrapnel or ammunition coming off. When the tank blew up, these guys started yelling and screaming at my soldiers, "Don't shoot us, don't shoot us," and one of my soldiers said, "Hey, we're from America; we don't shoot our prisoners." That sort of stuck with me.

2-82. The soldier's comment at the end of this story captures the essence of character. He said, "We're from America..." He defined, in a very simple way, the connection between who you are—your character—and what you do. This example illustrates character—shared values and attributes—telling soldiers what to do and what not to do. However, it's interesting for other reasons. Read it again: You can almost feel the soldiers' surprise when they realized what the Iraqi PWs were afraid of. You can picture the young soldier, nervous, hands on his weapon, but still managing to be a bit amused. The right thing, the ethical choice, was so deeply ingrained in those soldiers that it never occurred to them to do anything other than safeguard the PWs.

The Battle of the Bulge

In December 1944 the German Army launched its last major offensive on the Western Front of the ETO, sending massive infantry and armor formations into a lightly-held sector of the Allied line in Belgium. American units were overrun. Thousands of green troops, sent to that sector because it was quiet, were captured. For two desperate weeks the Allies fought to check the enemy advance. The 101st Airborne Division was sent to the town of Bastogne. The Germans needed to control the crossroads there to move equipment to the front; the 101st was there to stop them.

Outnumbered, surrounded, low on ammunition, out of medical supplies, and with wounded piling up, the 101st, elements of the 9th and 10th Armored Divisions, and a tank destroyer battalion fought off repeated attacks through some of the coldest weather Europe had seen in 50 years. Wounded men froze to death in their foxholes. Paratroopers fought tanks. Nonetheless, when the German commander demanded American surrender, BG Anthony C. McAuliffe, acting division commander, sent a one-word reply: "Nuts."

The Americans held. By the time the Allies regained control of the area and pushed the Germans back, Hitler's "Thousand Year Reich" had fewer than four months remaining.

2-83. BG McAuliffe spoke based on what he knew his soldiers were capable of, even in the most extreme circumstances. This kind of courage and toughness didn't develop overnight. Every Allied soldier brought a lifetime's worth of character to that battle; that character was the foundation for everything else that made them successful.

GA Eisenhower's Message

On 5 June 1944, the day before the D-Day invasion, with his hundreds of thousands of soldiers, sailors and airmen poised to invade France, GA Dwight D. Eisenhower took a few minutes to draft a message he hoped he would never deliver. It was a "statement he wrote out to have ready when the invasion was repulsed, his troops torn apart for nothing, his planes ripped and smashed to no end, his warships sunk, his reputation blasted."

In his handwritten statement, GA Eisenhower began, "Our landings in the Cherbourg-Havre area have failed to gain a satisfactory foothold and I have withdrawn the troops." Originally he had written, the "troops have been withdrawn," a use of the passive voice that conceals the actor. But he changed the wording to reflect his acceptance of full personal accountability.

GA Eisenhower went on, "My decision to attack at this time and place was based on the best information available." And after recognizing the courage and sacrifice of the troops he concluded, "If any blame or fault attaches to this attempt, it is mine alone."

2-84. GA Eisenhower, in command of the largest invasion force ever assembled and poised on the eve of a battle that would decide the fate of millions of people, was guided by the same values and attributes that shaped the actions of the soldiers in the Desert Storm example. His character allowed for nothing less than acceptance of total personal responsibility. If things went badly, he was ready to take the blame. When things went well, he gave credit to his subordinates. The Army values GA Eisenhower personified provide a powerful example for all members of the Army.

CHARACTER AND THE WARRIOR ETHOS

2-85. The *warrior ethos* refers to the professional attitudes and beliefs that characterize the American soldier. At its core, the warrior ethos grounds itself on the refusal to accept failure. The Army has forged the warrior ethos on training grounds from Valley Forge to the CTCs and honed it in battle from Bunker Hill to San Juan Hill, from the Meuse-Argonne to Omaha Beach, from Pork Chop Hill to the Ia Drang Valley, from Salinas Airfield to the Battle of 73 Easting. It derives from the unique realities of battle. It echoes through the precepts in the Code of Conduct. Developed through discipline, commitment to Army values, and knowledge of the Army's proud heritage, the warrior ethos makes clear that military service is much more than just another job: the purpose of winning the nation's wars calls for total commitment.

2-86. America has a proud tradition of winning. The ability to forge victory out of the chaos of battle includes overcoming fear, hunger, deprivation, and fatigue. The Army wins because it fights hard; it fights hard because it trains hard; and it trains hard because that's the way to *win*. Thus, the warrior ethos is about more than persevering under the worst of conditions; it fuels the fire to fight through those conditions to victory no matter how long it takes, no matter how much effort is required. It's one thing to make a snap decision to risk your life for a brief period of time. It's quite another to sustain the will to win when the situation looks hopeless and doesn't show any indications of getting better, when being away from home and family is a profound hardship. The soldier who jumps on a grenade to save his comrades is courageous, without question. That action requires great physical courage, but pursuing victory over time also requires a deep moral courage that concentrates on the mission.

2-87. The warrior ethos concerns character, shaping who you are and what you do. In that sense, it's clearly linked to Army values such as *personal courage, loyalty to comrades, and*

dedication to duty. Both loyalty and duty involve putting your life on the line, even when there's little chance of survival, for the good of a cause larger than yourself. That's the clearest example of *selfless service*. American soldiers never give up on their fellow soldiers, and they never compromise on doing their duty. *Integrity* underlies the character of the Army as well. The warrior ethos requires unrelenting and consistent determination to do what is right and to do it with pride, both in war and military operations other than war. Understanding what is right requires *respect* for both your comrades and other people involved in such complex arenas as peace operations and nation assistance. In such ambiguous situations, decisions to use lethal or nonlethal force severely test judgment and discipline. In whatever conditions Army leaders find themselves, they turn the personal warrior ethos into a collective commitment to win with *honor*.

2-88. The warrior ethos is crucial—and perishable—so the Army must continually affirm, develop, and sustain it. Its martial ethic connects American warriors today with those whose sacrifices have allowed our very existence. The Army's continuing drive to be the best, to triumph over all adversity, and to remain focused on mission accomplishment does more than preserve the Army's institutional culture; it sustains the nation.

2-89. Actions that safeguard the nation occur everywhere you find soldiers. The warrior ethos spurs the lead tank driver across a line of departure into uncertainty. It drives the bone-tired medic continually to put others first. It pushes the sweat-soaked gunner near muscle failure to keep up the fire. It drives the heavily loaded infantry soldier into an icy wind, steadily uphill to the objective. It presses the signaler through fatigue to provide communications. And the warrior ethos urges the truck driver across frozen roads bounded by minefields because fellow soldiers at an isolated outpost need supplies. Such tireless motivation comes in part from the comradeship that springs from the warrior ethos. Soldiers fight for each other; they would rather die than let their buddies down. That loyalty runs front to rear as well as left to right: mutual support marks Army culture regardless of who you are, where you are, or what you are doing.

2-90. That tight fabric of loyalty to one another and to collective victory reflects perhaps the noblest aspect of our American warrior ethos: the military's subordinate relationship to civilian authority. That subordination began in 1775, was reconfirmed at Newburgh, New York, in 1782, and continues to this day. It's established in the Constitution and makes possible the freedom all Americans enjoy. The Army sets out to achieve national objectives, not its own, for *selfless service* is an institutional as well as an individual value. And in the end, the Army returns its people back to the nation. America's sons and daughters return with their experience as part of a winning team and share that spirit as citizens. The traditions and values of the service derive from a commitment to excellent performance and operational success. They also point to the Army's unwavering commitment to the society we serve. Those characteristics serve America and its citizens—both in and out of uniform—well.

CHARACTER DEVELOPMENT

2-91. People come to the Army with a character formed by their background, religious or philosophical beliefs, education, and experience. Your job as an Army leader would be a great deal easier if you could check the values of a new DA civilian or soldier the way medics check teeth or run a blood test. You could figure out what values were missing by a quick glance at Figure 1-1 and administer the right combination, maybe with an injection or magic pill.

2-92. But character development is a complex, lifelong process. No scientist can point to a person and say, "This is when it all happens." However, there are a few things you can count on. You build character in subordinates by creating organizations in which Army values are not just words in a book but precepts for what their members do. You help build subordinates' character by acting the way you want them to act. You teach by example, and coach along the way. (Appendix E contains additional information on character development.) When you hold yourself and your subordinates to the highest standards, you reinforce the values those standards embody. They spread throughout the team, unit, or organization—throughout the Army—like the waves from a pebble dropped into a pond.

CHARACTER AND ETHICS

2-93. When you talk about character, you help your people answer the question, What kind of person should I be? You must not only embrace Army values and leader attributes but also use them to think, reason, and—after reflection—act. Acting in a situation that tests your character requires moral courage. Consider this example.

The Qualification Report

A battalion in a newly activated division had just spent a great deal of time and effort on weapons qualification. When the companies reported results, the battalion commander could not understand why B and C Companies had reported all machine gunners fully qualified while A Company had not. The A Company Commander said that he could not report his gunners qualified because they had only fired on the 10-meter range and the manual for qualification clearly stated that the gunners had to fire on the transition range as well. The battalion commander responded that since the transition range was not built yet, the gunners should be reported as qualified: "They fired on the only range we have. And besides, that's how we did it at Fort Braxton."

Some of the A Company NCOs, who had also been at Fort Braxton, tried to tell their company commander the same thing. But the captain insisted the A Company gunners were not fully qualified, and that's how the report went to the brigade commander.

The Qualification Report (continued)

The brigade commander asked for an explanation of the qualification scores. After hearing the A Company Commander's story, he agreed that the brigade would be doing itself no favors by reporting partially qualified gunners as fully qualified. The incident also sent a message to division: get that transition range built.

The A Company Commander's choice was not between loyalty to his battalion commander and honesty; doing the right thing here meant being loyal and honest. And the company commander had the moral courage to be both honest and loyal—loyal to the Army, loyal to his unit, and loyal to his soldiers.

2-94. The A Company Commander made his decision and submitted his report without knowing how it would turn out. He didn't know the brigade commander would back him up, but he reported his company's status relative to the published Army standard anyway. He insisted on reporting the truth—which took character—because it was the right thing to do.

2-95. Character is important in living a consistent and moral life, but character doesn't always provide the final answer to the specific question, What should I do now? Finding that answer can be called ethical reasoning. Chapter 4 outlines a process for ethical reasoning. When you read it, keep in mind that the process is much more complex than the steps indicate and that you must apply your own values, critical reasoning skills, and imagination to the situation. There are no formulas that will serve every time; sometimes you may not even come up with an answer that completely satisfies you. But if you embrace Army values and let them govern your actions, if you learn from your experiences and develop your skills over time, you're as prepared as you can be to face the tough calls.

2-96. Some people try to set different Army values against one another, saying a problem is about loyalty versus honesty or duty versus respect. Leadership is more complicated than that; the world isn't always black and white. If it were, leadership would be easy and anybody could do it. However, in the vast majority of cases, Army values are perfectly compatible; in fact, they reinforce each other.

CHARACTER AND ORDERS

2-97. Making the right choice and acting on it when faced with an ethical question can be difficult. Sometimes it means standing your ground. Sometimes it means telling your boss you think the boss is wrong, like the finance supervisor in Chapter 1 did. Situations like these test your character. But a situation in which you think you've received an illegal order can be even more difficult.

2-98. In Chapter 1 you read that a good leader executes the boss's decision with energy and enthusiasm. The only exception to this principle is your duty to disobey illegal orders. This isn't a privilege you can conveniently claim, but a duty you must perform. If you think an order is illegal, first be sure that you understand both the details of the order and its original intent. Seek clarification from the person who gave the order. This takes moral courage, but the question will be straightforward: Did you really mean for me to...steal the part...submit a false report...shoot the prisoners? If the question is complex or time permits, always seek legal counsel. However, if you must decide immediately—as may happen in the heat of combat make the best judgment possible based on Army values, your experience, and your previous study and reflection. You take a risk when you disobey what you believe to be an illegal order. It may be the most difficult decision you'll ever make, but that's what leaders do.

2-99. While you'll never be completely prepared for such a situation, spending time reflecting on Army values and leader attributes may help. Talk to your superiors, particularly those who have done what you aspire to do or what you think you'll be called on to do; providing counsel of this sort is an important part of mentoring (which Chapter 5 discusses). Obviously, you need to make time to do this before you're faced with a tough call. When you're in the middle of a firefight, you don't have time to reflect.

CHARACTER AND BELIEFS

2-100. What role do beliefs play in ethical matters? Beliefs are convictions people hold as true; they are based on their upbringing, culture, heritage, families, and traditions. As a result, different moral beliefs have been and will continue to be shaped by diverse religious and philosophical traditions. You serve a nation that takes very seriously the notion that people are free to choose their own beliefs and the basis for those beliefs. In fact, America's strength comes from that diversity. The Army respects different moral backgrounds and personal convictions—as long as they don't conflict with Army values.

2-101. Beliefs matter because they are the way people make sense of what they experience. Beliefs also provide the basis for personal values; values are moral beliefs that shape a person's behavior. Effective leaders are careful not to require their people to violate their beliefs by ordering or encouraging any illegal or unethical action.

2-102. The Constitution reflects our deepest national values. One of these values is the guarantee of freedom of religion. While religious beliefs and practices are left to individual conscience, Army leaders are responsible for ensuring their soldiers' right to freely practice their religion. Title 10 of the United States Code states, "Each commanding officer shall furnish facilities, including necessary transportation, to any chaplain assigned to his command, to assist the chaplain in performing his duties." What does this mean for Army leaders? The commander delegates staff responsibility to the chaplain for programs to enhance spiritual fitness since many people draw moral fortitude and inner strength from a spiritual foundation. At the same time, no leader may apply undue influence or coerce others in matters of religion—whether to practice or not to practice specific religious beliefs. (The first ten amendments to the Constitution are called the Bill of Rights. Freedom of religion is guaranteed by the First Amendment, an indication of how important the Founders considered it. You can read the Bill of Rights in Appendix F.)

2-103. Army leaders also recognize the role beliefs play in preparing soldiers for battle. Soldiers often fight and win over tremendous odds when they are convinced of the ideals (beliefs) for which they are fighting. Commitment to such beliefs as justice, liberty, freedom, and not letting down your fellow soldier can be essential ingredients in creating and sustaining the will to fight and prevail. A common theme expressed by American PWs during the Vietnam Conflict was the importance of values instilled by a common American culture. Those values helped them to withstand torture and the hardships of captivity.

SECTION II

COMPETENCE: WHAT A LEADER MUST KNOW

The American soldier...demands professional competence in his leaders. In battle, he wants to know that the job is going to be done right, with no unnecessary casualties. The noncommissioned officer wearing the chevron is supposed to be the best soldier in the platoon and he is supposed to know how to perform all the duties expected of him. The American soldier expects his sergeant to be able to teach him how to do his job. And he expects even more from his officers.

General of the Army Omar N. Bradley

2-104. Army values and leader attributes form the foundation of the character of soldiers and

DA civilians. Character, in turn, serves as the basis of knowing (competence) and doing (leadership). The self-discipline that leads to teamwork is rooted in character. In the Army, teamwork depends on the actions of competent leaders of proven character who know their profession and act to improve their organizations. The best Army leaders constantly strive to improve, to get better at what they do. Their self-discipline focuses on learning more about their profession and continually getting the team to perform better. They build competence in themselves and their subordinates. Leader skills increase in scope and complexity as one moves from direct leader positions to organizational and strategic leader positions. Chapters 4, 6, and 7 discuss in detail the different skills direct, organizational, and strategic leaders require.

2-105. Competence results from hard, realistic training. That's why Basic Training starts with simple skills, such as drill and marksmanship. Soldiers who master these skills have a couple of victories under their belts. The message from the drill sergeants—explicit or not—is, "You've learned how to do those things; now you're ready to take on something tougher." When you lead people through progressively more complex tasks this way, they develop the confidence and will—the inner drive—to take on the next, more difficult challenge.

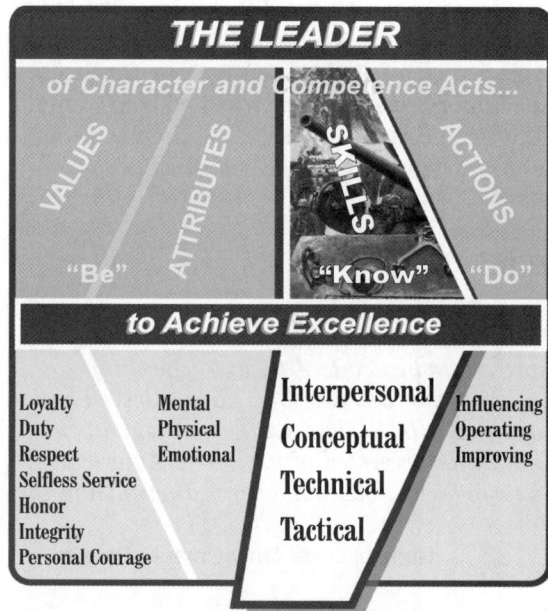

Figure 2-3. Leader Skills

2-106. For you as an Army leader, competence means much more than being well-trained. Competence links character (knowing the right thing to do) and leadership (doing or influencing your people to do the right thing). Leaders are responsible for being personally competent, but even that isn't enough: as a leader, you're responsible for your subordinates' competence as well.

2-107. Figure 2-3 highlights the four categories containing skills an Army leader must KNOW:

- **Interpersonal skills** affect how you deal with people. They include coaching, teaching, counseling, motivating, and empowering.

- **Conceptual skills** enable you to handle ideas. They require sound judgment as well as the ability to think creatively and reason analytically, critically, and ethically.

- **Technical skills** are job-related abilities. They include basic soldier skills. As an Army leader, you must possess the expertise necessary to accomplish all tasks and functions you're assigned.

- **Tactical skills** apply to solving tactical problems, that is, problems concerning employment of units in combat. You enhance tactical skills when you combine them with interpersonal, conceptual, and technical skills to accomplish a mission.

2-108. Leaders in combat combine interpersonal, conceptual, technical, and tactical skills to accomplish the mission. They use their interpersonal skills to communicate their intent effectively and motivate their soldiers. They apply their conceptual skills to determine viable concepts of operations, make the right decisions, and execute the tactics the operational environment requires. They capitalize on their technical skills to properly employ the techniques, procedures, fieldcraft, and equipment that fit the situation. Finally, combat leaders employ tactical skill, combining skills from the other skill categories with knowledge of the art of tactics appropriate to their level of responsibility and unit type to accomplish the mission. When plans go wrong and leadership must turn the tide, it is tactical skill, combined with character, that enables an Army leader to seize control of the situation and lead the unit to mission accomplishment.

2-109. The Army leadership framework draws a distinction between developing skills and performing actions. Army leaders who take their units to a combat training center (CTC) improve their skills by performing actions—by doing their jobs on the ground in the midst of intense simulated combat. But they don't wait until they arrive at the CTC to develop their skills; they practice ahead of time in command post exercises, in combat drills, on firing ranges, and even on the physical training (PT) field.

2-110. Your leader skills will improve as your experience broadens. A platoon sergeant gains valuable experience on the job that will help him be a better first sergeant. Army leaders take advantage of every chance to improve: they look for new learning opportunities, ask questions, seek training opportunities, and request performance critiques.

SECTION III

LEADERSHIP: WHAT A LEADER MUST DO

He gets his men to go along with him because they want to do it for him and they believe in him.

General of the Army Dwight D. Eisenhower

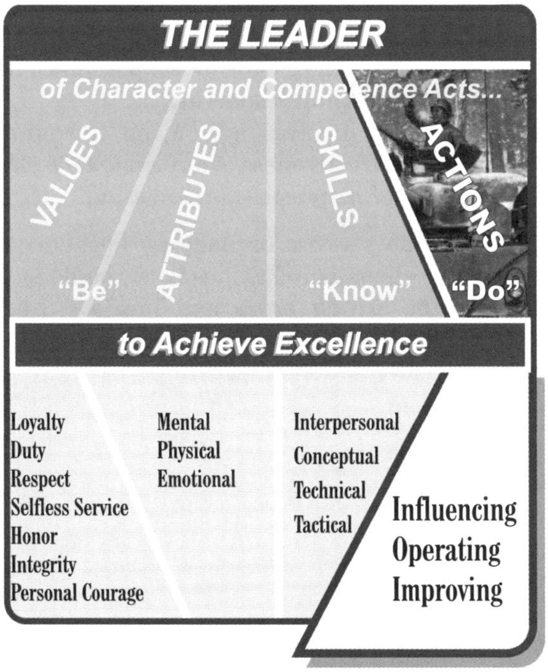

Figure 2-4. Leader Actions

2-111. Leaders act. They bring together everything they are, everything they believe, and everything they know how to do to provide purpose, direction, and motivation. Army leaders work to influence people, operate to accomplish the mission, and act to improve their organization. This section introduces leader actions. Chapters 5, 6, and 7 discuss them more fully. As with leader skills, leader actions increase in scope and complexity as you move from direct leader positions to organizational and strategic leader positions.

2-112. Developing the right values, attributes, and skills is only preparation to lead. Leadership doesn't begin until you act. Leaders who live up to Army values, who display leader attributes, who are competent, who act at all times as they would have their people act, will succeed. Leaders who talk a good game but can't back their words with actions will fail in the long run.

INFLUENCING

2-113. Army leaders use interpersonal skills to guide others toward a goal. Direct leaders most often influence subordinates face to face—such as when a team leader gives instructions, recognizes achievement, and encourages hard work. Organizational and strategic leaders also influence their immediate subordinates and staff face to face; however, they guide their organizations primarily by indirect influence. Squad leaders, for example, know what their division commander wants, not because the general has briefed each one personally, but because his intent is passed through the chain of command. Influencing actions fall into these categories:

- **Communicating** involves displaying good oral, written, and listening skills for individuals and groups.
- **Decision making** involves selecting the line of action intended to be followed as the one most favorable to the successful accomplishment of the mission. This involves using sound judgment, reasoning logically, and managing resources wisely.
- **Motivating** involves inspiring and guiding others toward mission accomplishment.

OPERATING

2-114. Operating is what you do to accomplish the immediate mission, to get the job done on time and to standard. Operating actions fall into these categories:

- **Planning and preparing** involve developing detailed, executable plans that are feasible, acceptable, and suitable; arranging unit support for the exercise or operation; and conducting rehearsals. During tactical operations, decision making and planning are enhanced by two methodologies: the military decision making process (MDMP) and the troop leading procedures (TLP). Battalion and higher echelons follow the MDMP. Company and lower echelons follow the TLP. (FM 101-5 discusses the MDMP.)
- **Executing** involves meeting mission standards, taking care of people, and efficiently managing resources.
- **Assessing** involves evaluating the efficiency and effectiveness of any system or plan in terms of its purpose and mission.

2-115. Leaders assess, or judge, performance so they can determine what needs to be done to sustain the strong areas and improve weak ones. This kind of forward thinking is linked to the last leader action, improving.

IMPROVING

2-116. Good leaders strive to leave an organization better than they found it. A child struggling to understand why it is better to put money in a piggy bank is learning what leaders know: plan and sacrifice now for the sake of the future. All leaders are tempted to focus on the short-term gain that makes them and their organizations look good today: "Why bother to fix it now? By the time next year rolls around, it will be someone else's problem." But that attitude doesn't serve either your subordinates or the Army well. When an organization sacrifices important training with long-term effects—say, training that leads to true marksmanship skill—and focuses exclusively on short-term appearances—such as qualification scores—the organization's capabilities suffers.

2-117. The results of shortsighted priorities may not appear immediately, but they will appear. Loyalty to your people as well as the Army as an institution demands you consider the long-term effects of your actions. Some of your people will remain in the organization after you've moved on. Some will still be in the Army after you're long gone. Soldiers and DA civilians tomorrow must live with problems leaders don't fix today.

2-118. Army leaders set priorities and balance competing demands. They focus their organizations' efforts on short- and long-term goals while continuing to meet requirements that may or may not contribute directly to achieving those goals. In the case of weapons proficiency, qualification is a requirement but true marksmanship skill is the goal. For battlefield success, soldiers need training that leads to understanding and mastery of technical and tactical skills that hold up under the stress of combat. Throw in all the other things vying for an organization's time and resources and your job becomes even more difficult. Guidance from higher headquarters may help, but you must make the tough calls. Improving actions fall into these categories:

- **Developing** involves investing adequate time and effort to develop individual subordinates as leaders. It includes mentoring.
- **Building** involves spending time and resources to improve teams, groups, and units and to foster an ethical climate.
- **Learning** involves seeking self-improvement and organizational growth. It includes envisioning, adapting, and leading change.

SUMMARY

2-119. As an Army leader, leadership in combat is your primary and most important challenge. It requires you to accept a set of values that contributes to a core of motivation and will. If you fail to accept and live these Army values, your soldiers may die unnecessarily and you may fail to accomplish your mission.

2-120. What must you, as an Army leader, BE, KNOW, and DO? You must have character, that combination of values and attributes that underlie your ability to see what needs to be done, decide to do it, and influence others to follow you. You must be competent, that is, possess the knowledge and skills required to do your job right. And you must lead, take the proper actions to accomplish the mission based

on what your character tells you is ethically right and appropriate for the situation.

2-121. Leadership in combat, the greatest challenge, requires a basis for your motivation and will. That foundation is Army values. In them are rooted the basis for the character and self-discipline that generate the will to succeed and the motivation to persevere. From this motivation derives the lifelong work of self-development in the skills that make a successful Army leader, one who walks the talk of BE, KNOW, DO. Chapter 3 examines the environment that surrounds your people and how what you do as a leader affects it. Understanding the human dimension is essential to mastering leader skills and performing leader actions.

Chapter 3

The Human Dimension

All soldiers are entitled to outstanding leadership; I will provide that leadership. I know my soldiers and I will always place their needs above my own. I will communicate consistently with my soldiers and never leave them uninformed.

Creed of the Noncommissioned Officer

3-1. Regardless of the level, keep in mind one important aspect of leadership: you lead people. In the words of former Army Chief of Staff Creighton W. Abrams,

The Army is not made up of people; the Army is people…living, breathing, serving human beings. They have needs and interests and desires. They have spirit and will, strengths and abilities. They have weaknesses and faults, and they have means. They are the heart of our preparedness…and this preparedness—as a nation and as an Army—depends upon the spirit of our soldiers. It is the spirit that gives the Army…life. Without it we cannot succeed.

3-2. GEN Abrams could not have been more clear about what's important. To fully appreciate the human dimension of leadership, you must understand two key elements: *leadership* itself and the *people* you lead. Leadership—what this manual is about—is far from an exact science; every person and organization is different. Not only that, the environment in which you lead is shaped first by who you are and what you know; second, by your people and what they know; and third, by everything that goes on around you.

3-3. This chapter examines this all-important human dimension. Later chapters discuss the levels of Army leadership and the skills and actions required of leaders at each level.

PEOPLE, THE TEAM, AND THE INSTITUTION

3-4. Former Army Chief of Staff John A. Wickham Jr. described the relationship between the people who are the Army and the Army as an institution this way:

The Army is an institution, not an occupation. Members take an oath of service to the nation and the Army, rather than simply accept a job…the Army has moral and ethical obligations to those who serve and their families; they, correspondingly, have responsibilities to the Army.

3-5. The Army has obligations to soldiers, DA civilians, and their families that most organizations don't have; in return, soldiers and DA civilians have responsibilities to the Army that far exceed those of an employee to most employers. This relationship, one of mutual obligation and responsibility, is at the very center of what

makes the Army a team, an institution rather than an occupation.

3-6. Chapter 2 discussed how the Army can't function except as a team. This team identity doesn't come about just because people take an oath or join an organization; you can't force a team to come together any more than you can force a plant to grow. Rather, the team identity comes out of mutual respect among its members and a trust between leaders and subordinates. That bond between leaders and subordinates likewise springs from mutual respect as well as from discipline. The highest form of discipline is the willing obedience of subordinates who trust their leaders, understand and believe in the mission's purpose, value the team and their place in it, and have the will to see the mission through. This form of discipline produces individuals and teams who—in the really tough moments—come up with solutions themselves.

Soldiers Are Our Credentials

In September 1944 on the Cotentin Peninsula in France, the commander of a German stronghold under siege by an American force sent word that he wanted to discuss surrender terms. German MG Hermann Ramcke was in his bunker when his staff escorted the assistant division commander of the US 8th Infantry Division down the concrete stairway to the underground headquarters. MG Ramcke addressed BG Charles D. W. Canham through an interpreter: "I am to surrender to you. Let me see your credentials." Pointing to the dirty, tired, disheveled—but victorious—American infantrymen who had accompanied him and were now crowding the dugout entrance, the American officer replied, "These are my credentials."

DISCIPLINE

I am confident that an army of strong individuals, held together by a sound discipline based on respect for personal initiative and rights and dignity of the individual, will never fail this nation in time of need.

General J. Lawton Collins
Former Army Chief of Staff

3-7. People are our most important resource; soldiers are in fact our "credentials." Part of knowing how to use this most precious resource is understanding the stresses and demands that influence people.

3-8. One sergeant major has described discipline as "a moral, mental, and physical state in which all ranks respond to the will of the [leader], whether he is there or not." Disciplined people take the right action, even if they don't feel like it. True discipline demands habitual and reasoned obedience, an obedience that preserves initiative and works, even when the leader isn't around. Soldiers and DA civilians who understand the purpose of the mission, trust the leader, and share Army values will do the right thing because they're truly committed to the organization.

3-9. Discipline doesn't just mean barking orders and demanding an instant response—it's more complex than that. You build discipline by training to standard, using rewards and punishment judiciously, instilling confidence in and building trust among team members, and creating a knowledgeable collective will. The confidence, trust, and collective will of a disciplined, cohesive unit is crucial in combat.

3-10. You can see the importance of these three characteristics in an example that occurred during the 3 October 1993 American raid in Somalia. One soldier kept fighting despite his wounds. His comrades remembered that he seemed to stop caring about himself, that he had to keep fighting because the other guys—his buddies—were all that mattered. When things go badly, soldiers draw strength from their own and their unit's discipline; they know that other members of the team are depending on them.

3-11. Soldiers—like those of Task Force Ranger in Somalia (which you'll read about later in this chapter) and SGT Alvin York (whose story is in Chapter 5)—persevere in tough situations. They fight through because they have confidence in themselves, their buddies, their leaders, their equipment, and their training—and because they have discipline and will. A young sergeant who participated in Operation Uphold Democracy in Haiti in 1994 asserted this fact when interviewed by the media. The soldier said that operations went well because his unit did things just the way they did them in training and that his training never let him down.

3-12. Even in the most complex operations, the performance of the Army comes down to the training and disciplined performance of individuals and teams on the ground. One example of this fact occurred when a detachment of American soldiers was sent to guard a television tower in Udrigovo, Bosnia-Herzegovina.

3-13. After the soldiers had assumed their posts, a crowd of about 100 people gathered, grew to about 300, and began throwing rocks at the Americans. However, the soldiers didn't overreact. They prevented damage to the tower without creating an international incident. There was no "Boston Massacre" in Udrigovo. The discipline of American soldiers sent into this and other highly volatile situations in Bosnia kept the lid on that operation. The bloody guerrilla war predicted by some didn't materialize. This is a testament to the professionalism of today's American soldiers—your soldiers—and the quality of their leaders—you.

MORALE

NSDQ [Night Stalkers Don't Quit]

Motto of the 160th Special Operations
Aviation Regiment, "The Night Stalkers"
Message sent by Chief Warrant Officer Mike Durant,
held by Somali guerrillas, to his wife, October 1993

3-14. When military historians discuss great armies, they write about weapons and equipment, training and the national cause. They may mention sheer numbers (Voltaire said, "God is always on the side of the heaviest battalions") and all sorts of other things that can be analyzed, measured, and compared. However, some also write about another factor equally important to success in battle, something that can't be measured: the emotional element called morale.

3-15. Morale is the human dimension's most important intangible element. It's a measure of how people feel about themselves, their team, and their leaders. High morale comes from good leadership, shared hardship, and mutual respect. It's an emotional bond that springs from common values like loyalty to fellow soldiers and a belief that the organization will care for families. High morale results in a cohesive team that enthusiastically strives to achieve common goals. Leaders know that morale, the essential human element, holds the team together and keeps it going in the face of the terrifying and dispiriting things that occur in war.

You have a comradeship, a rapport that you'll never have again...There's no competitiveness, no money values. You trust the man on your left and your right with your life.

Captain Audie Murphy
Medal of Honor recipient and most decorated
American soldier of World War II

TAKING CARE OF SOLDIERS

Readiness is the best way of truly taking care of soldiers.

Former Sergeant Major of the Army
Richard A. Kidd

3-16. Sending soldiers in harm's way, into places where they may be killed or wounded, might seem to contradict all the emphasis on taking care of soldiers. Does it? How can you truly care for your comrades and send them on missions that might get them killed? Consider this important and fundamental point as you read the next few paragraphs.

3-17. Whenever the talk turns to what leaders do, you'll almost certainly hear someone say, "Take care of your soldiers." And that's good advice. In fact, if you add one more clause, "Accomplish the mission _and_ take care of your soldiers," you have guidance for a career. But "taking care of soldiers" is one of those slippery phrases, like the word "honor," that lots of people talk about but few take the trouble to explain. So what does taking care of soldiers mean?

3-18. Taking care of soldiers means creating a disciplined environment where they can learn and grow. It means holding them to high standards, training them to do their jobs so they can function in peace and win in war. You take care of soldiers when you treat them fairly, refuse to cut corners, share their hardships, and set the example. Taking care of soldiers encompasses everything from making sure a soldier has time for an annual dental exam to visiting off-post housing to make sure it's adequate. It also means providing the family support that assures soldiers their families will be taken care of, whether the soldier is home or deployed. Family support means ensuring there's a support group in place, that even the most junior soldier and most inexperienced family members know where to turn for help when their soldier is deployed.

3-19. Taking care of soldiers also means demanding that soldiers do their duty, even at the risk of their lives. It doesn't mean coddling them or making training easy or comfortable. In fact, that kind of training can get soldiers killed. Training must be rigorous and as much like combat as is possible while being safe. Hard training is one way of preparing soldiers for the rigors of combat. Take care of soldiers by giving them the training, equipment, and support they need to keep them alive in combat.

3-20. In war, soldiers' comfort is important because it affects morale and combat effectiveness, but comfort takes a back seat to the mission. Consider this account of the 1944 landings on the island of Leyte in the Philippines, written more than 50 years later by Richard Gerhardt. Gerhardt, who was an 18-year-old rifleman in the 96th Infantry Division, survived two amphibious landings and months of close combat with the Japanese.

The 96th Division on Leyte

By the time we reached the beach, the smoke and dust created by the preparation fire had largely dissipated and we could see the terrain surrounding the landing area, which was flat and covered with some underbrush and palm trees. We were fortunate in that our sector of the beach was not heavily defended, and in going ashore there were few casualties in our platoon. Our company was engaged by small arms fire and a few mortar rounds, but we were able to move forward and secure the landing area in short order. Inland from the beach, however, the terrain turned into swamps, and as we moved ahead it was necessary to wade through muck and mud that was knee-deep at times....Roads in this part of the island were almost nonexistent, with the area being served by dirt trails around the swamps, connecting the villages....The Japanese had generally backed off the beaches and left them lightly defended, setting up their defense around certain villages which were at the junctions of the road system, as well as dug-in positions at points along the roads and trails. Our strategy was to...not use the roads and trails, but instead to move through the swamps and rice paddies and attack the enemy strong points from directions not as strongly defended. This was slow, dirty, and extremely fatiguing, but by this tactic we reduced our exposure to the enemy defensive plan, and to heavy fire from their strong points. It must be recognized that in combat the comfort of the front-line troops isn't part of the...planning process, but only what they can endure and still be effective. Conditions that seriously [affect] the combat efficiency of the troops then become a factor.

3-21. Gerhardt learned a lifetime's worth of lessons on physical hardship in the Pacific. Mud, tropical heat, monsoon rains, insects, malaria, Japanese snipers, and infiltrators—the details are still clear in his mind half a century later. Yet he knows—and he tells you—that soldiers must endure physical hardship when the best plan calls for it. In the Leyte campaign, the best plan was extremely difficult to execute, but it was tactically sound and it saved lives.

3-22. This concept doesn't mean that leaders sit at some safe, dry headquarters and make plans without seeing what their soldiers are going through, counting on them to tough out any situation. Leaders know that graphics on a map symbolize soldiers going forward to fight. Leaders get out with the soldiers to see and feel what they're experiencing as well as to influence the battle by their presence. (Gerhardt and numerous other front-line writers refer to the rear echelon as "anything behind my foxhole.") Leaders who stay a safe distance from the front jeopardize operations because they don't know what's going on. They risk destroying their soldiers' trust, not to mention their unit.

The K Company Visit

1LT Harold Leinbaugh, commander of K Company, 333d Infantry Regiment, 84th Division,related this experience from the ETO in January,1945, during the coldest winter in Europe in nearly 50 years:

On a front-line visit, the battalion commander criticized 1LT Leinbaugh and CPT Jay Prophet, the A Company Commander, for their own and their men's appearance. He said it looked like no one had shaved for a week. 1LT Leinbaugh replied that there was no hot water. Sensing a teaching moment, the colonel responded: "Now if you men would save some of your morning coffee it could be used for shaving." Stepping over to a snowbank, 1LT Leinbaugh picked up a five-gallon GI [general issue] coffee can brought up that morning, and shook it in the colonel's face. The frozen coffee produced a thunk. 1LT Leinbaugh shook it again.

"That's enough," said the colonel,"…I can hear."

3-23. This example illustrates three points:

- The importance of a leader going to where the action is to see and feel what's really going on.

- The importance of a first-line leader telling the boss something he doesn't want to hear.

- The importance of a leader accepting information that doesn't fit his preconceived notions.

3-24. Soldiers are extremely sensitive to situations where their leaders are not at risk, and they're not likely to forget a mistake by a leader they haven't seen. Leaders who are out with their soldiers—in the same rain or snow, under the same blazing sun or in the same dark night, under the same threat of enemy artillery or small arms fire—will not fall into the trap of ignorance. Those who lead from the front can better motivate their soldiers to carry on under extreme conditions.

3-25. Taking care of soldiers is every leader's business. A DA civilian engineering team chief volunteered to oversee the installation of six Force Provider troop life support systems in the vicinity of Tuzla, Bosnia-Herzogovina. Using organizational skills, motivational techniques, and careful supervision, the team chief ensured that the sites were properly laid out, integrated, and installed. As a result of thorough planning and the teamwork the DA civilian leader generated, the morale and quality of life of over 5,000 soldiers were significantly improved.

COMBAT STRESS

All men are frightened. The more intelligent they are, the more they are frightened. The courageous man is the man who forces himself, in spite of his fear, to carry on.

General George S. Patton Jr.
War As I Knew It

3-26. Leaders understand the human dimension and anticipate soldiers' reactions to stress, especially to the tremendous stress of combat. The answers may look simple as you sit somewhere safe and read this manual, but be sure easy answers don't come in combat. However, if you think about combat stress and its effects on you and your soldiers ahead of time, you'll be less surprised and better prepared to deal with and reduce its effects. It takes mental discipline to imagine the unthinkable—the plan going wrong, your soldiers wounded or dying, and the enemy coming after YOU. But in combat all of these things can happen, and your soldiers expect you, their leader, to have thought through each of them. Put yourself in the position of the squad leader in the following example.

Task Force Ranger in Somalia, 1993

"Sarge" was a company favorite, a big powerful kid from New Jersey who talked with his hands and played up his "Joy-zee" accent. He loved practical jokes. One of his favorites was to put those tiny charges in guys' cigarettes, the kind that would explode with a loud "POP!" about halfway through a smoke. If anyone else had done it, it would have been annoying; Sarge usually got everyone to laugh—even the guy whose cigarette he destroyed.

During the 3 October 1993 raid in Mogadishu, Sarge was manning his Humvee's .50 cal when he was hit and killed. The driver and some of the guys in back screamed, "He's dead! He's dead!" They panicked and were not responding as their squad leader tried to get someone else up and behind the gun. The squad leader had to yell at them, "Just calm down! We've got to keep fighting or none of us will get back alive."

3-27. Consider carefully what the squad leader did. First he told his squad to calm down. Then he told them why it was important: they had to continue the fight if they wanted to make it back to their base alive. In this way he jerked his soldiers back to a conditioned response, one that had been drilled during training and that took their minds off the loss. The squad leader demonstrated the calm, reasoned leadership under stress that's critical to mission success. In spite of the loss, the unit persevered.

WILL AND WINNING IN BATTLE

3-28. The Army's ultimate responsibility is to win the nation's wars. And what is it that carries soldiers through the terrible challenges of combat? It's the will to win, the ability to gut it out when things get really tough, even when things look hopeless. It's the will not only to persevere but also to find workable solutions to the toughest problems. This drive is part of the warrior ethos, the ability to forge victory out of the chaos of battle—to overcome fear, hunger, deprivation, and fatigue and accomplish the mission. And the will to win serves you just as well in peacetime, when it's easy to become discouraged, feel let down, and spend your energy complaining instead of using your talents to make things better. Discipline holds a team together; the warrior ethos motivates its members—you and your people—to continue the mission.

3-29. All soldiers are warriors: all need to develop and display the will to win—the desire to do their job well—to persevere, no matter what the circumstances. The Army is a team, and all members' contributions are essential to mission accomplishment. As an Army leader, you're responsible for developing this sense of belonging in your subordinates. Not only that; it's your job to inculcate in your people the winning spirit—the commitment to do their part to accomplish the mission, no matter when, no matter where, no matter what.

3-30. Army operations often involve danger and therefore fear. Battling the effects of fear has nothing to do with denying it and everything to do with recognizing fear and handling it. Leaders let their subordinates know, "You can expect to be afraid; here's what we'll do about it." The Army standard is to continue your mission to successful completion, as GEN Patton said, in spite of your fears. But saying this isn't going to make it happen. Army leaders expect fear to take hold when things go poorly, setbacks occur, the unit fails to complete a mission, or there are casualties. The sights and sounds of the modern battlefield are terrifying. So is fear of the unknown. Soldiers who see their buddies killed or wounded suddenly have a greater burden: they become aware of their own mortality. On top of all these obvious sources of fear is the insecurity before battle that many veterans have written about: "Will I perform well or will I let my buddies down?"

3-31. In the October 1993 fight in Somalia, one soldier who made it back to the safety of the American position was told to prepare to go back out; there were other soldiers in trouble. He had just run a gauntlet of fire, had just seen his friends killed and wounded, and was understandably afraid. "I can't go back out there," he told his sergeant. The leader reassured the soldier while reminding him of the mission and his responsibility to the team: "I know you're scared...I'm scared...I've never been in a situation like this, either. But we've got to go. It's our job. The difference between being a coward

and a man isn't whether you're scared; it's what you do while you're scared." That frightened soldier probably wasn't any less afraid, but he climbed back on the vehicle and went out to rescue the other American soldiers.

3-32. Will and a winning spirit apply in more situations than those requiring physical courage; sometimes you'll have to carry on for long periods in very difficult situations. The difficulties soldiers face may not be ones of physical danger, but of great physical, emotional, and mental strain. Physical courage allowed the soldier in the situation described above to return to the fight; will allowed his leader to say the right thing, to influence his frightened subordinate to do the right thing. Physical courage causes soldiers to charge a machine gun; will empowers them to fight on when they're hopelessly outnumbered, under appalling conditions, and without basic necessities.

STRESS IN TRAINING

When the bullets started flying...I never thought about half the things I was doing. I simply relied on my training and concentrated on the mission.

Captain Marie Bezubic
Operation Just Cause, Panama

3-33. Leaders must inject stress into training to prepare soldiers for stress in combat. However, creating a problem for subordinates and having them react to it doesn't induce the kind of stress required for combat training. A meaningful and productive mission, given with detailed constraints and limitations plus high standards of performance, does produce stress. Still, leaders must add unanticipated conditions to that stress to create a real learning environment. Sometimes, you don't even have to add stress; it just happens, as in this example.

Mix-up at the Crossroads

A young transportation section chief was leading a convoy of trucks on a night move to link up with several rifle companies. He was to transport the infantry to a new assembly area. When a sudden rainstorm dropped visibility to near zero, the section chief was especially glad that he had carefully briefed his drivers, issued strip maps, and made contingency plans. At a road intersection, his northbound convoy passed through an artillery battery moving east. When his convoy reached the rendezvous and the section chief got out to check his vehicles, he found he was missing two of his own trucks but had picked up three others towing howitzers. The tired and wet infantry commander was concerned that his unit would be late crossing the line of departure and forcefully expressed that concern to the section chief. The section chief now had to accomplish the same mission with fewer resources as well as run down his lost trucks and soldiers. There was certainly enough stress to go around.

After the section chief sent one of his most reliable soldiers with the artillery vehicles to find his missing trucks, he started shuttling the infantrymen to their destination. Later, after the mission was accomplished, the section chief and his drivers talked about what had happened. The leader admitted that he needed to supervise a convoy more closely under difficult conditions, and his soldiers recognized the need to follow the part of the unit SOP concerning reduced visibility operations.

3-34. The section chief fixed the immediate problem by starting to shuttle the infantry soldiers in the available trucks. During the AAR with the drivers, the leader admitted a mistake and figured out how to prevent similar errors in the future. The section chief also let the team know that sometimes, in spite of the best plans, things go wrong. A well-trained organization doesn't buckle under stress but deals with any setbacks and continues the mission.

THE STRESS OF CHANGE

3-35. Since the end of the Cold War, the Army has gone through tremendous change—dramatic decreases in the number of soldiers and DA civilians in all components, changes in assignment policies, base closings, and a host of other shifts that put stress on soldiers, DA civilians, and families. In those same years, the number of deployments to support missions such as peace operations and nation assistance has increased. And these changes have occurred in a peacetime Army. At the same time, Army leaders have had to prepare their soldiers for the stresses of combat, the ultimate crucible.

3-36. The stresses of combat you read about earlier in this chapter are classic: they've been the same for centuries. However, there's an aspect of the human dimension that has assumed an increasing importance: the effect of technological advances on organizations and people. Military leaders have always had to deal with the effect of technological changes. What's different today is the rate at which technology, to include warfighting technology, is changing. Rapid advances in new technologies are forcing the Army to change many aspects of the way it operates and are creating new leadership challenges.

TECHNOLOGY AND LEADERSHIP

3-37. Technology's presence challenges all Army leaders. Technology is here to stay and you, as an Army leader, need to continually learn how to manage it and make it work for you. The challenges come from many directions. Among them—

- You need to learn the strengths and vulnerabilities of the different technologies that support your team and its mission.

- You need to think through how your organization will operate with organizations that are less or more technologically complex. This situation may take the form of heavy and light Army units working together, operating with elements of another service, or cooperating with elements of another nation's armed forces.

- You need to consider the effect of technology on the time you have to analyze problems, make a decision, and act. Events happen faster today, and the stress you encounter as an Army leader is correspondingly greater.

Technological advances have the potential to permit better and more sustainable operations. However, as an Army leader you must remember the limitations of your people. No matter what technology you have or how it affects your mission, it's still your soldiers and DA civilians—their minds, hearts, courage, and talents—that will win the day.

3-38. Advances in electronic data processing let you handle large amounts of information easily. Today's desktop computer can do more, and do it faster, than the room-sized computers of only 20 years ago. Technology is a powerful tool—if you understand its potential uses and limitations. The challenge for all Army leaders is to overcome confusion on a fast-moving battlefield characterized by too much information coming in too fast.

3-39. Army leaders and staffs have always needed to determine mission-critical information, prioritize incoming reports, and process them quickly. The volume of information that current technology makes available makes this skill even more important than in the past. Sometimes something low-tech can divert the flood of technological help into channels the leader and staff can manage. For example, a well-understood commander's intent and thought-through commander's critical information requirements (CCIR) can help free leaders from nonessential information while pushing decisions to lower levels. As an Army leader, you must work hard to overcome the attractiveness and potential pitfalls of centralized decision making that access to information will appear to make practical.

3-40. Technology is also changing the size of the battlefield and the speed of battle. Instant global communications are increasing the pace of military actions. Global positioning systems and night vision capabilities mean the Army can fight at night and during periods of limited visibility—conditions that used to slow things down. Continuous operations increase the mental and physical stress on soldiers and leaders. Nonlinear operations make it more difficult for commanders to determine critical points on the battlefield. Effective leaders develop techniques to identify and manage stress well before actual conflict occurs. They also find ways to overcome the soldier's increased sense of isolation that comes with the greater breadth and depth of the modern battlefield. (FM 100-34 discusses continuous operations. FM 22-51 discusses combat stress control.)

3-41. Modern technology has also increased the number and complexity of skills the Army requires. Army leaders must carefully manage low-density specialties. They need to ensure that critical positions are filled and that their people maintain perishable skills. Army leaders must bring together leadership, personnel management, and training management to ensure their organizations are assigned people with the right specialties and that the entire organization is trained and ready. On top of this, the speed and lethality of modern battle have made mental agility and initiative even more necessary for fighting and winning. As in the past, Army leaders must develop these attributes in their subordinates.

3-42. To some, technology suggests a bloodless battlefield that resembles a computer war game more than the battlefields of the past. That isn't true now and it won't be true in the immediate future. Technology is still directed at answering the same basic questions that Civil War leaders tried to answer when they sent out a line of skirmishers: Where am I? Where are my buddies? Where is the enemy? How do I defeat him? Armed with this information, the soldiers and DA civilians of the Army will continue to accomplish the mission with character, using their technological edge to do the job better, faster, and smarter.

3-43. Modern digital technology can contribute a great deal to the Army leader's understanding of the battlefield; good leaders stay abreast of advances that enhance their tactical abilities. Digital technology has a lot to offer, but don't be fooled. A video image of a place, an action, or an organization can never substitute for the leader's getting down on the ground with the soldiers to find out what's going on. Technology can provide a great deal of information, but it may not present a completely accurate picture. The only way leaders can see the urgency in the faces of their soldiers is to get out and see them. As with any new weapon, the Army leader must know how to use technology without being seduced by it. Technology may be invaluable; however, effective leaders understand its limits.

3-44. Whatever their feeling regarding technology, today's leaders must contend more and more with an increased information flow and operational tempo (OPTEMPO). Pressures to make a decision increase, even as the time to verify and validate information decreases. Regardless of the crunch, Army leaders are responsible for the consequences of their decisions, so they gather, process, analyze, evaluate—and check —information. If they don't, the costs can be disastrous. (FM 100-34 discusses information management and decision making.)

"Superior Technology"

In the late fall of 1950, as United Nations (UN) forces pushed the North Korean People's Army northward, the People's Republic of China prepared to enter the conflict in support of its ally. The UN had air superiority, a marked advantage that had contributed significantly to the UN tactical and operational successes of the summer and early fall. Nonetheless, daily reconnaissance missions over the rugged North Korean interior failed to detect the Chinese People's Liberation Army's movement of nearly a quarter of a million ground troops across the border and into position in the North Korean mountains.

When the first reports of Chinese soldiers in North Korea arrived at Far East Command in Tokyo, intelligence analysts ignored them because they contradicted the information provided by the latest technology—aerial surveillance. Tactical commanders failed to send ground patrols into the mountains. They assumed the photos gave an accurate picture of the enemy situation when, in fact, the Chinese were practicing strict camouflage discipline. When the Chinese attacked in late November, UN forces were surprised, suffered heavy losses, and were driven from the Chinese border back to the 38th parallel.

When GEN Matthew B. Ridgway took over the UN forces in Korea in December, he immediately visited the headquarters of every regiment and many of the battalions on the front line. This gave GEN Ridgway an unfiltered look at the situation, and it sent a message to all his commanders: get out on the ground and find out what's going on.

3-45. The Chinese counterattack undid the results of the previous summer's campaign and denied UN forces the opportunity for a decisive victory that may have ended the war. The UN forces, under US leadership, enjoyed significant technological advantages over the Chinese. However, failure to verify the information provided by aerial photography set this advantage to zero. And this failure was one of leadership, not technology. Questioning good news provided by the latest "gee-whiz" system and ordering reconnaissance patrols to go out in lousy weather both require judgment and moral

courage: judgment as to when a doubt is reasonable and courage to order soldiers to risk their lives in cold, miserable weather. But Army leaders must make those judgments and give those orders. Technology has not changed that.

3-46. Technology and making the most of it will become increasingly important. Today's Army leaders require systems understanding and more technical and tactical skills. Technical skill: What does this system do? What does it not do? What are its strengths? What are its weaknesses? What must I check? Tactical skill:

How do this system's capabilities support my organization? How should I employ it to support this mission? What must I do if it fails? There's a fine line between a healthy questioning of new systems' capabilities and an unreasoning hostility that rejects the advantages technology offers. You, as an Army leader, must stay on the right side of that line, the side that allows you to maximize the advantages of technology. You need to remain aware of its capabilities and shortcomings, and you need to make sure your people do as well.

LEADERSHIP AND THE CHANGING THREAT

3-47. Another factor that will have a major impact on Army leadership in the near future is the changing nature of the threat. For the Army, the twenty-first century began in 1989 with the fall of the Berlin Wall and subsequent collapse of the Soviet Union. America no longer defines its security interests in terms of a single, major threat. Instead, it faces numerous, smaller threats and situations, any of which can quickly mushroom into a major security challenge.

3-48. The end of the Cold War has increased the frequency and variety of Army missions. Since 1989, the Army has fought a large-scale land war and been continually involved in many different kinds of stability operations and support operations. There has been a greater demand for special, joint, and multinational operations as well. Initiative at all levels is becoming more and more important. In many instances, Army leaders on the ground have had to invent ways of doing business for situations they could not have anticipated.

3-49. Not only that, the importance of direct leaders—NCOs and junior officers—making the right decisions in stressful situations has increased. Actions by direct-level leaders—sergeants, warrant officers, lieutenants, and captains—can have organizational- and strategic-level implications. Earlier in this chapter, you read about the disciplined soldiers and leaders who accomplished their mission of securing a television tower in Udrigovo, Bosnia-Herzegovina. In that case, the local population's perception of how American soldiers secured the tower was just as important as securing the tower itself. Had the American detachment created an international incident by using what could have been interpreted as excessive force, maintaining order throughout Bosnia Herzegovina would have become more difficult. The Army's organizational and strategic leaders count on direct leaders. It has always been important to accomplish the mission the right way the first time; today it's more important than ever.

3-50. The Army has handled change in the past. It will continue to do so in the future as long as Army leaders emphasize the constants—Army values, teamwork, and discipline—and help their people anticipate change by seeking always to improve. Army leaders explain, to the extent of their knowledge and in clear terms, what may happen and how the organization can effectively react if it does. Change is inevitable; trying to avoid it is futile. The disciplined, cohesive organization rides out the tough times and will emerge even better than it started. Leadership, in a very real sense, includes managing change and making it work for you. To do that, you must know what to change and what not to change.

3-51. FM 100-5 provides a doctrinal framework for coping with these challenges while executing operations. It gives Army leaders clues as to what they will face and what will be required of them, but as COL Chamberlain found on Little Round Top, no manual can cover all possibilities. The essence of leadership remains the same: Army leaders create a vision of what's necessary, communicate it in a way that makes their intent clear, and vigorously execute it to achieve success.

CLIMATE AND CULTURE

3-52. Climate and culture describe the environment in which you lead your people. Culture refers to the environment of the Army as an institution and of major elements or communities within it. Strategic leaders maintain the Army's institutional culture. (Chapter 7 discusses their role.) Climate refers to the environment of units and organizations. All organizational and direct leaders establish their organization's climate, whether purposefully or unwittingly. (Chapters 5 and 6 discuss their responsibilities.)

CLIMATE

3-53. Taking care of people and maximizing their performance also depends on the climate a leader creates in the organization. An organization's climate is the way its members feel about their organization. Climate comes from people's shared perceptions and attitudes, what they believe about the day-to-day functioning of their outfit. These things have a great impact on their motivation and the trust they feel for their team and their leaders. Climate is generally short-term: it depends on a network of the personalities in a small organization. As people come and go, the climate changes. When a soldier says "My last platoon sergeant was pretty good, but this new one is great," the soldier is talking about one of the many elements that affect organizational climate.

3-54. Although such a call seems subjective, some very definite things determine climate. The members' collective sense of the organization—its organizational climat —is directly attributable to the leader's values, skills, and actions. As an Army leader, you establish the climate of your organization, no matter how small it is or how large. Answering the following questions can help you describe an organization's climate:

- Does the leader set clear priorities and goals?
- Is there a system of recognition, rewards and punishments? Does it work?
- Do the leaders know what they're doing? Do they admit when they're wrong?
- Do leaders seek input from subordinates? Do they act on the feedback they're provided?
- In the absence of orders, do junior leaders have authority to make decisions that are consistent with the leader's intent?
- Are there high levels of internal stress and negative competition in the organization? If so, what's the leader doing to change that situation?
- Do the leaders behave the way they talk? Is that behavior consistent with Army values? Are they good role models?
- Do the leaders lead from the front, sharing hardship when things get tough?
- Do leaders talk to their organizations on a regular basis? Do they keep their people informed?

3-55. Army leaders who do the right things for the right reasons—even when it would be easier to do the wrong thing—create a healthy organizational climate. In fact, it's the leader's behavior that has the greatest effect on the organizational climate. That behavior signals to every member of the organization what the leader will and will not tolerate. Consider this example.

Changing a Unit Climate—The New Squad Leader

SSG Withers was having a tough week. He had just been promoted to squad leader in a different company; he had new responsibilities, new leaders, and new soldiers. Then, on his second day, his unit was alerted for a big inspection in two days. A quick check of the records let him know that the squad leader before him had let maintenance slip; the records were sloppy and a lot of the scheduled work had not been done. On top of that, SSG Withers was sure his new platoon sergeant didn't like him. SFC King was professional but gruff, a person of few words. The soldiers in SSG Withers' squad seemed a little afraid of the platoon sergeant.

After receiving the company commander's guidance about the inspection, the squad leaders briefed the platoon sergeant on their plans to get ready. SSG Withers had already determined that he and his soldiers would have to work late. He could have complained about his predecessor, but he thought it would be best just to stick to the facts and talk about what he had found in the squad. For all he knew, the old squad leader might have been a favorite of SFC King.

SFC King scowled as he asked, "You're going to work late?"

SSG Withers had checked his plan twice: "Yes, sergeant. I think it's necessary."

SFC King grunted, but the sound could have meant "okay" or it could have meant "You're being foolish." SSG Withers wasn't sure.

The next day SSG Withers told his soldiers what they would have to accomplish. One of the soldiers said that the old squad leader would have just fudged the paperwork. "No kidding," SSG Withers thought. He wondered if SFC King knew about it. Of course, there was a good chance he would fail the inspection if he didn't fudge the paperwork—and wouldn't *that* be a good introduction to the new company? But he told his squad that they would do it right: "We'll do the best we can. If we don't pass, we'll do better next time."

SSG Withers then asked his squad for their thoughts on how to get ready. He listened to their ideas and offered some of his own. One soldier suggested that they could beat the other squads by sneaking into the motor pool at night and lowering the oil levels in their vehicles. "SFC King gives a half day off to whatever squad does best," the soldier explained. SSG Withers didn't want to badmouth the previous squad leader; on the other hand, the squad was his responsibility now. "It'd be nice to win," SSG Withers said, "but we're not going to cheat."

The squad worked past 2200 hours the night before the inspection. At one point SSG Withers found one of the soldiers sleeping under a vehicle. "Don't you want to finish and go home to sleep?" he asked the soldier.

"I...uh...I didn't think you'd still be here," the soldier answered.

"Where else would I be?" replied the squad leader.

The next day, SFC King asked SSG Withers if he thought his squad's vehicle was going to pass the inspection.

"Not a chance," SSG Withers said.

SFC King gave another mysterious grunt.

Later, when the inspector was going over his vehicle, SSG Withers asked if his soldiers could follow along. "I want them to see how to do a thorough inspection," he told the inspector. As the soldiers followed the inspector around and learned how to look closely at the vehicle, one of them commented that the squad had never been around for any inspection up to that point. "We were always told to stay away," he said.

Later, when the company commander went over the results of the inspection, he looked up at SSG Withers as he read the failing grade. SSG Withers was about to say, "We'll try harder next time, sir," but he decided that sounded lame, so he said nothing. Then SFC King spoke up.

"First time that squad has ever failed an inspection," the platoon sergeant said, "but they're already better off than they were the day before yesterday, failing grade and all."

3-56. SFC King saw immediately that things had changed for the better in SSG Withers' squad. The failing grade was real; previous passing grades had not been. The new squad leader told the truth and expected his soldiers to do the same. He was there when his people were working late. He acted to improve the squad's ethical and performance standards (by clearly stating and enforcing them). He moved to teach his soldiers the skills and standards associated with vehicle maintenance (by asking the inspector to show them how to look at a vehicle). And not once did SSG Withers whine that the failing grade was not his fault; instead, he focused on how to make things better. SSG Withers knew how to motivate soldiers to perform to standard and had the strength of character to do the right thing. In addition, he trusted the chain of command to take the long-term view. Because of his decisive actions, based on his character and competence, SSG Withers was well on his way to creating a much healthier climate in his squad.

3-57. No matter how they complain about it, soldiers and DA civilians expect to be held to standard; in the long run they feel better about themselves when they do hard work successfully. They gain confidence in leaders who help them achieve standards and lose confidence in leaders who don't know the standards or who fail to demand performance.

CULTURE

When you're first sergeant, you're a role model whether you know it or not. You're a role model for the guy that will be in your job. Not next month or next year, but ten years from now. Every day soldiers are watching you and deciding if you are the kind of first sergeant they want to be.

An Army First Sergeant
1988

3-58. Culture is a longer lasting, more complex set of shared expectations than climate. While climate is how people feel about their organization right now, culture consists of the shared attitudes, values, goals, and practices that characterize the larger institution. It's deeply rooted in long-held beliefs, customs, and practices. For instance, the culture of the armed forces is different from that of the business world, and the culture of the Army is different from that of the Navy. Leaders must establish a climate consistent with the culture of the larger institution. They also use the culture to let their people know they're part of something bigger than just themselves, that they have responsibilities not only to the people around them but also to those who have gone before and those who will come after.

3-59. Soldiers draw strength from knowing they're part of a tradition. Most meaningful traditions have their roots in the institution's culture. Many of the Army's everyday customs and traditions are there to remind you that you're just the latest addition to a long line of American soldiers. Think of how much of your daily life connects you to the past and to American soldiers not yet born: the uniforms you wear, the martial music that punctuates your day, the way you salute, your title, your organization's history, and Army values such as selfless service. Reminders of your place in history surround you.

3-60. This sense of belonging is vitally important. Visit the Vietnam Memorial in Washington, DC, some Memorial Day weekend and you'll see dozens of veterans, many of them wearing bush hats or campaign ribbons or fatigue jackets decorated with unit patches. They're paying tribute to their comrades in this division or that company. They're also acknowledging what for many of them was the most intense experience of their lives.

3-61. Young soldiers want to belong to something bigger than themselves. Look at them off duty, wearing tee shirts with names of sports teams and famous athletes. It's not as if an 18-year-old who puts on a jacket with a professional sports team's logo thinks anyone will mistake him for a professional player; rather, that soldier wants to be associated with a winner. Advertising and mass media make heroes of rock stars, athletes, and actors. Unfortunately, it's easier to let some magazine or TV show tell you whom to admire than it is to dig up an organization's history and learn about heroes.

3-62. Soldiers want to have heroes. If they don't know about SGT Alvin York in World War I, about COL Joshua Chamberlain's 20th Maine during the Civil War, about MSG Gary Gordon and SFC Randall Shughart in the 1993 Somalia fight, then it's up to you, their leaders, to teach them. (The bibliography lists works you can use to learn more about your profession, its history, and the people who made it.)

3-63. When soldiers join the Army, they become part of a history: the Big Red One, the King of Battle, Sua Sponte. Teach them the history behind unit crests, behind greetings, behind decorations and badges. The Army's culture isn't something that exists apart from you; it's part of who you are, something you can use to give your soldiers pride in themselves and in what they're doing with their lives.

LEADERSHIP STYLES

3-64. You read in Chapter 2 that all people are shaped by what they've seen, what they've learned, and whom they've met. Who you are determines the way you work with other people. Some people are happy and smiling all the time; others are serious. Some leaders can wade into a room full of strangers and inside of five minutes have everyone there thinking, "How have I lived so long without meeting this person?" Other very competent leaders are uncomfortable in social situations. Most of us are somewhere in between. Although Army leadership doctrine describes at great length how you should interact with your subordinates and how you must strive to learn and improve your leadership skills, the Army recognizes that you must always be yourself; anything else comes across as fake and insincere.

3-65. Having said that, effective leaders are flexible enough to adjust their leadership style and techniques to the people they lead. Some subordinates respond best to coaxing, suggestions, or gentle prodding; others need, and even want at times, the verbal equivalent of a kick in the pants. Treating people fairly doesn't mean treating people as if they were clones of one another. In fact, if you treat everyone the same way, you're probably being unfair, because different people need different things from you.

3-66. Think of it this way: Say you must teach map reading to a large group of soldiers ranging in rank from private to senior NCO. The senior NCOs know a great deal about the subject, while the privates know very little. To meet all their needs, you must teach the privates more than you teach the senior NCOs. If you train

the privates only in the advanced skills the NCOs need, the privates will be lost. If you make the NCOs sit through training in the basic tasks the privates need, you'll waste the NCOs' time. You must fit the training to the experience of those being trained. In the same way, you must adjust your leadership style and techniques to the experience of your people and characteristics of your organization.

3-67. Obviously, you don't lead senior NCOs the same way you lead privates. But the easiest distinctions to make are those of rank and experience. You must also take into account personalities, self-confidence, self-esteem—all the elements of the complex mix of character traits that makes dealing with people so difficult and so rewarding. One of the many things that makes your job tough is that, in order to get their best performance, you must figure out what your subordinates need and what they're able to do—even when they don't know themselves.

3-68. When discussing leadership styles, many people focus on the extremes: autocratic and democratic. Autocratic leaders tell people what to do with no explanation; their message is, "I'm the boss; you'll do it because I said so." Democratic leaders use their personalities to persuade subordinates. There are many shades in between; the following paragraphs discuss five of them. However, bear in mind that competent leaders mix elements of all these styles to match to the place, task, and people involved. Using different leadership styles in different situations or elements of different styles in the same situation isn't inconsistent. The opposite is true: if you can use only one leadership style, you're inflexible and will have difficulty operating in situations where that style doesn't fit.

DIRECTING LEADERSHIP STYLE

3-69. The directing style is leader-centered. Leaders using this style don't solicit input from subordinates and give detailed instructions on how, when, and where they want a task performed. They then supervise its execution very closely.

3-70. The directing style may be appropriate when time is short and leaders don't have a chance to explain things. They may simply give orders: Do this. Go there. Move. In fast-paced operations or in combat, leaders may revert to the directing style, even with experienced subordinates. This is what the motor sergeant you read about in Chapter 1 did. If the leader has created a climate of trust, subordinates will assume the leader has switched to the directing style because of the circumstances.

3-71. The directing style is also appropriate when leading inexperienced teams or individuals who are not yet trained to operate on their own. In this kind of situation, the leader will probably remain close to the action to make sure things go smoothly.

3-72. Some people mistakenly believe the directing style means using abusive or demeaning language or includes threats and intimidation. This is wrong. If you're ever tempted to be abusive, whether because of pressure or stress or what seems like improper behavior by a subordinate, ask yourself these questions: Would I want to work for someone like me? Would I want my boss to see and hear me treat subordinates this way? Would I want to be treated this way?

PARTICIPATING LEADERSHIP STYLE

3-73. The participating style centers on both the leader and the team. Given a mission, leaders ask subordinates for input, information, and recommendations but make the final decision on what to do themselves. This style is especially appropriate for leaders who have time for such consultations or who are dealing with experienced subordinates.

3-74. The team-building approach lies behind the participating leadership style. When subordinates help create a plan, it becomes—at least in part—their plan. This ownership creates a strong incentive to invest the effort necessary to make the plan work. Asking for this kind of input is a sign of a leader's strength and self-confidence. But asking for advice doesn't mean the leader is obligated to follow it; the leader alone is always responsible for the quality of decisions and plans.

DELEGATING LEADERSHIP STYLE

3-75. The delegating style involves giving subordinates the authority to solve problems and make decisions without clearing them through the leader. Leaders with mature and experienced subordinates or who want to create a learning experience for subordinates often need only to give them authority to make decisions, the necessary resources, and a clear understanding of the mission's purpose. As always, the leader is ultimately responsible for what does or does not happen, but in the delegating leadership style, the leader holds subordinate leaders accountable for their actions. This is the style most often used by officers dealing with senior NCOs and by organizational and strategic leaders.

TRANSFORMATIONAL AND TRANSACTIONAL LEADERSHIP STYLES

A man does not have himself killed for a few halfpence a day or for a petty distinction. You must speak to the soul in order to electrify the man.

Napoleon Bonaparte

3-76. These words of a distinguished military leader capture the distinction between the transformational leadership style, which focuses on inspiration and change, and the transactional leadership style, which focuses on rewards and punishments. Of course Napoleon understood the importance of rewards and punishments. Nonetheless, he also understood that carrots and sticks alone don't inspire individuals to excellence.

Transformational Leadership Style

3-77. As the name suggests, the transformational style "transforms" subordinates by challenging them to rise above their immediate needs and self-interests. The transformational style is developmental: it emphasizes individual growth (both professional and personal) and organizational enhancement. Key features of the transformational style include empowering and mentally stimulating subordinates: you consider and motivate them first as individuals and then as a group. To use the transformational style, you must have the courage to communicate your intent and then step back and let your subordinates work. You must also be aware that immediate benefits are often delayed until the mission is accomplished.

3-78. The transformational style allows you to take advantage of the skills and knowledge of experienced subordinates who may have better ideas on how to accomplish a mission. Leaders who use this style communicate reasons for their decisions or actions and, in the process, build in subordinates a broader understanding and ability to exercise initiative and operate effectively. However, not all situations lend themselves to the transformational leadership style. The transformational style is most effective during periods that call for change or present new opportunities. It also works well when organizations face a crisis, instability, mediocrity, or disenchantment. It may not be effective when subordinates are inexperienced, when the mission allows little deviation from accepted procedures, or when subordinates are not motivated. Leaders who use only the transformational leadership style limit their ability to influence individuals in these and similar situations.

Transactional Leadership Style

3-79. In contrast, some leaders employ only the transactional leadership style. This style includes such techniques as—

- Motivating subordinates to work by offering rewards or threatening punishment.
- Prescribing task assignments in writing.
- Outlining all the conditions of task completion, the applicable rules and regula-

tions, the benefits of success, and the consequences—to include possible disciplinary actions—of failure.
- "Management-by-exception," where leaders focus on their subordinates' failures, showing up only when something goes wrong.

The leader who relies exclusively on the transactional style, rather than combining it with the transformational style, evokes only short-term commitment from his subordinates and discourages risk-taking and innovation.

3-80. There are situations where the transactional style is acceptable, if not preferred. For example, a leader who wants to emphasize safety could reward the organization with a three-day pass if the organization prevents any serious safety-related incidents over a two-month deployment. In this case, the leader's intent appears clear: unsafe acts are not tolerated and safe habits are rewarded.

3-81. However, using only the transactional style can make the leader's efforts appear self-serving. In this example, soldiers might interpret the leader's attempt to reward safe practices as an effort to look good by focusing on something that's unimportant but that has the boss's attention. Such perceptions can destroy the trust subordinates have in the leader. Using the transactional style alone can also deprive subordinates of opportunities to grow, because it leaves no room for honest mistakes.

3-82. The most effective leaders combine techniques from the transformational and transactional leadership styles to fit the situation. A strong base of transactional understanding supplemented by charisma, inspiration and individualized concern for each subordinate, produces the most enthusiastic and genuine response. Subordinates will be more committed, creative, and innovative. They will also be more likely to take calculated risks to accomplish their mission. Again referring to the safety example, leaders can avoid any misunderstanding of their intent by combining transformational techniques with transactional techniques. They can explain why safety is important (intellectual stimulation) and encourage their subordinates to take care of each other (individualized concern).

INTENDED AND UNINTENDED CONSEQUENCES

3-83. The actions you take as a leader will most likely have unintended as well as intended consequences. Like a chess player trying to anticipate an opponent's moves three or four turns in advance—if I do this, what will my opponent do; then what will I do next?—leaders think through what they can expect to happen as a result of a decision. Some decisions set off a chain of events; as far as possible, leaders must anticipate the second- and third-order effects of their actions. Even lower-level leaders' actions may have effects well beyond what they expect.

3-84. Consider the case of a sergeant whose team is manning a roadblock as part of a peace operation. The mission has received lots of media attention (Haiti and Bosnia come to mind), and millions of people back home are watching. Early one morning, a truckload of civilians appears, racing toward the roadblock. In the half-light, the sergeant can't tell if the things in the passengers' hands are weapons or farm tools, and the driver seems intent on smashing through the barricade. In the space of a few seconds, the sergeant must decide whether or not to order his team to fire on the truck.

3-85. If the sergeant orders his team to fire because he feels he and his soldiers are threatened, that decision will have international consequences. If he kills any civilians, chances are good that his chain of command from the president on down—not to mention the entire television audience of the developed world—will know about the incident in a few short hours. But the decision is tough for another reason: if the sergeant doesn't order his team to fire and the civilians turn out to be an armed gang, the team may take casualties that could have been avoided. If the only factor involved was avoiding civilian casualties, the choice is simple: don't shoot. But the sergeant must also consider the requirement to protect his force and accomplish the mission of preventing unauthorized traffic from passing the roadblock. So the sergeant must act; he's the leader, and he's in charge. Leaders who have thought through the consequences of possible actions, talked with their own leaders about the commander's intent and mission priorities, and trust their chain of command to support them are less likely to be paralyzed by this kind of pressure.

INTENDED CONSEQUENCES

3-86. Intended consequences are the anticipated results of a leader's decisions and actions. When a squad leader shows a team leader a better way to lead PT, that action will have intended consequences: the team leader will be better equipped to do the job. When leaders streamline procedures, help people work smarter, and get the resources to the right place at the right time, the intended consequences are good.

UNINTENDED CONSEQUENCES

3-87. Unintended consequences are the results of things a leader does that have an unplanned impact on the organization or accomplishment of the mission. Unintended consequences are often more lasting and harder to anticipate than intended consequences. Organizational and strategic leaders spend a good deal of energy considering possible unintended consequences of their actions. Their organizations are complex, so figuring out the effects today's decisions will have a few years in the future is difficult.

3-88. Unintended consequences are best described with an example, such as setting the morning PT formation time: Setting the formation time at 0600 hours results in soldiers standing in formation at 0600 hours, an intended consequence. To not be late, soldiers living off post may have to depart their homes at 0500 hours, a consequence that's probably also anticipated. However, since most junior enlisted soldiers with families probably own only one car, there will most likely be another consequence: entire families rising at 0430 hours. Spouses must drive their soldiers to post and children, who can't be left at home unattended, must accompany them. This is an unintended consequence.

SUMMARY

3-89. The human dimension of leadership, how the environment affects you and your people, affects how you lead. Stress is a major part of the environment, both in peace and war. Major sources of stress include the rapid pace of change and the increasing complexity of technology. As an Army leader, you must stay on top of both. Your character and skills—how you handle stress—and the morale and discipline you develop and your team are more important in establishing the climate in your organization than any external circumstances.

3-90. The organizational climate and the institutional culture define the environment in which you and your people work. Direct, organizational, and strategic leaders all have different responsibilities regarding climate and culture; what's important now is to realize that you, the leader, establish the climate of your organization. By action or inaction, you determine the environment in which your people work.

3-91. Leadership styles are different ways of approaching the DO of BE, KNOW, DO—the actual work of leading people. You've read about five leadership styles: directing, participating, delegating, transformational, and transactional. But remember that you must be able to adjust the leadership style you use to the situation and the people you're leading. Remember also that you're not limited to any one style in a given situation: you should use techniques from different styles if that will help you motivate your people to accomplish the mission. Your leader attributes of judgment, intelligence, cultural awareness, and self-control all play major roles in helping you choose the proper style and the appropriate techniques for the task at hand. That said, you must always be yourself.

3-92. All leader actions result in intended and unintended consequences. Two points to remember: think through your decisions and do your duty. It might not seem that the actions of one leader of one small unit matter in the big picture. But they do. In the words of Confederate COL William C. Oats, who faced COL Joshua Chamberlain at Little Round Top: "Great events sometimes turn on comparatively small affairs."

3-93. In spite of stress and changes, whether social or technological, leadership always involves shaping human emotions and behaviors. As they serve in more complex environments with wider-ranging consequences, Army leaders refine what they've known and done as well as develop new styles, skills, and actions. Parts Two and Three discuss the skills and actions required of leaders from team to Department of the Army level.

PART TWO

Direct Leadership

The first three chapters of this manual cover the constants of leadership. They focus primarily on what a leader must BE. Part Two examines what a direct leader must KNOW and DO. Note the distinction between a skill, *knowing* something, and an action, *doing* something. The reason for this distinction bears repeating: knowledge isn't enough. You can't be a leader until you apply what you know, until you act and DO what you must.

Army leaders are grounded in the heritage, values, and tradition of the Army. They embody the warrior ethos, value continuous learning, and demonstrate the ability to lead and train their subordinates. Army leaders lead by example, train from experience, and maintain and enforce standards. They do these things while taking care of their people and adapting to a changing world. Chapters 4 and 5 discuss these subjects in detail.

The *warrior ethos* is the will to win with honor. Despite a thinking enemy, despite adverse conditions, you accomplish your mission. You express your character— the BE of BE, KNOW, DO—when you and your people confront a difficult mission and persevere. The warrior ethos applies to all soldiers and DA civilians, not just those who close with and destroy the enemy. It's the will to meet mission demands no matter what, the drive to get the job done whatever the cost.

Continuous learning requires dedication to improving your technical and tactical skills through study and practice. It also includes learning about the world around you—mastering new technology, studying other cultures, staying aware of current events at home and abroad. All these things affect your job as a leader.

Continuous learning also means consciously developing your *character* through study and reflection. It means reflecting on Army values and developing leader attributes. Broad knowledge and strong character underlie the right decisions in hard times. Seek to learn as much as you can about your job, your people, and yourself. That way you'll be prepared when the time comes for tough decisions. You'll BE a leader of character, KNOW the necessary skills, and DO the right thing.

Army leaders train and lead people. Part of this responsibility is maintaining and enforcing standards. Your subordinates expect you to show them what the standard is and train them to it: they expect you to lead by example. In addition, as an Army leader you're required to take care of your people. You may have to call on them to do things that seem impossible. You may have to ask them to make extraordinary sacrifices to accomplish the mission. If you train your people to standard, inspire the warrior ethos in them, and consistently look after their interests, they'll be prepared to accomplish the mission—anytime, anywhere.

Chapter 4

Direct Leadership Skills

Never get so caught up in cutting wood that you forget to sharpen your ax.

First Sergeant James J. Karolchyk, 1986

4-1. The Army's direct leaders perform a huge array of functions in all kinds of places and under all kinds of conditions. Even as you read these pages, someone is in the field in a cold place, someone else in a hot place. There are people headed to a training exercise and others headed home. Somewhere a motor pool is buzzing, a medical ward operating, supplies moving. Somewhere a duty NCO is conducting inspections and a sergeant of the guard is making the rounds. In all these places, no matter what the conditions or the mission, direct leaders are guided by the same principles, using the same skills, and performing the same actions.

4-2. This chapter discusses the skills a direct leader must master and develop. It addresses the KNOW of BE, KNOW, and DO for direct leaders. The skills are organized under the four skill groups Chapter 1 introduced: interpersonal, conceptual, technical, and tactical. (Appendix B lists performance indicators for leader skills.)

INTERPERSONAL SKILLS

4-3. A DA civilian supervisor was in a frenzy because all the material needed for a project wasn't available. The branch chief took the supervisor aside and said, "You're worrying about *things*. Things are not important; things will or won't be there. Worry about working with the people who will get the job done."

4-4. Since leadership is about people, it's not surprising to find interpersonal skills, what some call "people skills," at the top of the list of what an Army leader must KNOW. Figure 4-1 (on page 4-3) identifies the direct leader interpersonal skills. All these skills—communicating, team building, supervising, and counseling—require communication. They're all closely related; you can hardly use one without using the others.

COMMUNICATING

4-5. Since leadership is about getting other people to do what you want them to do, it follows that communicating—transmitting information so that it's clearly understood—is an important skill. After all, if people can't understand you, how will you ever let them know what you want? The other interpersonal skills—supervising, team building, and counseling—also depend on your ability to communicate.

4-6. If you take a moment to think about all the training you've received under the heading "communication," you'll see that it probably falls into four broad categories: speaking, reading, writing, and listening. You begin practicing speech early; many children are using words by the age of one. The heavy emphasis on reading and writing begins in school, if not before. Yet how many times have you been taught how to listen? Of the four forms of communication,

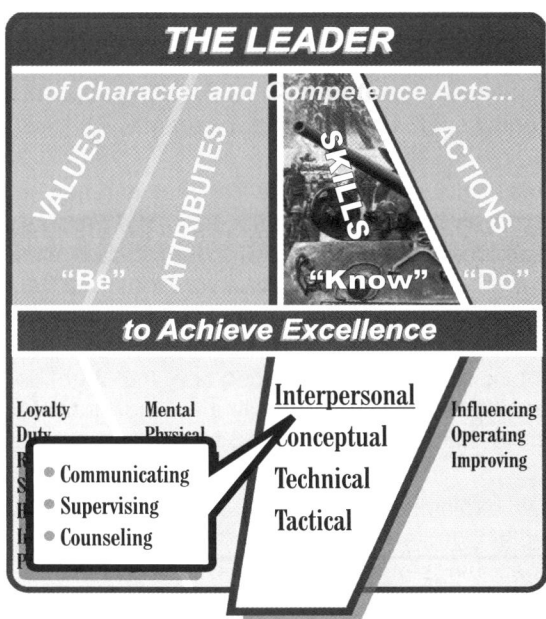

**Figure 4-1. Direct Leader
Skills—Interpersonal**

listening is the one in which most people receive the least amount of formal training. Yet for an Army leader, it's every bit as important as the others. It often comes first because you must listen and understand before you can decide what to say.

One-Way and Two-Way Communication

4-7. There are two common forms of one-way communication that are not necessarily the best way to exchange information: seeing and hearing. The key difference between one-way and two-way communication is that one-way communication—hearing or seeing something on television, reading a copy of a slide presentation, or even watching a training event unfold—may not give you a complete picture. You

may have unanswered questions or even walk away with the wrong concept of what has occurred. That's why two-way communication is preferred when time and resources permit.

Active Listening

4-8. An important form of two-way communication is active listening. When you practice active listening, you send signals to the speaker that say, "I'm paying attention." Nod your head every once in a while, as if to say, "Yes, I understand." When you agree with the speaker, you might use an occasional "uh-huh." Look the speaker in the eye. Give the speaker your full attention. Don't allow yourself to be distracted by looking out the window, checking your watch, playing with something on your desk, or trying to do more than one thing at a time. Avoid interrupting the speaker; that's the cardinal sin of active listening.

4-9. Be aware of barriers to listening. Don't form your response while the other person is still talking. Don't allow yourself to become distracted by the fact that you're angry, or that you have a problem with the speaker, or that you have lots of other things you need to be thinking about. If you give in to these temptations, you'll miss most of what's being said.

Nonverbal Communication

4-10. In face-to-face communication, even in the simplest conversation, there's a great deal going on that has almost nothing to do with the words being used. Nonverbal communication involves all the signals you send with your facial expressions, tone of voice, and body language. Effective leaders know that communication includes both verbal and nonverbal cues. Look for them in this example.

The Checking Account

A young soldier named PVT Bell, new to the unit, approaches his team leader, SGT Adams, and says, "I have a problem I'd like to talk to you about."

The team leader makes time—right then if possible—to listen. Stopping, looking the soldier in the eye, and asking, "What's up?" sends many signals: *I am concerned about your problem. You're part of the team, and we help each other. What can I do to help?* All these signals, by the way, reinforce Army values.

The Checking Account (continued)

PVT Bell sees the leader is paying attention and continues, "Well, I have this checking account, see, and it's the first time I've had one. I have lots of checks left, but for some reason the PX [post exchange] is saying they're no good."

SGT Adams has seen this problem before: PVT Bell thinks that checks are like cash and has no idea that there must be money in the bank to cover checks written against the account. SGT Adams, no matter how tempted, doesn't say anything that would make PVT Bell think that his difficulty was anything other than the most important problem in the world. He is careful to make sure that PVT Bell doesn't think that he's anyone other than the most important soldier in the world. Instead, SGT Adams remembers life as a young soldier and how many things were new and strange. What may seem like an obvious problem to an experienced person isn't so obvious to an inexperienced one. Although the soldier's problem may seem funny, SGT Adams doesn't laugh at the subordinate. And because nonverbal cues are important, SGT Adams is careful that his tone of voice and facial expressions don't convey contempt or disregard for the subordinate.

Instead, the leader listens patiently as PVT Bell explains the problem; then SGT Adams reassures PVT Bell that it can be fixed and carefully explains the solution. What's more, SGT Adams follows up later to make sure the soldier has straightened things out with the bank.

A few months later, a newly promoted PFC Bell realizes that this problem must have looked pretty silly to someone with SGT Adams' experience. But PFC Bell will always remember the example SGT Adams set. Future leaders are groomed every day and reflect their past leaders. By the simple act of listening and communicating, SGT Adams won the loyalty of PFC Bell. And when the next batch of new soldiers arrives, PFC Bell, now the old-timer, will say to them, "Yeah, in all my experience, I've got to say this is one of the best units in the Army. And SGT Adams is the best team leader around. Why, I remember a time..."

4-11. SGT Adams performed crisis counseling, a leader action Appendix C discusses. Look for the communicating skills in this example. SGT Adams listened actively and controlled his non-verbal communication. He gave PVT Bell his full attention and was careful not to signal indifference or a lack of concern. SGT Adams' ability to do this shows the mental attribute of self-discipline and the emotional attribute of self-control, which you read about in Chapter 2. The leader also displayed empathy, that is, sensitivity to the feelings, thoughts, and experiences of another person. It's an important quality for a counselor.

SUPERVISING

If a squad leader doesn't check, and the guy on point has no batteries for his night vision goggles, he has just degraded the effectiveness of the entire unit.

A Company Commander, Desert Storm

4-12. Direct leaders check and recheck things. Leaders strike a balance between checking too much and not checking enough. Training subordinates to act independently is important; that's why direct leaders give instructions or their intent and then allow subordinates to work without constantly looking over their shoulders. Accomplishing the mission is equally important; that's why leaders check things— especially conditions critical to the mission (fuel levels), details a soldier might forget (spare batteries for night vision goggles), or tasks at the limit of what a soldier has accomplished before (preparing a new version of a report).

4-13. Checking minimizes the chance of oversights, mistakes, or other circumstances that might derail a mission. Checking also gives leaders a chance to see and recognize subordinates who are doing things right or make on-the-spot corrections when necessary. Consider this example: A platoon sergeant delegates to the platoon's squad leaders the authority to get their squads ready for a tactical road march. The platoon sergeant oversees the activity but doesn't intervene unless errors, sloppy work, or lapses occur. The leader is there to answer questions or resolve problems that the squad leaders can't handle. This supervision ensures that the squads are prepared to standard and demonstrates to the squad leaders that the platoon sergeant cares about them and their people.

The Rusty Rifles Incident

While serving in the Republic of Vietnam, SFC Jackson was transferred from platoon sergeant of one platoon to platoon leader of another platoon in the same company. SFC Jackson quickly sized up the existing standards in the platoon. He wasn't pleased. One problem was that his soldiers were not keeping their weapons cleaned properly: rifles were dirty and rusty. He put out the word: weapons would be cleaned to standard each day, each squad leader would inspect each day, and he would inspect a sample of the weapons each day. He gave this order three days before the platoon was to go to the division rest and recuperation (R&R) area on the South China Sea.

The next day SFC Jackson checked several weapons in each squad. Most weapons were still unacceptable. He called the squad leaders together and explained the policy and his reasons for implementing it. SFC Jackson checked again the following day and still found dirty and rusty weapons. He decided there were two causes for the problem. First, the squad leaders were not doing their jobs. Second, the squad leaders and troops were bucking him—testing him to see who would really make the rules in the platoon. He sensed that, because he was new, they resisted his leadership. He knew he had a serious discipline problem he had to handle correctly. He called the squad leaders together again. Once again, he explained his standards clearly. He then said, "Tomorrow we are due to go on R&R for three days and I'll be inspecting rifles. We won't go on R&R until each weapon in this platoon meets the standard."

The next morning SFC Jackson inspected and found that most weapons in each squad were still below standard. He called the squad leaders together. With a determined look and a firm voice, he told them he would hold a formal in-ranks inspection at 1300 hours, even though the platoon was scheduled to board helicopters for R&R then. If every weapon didn't meet the standard, he would conduct another in-ranks inspection for squad leaders and troops with substandard weapons. He would continue inspections until all weapons met the standard.

At 1300 hours the platoon formed up, surly and angry with the new platoon leader, who was taking their hard-earned R&R time. The soldiers could hardly believe it, but his message was starting to sink in. This leader meant what he said. This time all weapons met the standard.

COUNSELING

Nothing will ever replace one person looking another in the eyes and telling the soldier his strengths and weaknesses. [Counseling] charts a path to success and diverts soldiers from heading down the wrong road.

Sergeant Major Randolph S. Hollingsworth

4-14. Counseling is subordinate-centered communication that produces a plan outlining actions necessary for subordinates to achieve individual or organizational goals. Effective counseling takes time, patience, and practice. As with everything else you do, you must develop your skills as a counselor. Seek feedback on how effective you are at counseling, study various counseling techniques, and make efforts to improve. (Appendix C discusses developmental counseling techniques.)

4-15. Proper counseling leads to a specific plan of action that the subordinate can use as a road map for improvement. Both parties, counselor and counseled, prepare this plan of action. The leader makes certain the subordinate understands and takes ownership of it. The best plan of action in the world does no good if the subordinate doesn't understand it, follow it, and believe in it. And once the plan of action is agreed upon, the leader must follow up with one-on-one sessions to ensure the subordinate stays on track.

4-16. Remember the Army values of loyalty, duty, and selfless service require you to counsel your subordinates. The values of honor, integrity, and personal courage require you to give them straightforward feedback. And the Army value of respect requires you to find the best way to communicate that feedback so that your subordinates understand it. These Army values all point to the requirement for you to become a proficient counselor. Effective counseling helps your subordinates develop personally and professionally.

4-17. One of the most important duties of all direct, organizational, and strategic leaders is to develop subordinates. Mentoring, which links the operating and improving leader actions, plays a major part in developing competent and confident future leaders. Counseling is an interpersonal skill essential to effective mentoring. (Chapters 5, 6, and 7 discuss the direct, organizational, and strategic leader mentoring actions.)

CONCEPTUAL SKILLS

4-18. Conceptual skills include competence in handling ideas, thoughts, and concepts. Figure 4-2 (on page 4-7) lists the direct leader conceptual skills.

CRITICAL REASONING

4-19. Critical reasoning helps you think through problems. It's the key to understanding situations, finding causes, arriving at justifiable conclusions, making good judgments, and learning from the experience—in short, solving problems. Critical reasoning is an essential part of effective counseling and underlies ethical reasoning, another conceptual skill. It's also a central aspect of decision making, which Chapter 5 discusses.

4-20. The word "critical" here doesn't mean finding fault; it doesn't have a negative meaning at all. It means getting past the surface of the problem and thinking about it in depth. It means looking at a problem from several points of view instead of just being satisfied with the first answer that comes to mind. Army leaders need this ability because many of the choices they face are complex and offer no easy solution.

4-21. Sometime during your schooling you probably ran across a multiple choice test, one that required you to "choose answer a, b, c, or d" or "choose one response from column a and two from column b." Your job as an Army leader would be a lot easier if the problems you faced were presented that way, but leadership is a lot more complex than that. Sometimes just figuring out the real problem presents a huge hurdle; at other times you have to sort through distracting multiple problems to get to the real difficulty. On some occasions you know what the problem is but have no clue as to what an answer might be. On others you can come up with two or three answers that all look pretty good.

Finding the Real Problem

A platoon sergeant directs the platoon's squad leaders to counsel their soldiers every month and keep written records. Three months later, the leader finds the records are sloppy or incomplete; in many cases, there's no record at all. The platoon sergeant's first instinct is to chew out the squad leaders for ignoring his instructions. It even occurs to him to write a counseling annex to the platoon SOP so he can point to it the next time the squad leaders fail to follow instructions.

Finding the Real Problem (continued)

But those are just knee-jerk reactions and the platoon sergeant knows it. Instead of venting his frustration, the leader does a little investigating and finds that two squad leaders have never really been taught how to do formal, written counseling. The third one has no idea why counseling is important. So what looked like a disciplinary problem—the squad leaders disobeying instructions—turns out to be a training shortfall. By thinking beyond the surface and by checking, the platoon sergeant was able to isolate the real problem: that the squad leaders had not been trained in counseling. The next step is to begin training and motivating subordinates to do the tasks.

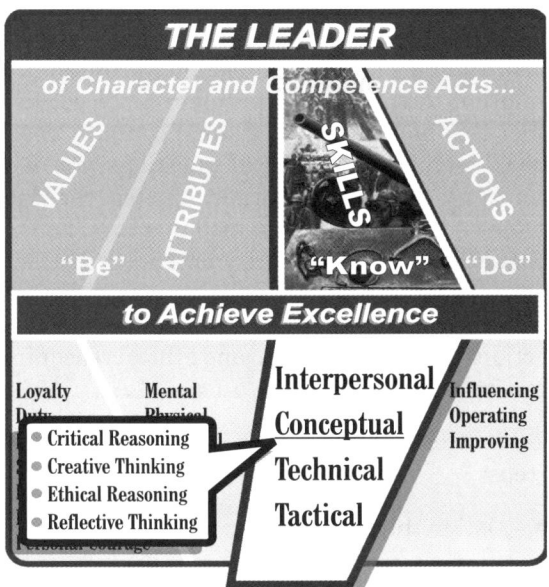

Figure 4-2. Direct Leader Skills—Conceptual

CREATIVE THINKING

4-22. Sometimes you run into a problem that you haven't seen before or an old problem that requires a new solution. Here you must apply imagination; a radical departure from the old way of doing things may be refreshing. Army leaders prevent complacency by finding ways to challenge subordinates with new approaches and ideas. In these cases, rely on your intuition, experience, and knowledge. Ask for input from your subordinates. Reinforce team building by making everybody responsible for, and shareholders in, the accomplishment of difficult tasks.

4-23. Creative thinking isn't some mysterious gift, nor does it have to be outlandish. It's not reserved for senior officers; all leaders think creatively. You employ it every day to solve small problems. A unit that deploys from a stateside post on a peace operation, for instance, may find itself in a small compound with limited athletic facilities and no room to run. Its leaders must devise new ways for their soldiers to maintain physical fitness. These may include sports and games, even games the local nationals play.

Pulling Dragons' Teeth

As American forces approached the Siegfried Line between Germany and France at the end of World War II, the armored advance was slowed by "dragons' teeth," concrete obstacles that looked like large, tightly spaced traffic cones. Engineers predicted it would take many days and tons of explosives to reduce the obstacles, which were heavily reinforced and deeply rooted. Then an NCO suggested using bulldozers to push dirt on top of the spikes, creating an earthen ramp to allow tanks to drive over the obstacles. This is but one example of the creative thinking by American soldiers of all ranks that contributed to victory in the ETO.

ETHICAL REASONING

4-24. Ethical leaders do the right things for the right reasons all the time, even when no one is watching. But figuring out what's the "right" thing is often, to put it mildly, a most difficult task. To fulfill your duty, maintain your integrity, and serve honorably, you must be able to reason ethically.

4-25. Occasionally, when there's little or no time, you'll have to make a snap decision based on your experience and intuition about what feels right. For Army leaders, such decisions are guided by Army values (discussed in Chapter 2), the institutional culture, and the organizational climate (discussed in Chapter 3). These shared values then serve as a basis for the whole team's buying into the leader's decision. But comfortable as this might be, you should not make all decisions on intuition.

4-26. When there's time to consider alternatives, ask for advice, and think things through, you can make a deliberate decision. First determine what's legally right by law and regulation. In gray areas requiring interpretation, apply Army values to the situation. Inside those boundaries, determine the best possible answer from among competing solutions, make your decision, and act on it.

4-27. The distinction between snap and deliberate decisions is important. In many decisions, you must think critically because your intuition—what feels right—may lead to the wrong answer. In combat especially, the intuitive response won't always work.

4-28. The moral application of force goes to the heart of military ethics. S. L. A. Marshall, a military historian as well as a brigadier general, has written that the typical soldier is often at a disadvantage in combat because he "comes from a civilization in which aggression, connected with the taking of a human life, is prohibited and unacceptable." Artist Jon Wolfe, an infantryman in Vietnam, once said that the first time he aimed his weapon at another human being, a "little voice" in the back of his mind asked, "Who gave you permission to do this?" That "little voice" comes, of course, from a lifetime of living within the law. You can

determine the right thing to do in these very unusual circumstances only when you apply ethical as well as critical reasoning.

4-29. The right action in the situation you face may not be in regulations or field manuals. Even the most exhaustive regulations can't predict every situation. They're designed for the routine, not the exceptional. One of the most difficult tasks facing you as an Army leader is determining when a rule or regulation simply doesn't apply because the situation you're facing falls outside the set of conditions envisioned by those who wrote the regulation. Remember COL Chamberlain on Little Round Top. The drill manuals he had studied didn't contain the solution to the tactical problem he faced; neither this nor any other manual contain "cookbook" solutions to ethical questions you will confront. COL Chamberlain *applied* the doctrine he learned from the drill manuals. So you should apply Army values, your knowledge, and your experience to any decision you make and be prepared to accept the consequences of your actions. Study, reflection, and ethical reasoning can help you do this.

4-30. Ethical reasoning takes you through these steps:

- Define the problem.
- Know the relevant rules.
- Develop and evaluate courses of action.
- Choose the course of action that best represents Army values.

4-31. These steps correspond to some of the steps of the decision making leadership action in Chapter 5. Thus, ethical reasoning isn't a separate process you trot out only when you think you're facing an ethical question. It should be part of the thought process you use to make any decision. Your subordinates count on you to do more than make tactically sound decisions. They rely on you to make decisions that are ethically sound as well. You should always consider ethical factors and, when necessary, use Army values to gauge what's right.

4-32. That said, not every decision is an ethical problem. In fact, most decisions are ethically neutral. But that doesn't mean you don't have to think about the ethical consequences of your actions. Only if you reflect on whether what you're asked to do or what you ask your people to do accords with Army values will you develop that sense of right and wrong that marks ethical people and great leaders. That sense of right and wrong alerts you to the presence of ethical aspects when you face a decision.

4-33. Ethical reasoning is an art, not a science, and sometimes the best answer is going to be hard to determine. Often, the hardest decisions are not between right and wrong, but between shades of right. Regulations may allow more than one choice. There may even be more than one good answer, or there may not be enough time to conduct a long review. In those cases, you must rely on your judgment.

Define the Problem

4-34. Defining the problem is the first step in making any decision. When you think a decision may have ethical aspects or effects, it's especially important to define it precisely. Know who said what—and what specifically was said, ordered, or demanded. Don't settle for secondhand information; get the details. Problems can be described in more than one way. This is the hardest step in solving any problem. It's especially difficult for decisions in the face of potential ethical conflicts. Too often some people come to rapid conclusions about the nature of a problem and end up applying solutions to what turn out to be only symptoms.

Know the Relevant Rules

4-35. This step is part of fact gathering, the second step in problem solving. Do your homework. Sometimes what looks like an ethical problem may stem from a misunderstanding of a regulation or policy, frustration, or overenthusiasm. Sometimes the person who gave an order or made a demand didn't check the regulation and a thorough reading may make the problem go away. Other times, a difficult situation results from trying to do something right in the wrong way. Also, some regulations leave room for interpretation; the problem then becomes a policy matter rather than an ethical one. If you do perceive an ethical problem, explain it to the person you think is causing it and try to come up with a better way to do the job.

Develop and Evaluate Courses of Action

4-36. Once you know the rules, lay out possible courses of action. As with the previous steps, you do this whenever you must make a decision. Next, consider these courses of action in view of Army values. Consider the consequences of your courses of action by asking yourself a few practical questions: Which course of action best upholds Army values? Do any of the courses of action compromise Army values? Does any course of action violate a principle, rule, or regulation identified in Step 2? Which course of action is in the best interest of the Army and of the nation? This part will feel like a juggling act; but with careful ethical reflection, you can reduce the chaos, determine the essentials, and choose the best course—even when that choice is the least bad of a set of undesirable options.

Choose the Course of Action That Best Represents Army Values

4-37. The last step in solving any problem is making a decision and acting on it. Leaders are paid to make decisions. As an Army leader, you're expected—by your bosses and your people—to make decisions that solve problems without violating Army values.

4-38. As a values-based organization, the Army uses expressed values—Army values—to provide its fundamental ethical framework. Army values lay out the ethical standards expected of soldiers and DA civilians. Taken together, Army values and ethical decision making provide a moral touchstone and a workable process that enable you to make sound ethical decisions and take right actions confidently.

4-39. The ethical aspects of some decisions are more obvious that those of others. This example contains an obvious ethical problem. The issues will seldom be so clear-cut; however, as you read the example, focus on the steps SGT Kirk follows as he moves toward an ethical decision. Follow the same steps when you seek to do the right thing.

The EFMB Test

SGT Kirk, who has already earned the Expert Field Medical Badge (EFMB), is assigned as a grader on the division's EFMB course. Sergeant Kirk's squad leader, SSG Michaels, passes through SGT Kirk's station and fails the task. Just before SGT Kirk records the score, SSG Michaels pulls him aside.

"I need my EFMB to get promoted," SSG Michaels says. "You can really help me out here; it's only a couple of points anyway. No big deal. Show a little loyalty."

SGT Kirk wants to help SSG Michaels, who's been an excellent squad leader and who's loyal to his subordinates. SSG Michaels even spent two Saturdays helping SGT Kirk prepare for his promotion board. If SGT Kirk wanted to make this easy on himself, he would say the choice is between honesty and loyalty. Then he could choose loyalty, falsify the score, and send everyone home happy. His life under SSG Michaels would probably be much easier too.

However, SGT Kirk would not have defined the problem correctly. (Remember, defining the problem is often the hardest step in ethical reasoning.) SGT Kirk knows the choice isn't between loyalty and honesty. Loyalty doesn't require that he lie. In fact, lying would be disloyal to the Army, himself, and the soldiers who met the standard. To falsify the score would also be a violation of the trust and confidence the Army placed in him when he was made an NCO and a grader. SGT Kirk knows that loyalty to the Army and the NCO corps comes first and that giving SSG Michaels a passing score would be granting the squad leader an unfair advantage. SGT Kirk knows it would be wrong to be a coward in the face of this ethical choice, just as it would be wrong to be a coward in battle. And if all that were not enough, when SGT Kirk imagines seeing the incident in the newspaper the next morning—Trusted NCO Lies to Help Boss—he knows what he must do.

4-40. When SGT Kirk stands his ground and does the right thing, it may cost him some pain in the short run, but the entire Army benefits. If he makes the wrong choice, he weakens the Army. Whether or not the Army lives by its values isn't just up to generals and colonels; it's up to each of the thousands of SGT Kirks, the Army leaders who must make tough calls when no one is watching, when the easy thing to do is the wrong thing to do.

REFLECTIVE THINKING

4-41. Leader development doesn't occur in a vacuum. All leaders must be open to feedback on their performance from multiple perspectives—seniors, peers, and subordinates. But being open to feedback is only one part of the equation. As a leader, you must also listen to and use the feedback: you must be able to reflect. Reflecting is the ability to take information, assess it, and apply it to behavior to explain why things did or did not go well. You can then use the resulting explanations to improve future behavior. Good leaders are always

striving to become better leaders. This means you need consistently to assess your strengths and weaknesses and reflect on what you can do to sustain your strengths and correct your weaknesses. To become a better leader, you must be willing to change.

4-42. For reasons discussed fully in Chapter 5, the Army often places a premium on doing—on the third element of BE, KNOW, DO. All Army leaders are busy dealing with what's on their plates and investing a lot of energy in accomplishing tasks. But how often do they take the time to STOP and really THINK about what they are doing? How often have you seen this sign on a leader's door: Do Not Disturb—Busy Reflecting? Not often. Well, good leaders need to take the time to think and reflect. Schedule it; start really exercising your capacity to get feedback. Then reflect on it and use it to improve. There's nothing wrong with making mistakes, but there's plenty wrong with not learning from those mistakes. Reflection is the means to that end.

TECHNICAL SKILLS

The first thing the senior NCOs had to do was to determine who wasn't qualified with his weapon, who didn't have his protective mask properly tested and sealed—just all the basic little things. Those things had to be determined real fast.

A Command Sergeant Major, Desert Storm

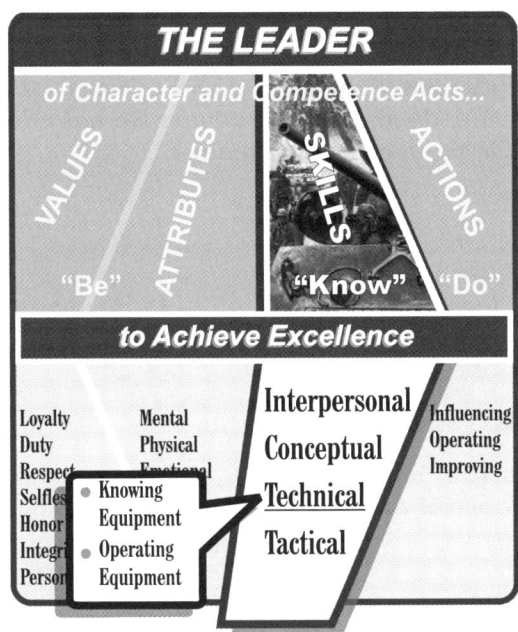

Figure 4-3. Direct Leader Skills—Technical

KNOWING EQUIPMENT

4-43. Technical skill is skill with things—equipment, weapons, systems—everything from the towing winch on the front of a vehicle to the computer that keeps track of corps personnel actions. Direct leaders must know their equipment and how to operate it. Figure 4-3 highlights direct leader technical skills. Technical manuals, training circulars, SOPs, and all the other publications necessary for efficient, effective performance explain specific skills more completely.

4-44. Direct leaders are closer to their equipment than organizational and strategic leaders. Thus, they have a greater need to know how it works and how to use it. In addition, direct leaders are the experts who are called upon to solve problems with the equipment, the ones who figure out how to make it work better, how to apply it, how to fix it—even how to modify it. Sergeants, junior officers, warrant officers, wage grade employees, and journeymen are the Army's technical experts and best teachers. Subordinates expect their first-line leaders to know their equipment and be experts in all the applicable technical skills.

OPERATING EQUIPMENT

4-45. Direct leaders know how to operate their equipment and make sure their people do as well. They set the example with a hands-on approach. When new equipment arrives, direct leaders find out how it works, learn how to use it themselves, and train their subordinates to do the same.

Technical Skill into Combat Power

Technical skill gave the Army a decided advantage in the 1944 battle for France. For example, the German Army had nothing like the US Army's maintenance battalions. Such an organization was a new idea, and a good one. These machine-age units were able to return almost half the battle-damaged tanks to action within two days. The job was done by young men who had been working at gas stations and body shops two years earlier and had brought their skill into the service of their country. Instead of fixing cars, they replaced damaged tank tracks, welded patches on the armor, and repaired engines. These combat supporters dragged tanks that were beyond repair to the rear and stripped them for parts. The Germans just left theirs in place.

I felt we had to get back to the basic soldier skills. The basics of setting up a training schedule for every soldier every day. We had to execute the standard field disciplines, such as NCOs checking weapons cleanliness and ensuring soldiers practiced personal hygiene daily. Our job is to go out there and kill the enemy. In order to do that, as Fehrenbach writes in [his study of the Korean Conflict entitled] This Kind of War, *we have to have disciplined teams; discipline brings pride to the unit. Discipline coupled with tough, realistic training is the key to high morale in units. Soldiers want to belong to good outfits, and our job as leaders is to give them the best outfit we can.*

A Company Commander, Desert Storm

4-46. This company commander is talking about two levels of skill. First is the individual level: soldiers are trained with their equipment and know how to do their jobs. Next is the collective level: leaders take these trained individuals and form them into teams. The result: a whole greater than the sum of its parts, a team that's more than just a collection of trained individuals, an organization that's capable of much more than any one of its elements. (FM 25-101 discusses how to integrate individual, collective, and leader training).

TACTICAL SKILLS

Man is and always will be the supreme element in combat, and upon the skill, the courage and endurance, and the fighting heart of the individual soldier the issue will ultimately depend.

General Matthew B. Ridgway
Former Army Chief of Staff

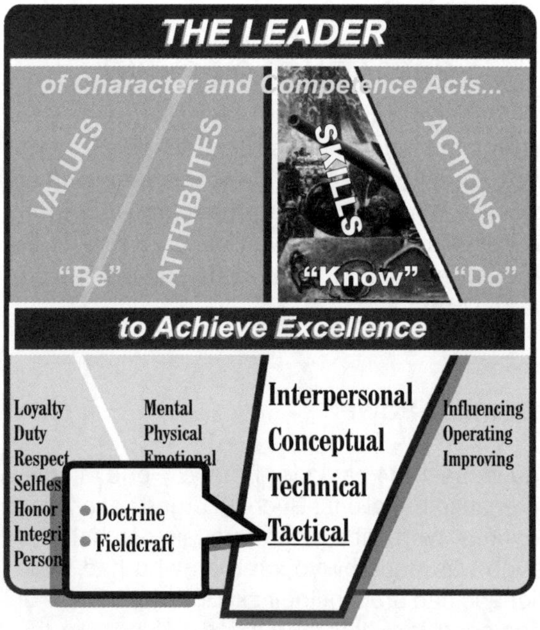

Figure 4-4. Direct Leader Skills—Tactical

DOCTRINE

4-47. Tactics is the art and science of employing available means to win battles and engagements. The science of tactics encompasses capabilities, techniques, and procedures that can be codified. The art of tactics includes the creative and flexible array of means to accomplish assigned missions, decision making when faced with an intelligent enemy, and the effects of combat on soldiers. Together, FM 100-34, FM 100-40, and branch-specific doctrinal manuals capture the tactical skills that are essential to mastering both the science and the art of tactics. Figure 4-4 highlights direct leader tactical skills.

FIELDCRAFT

4-48. Fieldcraft consists of the skills soldiers need to sustain themselves in the field. Proficiency in fieldcraft reduces the likelihood soldiers will become casualties. The requirement to be able to do one's job in a field environment distinguishes the soldier's profession from most civilian occupations. Likewise, the requirement that Army leaders make sure their soldiers take care of themselves and provide them with the means to do so is unique.

4-49. The Soldier's Manual of Common Tasks lists the individual skills all soldiers must master to operate effectively in the field. Those skills include everything from how to stay healthy, to how to pitch a tent, to how to run a heater. Some military occupational specialties (MOS) require proficiency in additional fieldcraft skills. Soldier's Manuals for these MOS list them.

4-50. Army leaders gain proficiency in fieldcraft through schooling, study, and practice. Once learned, few fieldcraft skills are difficult. However, they are easy to neglect during exercises, when everyone knows that the exercise will end at a specific time, sick and injured soldiers are always evacuated, and the adversary isn't using real ammunition. During peacetime, it's up to Army leaders to enforce tactical discipline, to make sure their soldiers practice the fieldcraft skills that will keep them from becoming casualties—battle or nonbattle—during operations.

TACTICAL SKILLS AND TRAINING

4-51. Direct leaders are the Army's primary tactical trainers, both for individuals and for teams. Practicing tactical skills is often challenging. The best way to improve individual and collective skills is to replicate operational conditions. Unfortunately, Army leaders can't always get the whole unit out in the field to practice maneuvers, so they make do with training parts of it separately. Sometimes they can't get the people, the time, and the money all together at the right time and the right place to train the entire team. There are always training distracters. There will always be a hundred excuses not to train together and one reason why such training must occur: units fight as they train. (FM 25-100 and FM 25-101 discuss training principles and techniques.)

4-52. Unfortunately, the Army has been caught unprepared for war more than once. In July 1950, American troops who had been on occupation duty in Japan were thrown into combat when North Korean forces invaded South Korea. Ill-trained, ill-equipped, and out of shape, they went into action and were overrun. However, that same conflict provides another example of how well things can go when a direct leader has tactical skill, the ability to pull people and things together into a team. Near the end of November 1950, American forces were chasing the remnants of the broken North Korean People's Army into the remote northern corners of the Korean Peninsula. Two American units pushed all the way to the Yalu River, which forms the boundary between North Korea and the People's Republic of China. One was the 17th Infantry Regiment. The other was a task force commanded by a 24-year-old first lieutenant named Joseph Kingston.

Task Force Kingston

1LT Joseph Kingston, a boyish-looking platoon leader in K Company, 3d Battalion, 32d Infantry, was the lead element for his battalion's move northward. The terrain was mountainous, the weather bitterly cold—the temperature often below zero—and the cornered enemy still dangerous. 1LT Kingston inched his way forward, with the battalion adding elements to his force. He had antiaircraft jeeps mounted with quad .50 caliber machine guns, a tank, a squad (later a platoon) of engineers, and an artillery forward observer. Some of these attachments were commanded by lieutenants who outranked him, as did a captain who was the tactical air controller. But 1LT Kingston remained in command, and battalion headquarters began referring to Task Force Kingston.

Bogged down in Yongsong-ni with casualties mounting, Task Force Kingston received reinforcements that brought the number of men to nearly 300. Despite tough fighting, the force continued to move northward. 1LT Kingston's battalion commander wanted him to remain in command, even though they sent several more officers who outranked 1LT Kingston. One of the

---

Task Force Kingston (continued)

attached units was a rifle company, commanded by a captain. But the arrangement worked, mostly because 1LT Kingston himself was an able leader. Hit while leading an assault on one enemy stronghold, he managed to toss a grenade just as a North Korean soldier shot him in the head. His helmet, badly grazed, saved his life. His personal courage inspired his men and the soldiers from the widely varied units who were under his control. Task Force Kingston was commanded by the soldier who showed, by courage and personal example, that he could handle the job.

4-53. 1LT Kingston made the task force work by applying skills at a level of responsibility far above what was normal for a soldier of his rank and experience. He knew how to shoot, move, and communicate. He knew the fundamentals of his profession. He employed the weapons under his command and controlled a rather unwieldy collection of combat assets. He understood small-unit tactics and applied his reasoning skills to make decisions. He fostered a sense of teamwork, even in this collection of units that had never trained together. Finally, he set the example with personal courage.

SUMMARY

4-54. Direct leadership is face-to-face, first-line leadership. It takes place in organizations where subordinates are used to seeing their leaders all the time: teams, squads, sections, platoons, companies, and battalions. To be effective, direct leaders must master many interpersonal, conceptual, technical, and tactical skills.

4-55. Direct leaders are first-line leaders. They apply the conceptual skills of critical reasoning and creative thinking to determine the best way to accomplish the mission. They use ethical reasoning to make sure their choice is the right thing to do, and they use reflective thinking to assess and improve team performance, their subordinates, and themselves. They employ the interpersonal skills of communicating and supervising to get the job done. They develop their people by mentoring and counseling and mold them into cohesive teams by training them to standard.

4-56. Direct leaders are the Army's technical experts and best teachers. Both their bosses and their people expect them to know their equipment and be experts in all the applicable technical skills. On top of that, direct leaders combine those skills with the tactical skills of doctrine, fieldcraft, and training to accomplish tactical missions.

4-57. Direct leaders use their competence to foster discipline in their units and to develop soldiers and DA civilians of character. They use their mastery of equipment and doctrine to train their subordinates to standard. They create and sustain teams with the skill, trust, and confidence to succeed—in peace and war.

Chapter 5

Direct Leadership Actions

5-1. Preparing to be a leader doesn't get the job done; the test of your character and competence comes when you act, when you DO those things required of a leader.

5-2. The three broad leader actions that Chapters 1 and 2 introduced—influencing, operating, and improving—contain other activities. As with the skills and attributes discussed previously, none of these exist alone. Most of what you do as a leader is a mix of these actions. This manual talks about them individually to explain them more clearly; in practice they're often too closely connected to sort out.

5-3. Remember that your actions say more about what kind of leader you are than anything else. Your people watch you all the time; you're always on duty. And if there's a disconnect between what you say and how you act, they'll make up their minds about you—and act accordingly—based on how you act. It's not good enough to talk the talk; you have to walk the walk.

> **The most important influence you have on your people is the example you set.**

INFLUENCING ACTIONS

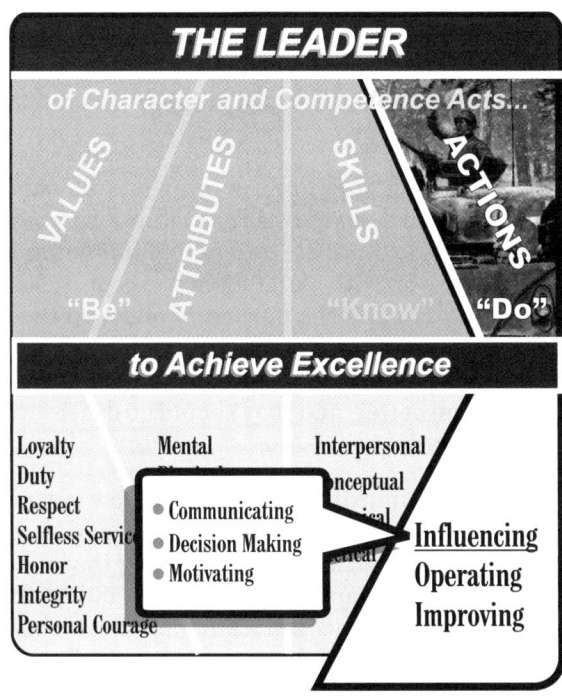

Figure 5-1. Direct Leader Actions—Influencing

5-4. Leadership is both art and science. It requires constant study, hard work, and frequent practice. Since you're dealing with people and their emotions, dreams, and fears, it also calls for imagination and a positive, upbeat approach.

5-5. Effective leaders act competently and confidently. Your attitude sets the tone for the entire unit, and you choose your attitude—day to day, task to task, even minute to minute. Remember that optimism, a positive outlook, and a sense of humor are infectious. This is especially true when you must make unpopular decisions and face the challenge of bringing the team on board.

5-6. Figure 5-1 shows that influencing consists of communicating, decision making and motivating. As a leader, you should be asking several questions: What's happening? What should be happening but isn't? Why are these things happening? Then ask yourself: How can I get this team moving toward the goal? (Appendix B lists leader performance indicators).

COMMUNICATING

You must talk to your soldiers...I don't just mean in formation or groups, but one-on-one. Take time (at least 15 to 30 minutes a day) to really talk to a soldier, one soldier a day.

Command Sergeant Major Daniel E. Wright

5-7. Leaders keep their subordinates informed because doing so shows trust, because sharing information can relieve stress, and because information allows subordinates to determine what they need to do to accomplish the mission when circumstances change. By informing them of a decision—and, as much as possible, the reasons for it—you show your subordinates they're important members of the team. Accurate information also relieves unnecessary stress and helps keep rumors under control. (Without an explanation for what's happening, your people will manufacture one—or several—of their own.) Finally, if something should happen to you, the next leader in the chain will be better prepared to take over and accomplish the mission if everyone knows what's going on. Subordinates must understand your intent. In a tactical setting, leaders must understand the intent of their commanders two levels up.

5-8. In other situations, leaders use a variety of means to keep people informed, from face-to-face talks to published memos and family newsletters. No matter what the method, keep two things in mind:

- As a leader, you are responsible for making sure your subordinates understand you.
- Communication isn't limited to your immediate superiors and subordinates.

5-9. The success or failure of any communication is the responsibility of the leader. If it appears your subordinates don't understand, check to make sure you've made yourself clear. In fact, even if you think your people understand, check anyway; ask for a back-brief.

5-10. Don't assume that communication begins or ends at the next level up or the next level down. If you're a team leader, listen carefully to what your supervisors, platoon sergeants, platoon leaders, and company commanders say. If you're a platoon sergeant, pass the word through your squad leaders or section chiefs, but also watch and *listen to the troops* to see if the information has made it all the way to where it needs to go. Listen carefully at least two levels up and two levels down.

5-11. In combat, subordinates may be out of contact with their leaders. Sometimes the plan falls apart because of something unexpected—weather, terrain, enemy action. Sometimes the leader may be killed or wounded. In those situations, subordinates who know the overall purpose of the mission and the commander's intent have the basic information they need to carry on. And if the leader has established a climate of trust, if the leader has trained the subordinate leaders in how and why decisions are made, one of these subordinates is more likely to step up and take charge.

5-12. To prepare your subordinates for such circumstances, create training situations where they must act on their own with a minimum of guidance—or with no guidance except a clear understanding of the purpose. Follow up these training situations with AARs so that subordinates learn what they did well, what they could have done better, and they should do differently next time.

5-13. Communicating also goes on from bottom to top. Leaders find out what their people are thinking, saying, and doing by using that most important communication tool: listening. By listening carefully, you can even hear those messages behind what a person is actually saying, the equivalent of reading between the lines. Practice "leadership by walking around." Get out and coach, listen, teach, and clarify; pass on what you learn to your superiors. They need to know what's going on to make good plans.

DECISION MAKING

A good leader must sometimes be stubborn. Armed with the courage of his convictions, he must often fight to defend them. When he has come to a decision after thorough analysis—and when he is sure he is right—he must stick to it even to the point of stubbornness.

General of the Army, Omar N. Bradley
Address to the US Army Command and
General Staff College, May 1967

5-14. A problem is an existing condition or situation in which what you want to happen is different from what actually is happening. Decision making is the process that begins to change that situation. Thus, decision making is knowing *whether* to decide, then *when* and *what* to decide. It includes understanding the consequences of your decisions.

5-15. Army leaders usually follow one of two decision-making processes. Leaders at company level and below follow the troop leading procedures (TLP). The TLP are designed to support solving tactical problems. Leaders at battalion level and above follow the military decision making process (MDMP). The MDMP, which FM 101-5 discusses, is designed for organizations with staffs. These established and proven methodologies combine elements of the planning operating action to save time and achieve parallel decision making and planning. Both follow the problem solving steps discussed below.

5-16. Every once in a while, you may come across a decision that's easy to make: yes or no, right or left, on or off. As you gain experience as a leader, some of the decisions you find difficult now will become easier. But there will always be difficult decisions that require imagination, that require rigorous thinking and analysis, or that require you to factor in your gut reaction. Those are the tough decisions, the ones you're getting paid to make. As an experienced first sergeant once said to a brand new company commander, "We get paid the big bucks to make the hard calls." The next several paragraphs explain the steps you should use to solve a problem; then you'll read about other factors that affect how you make those hard calls and the importance of setting priorities.

Problem Solving Steps

5-17. **Identify the problem**. Don't be distracted by the symptoms of the problem; get at its root cause. There may be more than one thing contributing to a problem, and you may run into a case where there are lots of contributing factors but no real "smoking gun." The issue you choose to address as the root cause becomes the mission (or restated mission for tactical problems). The mission must include a simple statement of who, what, when, where, and why. In addition, it should include your end state, how you want things to look when the mission is complete.

5-18. **Identify facts and assumptions**. Get whatever facts you can in the time you have. Facts are statements of what you know about the situation. Assumptions are statements of what you believe about the situation but don't have facts to support. Make only assumptions that are likely to be true and essential to generate alternatives. Some of the many sources of facts include regulations, policies, and doctrinal publications. Your organization's mission, goals, and objectives may also be a source. Sources of assumptions can be personal experiences, members of the organization, subject matter experts, or written observations. Analyze the facts and assumptions you identify to determine the scope of the problem. (FM 101-5 contains more information on facts and assumptions.)

5-19. **Generate alternatives**. Alternatives are ways to solve the problem. Develop more than one possible alternative. Don't be satisfied with the first thing that comes into your mind. That's lazy thinking; the third or fourth or twentieth alternative you come up with might be the best one. If you have time and experienced subordinates, include them in this step.

5-20. **Analyze the alternatives**. Identify intended and unintended consequences, resource or other constraints, and the advantages and disadvantages of each alternative. Be sure to consider all your alternatives. Don't prejudge the situation by favoring any one alternative over the others.

5-21. **Compare the alternatives**. Evaluate each alternative for its probability of success and its cost. Think past the immediate future. How will this decision change things tomorrow? Next week? Next year?

5-22. **Make and execute your decision**. Prepare a leader's plan of action, if necessary, and put it in motion. (Planning, an operating action, is covered later in this chapter. Appendix C discusses plans of action as part of developmental counseling. Appendix D contains an example of a leader's plan of action.)

5-23. **Assess the results**. Check constantly to see how the execution of your plan of action is going. Keep track of what happens. Adjust your plan, if necessary. Learn from the experience so you'll be better equipped next time. Follow up on results and make further adjustments as required.

Factors to Consider

5-24. All of this looks great on paper; and it's easy to talk about when things are calm, when there's plenty of time. But even when there isn't a great deal of time, you'll come up with the best solution if you follow this process to the extent that time allows.

5-25. Even following these steps, you may find that with some decisions you need to take into account your knowledge, your intuition, and your best judgment. Intuition tells you what feels right; it comes from accumulated experience, often referred to as "gut feeling." However, don't be fooled into relying only on intuition, even if it has worked in the past. A leader who says "Hey, I just do what feels right"

may be hiding a lack of competence or may just be too lazy to do the homework needed to make a reasoned, thought-out decision. Don't let that be you. Use your experience, listen to your instincts, but do your research as well. Get the facts and generate alternatives. Analyze and compare as many as time allows. Then make your decision and act.

5-26. Remember also that any decision you make must reflect Army values. Chapter 4 discusses ethical reasoning. Its steps match the problem solving steps outlined here. Most problems are not ethical problems, but many have ethical aspects. Taking leave for example, is a right soldiers and DA civilians enjoy, but leaders must balance mission requirements with their people's desires and their own. Reconciling such issues may require ethical reasoning. As a leader, your superiors and your people expect you to take ethical aspects into account and make decisions that are right as well as good.

Setting Priorities

5-27. Decisions are not often narrowly defined, as in "Do I choose A or B?" Leaders make decisions when they establish priorities and determine what's important, when they supervise, when they choose someone for a job, when they train.

5-28. As a leader, you must also set priorities. If you give your subordinates a list of things to do and say "They're all important," you may be trying to say something about urgency. But the message you actually send is "I can't decide which of these are most important, so I'll just lean on you and see what happens."

5-29. Sometimes all courses of action may appear equally good (or equally bad) and that any decision will be equally right (or equally wrong). Situations like that may tempt you to sit on the fence, to make no decision and let things work themselves out. Occasionally that may be appropriate; remember that decision making involves judgment, knowing _whether_ to decide. More often, things left to themselves go from bad to worse. In such situations, the decision you make may be less important than simply deciding to do something. Leaders must have the personal courage to say which tasks are more important than others. In the absence of a clear priority, you must set one; not everything can be a top priority, and you can't make progress without making decisions.

Solving a Training Problem

A rifle platoon gets a new platoon leader and a new platoon sergeant within days of a poor showing in the division's military operations on urbanized terrain (MOUT) exercise. The new leaders assume the platoon's poor showing is a problem. Feedback from the evaluators is general and vague. The platoon's squad and fire team leaders are angry and not much help in assessing what went wrong, so the new leaders begin investigating. In their fact-finding step they identify the following facts: (1) The soldiers are out of shape and unable to complete some of the physical tasks. (2) The fire team leaders don't know MOUT tactics, and some of the squad leaders are also weak. (3) Third Squad performed well, but didn't help the other squads. (4) The soldiers didn't have the right equipment at the training site.

Pushing a bit further to get at the root causes of these problems, the new leaders uncover the following: (1) Platoon PT emphasizes preparation for the APFT only. (2) Third Squad's leaders know MOUT techniques, and had even developed simple drills to help their soldiers learn, but because of unhealthy competition encouraged by the previous leaders, Third Squad didn't share the knowledge. (3) The company supply sergeant has the equipment the soldiers needed, but because the platoon had lost some equipment on the last field exercise, the supply sergeant didn't let the platoon sign out the equipment.

The new platoon leader and platoon sergeant set a goal of successfully meeting the exercise standard in two months. To generate alternatives, they meet with the squad leaders and ask for suggestions to improve training. They use all their available resources to develop solutions. Among the things suggested was to shuffle some of the team leaders to break up Third Squad's clique and spread some of the tactical knowledge around. When squad leaders complained, the platoon sergeant emphasized that they must think as a platoon, not just a collection of squads.

The platoon sergeant talks to the supply sergeant, who tells him the platoon's previous leadership had been lax about property accountability. Furthermore, the previous leaders didn't want to bother keeping track of equipment, so they often left it in garrison. The platoon sergeant teaches his squad leaders how to keep track of equipment and says that, in the future, soldiers who lose equipment will pay for it: "We wouldn't leave our stuff behind in war, so we're not going to do it in training."

Building on Third Squad's experience, the platoon leader works with the squad and fire team leaders to come up with some simple drills for the platoon's missions. He takes the leaders to the field and practices the drills with them so they'll be able to train their soldiers to the new standard.

The platoon sergeant also goes to the brigade's fitness trainers and, with their help, develops a PT program that emphasizes skills the soldiers need for their combat tasks. The new program includes rope climbing, running with weapons and equipment, and road marches. Finally, the leaders monitor how their plan is working. A few weeks before going through the course again, they decide to eliminate one of the battle drills because the squad leaders suggested that it wasn't necessary after all.

5-30. The platoon leader and platoon sergeant followed the problem solving steps you just read about. Given a problem (poor performance), they identified the facts surrounding it (poor PT practices, poor property accountability, and unhealthy competition), developed a plan of action, and executed it. Where appropriate, they analyzed and compared different alternatives (Third Squad's drills). They included their subordinates in the process, but had the moral courage to make unpopular decisions (breaking up the Third Squad clique). Will the platoon do better the next time out? Probably, but before then the new leaders will have to assess the results of their actions to make sure they're accomplishing what the leaders want. There may be other aspects of this problem that were not apparent at first. And following this or any process doesn't guarantee success. The process is only a framework that helps you make a plan and act. Success depends on your ability to apply your attributes and skills to influencing and operating actions.

5-31. Army leaders also make decisions when they evaluate subordinates, whether it's with a counseling statement, an evaluation report, or even on-the-spot encouragement. At an in-ranks inspection, a new squad leader takes a second look at a soldier's haircut—or lack of one. The squad leader's first reaction may be to ask, "Did you get your haircut lately?" But that avoids the problem. The soldier's haircut is either to standard or not—the NCO must decide. The squad leader either says—without apologizing or dancing around the subject—"You need a haircut" or else says nothing. Either way, the decision communicates the leader's standard. Looking a subordinate in the eye and making a necessary correction is a direct leader hallmark.

MOTIVATING

A unit with a high esprit de corps can accomplish its mission in spite of seemingly insurmountable odds.

FM 22-10, 1951

5-32. Recall from Chapter 1 that motivation involves using word and example to give your subordinates the will to accomplish the mission. Motivation grows out of people's confidence in themselves, their unit, and their leaders. This confidence is born in hard, realistic training; it's nurtured by constant reinforcement and through the kind of leadership—consistent, hard, and fair—that promotes trust. Remember that trust, like loyalty, is a gift your soldiers give you only when you demonstrate that you deserve it. Motivation also springs from the person's faith in the larger mission of the organization—a sense of being a part of the big picture.

Empowering People

5-33. People want to be recognized for the work they do and want to be empowered. You empower subordinates when you train them to do a job, give them the necessary resources and authority, get out of their way, and let them work. Not only is this a tremendous statement of the trust you have in your subordinates; it's one of the best ways to develop them as leaders. Coach and counsel them, both when they succeed and when they fail.

Positive Reinforcement

5-34. Part of empowering subordinates is finding out their needs. Talk to your people: find out what's important to them, what they want to accomplish, what their personal goals are. Give them feedback that lets them know how they're doing. Listen carefully so that you know what they mean, not just what they say. Use their feedback when it makes sense, and if you change something in the organization because of a subordinate's suggestion, let everyone know where the good idea came from. Remember, there's no limit to the amount of good you can do as long as you don't worry about who gets the credit. Give the credit to those who deserve it and you'll be amazed at the results.

5-35. You recognize subordinates when you give them credit for the work they do, from a pat on the back to a formal award or decoration. Don't underestimate the power of a few choice words of praise when a person has done a good job. Don't hesitate to give out awards—commendations, letters, certificates—when appropriate. (Use good judgment, however. If you give out a medal for every little thing, pretty soon the award becomes meaningless. Give an award for the wrong thing and you show you're out of touch.) Napoleon marveled at the motivational power of properly awarded ribbons and medals. He once said that if he had enough ribbon, he could rule the world.

5-36. When using rewards, you have many options. Here are some things to consider:

- Consult the leadership chain for recommendations.
- Choose a reward valued by the person receiving it, one that appeals to the individual's personal pride. This may be a locally approved award that's more respected than traditional DA awards.
- Use the established system of awards (certificates, medals, letters of commendation, driver and mechanic badges) when appropriate. These are recognized throughout the Army; when a soldier goes to a new unit, the reward will still be valuable.
- Present the award at an appropriate ceremony. Emphasize its importance. Let others see how hard work is rewarded.
- Give rewards promptly.
- Praise only good work or honest effort. Giving praise too freely cheapens its effect.
- Promote people who get the job done and who influence others to do better work.
- Recognize those who meet the standard and improve their performance. A soldier who works hard and raises his score on the APFT deserves some recognition, even if the soldier doesn't achieve the maximum score. Not everyone can be soldier of the quarter.

Negative Reinforcement

5-37. Of course, not everyone is going to perform to standard. In fact, some will require punishment. Using punishment to motivate a person away from an undesirable behavior is effective, but can be tricky. Sound judgment must guide you when administering punishment. Consider these guidelines:

- Before you punish a subordinate, make sure the subordinate understands the reason for the punishment. In most—although not all—cases, you'll want to try to change the subordinate's behavior by counseling or retraining before resulting to punishment.
- Consult your leader or supervisor before you punish a subordinate. They'll be aware of policies you need to consider and may be able to assist you in changing the subordinate's behavior.
- Avoid threatening a subordinate with punishment. Making a threat puts you in the position of having to deliver on that threat. In such a situation you may end up punishing because you said you would rather than because the behavior merits punishment. This undermines your standing as a leader.
- Avoid mass punishment. Correctly identify the problem, determine if an individual or individuals are responsible, and use an appropriate form of correction.
- With an open mind and without prejudging, listen to the subordinate's side of the story.
- Let the subordinate know that it's the behavior—not the individual—that is the problem. "You let the team down" works; "You're a loser" sends the wrong message.
- Since people tend to live up to their leader's expectations, tell them, "I know you can do better than that. I expect you to do better than that."
- Punish those who are able but *unwilling* to perform. Retrain a person who's *unable* to complete a task.
- Respond immediately to undesirable behavior. Investigate fully. Take prompt and prudent corrective action in accordance with established legal or regulatory procedures.
- Never humiliate a subordinate; avoid public reprimand.
- Ensure the person knows exactly what behavior got the person in trouble.
- Make sure the punishment isn't excessive or unreasonable. It's not only the severity of punishment that keeps subordinates in line; it's the certainty that they can't get away with undesirable behavior.
- Control your temper and hold no grudges. Don't let your personal feelings interfere; whether you like or dislike someone has nothing to do with good order and discipline.

5-38. If you were surprised to find a discussion of punishment under the section on motivation, consider this: good leaders are always on the lookout for opportunities to develop subordinates, even the ones who are being punished. Your people—even the ones who cause you problems—are still the most important resource you have. When a vehicle is broken, you don't throw it out; you fix it. If one of your people is performing poorly, don't just get rid of the person; try to help fix the problem.

OPERATING ACTIONS

Figure 5-2. Direct Leader Actions—Operating

5-39. You're operating when you act to achieve an immediate objective, when you're working to get today's job done. Although FM 25-100 is predominantly a training tool, its methodology applies to a unit's overall operational effectiveness. Because operating includes planning, preparing, executing, and assessing (see Figure 5-2), you can use the FM 25-100 principles as a model for operations other than training. Sometimes these elements are part of a cycle; other times they happen simultaneously.

5-40. You'll often find yourself influencing after you've moved on to operating. In practice, the nice, neat divisions in this manual are not clear-cut; you often must handle multiple tasks requiring different skills at the same time. (Appendix B lists operating actions and some indicators of effectiveness.)

PLANNING AND PREPARING

5-41. In peacetime training, in actual operations, and especially in combat, your job is to help your organization function effectively—accomplish the mission—in an environment that can be chaotic. That begins with a well thought-out plan and thorough preparation. A well-trained organization with a sound plan is much better prepared than one without a plan. Planning ahead reduces confusion, builds subordinates' confidence in themselves and the organization, and helps ensure success with a minimum of wasted effort—or in combat, the minimum number of casualties. (FM 101-5 discusses the different types of plans.)

5-42. A plan is a proposal for executing a command decision or project. Planning begins with a mission, specified or implied. A specified mission comes from your boss or from higher headquarters. An implied mission results when the leader, who may be you, sees something within his area of responsibility that needs to be done and, on his own initiative, develops a leader plan of action. (Remember that a problem exists when you're not satisfied with the way things are or the direction they're heading.) Either type of mission contains implied and specified tasks, actions that must be completed to accomplish the mission. (FM 101-5 discusses how the MDMP supports planning.)

Reverse Planning

5-43. When you begin with the goal in mind, you often will use the reverse planning method. Start with the question "Where do I want to end up?" and work backward from there until you reach "We are here right now."

5-44. Along the way, determine the basics of what's required: who, what, when, where, and why. You may also want to consider how to accomplish the task, although the "how" is usually not included in a mission to a subordinate. As you plan, consider the amount of time needed to coordinate and conduct each step. For instance, a tank platoon sergeant whose platoon has to spend part of a field exercise on the firing range might have to arrange, among other things, refueling at the range. No one explicitly said to refuel at the range, but the platoon sergeant knows what needs to happen. The platoon sergeant must think through the steps from the last to the first: (1) when the refueling must be complete, (2) how long the refueling will take, (3) how long it takes the refueling unit to get set up, and finally (4) when the refueling vehicles should report to the range.

5-45. After you have figured out what must happen on the way to the goal, put the tasks in sequence, set priorities, and determine a schedule. Look at the steps in the order they will occur. Make sure events are in logical order and you have allotted enough time for each one. As always, a good leader asks for input from subordinates when time allows. Getting input not only acts as a check on the your plan (you may have overlooked something), but also gets your people involved; involvement builds trust, self-confidence, and the will to succeed.

Preparing

5-46. While leaders plan, subordinates prepare. Leaders can develop a plan while their organization is preparing if they provide advance notice of the task or mission and initial guidance for preparation in a warning order. (Warning orders are part of the TLP and MDMP; however, any leader—uniformed or DA

civilian—can apply the principle of the warning order by giving subordinates advance notice of an impending requirement and how they'll be expected to contribute to it. FM 101-5 discusses warning orders.) Based on this guidance, subordinates can draw ammunition, rehearse key actions, inspect equipment, conduct security patrols, or begin movement while the leader completes the plan. In the case of a nontactical requirement, preparation may include making sure the necessary facilities and other resources are available to support it. In all cases, preparation includes coordinating with people and organizations that are involved or might be affected by the operation or project. (TC 25-30 discusses preparing for company- and platoon-level training).

5-47. Rehearsal is an important element of preparation. Rehearsing key combat actions lets subordinates see how things are supposed to work and builds confidence in the plan for both soldiers and leaders. Even a simple walk-through helps them visualize who's supposed to be where and do what when. Mobilization exercises provide a similar function for DA civilians and reserve component soldiers: they provide a chance to understand and rehearse mobilization and deployment support functions. Execution goes more smoothly because everyone has a mental picture of what's supposed to happen. Rehearsals help people remember their responsibilities. They also help leaders see how things might happen, what might go wrong, how the plan needs to be changed, and what things the leader didn't think of. (FM 101-5 contains an in-depth discussion of rehearsals.)

An Implied Mission and Leader Plan of Action

Not all missions originate with higher headquarters; sometimes the leader sees what's required and, exercising initiative, develops a leader plan of action.

Suppose a platoon sergeant's soldiers had trouble meeting minimum weapons qualification requirements. Since everyone qualified, no one has said to work on marksmanship. But the platoon sergeant knows the platoon made it through range week on sheer luck. The leader develops a training

---

An Implied Mission and Leader Plan of Action (continued)

plan to work on basic marksmanship, then goes to the platoon leader and presents it. Together the two leaders figure out a way to make sure their soldiers get a chance to train, even with all the other mission requirements. After they've talked it over, they bring in their subordinate leaders and involve them in the planning process. The platoon sergeant keeps track of the progress and effectiveness of the leader plan of action, making sure it accomplishes the intent and changing it when necessary. Later, the platoon leader and platoon sergeant meet to assess the plan's results and to decide if further action is required.

5-48. Leader plans of action can be used to reinforce positive behavior, improve performance, or even change an aspect of the organizational climate. A leader plan of action may also be personal—as when the leader decides "I need to improve my skills in this area."

Brief Solutions, Not Problems

Leaders develop their subordinates by requiring those subordinates to plan. A lieutenant, new to the battalion staff, ran into a problem getting all the resources the unit was going to need for an upcoming deployment. The officer studied the problem, talked to the people involved, checked his facts, and generally did a thorough analysis—of which he was very proud. Then he marched into the battalion executive officer's (XO's) office and laid it all out in a masterly fashion. The XO looked up from his desk and said, "Great. What are you going to do about it?"

The lieutenant was back in a half-hour with three possible solutions he had worked out with his NCOs. From that day on, the officer never presented a problem to any boss without offering some solutions as well. The lieutenant learned a useful technique from the XO. He learned it so well he began using it with his soldiers and became a better coach and mentor because of it.

5-49. No matter what your position is, part of your duty is making your boss's job easier. Just as you loyally provide resources and authority for your subordinates to do their jobs, you leave the boss free to do his. Ask only for decisions that fall outside your scope of authority—not those you want to avoid. Forward only problems you can't fix—not those whose solutions are just difficult. Ask for advice from others with more experience or seek clarification when you don't understand what's required. Do all that and exercise disciplined initiative within your boss's intent. (Appendix A discusses delegation of authority.)

EXECUTING

Soldiers do what they are told to do. It's leadership that's the key. Young men and women join the Army; if they're with competent,
confident, capable leaders they turn into good soldiers.

Sergeant Major of the Army Robert E. Hall

5-50. Executing means acting to accomplish the mission, moving to achieve the leader's goals as expressed in the leader's vision—to standard and on time—while taking care of your people.

5-51. Execution, the payoff, is based on all the work that has gone before. But planning and preparation alone can't guarantee success. Things will go wrong. Large chunks of the plan will go flying out the window. At times, it will seem as if everything is working against you. Then you must have the will to fight through, keeping in mind your higher leaders' intent and the mission's ultimate goal. You must adapt and improvise.

5-52. In a tactical setting, all leaders must know the intent of commanders two levels up. During execution, position yourself to best lead your people, initiate and control the action, get others to follow the plan, react to changes, keep your people focused, and work the team to accomplish the goal to standard. A well-trained organization accomplishes the mission, even when things go wrong.

5-53. Finally, leaders ensure they and their subordinate leaders are doing the right jobs. This goes hand in hand with empowerment. A company commander doesn't do a squad leader's job. A division chief doesn't do a branch chief's job. A supervisor doesn't do a team leader's job.

Maintaining Standards

5-54. The Army has established standards for all military activities. Standards are formal, detailed instructions that can be stated, measured, and achieved. They provide a performance baseline to evaluate how well a specific task has been executed. You must know, communicate and enforce standards. Explain the ones that apply to your organization and give your subordinate leaders the authority to enforce them. Then hold your subordinates responsible for achieving them.

5-55. Army leaders don't set the minimum standards as goals. However, everything can't be a number one priority. As an Army leader, you must exercise judgment concerning which tasks are most important. Organizations are required to perform many tasks that are not mission-related. While some of these are extremely important, others require only a minimum effort. Striving for excellence in every area, regardless of how trivial, quickly works an organization to death. On the other hand, the fact that a task isn't a first priority doesn't excuse a sloppy performance. Professional soldiers accomplish all tasks to standard. Competent leaders make sure the standard fits the task's importance.

Setting Goals

5-56. The leader's ultimate goal—your ultimate goal—is to train the organization to succeed in its wartime mission. Your daily work includes setting intermediate goals to get the organization ready. Involve your subordinates in goal setting. This kind of cooperation fosters trust and makes the best use of subordinates' talents. When developing goals, consider these points:

- Goals must be realistic, challenging, and attainable.
- Goals should lead to improved combat readiness.
- Subordinates ought to be involved in the goal setting.
- Leaders develop a plan of action to achieve each goal.

ASSESSING

Schools and their training offer better ways to do things, but only through experience are we able to capitalize on this learning. The process of profiting from mistakes becomes a milestone in learning to become a more efficient soldier.

Former Sergeant Major of the Army
William G. Bainbridge

5-57. Setting goals and maintaining standards are central to assessing mission accomplishment. Whenever you talk about accomplishing the mission, always include the phrase "to standard." When you set goals for your subordinates, make sure they know what the standards are. To use a simple example, the goal might be "All unit members will pass the APFT." The APFT standard tells you, for each exercise, how many repetitions are required in how much time, as well as describing a proper way to do the exercise.

5-58. Also central to assessing is spot checking. Army leaders check things: people, performance, equipment, resources. They check things to ensure the organization is meeting standards and moving toward the goals the leader has established. Look closely; do it early and often; do it both before and after the fact. Praise good performance and figure out how to fix poor performance. Watch good first sergeants or command sergeants major as they go through the mess line at the organizational dining facility. They pick up the silverware and run their fingers over it—almost unconsciously—checking for cleanliness. Good leaders supervise, inspect, and correct their subordinates. They don't waste time; they're always on duty.

5-59. Some assessments you make yourself. For others, you may want to involve subordinates. Involving subordinates in assessments and obtaining straightforward feedback from them become more important as your span of authority increases. Two techniques that involve your subordinates in assessing are in-process reviews (IPRs) and after-action reviews (AARs).

In-Process Reviews

5-60. Successful assessment begins with forming a picture of the organization's performance early. Anticipate which areas the organization might have trouble in; that way you know which areas to watch closely. Once the organization begins the mission, use IPRs to evaluate performance and give feedback. Think of an IPR as a checkpoint on the way to mission accomplishment.

5-61. Say you tell your driver to take you to division headquarters. If you recognize the landmarks, you decide your driver knows the way and probably say nothing. If you don't recognize the landmarks, you might ask where you are. And if you determine that the driver is lost or has made a wrong turn, you give instructions to get back to where you need to be. In more complex missions, IPRs give leaders and subordinates a chance to talk about what's going on. They can catch problems early and take steps to correct or avoid them.

After-Action Reviews

5-62. AARs fill a similar role at the end of the mission. Army leaders use AARs as opportunities to develop subordinates. During an AAR, give subordinates a chance to talk about how they saw things. Teach them how to look past a problem's symptoms to its root cause. Teach them how to give constructive, useful feedback. ("Here's what we did well; here's what we can do better.") When subordinates share in identifying reasons for success and failure, they become owners of a stake in how things get done. AARs also give you a chance to hear what's on your subordinates' minds—and

good leaders listen closely. (FM 25-101 and TC 25-20 discuss how to prepare, conduct, and follow up after AARs.)

5-63. Leaders base reviews on accurate observations and correct recording of those observations. If you're evaluating a ten-day field exercise, take good notes because you won't remember everything. Look at things in a systematic way; get out and see things firsthand. Don't neglect tasks that call for subjective judgment: evaluate unit cohesion, discipline, and morale. (FM 25-100 and FM 25-101 discuss training assessment.)

Initial Leader Assessments

5-64. Leaders often conduct an initial assessment before they take over a new position. How competent are your new subordinates? What's expected of you in your new job? Watch how people operate; this will give you clues about the organizational climate. (Remember SSG Withers and the vehicle inspection in Chapter 3?) Review the organization's SOP and any regulations that apply. Meet with the outgoing leader and listen to his assessment. (But don't take it as the absolute truth; everyone sees things through filters.) Review status reports and recent inspection results. Identify the key people outside the organization whose help you'll need to be successful. However, remember that your initial impression may be off-base. After you've been in the position for a while, take the necessary time to make an in-depth assessment.

5-65. And in the midst of all this checking and rechecking, don't forget to take a look at yourself. What kind of leader are you? Do you oversupervise? Undersupervise? How can you improve? What's your plan for working on your weak areas? What's the best way to make use of your strengths? Get feedback on yourself from as many sources as possible: your boss, your peers, even your subordinates. As Chapter 1 said in the discussion of character, make sure your own house is in order.

Assessment of Subordinates

5-66. Good leaders provide straightforward feedback to subordinates. Tell them where you see their strengths; let them know where they can improve. Have them come up with a plan of action for self-improvement; offer your help. Leader assessment should be a positive experience that your subordinates see as a chance for them to improve. They should see it as an opportunity to tap into your experience and knowledge for their benefit.

5-67. To assess your subordinate leaders, you must—

- Observe and record leadership actions. Figure 1-1 is a handy guide for organizing your thoughts.
- Compare what you see to the performance indicators in Appendix B or the appropriate reference.
- Determine if the performance meets, exceeds, or falls below standard.
- Tell your subordinates what you saw; give them a chance to assess themselves.
- Help your subordinate develop a plan of action to improve performance.

Leader Assessments and Plans of Action

5-68. Leader assessment won't help anyone improve unless it includes a plan of action designed to correct weaknesses and sustain strengths. Not only that, you and the subordinate must use the plan; it doesn't do anyone any good if you stick it in a drawer or file cabinet and never think about it again. Here is what you must do:

- Design the plan of action together; let your subordinate take the lead as much as possible.
- Agree on the actions necessary to improve leader performance; your subordinate must buy into this plan if it's going to work.
- Review the plan frequently, check progress, and change the plan if necessary.

(Appendix C discusses the relationship between a leader plan of action and developmental counseling.)

IMPROVING ACTIONS

How can you know if you've made a difference? Sometimes—rarely—the results are instant. Usually it takes much longer. You may see a soldier again as a seasoned NCO; you may get a call or a letter or see a name in the Army Times. *In most cases, you will never be sure how well you succeeded, but don't let that stop you.*

Command Sergeant Major John D. Woodyard, 1993

5-69. Improving actions are things leaders do to leave their organizations better than they found them. Improving actions fall into the categories highlighted in Figure 5-3: developing, building, and learning.

5-70. Developing refers to people: you improve your organization and the Army as an institution when you develop your subordinates.

5-71. Building refers to team building: as a direct leader, you improve your organization by building strong, cohesive teams that perform to standard, even in your absence.

5-72. Learning refers to you, your subordinates, and your organization as a whole. As a leader, you must model self-development for your people; you must constantly be learning. In addition, you must also encourage your subordinates to learn and reward their self-development efforts. Finally, you must establish an organizational climate that rewards collective learning and act to ensure your organization learns from its experiences.

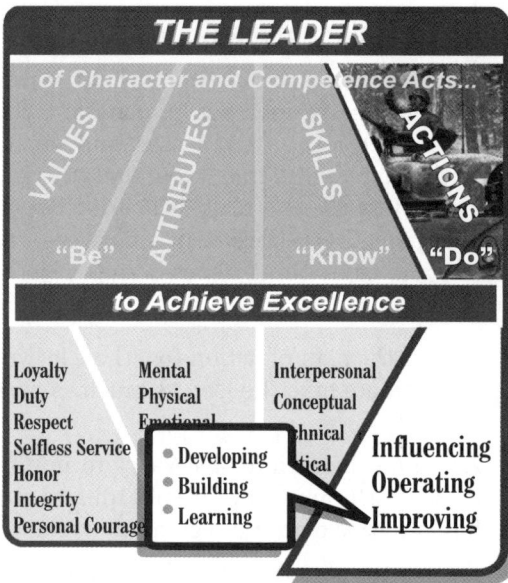

Figure 5-3. Direct Leader Actions
—Improving

DEVELOPING

I've reminded many NCOs that they wouldn't be where they are today if someone hadn't given them a little extra time. I know I wouldn't be where I am.

Former Sergeant Major of the Army Glenn E. Morrell

5-73. In the Army, developing means developing people. Your subordinates are the leaders of tomorrow's Army. You have a responsibility to train them, to be the kind of leader they deserve so that they'll see how leading is done. It's your duty to invest the time and energy it takes to help them reach their fullest potential. The driving principle behind Army leader development is that leaders must be prepared before assuming leadership positions; they must be competent and confident in their abilities. This principle applies to all ranks and levels, to soldiers and DA civilians, and to both the active and reserve components.

5-74. As Figure 5-4 shows, a trained and ready Army rests on effective leader development. In turn, leader development rests on a foundation of training and education, expectations and standards, and values and ethics. This foundation supports the three leader development

pillars: institutional training (schooling), operational assignments, and self-development. (DA Pam 350-58 discusses Army leader development.)

Institutional Training

5-75. The Army school system provides formal education and training for job-related and leadership skills. The American public education system is progressive; that is, children attend primary school before middle school or junior high and then go on to high school. Likewise, the Army school system is progressive. The main difference is that you can expect to go out and use your skills in an assignment before being considered for the next level of schooling. Institutional training is critical in developing leaders and preparing them for increased positions of responsibility throughout the Army.

Figure 5-4. Leader Development

Operational Assignments

5-76. When you take what you've learned in school into the field Army, you continue to learn through on-the-job experience and by watching your leaders, peers, and subordinates. Operational assignments provide opportunities to broaden your knowledge and refine skills you gain during institutional training and previous assignments. You gain and expand your experience base by performing a wide range of duties and tasks under a variety of frequently changing conditions and situations. Operational assignments provide a powerful resource for leader development—an opportunity to learn by doing.

Self-Development

5-77. Self-development is a process you should use to enhance previously acquired skills, knowledge, and experience. Its goal is to increase your readiness and potential for positions of greater responsibility. Effective self-development focuses on aspects of your character, knowledge, and capabilities that you believe need developing or improving. You can use the dimensions of the Army leadership framework to help you determine what areas to work on. Self-development is continuous: it takes place during institutional training and operational assignments.

5-78. Self-development is a joint effort involving you, your first-line leader, and your commander. Commanders establish and monitor self-development programs for their organizations. You and your first-line leader together establish goals to meet your individual needs and plan the actions you must take to meet them. You do this as part of developmental counseling, which is discussed below and in Appendix C. Finally, you must execute your plan of action. If you have subordinates, you monitor how well they're acting on their plans of action. You can't execute their plans for them, but you can give them advice, encouragement, and—when necessary and mission permits—time.

5-79. Self-development for junior personnel is very structured and generally narrow in focus. The focus broadens as individuals learn their strengths and weaknesses, determine their individual needs, and become more independent. Everyone's knowledge and perspective increases with age, experience, institutional training, and operational assignments. Specific, goal-oriented self-development actions can accelerate and broaden a person's skills and knowledge. As a member of the Army, you're obligated to develop your abilities to the greatest extent possible. As an Army leader, you're responsible to assist your subordinates in their self-development.

5-80. Civilian and military education is part of self-development. Army leaders never stop learning. They seek to educate and train themselves beyond what's offered in formal schooling or even in their duty assignments. Leaders look for educational opportunities to prepare themselves for their next job and future responsibilities. Look for Army off-duty education that interests you and will give you useful skills. Seek civilian education to broaden your outlook on life. Look for things to read that will develop your mind and help you build skills. Challenge yourself and apply the same initiative here as you do in your day-to-day duties.

5-81. Remember that Army leaders challenge themselves and take advantage of work done by others in such fields as leadership and military history as well as in their off-duty areas of interest. In the leadership area, you can begin with some of the books listed in the bibliography or go to any bookstore or library. You'll find hundreds of titles under the heading of leadership. The libraries of your post, nearby civilian communities, and colleges contain works on these topics. In addition, the Internet can also be a useful place for obtaining information on some areas. However, be careful. Some books contain more reliable and useful information than others; the same is true of Internet sites.

5-82. Figure 5-4 also shows that actions, skills, and attributes form the foundation of success in operational assignments. This is where you, the leader, fit into Army leader development. As a leader, you help your subordinates internalize Army values. You also assist them in developing the individual attributes, learning the skills, and mastering the actions required to become leaders of character and competence themselves. You do this through the action of mentoring.

Mentoring

Good NCOs are not just born—they are groomed and grown through a lot of hard work and strong leadership by senior NCOs.

Former Sergeant Major of the Army
William A. Connelly

> **Mentoring** (in the Army) is the proactive development of each subordinate through observing, assessing, coaching, teaching, developmental counseling, and evaluating that results in people being treated with fairness and equal opportunity. Mentoring is an inclusive process (not an exclusive one) for everyone under a leader's charge.

5-83. Mentoring is totally inclusive, real-life leader development for every subordinate. Because leaders don't know which of their subordinates today will be the most significant contributors and leaders in the future, they strive to provide all their subordinates with the knowledge and skills necessary to become the best they can be—for the Army and for themselves.

5-84. Mentoring begins with the leader setting the right example. As an Army leader, you mentor people every day in a positive or negative way, depending on how you live Army values and perform leader actions. Mentoring shows your subordinates a mature example of values, attributes, and skills in action. It encourages them to develop their own character and leader attributes accordingly.

5-85. Mentoring links operating leader actions to improving leader actions. When you mentor, you take the observing, assessing, and evaluating you do when you operate and apply these actions to developing individual subordinates. Mentoring techniques include teaching, developmental counseling, and coaching.

> **Teaching** gives knowledge or provides skills to others, causing them to learn by example or experience.

5-86. Teaching is passing on knowledge and skills to subordinates. It's a primary task for first-line leaders. Teaching focuses primarily on technical and tactical skills. Developmental counseling is better for improving interpersonal and conceptual skills. Technical

competence is critical to effective teaching. In order to develop subordinates, you must be able to demonstrate the technical and tactical skills you expect them to perform; otherwise they won't listen to you.

5-87. To be an Army leader, you must be a teacher. You give your subordinates knowledge and skills all the time, whether in formal, classroom settings or through your example. To be an effective teacher, you must first be professionally competent; then you must create conditions in which your subordinates can learn.

> *Soldiers learn to be good leaders from good leaders.*
> Former Sergeant Major of the Army Richard A. Kidd

5-88. The measure of how well you teach is how well your people learn. In most cases, your people will learn more by performing a skill than they will by watching you do it or by hearing you talk about how to do it. However, it's up to you to choose the teaching method that best fits the material. To make this choice, you need to understand the different ways people learn. People learn—

- Through the example of others (observing).
- By forming a picture in their minds of what they're trying to learn (thinking).
- By absorbing information (thinking).
- Through practice (hands-on experience).

5-89. Teaching is a complex art, one that you must learn in addition to the competencies you seek to teach. Just because you can pull the engine out of a tank doesn't mean you would be any good at teaching other people to do it. There are techniques and methods involved in teaching that have nothing to do with how good you are on the job; you must know both the skills related to the subject and another set of teaching skills. As an Army leader, you must develop these teaching skills as well. A subject matter expert who has acquired technical knowledge but is unable to teach that knowledge to others isn't improving the organization or the Army. (FM 25-101 addresses these and other areas related to conducting training.)

> **Developmental Counseling** is subordinate-centered communication that produces a plan outlining actions necessary for subordinates to achieve individual or organizational goals.

5-90. Developmental counseling is central to leader development. It's the means by which you prepare your subordinates of today to be the leaders of tomorrow. (Appendix C contains more details on developmental counseling.)

5-91. Developmental counseling isn't a time for war stories or for tales of how things were done way back when. It should focus on today's performance and problems and tomorrow's plans and solutions. Effective developmental counseling is centered on the subordinate, who is actively involved—listening, asking for more feedback, seeking elaboration of what the counselor has to say.

5-92. Developmental counseling isn't an occasional event that you do when you feel like it. It needs to be part of your program to develop your subordinates. It requires you to use all your counseling tools and skills. This means using counseling requirements such as those prescribed in the NCO Evaluation Reporting System (NCOERS), Officer Evaluation Reporting System (OERS), and Total Army Performance Evaluation System (TAPES, which is used to evaluate DA civilians) as more than paper drills. It means face-to-face counseling of individuals you rate. But more important, it means making time throughout the rating period to discuss performance objectives and provide meaningful assessments and feedback. No evaluation report—positive or negative—should be a surprise. A consistent developmental counseling program ensures your people know where they stand and what they should be doing to improve their performance and develop themselves. Your program should include all your people, not just the ones you think have the most potential. (The bibliography lists the evaluation and support forms prescribed by the OERS, NCOERS, and TAPES. Appendix C discusses

how to use support forms to assist you with developmental counseling.)

5-93. New direct leaders are sometimes uncomfortable confronting a subordinate who isn't performing to standard. However, remember that counseling isn't about how comfortable or uncomfortable you are; counseling is about correcting the performance or developing the character of a subordinate. Therefore, be honest and frank with your subordinates during developmental counseling. If you let your people get away with substandard behavior because you want them to like you or because you're afraid to make a hard call, you're sacrificing Army standards for your personal well-being—and you're not developing your subordinates.

5-94. This manual has emphasized throughout the importance of the example that you, as an Army leader, set for your subordinates. Your people look to you to see what kind of leader they want to be. The example you set in counseling is especially important. Army leaders at every level must ensure their subordinate leaders use counseling to develop their own subordinates. Setting the example is a powerful leadership tool: if you counsel your subordinates, your subordinate leaders will counsel theirs as well. The way you counsel is the way they'll counsel. Your people copy your behavior. The significance of your position as a role model can't be understated. It's a powerful teaching tool, for developmental counseling as well as other behaviors.

5-95. Although you're responsible for developing your subordinates, no leader can be all things to all people. In addition, the Army is already culturally diverse and is becoming increasingly technologically complex. In this environment, some of your subordinates may seek advice and counsel from informal relationships in addition to their leadership chain. Such relationships can be particularly important for women, minorities, and those in low-density specialties who have relatively few role models nearby.

5-96. This situation in no way relieves you, the leader, of any of your responsibilities regarding caring for and developing your people. Rather, being sensitive to your subordinates' professional development and cultural needs is part of the cultural awareness leader attribute. As an Army leader, you must know your people and take advantage of every resource available to help your subordinates develop as leaders. This includes other leaders who have skills or attributes different from your own.

> **Coaching** involves a leader's assessing performance based on observations, helping the subordinate develop an effective plan of action to sustain strengths and overcome weaknesses, and supporting the subordinate and the plan.

5-97. You can consider coaching to be both an operating and an improving leader action. It's less formal than teaching. When you're dealing with individuals, coaching is a form of specific instance counseling (which Appendix C discusses). When you're dealing with all or part of a team, it's generally associated with AARs (which you read about earlier in this chapter).

5-98. Coaching follows naturally from the assessing leader action. As you observe your subordinates at work, you'll see them perform some tasks to standard and some not to standard. Some of their plans will work; some won't. Your subordinates know when you're watching them. They expect you to tell them what they need to do to meet the standard, improve the team's performance, or develop themselves. You provide this sort of feedback through coaching. And don't limit your coaching to formal sessions. Use every opportunity to teach, counsel or coach from quarterly training briefings to AARs. Teaching moments and coaching opportunities occur all the time when you concentrate on developing leaders.

Mentoring and Developing Tomorrow's Army

5-99. Mentoring is demanding business, but the future of the Army depends on the trained and effective leaders whom you leave behind. Sometimes it requires you to set priorities, to balance short-term readiness with long-term leader development. The commitment to mentoring future leaders may require you to take risks. It requires you to give subordinates the opportunity to learn and develop themselves while using your experience to guide them without micromanaging. Mentoring will lead your subordinates to successes that build their confidence and skills for the future.

5-100. Mentoring isn't something new for the Army. Past successes and failures can often be traced to how seriously those in charge took the challenge of developing future leaders. As you consider the rapid pace of change in today's world, it's critical that you take the time to develop leaders capable of responding to that change. The success of the next generation of Army leaders depends on how well you accept the responsibility of mentoring your subordinates. Competent and confident leaders trained to meet tomorrow's challenges and fight and win future conflicts will be your legacy.

5-101. As you assume positions of greater responsibility, as the number of people for whom you are responsible increases, you need to do even more to develop your subordinates. More, in this case, means establishing a leader development program for your organization. It also means encouraging your subordinates to take actions to develop themselves personally and professionally. In addition, you may have to provide time for them to pursue self-development. (FM 25-101 discusses leader development programs.)

> **What have YOU done TODAY to develop the leaders of tomorrow's Army?**

BUILDING
Building Teams

5-102. You've heard—no doubt countless times—that the Army is a team. Just how important is it that people have a sense of the team? Very important. The national cause, the purpose of the mission, and all the larger concerns may not be visible from the battlefield.

Regardless of other issues, soldiers perform for the other people in the squad or section, for others in the team or crew, for the person on their right or left. This is a fundamental truth: soldiers perform because they don't want to let their buddies down.

5-103. If the leaders of the small teams that make up the Army are competent, and if their members trust one another, those teams and the larger team of teams will hang together and get the job done. People who belong to a successful team look at nearly everything in a positive light; their winners' attitudes are infectious, and they see problems as challenges rather than obstacles. Additionally, a cohesive team accomplishes the mission much more efficiently than a group of individuals. Just as a football team practices to win on the gridiron, so must a team of soldiers practice to be effective on the battlefield.

5-104. Training together builds collective competence; trust is a product of that competence. Subordinates learn to trust their leaders if the leaders know how to do their jobs and act consistently—if they say what they mean and mean what they say. Trust also springs from the collective competence of the team. As the team becomes more experienced and enjoys more successes, it becomes more cohesive.

Trust Earned

In a 1976 interview, Congressman Hamilton Fish of New York told of his experiences as a white officer with the 369th Infantry Regiment, an all-black unit in the segregated Army of 1917. Fish knew that his unit would function only if his soldiers trusted him; his soldiers, all of whom had volunteered for combat duty, deserved nothing less than a trustworthy leader. When a white regiment threatened to attack the black soldiers in training camp, Fish, his pistol drawn, alerted the leaders of that regiment and headed off a disaster.

"There was one thing they wanted above all from a white officer," [Fish recalled in an interview nearly 60 years later] "and that was fair treatment. You see, even in New York City [home of most of his soldiers] they really did not get a square deal most of the time. But if they felt you were on the level with them, they would go all out for you. And they seemed to have a sixth sense in realizing just how you felt. I sincerely wanted to lead them as real soldiers, and they knew it."

5-105. Developing teams takes hard work, patience, and quite a bit of interpersonal skill on the part of the leader, but it's a worthwhile investment. Good teams get the job done. People who are part of a good team complete the mission on time with the resources given them and a minimum of wasted effort; in combat, good teams are the most effective and take the fewest casualties.

5-106. Good teams—

- Work together to accomplish the mission.
- Execute tasks thoroughly and quickly.
- Meet or exceed the standard.
- Thrive on demanding challenges.
- Learn from their experiences and are proud of their accomplishments.

5-107. The Army is a team that includes members who are not soldiers but whose contributions are essential to mission success. The contributions made by almost 1,600 DA civilians in the Persian Gulf region were all but lost in the celebrations surrounding the military victory against Iraq and the homecoming celebration for the soldiers that followed. However, one safety specialist noted that these deployed DA civilians recognized the need for a team effort:

Patriotism was their drawing force for being there....We were part of the team supporting our soldiers! The focus is where it should be—on the military. They're here to do the job; we're here to help them.

5-108. People will do the most extraordinary things for their buddies. It's your job as an Army leader to pull each member into the team because you may someday ask that person for extraordinary effort. Team building involves applying interpersonal leader skills that transform individuals into productive teams. If you've done your work, the team member won't let you down.

5-109. Within a larger team, smaller teams may be at different stages of development. For instance, members of First Squad may be used to working together. They trust one another and get the job done—usually exceeding the standard—with no wasted motion. Second Squad in the same platoon just received three new soldiers and a team leader from another company. As a team, Second Squad is less mature; it will take them some time to get up to the level of First Squad. New team members have to learn how things work: they have to be brought on board and made to feel members of the team; they must learn the standards and the climate of their new unit; they'll have to demonstrate some competence before other members really accept them; and finally, they must practice working together. Leaders, who must oversee all this, are better equipped if they know what to expect. Make use of the information on the next few pages; learn what to look for—and stay flexible.

5-110. Figure 5-5 lists things you must do to pull a team together, get it going in the right direction, and keep it moving. And that list only hints at the work that lies ahead as you get your team to work together. Your subordinates must know—must truly believe—that they're a part of the team, that their contribution is important and valued. They must know that you'll train them and listen to them. They don't want you to let them get away with shoddy work or half-baked efforts; there's no pride in loafing. You must constantly observe, counsel, develop, listen; you must be every bit the team player you want your subordinates to be—and more.

5-111. Teams don't come together by accident; leaders must build and guide them through a series of developmental stages: formation, enrichment, and sustainment. This discussion may make the process seem more orderly than it actually is; as with so many things leaders do, the reality is more complicated than the explanation. Each team develops differently: the boundaries between stages are not hard and fast. As a leader, you must be sensitive to the characteristics of the team you're building and of its individual members—your people. Compare the characteristics of your team with the team building stage descriptions. The information that results can help you determine what to expect of your team and what you need to do to improve its capabilities.

Stages of Team Building

5-112. Teams, like individuals, have different personalities. As with individuals, the leader's job isn't to make teams that are clones of one another; the job is to make best use of the peculiar talents of the team, maximize the potential of the unit climate, and motivate aggressive execution.

5-113. **Formation stage.** Teams work best when new members are brought on board quickly, when they're made to feel a part of the team. The two steps—reception and orientation—are dramatically different in peace and war. In combat, this sponsorship process can literally mean life or death to new members and to the team.

5-114. Reception is the leader's welcome: the orientation begins with meeting other team members, learning the layout of the workplace, learning the schedule and other requirements, and generally getting to know the lay of the land. In combat, leaders may not have time to spend with new members. In this case, new arrivals are often assigned a buddy who will help them get oriented and keep them out of trouble until they learn their way around. Whatever technique you use, your soldiers should never encounter a situation similar to the one in the next example.

ESSENTIAL ARMY MANUAL
(345) ARMY LEADERSHIP

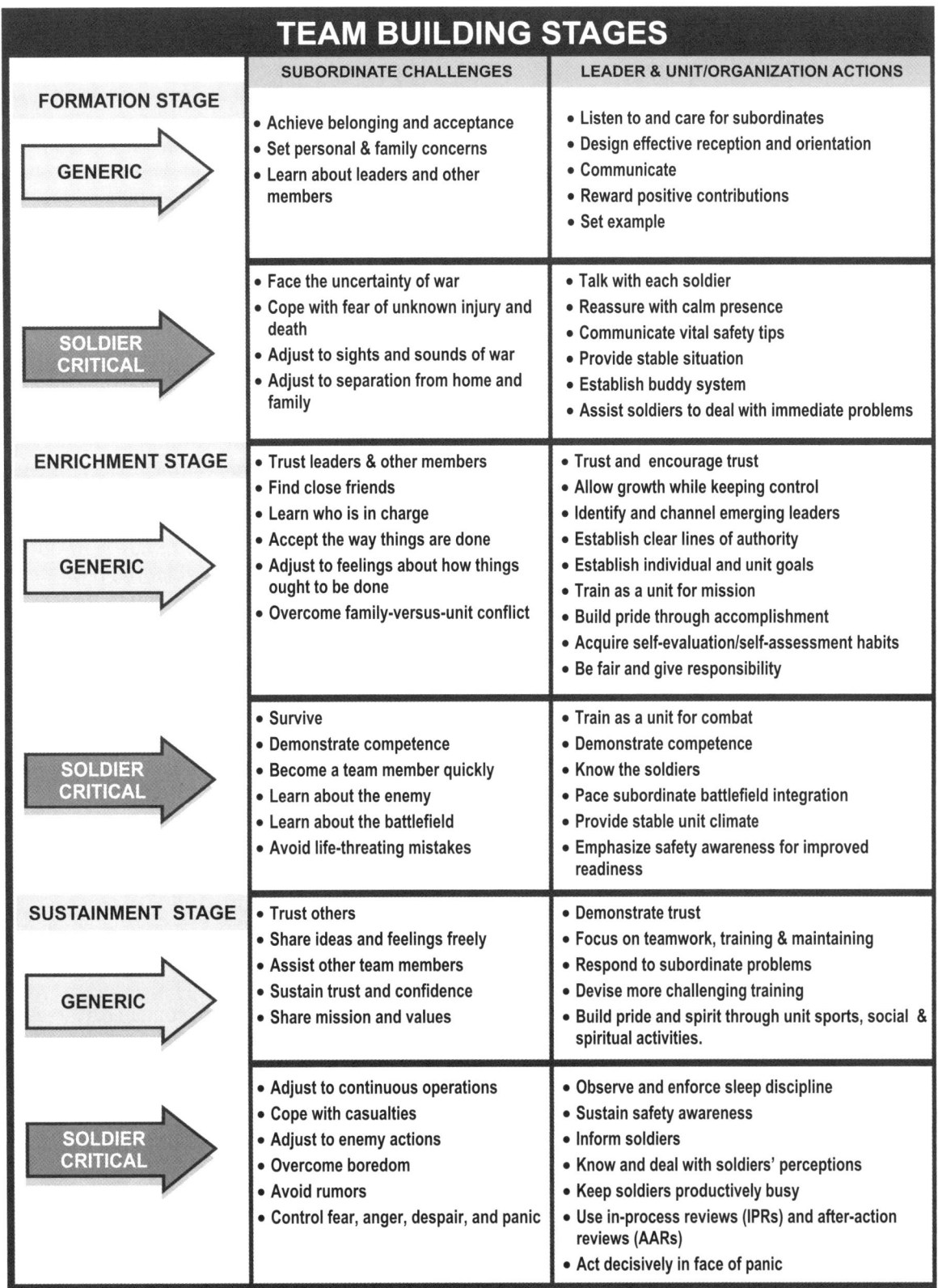

TEAM BUILDING STAGES

	SUBORDINATE CHALLENGES	LEADER & UNIT/ORGANIZATION ACTIONS
FORMATION STAGE GENERIC	• Achieve belonging and acceptance • Set personal & family concerns • Learn about leaders and other members	• Listen to and care for subordinates • Design effective reception and orientation • Communicate • Reward positive contributions • Set example
SOLDIER CRITICAL	• Face the uncertainty of war • Cope with fear of unknown injury and death • Adjust to sights and sounds of war • Adjust to separation from home and family	• Talk with each soldier • Reassure with calm presence • Communicate vital safety tips • Provide stable situation • Establish buddy system • Assist soldiers to deal with immediate problems
ENRICHMENT STAGE GENERIC	• Trust leaders & other members • Find close friends • Learn who is in charge • Accept the way things are done • Adjust to feelings about how things ought to be done • Overcome family-versus-unit conflict	• Trust and encourage trust • Allow growth while keeping control • Identify and channel emerging leaders • Establish clear lines of authority • Establish individual and unit goals • Train as a unit for mission • Build pride through accomplishment • Acquire self-evaluation/self-assessment habits • Be fair and give responsibility
SOLDIER CRITICAL	• Survive • Demonstrate competence • Become a team member quickly • Learn about the enemy • Learn about the battlefield • Avoid life-threating mistakes	• Train as a unit for combat • Demonstrate competence • Know the soldiers • Pace subordinate battlefield integration • Provide stable unit climate • Emphasize safety awareness for improved readiness
SUSTAINMENT STAGE GENERIC	• Trust others • Share ideas and feelings freely • Assist other team members • Sustain trust and confidence • Share mission and values	• Demonstrate trust • Focus on teamwork, training & maintaining • Respond to subordinate problems • Devise more challenging training • Build pride and spirit through unit sports, social & spiritual activities.
SOLDIER CRITICAL	• Adjust to continuous operations • Cope with casualties • Adjust to enemy actions • Overcome boredom • Avoid rumors • Control fear, anger, despair, and panic	• Observe and enforce sleep discipline • Sustain safety awareness • Inform soldiers • Know and deal with soldiers' perceptions • Keep soldiers productively busy • Use in-process reviews (IPRs) and after-action reviews (AARs) • Act decisively in face of panic

Figure 5-5. Team Building Stages

Replacements in the ETO

Most historians writing about World War II agree that the replacement system that fed new soldiers into the line units was seriously flawed, especially in the ETO, and did tremendous harm to the soldiers and the Army. Troops fresh from stateside posts were shuffled about in tent cities where they were just numbers. 1LT George Wilson, an infantry company commander who received one hundred replacements on December 29, 1944, in the midst of the Battle of the Bulge, remembers the results: "We discovered that these men had been on a rifle range only once; they had never thrown a grenade or fired a bazooka [antitank rocket], mortar or machine gun."

PVT Morris Dunn, another soldier who ended up with the 84th Division after weeks in a replacement depot recalls how the new soldiers felt: "We were just numbers, we didn't know anybody, and I've never felt so alone and miserable and helpless in my entire life—we'd been herded around like cattle at roundup time....On the ride to the front it was cold and raining with the artillery fire louder every mile, and finally we were dumped out in the middle of a heavily damaged town."

5-115. In combat, Army leaders have countless things to worry about; the mental state of new arrivals might seem low on the list. But if those soldiers can't fight, the unit will suffer needless casualties and may fail to complete the mission.

5-116. Discipline and shared hardship pull people together in powerful ways. SGT Alvin C. York, who won the Medal of Honor in an action you'll read about later in this chapter, talked about cohesion this way:

War brings out the worst in you. It turns you into a mad, fighting animal, but it also brings out something else, something I just don't know how to describe, a sort of tenderness and love for the fellow fighting with you.

5-117. However, the emotions SGT York mentions don't emerge automatically in combat. One way to ensure cohesion is to build it during peacetime. Team building begins with receiving new members; you know how important first impressions are when you meet someone new. The same thing is true of teams; the new member's reception and orientation creates that crucial first impression that colors the person's opinion of the team for a long time. A good experience joining the organization will make it easier for the new member to fit in and contribute. Even in peacetime, the way a person is received into an organization can have long-lasting effects—good or bad—on the individual and the team. (Appendix C discusses reception and integration counseling.)

Reception on Christmas Eve

An assistant division commander of the 25th Infantry Division told this story as part of his farewell speech:

"I ran across some new soldiers and asked them about their arrival on the island [of Oahu]. They said they got in on Christmas Eve, and I thought to myself, 'Can't we do a better job when we ship these kids out, so they're not sitting in some airport on their first big holiday away from home?' I mean, I really felt sorry for them. So I said, 'Must have been pretty lonesome sitting in a new barracks where you didn't know anyone.' And one of them said, 'No, sir. We weren't there a half-hour before the CQ [charge of quarters] came up and told us to get into class B's and be standing out front of the company in 15 minutes. Then this civilian drives up, a teenager, and the CQ orders us into the car. Turns out the kid was the first sergeant's son; his father had sent him over to police up anybody who was hanging around the barracks. We went over to the first sergeant's house to a big luau [party] with his family and a bunch of their neighbors and friends.'

Reception on Christmas Eve (continued)

"My guess is that those soldiers will not only do anything and everything that first sergeant wants, but they are going to tell anyone who will listen that they belong to the best outfit in the Army."

5-118. **Enrichment stage.** New teams and new team members gradually move from questioning everything to trusting themselves, their peers, and their leaders. Leaders earn that trust by listening, following up on what they hear, establishing clear lines of authority, and setting standards. By far the most important thing a leader does to strengthen the team is training. Training takes a group of individuals and molds them into a team while preparing them to accomplish their missions. Training occurs during all three team building stages, but is particularly important during enrichment; it's at this point that the team is building collective proficiency.

5-119. **Sustainment stage.** When a team reaches this stage, its members think of the team as "their team." They own it, have pride in it, and want the team to succeed. At this stage, team members will do what needs to be done without being told. Every new mission gives the leader a chance to make the bonds even stronger, to challenge the team to reach for new heights. The leader develops his subordinates because they're tomorrow's team leaders. He continues to train the team so that it maintains proficiency in the collective and individual tasks it must perform to accomplish its missions. Finally, the leader works to keep the team going in spite of the stresses and losses of combat.

Building the Ethical Climate

5-120. As an Army leader, you are the ethical standard bearer for your organization. You're responsible for building an ethical climate that demands and rewards behavior consistent with Army values. The primary factor affecting an organization's ethical climate is its leader's ethical standard. Leaders can look to other organizational or installation personnel—for example, the chaplain, staff judge advocate, inspector general, and equal employment opportunity manager—to assist them in building and assessing their organization's ethical climate, but the ultimate responsibility belongs to the leader—period.

5-121. Setting a good ethical example doesn't necessarily mean subordinates will follow it. Some of them may feel that circumstances justify unethical behavior. (See, for example, the situation portrayed in Appendix D.) Therefore, you must constantly seek to maintain a feel for your organization's current ethical climate and take prompt action to correct any discrepancies between the climate and the standard. One tool to help you is the Ethical Climate Assessment Survey (ECAS), which is discussed in Appendix D. You can also use some of the resources listed above to help you get a feel for your organization's ethical climate. After analyzing the information gathered from the survey or other sources, a focus group may be a part of your plan of action to improve the ethical climate. Your abilities to listen and decide are the most important tools you have for this job.

5-122. It's important for subordinates to have confidence in the organization's ethical environment because much of what is necessary in war goes against the grain of the societal values individuals bring into the Army. You read in the part of Chapter 4 that discusses ethical reasoning that a soldier's conscience may tell him it's wrong to take human life while the mission of the unit calls for exactly that. Unless you've established a strong ethical climate that lets that soldier know his duty, the conflict of values may sap the soldier's will to fight.

SGT York

A conscientious objector from the Tennessee hills, Alvin C. York was drafted after America's entry into World War I and assigned to the 328th Infantry Regiment of the 82d Division, the "All Americans." PVT York, a devout Christian, told his commander, CPT E. C. B. Danforth, that he would bear arms against the enemy but didn't believe in killing. Recognizing PVT York as a potential leader but unable to sway him from his convictions, CPT Danforth consulted his battalion commander, MAJ George E. Buxton, about how to handle the situation.

MAJ Buxton was also deeply religious and knew the Bible as well as PVT York did. He had CPT Danforth bring PVT York to him, and they talked at length about the Scriptures, about God's teachings, about right and wrong, about just wars. Then MAJ Buxton sent PVT York home on leave to ponder and pray over the dilemma. The battalion commander promised to release him from the Army if PVT York decided he could not serve his country without sacrificing his integrity. After two weeks of reflection and deep soul-searching, PVT York returned, having reconciled his personal values with those of the Army. PVT York's decision had great consequences for both himself and his unit.

Alvin York performed an exploit of almost unbelievable heroism in the morning hours of 8 October 1918 in France's Argonne Forest. He was now a corporal (CPL), having won his stripes during combat in the Lorraine. That morning CPL York's battalion was moving across a valley to seize a German-held rail point when a German infantry battalion, hidden on a wooded ridge overlooking the valley, opened up with machine gun fire. The American battalion dived for cover, and the attack stalled. CPL York's platoon, already reduced to 16 men, was sent to flank the enemy machine guns.

As the platoon advanced through the woods to the rear of the German outfit, it surprised a group of about 25 German soldiers. The shocked enemy offered only token resistance, but then more hidden machine guns swept the clearing with fire. The Germans dropped safely to the ground, but nine Americans, including the platoon leader and the other two corporals, fell dead or wounded. CPL York was the only unwounded leader remaining.

CPL York found his platoon trapped and under fire within 25 yards of the enemy's machine gun pits. Nonetheless, he didn't panic. Instead, he began firing into the nearest enemy position, aware that the Germans would have to expose themselves to get an aimed shot at him. An expert marksman, CPL York was able to hit every enemy soldier who popped his head over the parapet.

After he had shot more than a dozen enemy, six German soldiers charged him with fixed bayonets. As the Germans ran toward him, CPL York once again drew on the instincts of a Tennessee hunter and shot the last man first (so the ones in front wouldn't see the ones he shot fall), then the fifth, and so on. After he had shot all the assaulting Germans, CPL York again turned his attention to the machine gun pits. In between shots, he called for the Germans to give up. It may have initially seemed ludicrous for a lone soldier in the open to call on a well-entrenched enemy to surrender, but their situation looked desperate to the German battalion commander, who had seen over 20 of his soldiers killed by this one American. The commander advanced and offered to surrender if CPL York would stop shooting.

CPL York now faced a daunting task. His platoon, now numbering seven unwounded soldiers, was isolated behind enemy lines with several dozen prisoners. However, when one American said their predicament was hopeless, CPL York told him to be quiet and began organizing the prisoners for a movement. CPL York moved his unit and prisoners toward American lines, encountering other German positions and forcing their surrender. By the time the platoon reached the edge of the valley they had left just a few hours before, the hill was clear of German machine guns. The fire on the Americans in the valley was substantially reduced and their advance began again.

SGT York (continued)

CPL York returned to American lines, having taken a total of 132 prisoners and putting 35 machine guns out of action. He left the prisoners and headed back to his own outfit. Intelligence officers questioned the prisoners and learned from their testimony the incredible story of how a fighting battalion was destroyed by one determined soldier armed only with a rifle and pistol. Alvin C. York was promoted to sergeant and awarded the Medal of Honor for this action. His character, physical courage, technical competence, and leadership enabled him to destroy the morale and effectiveness of an entire enemy infantry battalion.

5-123. CPT Danforth and MAJ Buxton could have ordered SGT York to go to war, or they might have shipped him out to a job that would take him away from the fight. Instead, these leaders carefully addressed the soldier's ethical concerns. MAJ Buxton, in particular, established the ethical climate by showing that he, too, had wrestled with the very questions that troubled SGT York. The climate these leaders created held that every person's beliefs are important and should be considered. MAJ Buxton demonstrated that a soldier's duties could be consistent with the ethical framework established by his religious beliefs. Leaders who create a healthy ethical environment inspire confidence in their subordinates; that confidence and the trust it engenders builds the unit's will. They create an environment where soldiers can truly "be all they can be."

LEARNING

For most men, the matter of learning is one of personal preference. But for Army [leaders], the obligation to learn, to grow in their profession, is clearly a public duty.

General of the Army Omar N. Bradley

5-124. The Army is a learning organization, one that harnesses the experience of its people and organizations to improve the way it does business. Based on their experiences, learning organizations adopt new techniques and procedures that get the job done more efficiently or effectively. Likewise, they discard techniques and procedures that have outlived their purpose. However, you must remain flexible when trying to make sense of your experiences. The leader who works day after day after day and never stops to ask "How can I do this better?" is never going to learn and won't improve the team.

5-125. Leaders who learn look at their experience and find better ways of doing things. Don't be afraid to challenge how you and your subordinates operate. When you ask "Why do we do it that way?" and the only answer you get is "Because we've always done it that way," it's time for a closer look. Teams that have found a way that works still may not be doing things the best way. Unless leaders are willing to question how things are, no one will ever know what can be.

"Zero Defects" and Learning

5-126. There's no room for the "zero-defects" mentality in a learning organization. Leaders willing to learn welcome new ways of looking at things, examine what's going well, and are not afraid to look at what's going poorly. When direct leaders stop receiving feedback from subordinates, it's a good indication that something is wrong. If the message you hammer home is "There will be no mistakes," or if you lose your temper and "shoot the messenger" every time there's bad news, eventually your people will just stop telling you when things go wrong or suggesting how to make things go right. Then there will be some unpleasant surprises in store. Any time you have human beings in a complex organization doing difficult jobs, often under pressure, there are going to be problems. Effective leaders use those mistakes to figure out how to do things better and share what they have learned with other leaders in the organization, both peers and superiors.

5-127. That being said, all environments are not learning environments; a standard of "zero-defects" is acceptable, if not mandatory, in some circumstances. A parachute rigger is charged with a "zero-defect" standard. If a rigger makes a mistake, a parachutist will die. Helicopter repairers live in a "zero-defect" environment as well. They can't allow aircraft to be mechanically unstable during flight. In these and similar work environments, safety concerns mandate a "zero-defects" mentality. Of course, organizations and people make mistakes; mistakes are part of training and may be the price of taking action. Leaders must make their intent clear and ensure their people understand the sorts of mistakes that are acceptable and those that are not.

5-128. Leaders can create a "zero-defects" environment without realizing it. Good leaders want their organizations to excel. But an organizational "standard" of excellence can quickly slide into "zero defects" if the leader isn't careful. For example, the published minimum standard for passing the APFT is 180 points—60 points per event. However, in units that are routinely assigned missions requiring highly strenuous physical activities, leaders need to train their people to a higher-than-average level of physical fitness. If leaders use APFT scores as the primary means of gauging physical fitness, their soldiers will focus on the test rather than the need for physical fitness. A better course would be for leaders to train their people on mission-related skills that require the higher level of physical readiness while at the same time motivating them to strive for their personal best on the APFT.

Barriers to Learning

5-129. Fear of mistakes isn't the only thing that can get in the way of learning; so can rigid, lockstep thinking and plain mental laziness. These habits can become learning barriers leaders are so used to that they don't even notice them. Fight this tendency. Challenge yourself. Use your imagination. Ask how other people do things. Listen to subordinates.

Helping People Learn

5-130. Certain conditions help people learn. First, you must motivate the person to learn. Explain to the subordinate why the subject is important or show how it will help the individual perform better. Second, involve the subordinate in the learning process; make it active. For instance, you would never try to teach someone how to drive a vehicle with classroom instruction alone; you have to get the person behind the wheel. That same approach applies to much more complex tasks; keep the lecture to a minimum and maximize the hands-on time.

5-131. Learning from experience isn't enough; you can't have every kind of experience. But if you take advantage of what others have learned, you get the benefit without having the experience. An obvious example is when combat veterans in a unit share their experiences with soldiers who haven't been to war. A less obvious, but no less important, example is when leaders share their experience with subordinates during developmental counseling.

After-Action Reviews and Learning

5-132. Individuals benefit when the group learns together. The AAR is one tool good leaders use to help their organizations learn as a group. Properly conducted, an AAR is a professional discussion of an event, focused on performance standards, that enables people to discover for themselves what happened, why it happened, and how to sustain strengths and improve on weaknesses. Like warning orders and rehearsals, the AAR is a technique that all leaders—military or DA civilian—can use in garrison as well as field environments. (FM 25-101 and TC 25-20 discuss how to prepare for, conduct, and follow up after an AAR.) When your team sits down for an AAR, make sure everyone participates and all understand what's being said. With input from the whole team, your people will learn more than if they just think about the experience by themselves.

Organizational Climate and Learning

5-133. It takes courage to create a learning environment. When you try new things or try things in different ways, you're bound to make mistakes. Learn from your mistakes and the mistakes of others. Pick your team and yourself up, determine what went right and wrong, and continue the mission. Be confident in your abilities. Theodore Roosevelt, a colonel during the Spanish-American War and twenty-sixth President of the United States, put it this way:

Whenever you are asked if you can do a job, tell 'em, Certainly I can!—and get busy and find out how to do it.

5-134. Your actions as a direct leader move the Army forward. How you influence your subordinates and the people you work for, how you operate to get the job done, how you improve the organization for a better future, all determine the Army's success or failure.

SUMMARY

5-135. Direct leaders influence their subordinates face-to-face as they operate to accomplish the mission and improve the organization. Because their leadership is face-to-face, direct leaders see the outcomes of their actions almost immediately. This is partly because they receive immediate feedback on the results of their actions.

5-136. Direct leaders influence by determining their purpose and direction from the boss's intent and concept of the operation. They motivate subordinates by completing tasks that reinforce this intent and concept. They continually acquire and assess outcomes and motivate

their subordinates through face-to-face contact and personal example.

5-137. Direct leaders operate by focusing their subordinates' activities toward the organization's objective and achieving it. Direct leaders plan, prepare, execute, and assess as they operate. These functions sometimes occur simultaneously.

5-138. Direct leaders improve by living Army values and providing the proper role model for subordinates. Leaders must develop all subordinates as they build strong, cohesive teams and establish an effective learning environment.

PART THREE

Organizational and Strategic Leadership

As they mature and assume greater responsibilities, Army leaders must also learn new skills, develop new abilities, and act in more complex environments. Organizational and strategic leaders maintain their own personalities and propensities, but they also expand what they know and refine what they do.

Chapters 6 and 7 describe (rather than prescribe or mandate) skills and actions required of organizational and strategic leaders. The chapters discuss much of what developing leaders often sense and explore some concepts that may seem foreign to them. Neither chapter outlines exhaustively what leaders know and do at higher levels; they simply introduce what's different.

The audience for Chapters 6 and 7 is only in part organizational and strategic leaders, who have prepared to serve in those positions by career-long experience and study. Primarily, these chapters offer staffs and subordinates who work for those leaders insight into the additional concerns and activities of organizational and strategic leadership.

Chapter 6

Organizational Leadership

6-1. During the Battle of the Bulge, with the Germans bearing down on retreating US forces, PFC Vernon L. Haught dug in and told a sergeant in a tank destroyer, "Just pull your vehicle behind me . . . I'm the 82d Airborne, and this is as far as the bastards are going." He knew his division commander's intent. Despite desperate odds, he had confidence in himself and his unit and knew they would make the difference. Faced with a fluid situation, he knew where the line had to be drawn; he had the will to act and he didn't hesitate to do what he thought was right.

6-2. Whether for key terrain in combat or for results in peacetime training, leaders in units and organizations translate strategy into policy and practice. They develop programs, plans, and systems that allow soldiers in teams, like the infantryman in the All-American Division, to turn plans and orders into fire and maneuver that seize victory at the least possible cost in sweat and blood. By force of will and application of their leadership skills, organizational leaders build teams with discipline, cohesion, trust, and proficiency. They clarify missions throughout the ranks by producing an intent, concept, and systematic approach to execution.

6-3. Organizational leadership builds on direct leader actions. Organizational leaders apply direct leader skills in their daily work with their command and staff teams and, with soldiers and subordinate leaders, they influence during their contacts with units. But to lead complex organizations like brigades, divisions, and corps at today's OPTEMPO and under the stresses of training, contingency operations, and combat, organizational leaders must add a whole new set of skills and actions to their leadership arsenal. They must practice direct and organizational leadership simultaneously.

6-4. Communicating to NCOs, like the airborne soldier at the Battle of the Bulge, occurs through individual subordinates, the staff, and the chain of command. Organizational leaders divide their attention between the concerns of the larger organization and their staffs and those of their subordinate leaders, units, and individuals. This tradeoff requires them to apply interpersonal and conceptual skills differently when exercising organizational leadership than when exercising direct leadership.

6-5. Organizational leaders rely heavily on mentoring subordinates and empowering them to execute their assigned responsibilities and missions. They stay mentally and emotionally detached from their immediate surroundings so they can visualize the larger impact on the organization and mission. Soldiers and subordinate leaders look to their organizational leaders

to establish standards for mission accomplishment and provide resources (conditions) to achieve that goal. Organizational leaders provide direction and programs for training and execution that focus efforts on mission success.

6-6. Due to the indirect nature of their influence, organizational leaders assess interrelated systems and design long-term plans to accomplish the mission. They must sharpen their abilities to assess their environments, their organization, and their subordinates. Organizational leaders determine the cause and effect of shortcomings, translate these new understandings into plans and programs, and allow their subordinate leaders latitude to execute and get the job done.

6-7. Organizational demands also differ as leaders develop a systems perspective. At the strategic level, the Army has identified six imperatives: quality people, training, force mix, doctrine, modern equipment, and leader development. In organizations these imperatives translate into doctrine, training, leader development, organization, materiel, and soldiers—commonly called DTLOMS. Together with Army values, these systems provide the framework for influencing people and organizations at all levels, conducting a wide variety of operations, and continually improving the force. Doctrine includes techniques to drive the functional systems in Army organizations. FMs 25-100, 25-101, and 101-5 lay out procedures for training management and military decision making that enable and focus execution. The training management and military decision-making processes provide a ready-made, systemic approach to planning, preparing, executing, and assessing.

SECTION I

WHAT IT TAKES TO LEAD ORGANIZATIONS—SKILLS

6-8. Organizational leaders continue to use the direct leader skills discussed in Chapter 4. However their larger organizations and spans of authority require them to master additional skills. As with direct leader skills, these span four areas: interpersonal, conceptual, technical, and tactical.

INTERPERSONAL SKILLS

To get the best out of your men, they must feel that you are their real leader and must know that they can depend upon you.

General of the Armies John J. Pershing

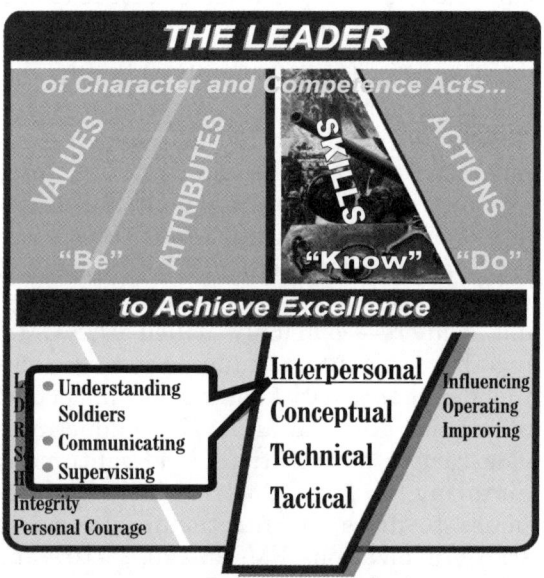

Figure 6-1. Organizational Leader Skills— Interpersonal

UNDERSTANDING SOLDIERS

6-9. Good organizational leaders understand the human dimension, which Chapter 3 discussed. They use that understanding to motivate subordinates and to encourage initiative. Chapter 5 explained that motivation means much more than an individual willingness to do what's directed. It imparts a desire on the part of individuals and organizations to do what's needed *without* being directed. This collective desire to accomplish the mission underlies good organizational discipline: good soldiers and competent DA civilians adhere to standards because they understand that doing so, even when it's a nuisance or hardship, leads to success.

6-10. This understanding, along with Army values, forms the foundation of great units. Units that have solid discipline can take tremendous stress and friction yet persevere, fight through, and win. Fostering initiative builds on motivation and discipline. It requires subordinates' confidence that in an uncertain situation, when they know the commander's intent and develop a competent solution, the commander will underwrite the risk they take. While this principle applies to both direct and organizational leaders, the stakes are usually higher in larger, more complex organizations. Additionally, organizational leaders may be more remote in time and distance and subordinates' ability to check back with them is diminished. Therefore, organizational leaders' understanding must develop beyond what they can immediately and personally observe.

COMMUNICATING

6-11. Persuasion is a communication skill important to organizational leaders. Well-developed skills of persuasion and an openness to working through controversy in a positive way help organizational leaders overcome resistance and build support. These characteristics are particularly important in dealing with other organizational leaders. By reducing grounds for misunderstanding, persuasion reduces time wasted in overcoming unimportant issues. It also ensures involvement of others, opens communication with them, and places value on their opinions—all team-building actions. Openness to discussing one's position and a positive attitude toward a dissenting view often diffuses tension and saves time and resistance in the long run. By demonstrating these traits, organizational leaders also provide an example that subordinates can use in self-development.

6-12. In some circumstances, persuasion may be inappropriate. In combat, all leaders make decisions quickly, modifying the decision-making process to fit the circumstances. But this practice of using the directing leadership styles as opposed to more participatory ones should occur when situations are in doubt, risks are high, and time is short—circumstances that often appear in combat. No exact blueprints exist for success in every context; leadership and the ability to adapt to the situation will carry the day. Appropriate style, seasoned instinct, and the realities of the situation must prevail.

SUPERVISING

6-13. Organizations pay attention to things leaders check. Feedback and coaching enhance motivation and improve performance by showing subordinates how to succeed. But how much should you check and how much is too much? When are statistics and reports adequate indicators and when must you visit your front-line organizations, talk to your soldiers and DA civilians and see what's going on yourself?

6-14. Overcentralized authority and oversupervising undermine trust and empowerment. Undersupervising can lead to failure, especially in cases where the leader's intent wasn't fully understood or where subordinate organizations lack the training for the task. Different subordinate commanders need different levels of supervision: some need a great deal of coaching and encouragement, though most would just as soon be left alone. As always, a good leader knows his subordinates and has the skill to supervise at the appropriate level.

Knowing Your People

This General said, "Each of our three regimental commanders must be handled differently. Colonel 'A' does not want an order. He wants to do everything himself and always does well. Colonel 'B' executes every order, but has no initiative. Colonel 'C' opposes everything he is told to do and wants to do the contrary."

A few days later the troops confronted a well-entrenched enemy whose position would have to be attacked. The General issued the following orders:

To Colonel "A" (who wants to do everything himself): "My dear Colonel "A", I think we will attack. Your regiment will have to carry the burden of the attack. I have, however, selected you for this reason. The boundaries of your regiment are so-and-so. Attack at X-hour. I don't have to tell you anything more."

To Colonel "C" (who opposes everything): "We have met a very strong enemy. I am afraid we will not be able to attack with the forces at our disposal." "Oh, General, certainly we will attack. Just give my regiment the time of attack and you will see that we are successful," replied Colonel "C." "Go then, we will try it," said the General, giving him the order for the attack, which he had prepared some time previously.

To Colonel "B" (who always must have detailed orders) the attack order was merely sent with additional details.

All three regiments attacked splendidly.

Adolph von Schell
German liaison to the Infantry School between the World Wars

CONCEPTUAL SKILLS

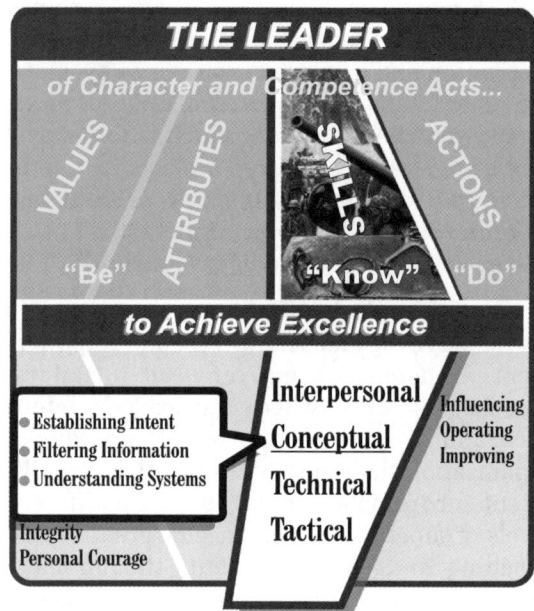

Figure 6-2. Organizational Leader Skills—Conceptual

6-15. The complexity of the organizational leader's environment requires patience, the willingness to think before acting. Furthermore, the importance of conceptual and analytical skills increases as an organizational leader moves into positions of greater responsibility. Organizational environments with multiple dimensions offer problems that become more abstract, complex, and uncertain.

6-16. Figure 6-2 identifies the conceptual skills required of organizational leaders. For organizational leaders, reasoning skills are crucial for developing intent and direction toward common goals. Critical thinking at the organizational level requires understanding systems and an increased ability to filter information, that is, to identify quickly information that applies to the task at hand and separate the important from the unimportant. Organizational leaders use this analytical ability to assess ambiguous environments and to calculate and manage risk. Their experience may allow them to see and define problems more easily—but not necessarily *fix* them quickly. Therefore, they

also dedicate time to think and generate alternative ways of organizing their organizations and resources for maximum effect. It's important for organizational leaders to encourage critical thinking in subordinates because subordinates also assess organizational challenges, analyze indicators, and recommend courses of action. It's also important, time and mission permitting, to allow subordinates' solutions to bear fruit.

ESTABLISHING INTENT

In an organization like ours, you have to think through what it is that you are becoming. Like a marathon runner, you have to get out in front, mentally, and pull the organization to you. You have to visualize the finish line—to see yourself there—and pull yourself along—not push—pull yourself to the future.

General Gordon R. Sullivan
Former Army Chief of Staff

6-17. Intent is the leader's personal expression of a mission's end state and the key tasks the organization must accomplish to achieve it. During operations and field training, it's a clear, concise statement of what the force must do to succeed with respect to the enemy and the terrain and to the desired end state. It provides the link between the mission and the concept of operations. By describing their intent, organizational leaders highlight the key tasks that, along with the mission, are the basis for subordinates to exercise initiative when unanticipated opportunities arise or when the original concept of operations no longer applies. Clear and concise, the leader's intent includes a mission's overall purpose and expected results. It provides purpose, motivation, and direction, whether the leader is commanding a division or running a staff directorate. An organizational leader visualizes the sequence of activities that will move the organization from its current state to the desired end state and expresses it as simply and clearly as possible. (FM 101-5-1 contains a complete definition of commander's intent. FM 100-34 discusses the relationship of intent and visualization to command and control. FM 101-5 discusses the development of intent in the MDMP.)

6-18. After establishing a clear and valid intent, the art of organizational leadership lies in having subordinates take actions on their own to transform that intent into reality. Since organizational leaders are likely to be farther away from the point of execution in time and space, they must describe the collective goal rather than list tasks for individual subordinates. With clearly communicated purpose and direction, subordinates can then determine what they must do and why. Within that broad framework, leaders empower subordinates, delegating authority to act within the intent: "Here's where we're headed, why we're going

there, and how we're going to get there." Purpose and direction align the efforts of subordinates working toward common goals.

6-19. A former division commander has said, "You must be seen to be heard." There's a great temptation for organizational leaders to rely exclusively on indirect leadership, to spread intent by passing orders through subordinates or communicating electronically with troops scattered far and wide. However, nothing can take the place of face-to-face contact. Organizational leaders make every effort to get out among the troops. There they can spot-check intent to see that it's disseminated and understood among those who must execute it.

GEN Grant and the End of the Civil War

I propose to fight it out on this line if it takes all summer.

General Ulysses S. Grant
Dispatch, May 11, 1864

GEN Ulysses S. Grant penned those words at Spotsylvania, Virginia, after being appointed general in chief of the Union Army and stationing himself forward with the Army of the Potomac. After fighting a bloody draw at the Wilderness, the Army of the Potomac had moved aggressively to outflank GEN Robert E. Lee's Army of Northern Virginia but once more faced a dug-in Confederate Army. However, where previous Union commanders had turned away, GEN Grant would not relent. His intent was clear. It was reinforced in the sentiment of his army, which wanted to finish the war.

In a series of determined attacks, the new Union commander broke GEN Lee's defense and used a turning movement to force his opponent out of position. Again they met at Cold Harbor, where GEN Grant attacked frontally and failed. Prior to his attack, Union soldiers, literally days from the end of their enlistment, were seen writing notes to their families and pinning them on the backs of their shirts so one last message would get home if they were killed. Their resolution demonstrates the power of their commitment to a shared intent.

After his bloody repulse at Cold Harbor, GEN Grant again moved south, maintaining the initiative, always pressing, always threatening to turn the Confederate flank and expose Richmond, the capital of Virginia. GEN Lee, in turn, was forced to block and defend Petersburg, Richmond's railroad hub. Uncovering it would have isolated Richmond as well as his army's rail-based lines of communication. GEN Grant had his opponent pinned to a critical strategic resource. Doing this denied the Army of Northern Virginia its greatest asset, its excellent ability to maneuver. It would not escape.

GEN Grant was unstinting in his resolve to totally defeat GEN Lee. He was a familiar figure to his soldiers, riding among them with his slouch hat in his private's uniform with general's rank. He drove his subordinates, who themselves wanted to finish off their old foe; despite casualties, he unflinchingly resisted pressure to back away from his intent.

6-20. Having established an azimuth, organizational leaders assist their subordinates' efforts to build and train their organizations on those tasks necessary for success. Finally, they act to motivate subordinate leaders and organizations to meet the operational standards upon which discipline depends.

FILTERING INFORMATION

Leaders at all levels, but particularly those at higher levels who lack recent personal observations, can only make decisions based on the information given to them. What sets senior leaders apart is their ability to sort through great amounts of information, key in on what is significant, and then make decisions. But, these decisions are only as good as the information provided.

A Former Battalion Commander

6-21. Organizational leaders deal with a tremendous amount of information. Some information will make sense only to someone with a broad perspective and an understanding of the entire situation. Organizational leaders communicate clearly to their staffs what information they need and then hold the staff accountable for providing it. Then, they judge—based on their education, training, and experience—what's important and make well-informed, timely decisions.

6-22. Analysis and synthesis are essential to effective decision-making and program development. Analysis breaks a problem into its component parts. Synthesis assembles complex and disorganized data into a solution. Often, data must be processed before it fits into place.

6-23. Commander's critical information requirements (CCIR) are the commander's most important information filters. Commanders must know the environment, the situation, their organizations, and themselves well enough to articulate what they need to know to control their organizations and accomplish their missions. They must also ensure they have thought through the feedback systems necessary to supervise execution. Organizational-level commanders must not only establish CCIR but also train their staffs to battle drill proficiency in information filtering. (FM 101-5 discusses CCIR, mission analysis, information management, and other staff operations.)

UNDERSTANDING SYSTEMS

6-24. Organizational leaders think about systems in their organization: how they work together, how using one affects the others, and how to get the best performance from the whole. They think beyond their own organizations to how what their organization does affects other organizations and the team as a whole. Whether coordinating fires among different units or improving sponsorship of new personnel, organizational leaders use a systems perspective. While direct leaders think about tasks, organizational leaders integrate, synchronize, and fine-tune systems and monitor outcomes. If organizational leaders can't get something done, the flaw or failure is more likely systemic than human. Being able to understand and leverage systems increases a leader's ability to achieve organizational goals and objectives.

6-25. Organizational leaders also know how effectively apply all available systems to achieve mission success. They constantly make sure that the systems for personnel, administration, logistical support, resourcing, and training work effectively. They know where to look to see if the critical parts of the system are functioning properly.

DA Civilian Support to Desert Shield

During Operation Desert Shield, a contingent of DA civilians deployed to a depot in the combat theater to provide warfighting supplies and operational equipment to Third (US) Army. These DA civilians were under the supervision of DA civilian supervisors, who motivated their employees in spite of the harsh conditions in the region: hot weather, a dismal environment, and the constant threat of Iraqi missile and chemical attacks. It turned out that the uplifting organizational climate these leaders provided overcame the physical deprivation.

Two senior DA civilian leaders, the depot's deputy director of maintenance and the chief of the vehicle branch, developed a plan to replace arriving units' M1 tanks with M1A1s, which boasted greater firepower, better armor, and a more advanced nuclear, biological, and chemical protective system. They also developed systems for performing semiannual and annual maintenance checks, quickly resolving problems, applying modifications such as additional armor, and repainting the tanks in the desert camouflage pattern.

Although similar programs normally take 18 to 24 months to complete, the two leaders set an ambitious objective of 6 months. Many experts thought the goal could not be met, but the tenacious leaders never wavered in their resolve. After 24-hour-a-day, 7-day-a week operations, their inspired team of teams completed the project in 2 months. These DA civilian leaders with clear intent, firm objectives, and unrelenting will motivated their team and provided modern, lethal weapons to the soldiers who needed them when they were needed.

6-26. Organizational leaders analyze systems and results to determine why things happened the way they did. Performance indicators and standards for systems assist them in their analysis. Equipment failure rates, unit status reports (USR) Standard Installation/Division Personnel System (SIDPERS), Defense Civilian Personnel Data System (DCPDS) data, and evaluation report timeliness all show the health of systems. Once an assessment is complete and causes of a problem known, organizational leaders develop appropriate solutions that address the problem's root cause.

6-27. Isolating why things go wrong and where systems break down usually requires giving subordinates time and encouragement to ferret out what's really happening. The dilemma for organizational leaders occurs when circumstances and mission pressures require immediate remedial action and preclude gathering more data. It's then that they must fall back on their experience and that of their subordinates, make a judgment, and act.

Innovative Reorganization

Facing a long-term downsizing of his organization, a DA civilian director didn't simply shrink its size. Instead, the director creatively flattened the organization by reducing the number of deputy executives, managers, and supervisors. The director increased responsibilities of those in leadership positions and returned to a technical focus those managers and supervisors with dominant mission skills. The result was a better leader-to-led ratio, a reduced number of administrative and clerical positions, and a smoother transition to multidisciplined team operations. The director's systems understanding led him to tailor inputs that maintained healthy systems and improved outputs.

TECHNICAL SKILLS

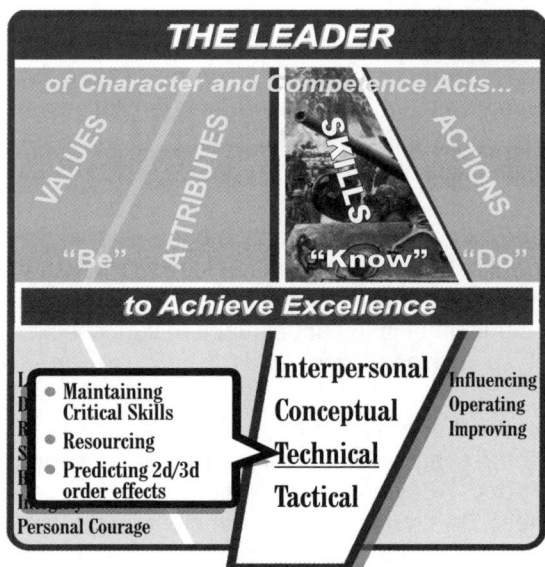

Figure 6-3. Organizational Leader Skills—Technical

6-28. The external responsibilities of organizational leaders are greater than those of direct leaders, both vertically and horizontally. Their organizations have more levels than direct-level organizations and, depending on the organization's role, command interest may reach to the CINC or national command authority. There are more requirements to coordinate with other organizations, which may include agencies outside the Department of Defense (DOD). To make full use of their organizations' capabilities, organizational leaders must continue to master technical skills outside their original area of expertise.

MAINTAINING CRITICAL SKILLS

6-29. Organizational leaders have fewer opportunities to practice many of the technical skills they mastered as direct leaders. However, that doesn't mean they can forget about them. In every organization there are certain skills in which all members must be proficient. Soldiers know what they are and expect their leaders to be able to perform them. This doesn't mean that organizational leaders must be able to perform every specialty-related skill as well as an individual holding that specialty. The Army is

too complex for that. It does, however, mean that organizational leaders must identify and be proficient in those critical, direct-leader skills they need to assess tactical training and set the example.

6-30. One organizational leader who set the example by drawing on deeply embedded technical skills was COL Marian Tierney. In her final military assignment, COL Tierney was responsible for nursing operations at 38 hospitals with 2500 nurses in the Republic of Vietnam. In 1966 she had to call on the basic medical skills and personal character she had honed throughout a career in places like Omaha Beach during the Normandy invasion. That day, 22 years after D-Day, the aircraft on which she was a passenger crashed, leaving many injured and panicked survivors. Ignoring her own injuries COL Tierney treated her comrades and took charge of evacuating the scene. For her heroism she received the Soldier's Medal. Her actions demonstrate that courageous leaders of character and competence serve at all levels.

RESOURCING

6-31. In addition to using the technical skills they learned as direct leaders, organizational leaders must also master the skill of resourcing. Resources—which include time, equipment, facilities, budgets, and people—are required to achieve organizational goals. Organizational leaders must aggressively manage the resources at their disposal to ensure their organizations' readiness. The leader's job grows more difficult when unprogrammed costs—such as an emergency deployment—shift priorities.

6-32. Organizational leaders are stewards of their people's time and energy and their own will and tenacity. They don't waste these resources but skillfully evaluate objectives, anticipate resource requirements, and efficiently allocate what's available. They balance available resources with organizational requirements and distribute them in a way that best achieves organizational goals—in combat as well as peacetime. For instance, when a cavalry squadron acting as the division flank guard makes contact, its commander asks for priority of fires. The division commander considers the needs of the squadron but must weigh it against the overall requirements of the current and future missions.

PREDICTING SECOND- AND THIRD-ORDER EFFECTS

6-33. Because the decisions of organizational leaders have wider-ranging effects than those of direct leaders, organizational leaders must be more sensitive to how their own actions affect the organization's climate. These actions may be conscious, as in the case of orders and policies, or unconscious, such as requirements for routine or unscheduled reports and meetings. The ability to discern and predict second- and third-order effects helps organizational leaders assess the health of the organizational climate and provide constructive feedback to subordinates. It can also result in identifying resource requirements and changes to organizations and procedures. (The ECAS process

illustrated in Appendix D or a similar one can be applied by organizational as well as direct leaders.)

6-34. For instance, when the Army Chief of Staff approved a separate military occupational specialty code for mechanized infantry soldiers, the consequences were wide-ranging. Second-order effects included more specialized schooling for infantry NCOs, a revised promotion system to accommodate different infantry NCO career patterns, and more doctrinal and training material to support the new specialty. Third-order effects included resource requirements for developing the training material and adding additional instructor positions at the Infantry Center and School. Organizational leaders are responsible for anticipating the consequences of any action they take or direct. Requiring thorough staff work can help. However, proper anticipation also requires imagination and vision as well as an appreciation for other people and organizations.

TACTICAL SKILLS

Soldiers need leaders who know how to fight and how to make the right decisions.

General Carl F. Vuono
Former Army Chief of Staff

Figure 6-4. Organizational Leader Skills—Tactical

6-35. Organizational leaders must master the tactical skills of synchronization and orchestration. Synchronization applies at the tactical level of war; orchestration is an operational-level term. Synchronization arranges activities in time, space, and purpose to focus maximum relative military power at a decisive point in space and time. Organizational leaders synchronize battles, each of which may comprise several synchronized engagements. (FM 100-40 discusses synchronization. FM 100-5 discusses orchestration.)

6-36. Organizational leaders at corps and higher levels orchestrate by applying the complementary and reinforcing effects of all military and nonmilitary assets to overwhelm opponents at one or more decisive points. Both synchronization and orchestration require leaders to put together technical, interpersonal, and conceptual skills and apply them to warfighting tasks.

6-37. Tactical skill for direct leaders involves employing individuals and teams of company size and smaller. In contrast, tactical skill for organizational leaders entails employing *units* of battalion size and larger. Organizational leaders get divisions, brigades, and battalions to the right place, at the right time, and in the right combination to fight and win battles and engagements. (FM 100-40 discusses battles and engagements.) They project the effects of their decisions further out—in time and distance—than do direct leaders.

6-38. The operational skill of orchestrating a series of tactical events is also more demanding and far-reaching. Time horizons are longer. Effects take more time to unfold. Decision sets are more intricate. GEN Grant's Vicksburg campaign in the spring of 1863, which split the Confederacy and opened the Mississippi River to Union use, is a classic example of an organizational leader orchestrating the efforts of subordinate forces.

GEN Grant at Vicksburg

After failing to capture Vicksburg by attacking from the north, GEN Ulysses S. Grant moved along the west bank of the Mississippi River to a point south of the city. He masked his movement and intentions by sending COL Benjamin Grierson's cavalry deep into Mississippi to conduct a series of raids. The Union commander also synchronized the daring dispatch of US Navy gunboats through Confederate shore batteries to link up with his army south of Vicksburg. Using Admiral (ADM) David D. Porter's gunboats, the Union Army crossed to the east bank of the Mississippi while MG William T. Sherman conducted a diversionary attack on the northern approaches to Vicksburg.

Once across the Mississippi, GEN Grant bypassed Vicksburg, used the Big Black River to protect his flank, and maneuvered east toward Jackson, Mississippi. By threatening both Jackson and Vicksburg, GEN Grant prevented Confederate forces from uniting against him. After a rapid series of engagements, the Union Army forced the enemy out of Jackson, blocking Vicksburg's main line of supply. It then turned west for an assault of Vicksburg, the key to control of the Mississippi. With supply lines severed and Union forces surrounding the city, Confederate forces at Vicksburg capitulated on 4 July 1863.

GEN Grant's Vicksburg campaign demonstrates the orchestration of a series of subordinate unit actions. In a succession of calculated moves, he defeated the Confederate forces under the command of Generals Joseph E. Johnston and John C. Pemberton, gained control of the Mississippi River, and divided the Confederacy.

6-39. Organizational leaders know doctrine, tactics, techniques, and procedures. Their refined tactical skills allow them to understand, integrate, and synchronize the activities of systems, bringing all resources and systems to bear on warfighting tasks.

SECTION II

WHAT IT TAKES TO LEAD ORGANIZATIONS—ACTIONS

Making decisions, exercising command, managing, administering—those are the dynamics of our calling. Responsibility is its core.

General Harold K. Johnson
Former Army Chief of Staff

6-40. Actions by organizational leaders have far greater consequences for more people over a longer time than those of direct leaders. Because the connections between action and effect are sometimes more remote and difficult to see, organizational leaders spend more time thinking about what they're doing and how they're doing it than direct leaders do. When organizational leaders act, they must translate their intent into action through the larger number of people working for them.

6-41. Knowledge of subordinates is crucial to success. To maximize and focus the energy of their staffs, organizational leaders ensure that subordinates know what must be done and why. In addition, they ensure that work being done is moving the organization in the right direction. They develop concepts for operations and policies and procedures to control and monitor their execution. Since the challenges they face are varied and complicated, no manual can possibly address them all. However, the following section provides a framework for examining, explaining, and reflecting on organizational leader actions.

INFLUENCING ACTIONS

A soldier may not always believe what you say, but he will never doubt what you do.

The Battalion Commander's Handbook

6-42. As Figure 6-5 shows, influencing is achieved through communicating, decision making, and motivating. At the organizational level, influencing means not only getting the order or concept out; it means marshaling the activities of the staff and subordinate leaders to move towards the organization's objective. Influencing involves continuing to reinforce the intent and concept, continually acquiring and assessing available feedback, and inspiring subordinates with the leader's own presence and encouragement.

6-43. The chain of command provides the initial tool for getting the word out from, and returning feedback to, the commander. In training, commanders must constantly improve its functioning. They must stress it in training situations, pushing it to the point of failure. Combat training centers (CTCs) offer tremendous opportunities to exercise and assess the chain of command in their

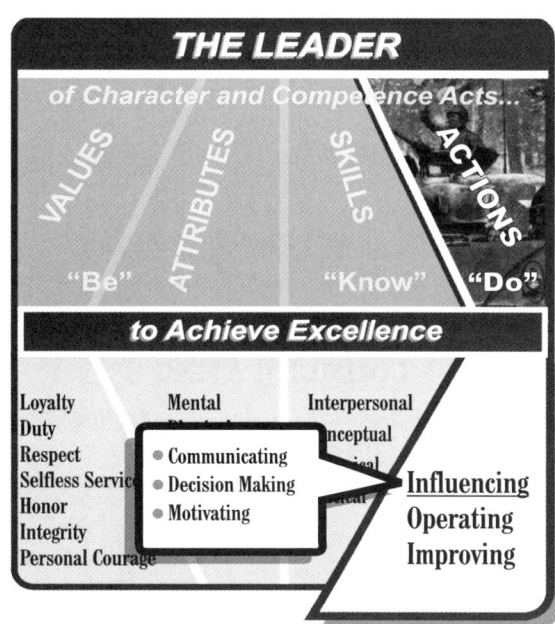

Figure 6-5. Organizational Leader Actions—
Influencing

communicating and monitoring tasks. Programs for officer and NCO professional development based on either terrain walks or seminars can reinforce chain of command functioning. Checking organizational functions daily ("leading by walking around") can reveal whether the commander's intent is getting to the lowest level.

6-44. Communication becomes more complex for organizational leaders because of their increased span of control and separation from elements actually executing the mission. Whatever organizational leaders ask for, explicitly or implicitly, causes ripples throughout the organization. Therefore, they must consider how subordinates might interpret their wishes. Directives and actions must be clear and issued in a manner that discourages overreaction. The installation commander who remarks out loud about bland walls may cause an entire organization's soldiers to paint all weekend (it has happened).

6-45. Organizational leaders also lose the right to complain in public; for example, their opinion of a support agency affects the attitude of hundreds or thousands of people. Where the leader is, how the leader looks, and what the leader does and says influence routine leadership actions throughout the organization. Like direct leaders, organizational leaders are always on display, and their demeanor and presence set the tone and climate for subordinate organizations. However, the position of organizational leaders makes them more prominent, and they must remain aware of how their behavior affects their organization. A bad day for the leader should not have to be a bad day for everyone else.

COMMUNICATING

Too often we place the burden of comprehension on [those at a different level from] us, assuming both the existence of a common language and motivation.

General Edward C. Meyer
Former Army Chief of Staff

6-46. Ironically, organizational leaders' face-to-face communication must be more powerful,

more focused, and more unequivocal than direct leaders' communication. Because organizational leaders move quickly from one project to another and one part of the organization to the other, they must be careful that the right message goes out the first time. Poor communication can have tremendously negative consequences.

Know Yourself

6-47. Even before assuming an organizational, leadership position, leaders must assess themselves, understand their strengths and weaknesses, and commit to an appropriate leadership philosophy. Organizational leaders must realize that some techniques that worked in direct-level positions may no longer work at the organizational level. They must resist the temptation to revert to their old role and thus preempt their subordinates by making decisions for them.

6-48. That said, personal qualities that contributed to their previous success are still important for organizational leaders. They must be themselves. They must know their biases, frustrations, and desires and try to keep these factors from negatively influencing their communication. It's not enough to be careful about what they say. Nonverbal communication is so powerful that organizational leaders need to be aware of personal mannerisms, behavioral quirks, and demeanor that reinforce or contradict a spoken message.

Know the Purpose

6-49. Organizational leaders know themselves, the mission, and the message. They owe it to their organization and their people to share as much as possible. People have to know what to do and why. At the most basic level, communication provides the primary way that organizational leaders show they care. If subordinates are to succeed and the organization is to move forward, then the organizational leader must work hard at maintaining positive communication. Encouraging open dialogue, actively listening to all perspectives, and ensuring that subordinate leaders and staffs can have a forthright, open, and honest voice in the organization without fear of negative consequences greatly fosters communication at all levels. Organizational leaders who communicate openly and genuinely reinforce team values, send a message of trust to subordinates, and benefit from subordinates' good ideas.

The Commander's Notebook

A brigade commander met with his subordinate leaders and outlined his goals for an upcoming training exercise. In the days following, while the brigade staff worked on the formal orders and requirements, the commander spent time visiting subordinate units as they trained. As a part of each visit, he asked his subordinate leaders for specific feedback on his intent. Was it clear? Could they repeat the three main points he had tried to make? What would they add to the unit's goals for the training? He listened, asked his own questions, and allowed them to question him. It turned out that most of the people he spoke to missed one of his three main points, which led the commander to believe that he hadn't made himself clear the first time. Eventually, he started the conversation by saying, "There are a couple of points I tried to make in my talk; apparently, I dropped the ball on at least one of them. Let me take another shot at it." Then he explained the point again.

Whenever subordinate leaders offered suggestions about the upcoming exercise, the brigade commander took out a pocket notebook and wrote some notes. Even when suggestions sounded lame, he wrote them down. That way, he signaled to the speaker, "Yes, your opinion counts, too." Secondly, by writing down the ideas, the commander guaranteed himself a chance to look at the comments later. He knew from experience that sometimes the ones that don't seem to make sense at first turn out to be quite useful later. Many of the direct leaders remarked that they had never seen a brigade commander do anything like that before. They were even more astonished when they got feedback on the suggestion. The brigade adjutant even explained to one company commander why his suggestion wasn't implemented. On a Saturday morning the brigade commander was standing in line at the PX when a platoon sergeant engaged him in conversation. "I wasn't around the day you visited my company last week, sir," the NCO said, "but I heard the other folks had a few suggestions for you. I wonder if I could add something?"

Know the Environment

6-50. Before organizational leaders can effectively communicate, they must assess the environment—people, events, and systems—and tailor their message to the target audience. Organizational leaders constantly communicate by persuading and conveying intent, standards, goals, and priorities at four levels within the Army: their people, their own and higher staffs, their subordinate leaders and commanders, and their superiors. There may also be occasions that require organizational leaders to speak to audiences outside the Army such as the media or community groups. They may have to repeat the message to different audiences and retune it for different echelons, but only leaders can reinforce their true intent.

Know the Boss

6-51. Working to communicate consistently with the boss is especially important for organizational leaders. Organizational leaders have to figure out how to reach the boss. They must assess how the boss communicates and how the boss receives information. For some leaders, direct and personal contact is best; others may be more comfortable with weekly meetings, e-mail, or letters. Knowing the boss's intent, priorities, and thought processes greatly enhances organizational success. An organizational leader who communicates well with the boss minimizes friction between the organization and the higher headquarters and improves the overall organizational climate.

Know the Subordinates

If it's dumb it's not our policy.

Lieutenant General Walter F. Ulmer Jr.
Former Commanding General, III Corps

6-52. The mere presence of an organizational leader somewhere communicates the leader's character and what the leader values. The organizational leader who hurries through a talk about caring for subordinates, then passes up an opportunity to speak face-to-face with some soldiers, does more than negate the message; he undercuts whatever trust his subordinates may have had.

6-53. Because organizational leaders know themselves, they also know that others bring the sum total of their experience to their duties. They analyze interpersonal contact to gather meaning; they look for the message behind the words. In this way, they gain a greater understanding of peers, subordinates, and superiors. Improving communication skills becomes a major self-development challenge for them. By stating their intent openly, organizational leaders give subordinates an open door for feedback on unintended consequences or just bad policy: "Hey sir, did you really mean it when you said, 'If it's dumb it's not our policy?' OK, well what about...?" Leaders must be seen to be heard.

Know the Staff

6-54. Organizational leaders must understand what's going on within their own and the next-higher echelon staff. Networking allows them to improve communication and mission accomplishment by giving them a better understanding of the overall environment. Networking requires leaders to constantly interact and share thoughts, ideas, and priorities. Informed staffs can then turn policies, plans, and programs into realities.

6-55. Organizational leaders must also know the focus of the next-higher staff and commander. Through taking time to interact with the next-higher staff, organizational leaders gain a greater understanding of the boss's priorities and also help set the conditions for their own requirements. Constantly sensing—observing, talking, questioning, and actively listening

to—what's going on helps organizational leaders better identify and solve potential problems and allows them to anticipate decisions and put their outfit in the best possible position to execute.

Know the Best Method

6-56. To disseminate information accurately and rapidly, organizational leaders must also develop an effective communications network. Some of these networks—such as the chain of command, the family support network, the NCO support channel, and staff relationships—simply need to be recognized and exploited. Other informal chains must be developed. Different actions may require different networks.

6-57. The more adept organizational leaders become in recognizing, establishing, and using these networks, the more successful the outcome, especially as they become comfortable using a wider range of communications forums. Memorandums, notes, and e-mail as well as formal and informal meetings, interactions, and publications are tools of an effective communicator. Organizational leaders must know the audiences these methods reach and use them accordingly.

DECISION MAKING

The key is not to make quick decisions, but to make timely decisions.

General Colin Powell
Former Chairman, Joint Chiefs of Staff

6-58. Organizational leaders are far more likely than direct leaders to be required to make decisions with incomplete information. They determine whether they have to decide at all, which decisions to make themselves, and which ones to push down to lower levels. To determine the right course of action, they consider possible second- and third-order effects and think farther into the future—months, or even years, out in the case of some directorates.

6-59. Organizational leaders identify the problem, collect input from all levels, synthesize that input into solutions, and then choose and execute the best solution in time to make a difference. To maximize the use of resources and have the greatest effect on developing an effective organization, organizational leaders move beyond a reacting, problem-*solving* approach to an anticipating, problem-*preemption* method. While there will always be emergencies and unforeseen circumstances, organizational leaders focus on anticipating future events and making decisions about the systems and people necessary to minimize crises. Vision is essential for organizational leaders.

6-60. During operations, the pace and stress of action increase over those of training. Organizational leaders use the MDMP to make tactical decisions; however, they must add their conceptual skill of systems understanding to their knowledge of tactics when considering courses of action. Organizational leaders may be tempted—because of pressure, the threat, fear, or fatigue—to abandon sound decision making by reacting to short-term demands. The same impulses may result in focusing too narrowly on specific events and losing their sense of time and timing. But there's no reason for organizational leaders to abandon proven decision-making processes in crises, although they shouldn't hesitate to modify a process to fit the situation. In combat, success comes from creative, flexible decision making by leaders who quickly analyze a problem, anticipate enemy actions, and rapidly execute their decisions. (Remember GEN Grant's actions at Vicksburg.) Leaders who delay or attempt to avoid a decision may cause unnecessary casualties and even mission failure.

6-61. Effective and timely decision making—both the commander's and subordinates'—is crucial to success. As part of decision making, organizational leaders establish responsibility and accountability among their subordinates. They delegate decision making authority as far as it will go, empowering and encouraging subordinates to make decisions that affect their areas of responsibility or to further delegate that authority to their own subordinates.

6-62. Effective organizational leaders encourage initiative and risk-taking. They remember that they are training leaders and soldiers; the goal is a better-trained team, not some ideal outcome. When necessary, they support subordinates' bad decisions, but only those made attempting to follow the commander's intent. Failing through want of experience or luck is forgivable. Negligence, indecision, or attempts to take an easy route should never be tolerated.

6-63. As GEN Powell's comment makes clear, a decision's timeliness is as important as the speed at which it is made. Just as for direct leaders, a good decision now is better than a perfect one too late. Leaders who are good at handling the decision-making process will perform better when the OPTEMPO speeds up. Leaders who don't deliver timely decisions leave their subordinates scrambling and trying to make up for lost time. Better to launch the operation with a good concept and let empowered subordinates develop subsequent changes to the plan than to court failure by waiting too long for the perfect plan.

6-64. In tough moments, organizational leaders may need the support of key subordinates to close an issue. Consider MG George G. Meade's position at Gettysburg. In command of the Army of the Potomac for only a few days, MG Meade met with his subordinates on the night of 2 July 1863 after two days of tough fighting. Uncertainty hung heavy in the air. MG Meade's decision to stand and fight, made with the support of his corps commanders, influenced the outcome of the battle and became a turning point of the Civil War.

6-65. Coping with uncertainty is normal for all leaders, increasingly so for organizational leaders. Given today's information technology, the dangerous temptation to wait for all available information before making a decision will persist. Even though this same technology may also bring the unwanted attention of a superior, leaders should not allow it to unduly influence their decisions. Organizational leaders are where they are because of their experience, intuition, initiative, and judgment. Events move quickly, and it's more important for decisive organizational leaders to recognize and seize opportunities, thereby creating success, than to wait for all the facts and risk failure.

6-66. In the end, leaders bear ultimate responsibility for their organizations' success or failure. If the mission fails, they can't lay the blame elsewhere but must take full responsibility. If the mission succeeds, good leaders give credit to their subordinates. While organizational leaders can't ensure success by being all-knowing or present everywhere, they can assert themselves throughout the organization by being decisive in times of crisis and quick to seize opportunities. In combat, leaders take advantage of fleeting windows of opportunity: they see challenges rather than obstacles; they seek solutions rather than excuses; they fight through uncertainty to victory.

MOTIVATING

It is not enough to fight. It is the spirit which we bring to the fight that decides the issue. It is morale that wins the victory.

General of the Army George C. Marshall

6-67. Interpersonal skills involved in creating and sustaining ethical and supportive climates are required at the organizational as well as the direct leadership level. As Chapter 3 explains, the organizational, unit, or command climate describes the environment in which subordinates work. Chapter 5 discusses how direct leaders focus their motivational skills on individuals or small groups of subordinates. While direct leaders are responsible for their organizations' climate, their efforts are constrained (or reinforced) by the larger organization's climate. Organizational leaders shape that larger environment. Their primary motivational responsibility is to establish and maintain the climate of their entire organization.

6-68. Disciplined organizations evolve within a positive organizational climate. An organization's climate springs from its leader's attitudes, actions, and priorities. Organizational leaders set the tone most powerfully through a personal example that brings Army values to life. Upon assuming an organizational leadership position, a leader determines the organizational climate by assessing the organization from the bottom up. Once this assessment is complete, the leader can provide the guidance and focus (purpose, direction, and motivation) required to move the organizational climate to the desired end state.

6-69. A climate that promotes Army values and fosters the warrior ethos encourages learning and promotes creative performance. The foundation for a positive organizational climate is a healthy ethical climate, but that alone is insufficient. Characteristics of successful organizational climates include a clear, widely known intent; well-trained and confident soldiers; disciplined, cohesive teams; and trusted, competent leadership.

6-70. To create such a climate, organizational leaders recognize mistakes as opportunities to learn, create cohesive teams, and reward leaders of character and competence. Organizational leaders value honest feedback and constantly use all available means to maintain a feel for the environment. Staff members who may be good sources for straightforward feedback may include equal opportunity advisors and chaplains. Methods may include town hall meetings, surveys, and councils. And of course, personal observation—getting out and talking to DA civilians, soldiers, and family members—brings organizational leaders face-to-face with the people affected by their decisions and policies. Organizational leaders' consistent, sincere effort to see what's really going on and fix things that are not working right can result in mutual respect throughout their organizations. They must know the intricacies of the job, trust their people, develop trust among them, and support their subordinates.

6-71. Organizational leaders who are positive, fair, and honest in their dealings and who are not afraid of constructive criticism encourage an atmosphere of openness and trust. Their people willingly share ideas and take risks to get the job done well because their leaders strive for more than compliance; they seek to develop subordinates with good judgment.

6-72. Good judgment doesn't mean lockstep thinking. Thinking "outside the box" isn't the same as indiscipline. In fact, a disciplined organization systematically encourages creativity and taking prudent risks. The leader convinces subordinates that anything they break can be fixed, except life or limb. Effective organizational leaders actively listen to, support, and reward subordinates who show disciplined initiative. All these things create opportunities for subordinates to succeed and thereby build their confidence and motivation.

6-73. However, it's not enough that individuals can perform. When people are part of a disciplined and cohesive team, they gain proficiency, are motivated, and willingly subordinate themselves to organizational needs. People who sense they're part of a competent, well-trained team act on what the team needs; they're confident in themselves and feel a part of something important and compelling. These team members know that what they do matters and discipline themselves.

The 505th Parachute Infantry Regiment at Normandy

On 7 June 1944, the day after D-Day, nearly 600 paratroopers of the 505th Parachute Infantry Regiment were in position in the town of Ste. Mère Église in Normandy to block any German counterattack of the Allied invasion force. Although outnumbered by an enemy force of over 6,000 soldiers, the paratroopers attacked the German flank and prevented the enemy's assault. The paratroopers were motivated and well-trained, and they all understood the absolute necessity of preventing the German counterattack. Even in the fog of war, they did what needed to be done to achieve victory. Their feat is especially noteworthy since many landed outside their planned drop zones and had to find their units on their own. They did so quickly and efficiently in the face of the enemy.

The 505th combined shared purpose, a positive and ethical climate, and cohesive, disciplined teams to build the confidence and motivation necessary to fight and win in the face of uncertainty and adversity. Both leaders and soldiers understood that no plan remains intact after a unit crosses the line of departure. The leaders' initiative allowed the disciplined units to execute the mission by following the commander's intent, even when the conditions on the battlefield changed.

OPERATING ACTIONS

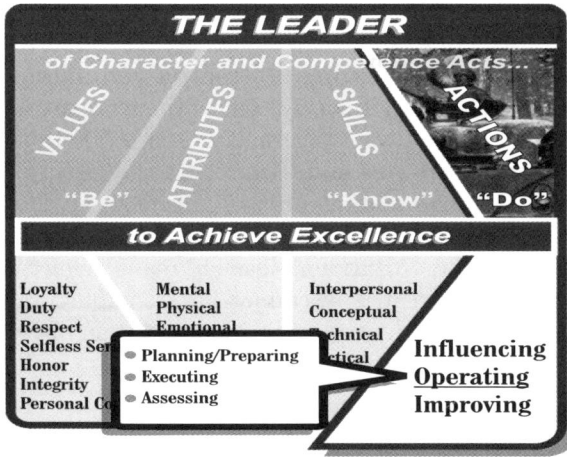

Figure 6-6. Organizational Leader Actions— Operating

6-74. Organizational leaders see, decide, and act as they perform the operating actions shown in Figure 6-6. They emphasize teamwork and cooperation over competition. They provide their intent so subordinates can accomplish the mission, no matter what happens to the original plan. Because organizational leaders primarily work through subordinates, empowerment and delegation are indispensable. As a result of communicating with subordinates, listening to their responses, and obtaining feedback from their assessments, organizational leaders are better equipped to make decisions.

6-75. Organizational-level commanders usually use the MDMP for tactical decision making and planning. However, those who command in the joint environment must use the Joint Operation Planning and Execution System (JOPES) methodology. Both the MDMP and JOPES allow organizational commanders to apply the factors discussed in this chapter.

SYSTEMS PLANNING AND PREPARING

[A] good plan violently executed now *is better than a perfect plan next week.*

General George S. Patton Jr.
War As I Knew It

6-76. GEN Patton wasn't belittling the importance of planning; he was emphasizing the balance necessary for successful operations. Planning, getting ready for the future by laying out how leaders aim to turn their intent into reality, is something leaders do every day and something the Army does very well. However, organizational leaders plan for the systems that support training and operations as well as for the actual training event or operation. Systems planning involves seven steps:

- Establish intent.
- Set goals.
- Determine objectives.
- Determine tasks.
- Establish priorities.
- Follow up.

Establish Intent

6-77. The first step in systems planning is for the organizational leader to have a clear intent for what he wants the organization to be. What will it look like at some future point? Spending extra time visualizing the end state up front is more important than quickly jumping into the mechanics of planning. Obviously, the actual mission is critical in determining this end state. The organizational leader's intent should be announced at the earliest practicable time after it has been formulated so the staff and subordinate commanders can have maximum time to plan. For a division, the intent might be—

- The best infantry division in the world.
- Supported by the finest installation in the Army.
- Trained and ready to deploy anywhere in the world to fight and win.
- But flexible enough to accomplish any other mission the nation asks us to perform.
- A values-based organization that takes care of its soldiers, DA civilians, and families.

6-78. Organizational leaders must determine how this intent affects the various systems for which they are responsible. By their actions and those of their subordinates and by using their presence to be heard, organizational leaders bring meaning for their intent to their people.

Set Goals

6-79. Once they have established their intent, organizational leaders, with the help of their team of subordinate leaders and staffs, set specific goals for their organizations. Goals frame the organizational leader's intent. For instance, the goal "Improve fire control and killing power" could support that part of the intent that states the division will be "trained and ready to deploy anywhere in the world to fight and win." Organizational leaders are personally involved in setting goals and priorities to execute their intent and are aware that unrealistic goals punish subordinates.

Determine Objectives

6-80. In the third step, organizational leaders establish objectives that are specific and measurable. For example, an objective that supports the goal of improving fire control and killing power could be "Fifty percent of the force must fire expert on their personal weapons." Establishing objectives is difficult because the process requires making precise calls from a wide variety of options. Since time and resources are limited, organizational leaders make choices about what can and cannot be accomplished. They check key system nodes to monitor subsystem functions.

Determine Tasks

6-81. The fourth step involves determining the measurable, concrete steps that must be taken on the way to the objective. For example, the commander of a forward-stationed division might ensure family readiness by ordering that any newly arriving soldier with a family may not be deployed without having a vehicle in country and household goods delivered.

Establish Priorities

6-82. The fifth step is to establish a priority for the tasks. This crucial step lets subordinates know how to spend one of their most critical resources: time.

6-83. This system of establishing priorities is important for the organization; organizational leaders must also practice it personally. In fact, a highly developed system of time management may be the only way for organizational leaders to handle all the demands upon them. There's rarely enough time to do everything, yet they must make the time to assess and synthesize information and make timely decisions. Leaders who recognize distractions are better equipped to handle their time well.

Prepare

6-84. Though organizational leaders have more complex missions than direct leaders,

they also have more assets: a staff and additional subordinate leaders, specialists, and equipment allow their preparation to be diverse and complete. Direct leaders prepare by getting individuals moving in the right direction; organizational leaders take a step back and check to make sure the systems necessary to support the mission are in place and functioning correctly.

Follow Up

6-85. The final step in systems planning is to follow up: Does the team understand the tasks? Is the team taking the necessary actions to complete them? Check the chain of command again: does everyone have the word? Organizational leader involvement in this follow up validates the priorities and demonstrates that the leader is serious about seeing the mission completed. Organizational leaders who fail to follow up send a message that the priorities are not really that important and that their orders are not really binding.

The "Paperwork Purge"

The division's new chief of staff was surprised at how much time subordinates spent at meetings; it seemed they had time for little else. After observing the way things worked for two weeks, the chief did away with most of the scheduled meetings, telling the staff, "We'll meet when we need to meet, and not just because it's Friday morning." What's more, the chief required an agenda for each meeting ahead of time: "That way, people can do their homework and see who needs to be there and who doesn't." The chief was always on time for meetings and started at the time specified on the agenda. There were no interruptions of whomever had the floor, and the long, meandering speeches that had marked previous meetings were cut short.

The chief put a one-page limit on briefing papers for the boss. This meant subordinates learned to write concisely. Each staff section did a top-to-bottom review of procedures that had been in place as long as anyone could remember. Anything that couldn't be justified was thrown out. The chief handled most of the correspondence that came across his desk with a quick note written on the original and told the staff to do the same.

The chief made the staff justify requirements they sent to subordinate organizations, with the comment, "If you can't tell them why it's important, then maybe it's not important." The explanation also helped subordinate elements determine their own priorities: "You can't keep sending stuff down saying, 'This is critical!' It gets to be like the boy who cried wolf."

The "Paperwork Purge" (continued)

Of course, the staff didn't take the new chief quite seriously at first, and after a week of reviewing old policy letters, some staff sections let the requirement slide. Then the chief showed up one day and had them give him a rundown on all the policies left after what everyone was calling "the big paperwork purge." A few more outdated requirements fell by the wayside that afternoon. More important, the staff got the message that the chief followed through on decisions.

Finally, and most startling, the chief told staff members that now and then they should sit quietly and stare into space: "You're getting paid to think, and every once in a while you've got to stop moving to do that well."

6-86. Keeping their intent in mind, organizational leaders fight distracters, make time to reflect, and seek to work more efficiently. Despite the pressure of too much to do in too little time, they keep their sense of humor and help those around them do the same.

THE CREATIVE STAFF PROCESS

None of us is as smart as all of us.

A Former Brigade Commander

6-87. The size and complexity of the organizations led by organizational leaders requires well-trained, competent staffs. Training these staffs is a major responsibility of organizational leaders. The chief of staff or executive officer is the organizational leader's right hand in that effort.

6-88. In the 100 days leading to the Battle of Waterloo, Napoleon had to campaign without his intensely loyal and untiring chief of staff, Berthier. In all his other campaigns, Berthier had transformed Napoleon's orders into instructions to the marshals, usually in quadruplicate with different riders carrying four copies to the same marshal over different routes. Berthier's genius for translating Napoleon's intent into tasks for each corps underlay the French Army's versatile, fluid maneuver style. Without Berthier and with an increasingly rigid Napoleon disdaining advice from any source, Napoleon's formations lost a good deal of their flexibility and speed.

6-89. Great staffs work in concert with the leader to turn intent into reality. A single leader in isolation has no doubt done great things and made good decisions. However, the organizational leader alone can't consistently make the right decisions in an environment where operational momentum never stops.

6-90. Building a creative, thinking staff requires the commander's time, maturity, wisdom, and patience. Although managing information is important, the organizational leader needs to invest in both quality people and in training them to think rather than just process information. Several factors contribute to building a creative, thinking staff.

The Right People

6-91. A high-performing staff starts with putting the right people in the right places. Organizational leaders are limited to their organization's resources, but have many choices about how to use them. They assemble, from throughout their organizations, people who think creatively, possess a vast array of technical skills, are trained to solve problems, and can work together. They take the time to evaluate the staff and implement a training program to improve it as a whole. They avoid micromanaging the staff, instead trusting and empowering it to think creatively and provide answers.

The Chief of Staff

6-92. The staff needs its own leader to take charge—someone who can focus it, work with it, inspire it, and move it to decisive action in the absence of the commander. The sections of the staff work as equals, yet without superb leadership they won't perform exceptionally.

To make a staff a true team, an empowered deputy must be worthy of the staff and have its respect. The chief of staff must have the courage to anticipate and demand the best possible quality. On the other hand, the chief must take care of the hardworking people who make up the staff and create an environment that fosters individual initiative and develops potential. (FM 100-34 discusses the role of the chief of staff.)

Challenging Problems

6-93. A staff constantly needs challenging problems to solve if it's to build the attitude that it can overcome any obstacle. Tackling problems with restricted time and resources improves the staff members' confidence and proficiency, as long as they get an opportunity to celebrate successes and to recharge their batteries. Great confidence comes from training under conditions more strenuous than they would likely face otherwise.

Clear Guidance

6-94. The commander constantly shares thoughts and guidance with the staff. Well-trained staffs can then synthesize data according to those guidelines. Computers, because of their ability to handle large amounts of data, are useful analytical tools, but they can do only limited, low-order synthesis. There's no substitute for a clear commander's intent, clearly understood by every member of the staff.

EXECUTING

The American soldier demonstrated that, properly equipped, trained, and led, he has no superior among all the armies of the world.

General Lucian K. Truscott
Former Commanding General, 5th Army

6-95. Planning and preparation for branches and sequels of a plan and contingencies for future operations may continue, even during execution. However, execution is the purpose for which the other operating actions occur; at some point, the organizational leader commits to action, spurs his organization forward, and sees the job through to the end. (FM 100-34 and FM 101-5 discuss branches and sequels.)

6-96. In combat, organizational leaders integrate and synchronize all available elements of the combined arms team, empower subordinates, and assign tasks to accomplish the mission. But the essence of warfighting for organizational leaders is their will. They must persevere despite limitations, setbacks, physical exhaustion, and declining mental and emotional reserves. They then directly and indirectly energize their units—commanders and soldiers—to push through confusion and hardship to victory.

6-97. Whether they're officers, NCOs, or DA civilians, the ultimate responsibility of organizational-level leaders is to accomplish the mission. To this end, they must mass the effects of available forces on the battlefield, to include supporting assets from other services. The process starts before the fight as leaders align forces, resources, training, and other supporting systems.

Combined Arms and Joint Warfighting

6-98. Brigades and battalions usually conduct single-service operations supported by assets from other services. In contrast, the large areas of responsibility in which divisions and corps operate make division and corps fights joint by nature. Joint task forces (JTFs) are also organizational-level formations. Therefore, organizational leaders and their staffs at division-level and higher must understand joint procedures and concerns at least as well as they understand Army procedures and concerns. In addition, it's not unusual for a corps to control forces of another nation; divisions do also, but not as frequently. This means that corps and division headquarters include liaison officers from other nations. In some cases, these staffs may have members of other nations permanently assigned: such a staff is truly multinational.

6-99. Today's operations present all Army leaders—but particularly organizational leaders—with a nonlinear, dynamic environment ranging the full spectrum of continuous operations. These dispersed conditions create an information-intense environment that challenges leaders to synchronize their efforts with nonmilitary and often nongovernmental agencies.

Empowering

Never tell people how to do things. Tell them what to do and they will surprise you with their ingenuity.

General George S. Patton Jr.
War As I Knew It

6-100. To increase the effects of their will, organizational leaders must encourage initiative in their subordinates. Although unity of command is a principle of war, at some level a single leader alone can no longer control all elements of an organization and personally direct the accomplishment of every aspect of its mission. As leaders approach the brigade or directorate level, hard work and force of personality alone cannot carry the organization. Effective organizational leaders delegate authority and support their subordinates' decisions, while holding subordinates accountable for their actions.

6-101. Delegating successfully involves convincing subordinates that they're empowered, that they indeed have the freedom to act independently. Empowered subordinates have, and know they have, more than the responsibility to get the job done. They have the authority to operate in the way they see fit and are limited only by the leader's intent.

6-102. To do that, the organizational leader gives subordinates the mission, motivates them, and lets them go. Subordinates know that the boss trusts them to make the right things happen; this security motivates them, in turn, to lead their people with determination. They know the boss will underwrite honest mistakes, well-intentioned mistakes—not stupid, careless, or repeated ones. So for the boss, empowering subordinates means building the systems and establishing the climate that gives subordinates the rein to do the job within the bounds of acceptable risk. It means setting organizational objectives and delegating tasks to ensure parallel, synchronized progress.

6-103. Delegation is a critical task: Which subordinates can be trusted with independent action? Which need a short rein? In fluid situations—especially in combat, where circumstances can change rapidly or where leaders may be out of touch or become casualties—empowered subordinates will pursue the commander's intent as the situation develops and react correctly to changes that previous orders failed to anticipate. However, as important as delegation is to the success of organizations, it does not imply in any way a reduction of the commander's responsibility for the outcome. Only the commander is accountable for the overall outcome, the success or failure, of the mission.

ASSESSING

6-104. The ability to assess a situation accurately and reliably—a critical tool in the leader's arsenal—requires instinct and intuition based on experience and learning. It also requires a feel for the reliability and validity of information and its sources. Organizational assessment is necessary to determine organizational weaknesses and preempt mishaps. Accurately determining causes is essential to training management, developing subordinate leadership, and process improvement.

6-105. There are several different ways to gather information: asking subordinates questions to find out if the word is getting to them, meeting people, and checking for synchronized plans are a few. Assessing may also involve delving into the electronic databases upon which situational understanding depends. Assessment techniques are more than measurement tools; in fact, the way a leader assesses something can influence the process being assessed. The techniques used may produce high quality, useful feedback; however, in a dysfunctional command climate, they can backfire and send the wrong message about priorities.

6-106. Staff and subordinates manage and process information for a leader, but this doesn't relieve the leader from the responsibility of analyzing information as part of the decision-making process. Leaders obtain information from various sources so they can compare and make judgments about the accuracy of sources.

6-107. As Third Army commander during World War II, GEN George Patton did this continuously. Third Army staff officers visited front-line units daily to gather the latest available information. In addition, the 6th Cavalry Group, the so-called "Household Cavalry" monitored subordinate unit reconnaissance nets and sent liaison patrols to visit command and observation posts of units in contact. These liaison patrols would exchange information with subordinate unit G2s and G3s and report tactical and operational information directly to the Third Army forward headquarters (after clearing it with the operations section of the unit they were visiting).

6-108. In addition to providing timely combat information, the Household Cavalry and staff visits reduced the number of reports Third Army headquarters required and created a sense of cohesiveness and understanding not found in other field armies. Other organizational leaders have accomplished the same thing using liaison officers grounded in their commander's intent. Whatever the method they choose, organizational leaders must be aware of the second- and third-order effects of having "another set of eyes."

6-109. In the world of digital command and control, commanders may set screens on various command and control systems to monitor the status of key units, selected enemy parameters, and critical planning and execution timelines. They may establish prompts in the command and control terminal that warn of imminent selected events, such as low fuel levels in maneuver units, tight fighter management timelines among aviation crews, or massing enemy artillery.

6-110. A leader's preconceived notions and opinions (such as "technology undermines basic skills" or "technology is the answer") can interfere with objective analysis. It's also possible to be too analytical, especially with limited amounts of information and time. Therefore, when analyzing information, organizational leaders guard against dogmatism, impatience, or overconfidence that may bias their analysis.

6-111. The first step in designing an assessment system is to determine the purpose of the assessment. While purposes vary, most fall into one of the following categories:

- Evaluate progress toward organizational goals (using an emergency deployment readiness exercise to check unit readiness or monitoring progress of units through stages of reception, staging, onward movement, and integration).

- Evaluate the efficiency of a system, that is, the ratio of the resources expended to the results gained (comparing the amount of time spent performing maintenance to the organization's readiness rate).

- Evaluate the effectiveness of a system, that is, the quality of the results it produces (analyzing the variation in Bradley gunnery scores).

- Compare the relative efficiency or effectiveness against standards.

- Compare the behavior of individuals in a group with the prescribed standards (APFT or gunnery scores).

- Evaluate systems supporting the organization (following up "no pay dues" to see what the NCO support channel did about them).

6-112. Organizational leaders consider the direct and indirect costs of assessing. Objective costs include the manpower required to design and administer the system and to collect, analyze, and report the data. Costs may also include machine-processing times and expenses related to communicating the data. Subjective costs include possible confusion regarding organizational priorities and philosophies, misperceptions regarding trust and decentralization, fears over unfair use of collected data, and the energy expended to collect and refine the data.

6-113. Organizational leaders ask themselves these questions: What's the standard? Does the standard make sense to all concerned? Did we meet it? What system measures it? Who's responsible for the system? How do we reinforce or correct our findings? One of the greatest contributions organizational leaders can make to their organizations is to assess their own leadership actions: Are you doing things the way you would to support the nation at war? Will your current systems serve equally well under the stress and strain of continuous fighting? If not, why not?

6-114. It follows that organizational leaders who make those evaluations every day will also hold their organizations to the highest standards. When asked, their closest subordinates will give them informal AARs of their leadership behaviors in the critical situations. When they arrange to be part of official AARs, they can invite subordinates to comment on how they could have made things go better. Organizational leader errors are very visible; their results are probably observed and felt by many subordinates. Thus, there's no sense in not admitting, analyzing. and learning from these errors. A bit of reflection in peacetime may lead to greater effectiveness in war.

6-115. The 1991 ground war in the Iraqi desert lasted only 100 hours, but it was won through hard work over a period of years, in countless field exercises on ranges and at the combat training centers. The continual assessment process allowed organizational leaders to trade long hours of hard work in peacetime for operations in war.

6-116. Organizational leaders are personally dedicated to providing tough, battle-focused training so that the scrimmage is always harder than the game. They must ensure that in training, to the extent that resources and risks allow, nothing is simulated. Constant assessments refine training challenges, forge confidence, and foster the quiet, calculating, and deadly warrior ethos that wins battles and campaigns. (FM 25-100 and FM 25-101 discuss battle focus and training assessment.)

IMPROVING ACTIONS

The creative leader is one who will rewrite doctrine, employ new weapons systems, develop new tactics and who pushes the state of the art.

John O. Marsh Jr.
Former Secretary of the Army

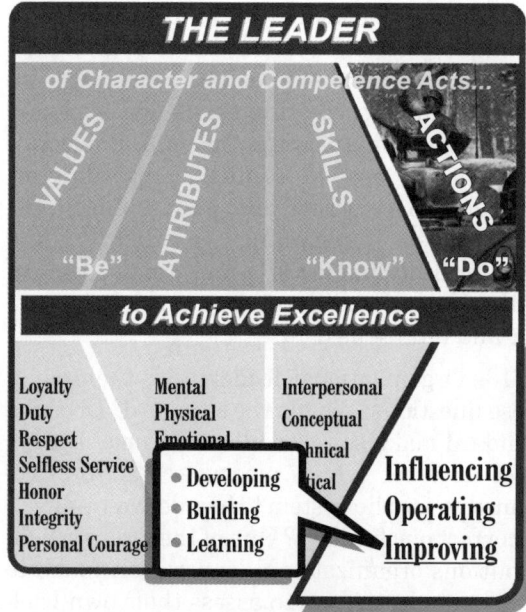

Figure 6-7. Organizational Leader Actions—Improving

6-117. Improving actions are what all leaders do today to make their organization and subordinates better tomorrow, next month, next year, five years from now. The responsibility for how the Army fights the next war lies with today's leaders; the work to improve the organization in the long term never ends. Leaders teaching subordinates to do the leader's job in combat is the hallmark of the profession of arms.

6-118. The payoff for improving actions might not be evident for years. In fact, leaders at all levels may never see the benefit of developing subordinates, since those subordinates go on to work for someone else. But this doesn't stop them from taking pride in their subordinates' development and performance; a subordinate's success is a great measure of a leader's success. Further, it's often difficult to draw a cause-and-effect line from what leaders do today to how it pays off tomorrow. Precisely because of these difficulties, organizational leaders ensure the goals they establish include improving people and organizations. They also make sure they communicate this to their subordinates.

6-119. The developing, building, and learning actions may be more difficult at the organizational level because the leaders themselves must rely more on indirect leadership methods. The challenge is greater because of the size of the organization, but the rewards increase as well: organizational leaders can influence large numbers of people and improve large segments of the Army.

DEVELOPING

Let us set for ourselves a standard so high that it will be a glory to live up to it, and then let us live up to it and add a new laurel to the crown of America.

Woodrow Wilson
28th President of the United States

6-120. Just as leadership begins at the top, so does developing. Organizational leaders keep a focus on where the organization needs to go and what all leaders must be capable of accomplishing. They continually develop themselves and mentor their subordinate leaders. As discussed in Chapter 3, leaders search for and take advantage of opportunities to mentor their subordinates. At the organizational level, commanders ensure that systems and conditions are in place for the mentoring of all organizational members.

6-121. Effective organizational leaders grow leaders at all levels of their organization. Just as they prepare their units for in-stride breaches, for example, they combine existing opportunities into a coherent plan for leadership development. Leaders get much of their development when they practice what they've learned and receive straightforward feedback in rigorously honest AARs. Feedback also comes from self-assessments as well as from peers, subordinates, and supervisors.

6-122. Organizational leaders design and integrate leader development programs into everyday training. They aim to capture learning in common duties, ensure timely feedback, and allow reflection and analysis. As Frederick the Great said, "What good is experience if you do not reflect?" Simply scheduling officer and NCO professional development sessions isn't enough for genuine, lasting leader development. Letting "operating" overwhelm "improving" threatens the future.

6-123. Leadership development is purposeful, not accidental. Everyday mission requirements are opportunities to grow leaders. Based on assessment of their subordinate leaders, organizational leaders describe how they intend to deliberately influence leader development through a comprehensive leadership development program that captures and harnesses what's already occurring in the organization. A leadership development program must provide for learning skills, practicing actions, and receiving feedback.

6-124. Organizational leaders assess their organizations to determine organization-specific developmental needs. They analyze their mission, equipment, and long-term schedule as well as the experience and competence of their subordinate leaders to determine leadership requirements. In addition to preparing their immediate subordinates to take their place, organizational leaders must also prepare subordinate leaders selected for specific duties to actually execute them.

6-125. Based on their assessment, organizational leaders define and clearly articulate their goals and objectives for leadership development within the organization. They create program goals and objectives to support their focus as well as to communicate specific responsibilities for subordinate leaders. These subordinate leaders help bring leadership development to life through constant mentoring and experiential learning opportunities. Leadership development is an important responsibility shared by leaders at every level. It becomes their greatest contribution—their legacy.

6-126. The development technique used depends on the leaders involved. Learning by making mistakes is possible, but having subordinates develop habits of succeeding is better for instilling self-confidence and initiative. Newly assigned assistant operations officers may need time to visit remote sites over which they have day-to-day control. They may need time to visit higher headquarters to establish rapport with those action officers they will have to deal with under the pressure of a tense operational situation. They may have to see the tactical operations center set up in the field and get a chance to review its SOPs before a big exercise. These activities can only happen if the organizational leader supports the leader development program and demands it take place in spite of the pressures of daily (routine) business.

6-127. There are many ways to tackle leader development. For example, instead of pursuing long-term training programs for his career civilians, one DA civilian leader enrolled both his DA civilian managers and military officers in graduate programs while they continued to work full-time. This approach allowed the directorate to provide development opportunities to four times as many personnel at one-third the cost of long-term training. In addition, the students were able to apply what they were learning directly to their jobs, thus providing immediate benefit to the organization.

6-128. In addition to educational programs, innovative interagency exchange assignments can cross-level the knowledge, skills, and experience of DA civilian leaders. Whether taking on new interns or expanding the perspectives of seasoned managers, the DA civilian component mirrors the uniformed components in its approach to broad-based leadership opportunities.

6-129. Often developmental programs involve historical events similar to current operational challenges. Such situations allow all to share a sense of what works and what does not from what worked before and what did not. This analysis can also be applied to recent organizational experiences. For example, in preparation for a CTC rotation, leaders review their own as well as others' experiences to determine valuable lessons learned. They master the individual and collective tasks through a training program that sets up soldiers and leaders for success. Based on internal AARs, they continue

to learn, practice, and assess. CTCs also provide individual leaders with invaluable experience in operating under harsh conditions. Organizations execute missions, receive candid feedback and coaching to facilitate lessons learned, then execute again.

6-130. Commanders must take the time to ensure they do developmental counseling. Nothing can replace the face-to-face contribution made by a commander mentoring a subordinate. Developing the most talented and (often) those with the greatest challenges requires a great amount of time and energy, but it's an essential investment for the future.

BUILDING

Building Combat Power

6-131. Emphasis on winning can't waver during training, deploying, and fighting. By developing the right systems and formulating appropriate contingency plans, organizational leaders ensure that the organization is prepared for a variety of conditions and uncertainties. In wartime, building combat power derives from task organization, resourcing, and preparing for execution while still meeting the human needs of the organization. Commanders must preserve and recycle organizational energy throughout the campaign. In peacetime, the main component of potential combat power is embedded collective skill and organizational readiness stemming from hard, continuous, and challenging training to standard.

Building Teams

All United States military doctrine is based upon reliance on the ingenuity of the individual working on his own initiative as a member of a team and using the most modern weapons and equipment which can be provided him.

General Manton S. Eddy
Commanding General, XII Corps, World War II

6-132. Organizational leaders rely on others to follow and execute their intent. Turning a battlefield vision or training goals into reality takes the combined efforts of many teams inside and outside of the leader's organization.

Organizational leaders build solid, effective teams by developing and training them and sustain those teams by creating healthy organizational climates.

6-133. Organizational leaders work consistently to create individual and team ownership of organizational goals. By knowing their subordinates—their aspirations, fears, and concerns—organizational leaders can ensure their subordinate organizations and leaders work together. Taking time to allow subordinates to develop ways to meet organizational missions fosters ownership of a plan. The FM 25-100 training management process, in which subordinate organizations define supporting tasks and suggest the training required to gain and maintain proficiency, is an example of a process that encourages collective investment in training. That investment leads to a commitment that not only supports execution but also reduces the chances of internal conflict.

6-134. Subordinates work hard and fight tenaciously when they're well-trained and feel they're part of a good team. Collective confidence comes from winning under challenging and stressful conditions. People's sense of belonging comes from technical and tactical proficiency—as individuals and then collectively as a team—and the confidence they have in their peers and their leaders. As cohesive teams combine into a network, a team of teams, organizations work in harness with those on the left and right to fight as a whole. The balance among three good battalions is more important than having a single outstanding one. Following that philosophy necessarily affects resource allocation and task assignment.

6-135. Organizational leaders build cohesive organizations. They overcome, and even capitalize on, diversity of background and experience to create the energy necessary to achieve organizational goals. They resolve conflicts among subordinate leaders as well as any conflicts between their own organization and others.

6-136. For example, subordinate leaders may compete for limited resources while pursuing their individual organization's goals. Two battalion commanders may both want and need a certain maneuver training area to prepare for deployment, so they both present the issue professionally and creatively to their commander. The brigade commander must then weigh and decide between the different unit requirements, balancing their competing demands with the greater good of the entire organization and the Army. An even better situation would be if the organizational climate facilitates teamwork and cooperation that results in the subordinate commanders themselves producing a satisfactory solution.

6-137. Similarly, the brigade commander's own interests may at times conflict with that of other organizations. He must maintain a broad perspective and develop sensible solutions for positive resolution with his contemporaries. In both these cases, subordinates observe the actions of leaders and pattern their attitudes and actions after them. Everyone, even experienced leaders, looks up the chain of command for the example of "how we do it here," how to do it right. Organizational leaders empower their subordinates with a powerful personal example.

6-138. Like direct leaders, organizational leaders build teams by keeping team members informed. They share information, solicit input, and consider suggestions. This give-and-take also allows subordinates a glimpse into the mind of their leaders, which helps them prepare for the day when they fill that job. The leader who sends these messages—"I value your opinion; I'm preparing you for greater responsibilities; you're part of the team"—strengthens the bonds that hold the team together.

6-139. Team building produces trust. Trust begins with action, when leaders demonstrate discipline and competence. Over time, subordinates learn that leaders do what they say they'll do. Likewise leaders learn to trust their subordinates. That connection, that mutual assurance, is the link that helps organizations accomplish the most difficult tasks. (FM 100-34 discusses the importance of building trust for command and control.)

LEARNING

6-140. Organizational leaders create an environment that supports people within their organizations learning from their own experiences and the experiences of others. How leaders react to failure and encourage success now is critical to reaching excellence in the future. Subordinates who feel they need to hide mistakes deprive others of valuable lessons. Organizational leaders set the tone for this honest sharing of experiences by acknowledging that not all experiences (even their own) are successful. They encourage subordinates to examine their experiences, and make it easy for them to share what they learn.

6-141. Learning is continuous and occurs throughout an organization: someone is always experiencing something from which a lesson can be drawn. For this reason, organizational leaders ensure continual teaching at all levels; the organization as a whole shares knowledge and applies relevant lessons. They have systems in place to collect and disseminate those lessons so that individual mistakes become organizational tools. This commitment improves organizational programs, processes, and performances.

SECTION III

A HISTORICAL PERSPECTIVE OF ORGANIZATIONAL LEADERSHIP—GENERAL RIDGWAY IN KOREA

6-142. Few leaders have better exemplified effective organizational leadership in combat than GEN Matthew B. Ridgway. GEN Ridgway successfully led the 82d Airborne Division and XVIII Airborne Corps in the ETO during World War II and Eighth (US) Army during the Korean War. His actions during four months in command of Eighth Army prior to his appointment as UN Supreme Commander bring to life the skills and actions described throughout this chapter.

6-143. At the outbreak of the Korean War in June 1950, GEN Ridgway was assigned as the Army Deputy Chief of Staff, Operations. In an agreement between the Army Chief of Staff, GEN J. Lawton Collins, and the UN Supreme Commander, GA Douglas MacArthur, GEN Ridgway was identified early as the replacement for the Eighth Army commander, GEN Walton H. Walker, in the event GEN Walker was killed in combat.

6-144. That year, on 23 December, GEN Walker died in a jeep accident. Following approval by Secretary of Defense George C. Marshall and President Truman, GEN Ridgway was ordered to take command of Eighth Army. At that time, Eighth Army was defending near the 38th parallel, having completed a 300-mile retreat after the Chinese intervention and stunning victory on the Chongchin River.

6-145. The UN defeat had left its forces in serious disarray. One of Eighth Army's four American divisions, the 2d, needed extensive replacements and reorganization. Two other divisions, the 25th and 1st Cavalry, were seriously battered. Of the Republic of Korea divisions, only the 1st was in good fighting shape. A British brigade was combat ready, but it too had suffered substantial losses in helping cover the retreat.

6-146. Within 24 hours of GEN Walker's death, GEN Ridgway was bound for Korea. During the long flight from Washington, DC, to GA MacArthur's headquarters in Japan, GEN Ridgway had an opportunity to reflect on what lay ahead. He felt this problem was like so many others he had experienced: "Here's the situation—what's your solution?" He began to formulate his plan of action. He determined each step based on his assessment of the enemy's strengths and capabilities as well as his own command's strengths and capabilities.

6-147. The necessary steps seemed clear: gain an appreciation for the immediate situation from GA MacArthur's staff, establish his presence as Eighth Army commander by sending a statement of his confidence in them, and then meet with his own staff to establish his priorities. His first message to his new command was straight to the point: "You will have my utmost. I shall expect yours."

6-148. During the flight from Japan to his forward command post, GEN Ridgway carefully looked at the terrain upon which he was to fight. The battered Eighth Army had to cover a rugged, 100-mile-long front that restricted both maneuver and resupply. Poor morale presented a further problem. Many military observers felt that Eighth Army lacked spirit and possessed little stomach for continuing the bruising battle with the Chinese.

6-149. For three days GEN Ridgway traveled the army area by jeep, talking with commanders who had faced the enemy beyond the Han River. GEN Ridgway wrote later,

I held to the old-fashioned idea that it helped the spirits of the men to see the Old Man up there, in the snow and the sleet and the mud, sharing the same cold, miserable existence they had to endure.

6-150. GEN Ridgway believed a commander should publicly show a personal interest in the well-being of his soldiers. He needed to do something to attract notice and display his concern for the front-line fighters. Finding that one of his units was still short of some winter equipment, GEN Ridgway dramatically ordered that the equipment be delivered within 24 hours. In response, the logistical command made a massive effort to comply, flying equipment from Pusan to the front lines. Everyone noticed. He also ordered—and made sure the order was known—that the troops be served hot meals, with any failures to comply reported directly to him.

6-151. GEN Ridgway was candid, criticizing the spirit of both the commanders and soldiers of Eighth Army. He talked with riflemen and generals, from front-line foxholes to corps command posts. He was appalled at American infantrymen who didn't patrol, who had no knowledge of the terrain in which they fought, and who failed to know the whereabouts of their enemy. Moreover, this army was road-bound and failed to occupy commanding terrain overlooking its positions and supply lines. GEN Ridgway also sensed that Eighth Army—particularly the commanders and their staffs—kept looking over their shoulders for the best route to the rear and planned only for retreat. In short, he found his army immobilized and demoralized.

6-152. An important part of GEN Ridgway's effort to instill fighting spirit in Eighth Army was to order units to close up their flanks and tie in with other units. He said he wanted no units cut off and abandoned, as had happened to the 3d Battalion, 8th Cavalry at Unsan, Task Force Faith at Chosin Reservoir, and the 2d Division at Kuni-ri. GEN Ridgway felt that it was essential for soldiers to know they would not be left to fend for themselves if cut off. He believed that soldiers would be persuaded to stand and fight only if they realized help would come. Without that confidence in the command and their fellow soldiers, they would pull out, fearing to be left behind.

6-153. As he visited their headquarters, GEN Ridgway spoke to commanders and their staffs. These talks contained many of his ideas about proper combat leadership. He told his commanders to get out of their command posts and up to the front. When commanders reported on terrain, GEN Ridgway demanded that they base their information on personal knowledge and that it be correct.

6-154. Furthermore, he urged commanders to conduct intensive training in night fighting and make full use of their firepower. He also required commanders to personally check that their men had adequate winter clothing, warming tents, and writing materials. In addition, he encouraged commanders to locate wounded who had been evacuated and make every effort to return them to their old units. Finally, the army commander ordered his officers to stop wasting resources, calling for punishment of those who lost government equipment.

6-155. During its first battle under GEN Ridgway's command in early January 1951, Eighth Army fell back another 70 miles and lost Seoul, South Korea's capital. Major commanders didn't carry out orders to fall back in an orderly fashion, use field artillery to inflict the heaviest possible enemy casualties, and counterattack in force during daylight hours. Eighth Army's morale and sense of purpose reached their lowest point ever.

6-156. Eighth Army had only two choices: substantially improve its fighting spirit or get out of Korea. GEN Ridgway began to restore his men's fighting spirit by ordering aggressive patrolling into areas just lost. When patrols found the enemy few in number and not aggressive, the army commander increased the number and size of patrols. His army discovered it could drive back the Chinese without suffering overwhelming casualties. Buoyed by these successes, GEN Ridgway ordered a general advance along Korea's west coast, where the terrain was more open and his forces could take advantage of its tanks, artillery, and aircraft.

6-157. During this advance, GEN Ridgway also attempted to tell the men of Eighth Army why they were fighting in Korea. He sought to build a fighting spirit in his men based on unit and soldier pride. In addition, he called on them to defend Western Civilization from Communist degradation, saying:

In the final analysis, the issue now joined right here in Korea is whether Communism or individual freedom shall prevail; whether the flight of the fear-driven people we have witnessed here shall be checked, or shall at some future time, however distant, engulf our own loved ones in all its misery and despair.

6-158. In mid-February of 1951, the Chinese and North Koreans launched yet another offensive in the central area of Korea, where US tanks could not maneuver as readily and artillery could be trapped on narrow roads in mountainous terrain. In heavy fights at Chipyon-ni and Wonju, Eighth Army, for the first time, re-pulsed the Communist attacks. Eighth Army's offensive spirit soared as GEN Ridgway quickly followed up with a renewed attack that took Seoul and regained roughly the same positions Eight Army had held when he first took command. In late March, Eighth Army pushed the Communist forces north of the 38th parallel.

6-159. GEN Ridgway's actions superbly exemplify those expected of organizational leaders. His knowledge of American soldiers, units, and the Korean situation led him to certain expectations. Those expectations gave him a baseline from which to assess his command once he arrived. He continually visited units throughout the army area, talked with soldiers and their commanders, assessed command climate, and took action to mold attitudes with clear intent, supreme confidence, and unyielding tactical discipline.

6-160. He sought to develop subordinate commanders and their staffs by sharing his thoughts and expectations of combat leadership. He felt the pulse of the men on the front, shared their hardships, and demanded they be taken care of. He pushed the logistical systems to provide creature comforts as well as the supplies of war. He eliminated the skepticism of purpose, gave soldiers cause to fight, and helped them gain confidence by winning small victories. Most of all, he led by example.

6-161. In April GEN Ridgway turned Eighth Army over to GEN James A. Van Fleet. In under four months, a dynamic, aggressive commander had revitalized and transformed a traumatized and desperate army into a proud, determined fighting force. GA Omar N. Bradley, Chairman of the Joint Chiefs of Staff, summed up GEN Ridgway's contributions:

It is not often that a single battlefield commander can make a decisive difference. But in Korea Ridgway would prove to be that exception. His brilliant, driving, uncompromising leadership would turn the tide of battle like no other general's in our military history.

SUMMARY

6-162. This chapter has covered how organizational leaders train and lead staffs, subordinate leaders, and entire organizations. The influence of organizational leaders is primarily indirect: they communicate and motivate through staffs and subordinate commanders. Because their leadership is much more indirect, the eventual outcomes of their actions are often difficult to foresee. Nor do organizational leaders receive the immediate feedback that direct leaders do.

6-163. Still, as demonstrated by GEN Ridgway in Korea, the presence of commanders at the critical time and place boosts confidence and performance. Regardless of the type of organization they head, organizational leaders direct operations by setting the example, empowering their subordinates and organizations and supervising them appropriately. Organizational leaders concern themselves with combat power—how to build, maintain, and recover it. That includes developing systems that will provide the organization and the Army with its next generation of leaders. They also improve conditions by sustaining an ethical and supportive climate, building strong cohesive teams and organizations, and improving the processes that work within the organization.

6-164. Strategic leaders provide leadership at the highest levels of the Army. Their influence is even more indirect and the consequences of their actions more delayed than those of organizational leaders. Because of this, strategic leaders must develop additional skills based on those they've mastered as direct and organizational leaders. Chapter 7 discusses these and other aspects of strategic leadership.

Chapter 7

Strategic Leadership

It became clear to me that at the age of 58 I would have to learn new tricks that were not taught in the military manuals or on the battlefield. In this position I am a political soldier and will have to put my training in rapping-out orders and making snap decisions on the back burner, and have to learn the arts of persuasion and guile. I must become an expert in a whole new set of skills.

General of the Army George C. Marshall

7-1. Strategic leaders are the Army's highest-level thinkers, warfighters, and political-military experts. Some work in an institutional setting within the United States; others work in strategic regions around the world. They simultaneously sustain the Army's culture, envision the future, convey that vision to a wide audience, and personally lead change. Strategic leaders look at the environment outside the Army today to understand the context for the institution's future role. They also use their knowledge of the current force to anchor their vision in reality. This chapter outlines strategic leadership for audiences other than the general officers and Senior Executive Service DA civilians who actually lead there. Those who support strategic leaders need to understand the distinct environment in which these leaders work and the special considerations it requires.

7-2. Strategic leadership requires significantly different techniques in both scope and skill from direct and organizational leadership. In an environment of extreme uncertainty, complexity, ambiguity, and volatility, strategic leaders think in multiple time domains and operate flexibly to manage change. Moreover, strategic leaders often interact with other leaders over whom they have minimal authority.

7-3. Strategic leaders are not only experts in their own domain—warfighting and leading large military organizations—but also are astute in the departmental and political environments of the nation's decision-making process. They're expected to deal competently with the public sector, the executive branch, and the legislature. The complex national security

environment requires an in-depth knowledge of the political, economic, informational, and military elements of national power as well as the interrelationship among them. In short, strategic leaders not only know themselves and their own organizations but also understand a host of different players, rules, and conditions.

7-4. Because strategic leaders implement the National Military Strategy, they deal with the elements that shape that strategy. The most important of these are Presidential Decision Memorandums, Department of State Policies, the will of the American people, US national security interests, and the collective strategies—theater and functional—of the combatant commanders (CINCs). Strategic leaders operate in intricate networks of competing constituencies and cooperate in endeavors extending beyond their establishments. As institutional leaders, they represent their organizations to soldiers, DA civilians, citizens, statesmen, and the media, as well as to other services and nations. Communicating effectively with these different audiences is vital to the organization's success.

7-5. Strategic leaders are keenly aware of the complexities of the national security environment. Their decisions take into account factors such as congressional hearings, Army budget constraints, reserve component issues, new systems acquisition, DA civilian programs, research, development, and interservice cooperation. Strategic leaders process information from these areas quickly, assess alternatives based on incomplete data, make decisions, and garner support. Often, highly developed interpersonal skills are essential to building consensus among civilian and military policy makers. Limited interpersonal skills can limit the effect of other skills.

7-6. While direct and organizational leaders have a short-term focus, strategic leaders have a "future focus." Strategic leaders spend much of their time looking toward the mid-term and positioning their establishments for long-term success, even as they contend with immediate issues. With that perspective, strategic leaders seldom see the whole life span of their ideas; initiatives at this level may take years to come to fruition. Strategic leaders think, therefore, in terms of strategic systems that will operate over extended time periods. They ensure these systems are built in accord with the six imperatives mentioned in Chapter 6—quality people, training, force mix, doctrine, modern equipment, and leader development—and they ensure that programs and resources are in place to sustain them. This systems approach sharpens strategic leaders' "future focus" and helps align separate actions, reduce conflict, and improve cooperation.

SECTION I
STRATEGIC LEADERSHIP SKILLS

7-7. The values and attributes demanded of Army leaders are the same at all leadership levels. Strategic leaders live by Army values and set the example just as much as direct and organizational leaders, but they face additional challenges. Strategic leaders affect the culture of the entire Army and may find themselves involved in political decision making at the highest national or even global levels. Therefore, nearly any task strategic leaders set out to accomplish requires more coordination, takes longer, has a wider impact, and produces longer-term effects than a similar organizational-level task.

7-8. Strategic leaders understand, embody, and execute values-based leadership. The political and long-term nature of their decisions doesn't release strategic leaders from the current demands of training, readiness, and unforeseen crises; they are responsible to continue to work toward the ultimate goals of the force, despite the burden of those events. Army values provide the constant reference for actions in the stressful environment of strategic leaders. Strategic leaders understand, embody, and execute leadership based on Army values.

INTERPERSONAL SKILLS

7-9. Strategic leaders continue to use interpersonal skills developed as direct and organizational leaders, but the scope, responsibilities, and authority of strategic positions require leaders with unusually sophisticated interpersonal skills. Internally, there are more levels of people to deal with; externally, there are more interactions with outside agencies, with the media, even with foreign governments. Knowing the Army's needs and goals, strategic leaders patiently but tenaciously labor to convince the proper people about what the Army must have and become. Figure 7-1 lists strategic leader interpersonal skills.

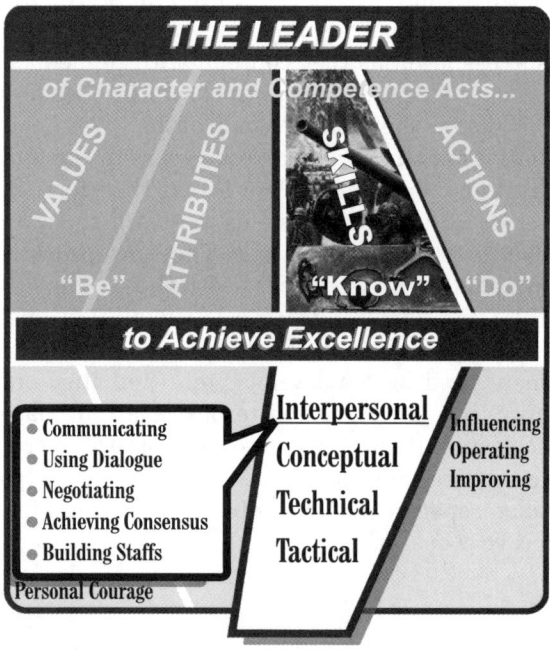

Figure 7-1. Strategic Leader Skills— Interpersonal

7-10. Strategic leaders and their staffs develop networks of knowledgeable individuals in organizations and agencies that influence their own organizations. Through penetrating assessments, these leaders seek to understand the personal strengths and weaknesses of all the main players on a particular issue. Strategic leaders are adept at reading other people, and they work to completely control their own actions and reactions. Armed with improved knowledge of others, self-control, and established networks, strategic leaders influence external events by providing leadership, timely and relevant information, and access to the right people and agencies.

COMMUNICATING

7-11. Communication at the strategic level is complicated by the wide array of staff, functional, and operational components interacting with each other and with external agencies. These complex relationships require strategic leaders to employ comprehensive

communications skills as they represent their organizations. One of the most prominent differences between strategic leaders and leaders at other levels is the greater importance of symbolic communication. The example strategic leaders set, their decisions, and their actions have meaning beyond their immediate consequences to a much greater extent than those of direct and organizational leaders.

7-12. Thus, strategic leaders identify those actions that send messages. Then they use their positions to send the desired messages to their organizations and ensure that the right audiences hear them. The messages strategic leaders send set the example in the largest sense. For instance, messages that support traditions, Army values, or a particular program indicate the strategic leader's priorities.

7-13. Thus, strategic leaders communicate not only to the organization but also to a large external audience that includes the political leadership, media, and the American people. To influence those audiences, strategic leaders seek to convey integrity and win trust. As GA Marshall noted, they become expert in "the art of persuasion."

7-14. Strategic leaders commit to a few common, powerful, and consistent messages and repeat them over and over in different forms and settings. They devise a communications campaign plan, written or conceptual, that outlines how to deal with each target group. When preparing to address a specific audience, they determine its composition and agenda so they know how best to reach its members. Finding some apparent success with the medium, frequency, and words of the message, strategic leaders determine the best way to measure the message's effectiveness and continually scan and assess the environment to make sure that the message is going to all the right groups.

USING DIALOGUE

7-15. One of the forms of communication that strategic leaders use to persuade individuals, rather than groups, is dialogue. Dialogue is a conversation between two or more people. It requires not only active listening, but carefully considering what's said (and not said), logically assessing it without personal bias, and specifying issues that are not understood or don't make sense within the strategic leader's frame of reference. By using dialogue to thoroughly exchange points of view, assumptions, and concepts, strategic leaders gather information, clarify issues, and enlist support of subordinates and peers.

NEGOTIATING

7-16. Many relationships between strategic-level organizations are lateral and without clear subordination. Often, strategic leaders rely heavily on negotiating skills to obtain the cooperation and support necessary to accomplish a mission or meet the command's needs. For example, commanders of the national contingents that made up the North Atlantic Treaty Organization (NATO) implementation force (IFOR) sent to Bosnia to support the 1995 Dayton peace accords all had limitations imposed on the extent of their participation. In addition, they all had direct lines to their home governments, which they used when they believed IFOR commanders exceeded those limits. NATO strategic leaders had to negotiate some actions that ordinarily would have required only issuing orders. They often had to interpret a requirement to the satisfaction of one or more foreign governments.

7-17. Successful negotiation requires a range of interpersonal skills. Good negotiators are also able to visualize several possible end states while maintaining a clear idea of the best end state from the command's perspective. One of the most important skills is the ability to stand firm on nonnegotiable points while simultaneously communicating respect for other participants and their negotiating limits. In international forums, firmness and respect demonstrate that the negotiator knows and understands US interests. That understanding can help the negotiator persuade others of the validity of US interests and convince others that the United States understands and respects the interests of other states.

7-18. A good negotiator is particularly skilled in active listening. Other essential personal characteristics include perceptiveness and objectivity. Negotiators must be able to diagnose

unspoken agendas and detach themselves from the negotiation process. Successful negotiating involves communicating a clear position on all issues while still conveying willingness to bargain on negotiable issues, recognizing what's acceptable to all concerned, and achieving a compromise that meets the needs of all participants to the greatest extent possible.

7-19. Sometimes strategic leaders to put out a proposal early so the interchange and ultimate solution revolve around factors important to the Army. However, they are confident enough to resist the impulse to leave their thumbprints on final products. Strategic leaders don't have to claim every good idea because they know they will have more. Their understanding of selfless service allows them to subordinate personal recognition to negotiated settlements that produce the greatest good for their establishment, the Army, and the nation or coalition.

ACHIEVING CONSENSUS

7-20. Strategic leaders are skilled at reaching consensus and building and sustaining coalitions. They may apply these skills to tasks as diverse as designing combatant commands, JTFs, and policy working groups or determining the direction of a major command or the Army as an institution. Strategic leaders routinely weld people together for missions lasting from months to years. Using peer leadership rather than strict positional authority, strategic leaders oversee progress toward their visualized end state and monitor the health of the relationships necessary to achieve it. Interpersonal contact sets the tone for professional relations: strategic leaders are tactful and discreet.

7-21. GA Eisenhower's creation of SHAEF during World War II (which was mentioned in Chapter 2) is an outstanding example of coalition building and sustainment. GA Eisenhower insisted on unity of command over the forces assigned to him. He received this authority from both the British and US governments but exercised it through an integrated command and staff structure that related influence roughly to the contribution of the nations involved. The sections within SHAEF all had chiefs of one nationality and deputies of another.

7-22. GA Eisenhower also insisted that military, rather than political, criteria would predominate in his operational and strategic decisions as Supreme Allied Commander. His most controversial decisions, adoption of the so-called broad-front strategy and the refusal to race the Soviet forces to Berlin, rested on his belief that maintaining the Anglo-American alliance was a national interest and his personal responsibility. Many historians argue that this feat of getting the Allies to work together was his most important contribution to the war.

Allied Command During the Battle of the Bulge

A pivotal moment in the history of the Western Alliance arrived on 16 December 1944, when the German Army launched a massive offensive in a lightly held-sector of the American line in the Ardennes Forest. This offensive, which became known as the Battle of the Bulge, split GEN Omar Bradley's Twelfth Army Group. North of the salient, British GEN Bernard Montgomery commanded most of the Allied forces, so GA Eisenhower shifted command of the US forces there to GEN Montgomery rather than have one US command straddle the gap. GEN Bradley, the Supreme Allied Commander reasoned, could not effectively control forces both north and south of the penetration. It made more sense for GEN Montgomery to command all Allied forces on the northern shoulder and GEN Bradley all those on its southern shoulder. GA Eisenhower personally telephoned GEN Bradley to tell his old comrade of the decision. With the SHAEF staff still present, GA Eisenhower passed the order to his reluctant subordinate, listened to GEN Bradley's protests, and then said sharply, "Well, Brad, those are my orders."

According to historian J.D. Morelock, this short conversation, more than any other action taken by GA Eisenhower and the SHAEF staff during the battle, "discredited the German assumption that nationalistic fears and rivalries would inhibit prompt and effective steps to meet the German challenge." It demonstrated GA Eisenhower's "firm grasp of the true nature of an allied command" and it meant that Hitler's gamble to win the war had failed.

7-23. Across the Atlantic Ocean, GA George C. Marshall, the Army Chief of Staff, also had to seek consensus with demanding peers, none more so than ADM Ernest J. King, Commander in Chief, US Fleet and Chief of Naval Operations. GA Marshall expended great personal energy to ensure that interservice feuding at the top didn't mar the US war effort. ADM King, a forceful leader with strong and often differing views, responded in kind. Because of the ability of these two strategic leaders to work in harmony, President Franklin D. Roosevelt had few issues of major consequence to resolve once he had issued a decision and guidance.

7-24. Opportunities for strategic leadership may come at surprising moments. For instance, Joshua Chamberlain's greatest contribution to our nation may have been not at Gettysburg or Petersburg, but at Appomattox. By that time a major general, Chamberlain was chosen to command the parade at which GEN Lee's Army of Northern Virginia laid down its arms and colors. GEN Grant had directed a simple ceremony that recognized the Union victory without humiliating the Confederates.

7-25. However, MG Chamberlain sensed the need for something even greater. Instead of gloating as the vanquished army passed, he directed his bugler to sound the commands for attention and present arms. His units came to attention and rendered a salute, following his order out of respect for their commander, certainly not out of sudden warmth for recent enemies. That act set the tone for reconciliation and reconstruction and marks a brilliant leader, brave in battle and respectful in peace, who knew when, where, and how to lead.

BUILDING STAFFS

The best executive is the one who has sense enough to pick good men to do what he wants done, and self-restraint enough to keep from meddling with them while they do it.

Theodore Roosevelt
26th President of the United States

7-26. Until Army leaders reach the highest levels, they cannot staff positions and projects as they prefer. Strategic leaders have not only the authority but also the responsibility to pick the best people for their staffs. They seek to put the right people in the right places, balancing strengths and weaknesses for the good of the nation. They mold staffs able to package concise, unbiased information and build networks across organizational lines. Strategic leaders make so many wide-ranging, interrelated decisions that they must have imaginative staff members who know the environment, foresee consequences of various courses of action, and identify crucial information accordingly.

7-27. With their understanding of the strategic environment and vision for the future, strategic leaders seek to build staffs that compensate for their weaknesses, reinforce their vision, and ensure institutional success. Strategic leaders can't afford to be surrounded by staffs that blindly agree with everything they say. Not only do they avoid surrounding themselves with "yes-men," they also reward staff members for speaking the truth. Strategic leaders encourage their staffs to participate in dialogue with them, discuss alternative points of view, and explore all facts, assumptions, and implications. Such dialogue assists strategic leaders to fully assess all aspects of an issue and helps clarify their intent and guidance.

7-28. As strategic leaders build and use their staffs, they continually seek honesty and competence. Strategic-level staffs must be able to discern what the "truth" is. During World War II, GA Marshall's ability to fill his staff and commands with excellent officers made a difference in how quickly the Army could create a wartime force able to mobilize, deploy, fight, and win. Today's strategic leaders face an environment more complex than the one GA Marshall faced. They often have less time than GA Marshall had to assess situations, make plans, prepare an appropriate response, and execute. The importance of building courageous, honest, and competent staffs has correspondingly increased.

CONCEPTUAL SKILLS

From an intellectual standpoint, Princeton was a world-shaking experience. It fundamentally changed my approach to life. The basic thrust of the curriculum was to give students an appreciation of how complex and diverse various political systems and issues are....The bottom line was that answers had to be sought in terms of the shifting relationships of groups and individuals, that politics pervades all human activity, a truth not to be condemned but appreciated and put to use.

Admiral William Crowe
Former Chairman, Joint Chiefs of Staff

Figure 7-2. Strategic Leader Skills— Conceptual

7-29. Strategic leaders, more than direct and organizational leaders, draw on their conceptual skills to comprehend national, national security, and theater strategies, operate in the strategic and theater contexts, and improve their vast, complex organizations. The variety and scope of their concerns demand the application of more sophisticated concepts.

7-30. Strategic leaders need wisdom—and wisdom isn't just knowledge. They routinely deal with diversity, complexity, ambiguity, change, uncertainty, and conflicting policies. They are responsible for developing well-reasoned positions and providing their views and advice to our nation's highest leaders. For the good of the Army and the nation, strategic leaders seek to determine what's important now and what will be important in the future. They develop the necessary wisdom by freeing themselves to stay in touch with the force and spending time thinking, simply thinking.

ENVISIONING

It is in the minds of the commanders that the issue of battle is really decided.

Sir Basil H. Liddell Hart

7-31. Strategic leaders design compelling visions for their organizations and inspire a collaborative effort to articulate the vision in detail. They then communicate that vision clearly and use it to create a plan, gain support, and focus subordinates' work. Strategic leaders have the further responsibility of defining for their diverse organizations what counts as success in achieving the vision. They monitor their progress by drawing on personal observations, review and analysis, strategic management plans, and informal discussions with soldiers and DA civilians.

7-32. Strategic leaders look realistically at what the future may hold. They consider things they know and things they can anticipate. They incorporate new ideas, new technologies, and new capabilities. The National Security Strategy and National Military Strategy guide strategic leaders as they develop visions for their organizations. From a complicated mixture of ideas, facts, conjecture, and personal experience they create an image of what their organizations need to be.

7-33. Once strategic leaders have developed a vision, they create a plan to reach that end state. They consider objectives, courses of action to take the organization there, and resources needed to do the job. The word "vision" implies that strategic leaders create a conceptual model of what they want. Subordinates will be more involved in moving the organization forward if they can "see" what the leader has in mind. And because moving a large organization is often a long haul, subordinates need some sign that they're making progress. Strategic leaders therefore provide intermediate objectives that act as milestones for their subordinates in checking their direction and measuring their progress.

7-34. The strategic leader's vision provides the ultimate sense of purpose, direction, and motivation for everyone in the organization. It is at once the starting point for developing specific goals and plans, a yardstick for measuring what the organization accomplishes, and a check on organizational values. Ordinarily, a strategic leader's vision for the organization may have a time horizon of years, or even decades. In combat, the horizon is much closer, but strategic leaders still focus far beyond the immediate actions.

7-35. The strategic leader's vision is a goal, something the organization strives for (even though some goals may always be just out of reach). When members understand the vision, they can see it as clearly as the strategic leader can. When they see it as worthwhile and accept it, the vision creates energy, inspiration, commitment, and a sense of belonging.

7-36. Strategic leaders set the vision for their entire organization. They seek to keep the vision consistent with the external environment, alliance or coalition goals, the National Security Strategy, and the National Military Strategy. Subordinate leaders align their visions and intent with their strategic leader's vision. A strategic leader's vision may be expressed in everything from small acts to formal, written policy statements.

7-37. Joint Vision 2010 and Army Vision 2010, which is derived from it, are not based on formal organizations; rather they array future technologies and force structure against emerging threats. While no one can yet see exactly what that force will look like, the concepts themselves provide an azimuth and a point on the horizon. Achieving well-publicized milepost initiatives shows that the Army as an institution is progressing toward the end state visualized by its strategic leaders.

DEVELOPING FRAMES OF REFERENCE

7-38. All Army leaders build a personal frame of reference from schooling, experience, self-study, and reflection on current events and history. Strategic leaders create a comprehensive frame of reference that encompasses their organization and places it in the strategic environment. To construct a useful frame, strategic leaders are open to new experiences and to comments from others, including subordinates. Strategic leaders are reflective, thoughtful, and unafraid to rethink past experiences and to learn from them. They are comfortable with the abstractions and concepts common in the strategic environment. Moreover, they understand the circumstances surrounding them, their organization, and the nation.

7-39. Much like intelligence analysts, strategic leaders look at events and see patterns that others often miss. These leaders are likely to identify and understand a strategic situation and, more important, infer the outcome of interventions or the absence of interventions. A strategic leader's frame of reference helps identify the information most relevant to a strategic situation so that the leader can go to the heart of a matter without being distracted. In the new information environment, that talent is more important than ever. Cosmopolitan strategic leaders, those with comprehensive frames of reference and the wisdom that comes from thought and reflection, are well equipped to deal with events having complex causes and to envision creative solutions.

7-40. A well-developed frame of reference also gives strategic leaders a thorough understanding of organizational subsystems and their interacting processes. Cognizant of the relationships among systems, strategic leaders foresee the possible effects on one system of actions in others. Their vision helps them anticipate and avoid problems.

DEALING WITH UNCERTAINTY AND AMBIGUITY

True genius resides in the capacity for evaluation of uncertain, hazardous, and conflicting information.

Sir Winston Churchill
Prime Minister of Great Britain, World War II

7-41. Strategic leaders operate in an environment of increased volatility, uncertainty, complexity, and ambiguity. Change at this level may arrive suddenly and unannounced. As they plan for contingencies, strategic leaders prepare intellectually for a range of uncertain threats and scenarios. Since even great planning and foresight can't predict or influence all future events, strategic leaders work to shape the future on terms they can control, using diplomatic, informational, military, and economic instruments of national power.

7-42. Strategic leaders fight complexity by encompassing it. They must be more complex than the situations they face. This means they're able to expand their frame of reference to fit a situation rather than reducing a situation to fit their preconceptions. They don't lose sight of Army values and force capabilities as they focus on national policy. Because of their maturity and wisdom, they tolerate ambiguity, knowing they will never have all the information they want. Instead, they carefully analyze events and decide when to make a decision, realizing that they must innovate and accept some risk. Once they make decisions, strategic leaders then explain them to the Army and the nation, in the process imposing order on the uncertainty and ambiguity of the situation. Strategic leaders not only understand the environment themselves; they also translate their understanding to others.

Strategic Flexibility in Haiti

Operation Uphold Democracy, the 1994 US intervention in Haiti conducted under UN auspices, provides an example of strategic leaders achieving success in spite of extreme uncertainty and ambiguity. Prior to the order to enter Haiti, strategic leaders didn't know either D-day or the available forces. Neither did they know whether the operation would be an invitation (permissive entry), an invasion (forced entry), or something in between. To complicate the actual military execution, former President Jimmy Carter, retired GEN Colin Powell, and Senator Sam Nunn were negotiating with LTG Raoul Cedras, commander in chief of the Haitian armed forces, in the Haitian capital even as paratroopers, ready for a combat jump, were inbound.

When LTG Cedras agreed to hand over power, the mission of the inbound JTF changed from a forced to a permissive entry. The basis for the operation wound up being an operation plan based on an "in-between" course of action inferred by the JTF staff during planning. The ability of the strategic leaders involved to change their focus so dramatically and quickly provides an outstanding example of strategic flexibility during a crisis. The ability of the soldiers, sailors, airmen, and Marines of the JTF to execute the new mission on short notice is a credit to them and their leaders at all levels.

7-43. In addition to demonstrating the flexibility required to handle competing demands, strategic leaders understand complex cause-and-effect relationships and anticipate the second- and third-order effects of their decisions throughout the organization. The highly volatile nature of the strategic environment may tempt them to concentrate on the short term, but strategic leaders don't allow the crisis of the moment absorb them completely. They remain focused on their responsibility to shape an organization or policies that will perform successfully over the next 10 to 20 years. Some second- and third-order effects are desirable; leaders can design and pursue actions to achieve them. For example, strategic leaders who continually send—through their actions—messages of trust to subordinates inspire trust in themselves. The third-order effect may be to enhance subordinates' initiative.

TECHNICAL SKILLS

The crucial difference (apart from levels of innate ability) between Washington and the commanders who opposed him was they were sure they knew all the answers, while Washington tried every day and every hour to learn.

James Thomas Flexner
George Washington in the American Revolution

Figure 7-3. Strategic Leader Skills— Technical

7-44. Strategic leaders create their work on a broad canvas that requires broad technical skills of the sort named in Figure 7-3.

STRATEGIC ART

7-45. The strategic art, broadly defined, is the skillful formulation, coordination, and application of ends, ways, and means to promote and defend the national interest. Masters of the strategic art competently integrate the three roles performed by the complete strategist: strategic leader, strategic practitioner, and strategic theorist.

7-46. Using their understanding of the systems within their own organizations, strategic leaders

work through the complexity and uncertainty of the strategic environment and translate abstract concepts into concrete actions. Proficiency in the science of leadership—programs, schedules, and systems, for example—can bring direct or organizational leaders success. For strategic leaders, however, the intangible qualities of leadership draw on their long and varied experience to produce a rare art.

7-47. Strategic leaders do more than imagine and accurately predict the future; they shape it by moving out of the conceptual realm into practical execution. Although strategic leaders never lose touch with soldiers and their technical skills, some practical activities are unique to this level.

7-48. By reconciling political and economic constraints with the Army's needs, strategic leaders navigate to move the force forward using the strategy and budget processes. They spend a great deal of time obtaining and allocating resources and determining conceptual directions, especially those judged critical for future strategic positioning and necessary to prevent readiness shortfalls. They're also charged with overseeing of the Army's responsibilities under Title 10 of the United States Code.

7-49. Strategic leaders focus not so much on internal processes as on how the organization fits into the DOD and the international arena: What are the relationships among external organizations? What are the broad political and social systems in which the organization and the Army must operate? Because of the complex reporting and coordinating relationships, strategic leaders fully understand their roles, the boundaries of these roles, and the expectations of other departments and agencies. Understanding those interdependencies outside the Army helps strategic leaders do the right thing for the programs, systems, and people within the Army as well as for the nation.

7-50. Theater CINCs, with their service component commanders, seek to shape their environments and accomplish long-term national security policy goals within their theaters. They operate through congressional testimony, creative use of assigned and attached military forces, imaginative bilateral and multilateral programs, treaty obligations, person-to-person contacts with regional leaders, and various joint processes. These actions require strategic leaders to apply the strategic art just as much as does designing and employing force packages to achieve military end states.

7-51. GA Douglas MacArthur, a theater CINC during World War II, became military governor of occupied Japan after the Japanese surrender. His former enemies became his responsibility; he had to deal diplomatically with the defeated nation as well as the directives of American civil authorities and the interests of the former Allies. GA MacArthur understood the difference between preliminary (often called military) end state conditions and the broader set of end state conditions that are necessary for the transition from war to peace. Once a war has ended, military force can no longer be the principal means of achieving strategic aims. Thus, a strategic leader's end state vision must include diplomatic, economic, and informational—as well as military—aspects. GA MacArthur's vision, and the actions he took to achieve it, helped establish the framework that preserved peace in the Pacific Ocean and rebuilt a nation that would become a trusted ally.

7-52. A similar institutional example occurred in the summer of 1990. Then, while the Army was in the midst of the most precisely planned force "build-down" in history, Army Chief of Staff Carl Vuono had to halt the process to meet a crisis in the Persian Gulf. GEN Vuono was required to call up, mobilize and deploy forces necessary to meet the immediate crisis while

retaining adequate capabilities in other theaters. The following year he redeployed Third (US) Army, demobilized the activated reserves, and resumed downsizing toward the smallest active force since the 1930s. GEN Vuono demonstrated the technical skill of the strategic art and proved himself a leader of character and competence motivated by Army values.

LEVERAGING TECHNOLOGY

7-53. Leveraging technology—that is, applying technological capabilities to obtain a decisive military advantage—has given strategic leaders advantages in force projection, in command and control, and in the generation of overwhelming combat power. Leveraging technology has also increased the tempo of operations, the speed of maneuver, the precision of firepower, and the pace at which information is processed. Ideally, information technology, in particular, enhances not only communications, but also situational understanding. With all these advantages, of course, comes increasing complexity: it's harder to control large organizations that are moving quickly. Strategic leaders seek to understand emerging military technologies and apply that understanding to resourcing, allocating, and exploiting the many systems under their control.

7-54. Emerging combat, combat support, and combat service support technologies bring more than changes to doctrine. Technological change allows organizations to do the things they do now better and faster, but it also enables them to do things that were not possible before. So a part of leveraging technology is envisioning the future capability that could be exploited by developing a technology. Another aspect is rethinking the form the organization ought to take in order to exploit new processes that previously were not available. This is why strategic leaders take time to think "out of the box."

TRANSLATING POLITICAL GOALS INTO MILITARY OBJECTIVES

7-55. Leveraging technology takes more than understanding; it takes money. Strategic leaders call on their understanding and their knowledge of the budgetary process to determine which combat, combat support, and combat service support technologies will provide the leap-ahead capability commensurate with the cost. Wise Army leaders in the 1970s and 1980s realized that superior night systems and greater standoff ranges could expose fewer Americans to danger yet kill more of the enemy. Those leaders committed money to developing and procuring appropriate weapons systems and equipment. Operation Desert Storm validated these decisions when, for example, M1 tanks destroyed Soviet-style equipment before it could close within its maximum effective range. However strategic leaders are always in the position of balancing budget constraints, technological improvements, and current force readiness against potential threats as they shape the force for the future.

7-56. Strategic leaders identify military conditions necessary to satisfy political ends desired by America's civilian leadership. They must synchronize the efforts of the Army with those of the other services and government agencies to attain those conditions and achieve the end state envisioned by America's political leaders. To operate on the world stage, often in conjunction with allies, strategic leaders call on their international perspective and relationships with policy makers in other countries.

Show of Force in the Philippines

At the end of November 1989, 1,000 rebels seized two Filipino air bases in an attempt to overthrow the government of the Philippines. There had been rumors that someone was plotting a coup to end Philippine President Corazon Aquino's rule. Now rebel aircraft from the captured airfields had bombed and strafed the presidential palace. President Aquino requested that the United States help suppress the coup attempt by destroying the captured airfields. Vice President Dan Quayle and Deputy Secretary of State Lawrence Eagleburger favored US intervention to support the Philippine government. As the principal military advisor to the president, Joint Chiefs of Staff Chairman Colin Powell was asked to recommend a response to President Aquino's request.

GEN Powell applied critical reasoning to this request for US military power in support of a foreign government. He first asked the purpose of the proposed intervention. The State Department and White House answered that the United States needed to demonstrate support for President Aquino and keep her in power. GEN Powell then asked the purpose of bombing the airfields. To prevent aircraft from supporting the coup, was the reply. Once GEN Powell understood the political goal, he recommended a military response to support it.

The chairman recommended to the White House that American jets fly menacing runs over the captured airfields. The goal would be to prevent takeoffs from the airfields by intimidating the rebel pilots rather than destroying rebel aircraft and facilities. This course of action was approved by President George Bush and achieved the desired political goal: it deterred the rebel pilots from supporting the coup attempt. By understanding the political goal and properly defining the military objective, GEN Powell was able to recommend a course of action that applied a measured military response to what was, from the United States' perspective, a diplomatic problem. By electing to conduct a show of force rather than an attack, the United States avoided unnecessary casualties and damage to the Philippine infrastructure.

7-57. Since the end of the Cold War, the international stage has become more confused. Threats to US national security may come from a number of quarters: regional instability, insurgencies, terrorism, and proliferation of weapons of mass destruction to name a few. International drug traffickers and other transnational groups are also potential adversaries. To counter such diverse threats, the nation needs a force flexible enough to execute a wide array of missions, from warfighting to peace operations to humanitarian assistance. And of course, the nation needs strategic leaders with the sound perspective that allows them to understand the

nation's political goals in the complex international environment and to shape military objectives appropriate to the various threats.

SECTION II

STRATEGIC LEADERSHIP ACTIONS

Leadership is understanding people and involving them to help you do a job. That takes all of the good characteristics, like integrity, dedication of purpose, selflessness, knowledge, skill, implacability, as well as determination not to accept failure.

Admiral Arleigh A. Burke
Naval Leadership: Voices of Experience

7-58. Operating at the highest levels of the Army, the DOD, and the national security establishment, military and DA civilian strategic leaders face highly complex demands from inside and outside the Army. Constantly changing global conditions challenge their decision-making abilities. Strategic leaders tell the Army story, make long-range decisions, and shape the Army culture to influence the force and its partners inside and outside the United

States. They plan for contingencies across the range of military operations and allocate resources to prepare for them, all the while assessing the threat and the force's readiness. Steadily improving the Army, strategic leaders develop their successors, lead changes in the force, and optimize systems and operations. This section addresses the influencing, operating, and improving actions they use.

INFLUENCING ACTIONS

7-59. Strategic leaders act to influence both their organization and its outside environment. Like direct and organizational leaders, strategic leaders influence through personal example as well as by communicating, making decisions, and motivating.

7-60. Because the external environment is diverse and complex, it's sometimes difficult for strategic leaders to identify and influence the origins of factors affecting the organization. This difficulty applies particularly to fast-paced situations like theater campaigns. Strategic leaders meet this challenge by becoming masters of information, influence, and vision.

7-61. Strategic leaders also seek to control the information environment, consistent with US law and Army values. Action in this area can range from psychological operations campaigns to managing media relationships. Strategic leaders who know what's happening with present and future requirements, both inside and

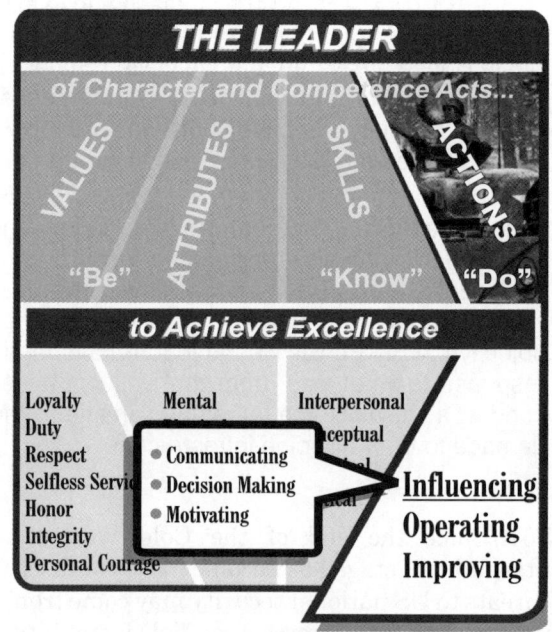

Figure 7-4. Strategic Leader Actions—Influencing

outside the organization, are in a position to influence events, take advantage of opportunities, and move the organization toward its goals.

7-62. As noted earlier, strategic leaders develop the wisdom and frames of reference necessary to identify the information relevant to the situation at hand. In addition, they use interpersonal skills to develop a network of knowledgeable people, especially in those organizations that can influence their own. They encourage staff members to develop similar networks. Through these networks, strategic leaders actively seek information relevant to their organizations and subject matter experts they can call on to assist themselves and their staffs. Strategic leaders can often call on the nation's best minds and information sources and may face situations where nothing less will do.

COMMUNICATING

Moving our Army into the next century is a journey, not a destination; we know where we are going and we are moving out.

General Gordon R. Sullivan
Former Army Chief of Staff

Communicating a Vision

7-63. The skill of envisioning is vital to the strategic leader. But forming a vision is pointless unless the leader shares it with a broad audience, gains widespread support, and uses it

as a compass to guide the organization. For the vision to provide purpose, direction, and motivation, the strategic leader must personally commit to it, gain commitment from the organization as a whole, and persistently pursue the goals and objectives that will spread the vision throughout the organization and make it a reality.

7-64. Strategic leaders identify trends, opportunities, and threats that could affect the Army's future and move vigorously to mobilize the talent that will help create strategic vision. In 1991 Army Chief of Staff Gordon R. Sullivan formed a study group of two dozen people to help craft his vision for the Army. In this process, GEN Sullivan considered authorship less important than shared vision:

Once a vision has been articulated and the process of buy-in has begun, the vision must be continually interpreted. In some cases, the vision may be immediately understandable at every level. In other cases, it must be translated—put into more appropriate language—for each part of the organization. In still other cases, it may be possible to find symbols that come to represent the vision.

7-65. Strategic leaders are open to ideas from a variety of sources, not just their own organizations. Some ideas will work; some won't. Some will have few, if any, long-lasting effects; others, like the one in this example, will have effects few will foresee.

Combat Power from a Good Idea

In 1941, as the American military was preparing for war, Congresswoman Edith Nourse Rogers correctly anticipated manpower shortages in industry and in the armed forces as the military grew. To meet this need, she proposed creation of a Women's Army Auxiliary Corps (WAAC) of 25,000 women to fill administrative jobs and free men for service with combat units. After the United States entered the war, when the size of the effort needed became clearer, Congresswoman Rogers introduced another bill for a WAAC of some 150,000 women. Although the bill met stiff opposition in some quarters, a version passed and eventually the Women's Army Corps was born. Congresswoman Rogers' vision of how to best get the job done in the face of vast demands on manpower contributed a great deal to the war effort.

Telling the Army Story

If you have an important point to make, don't try to be subtle or clever. Use a pile-driver. Hit the point once. Then come back and hit it a second time—a tremendous whack!

Sir Winston Churchill
Prime Minister of Great Britain, World War II

7-66. Whether by nuance or overt presentation, strategic leaders vigorously and constantly represent who Army is, what it's doing, and where it's going. The audience is the Army itself as well as the rest of the world. There's an especially powerful responsibility to explain things to the American people, who support their Army with money and lives. Whether working with other branches of government, federal agencies, the media, other militaries, the other services, or their own organizations, strategic leaders rely increasingly on writing and public speaking (conferences and press briefings) to reinforce the Army's central messages. Because so much of this communication is directed at outside agencies, strategic leaders avoid parochial language and remain sensitive to the Army's image.

7-67. Strategic leaders of all times have determined and reinforced the message that speaks to the soul of the nation and unifies the force. In 1973 Army leaders at all levels knew about "The Big Five," the weapons systems that would transform the Army (a new tank, an infantry fighting vehicle, an advanced attack helicopter, a new utility helicopter, and an air defense system). Those programs yielded the M1 Abrams, the M2/M3 Bradley, the AH-64 Apache, the UH-60 Blackhawk, and the Patriot. But those initiatives were more than sales pitches for newer hardware; they were linked to concepts about how to fight and win against a massive Soviet-style force. As a result, fielding the new equipment gave physical form to the new ideas being adopted at the same time. Soldiers could see improvements as well as read about them. The synergism of new equipment, new ideas, and good leadership resulted in the Army of Excellence.

7-68. Today, given the rapid growth of technology, unpredictable threats, and newly emerging roles, Army leaders can't cling to new hardware as the key to the Army's vision. Instead, today's strategic leaders emphasize the Army's core strength: Army values and the timeless character of the American soldier. The Army—trained, ready, and led by leaders of character and competence at all levels—has met and will continue to meet the nation's security needs. That's the message of today's Army to the nation it serves.

7-69. A recent example of successfully telling the Army story occurred during Operation Desert Shield. During the deployment phase, strategic leaders decided to get local reporters to the theater of war to report on mobilized reserve component units from their communities. That decision had several effects. The first-order effect was to get the Army story to the citizens of hometown America. That publicity resulted in an unintended second-order effect: a flood of mail that the nation sent to its deployed soldiers. That mail, in turn, produced a third-order effect felt by American soldiers: a new pride in themselves.

STRATEGIC DECISION MAKING

When I am faced with a decision—picking somebody for a post, or choosing a course of action—I dredge up every scrap of knowledge I can. I call in people. I telephone them. I read whatever I can get my hands on. I use my intellect to inform my instinct. I then use my instinct to test all this data. "Hey, instinct, does this sound right? Does it smell right, feel right, fit right?"

General Colin Powell
Former Chairman, Joint Chiefs of Staff

7-70. Strategic leaders have great conceptual resources; they have a collegial network to share thoughts and plan for the institution's continued success and well being. Even when there's consensual decision making, however, everyone knows who the boss is. Decisions made by strategic leaders—whether CINCs deploying forces or service chiefs initiating budget programs—often result in a major commitment of resources. They're expensive and tough to reverse. Therefore, strategic leaders rely on timely feedback throughout the decision-making process in order to avoid making a decision based on inadequate or faulty information. Their purpose, direction, and motivation flow down; information and recommendations surface from below.

7-71. Strategic leaders use the processes of the DOD, Joint Staff, and Army strategic planning systems to provide purpose and direction to subordinate leaders. These systems include the Joint Strategic Planning System (JSPS), the Joint Operation Planning and Execution System (JOPES), and the Planning, Programming and Budgeting System (PPBS). However, no matter how many systems are involved and no matter how complex they are, providing motivation remains the province of the individual strategic leader.

7-72. Because strategic leaders are constantly involved in this sort of planning and because decisions at this level are so complex and depend on so many variables, there's a temptation to analyze things endlessly. There's always new information; there's always a reason to wait for the next batch of reports or the next dispatch. Strategic leaders' perspective, wisdom, courage, and sense of timing help them know when to decide. In peacetime the products of those decisions may not see completion for 10 to 20 years and may require leaders to constantly adjust them along the way. By contrast, a strategic leader's decision at a critical moment in combat can rapidly alter the course of the war, as did the one in this example.

The D-Day Decision

On 4 June 1944 the largest invasion armada ever assembled was poised to strike the Normandy region of France. Weather delays had already caused a 24-hour postponement and another front of bad weather was heading for the area. If the Allies didn't make the landings on 6 June, they would miss the combination of favorable tides, clear flying weather, and moonlight needed for the assault. In addition to his concerns about the weather, GA Dwight D. Eisenhower, the Supreme Allied Commander, worried about his soldiers. Every hour they spent jammed aboard crowded ships, tossed about and seasick, degraded their fighting ability.

The next possible invasion date was 19 June; however the optimal tide and visibility conditions would not recur until mid-July. GA Eisenhower was ever mindful that the longer he delayed, the greater chance German intelligence had to discover the Allied plan. The Germans would use any additional time to improve the already formidable coastal defenses.

On the evening of 4 June GA Eisenhower and his staff received word that there would be a window of clear weather on the next night, the night of 5-6 June. If the meteorologists were wrong, GA Eisenhower would be sending seasick men ashore with no air cover or accurate naval gunfire. GA Eisenhower was concerned for his soldiers.

"Don't forget," GA Eisenhower said in an interview 20 years later, "some hundreds of thousands of men were down here around Portsmouth, and many of them had already been loaded for some time, particularly those who were going to make the initial assault. Those people in the ships and ready to go were in cages, you might say. You couldn't call them anything else. They were fenced in. They were crowded up, and everybody was unhappy."

GA Eisenhower continued, "Goodness knows, those fellows meant a lot to me. But these are the decisions that have to be made when you're in a war. You say to yourself, I'm going to do something that will be to my country's advantage for the least cost. You can't say without any cost. You know you're going to lose some of them, and it's very difficult."

A failed invasion would delay the end of a war that had already dragged on for nearly five years. GA Eisenhower paced back and forth as a storm rattled the windows. There were no guarantees, but the time had come to act.

He stopped pacing and, facing his subordinates, said quietly but clearly, "OK, let's go."

MOTIVATING

It is the morale of armies, as well as of nations, more than anything else, which makes victories and their results decisive.

Baron Antoine-Henri de Jomini
Precis de l'Art de Guerre, 1838

Shaping Culture

7-73. Strategic leaders inspire great effort. To mold morale and motivate the entire Army, strategic leaders cultivate a challenging, supportive, and respectful environment for soldiers and DA civilians to operate in. An institution with a history has a mature, well-established culture—a shared set of values and assumptions that members hold about it. At the same time, large and complex institutions like the Army are diverse; they have many subcultures, such as those that exist in the civilian and reserve components, heavy and light forces, and special operations forces. Gender, ethnic, religious, occupational, and regional differences also define groups within the force.

Culture and Values

7-74. The challenge for strategic leaders is to ensure that all these subcultures are part of the larger Army culture and that they all share Army values. Strategic leaders do this by working with the best that each subculture has to offer and ensuring that subcultures don't foster unhealthy competition with each other, outside agencies, or the rest of the Army. Rather, these various subcultures must complement each other and the Army's institutional culture. Strategic leaders appreciate the differences that characterize these subcultures and treat all members of all components with dignity and respect. They're responsible for creating an environment that fosters mutual understanding so that soldiers and DA civilians treat one another as they should.

7-75. Army values form the foundation on which the Army's institutional culture stands. Army values also form the basis for Army policies and procedures. But written values are of little use unless they are practiced. Strategic leaders help subordinates adopt these values by making sure that their experience validates them. In this, strategic leaders support the efforts all Army leaders make to develop the their subordinates' character. This character development effort (discussed in Appendix E) strives to have all soldiers and DA civilians adopt Army values, incorporate them into a personal code, and act according to them.

7-76. Like organizational and direct leaders, strategic leaders model character by their actions. Only experience can validate Army values: subordinates will hear of Army values, then look to see if they are being lived around them. If they are, the Army's institutional culture is strengthened; if they are not, the Army's institutional culture begins to weaken. Strategic leaders ensure Army values remain fundamental to the Army's institutional culture.

7-77. Over time, an institution's culture becomes so embedded in its members that they may not even notice how it affects their attitudes. The institutional culture becomes second nature and influences the way people think, the way they act in relation to each other and outside agencies, and the way they approach the mission. Institutional culture helps define the boundaries of acceptable behavior, ranging from how to wear the uniform to how to interact with foreign nationals. It helps determine how people approach problems, make judgments, determine right from wrong, and establish priorities. Culture shapes Army customs and traditions through doctrine, policies and regulations, and the philosophy that guides the institution. Professional journals, historical works, ceremonies—even the folklore of the organization—all contain evidence of the Army's institutional culture.

Culture and Leadership

7-78. A healthy culture is a powerful leadership tool strategic leaders use to help them guide their large diverse organizations. Strategic leaders seek to shape the culture to support their vision, accomplish the mission, and improve the organization. A cohesive culture molds the organization's morale, reinforcing an ethical climate built on Army values, especially respect. As leaders initiate changes for long-range improvements, soldiers and DA civilians must feel that they're valued as persons, not just as workers or program supporters.

7-79. One way the Army's institutional culture affirms the importance of individuals is through its commitment to leader development: in essence, this commitment declares that people are the Army's future. By committing to broad-based leader development, the Army has redefined what it means to be a soldier. In fact, Army leaders have even changed the appearance of American soldiers and the way they perform. Introducing height and weight standards, raising PT standards, emphasizing training and education, and deglamorizing alcohol have all fundamentally changed the Army's institutional culture.

OPERATING ACTIONS

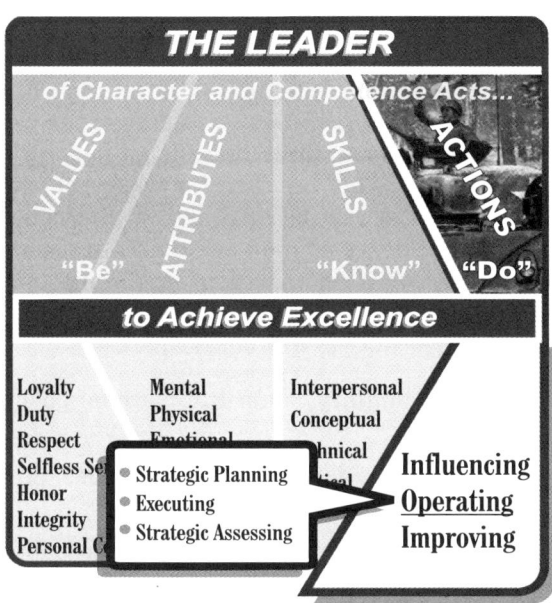

Figure 7-5. Strategic Leader Actions—Operating

leaders, strategic leaders also manage joint, multinational, and interagency relationships. For strategic leaders, planning, preparing, executing, and assessing are nearly continuous, more so than at the other leadership levels, because the larger organizations they lead have continuing missions. In addition, the preparing action takes on a more comprehensive meaning at the strategic leadership level. Leaders at all levels keep one eye on tomorrow. Strategic leaders, to a greater extent than leaders at other levels, must coordinate their organizations' actions, positioning them to accomplish the current mission in a way that will feed seamlessly into the next one. The Army doesn't stop at the end of a field exercise—or even after recovering from a major deployment; there's always another mission about to start and still another one on the drawing board.

7-80. Operating at the strategic level can involve both short-term and long-term actions. The most agile organizations have standing procedures and policies to take the guesswork out of routine actions and allow leaders to concentrate their imagination and energy on the most difficult tasks. Strategic leaders coordinate their organizations' actions to accomplish near-term missions, often without the benefit of direct guidance. Strategic leaders receive general guidance—frequently from several sources, including the national command authority.

7-81. Although they perform many of the same operating actions as organizational and direct

STRATEGIC PLANNING

7-82. Strategic-level plans must balance competing demands across the vast structure of the DOD, but the fundamental requirements for strategic-level planning are the same as for direct- and organizational-level planning. At all levels, leaders establish priorities and communicate decisions; however, at the strategic level, the sheer number of players who can influence the organization means that strategic leaders must stay on top of multiple demands. To plan coherently and comprehensively, they look at the mission from other players' points of view. Strategic planning depends heavily on wisely applying interpersonal and conceptual skills. Strategic leaders ask, What will these people want? How will they see things? Have I justified the mission? The interaction among strategic leaders' interpersonal and conceptual skills and their operating actions is highly complex.

7-83. Interpersonal and conceptual understanding helped the Army during Operation Uphold Democracy, the US intervention in Haiti. The success of the plan to collect and disarm former Haitian police and military officials, investigate them, remove them (if required), or retrain them owed much to recognizing the special demands of the Haitian psyche. The population needed a secure and stable environment and a way to know when that condition was in place. The Haitians feared the resurgence of government terror, and any long-term solution had to address their concerns. Strategic planners maintained a focus on the desired end state: US disengagement and a return to a peaceful, self-governing Haiti. In the end, the United States forced Haitian leaders to cooperate, restored the elected president, Jean-Bertrand Aristide, and made provisions for returning control of affairs to the Haitians themselves.

EXECUTING

There are no victories at bargain prices.

General of the Army Dwight D. Eisenhower

Allocating Resources

7-84. Because lives are precious and materiel is scarce, strategic leaders make tough decisions about priorities. Their goal is a capable, prepared, and victorious force. In peacetime, strategic leaders decide which programs get funded and consider the implications of those choices. Allocating resources isn't simply a matter of choosing helicopters, tanks, and missiles for the future Army. Strategic resourcing affects how the Army will operate and fight tomorrow. For example, strategic leaders determine how much equipment can be pre-positioned for contingencies without degrading current operational capabilities.

Managing Joint, Interagency, and Multinational Relationships

7-85. Strategic leaders oversee the relationship between their organizations, as part of the nation's total defense force, and the national policy apparatus. They use their knowledge of how things work at the national and international levels to influence opinion and build consensus for the organization's missions, gathering support of diverse players to achieve their vision. Among their duties, strategic leaders—

* Provide military counsel in national policy forums.
* Interpret national policy guidelines and directions.
* Plan for and maintain the military capability required to implement national policy.
* Present the organization's resource requirements.

Multinational Resource Allocation

Following the breakout and pursuit after the Normandy landings, Allied logistics systems became seriously overstretched. GA Eisenhower, the Supreme Allied Commander, had to make a number of decisions on resource allocation among his three army groups. These decisions had serious implications for the conduct of the war in the ETO. Both GEN George Patton, Commander of the Third (US) Army in GEN Omar Bradley's Twelfth Army Group, and British GEN Bernard Montgomery, the Twenty-first Army Group Commander, argued that sole priority for their single thrusts into the German homeland could win the war. GA Eisenhower, dedicated to preserving the alliance with an Allied success in the West, gave GEN Montgomery only a limited priority for a risky attempt to gain a Rhine bridgehead, and at the same time, slowed GEN Patton's effort to what was logistically feasible under the circumstances. The Supreme Allied Commander's decision was undoubtedly unpopular with his longtime colleague, GEN Patton, but it contributed to alliance solidarity, sent a message to the Soviets, and ensured a final success that did not rely on the still highly uncertain collapse of German defenses.

- Develop strategies to support national objectives.
- Bridge the gap between political decisions made as part of national strategy and the individuals and organizations that must carry out those decisions.

7-86. As part of this last requirement, strategic leaders clarify national policy for subordinates and explain the perspectives that contribute to that national policy. They develop policies reflecting national security objectives and prepare their organizations to respond to missions across the spectrum of military actions.

7-87. Just as direct and organizational leaders consider their sister units and agencies, strategic leaders consider and work with other armed services and government agencies. How important is this joint perspective? Most of the Army's four-star billets are joint or multinational. Almost half of the lieutenant generals hold similar positions on the Joint Staff, with the DOD, or in combatant commands. While the remaining strategic leaders are assigned to organizations that are nominally single service (Forces Command, Training and Doctrine Command, Army Materiel Command), they frequently work outside Army channels. In addition, many DA civilian strategic leaders hold positions that require a joint perspective.

7-88. The complexity of the work created by joint and multinational requirements is twofold. First, communication is more complicated because of the different interests, cultures, and languages of the participants. Even the cultures and jargon of the various US armed services differ dramatically. Second, subordinates may not be subordinate in the same sense as they are in a purely Army organization. Strategic leaders and their forces may fall under international operational control but retain their allegiances and lines of authority to their own national commanders. UN and NATO commands, such as the IFOR, discussed earlier are examples of this kind of arrangement.

7-89. To operate effectively in a joint or multinational environment, strategic leaders exercise a heightened multiservice and international sensitivity developed over their years of experience. A joint perspective results from shared experiences and interactions with leaders of other services, complemented by the leader's habitual introspection. Similar elements in the international arena inform an international perspective. Combing those perspectives with their own Army and national perspectives, strategic leaders—

- Influence the opinions of those outside the Army and help them understand Army needs.
- Interpret the outside environment for people on the inside, especially in the formulation of plans and policies.

Most Army leaders will have several opportunities to serve abroad, sometimes with forces of other nations. Perceptive leaders turn such service into opportunities for self-development and personal broadening.

7-90. Chapter 2 describes building a "third culture," that is, a hybrid culture that bridges the gap between partners in multinational operations. Strategic leaders take the time to learn about their partners' cultures—including political, social and economic aspects—so that they understand how and why the partners think and act as they do. Strategic leaders are also aware that the successful conduct of multinational operations requires a particular sensitivity to the effect that deploying US forces may have on the laws, traditions, and customs of a third country.

7-91. Strategic leaders understand American and Army culture. This allows them to see their own culturally-based actions from the viewpoint of another culture—civilian, military, or foreign. Effective testimony before Congress requires an understanding of how Congress works and how its members think. The same is true concerning dealings with other federal and state agencies, non-governmental organizations, local political leaders, the media, and other people who shape public opinion and national attitudes toward the military. Awareness of the audience helps strategic leaders represent their organizations to outside agencies. Understanding societal values—those values people bring into the Army—helps strategic leaders motivate subordinates to live Army values.

7-92. When the Army's immediate needs conflict with the objectives of other agencies, strategic leaders work to reconcile the differences. Reconciliation begins with a clear understanding of the other agency's position. Understanding the other side's position is the first step in identifying shared interests, which may permit a new outcome better for both parties. There will be times when strategic leaders decide to stick to their course; there will be other times when Army leaders bend to accommodate other organizations. Continued disagreement can impair the Army's ability to serve the nation; therefore, strategic leaders must work to devise Army courses of action that reflect national policy objectives and take into account the interests of other organizations and agencies.

7-93. Joint and multinational task force commanders may be strategic leaders. In certain operations they will work for a CINC but receive guidance directly from the Chairman of the Joint Chiefs of Staff, the Secretary of Defense, the State Department, or the UN. Besides establishing professional relationships within the DOD and US government, such strategic leaders must build personal rapport with officials from other countries and military establishments.

Military Actions Across the Spectrum

7-94. Since the character of the next war has not been clearly defined for them, today's strategic leaders rely on hints in the international environment to provide information on what sort of force to prepare. Questions they consider include these: Where is the next threat? Will we have allies or contend alone? What will our national and military goals be? What will the exit strategy be? Strategic leaders address the technological, leadership, and moral considerations associated with fighting on an asymmetrical battlefield. They're at the center of the tension between traditional warfare and the newer kinds of multiparty conflict emerging outside the

industrialized world. Recent actions like those in Bosnia, Somalia, Haiti, Grenada, and the Persian Gulf suggest the range of possible military contingencies. Strategic leaders struggle with the ramifications of switching repeatedly among the different types of military actions required under a strategy of engagement.

7-95. The variety of potential missions calls for the ability to quickly build temporary organizations able to perform specific tasks. As they design future joint organizations, strategic leaders must also determine how to engineer both cohesion and proficiency in modular units that are constantly forming and reforming.

STRATEGIC ASSESSING

7-96. There are many elements of their environment that strategic leaders must assess. Like leaders at other levels, they must first assess themselves: their leadership style, strengths and weaknesses, and their fields of excellence. They must also understand the present operational environment—to include the will of the American people, expressed in part through law, policy, and their leaders. Finally, strategic leaders must survey the political landscape and the international environment, for these affect the organization and shape the future strategic requirements.

7-97. Strategic leaders also cast a wide net to assess their own organizations. They develop performance indicators to signal how well they're communicating to all levels of their commands and how well established systems and processes are balancing the six imperatives. Assessment starts early in each mission and continues through its end. It may include monitoring such diverse areas as resource use, development of subordinates, efficiency, effects of stress and fatigue, morale, and mission accomplishment. Such assessments generate huge amounts of data; strategic leaders must make clear what they're looking for so their staffs can filter information for them. They must also guard against misuse of assessment data.

World War II Strategic Assessment

Pursuing a "Germany first" strategy in World War II was a deliberate decision based on a strategic and political assessment of the global situation. Military planners, particularly Army Chief of Staff George C. Marshall, worried that US troops might be dispersed and used piecemeal. Strategic leaders heeded Frederick the Great's adage, "He who defends everywhere defends nowhere." The greatest threat to US interests was a total German success in Europe: a defeated Russia and neutralized Great Britain. Still, the Japanese attacks on Pearl Harbor and the Philippines and the threat to the line of communications with Australia tugged forces toward the opposite hemisphere. Indeed, throughout the first months of 1942, more forces headed for the Pacific Theater than across the Atlantic.

However, before the US was at war with anyone, President Franklin Roosevelt had agreed with British Prime Minister Winston Churchill to a "Germany first" strategy. The 1942 invasion of North Africa restored this focus. While the US military buildup took hold and forces flowed into Great Britain for the Normandy invasion in 1944, operations in secondary theaters could and did continue. However, they were resourced only after measuring their impact on the planned cross-channel attack.

IMPROVING ACTIONS

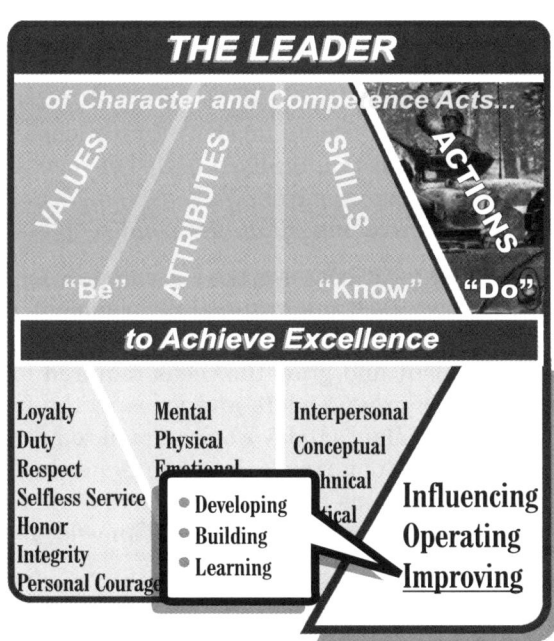

Figure 7-6. Strategic Leader Actions— Improving

7-98. Improving is institutional investment for the long haul, refining the things we do today for a better organization tomorrow. A fundamental goal of strategic leaders is to leave the Army better than they found it. Improving at this level calls for experimentation and innovation; however, because strategic-level organizations are so complex, quantifying the results of changes may be difficult.

7-99. Improving the institution and organizations involves an ongoing tradeoff between today and tomorrow. Wisdom and a refined frame of reference are tools to understand what improvement is and what change is needed. Knowing when and what to change is a constant challenge: what traditions should remain stable, and which long-standing methods need to evolve? Strategic leaders set the conditions for long-term success of the organization by developing subordinates, leading change, building the culture and teams, and creating a learning environment.

7-100. One technique for the Army as a learning institution is to decentralize the learning and other improving actions to some extent. That technique raises the questions of how to share good ideas across the entire institution and how to incorporate the best ideas into doctrine (thus establishing an Army-wide standard) without discouraging the decentralized learning process that generated the ideas in the first place. Those and other questions face the strategic leaders of the learning organization the Army seeks to become.

DEVELOPING

George C. Marshall learned leadership from John J. Pershing, and Marshall's followers became great captains themselves: Dwight D. Eisenhower, Omar N. Bradley...among them. Pershing and Marshall each taught their subordinates their profession; and, more importantly, they gave them room to grow.

General Gordon R. Sullivan
Former Army Chief of Staff

Mentoring

7-101. Strategic leaders develop subordinates by sharing the benefit of their perspective and experience. People arriving at the Pentagon know how the Army works in the field, but regardless of what they may have read, they don't really know how the institutional Army works. Strategic leaders act as a kind of sponsor by introducing them to the important players and pointing out the important places and activities. But strategic leaders actually become mentors as they, in effect, underwrite the learning, efforts, projects, and ideas of rising leaders. The moral responsibility associated with mentoring is compelling for all leaders; for strategic leaders, the potential significance is enormous.

7-102. More than a matter of required forms and sessions, mentoring by strategic leaders means giving the right people an intellectual boost so that they make the leap to operations and thinking at the highest levels. Because those being groomed for strategic leadership positions are among the most talented Army leaders, the manner in which leaders and subordinates interact also changes. Strategic leaders aim not only to pass on knowledge but also to grow wisdom in those they mentor.

7-103. Since few formal leader development programs exist beyond the senior service colleges, strategic leaders pay special attention to their subordinates' self-development, showing them what to study, where to focus, whom to watch, and how to proceed. They speak to audiences at service schools about what goes on "at

the top" and spend time sharing their perspectives with those who haven't yet reached the highest levels of Army leadership. Today's subordinates will, after all, become the next generation of strategic leaders. Strategic leaders are continually concerned that the Army institutional culture and the climates in subordinate organizations encourage mentoring by others so that growth opportunities are available from the earliest days of a soldier or DA civilian's careers.

Developing Intellectual Capital

7-104. What strategic leaders do for individuals they personally mentor, they also seek to provide to the force at large. They invest in the future of the force in several ways. Committing money to programs and projects and investing more time and resources in some actions than others are obvious ways strategic leaders choose what's important. They also value people and ideas as investments in the future. The concepts that shape the thinking of strategic leaders become the intellectual currency of the coming era; the soldiers and DA civilians who develop those ideas become trusted assets themselves. Strategic leaders must choose wisely the ideas that bridge the gap between today and tomorrow and skillfully determine how best to resource important ideas and people.

7-105. Strategic leaders make difficult decisions about how much institutional development is enough. They calculate how much time it will take to plant and grow the seeds required for the Army's great leaders and ideas in the future. They balance today's operational requirements with tomorrow's leadership needs to produce programs that develop a core of Army leaders with the required skills and knowledge.

7-106. Programs like training with industry, advanced civil schooling, and foreign area officer education complement the training and education available in Army schools and contribute to shaping the people who will shape the Army's future. Strategic leaders develop the institution using Army resources when they are available and those of other services or the public sector when they are not.

7-107. After Vietnam the Army's leadership thought investing in officer development so important that new courses were instituted to revitalize professional education for the force. The establishment of the Training and Doctrine Command revived Army doctrine as a central intellectual pillar of the entire service. The Goldwater-Nichols Act of 1986 provided similar attention and invigoration to professional joint education and joint doctrine.

7-108. Likewise, there has been a huge investment in and payoff from developing the NCO corps. The Army has the world's finest noncommissioned officers, in part because they get the world's best professional development. The strategic decision to resource a robust NCO education system signaled the Army's investment in developing the whole person—not just the technical skills—of its first-line leaders.

7-109. The Army Civilian Training and Education Development System is the Army's program for developing DA civilian leaders. Like the NCO education system, it continues throughout an individual's career. The first course integrates interns into the Army by explaining Army values, culture, customs, and policies. The Leadership Education and Development Course helps prepare leaders for supervisory demands with training in communication, counseling, team building, problem solving, and group development. For organizational managers, the Organizational Leadership for Executives course adds higher-order study on topics such as strategic planning, change management, climate, and culture. DA civilians in the Senior Executive Service have a variety of leadership education options that deal with leadership in both the military and civilian contexts. Together, these programs highlight ways that leadership development of DA civilians parallels that of soldiers.

BUILDING

The higher up the chain of command, the greater is the need for boldness to be supported by a reflective mind, so that boldness does not degenerate into purposeless bursts of blind passion.

Carl von Clausewitz

Building Amid Change

7-110. The Army has no choice but to face change. It's in a nearly constant state of flux, with new people, new missions, new technologies, new equipment, and new information. At the same time, the Army, inspired by strategic leaders, must innovate and create change. The Army's customs, procedures, hierarchical structure, and sheer size make change especially daunting and stressful. Nonetheless, the Army must be flexible enough to produce and respond to change, even as it preserves the core of traditions that tie it to the nation, its heritage and its values.

7-111. Strategic leaders deal with change by being proactive, not reactive. They anticipate change even as they shield their organizations from unimportant and bothersome influences; they use the "change-drivers" of technology, education, doctrine, equipment, and organization to control the direction and pace of change. Many agencies and corporations have "futures" groups charged with thinking about tomorrow; strategic leaders and their advisory teams are the Army's "futures people."

Leading Change

7-112. Strategic leaders lead change by—

- Identifying the force capabilities necessary to accomplish the National Military Strategy.
- Assigning strategic and operational missions, including priorities for allocating resources.
- Preparing plans for using military forces across the spectrum of operations.
- Creating, resourcing, and sustaining organizational systems, including—
  - Force modernization programs.
  - Requisite personnel and equipment.
  - Essential command, control, communications, computers, and intelligence systems.
- Developing and improving doctrine and the training methods to support it.

- Planning for the second- and third-order effects of change.

- Maintaining an effective leader development program and other human resource initiatives.

Change After Vietnam

The history of the post-Vietnam Army provides an example of how strategic leaders' commitment can shape the environment and harness change to improve the institution while continuing to operate.

The Army began seeking only volunteers in the early 1970s. With the all-volunteer force came a tremendous emphasis on doctrinal, personnel, and training initiatives that took years to mature. The Army tackled problems in drug abuse, racial tensions, and education with ambitious, long-range plans and aggressive leader actions. Strategic leaders overhauled doctrine and created an environment that improved training at all levels; the CTC program provided a uniform, rock-solid foundation of a single, well-understood warfighting doctrine upon which to build a trained and ready Army. Simultaneously, new equipment, weapons, vehicles, and uniforms were introduced. The result was the Army of Desert Storm, which differed greatly from the force of 15 years earlier.

None of these changes happened by chance or through evolution. Change depended on the hard work of direct and organizational leaders who developed systematically in an environment directed, engineered, and led by strategic leaders.

7-113. Strategic leaders must guide their organizations through eight stages if their initiatives for change are to make lasting progress. Skipping a step or moving forward prematurely subverts the process and compromises success. Strategic leaders (1) demonstrate a sense of urgency by showing not only the benefits of but the necessity for change. They (2) form guiding coalitions to work the process all the way from concept through implementation. With those groups they (3) develop a vision of the future and strategy for achieving it. Because change is most effective when members embrace it, strategic leaders (4) communicate the vision throughout the institution or organization, and then (5) empower subordinates at all levels for widespread, parallel efforts. They (6) plan for short-term successes to validate the programs and keep the vision credible and (7) consolidate those wins and produce further change. Finally, the leader (8) preserves the change culturally. The result is an institution that constantly prepares for and even shapes the future environment. Strategic leaders seek to sustain the Army as that kind of institution.

LEARNING

A good soldier, whether he leads a platoon or an army, is expected to look backward as well as forward; but he must think only forward.

General of the Army Douglas MacArthur

7-114. The nation expects military professionals as individuals and the Army as an institution to learn from the experience of others and apply that learning to understanding the present and preparing for the future. Such learning requires both individual and institutional commitments. Each military professional must be committed to self-development, part of which is studying military history and other disciplines related to military operations. The Army as an institution must be committed to conducting technical research, monitoring emerging threats, and developing leaders for the next generation. Strategic leaders, by their example and resourcing decisions, sustain the culture and policies that encourage both the individual and the Army to learn.

7-115. Strategic leaders promote learning by emplacing systems for studying the force and the future. Strategic leaders must resource a structure that constantly reflects on how the Army fights and what victory may cost. All that means constantly assessing the culture and deliberately encouraging creativity and learning.

7-116. The notion of the Army as a "learning organization" is epitomized by the AAR concept, which was developed as part of the REALTRAIN project, the first version of engagement simulation. Since then, it has been part of a cultural change, in which realistic "hot washes," such as following tough engagements at CTCs, are now embedded in all training. Twenty years ago, anything like today's AARs would have been rare.

7-117. Efficient and effective operations require aligning various initiatives so that different factions are not working at cross-purposes. Strategic leaders focus research and development efforts on achieving combined arms success. They deal with questions such as: Can these new systems from various sources communicate with one another? What happens during digitization lapses—what's our residual

combat capability? Strategic leaders coordinate time lines and budgets so that compatible systems are fielded together. However, they are also concerned that the force have optimal capability across time; therefore, they prepare plans that integrate new equipment and concepts into the force as they're developed, rather than waiting for all elements of a system to be ready before fielding it. Finally, learning what the force should be means developing the structure, training, and leaders those future systems will support and studying the variety of threats they may face.

7-118. The Louisiana Maneuvers in 1941 taught the Army what mechanized warfare would look like and what was needed to prepare for it. The study of the same name 50 years later helped produce the conceptual Force XXI and the first digitized division. Strategic leaders commissioned these projects because the Army is dedicated to learning about operations in new environments, against different threats. The projects were strategic counterparts to the rehearsals that direct and organizational leaders conduct to prepare for upcoming missions.

SECTION III
A HISTORICAL PERSPECTIVE OF STRATEGIC LEADERSHIP—
GENERAL OF THE ARMY MARSHALL DURING WORLD WAR II

7-119. GA George C. Marshall was one of the greatest strategic leaders of World War II, of this century, of our nation's history. His example over many years demonstrates the skills and actions this chapter has identified as the hallmarks of strategic leadership.

7-120. Chosen over 34 officers senior to him, GA Marshall became Army Chief of Staff in 1939, a time of great uncertainty about the future of the free world. Part of his appeal for President Roosevelt was his strength of character and personal integrity. The honesty and candor that GA Marshall displayed early in their relationship were qualities the president knew he and the nation would need in the difficult times ahead.

7-121. The new Army Chief of Staff knew he had to wake the Army from its interwar slumber and grow it beyond its 174,000 soldiers—a size that ranked it seventeenth internationally, behind Bulgaria and Portugal. By 1941 he had begun to move the Army toward his vision of what it needed to become: a world-changing force of 8,795,000 soldiers and airmen. His vision was remarkably accurate: by the end of the war, 89 divisions and over 8,200,000 soldiers in Army uniforms had made history.

7-122. GA Marshall reached deep within the Army for leaders capable of the conceptual leaps necessary to fight the impending war. He demanded leaders ready for the huge tasks ahead, and he accepted no excuses. As he found colonels, lieutenant colonels, and even majors who seemed ready for the biggest challenge of their lives, he promoted them ahead of those more senior but less capable and made many of them generals. He knew firsthand that such jumps could be productive. As a lieutenant in the Philippines, he had commanded 5,000 soldiers during an exercise. For generals who could not adjust to the sweeping changes in the Army, he made career shifts as well: he retired them. His loyalty to the institution and the nation came before any personal relationships.

7-123. Merely assembling the required number of soldiers would not be enough. The mass Army that was forming required a new structure to manage the forces and resources the nation was mobilizing for the war effort. Realizing this, GA Marshall reorganized the Army into the Army Ground Forces, Army Air Forces, and Army Service Forces. His foresight organized the Army for the evolving nature of warfare.

7-124. Preparing for combat required more than manning the force. GA Marshall understood that World War I had presented confusing lessons about the future of warfare. His in-theater experience during that war and later reflection distilled a vision of the future. He believed that maneuver of motorized formations spearheaded by tanks and supported logistically by trucks (instead of horse-drawn wagons) would replace the almost siege-like battles of World War I. So while the French trusted the Maginot Line, GA Marshall emphasized the new technologies that would heighten the speed and complexity of the coming conflict.

7-125. Further, GA Marshall championed the common sense training to prepare soldiers to go overseas ready to fight and win. By having new units spend sufficient time on marksmanship, fitness, drill, and fieldcraft, GA Marshall ensured that soldiers and leaders had the requisite competence and confidence to face an experienced enemy.

7-126. Before and during the war, GA Marshall showed a gift for communicating with the American public. He worked closely with the press, frequently confiding in senior newsmen so they would know about the Army's activities

and the progress of the war. They responded to his trust by not printing damaging or premature stories. His relaxed manner and complete command of pertinent facts reassured the press, and through it the nation, that America's youth were entrusted to the right person.

7-127. He was equally successful with Congress. GA Marshall understood that getting what he wanted meant asking, not demanding. His humble and respectful approach with lawmakers won his troops what they needed; arrogant demands would have never worked. Because he never sought anything for himself (his five-star rank was awarded over his objections), his credibility soared.

7-128. However, GA Marshall knew how to shift his approach depending on the audience, the environment, and the situation. He refused to be intimidated by leaders such as Prime Minister Winston Churchill, Secretary of War Henry Stimson, or even the president. Though he was always respectful, his integrity demanded that he stand up for his deeply held convictions—and he did, without exception.

7-129. The US role in Europe was to open a major second front to relieve pressure on the Soviet Union and ensure the Allied victory over Germany. GA Marshall had spent years preparing the Army for Operation Overlord, the D-Day invasion that would become the main effort by the Western Allies and the one expected to lead to final victory over Nazi Germany. Many assumed GA Marshall would command it. President Roosevelt might have felt obligated to reward the general's faithful and towering service, but GA Marshall never raised the subject. Ultimately, the president told GA Marshall that it was more important that he lead global resourcing than command a theater of war. GA Eisenhower got the command, while GA Marshall continued to serve on staff.

7-130. GA Marshall didn't request the command that would have placed him alongside immortal combat commanders like Washington, Grant, and Lee. His decision reflects the value of selfless service that kept him laboring for decades without the recognition that came to some of his associates. GA Marshall never attempted to be anyplace but where his country needed him. And there, finally as Army Chief of Staff, GA Marshall served with unsurpassed vision and brilliance, engineering the greatest victory in our nation's history and setting an extraordinary example for those who came after him.

SUMMARY

7-131. Just as GA Marshall prepared for the coming war, strategic leaders today ready the Army for the next conflict. They may not have years before the next D-Day; it could be just hours away. Strategic leaders operate between extremes, balancing a constant awareness of the current national and global situation with a steady focus on the Army's long-term mission and goals.

7-132. Since the nature of future military operations is so unclear, the vision of the Army's strategic leaders is especially crucial. Identifying what's important among the concerns of mission, soldiers, weapons, logistics, and technology produces decisions that determine the structure and capability of tomorrow's Army.

7-133. Within the institution, strategic leaders build support for the end state they desire. That means building a staff that can take broad guidance and turn it into initiatives that move the Army forward. To obtain the required support, strategic leaders also seek to achieve consensus beyond the Army, working with Congress and the other services on budget, force structure, and strategy issues and working with other countries and militaries on shared interests. The way strategic leaders communicate direction to soldiers, DA civilians, and citizens determines the understanding and support for the new ideas.

7-134. Like GA Marshall, today's strategic leaders are deciding how to transform today's force into tomorrow's. These leaders have little guidance. Still, they know that they work to develop the next generation of Army leaders, build the organizations of the future, and resource the systems that will help gain the next success. The way strategic leaders communicate direction to soldiers, DA civilians, and citizens determines the understanding and support for the new ideas. To communicate with these diverse audiences, strategic leaders work through multiple media, adjust the message when necessary, and constantly reinforce Army themes.

7-135. To lead change personally and move the Army establishment toward their concept of the future, strategic leaders transform political and conceptual programs into practical and concrete initiatives. That process increasingly involves leveraging technology and shaping the culture. By knowing themselves, the strategic players, the operational requirements, the geopolitical situation, and the American public, strategic leaders position the force and the nation for success. Because there may be no time for a World War II or Desert Storm sort of buildup, success for Army strategic leaders means being ready to win a variety of conflicts now and remaining ready in the uncertain years ahead.

7-136. Strategic leaders prepare the Army for the future through their leadership. That means **influencing** people—members of the Army, members of other government agencies, and the people of the nation the Army serves—by providing purpose, direction, and motivation. It means **operating** to accomplish today's missions, foreign and domestic. And it means **improving** the institution—making sure its people are trained and that its equipment and organizations are ready for tomorrow's missions, anytime, anywhere.

Appendix A
Roles and Relationships

A-1. When the Army speaks of soldiers, it refers to commissioned officers, warrant officers, noncommissioned officers (NCOs), and enlisted personnel—both men and women. The terms commissioned officer and warrant officer are used when it is necessary to specifically address or refer to a particular group of officers. All Army leaders—soldiers and DA civilians—share the same goal: to accomplish their organization's mission. The roles and responsibilities of Army leaders—commissioned, warrant, noncommissioned, and DA civilian—overlap. Figure A-1 summarizes them.

A-2. Commissioned officers are direct representatives of the President of the United States. Commissions are legal instruments the president uses to appoint and exercise direct control over qualified people to act as his legal agents and help him carry out his duties. The Army retains this direct-agent relationship with the president through its commissioned officers. The commission serves as the basis for a commissioned officer's legal authority. Commissioned officers command, establish policy, and manage Army resources. They are normally generalists who assume progressively broader responsibilities over the course of a career.

A-3. Warrant officers are highly specialized, single-track specialty officers who receive their authority from the Secretary of the Army upon their initial appointment. However, Title 10 USC authorizes the commissioning of warrant officers (WO1) upon promotion to chief warrant officer (CW2). These commissioned warrant officers are direct representatives of the president of the United States. They derive their authority from the same source as commissioned officers but remain specialists, in contrast to commissioned officers, who are generalists. Warrant officers can and do command detachments, units, activities, and vessels as well as lead, coach, train, and counsel subordinates. As leaders and technical experts, they provide valuable skills, guidance, and expertise to commanders and organizations in their particular field.

A-4. NCOs, the backbone of the Army, train, lead, and take care of enlisted soldiers. They receive their authority from their oaths of office, law, rank structure, traditions, and regulations. This authority allows them to direct soldiers, take actions required to accomplish the mission, and enforce good order and discipline. NCOs represent officer, and sometimes DA civilian, leaders. They ensure their subordinates, along with their personal equipment, are prepared to function as effective unit and team members. While commissioned officers command, establish policy, and manage resources, NCOs conduct the Army's daily business.

A-5. As members of the executive branch of the federal government, DA civilians are part of the Army. They derive their authority from a variety of sources, such as commanders, supervisors, Army regulations, and Title 5 USC. DA civilians' authority is job-related: they normally exercise authority related to their positions. DA civilians fill positions in staff and base sustaining operations that would otherwise have to be filled by officers and NCOs. Senior DA civilians establish policy and manage Army resources, but they do not have the authority to command.

A-6. The complementary relationship and mutual respect between the military and civilian members of the Army is a long-standing tradition. Since the Army's beginning in 1775, military and DA civilian duties have stayed separate, yet necessarily related. Taken in combination, traditions, functions, and laws serve to delineate the particular duties of military and civilian members of the Army.

THE COMMISSIONED OFFICER

- Commands, establishes policy, and manages Army resources.
- Integrates collective, leader, and soldier training to accomplish missions.
- Deals primarily with units and unit operations.
- Concentrates on unit effectiveness and readiness.

THE WARRANT OFFICER

- Provides quality advice, counsel, and solutions to support the command.
- Executes policy and manages the Army's systems.
- Commands special-purpose units and task-organized operational elements.
- Focuses on collective, leader, and individual training.
- Operates, maintains, administers, and manages the Army's equipment, support activities, and technical systems.
- Concentrates on unit effectiveness and readiness.

THE NONCOMMISSIONED OFFICER

- Trains soldiers and conducts the daily business of the Army within established policy.
- Focuses on individual soldier training.
- Deals primarily with individual soldier training and team leading.
- Ensures that subordinate teams, NCOs, and soldiers are prepared to function as effective unit and team members.

THE DEPARTMENT OF THE ARMY CIVILIAN

- Establishes and executes policy, leads people, and manages programs, projects, and Army systems.
- Focuses on integrating collective, leader, and individual training.
- Operates, maintains, administers, and manages Army equipment and support, research, and technical activities.
- Concentrates on DA civilian individual and organizational effectiveness and readiness.

Figure A-1. Roles and Responsibilities of Commissioned, Warrant, Noncommissioned, and DA Civilian Leaders

AUTHORITY

A-7. Authority is the legitimate power of leaders to direct subordinates or to take action within the scope of their positions. Military authority begins with the Constitution, which divides it between Congress and the president. (The Constitution appears in Appendix F.) Congress has the authority to make laws that govern the Army. The president, as commander in chief, commands the armed forces, including the Army. Two types of military authority exist: command and general military.

Command Authority

A-8. Command is the authority that a commander in the armed forces lawfully exercises over subordinates by virtue of rank or assignment. Command includes the authority and responsibility for effectively using available resources to organize, direct, coordinate, employ, and control military forces so that they accomplish assigned missions. It also includes responsibility for health, welfare, morale, and discipline of assigned personnel.

A-9. Command authority originates with the president and may be supplemented by law or regulation. It is the authority that a commander lawfully exercises over subordinates by virtue of rank or assignment. Only commissioned and warrant officers may command Army units and installations. DA civilians may exercise general supervision over an Army installation or activity; however, they act under the authority of a military supervisor. DA civilians do not command. (AR 600-20 addresses command authority in more detail.)

A-10. Army leaders are granted command authority when they fill command-designated positions. These normally involve the direction and control of other soldiers and DA civilians. Leaders in command-designated positions have the inherent authority to issue orders, carry out the unit mission, and care for both military members and DA civilians within the leader's scope of responsibility.

General Military Authority

A-11. General military authority originates in oaths of office, law, rank structure, traditions,

and regulations. This broad-based authority also allows leaders to take appropriate corrective actions whenever a member of any armed service, anywhere, commits an act involving a breach of good order or discipline. AR 600-20, paragraph 4-5, states this specifically, giving commissioned, warrant, and noncommissioned officers authority to "quell all quarrels, frays, and disorders among persons subject to military law"—in other words, to maintain good order and discipline.

A-12. All enlisted leaders have general military authority. For example, dining facility managers, platoon sergeants, squad leaders, and tank commanders all use general military authority when they issue orders to direct and control their subordinates. Army leaders may exercise general military authority over soldiers from different units.

A-13. For NCOs, another source of general military authority stems from the combination of the chain of command and the NCO support channel. The chain of command passes orders and policies through the NCO support channel to provide authority for NCOs to do their job.

Delegation of Authority

A-14. Just as Congress and the president cannot participate in every aspect of armed forces operations, most leaders cannot handle every action directly. To meet the organization's goals, officers delegate authority to NCOs and, when appropriate, to DA civilians. These leaders, in turn, may further delegate that authority.

A-15. Unless restricted by law, regulation, or a superior, leaders may delegate any or all of their authority to their subordinate leaders. However, such delegation must fall within the leader's scope of authority. Leaders cannot delegate authority they do not have and subordinate leaders may not assume authority that their superiors do not have, cannot delegate, or have retained. The task or duty to be performed limits the authority of the leader to which it is assigned.

A-16. When a leader is assigned a task or duty, the authority necessary to accomplish it accompanies the assignment. When a leader delegates a task or duty to a subordinate, he delegates the requisite authority as well. However, leaders always retain responsibility for the outcome of any tasks they assign. They must answer for any actions or omissions related to them.

RESPONSIBILITY AND ACCOUNTABILITY

A-17. No definitive lines separate officer, NCO, and DA civilian responsibilities. Officers, NCOs, and DA civilians lead other officers, NCOs, and DA civilians and help them carry out their responsibilities. Commanders set overall policies and standards, but all leaders must provide the guidance, resources, assistance, and supervision necessary for subordinates to perform their duties. Similarly, subordinates must assist and advise their leaders. Mission accomplishment demands that officers, NCOs, and DA civilians work together to advise, assist, and learn from each other. Responsibilities fall into two categories: command and individual.

Command Responsibility

A-18. Command responsibility refers to collective or organizational accountability and includes how well units perform their missions.

For example, a company commander is responsible for all the tasks and missions assigned to his company; his leaders hold him accountable for completing them. Military and DA civilian leaders have responsibility for what their sections, units, or organizations do or fail to do.

Individual Responsibility

A-19. All soldiers and DA civilians must account for their personal conduct. Commissioned officers, warrant officers, and DA civilians assume personal responsibility when they take their oath. DA civilians take the same oath as commissioned officers. Soldiers take their initial oath of enlistment. Members of the Army account for their actions to their fellow soldiers or coworkers, the appointed leader, their unit or organization, the Army, and the American people.

COMMUNICATIONS AND THE CHAIN OF COMMAND

A-20. Communication among individuals, teams, units, and organizations is essential to efficient and effective mission accomplishment. As Chapter 4 discusses, two-way communication is more effective than one-way communication. Mission accomplishment depends on information passing accurately to and from subordinates and leaders, up and down the chain of command and NCO support channel, and laterally among adjacent organizations or activities. In garrison operations, organizations working on the same mission or project should be considered "adjacent."

A-21. The Army has only one chain of command. Through this chain of command, leaders issue orders and instructions and convey policies. A healthy chain of command is a two-way communications channel. Its members do more than transmit orders; they carry information from within the unit or organization back up to its leader. They furnish information about how things are developing, notify the leader of problems, and provide requests for clarification and help. Leaders at all levels use the chain of command—their subordinate leaders—to keep their people informed and render assistance. They continually facilitate the process of gaining the necessary clarification and solving problems.

A-22. Beyond conducting their normal duties, NCOs train soldiers and advise commanders on individual soldier readiness and the training needed to ensure unit readiness. Officers and DA civilian leaders should consult their command sergeant major, first sergeant, or NCO assistant, before implementing policy. Commanders, commissioned and warrant officers, DA civilian leaders, and NCOs must continually communicate to avoid duplicating instructions or issuing conflicting orders. Continuous and open lines of communication enable commanders and DA civilian leaders to freely plan, make decisions, and program future training and operations.

THE NONCOMMISSIONED OFFICER SUPPORT CHANNEL

A-23. The NCO support channel parallels and reinforces the chain of command. NCO leaders work with and support the commissioned and warrant officers of their chain of command. For the chain of command to work efficiently, the NCO support channel must operate effectively. At battalion level and higher, the NCO support channel begins with the command sergeant major, extends through first sergeants and platoon sergeants, and ends with section chiefs, squad leaders, or team leaders. (TC 22-6 discusses the NCO support channel.)

A-24. The connection between the chain of command and NCO support channel is the senior NCO. Commanders issue orders through the chain of command, but senior NCOs must know and understand the orders to issue effective implementing instructions through the NCO support channel. Although the first sergeant and command sergeant major are not part of the formal chain of command, leaders should consult them on all individual soldier matters.

A-25. Successful leaders have a good relationship with their senior NCOs. Successful commanders have a good leader-NCO relationship with their first sergeants and command sergeants major. The need for such a relationship applies to platoon leaders and platoon sergeants as well as to staff officers and NCOs. Senior NCOs have extensive experience in successfully completing missions and dealing with enlisted soldier issues. Also, senior NCOs can monitor organizational activities at all levels, take corrective action to keep the organization within the boundaries of the commander's intent, or report situations that require the attention of the officer leadership. A positive relationship between officers and NCOs creates conditions for success.

DA CIVILIAN SUPPORT

A-26. The Army employs DA civilians because they possess or develop technical skills that are necessary to accomplish some missions. The specialized skills of DA civilians are essential to victory but, for a variety of reasons, they are difficult to maintain in the uniformed components. The Army expects DA civilian leaders to be more than specialists: they are expected to apply technical, conceptual, and interpersonal skills together to accomplish missions—in a combat theater, if necessary.

A-27. While the command sergeant major is the advocate in a unit for soldier issues, DA civilians have no single advocate. Rather, their own leaders, civilian personnel advisory center, or civilian personnel operations center represent them and their issues to the chain of command. Often the senior DA civilian in an organization or the senior DA civilian in a particular career field has the additional duty of advising and counseling junior DA civilians on job-related issues and career development.

Appendix B

Performance Indicators

B-1. Appendix B is organized around the leadership dimensions that Chapters 1 through 7 discuss and that Figure B-1 shows. This appendix lists indicators for you to use to assess the leadership of yourself and others based on these leadership dimensions. Use it as an assessment and counseling tool, not as a source of phrases for evaluation reports. When you prepare an evaluation, make comments that apply specifically to the individual you are evaluating. Do not limit yourself to the general indicators listed here. Be specific; be precise; be objective; be fair.

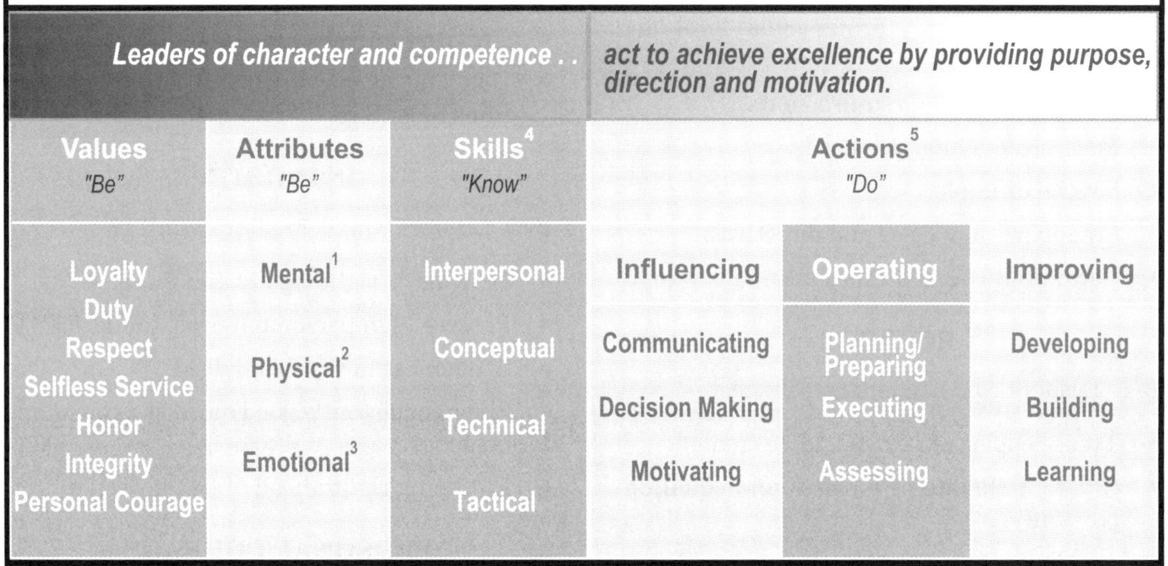

Leaders of character and competence . .			act to achieve excellence by providing purpose, direction and motivation.		
Values "Be"	Attributes "Be"	Skills[4] "Know"	Actions[5] "Do"		
Loyalty Duty Respect Selfless Service Honor Integrity Personal Courage	Mental[1] Physical[2] Emotional[3]	Interpersonal Conceptual Technical Tactical	Influencing Communicating Decision Making Motivating	Operating Planning/ Preparing Executing Assessing	Improving Developing Building Learning

[1] The mental attributes of an Army leader are will, self-discipline, initiative, judgment, self-confidence, intelligence, and cultural awareness.

[2] The physical attributes of an Army leader are health fitness, physical fitness, and military and professional bearing.

[3] The emotional attributes of an Army leader are self-control, balance, and stability.

[4]The interpersonal, conceptual, technical, and tactical skills are different for direct, organizational, and strategic leaders.

[5]The influencing, operating, and improving actions are different for direct, organizational, and strategic leaders.

Figure B-1. Leadership Dimensions

VALUES

LOYALTY

B-2. Leaders who demonstrate loyalty—

- Bear true faith and allegiance in the correct order to the Constitution, the Army, and the organization.
- Observe higher headquarters' priorities.
- Work within the system without manipulating it for personal gain.

DUTY

B-3. Leaders who demonstrate devotion to duty—

- Fulfill obligations—professional, legal, and moral.
- Carry out mission requirements.
- Meet professional standards.
- Set the example.
- Comply with policies and directives.
- Continually pursue excellence.

RESPECT

B-4. Leaders who demonstrate respect—

- Treat people as they should be treated.
- Create a climate of fairness and equal opportunity.
- Are discreet and tactful when correcting or questioning others.
- Show concern for and make an effort to check on the safety and well-being of others.
- Are courteous.
- Don't take advantage of positions of authority.

SELFLESS SERVICE

B-5. Leaders who demonstrate selfless service—

- Put the welfare of the nation, the Army, and subordinates before their own.
- Sustain team morale.
- Share subordinates' hardships.
- Give credit for success to others and accept responsibility for failure themselves.

HONOR

B-6. Leaders who demonstrate honor—

- Live up to Army values.
- Don't lie, cheat, steal, or tolerate those actions by others.

INTEGRITY

B-7. Leaders who demonstrate integrity—

- Do what is right legally and morally.
- Possess high personal moral standards.
- Are honest in word and deed.
- Show consistently good moral judgment and behavior.
- Put being right ahead of being popular.

PERSONAL COURAGE

B-8. Leaders who demonstrate personal courage—

- Show physical and moral bravery.
- Take responsibility for decisions and actions.
- Accept responsibility for mistakes and shortcomings.

ATTRIBUTES

MENTAL ATTRIBUTES

B-9. Leaders who demonstrate desirable mental attributes—

- Possess and display will, self-discipline, initiative, judgment, self-confidence, intelligence, common sense, and cultural awareness.
- Think and act quickly and logically, even when there are no clear instructions or the plan falls apart.
- Analyze situations.
- Combine complex ideas to generate feasible courses of action.
- Balance resolve and flexibility.
- Show a desire to succeed; do not quit in the face of adversity.
- Do their fair share.
- Balance competing demands.
- Embrace and use the talents of all members to build team cohesion.

PHYSICAL ATTRIBUTES

B-10. Leaders who demonstrate desirable physical attributes—

- Maintain an appropriate level of physical fitness and military bearing.
- Present a neat and professional appearance.

- Meet established norms of personal hygiene, grooming, and cleanliness.
- Maintain Army height and weight standards (not applicable to DA civilians).
- Render appropriate military and civilian courtesies.
- Demonstrate nonverbal expressions and gestures appropriate to the situation.
- Are personally energetic.
- Cope with hardship.
- Complete physically demanding endeavors.
- Continue to function under adverse conditions.
- Lead by example in performance, fitness, and appearance.

EMOTIONAL ATTRIBUTES

B-11. Leaders who demonstrate appropriate emotional attributes—

- Show self-confidence.
- Remain calm during conditions of stress, chaos, and rapid change.
- Exercise self-control, balance, and stability.
- Maintain a positive attitude.
- Demonstrate mature, responsible behavior that inspires trust and earns respect.

SKILLS

INTERPERSONAL SKILLS

B-12. Leaders who demonstrate interpersonal skills—

- Coach, teach, counsel, motivate, and empower subordinates.
- Readily interact with others.
- Earn trust and respect.
- Actively contribute to problem solving and decision making.
- Are sought out by peers for expertise and counsel

CONCEPTUAL SKILLS

B-13. Leaders who demonstrate conceptual skills—

- Reason critically and ethically.
- Think creatively.
- Anticipate requirements and contingencies.
- Improvise within the commander's intent.
- Use appropriate reference materials.
- Pay attention to details.

TECHNICAL SKILLS

B-14. Leaders who demonstrate technical skills—

- Possess or develop the expertise necessary to accomplish all assigned tasks and functions.
- Know standards for task accomplishment.
- Know the small unit tactics, techniques, and procedures that support the organization's mission.
- Know the drills that support the organization's mission.
- Prepare clear, concise operation orders.
- Understand how to apply the factors of mission, enemy, terrain and weather, troops, time available, and civil considerations (METT-TC) to mission analysis.
- Master basic soldier skills.
- Know how to use and maintain equipment.

- Know how and what to inspect or check.
- Use technology, especially information technology, to enhance communication.

TACTICAL SKILLS

B-15. Leaders who demonstrate tactical skills—

- Know how to apply warfighting doctrine within the commander's intent.
- Apply their professional knowledge, judgment, and warfighting skill at the appropriate leadership level.
- Combine and apply skill with people, ideas, and things to accomplish short-term missions.
- Apply skill with people, ideas, and things to train for, plan, prepare, execute and assess offensive, defensive, stability, and support actions.

ACTIONS

INFLUENCING

B-16. Leaders who influence—

- Use appropriate methods to reach goals while operating and improving.
- Motivate subordinates to accomplish assigned tasks and missions.
- Set the example by demonstrating enthusiasm for—and, if necessary, methods of—accomplishing assigned tasks.
- Make themselves available to assist peers and subordinates.
- Share information with subordinates.
- Encourage subordinates and peers to express candid opinions.
- Actively listen to feedback and act appropriately based on it.
- Mediate peer conflicts and disagreements.
- Tactfully confront and correct others when necessary.
- Earn respect and obtain willing cooperation of peers, subordinates, and superiors.
- Challenge others to match their example.

- Take care of subordinates and their families, providing for their health, welfare, morale, and training.
- Are persuasive in peer discussions and prudently rally peer pressure against peers when required.
- Provide a team vision for the future.
- Shape the organizational climate by setting, sustaining, and ensuring a values-based environment.

Communicating

B-17. Leaders who communicate effectively—

- Display good oral, written, and listening skills.
- Persuade others.
- Express thoughts and ideas clearly to individuals and groups.

B-18. **Oral Communication.** Leaders who effectively communicate orally—

- Speak clearly and concisely.
- Speak enthusiastically and maintain listeners' interest and involvement.

- Make appropriate eye contact when speaking.
- Use gestures that are appropriate but not distracting.
- Convey ideas, feelings, sincerity, and conviction.
- Express well-thought-out and well-organized ideas.
- Use grammatically and doctrinally correct terms and phrases.
- Use appropriate visual aids.
- Act to determine, recognize and resolve misunderstandings.
- Listen and watch attentively; make appropriate notes; convey the essence of what was said or done to others.
- React appropriately to verbal and nonverbal feedback.
- Keep conversations on track.

B-19. **Written Communication.** Leaders who effectively communicate in writing—

- Are understood in a single rapid reading by the intended audience.
- Use correct grammar, spelling, and punctuation.
- Have legible handwriting.
- Put the "bottom line up front."
- Use the active voice.
- Use an appropriate format, a clear organization, and a reasonably simple style.
- Use only essential acronyms and spell out those used.
- Stay on topic.
- Correctly use facts and data.

(DA Pam 600-67 discusses techniques for writing effectively.)

Decision Making

B-20. Leaders who make effective, timely decisions—

- Employ sound judgment and logical reasoning.
- Gather and analyze relevant information about changing situations to recognize and define emerging problems.

- Make logical assumptions in the absence of facts.
- Uncover critical issues to use as a guide in both making decisions and taking advantage of opportunities.
- Keep informed about developments and policy changes inside and outside the organization.
- Recognize and generate innovative solutions.
- Develop alternative courses of action and choose the best course of action based on analysis of their relative costs and benefits.
- Anticipate needs for action.
- Relate and compare information from different sources to identify possible cause-and-effect relationships.
- Consider the impact and implications of decisions on others and on situations.
- Involve others in decisions and keep them informed of consequences that affect them.
- Take charge when in charge.
- Define intent.
- Consider contingencies and their consequences.
- Remain decisive after discovering a mistake.
- Act in the absence of guidance.
- Improvise within commander's intent; handle a fluid environment.

Motivating

B-21. Leaders who effectively motivate—

- Inspire, encourage, and guide others toward mission accomplishment.
- Don't show discouragement when facing setbacks.
- Attempt to satisfy subordinates' needs.
- Give subordinates the reason for tasks.
- Provide accurate, timely, and (where appropriate) positive feedback.
- Actively listen for feedback from subordinates.
- Use feedback to modify duties, tasks, requirements, and goals when appropriate.

- Recognize individual and team accomplishments and reward them appropriately.
- Recognize poor performance and address it appropriately.
- Justly apply disciplinary measures.
- Keep subordinates informed.
- Clearly articulate expectations.
- Consider duty positions, capabilities, and developmental needs when assigning tasks.
- Provide early warning to subordinate leaders of tasks they will be responsible for.
- Define requirements by issuing clear and concise orders or guidance.
- Allocate as much time as possible for task completion.
- Accept responsibility for organizational performance. Credit subordinates for good performance. Take responsibility for and correct poor performance.

OPERATING

B-22. Leaders who effectively operate—

- Accomplish short-term missions.
- Demonstrate tactical and technical competency appropriate to their rank and position.
- Complete individual and unit tasks to standard, on time, and within the commander's intent.

Planning and Preparing

B-23. Leaders who effectively plan—

- Develop feasible and acceptable plans for themselves and others that accomplish the mission while expending minimum resources and posturing the organization for future missions.
- Use forward planning to ensure each course of action achieves the desired outcome.
- Use reverse planning to ensure that all tasks can be executed in the time available and that tasks depending on other tasks are executed in the correct sequence.
- Determine specified and implied tasks and restate the higher headquarters' mission in terms appropriate to the organization.

- Incorporate adequate controls such as time phasing; ensure others understand when actions should begin or end.
- Adhere to the "1/3–2/3 Rule"; give subordinates time to plan.
- Allocate time to prepare and conduct rehearsals.
- Ensure all courses of action accomplish the mission within the commander's intent.
- Allocate available resources to competing demands by setting task priorities based on the relative importance of each task.
- Address likely contingencies.
- Remain flexible.
- Consider SOPs, the factors of METT-TC, and the military aspects of terrain (OCOKA).
- Coordinate plans with higher, lower, adjacent, and affected organizations.
- Personally arrive on time and meet deadlines; require subordinates and their organizations to accomplish tasks on time.
- Delegate all tasks except those they are required to do personally.
- Schedule activities so the organization meets all commitments in critical performance areas.
- Recognize and resolve scheduling conflicts.
- Notify peers and subordinates as far in advance as possible when their support is required.
- Use some form of a personal planning calendar to organize requirements.

Executing

B-24. Leaders who effectively execute—

- Use technical and tactical skills to meet mission standards, take care of people, and accomplish the mission with available resources.
- Perform individual and collective tasks to standard.
- Execute plans, adjusting when necessary, to accomplish the mission.
- Encourage initiative.
- Keep higher and lower headquarters, superiors, and subordinates informed.

- Keep track of people and equipment.
- Make necessary on-the-spot corrections.
- Adapt to and handle fluid environments.
- Fight through obstacles, difficulties, and hardships to accomplish the mission.
- Keep track of task assignments and suspenses; adjust assignments, if necessary; follow up.

Assessing

B-25. Leaders who effectively assess—

- Use assessment techniques and evaluation tools (especially AARs) to identify lessons learned and facilitate consistent improvement.
- Establish and employ procedures for monitoring, coordinating, and regulating subordinates' actions and activities.
- Conduct initial assessments when beginning a new task or assuming a new position.
- Conduct IPRs.
- Analyze activities to determine how desired end states are achieved or affected.
- Seek sustainment in areas when the organization meets the standard.
- Observe and assess actions in progress without oversupervising.
- Judge results based on standards.
- Sort out important actual and potential problems.
- Conduct and facilitate AARs; identify lessons.
- Determine causes, effects, and contributing factors for problems.
- Analyze activities to determine how desired end states can be achieved ethically.

IMPROVING

B-26. Leaders who effectively improve the organization—

- Sustain skills and actions that benefit themselves and each of their people for the future.
- Sustain and renew the organization for the future by managing change and exploiting individual and institutional learning capabilities.

- Create and sustain an environment where all leaders, subordinates, and organizations can reach their full potential.

Developing

B-27. Leaders who effectively develop—

- Strive to improve themselves, subordinates, and the organization.
- Mentor by investing adequate time and effort in counseling, coaching, and teaching their individual subordinates and subordinate leaders.
- Set the example by displaying high standards of duty performance, personal appearance, military and professional bearing, and ethics.
- Create a climate that expects good performance, recognizes superior performance, and doesn't accept poor performance.
- Design tasks to provide practice in areas of subordinate leaders' weaknesses.
- Clearly articulate tasks and expectations and set realistic standards.
- Guide subordinate leaders in thinking through problems for themselves.
- Anticipate mistakes and freely offer assistance without being overbearing.
- Observe, assess, counsel, coach, and evaluate subordinate leaders.
- Motivate subordinates to develop themselves.
- Arrange training opportunities that help subordinates achieve insight, self-awareness, self-esteem, and effectiveness.
- Balance the organization's tasks, goals, and objectives with subordinates' personal and professional needs.
- Develop subordinate leaders who demonstrate respect for natural resources and the environment.
- Act to expand and enhance subordinates' competence and self-confidence.
- Encourage initiative.
- Create and contribute to a positive organizational climate.
- Build on successes.
- Improve weaknesses.

Building

B-28. Leaders who effectively build—

- Spend time and resources improving the organization.
- Foster a healthy ethical climate.
- Act to improve the organization's collective performance.
- Comply with and support organizational goals.
- Encourage people to work effectively with each other.
- Promote teamwork and team achievement.
- Are examples of team players.
- Offer suggestions, but properly execute decisions of the chain of command and NCO support channel—even unpopular ones—as if they were their own.
- Accept and act on assigned tasks.
- Volunteer in useful ways.
- Remain positive when the situation becomes confused or changes.
- Use the chain of command and NCO support channel to solve problems.
- Support equal opportunity.
- Prevent sexual harassment.
- Participate in organizational activities and functions.
- Participate in team tasks and missions without being requested to do so.
- Establish an organizational climate that demonstrates respect for the environment and stewards natural resources.

Learning

B-29. Leaders who effectively learn—

- Seek self-improvement in weak areas.
- Encourage organizational growth.
- Envision, adapt, and lead change.
- Act to expand and enhance personal and organizational knowledge and capabilities.
- Apply lessons learned.
- Ask incisive questions.
- Envision ways to improve.
- Design ways to practice.
- Endeavor to broaden their understanding.
- Transform experience into knowledge and use it to improve future performance.
- Make knowledge accessible to the entire organization.
- Exhibit reasonable self-awareness.
- Take time off to grow and recreate.
- Embrace and manage change; adopt a future orientation.
- Use experience to improve themselves and the organization.

ESSENTIAL ARMY MANUAL
(427) ARMY LEADERSHIP

ESSENTIAL ARMY MANUAL
(429) ARMY LEADERSHIP

Developmental Counseling

C-1. Subordinate leadership development is one of the most important responsibilities of every Army leader. Developing the leaders who will come after you should be one of your highest priorities. Your legacy and the Army's future rests on the shoulders of those you prepare for greater responsibility.

C-2. Leadership development reviews are a means to focus the growing of tomorrow's leaders. Think of them as AARs with a focus of making leaders more effective every day. These important reviews are not necessarily limited to internal counseling sessions; leadership feedback mechanisms also apply in operational settings such as the CTCs.

C-3. Just as training includes AARs and training strategies to fix shortcomings, leadership development includes performance reviews. These reviews result in agreements between leader and subordinate on a development strategy or plan of action that builds on the subordinate's strengths and establishes goals to improve on weaknesses. Leaders conduct performance reviews and create plans of action during developmental counseling.

C-4. Leadership development reviews are a component of the broader concept of developmental counseling. Developmental counseling is subordinate-centered communication that produces a plan outlining actions that subordinates must take to achieve individual and organizational goals. During developmental counseling, subordinates are not merely passive listeners; they're actively involved in the process. The Developmental Counseling Form (DA Form 4856-E, which is discussed at the end of this appendix) provides a useful framework to prepare for almost any type of counseling. Use it to help you mentally organize issues and isolate important, relevant items to cover during counseling sessions.

C-5. Developmental counseling is a shared effort. As a leader, you assist your subordinates in identifying strengths and weaknesses and creating plans of action. Then you support them throughout the plan implementation and assessment. However, to achieve success, your subordinates must be forthright in their commitment to improve and candid in their own assessment and goal setting.

THE LEADER'S RESPONSIBILITIES

C-6. Organizational readiness and mission accomplishment depend on every member's ability to perform to established standards. Supervisors must mentor their subordinates through teaching, coaching, and counseling. Leaders coach subordinates the same way sports coaches improve their teams: by identifying weaknesses, setting goals, developing and implementing plans of action, and providing oversight and motivation throughout the process. To be effective coaches, leaders must thoroughly understand the strengths, weaknesses, and professional goals of their subordinates. (Chapter 5 discusses coaching.)

C-7. Army leaders evaluate DA civilians using procedures prescribed under the Total Army Performance Evaluation System (TAPES). Although TAPES doesn't address developmental counseling, you can use DA Form 4856-E to counsel DA civilians concerning professional growth and career goals. DA Form 4856-E is not appropriate for documenting counseling concerning DA civilian misconduct or poor performance. The servicing civilian personnel office can provide guidance for such situations.

C-8. Soldiers and DA civilians often perceive counseling as an adverse action. Effective leaders who counsel properly can change that perception. Army leaders conduct counseling to help subordinates become better members of the team, maintain or improve performance, and prepare for the future. Just as no easy answers exist for exactly what to do in all leadership situations, no easy answers exist for exactly what to do in all counseling situations. However, to conduct effective counseling, you should develop a counseling style with the characteristics listed in Figure C-1.

* **Purpose**: Clearly define the purpose of the counseling.

* **Flexibility**: Fit the counseling style to the character of each subordinate and to the relationship desired.

* **Respect**: View subordinates as unique, complex individuals, each with a distinct set of values, beliefs, and attitudes.

* **Communication**: Establish open, two-way communication with subordinates using spoken language, nonverbal actions, gestures, and body language. Effective counselors listen more than they speak.

* **Support**: Encourage subordinates through actions while guiding them through their problems.

Figure C-1. Characteristics of Effective Counseling

THE LEADER AS A COUNSELOR

C-9. Army leaders must demonstrate certain qualities to be effective counselors. These qualities include respect for subordinates, self-awareness and cultural awareness, empathy, and credibility.

RESPECT FOR SUBORDINATES

C-10. As an Army leader, you show respect for subordinates when you allow them to take responsibility for their own ideas and actions. Respecting subordinates helps create mutual respect in the leader-subordinate relationship. Mutual respect improves the chances of changing (or maintaining) behavior and achieving goals.

SELF AWARENESS AND CULTURAL AWARENESS

C-11. As an Army leader, you must be fully aware of your own values, needs, and biases prior to counseling subordinates. Self-aware leaders are less likely to project their biases onto subordinates. Also, aware leaders are more likely to act consistently with their own values and actions.

C-12. Cultural awareness, as discussed in Chapter 2, is a mental attribute. As an Army leader, you need to be aware of the similarities and differences between individuals of different cultural backgrounds and how these factors may influence values, perspectives, and actions. Don't let unfamiliarity with cultural backgrounds hinder you in addressing cultural issues, especially if they generate concerns within the organization or hinder team-building. Cultural awareness enhances your ability to display empathy.

EMPATHY

C-13. Empathy is the action of being understanding of and sensitive to the feelings, thoughts, and experiences of another person to the point that you can almost feel or experience them yourself. Leaders with empathy can put themselves in their subordinate's shoes; they can see a situation from the other person's perspective. By understanding the subordinate's position, you can help a subordinate develop a plan of action that fits the subordinate's personality and needs, one that works for the subordinate. If you don't fully comprehend a situation from your subordinate's point of view, you have less credibility and influence and your

subordinate is less likely to commit to the agreed upon plan of action.

CREDIBILITY

C-14. Leaders achieve credibility by being honest and consistent in their statements and actions. To be credible, use a straightforward style with your subordinates. Behave in a manner that your subordinates respect and trust. You can earn credibility by repeatedly demonstrating your willingness to assist a subordinate and being consistent in what you say and do. If you lack credibility with your subordinates you'll find it difficult to influence them.

LEADER COUNSELING SKILLS

C-15. One challenging aspect of counseling is selecting the proper approach to a specific situation. To counsel effectively, the technique you use must fit the situation, your capabilities, and your subordinate's expectations. In some cases, you may only need to give information or listen. A subordinate's improvement may call for just a brief word of praise. Other situations may require structured counseling followed by definite actions.

C-16. All leaders should seek to develop and improve their own counseling abilities. You can improve your counseling techniques by studying human behavior, learning the kinds of problems that affect your subordinates, and developing your interpersonal skills. The techniques needed to provide effective counseling will vary from person to person and session to session. However, general skills that you'll need in almost every situation include active listening, responding, and questioning.

ACTIVE LISTENING

C-17. During counseling, you must actively listen to your subordinate. When you're actively listening, you communicate verbally and nonverbally that you've received the subordinate's message. To fully understand a subordinate's message, you must listen to the words and observe the subordinate's manners. Elements of active listening you should consider include—

- **Eye contact**. Maintaining eye contact without staring helps show sincere interest. Occasional breaks of contact are normal and acceptable. Subordinates may perceive excessive breaks of eye contact, paper shuffling, and clock-watching as a lack of interest or concern. These are guidelines only. Based on cultural background, participants in a particular counseling session may have different ideas about what proper eye contact is.
- **Body posture**. Being relaxed and comfortable will help put the subordinate at ease. However, a too-relaxed position or slouching may be interpreted as a lack of interest.
- **Head nods**. Occasionally nodding your head shows you're paying attention and encourages the subordinate to continue.
- **Facial expressions**. Keep your facial expressions natural and relaxed. A blank look or fixed expression may disturb the subordinate. Smiling too much or frowning may discourage the subordinate from continuing.
- **Verbal expressions**. Refrain from talking too much and avoid interrupting. Let the subordinate do the talking while keeping the discussion on the counseling subject. Speaking only when necessary reinforces the importance of what the subordinate is saying and encourages the subordinate to

continue. Silence can also do this, but be careful. Occasional silence may indicate to the subordinate that it's okay to continue talking, but a long silence can sometimes be distracting and make the subordinate feel uncomfortable.

C-18. Active listening also means listening thoughtfully and deliberately to the way a subordinate says things. Stay alert for common themes. A subordinate's opening and closing statements as well as recurring references may indicate the subordinate's priorities. Inconsistencies and gaps may indicate a subordinate's avoidance of the real issue. This confusion and uncertainty may suggest additional questions.

C-19. While listening, pay attention to the subordinate's gestures. These actions complete the total message. By watching the subordinate's actions, you can "see" the feelings behind the words. Not all actions are proof of a subordinate's feelings, but they should be taken into consideration. Note differences between what the subordinate says and does. Nonverbal indicators of a subordinate's attitude include—

- **Boredom.** Drumming on the table, doodling, clicking a ball-point pen, or resting the head in the palm of the hand.
- **Self-confidence.** Standing tall, leaning back with hands behind the head, and maintaining steady eye contact.
- **Defensiveness.** Pushing deeply into a chair, glaring at the leader, and making sarcastic comments as well as crossing or folding arms in front of the chest.
- **Frustration.** Rubbing eyes, pulling on an ear, taking short breaths, wringing the hands, or frequently changing total body position.
- **Interest, friendliness, and openness.** Moving toward the leader while sitting.
- **Openness or anxiety.** Sitting on the edge of the chair with arms uncrossed and hands open.

C-20. Consider these indicators carefully. Although each indicator may show something about the subordinate, don't assume a particular behavior absolutely means something. Ask the subordinate about the indicator so you can better understand the behavior and allow the subordinate to take responsibility for it.

RESPONDING

C-21. Responding skills follow-up on active listening skills. A leader responds to communicate that the leader understands the subordinate. From time to time, check your understanding: clarify and confirm what has been said. Respond to subordinates both verbally and nonverbally. Verbal responses consist of summarizing, interpreting, and clarifying the subordinate's message. Nonverbal responses include eye contact and occasional gestures such as a head nod.

QUESTIONING

C-22. Although questioning is a necessary skill, you must use it with caution. Too many questions can aggravate the power differential between a leader and a subordinate and place the subordinate in a passive mode. The subordinate may also react to excessive questioning as an intrusion of privacy and become defensive. During a leadership development review, ask questions to obtain information or to get the subordinate to think about a particular situation. Generally, the questions should be open-ended so as to evoke more than a yes or no answer. Well-posed questions may help to verify understanding, encourage further explanation, or help the subordinate move through the stages of the counseling session.

COUNSELING ERRORS

C-23. Effective leaders avoid common counseling mistakes. Dominating the counseling by talking too much, giving unnecessary or inappropriate "advice," not truly listening, and projecting personal likes, dislikes, biases, and prejudices all interfere with effective counseling. You should also avoid other common mistakes such as rash judgments, stereotypes, loss of emotional control, inflexible methods of counseling and improper follow-up. To improve your counseling skills, follow the guidelines in Figure C-2.

ESSENTIAL ARMY MANUAL
(433) ARMY LEADERSHIP

- Determine the subordinate's role in the situation and what the subordinate has done to resolve the problem or improve performance.
- Draw conclusions based on more than the subordinate's statement.
- Try to understand what the subordinate says and feels; listen to what the subordinate says and how the subordinate says it.
- Show empathy when discussing the problem.
- When asking questions, be sure that you need the information.
- Keep the conversation open-ended; avoid interrupting.
- Give the subordinate your full attention.
- Be receptive to the subordinate's feelings without feeling responsible to save the subordinate from hurting.
- Encourage the subordinate to take the initiative and to say what the subordinate wants to say.
- Avoid interrogating.
- Keep your personal experiences out of the counseling session unless you believe your experiences will really help.
- Listen more; talk less.
- Remain objective.
- Avoid confirming a subordinate's prejudices.
- Help the subordinate help himself.
- Know what information to keep confidential and what to present to the chain of command.

Figure C-2. Guidelines to Improve Counseling

THE LEADER'S LIMITATIONS

C-24. Army leaders can't help everyone in every situation. Even professional counselors can't provide all the help that a person might need. You must recognize your limitations and, when the situation calls for it, refer a subordinate to a person or agency more qualified to help.

C-25. These agencies Figure C-3 lists can help you and your people resolve problems. Although it's generally in an individual's best interest to seek help first from their first-line leaders, leaders must always respect an individual's right to contact most of these agencies on their own.

Activity	Description
Adjutant General	Provides personnel and administrative services support such as orders, ID cards, retirement assistance, deferments, and in- and out-processing.
American Red Cross	Provides communications support between soldiers and families and assistance during or after emergency or compassionate situations.
Army Community Service	Assists military families through their information and referral services, budget and indebtedness counseling, household item loan closet, information on other military posts, and welcome packets for new arrivals.
Army Substance Abuse Program	Provides alcohol and drug abuse prevention and control programs for DA civilians.
Better Opportunities for Single Soldiers (BOSS)	Serves as a liaison between upper levels of command on the installation and single soldiers.
Army Education Center	Provides services for continuing education and individual learning services support.
Army Emergency Relief	Provides financial assistance and personal budget counseling; coordinates student loans through Army Emergency Relief education loan programs.
Career Counselor	Explains reenlistment options and provides current information on prerequisites for reenlistment and selective reenlistment bonuses.
Chaplain	Provides spiritual and humanitarian counseling to soldiers and DA civilians.
Claims Section, SJA	Handles claims for and against the government, most often those for the loss and damage of household goods.
Legal Assistance Office	Provides legal information or assistance on matters of contracts, citizenship, adoption, marital problems, taxes, wills, and powers of attorney.
Community Counseling Center	Provides alcohol and drug abuse prevention and control programs for soldiers.
Community Health Nurse	Provides preventive health care services.
Community Mental Health Service	Provides assistance and counseling for mental health problems.
Employee Assistance Program	Provides health nurse, mental health service, and social work services for DA civilians.
Equal Opportunity Staff Office and Equal Employment Opportunity Office	Provides assistance for matters involving discrimination in race, color, national origin, gender, and religion. Provides, information on procedures for initiating complaints and resolving complaints informally.
Family Advocacy Officer	Coordinates programs supporting children and families including abuse and neglect investigation, counseling, and educational programs.
Finance and Accounting Office	Handles inquiries for pay, allowances, and allotments.
Housing Referral Office	Provides assistance with housing on and off post.
Inspector General	Renders assistance to soldiers and DA civilians. Corrects injustices affecting individuals and eliminates conditions determined to be detrimental to the efficiency, economy, morale, and reputation of the Army. Investigates matters involving fraud, waste, and abuse.
Social Work Office	Provides services dealing with social problems to include crisis intervention, family therapy, marital counseling, and parent or child management assistance.
Transition Office	Provides assistance and information on separation from the Army.

Figure C-3. Support Activities

TYPES OF DEVELOPMENTAL COUNSELING

C-26. You can often categorize developmental counseling based on the topic of the session. The two major categories of counseling are event-oriented and performance/professional growth.

EVENT-ORIENTED COUNSELING

C-27. Event-oriented counseling involves a specific event or situation. It may precede events, such as going to a promotion board or attending a school; or it may follow events, such as a noteworthy duty performance, a problem with performance or mission accomplishment, or a personal problem. Examples of event-oriented counseling include, but are not limited to—

- Specific instances of superior or substandard performance.
- Reception and integration counseling.
- Crisis counseling.
- Referral counseling.
- Promotion counseling.
- Separation counseling.

Counseling for Specific Instances

C-28. Sometimes counseling is tied to specific instances of superior or substandard duty performance. You tell your subordinate whether or not the performance met the standard and what the subordinate did right or wrong. The key to successful counseling for specific performance is to conduct it as close to the event as possible.

C-29. Many leaders focus counseling for specific instances on poor performance and miss, or at least fail to acknowledge, excellent performance. You should counsel subordinates for specific examples of superior as well as substandard duty performance. To measure your own performance and counseling emphasis, you can note how often you document counseling for superior versus substandard performance.

C-30. You should counsel subordinates who don't meet the standard. If the subordinate's performance is unsatisfactory because of a lack of knowledge or ability, you and the subordinate should develop a plan to improve the subordinate's skills. Corrective training may be required at times to ensure the subordinate knows and achieves the standard. Once the subordinate can achieve the standard, you should end the corrective training.

C-31. When counseling a subordinate for a specific performance, take the following actions:

- Tell the subordinate the purpose of the counseling, what was expected, and how the subordinate failed to meet the standard.
- Address the specific unacceptable behavior or action, not the person's character.
- Tell the subordinate the effect of the behavior, action, or performance on the rest of the organization.
- Actively listen to the subordinate's response.
- Remain unemotional.
- Teach the subordinate how to meet the standard.
- Be prepared to do some personal counseling, since a failure to meet the standard may be related to or the result of an unresolved personal problem.
- Explain to the subordinate what will be done to improve performance (plan of action). Identify your responsibilities in implementing the plan of action; continue to assess and follow up on the subordinate's progress. Adjust plan of action as necessary.

Reception and Integration Counseling

C-32. As the leader, you must counsel new team members when they arrive at your organization. This reception and integration counseling serves two purposes. First, it identifies and helps fix any problems or concerns that new members may have, especially any issues resulting from the new duty assignment. Second, it lets them know the organizational standards and how they fit into the team. It clarifies job titles and sends the message that the chain of command cares. Reception and integration counseling should begin immediately upon arrival so new team members can quickly become integrated into the organization. (Figure C-4 gives some possible discussion points.)

- Organizational standards.
- Chain of command.
- NCO support channel (who and how used).
- On-and-off duty conduct.
- Personnel/personal affairs/initial clothing issue.
- Organizational history, organization, and mission.
- Soldier programs within the organization, such as soldier of the month/quarter/year and Audie Murphy.
- Off limits and danger areas.
- Functions and locations of support activities (see Figure C-3).
- On- and off-post recreational, educational, cultural, and historical opportunities.
- Foreign nation or host nation orientation.
- Other areas the individual should be aware of, as determined by the leader.

Figure C-4. Reception and Integration Counseling Points

Crisis Counseling

C-33. You may conduct crisis counseling to get a subordinate through the initial shock after receiving negative news, such as notification of the death of a loved one. You may assist the subordinate by listening and, as appropriate, providing assistance. Assistance may include referring the subordinate to a support activity or coordinating external agency support. Crisis counseling focuses on the subordinate's immediate, short-term needs.

Referral Counseling

C-34. Referral counseling helps subordinates work through a personal situation and may or may not follow crisis counseling. Referral counseling may also act as preventative counseling before the situation becomes a problem. Usually, the leader assists the subordinate in identifying the problem and refers the subordinate to the appropriate resource, such as Army Community Services, a chaplain, or an alcohol and drug counselor. (Figure C-3 lists support activities.)

Promotion Counseling

C-35. Leaders must conduct promotion counseling for all specialists and sergeants who are eligible for advancement without waivers but not recommended for promotion to the next higher grade. Army regulations require that soldiers within this category receive initial (event-oriented) counseling when they attain full eligibility and then periodic (performance/personal growth) counseling thereafter.

Adverse Separation Counseling

C-36. Adverse separation counseling may involve informing the soldier of the administrative actions available to the commander in the event substandard performance continues and of the consequences associated with those administrative actions (see AR 635-200).

C-37. Developmental counseling may not apply when an individual has engaged in more serious acts of misconduct. In those situations, you should refer the matter to the commander and the servicing staff judge advocate. When the leader's rehabilitative efforts fail, counseling with a view towards separation fills an administrative prerequisite to many administrative discharges and serves as a final warning to the soldier to improve performance or face discharge. In many situations, it may be beneficial to involve the chain of command as soon as you determine that adverse separation counseling might be required. A unit first sergeant or commander should be the person who informs the soldier of the notification requirements outlined in AR 635-200.

PERFORMANCE AND PROFESSIONAL GROWTH COUNSELING

Performance Counseling

C-38. During performance counseling, you conduct a review of a subordinate's duty performance during a certain period. You and the subordinate jointly establish performance objectives and standards for the next period. Rather than dwelling on the past, you should focus the session on the subordinate's strengths, areas needing improvement, and potential.

C-39. Performance counseling is required under the officer, NCO, and DA civilian evaluation reporting systems. The OER process requires periodic performance counseling as part of the OER Support Form requirements. Mandatory, face-to-face performance counseling between the rater and the rated NCO is required under the NCOERS. TAPES includes a combination of both of these requirements.

C-40. Counseling at the beginning of and during the evaluation period facilitates a subordinate's involvement in the evaluation process. Performance counseling communicates standards and is an opportunity for leaders to establish and clarify the expected values, attributes, skills, and actions. Part IVb (Leader Attributes/Skills/Actions) of the OER Support Form (DA Form 67-9-1) serves as an excellent tool for leaders doing performance counseling. For lieutenants and warrant officers one, the major performance objectives on the OER Support Form are used as the basis for determining the developmental tasks on the Junior Officer Developmental Support Form (DA Form 67-9-1a). Quarterly face-to-face performance and developmental counseling is required for these junior officers as outlined in AR 623-105.

C-41. As an Army leader, you must ensure you've tied your expectations to performance objectives and appropriate standards. You must establish standards that your subordinates can work towards and must teach them how to achieve the standards if they are to develop.

Professional Growth Counseling

C-42. Professional growth counseling includes planning for the accomplishment of individual and professional goals. You conduct this counseling to assist subordinates in achieving organizational and individual goals. During the counseling, you and your subordinate conduct a review to identify and discuss the subordinate's strengths and weaknesses and create a plan of action to build upon strengths and overcome weaknesses. This counseling isn't normally event-driven.

C-43. As part of professional growth counseling, you may choose to discuss and develop a "pathway to success" with the subordinate. This future-oriented counseling establishes short- and long-term goals and objectives. The discussion may include opportunities for civilian or military schooling, future duty assignments, special programs, and reenlistment options. Every person's needs are different, and leaders must apply specific courses of action tailored to each individual.

C-44. Career field counseling is required for lieutenants and captains before they're considered for promotion to major. Raters and senior raters, in conjunction with the rated officer, need to determine where the officer's skills best fit the needs of the Army. During career field counseling, consideration must be given to the rated officer's preference and his abilities (both performance and academic). The rater and senior rater should discuss career field designation with the officer prior to making a recommendation on the rated officer's OER.

C-45. While these categories can help you organize and focus counseling sessions, they should not be viewed as separate, distinct, or exhaustive. For example, a counseling session that focuses on resolving a problem may also address improving duty performance. A session focused on performance may also include a discussion on opportunities for professional growth. Regardless of the topic of the counseling session, leaders should follow the same basic format to prepare for and conduct it.

ESSENTIAL ARMY MANUAL
(439) ARMY LEADERSHIP

APPROACHES TO COUNSELING

C-46. An effective leader approaches each subordinate as an individual. Different people and different situations require different counseling approaches. Three approaches to counseling include nondirective, directive, and combined. These approaches differ in the techniques used, but they all fit the definition of counseling and contribute to its overall purpose. The major difference between the approaches is the degree to which the subordinate participates and interacts during a counseling session. Figure C-5 summarizes the advantages and disadvantages of each approach.

NONDIRECTIVE

C-47. The nondirective approach is preferred for most counseling sessions. Leaders use their experienced insight and judgment to assist subordinates in developing solutions. You should partially structure this type of counseling by telling the subordinate about the counseling process and explaining what you expect.

C-48. During the counseling session, listen rather than make decisions or give advice. Clarify what's said. Cause the subordinate to bring out important points, so as to better understand the situation. When appropriate, summarize the discussion. Avoid providing solutions or rendering opinions; instead, maintain a focus on individual and organizational goals and objectives. Ensure the subordinate's plan of action supports those goals and objectives.

DIRECTIVE

C-49. The directive approach works best to correct simple problems, make on-the-spot corrections, and correct aspects of duty performance. The leader using the directive style does most of the talking and tells the subordinate what to do and when to do it. In contrast to the nondirective approach, the leader directs a course of action for the subordinate.

C-50. Choose this approach when time is short, when you alone know what to do, or if a subordinate has limited problem-solving skills. It's also appropriate when a subordinate needs guidance, is immature, or is insecure.

COMBINED

C-51. In the combined approach, the leader uses techniques from both the directive and nondirective approaches, adjusting them to articulate what's best for the subordinate. The combined approach emphasizes the subordinate's planning and decision-making responsibilities.

C-52. With your assistance, the subordinate develops the subordinate's own plan of action. You should listen, suggest possible courses, and help analyze each possible solution to determine its good and bad points. You should then help the subordinate fully understand all aspects of the situation and encourage the subordinate to decide which solution is best.

	Advantages	Disadvantages
Nondirective	• Encourages maturity. • Encourages open communication. • Develops personal responsibility.	• More time-consuming. • Requires greatest counselor skill.
Directive	• Quickest method. • Good for people who need clear, concise direction. • Allows counselors to actively use their experience.	• Doesn't encourage subordinates to be part of the solution. • Tends to treat symptoms, not problems. • Tends to discourage subordinates from talking freely. • Solution is the counselor's, not the subordinate's.
Combined	• Moderately quick. • Encourages maturity. • Encourages open communication. • Allows counselors to actively use their experience.	• May take too much time for some situations.

Figure C-5. Counseling Approach Summary Chart

COUNSELING TECHNIQUES

C-53. As an Army leader, you may select from a variety of techniques when counseling subordinates. These counseling techniques, when appropriately used, cause subordinates to do things or improve upon their performance. You can use these methods during scheduled counseling sessions or while simply coaching a subordinate. Counseling techniques you can use during the nondirective or combined approaches include—

- **Suggesting alternatives**. Discuss alternative actions that the subordinate may take, but both you and the subordinate decide which course of action is most appropriate.

- **Recommending**. Recommend one course of action, but leave the decision to accept the recommended action to the subordinate.

- **Persuading**. Persuade the subordinate that a given course of action is best, but

leave the decision to the subordinate. Successful persuasion depends on the leader's credibility, the subordinate's willingness to listen, and their mutual trust.

- **Advising**. Advise the subordinate that a given course of action is best. This is the strongest form of influence not involving a command.

C-54. Some techniques you can use during the directive approach to counseling include—

- **Corrective training**. Teach and assist the subordinate in attaining and maintaining the standards. The subordinate completes corrective training when the subordinate attains the standard.

- **Commanding**. Order the subordinate to take a given course of action in clear, exact words. The subordinate understands that he has been given a command and will face the consequences for failing to carry it out.

THE COUNSELING PROCESS

C-55. Effective leaders use the counseling process. It consists of four stages:

- Identify the need for counseling.
- Prepare for counseling.
- Conduct counseling.
- Follow up.

IDENTIFY THE NEED FOR COUNSELING

C-56. Quite often organizational policies, such as counseling associated with an evaluation or counseling required by the command, focus a counseling session. However, you may conduct developmental counseling whenever the need arises for focused, two-way communication aimed at subordinate development. Developing subordinates consists of observing the subordinate's performance, comparing it to the standard, and then providing feedback to the subordinate in the form of counseling.

PREPARE FOR COUNSELING

C-57. Successful counseling requires preparation. To prepare for counseling, do the following:

- Select a suitable place.
- Schedule the time.
- Notify the subordinate well in advance.
- Organize information.
- Outline the counseling session components.
- Plan your counseling strategy.
- Establish the right atmosphere.

Select a Suitable Place

C-58. Schedule counseling in an environment that minimizes interruptions and is free from distracting sights and sounds.

Schedule the Time

C-59. When possible, counsel a subordinate during the duty day. Counseling after duty hours may be rushed or perceived as unfavorable. The length of time required for counseling depends on the complexity of the issue. Generally a counseling session should last less than an hour. If you need more time, schedule a second session. Additionally, select a time free from competition with other activities and consider what has been planned after the counseling session. Important events can distract a subordinate from concentrating on the counseling.

Notify the Subordinate Well in Advance

C-60. For a counseling session to be a subordinate-centered, two-person effort, the subordinate must have time to prepare for it. The subordinate should know why, where, and when the counseling will take place. Counseling following a specific event should happen as close to the event as possible. However, for performance or professional development counseling, subordinates may need a week or more to prepare or review specific products, such as support forms or counseling records.

Organize Information

C-61. Solid preparation is essential to effective counseling. Review all pertinent information. This includes the purpose of the counseling, facts and observations about the subordinate, identification of possible problems, main points of discussion, and the development of a plan of action. Focus on specific and objective behaviors that the subordinate must maintain or improve as well as a plan of action with clear, obtainable goals.

Outline the Components of the Counseling Session

C-62. Using the information obtained, determine what to discuss during the counseling session. Note what prompted the counseling, what you aim to achieve, and what your role as a counselor is. Identify possible comments or questions to help you keep the counseling session subordinate-centered and help the subordinate progress through its stages. Although you never know what a subordinate will say or do during counseling, a written outline helps organize the session and enhances the chance of positive results. (Figure C-6 is one example of a counseling outline prepared by a platoon leader about to conduct an initial NCOER counseling session with a platoon sergeant.)

Type of counseling: Initial NCOER counseling for SFC Taylor, a recently promoted new arrival to the unit.
Place and time: The platoon office, 1500 hours, 9 October.
Time to notify the subordinate: Notify SFC Taylor one week in advance of the scheduled counseling session.
Subordinate preparation: Have SFC Taylor put together a list of goals and objectives he would like to complete over the next 90 to 180 days. Review the values, attributes, skills, and actions from FM 22-100.
Counselor preparation:

- Review the NCO Counseling Checklist/Record (DA Form 2166-8-1).

- Update or review SFC Taylor's duty description and fill out the rating chain and duty description on the working copy of the NCOER (DA Form 2166-8, Parts II and III).

- Review each of the values and responsibilities in Part IV of the NCOER and the values, attributes, skills and actions in FM 22-100. Think of how each applies to SFC Taylor and the platoon sergeant position.

- Review the actions you consider necessary for a success or excellence in each value and responsibility.

- Make notes in blank spaces in Part IV of the NCOER to assist when counseling.
Role as counselor: Help SFC Taylor to understand the expectations and standards associated with the platoon sergeant position. Assist SFC Taylor in developing the values, attributes, skills, and actions that will enable him to achieve his performance objectives, consistent with those of the platoon and company. Resolve any aspects of the job that aren't clearly understood.
Session outline: Complete an outline following the counseling session components in Figure C-7 and based on the draft duty description on the NCOER, ideally at least two to three days prior to the actual counseling session.

Figure C-6. Example of a Counseling Outline

Plan Counseling Strategy

C-63. As many approaches to counseling exist as there are leaders. The directive, nondirective, and combined approaches to counseling were addressed earlier. Use a strategy that suits your subordinates and the situation.

Establish the Right Atmosphere

C-64. The right atmosphere promotes two-way communication between a leader and subordinate. To establish a relaxed atmosphere, you may offer the subordinate a seat or a cup of coffee. You may want to sit in a chair facing the subordinate since a desk can act as a barrier.

C-65. Some situations make an informal atmosphere inappropriate. For example, during counseling to correct substandard performance, you may direct the subordinate to remain standing while you remain seated behind a desk. This formal atmosphere,

normally used to give specific guidance, reinforces the leader's rank, position in the chain of command, and authority.

CONDUCT THE COUNSELING SESSION

C-66. Be flexible when conducting a counseling session. Often counseling for a specific incident occurs spontaneously as leaders encounter subordinates in their daily activities. Such counseling can occur in the field, motor pool, barracks—wherever subordinates perform their duties. Good leaders take advantage of naturally occurring events to provide subordinates with feedback.

C-67. Even when you haven't prepared for formal counseling, you should address the four basic components of a counseling session. Their purpose is to guide effective counseling rather than mandate a series of rigid steps. Counseling sessions consist of—

- Opening the session.
- Discussing the issues.
- Developing the plan of action.
- Recording and closing the session.

Ideally, a counseling session results in a subordinate's commitment to a plan of action. Assessment of the plan of action (discussed below) becomes the starting point for follow-up counseling. (Figure C-7 is an example of a counseling session.)

Open the Session

C-68. In the session opening, state the purpose of the session and establish a subordinate-centered setting. Establish the preferred setting early in the session by inviting the subordinate to speak. The best way to open a counseling session is to clearly state its purpose. For example, an appropriate purpose statement might be: "The purpose of this counseling is to discuss your duty performance over the past month and to create a plan to enhance performance and attain performance goals." If applicable, start the counseling session by reviewing the status of the previous plan of action.

C-69. You and the subordinate should attempt to develop a mutual understanding of the issues. You can best develop this by letting the subordinate do most of the talking. Use active listening; respond, and question without dominating the conversation. Aim to help the subordinate better understand the subject of the counseling, for example, duty performance, a problem situation and its impact, or potential areas for growth.

C-70. Both you and the subordinate should provide examples or cite specific observations to reduce the perception that either is unnecessarily biased or judgmental. However, when the issue is substandard performance, you should make clear how the performance didn't meet the standard. The conversation, which should be two-way, then addresses what the subordinate needs to do to meet the standard. It's important that you define the issue as substandard performance and don't allow the

subordinate to define the issue as an unreasonable standard—unless you consider the standard negotiable or are willing to alter the conditions under which the subordinate must meet the standard.

Develop a Plan of Action

C-71. A plan of action identifies a method for achieving a desired result. It specifies what the subordinate must do to reach the goals set during the counseling session. The plan of action must be specific: it should show the subordinate how to modify or maintain his behavior. It should avoid vague intentions such as "Next month I want you to improve your land navigation skills." The plan must use concrete and direct terms. For example, you might say: "Next week you'll attend the map reading class with 1st Platoon. After the class, SGT Dixon will coach you through the land navigation course. He will help you develop your skill with the compass. I will observe you going through the course with SGT Dixon, and then I will talk to you again and determine where and if you still need additional training." A specific and achievable plan of action sets the stage for successful development.

Record and Close the Session

C-72. Although requirements to record counseling sessions vary, a leader always benefits by documenting the main points of a counseling session. Documentation serves as a reference to the agreed upon plan of action and the subordinate's accomplishments, improvements, personal preferences, or problems. A complete record of counseling aids in making recommendations for professional development, schools, promotions, and evaluation reports.

C-73. Additionally, Army regulations require written records of counseling for certain personnel actions, such as a barring a soldier from reenlisting, processing a soldier for administrative separation, or placing a soldier in the overweight program. When a soldier faces involuntary separation, the leader must take special care to maintain accurate counseling records. Documentation of substandard actions conveys a strong corrective message to subordinates.

C-74. To close the session, summarize its key points and ask if the subordinate understands the plan of action. Invite the subordinate to review the plan of action and what's expected of you, the leader. With the subordinate, establish any follow-up measures necessary to support the successful implementation of the plan of action. These may include providing the subordinate with resources and time, periodically assessing the plan, and following through on referrals. Schedule any future meetings, at least tentatively, before dismissing the subordinate.

FOLLOW UP
Leader's Responsibilities

C-75. The counseling process doesn't end with the counseling session. It continues through implementation of the plan of action and evaluation of results. After counseling, you must support subordinates as they implement their plans of action. Support may include

teaching, coaching, or providing time and resources. You must observe and assess this process and possibly modify the plan to meet its goals. Appropriate measures after counseling include follow-up counseling, making referrals, informing the chain of command, and taking corrective measures.

Assess the Plan of Action

C-76. The purpose of counseling is to develop subordinates who are better able to achieve personal, professional, and organizational goals. During the assessment, review the plan of action with the subordinate to determine if the desired results were achieved. You and the subordinate should determine the date for this assessment during the initial counseling session. The assessment of the plan of action provides useful information for future follow-up counseling sessions.

Open the Session
- Establish a relaxed environment. Explain to SFC Taylor that the more one discusses and understands Army values and leader attributes, skills, and actions, the easier it is to develop and incorporate them into an individual leadership style.
- State the purpose of the counseling session. Explain that the initial counseling is based on leader actions (what SFC Taylor needs to do to be a successful platoon sergeant) and not on professional developmental needs (what SFC Taylor needs to do to develop further as an NCO).
- Come to an agreement on the duty description, the meaning of each value and responsibility, and the standards for success and excellence for each value and responsibility. Explain that subsequent counseling will focus on SFC Taylor's developmental needs as well as how well SFC Taylor is meeting the jointly agreed upon performance objectives. Instruct SFC Taylor to perform a self-assessment during the next quarter to identify his developmental needs.
- Ensure SFC Taylor knows the rating chain. Resolve any questions that SFC Taylor has about the job. Discuss the team relationship that exists between a platoon leader and a platoon sergeant and the importance of two-way communication between them.

Discuss the Issue
- Jointly review the duty description on the NCOER, including the maintenance, training, and taking care of soldiers responsibilities. Mention that the duty description can be revised as necessary. Highlight areas of special emphasis and appointed duties.
- Discuss the meaning of each value and responsibility on the NCOER. Discuss the values, attributes, skills, and actions outlined in FM 22-100. Ask open-ended questions to see if SFC Taylor can relate these items to his role as a platoon sergeant.

Figure C-7. Example of a Counseling Session

- Explain that even though the developmental tasks focus on developing leader actions, character development forms the basis for leadership development. Character and actions can't be viewed as separate; they're closely linked. In formulating the plan of action to accomplish major performance objectives, the proper values, attributes, and skills form the basis for the plan. As such, character development must be incorporated into the plan of action.

Assist in Developing a Plan of Action (During the Counseling Session)

- Ask SFC Taylor to identify actions that will facilitate the accomplishment of the major performance objectives. Categorize each action into one of the values or responsibilities listed on the NCOER.
- Discuss how each value and responsibility applies to the platoon sergeant position. Discuss specific examples of success and excellence in each value and responsibility block. Ask SFC Taylor for suggestions to make the goals more objective, specific, and measurable.
- Ensure that SFC Taylor has at least one example of a success or excellence bullet listed under each value and responsibility.
- Discuss SFC Taylor's promotion goals and ask him what he considers to be his strengths and weakness. Obtain copies of the last two master sergeant selection board results and match his goals and objectives to these.

Close the Session

- Check SFC Taylor's understanding of the duty description and performance objectives.
- Stress the importance of teamwork and two-way communication.
- Ensure SFC Taylor understands that you expect him to assist in your development as a platoon leader. This means that both of you have the role of teacher and coach.
- Remind SFC Taylor to perform a self-assessment during the next quarter.
- Set a tentative date during the next quarter for the routinely scheduled follow-up counseling.

Notes on Strategy

- Facilitate answering any questions SFC Taylor may have.
- Expect SFC Taylor to be uncomfortable with the terms and the developmental process; respond in a way that encourages participation throughout the session.

Figure C-7. Example of a Counseling Session (continued)

SUMMARY

C-77. This appendix has discussed developmental counseling. Developmental counseling is subordinate-centered communication that outlines actions necessary for subordinates to achieve individual and organizational goals and objectives. It can be either event-oriented or focused on personal and professional development. Figure C-8 summarizes the major aspects of developmental counseling and the counseling process.

Leaders must demonstrate these qualities to counsel effectively:
 	
• Respect for subordinates.
• Self and cultural awareness.
• Credibility.
• Empathy.
 	
Leaders must possess these counseling skills:
 	
• Active listening.
• Responding.
• Questioning.
 	
Effective leaders avoid common counseling mistakes. Leaders should avoid the influence of—
 	
• Personal bias.
• Rash judgments.
• Stereotyping.
• Losing emotional control.
• Inflexible counseling methods.
• Improper follow up.
	The Counseling Process
	
	1. **Identify the need for counseling.**
	
	2. **Prepare for counseling.**
	
	    • Select a suitable place.
	    • Schedule the time.
	    • Notify the subordinate well in advance.
	    • Organize information.
	    • Outline the components of the counseling session.
	    • Plan counseling strategy.
	    • Establish the right atmosphere.
	
	3. **Conduct the counseling session.**
	
	    • Open the session.
	    • Discuss the issue.
	    • Develop a plan of action (to include the leader's responsibilities).
	    • Record and close the session.
	
	4. **Follow up.**
	
	    • Support plan of action implementation
	    • Assess the plan of action.

Figure C-8. A Summary of Developmental Counseling

THE DEVELOPMENTAL COUNSELING FORM

C-78. The Developmental Counseling Form (DA Form 4856-E) is designed to help Army leaders conduct and record counseling sessions. Figure C-9 shows a completed DA Form 4856-E documenting the counseling of a young soldier with financial problems. While this is an example of a derogatory counseling , you can see that it is still developmental. Leaders must decide when counseling, additional training, rehabilitation, reassignment, or other developmental options have been exhausted. If the purpose of a counseling session is not developmental, refer to paragraphs C-36 and C-37. Figure C-10 shows a routine performance/professional growth counseling for a unit first sergeant. Figure C-11 shows a blank form with instructions on how to complete each block.

DEVELOPMENTAL COUNSELING FORM
For use of this form see FM 22-100

DATA REQUIRED BY THE PRIVACY ACT OF 1974

AUTHORITY: 5 USC 301, Departmental Regulations; 10 USC 3013, Secretary of the Army and E.O. 9397 (SSN)
PRINCIPAL PURPOSE: To assist leaders in conducting and recording counseling data pertaining to subordinates.
ROUTINE USES: For subordinate leader development IAW FM 22-100. Leaders should use this form as necessary.
DISCLOSURE: Disclosure is voluntary.

PART I - ADMINISTRATIVE DATA

Name (Last, First, MI) *Lloyd, Andrew*	Rank / Grade *PFC*	Social Security No. *123-45-6789*	Date of Counseling *28 March 1997*
Organization *2nd Platoon, B Battery, 1 - 1 ADA Bn*		Name and Title of Counselor *SGT Mark Levy, Squad Leader*	

PART II - BACKGROUND INFORMATION

Purpose of Counseling: (Leader states the reason for the counseling, e.g. performance/professional or event-oriented counseling and includes the leader's facts and observations prior to the counseling):

The purpose of this counseling is to inform PFC Lloyd of his responsibility to manage his financial affairs and the potential consequences of poorly managing finances and to help PFC Lloyd develop a plan of action to resolve his financial problems.

Facts: The battery commander received notice of delinquent payment on PFC Lloyd's Deferred Payment Plan (DPP).

A payment of $86.00 is 45 days delinquent

PART III - SUMMARY OF COUNSELING
Complete this section during or immediately subsequent to counseling.

Key Points of Discussion:

PFC Lloyd, late payments on a DPP account reflect a lack of responsibility and poor managing of finances. You should know that the letter of lateness has been brought to the attention of the battery commander, the first sergeant, and the platoon sergeant. They're all questioning your ability to manage your personal affairs. I also remind you that promotions and awards are based more than on just performing MOS-related duties; soldiers must act professionally and responsibly in all areas. Per conversation with PFC Lloyd, the following information was obtained:

He didn't make the DPP payment due to a lack of funds in his checking account. His most recent long distance phone bill was over $220 due to calling his house concerning his grandmother's failing health. PFC Lloyd stated that he wanted to pay for the phone calls himself in order not to burden his parents with the expense of collect calls. He also stated that his calling had tapered down considerably and he expects this month's phone bill to be approximately $50. We made an appointment at ACS and ACS came up with the following information:

PFC Lloyd's monthly obligations: Car payment: $330
    Car insurance: $138
    Rent including utilities: $400
    Other credit cards: $0
Total monthly obligations: $868.00
Monthly take-home pay: $1232.63

We discussed that with approximately $364 available for monthly living expenses, a phone bill in excess of $200 will severely affect PFC Lloyd's financial stability and can't continue. We discussed the need for PFC Lloyd to establish a savings account to help cover emergency expenses. PFC Lloyd agreed that his expensive phone bill and his inability to make the DPP payment is not responsible behavior. He confirmed that he wants to get his finances back on track and begin building a savings account.

OTHER INSTRUCTIONS
This form will be destroyed upon: reassignment (other than rehabilitative transfers), separation at ETS, or retirement. For separation requirements and notification of loss of benefits/consequences, see local directives and AR 635-200.

DA FORM 4856-E, JUN 99   EDITION OF JUN 85 IS OBSOLETE

Figure C-9. Example of a Developmental Counseling Form—Event Counseling

Plan of Action: (Outlines actions that the subordinate will do after the counseling session to reach the agreed upon goals(s). The actions must be specific enough to modify or maintain the subordinate's behavior and include a specific time line for implementation and assessment (Part IV below)).

Based on our discussion, PFC Lloyd will be able to resume normal payment on his DPP account next month (assuming that his phone bill is approximately $50). PFC Lloyd agreed to contact the DPP office and provide a partial payment of $20 immediately. He agreed to exercise self-restraint and not make long distance calls as frequently. He decided that his goal is to make one ten-minute phone call every two weeks. He will write letters to express concern over his grandmother's condition and ask his parents to do the same to keep him informed. His long-term goal is to establish a savings account with a goal of contributing $50 a month.

PFC Lloyd also agreed to attend the check cashing class at ACS on 2, 9, and 16 April.

Assessment date: 27 June

Session Closing: (The leader summarizes the key points of the session and checks if the subordinate understands the plan of action. The subordinate agrees/disagrees and provides remarks if appropriate).

Individual counseled: I agree/ ~~disagree~~ with the information above

Individual counseled remarks:

Signature of Individual Counseled: *Andrew Lloyd* _____ Date: *28 March 1997* _____

Leader Responsibilities: (Leader's responsibilities in implementing the plan of action).

PFC Lloyd will visit the DPP office to make an immediate partial payment of $20 and will give me a copy of the receipt as soon as the payment is made. PFC Lloyd will also provide me with a copy of the next month's phone bill and DPP payment receipt.
PFC Lloyd's finances will be a key topic of discussion at his next monthly counseling session.

Signature of Counselor: *Mark Levy* _____ Date: *28 March 1997* _____

PART IV - ASSESSMENT OF THE PLAN OF ACTION

Assessment (Did the plan of action achieve the desired results? This section is completed by both the leader and the individual counseled and provides useful information for follow-up counseling):

Counselor: _____ Individual Counseled: _____ Date of Assessment: _____

Note: Both the counselor and the individual counseled should retain a record of the counseling.

DA FORM 4856-E (Reverse)

Figure C-9 (continued). Example of a Developmental Counseling Form—Event Counseling

DEVELOPMENTAL COUNSELING FORM
For use of this form see FM 22-100

DATA REQUIRED BY THE PRIVACY ACT OF 1974

AUTHORITY: 5 USC 301, Departmental Regulations; 10 USC 3013, Secretary of the Army and E.O. 9397 (SSN)
PRINCIPAL PURPOSE: To assist leaders in conducting and recording counseling data pertaining to subordinates.
ROUTINE USES: For subordinate leader development IAW FM 22-100. Leaders should use this form as necessary.
DISCLOSURE: Disclosure is voluntary.

PART I - ADMINISTRATIVE DATA

Name (Last, First, MI) *McDonald, Stephen*	Rank / Grade *1SG*	Social Security No. *333-33-3333*	Date of Counseling *13 March 1998*
Organization *D Company, 3–95th IN*		Name and Title of Counselor *CPT Peterson, Company Commander*	

PART II - BACKGROUND INFORMATION

Purpose of Counseling: (Leader states the reason for the counseling, e.g. performance/professional or event-oriented counseling and includes the leader's facts and observations prior to the counseling):

- *To discuss duty performance for the period 19 Dec 97 to 11 March 1998.*

- *To discuss short-range professional growth goals/plan for next year.*

- *Talk about long-range professional growth (2-5 years) goals.*

PART III - SUMMARY OF COUNSELING
Complete this section during or immediately subsequent to counseling.

Key Points of Discussion:

- *Performance (sustain):*

- *Emphasized safety and knowledge of demolition, tactical proficiency on the Platoon Live Fire Exercises.*

- *Took charge of company defense during the last major field training exercise; outstanding integration and use of engineer, heavy weapons, and air defense artillery assets. Superb execution of defense preparations and execution.*

- *No dropped white cycle taskings.*

- *Good job coordinating with battalion adjutant on legal and personnel issues.*

- *Continue to take care of soldiers, keep the commander abreast of problems.*

- *Focused on subordinate NCO development; right man for the right job.*

Improve:

- *Get NCODPs on the calendar.*

- *Hold NCOs to standard on sergeants time training.*

OTHER INSTRUCTIONS
This form will be destroyed upon: reassignment (other than rehabilitative transfers), separation at ETS, or upon retirement. For separation requirements and notification of loss of benefits/consequences see local directives and AR 635-200.

DA FORM 4856-E, JUN 99   EDITION OF JUN 85 IS OBSOLETE

**Figure C-10. Example of a Developmental Counseling Form—
Performance/Professional Growth Counseling**

Plan of Action: (Outlines actions that the subordinate will do after the counseling session to reach the agreed upon goals(s). The actions must be specific enough to modify or maintain the subordinate's behavior and include a specific time line for implementation and assessment (Part IV below)).

- *Developmental Plan (next year):*
- *Develop a yearlong plan for NCODPs; coordinate to place on the calendar and training schedules.*
- *Resume civilian education; correspondence courses.*
- *Develop a company soldier of the month competition.*
- *Assist the company XO in modularizing the supply room for quick, efficient load-outs.*
- *Put in place a program to develop Ranger School candidates.*

Long-range goals (2 to 5 years):
- *Earn bachelor's degree.*
- *Attend and graduate the Sergeant Majors Academy.*

Session Closing: (The leader summarizes the key points of the session and checks if the subordinate understands the plan of action. The subordinate agrees/disagrees and provides remarks if appropriate).

Individual counseled: I agree/ ~~disagree~~ with the information above

Individual counseled remarks:

Signature of Individual Counseled: 1SG McDonald   Date: *13 March 1998*

Leader Responsibilities: (Leader's responsibilities in implementing the plan of action).

- *Coordinate with the 1SG on scheduling of NCODPs and soldier of the month boards.*
- *Have the XO meet with the 1SG on developing a plan for modularizing and improving the supply room.*
- *Provide time for Ranger candidate program.*

Signature of Counselor: *Mark Levy*   Date: *28 March 1997*

PART IV - ASSESSMENT OF THE PLAN OF ACTION

Assessment (Did the plan of action achieve the desired results? This section is completed by both the leader and the individual counseled and provides useful information for follow-up counseling):

1SG McDonald has enrolled in an associates degree program at the University of Kentucky. The supply room received all green evaluations during the last command inspection. Five of seven Ranger applicants successfully completed Ranger School, exceeding the overall course completion rate of 39%. Monthly soldier of the month boards proved to be impractical because of the OPTEMPO; however, the company does now hold quarterly boards during the white cycle. Brigade command sergeant major commented favorably on the last company NCODP he attended and gave the instructor a brigade coin.

Counselor: *CPT Peterson*   Individual Counseled: *1SG McDonald*   Date of Assessment: *1 Aug 98*

Note: Both the counselor and the individual counseled should retain a record of the counseling.

DA FORM 4856-E (Reverse)

**Figure C-10 (continued). Example of a Developmental Counseling Form—
Performance/Professional Growth Counseling**

DEVELOPMENTAL COUNSELING FORM
For use of this form see FM 22-100

DATA REQUIRED BY THE PRIVACY ACT OF 1974

AUTHORITY: 5 USC 301, Departmental Regulations; 10 USC 3013, Secretary of the Army and E.O. 9397 (SSN)
PRINCIPAL PURPOSE: To assist leaders in conducting and recording counseling data pertaining to subordinates.
ROUTINE USES: For subordinate leader development IAW FM 22-100. Leaders should use this form as necessary.
DISCLOSURE: Disclosure is voluntary.

PART I - ADMINISTRATIVE DATA

Name (Last, First, MI)	Rank / Grade	Social Security No.	Date of Counseling
Organization		Name and Title of Counselor	

PART II - BACKGROUND INFORMATION

Purpose of Counseling: (Leader states the reason for the counseling, e.g. performance/professional or event-oriented counseling and includes the leader's facts and observations prior to the counseling):

See paragraph C-68, Open the Session

The leader should annotate pertinent, specific, and objective facts and observations made. If applicable, the leader and subordinate start the counseling session by reviewing the status of the previous plan of action.

PART III - SUMMARY OF COUNSELING
Complete this section during or immediately subsequent to counseling.

Key Points of Discussion:

See paragraphs C-69 and C-70, Discuss the Issues.

The leader and subordinate should attempt to develop a mutual understanding of the issues. Both the leader and the subordinate should provide examples or cite specific observations to reduce the perception that either is unnecessarily biased or judgmental.

OTHER INSTRUCTIONS
This form will be destroyed upon: reassignment (other than rehabilitative transfers), separation at ETS, or upon retirement. For separation requirements and notification of loss of benefits/consequences see local directives and AR 635-200.

DA FORM 4856-E, JUN 99   EDITION OF JUN 85 IS OBSOLETE

Figure C-11. Guidelines on Completing a Developmental Counseling Form

Plan of Action: (Outlines actions that the subordinate will do after the counseling session to reach the agreed upon goals(s). The actions must be specific enough to modify or maintain the subordinate's behavior and include a specific time line for implementation and assessment (Part IV below)).

See paragraph C-71, Develop a Plan of Action

The plan of action specifies what the subordinate must do to reach the goals set during the counseling session. The plan of action must be specific and should contain the outline, guideline(s), and time line that the subordinate follows. A specific and achievable plan of action sets the stage for successful subordinate development.

Remember, event-oriented counseling with corrective training as part of the plan of action can't be tied to a specified time frame. Corrective training is complete once the subordinate attains the standard.

Session Closing: (The leader summarizes the key points of the session and checks if the subordinate understands the plan of action. The subordinate agrees/disagrees and provides remarks if appropriate).

Individual counseled: I agree/ disagree with the information above

Individual counseled remarks:

See paragraph C-72 through C-74, Close the Session

Signature of Individual Counseled:_____ Date: _____

Leader Responsibilities: (Leader's responsibilities in implementing the plan of action).

See paragraph C76, Leader's Responsibilities

To accomplish the plan of action, the leader must list the resources necessary and commit to providing them to the soldier.

Signature of Counselor: _____ Date: _____

PART IV - ASSESSMENT OF THE PLAN OF ACTION

Assessment (Did the plan of action achieve the desired results? This section is completed by both the leader and the individual counseled and provides useful information for follow-up counseling):

The assessment of the plan of action provides useful information for future follow-up counseling. This block should be completed prior to the start of a follow-up counseling session. During an event-oriented counseling session, the counseling session is not complete until this block is completed.

During performance/professional growth counseling, this block serves as the starting point for future counseling sessions. Leaders must remember to conduct this assessment based on resolution of the situation or the established time line discussed in the plan of action block above.

Counselor:_____ Individual Counseled: _____ Date of Assessment: _____

Note: Both the counselor and the individual counseled should retain a record of the counseling.

DA FORM 4856-E (Reverse)

Figure C-11 (continued). Guidelines on Completing a Developmental Counseling Form

Appendix D

A Leader Plan of Action and the ECAS

D-1. By completing a set of tasks (shown in Figure D-1), leaders can improve, sustain, or reinforce a standard of performance within their organizations. Leaders may complete some or all of the sub-tasks shown in Figure D-1, depending on the situation.

D-2. A leader plan of action (developed in step 3) identifies specific leader actions necessary to achieve improvement. It is similar to the individual plan of action that Appendix C discusses.

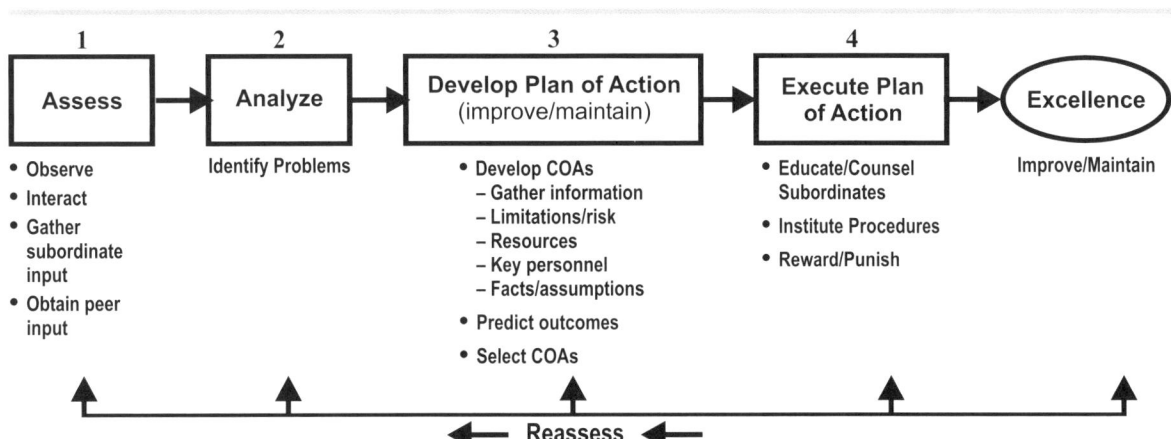

Figure D-1. The Leader Plan of Action Development Process

D-3. Begin your plan of action by assessing your unit (Step 1). Observe, interact, and gather feedback from others; or conduct formal assessments of the workplace. Then analyze the information you gathered to identify what needs improvement (Step 2). Once you have identified what needs improvement, begin to develop courses of action to make the improvements.

D-4. In Step 3, you develop your plan of action. First, develop and consider several possible courses of action to correct the weaknesses you identified. Gather important information, assess the limitations and risks associated with the various courses, identify available key personnel and resources, and verify facts and assumptions. Attempt to predict the outcome for each possible course of action. Based on your predictions, select several leader actions to deal with the problems.

D-5. Execute your plan of action (Step 4) by educating, training, or counseling your subordinates; instituting new policies or procedures; and revising or enforcing proper systems of rewards and punishment. Your organization moves towards excellence by improving substandard or weak areas and maintaining conditions that meet or exceed the standard. Finally, periodically reassesses your unit to identify new matters of concern or to evaluate the effectiveness of the leader actions.

D-6. You can use this process for many areas of interest within your organization. A case study demonstrating how to use an ECAS to prepare a leader plan of action follows. It includes a description of how one leader gathered information to complete the survey. (You can obtain the form used to conduct an ECAS through Training Support Centers by ordering GTA 22-6-1.)

PREPARATION OF AN ECAS

D-7. 2LT Christina Ortega has been a military police platoon leader for almost eight months. When she first came to the platoon, it was a well-trained, cohesive group. Within two months of her taking charge, she and her platoon deployed on a six-month rotation to support operations in Bosnia. The unit performed well, and she quickly earned a reputation as a leader with high standards for herself and her unit. Now redeployed, she must have her platoon ready in two months for a rotation at the Combat Maneuver Training Center (CMTC). She realizes that within that time she must get the unit's equipment ready for deployment, train her soldiers on different missions they will encounter at the CMTC, and provide them some much needed and deserved time off.

D-8. As 2LT Ortega reflects on her first eight months of leadership, she remembers how she took charge of the platoon. She spoke individually with the leaders in the platoon about her expectations and gathered information about her subordinates. She stayed up all night completing the leadership philosophy memorandum that she gave to every member of her platoon. After getting her feet on the ground and getting to know her soldiers, she assessed the platoon's ethical climate using the ECAS. Her unit's overall ECAS score was very good. She committed herself to maintaining that positive ethical climate by continuing the established policies and by monitoring the climate periodically.

D-9. Having completed a major deployment and received a recent influx of some new soldiers, 2LT Ortega decides to complete another ECAS. She heads to the unit motor pool to observe her soldiers preparing for the next day's training exercise. The platoon is deploying to the local training area for the "best squad" competition prior to the ARTEP evaluation at the CMTC. "The best squad competition has really become a big deal in the company," she thinks. "Squad rivalry is fierce, and the squad leaders seem to be looking for an edge so they can come out on top and win the weekend pass that goes to the winning squad."

D-10. She talks to as many of her soldiers as she can, paying particular attention to the newest members of the unit. One new soldier, a vehicle driver for SSG Smith, the 2nd Squad Leader, appears very nervous and anxious. During her conversation with the soldier, 2LT Ortega discovers some disturbing information.

D-11. The new soldier, PFC O'Brien, worries about his vehicle's maintenance and readiness for the next day. His squad leader has told him to "get the parts no matter what." PFC O'Brien says that he admires SSG Smith because he realizes that SSG Smith just wants to perform well and keep up the high standards of his previous driver. He recounts that SSG Smith has vowed to win the next day's land navigation competition. "SSG Smith even went so far as to say that he knows we'll win because he already knows the location of the points for the course. He saw them on the XO's desk last night and wrote them on his map."

D-12. 2LT Ortega thanks the soldier for talking honestly with her and immediately sets him straight on the proper and improper way to get repair parts. By the time she leaves, PFC O'Brien knows that 2LT Ortega has high standards and will not tolerate improper means of meeting them. Meanwhile, 2LT Ortega heads back toward the company headquarters to find the XO.

D-13. She finds the XO busily scribbling numbers and dates on pieces of paper. He is obviously involved and frantic. He looks up at her and manages a quick "Hi, Christina," before returning to his task. The battalion XO apparently did not like the way the unit status report (USR) portrayed the status of the maintenance in the battalion and refused to send that report forward. Not completely familiar with the USR, 2LT Ortega goes to the battalion motor officer to get some more information. After talking to a few more people in her platoon, 2LT Ortega completes the ECAS shown in Figure D-2.

GTA 22-6-1

Ethical **C**limate **A**ssessment **S**urvey

An ethical climate is one in which our stated Army values are routinely articulated, supported, practiced and respected. The Ethical Climate of an organization is determined by a variety of factors, including the *individual character* of unit members, the *policies and practices* within the organization, the *actions of unit leaders*, and *environmental and mission factors*. Leaders should periodically assess their unit's ethical climate and take appropriate actions to maintain the high ethical standards expected of all Army organizations. This survey will assist you in making these assessments and in identifying the actions necessary to accomplish this vital leader function. FM 22-100, <u>Army Leadership</u>, provides specific leader actions necessary to sustain or improve your ethical climate, as necessary.

_____

E. We maintain an organizational creed, motto, and/or philosophy that is consistent with Army values. `4`

F. We submit unit reports that reflect accurate information. `3`

G. We ensure unit members are aware of, and are comfortable using, the various channels available to report unethical behavior. `4`

H. We treat fairly those individuals in our unit who report unethical behavior. `5`

I. We hold accountable (i.e., report and/or punish) members of our organization who behave unethically. `4`

Section II Total `31`

*Use the following scale for questions in Section III.

Never	Hardly Ever	Sometimes	Almost Always	Always
1	2	3	4	5

III. Unit Leader Actions - *"What do I do?"* This section focuses on what you do as the leader of your organization to encourage an ethical climate.

A. I discuss Army values in orientation programs when I welcome new members to my organization. `5`

B. I routinely assess the ethical climate of my unit (i.e., sensing sessions, climate surveys, etc.). `5`

C. I communicate my expectations regarding ethical behavior in my unit, and require subordinates to perform tasks in an ethical manner. `5`

D. I encourage discussions of ethical issues in After Action Reviews, training meetings, seminars, and workshops. `3`

E. I encourage unit members to raise ethical questions and concerns to the chain of command or other individuals, if needed (i.e., chaplain, IG, etc.). `5`

F. I consider ethical behavior in performance evaluations, award and promotion recommendations, and adverse personnel actions. `4`

G. I include maintaining a strong ethical climate as one of my unit's goals and objectives. `5`

Section III Total `32`

INSTRUCTIONS

Answer the questions in this survey according to how you currently perceive your unit and your own leader actions, NOT according to how you would prefer them to be or how you think they should be. This information is for your use, (not your chain of command's) to determine if you need to take action to improve the Ethical Climate in your organization. Use the following scale for all questions in Sections I and II.

Strongly Disagree	Disagree	Neither Agree nor Disagree	Agree	Strongly Agree
1	2	3	4	5

I. Individual Character - *"Who are we?"* This section focuses on your organization's members' commitment to Army values. Please answer the following questions based on your observations of the ethical commitment in your unit. (This means your *immediate* unit. If you are a squad leader, it means you and your squad. If you are a civilian supervisor, it means you and your section.)

A. In general, the members of my unit demonstrate a commitment to Army values (honor, selfless service, integrity, loyalty, courage, duty and respect). `4`

B. The members of my unit typically accomplish a mission by "doing the right thing" rather than compromising Army values. `2`

C. I understand, and I am committed to, the Army's values as outlined in FM 22-100, <u>Army Leadership</u>. `5`

Section I Total `11`

II. Unit/Workplace Policies & Practices - *"What do we do?"*
This section focuses on what you, and the leaders who report to you, do to maintain an ethical climate in your workplace. (This does **not** mean your superiors. Their actions will be addressed in Section IV).

A. We provide clear instructions which help prevent unethical behavior. `2`

B. We promote an environment in which subordinates can learn from their mistakes. `5`

C. We maintain appropriate, not dysfunctional, levels of stress and competition in our unit. `1`

D. We discuss ethical behavior and issues during regular counseling sessions. `3`

IV. Environmental/Mission Factors - *"What surrounds us?"*
This section focuses on the external environment surrounding your organization. Answer the following questions to assess the impact of these factors on the ethical behavior in your organization.

Use the following scale for all questions in Section IV. ***Note: the scale is reversed for this section (Strongly Agree is scored as a "1", not a "5") ***

Strongly Agree	Agree	Neither Agree nor Disagree	Disagree	Strongly Disagree
1	2	3	4	5

A. My unit is currently under an excessive amount of stress (i.e., inspections, limited resources, frequent deployments, training events, deadlines, etc.). `1`

B. My higher unit leaders foster a 'zero defects' outlook on performance, such that they do not tolerate mistakes. `1`

C. My higher unit leaders over-emphasize competition between units. `1`

D. My higher unit leaders appear to be unconcerned with unethical behavior as long as the mission is accomplished. `2`

E. I do not feel comfortable bringing up ethical issues with my supervisors. `5`

F. My peers in my unit do not seem to take ethical behavior very seriously. `1`

Section IV Total `11`

Place the Total Score from each section in the spaces below:
(A score of 1 or 2 on any question requires some immediate leader action.)

Section I - Individual Character Total Score	11
Section II - Leader Action Total Score	31
Section III - Unit Policies and Procedures Total Score	32
Section IV - Environmental/Mission Factors Total Score	11

ECAS Total Socre (I + II + III + IV)	85

25 - 75	76-100	101 - 125
Take *Immediate* Action to Improve Ethical Climate	Take Actions to Improve Ethical Climate	Maintain a Healthy Ethical Climate

Figure D-2. Example of an Ethical Climate Assessment Survey

PREPARATION OF A LEADER PLAN OF ACTION

D-14. 2LT Ortega looks at her ECAS score and determines that she needs to take action to improve the ethical climate in her platoon. To help determine where she should begin, 2LT Ortega looks at the scores for each question. She knows that any question receiving a "1" or "2" must be addressed immediately in her plan of action. As 2LT Ortega reviews the rest of the scores for her unit, she identifies additional problems to correct. Furthermore, she decides to look at a few actions in which her unit excels and to describe ways to sustain the performance. As she continues to develop the leader plan of action, she looks at each subject she has identified. She next develops the plan shown in Figure D-3 to correct the deficiencies. At the bottom of the form, she lists at least two actions she plans to take to maintain the positive aspects of her platoon's ethical climate.

D-15. 2LT Ortega has already completed the first three steps (assess, analyze, and develop a plan of action) specified in Figure D-1. When she takes action to implement the plan she will have completed the process. She must then follow up to ensure her actions have the effects she intended.

Actions to *correct* negative aspects of the ethical climate in the organization

Problem: Dysfunctional competition/stress in the unit (the competition is causing some members of the unit to seek ways to gain an unfair advantage over others) [ECAS question # II.C., IV.A. & IV.C.]
Action:

- Postpone the platoon competition; focus on the readiness of equipment and soldier preparation rather than competition.
- Build some time in the long-range calendar to allow soldiers time to get away from work and relax.
- Focus on the group's accomplishment of the mission (unit excellence). Reward the platoon, not squads, for excellent performance. Reward teamwork.

Problem: Battalion XO "ordering" the changing of reports [IV B., D. & F.]
Action:

- Go see the company XO first and discuss what he should do.
- If the XO won't deal with it, see the commander myself to raise the issue.

Problem: Squad leader's unethical behavior [I.B. & II.A.]
Action:

- Reprimand the squad leader for getting the land navigation points unfairly.
- Counsel the squad leader on appropriate ways to give instructions and accomplish the mission without compromising values.

Problem: Unclear instructions given by the squad leader ("get the parts no matter what") [II.A.]
Action:

- Have the platoon sergeant give a class (NCODP) on proper guidelines for giving instructions and appropriate ethical considerations when asking subordinates to complete a task.
- Have the platoon sergeant counsel the squad leader(s) on the importance of using proper supply procedures.

Problem: Company XO "changing report" to meet battalion XO's needs [IV.B. & F.]
Action:

- Have an informal discussion with the company XO about correct reporting or see the company commander to raise the issue about the battalion XO.

Actions to *maintain* positive aspects of the ethical climate in the organization

Maintain: Continue to hold feedback (sensing) sessions and conduct ECAS assessments to maintain a feel for how the platoon is accomplishing its mission. [II.D. & G.; III.A. & B.]

Maintain: Continue to reward people who perform to high standards without compromising values. Punish those caught compromising them. [III.E. & F.]

Figure D-3. Example of a Leader Plan of Action

Appendix E

Character Development

E-1. Everyone who becomes part of America's Army, soldier or DA civilian, has character. On the day a person joins the Army, leaders begin building on that character. Army values emphasize the relationship between character and competence. Although competence is a fundamental attribute of Army leaders, character is even more critical. This appendix discusses the actions Army leaders take to develop their subordinates' character.

E-2. Army leaders are responsible for refining the character of soldiers and DA civilians. How does the Army as an institution ensure proper character development? What should leaders do to inculcate Army values in their subordinates?

E-3. Leaders teach Army values to every new member of the Army. Together with the leader attributes described in Chapter 2, Army values establish the foundation of leaders of character. Once members learn these values, their leaders ensure adherence. Adhering to the principles Army values embody is essential, for the Army cannot tolerate unethical behavior. Unethical behavior destroys morale and cohesion; it undermines the trust and confidence essential to teamwork and mission accomplishment.

E-4. Ethical conduct must reflect beliefs and convictions, not just fear of punishment. Over time, soldiers and DA civilians adhere to Army values because they want to live ethically and profess the values because they know it's right to do so. Once people believe and demonstrate Army values, they are persons of character. Ultimately, Army leaders are charged with the with the essential role of developing character in others. Figure E-1 shows the leader actions that support character development.

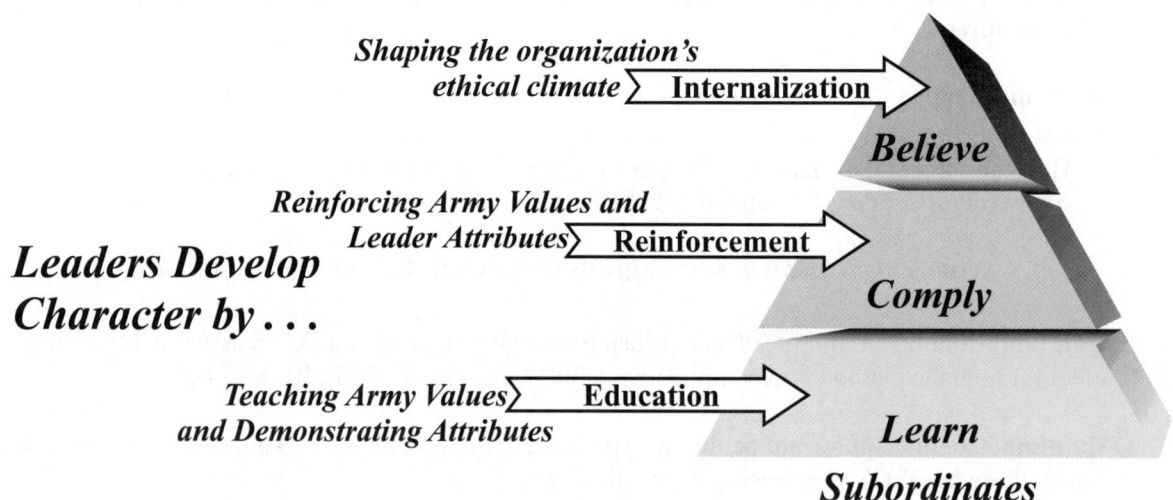

Figure E-1. Character Development

LEADERS TEACH VALUES; SUBORDINATES LEARN THE CULTURE

E-5. Army leaders must teach their subordinates moral principles, ethical theory, Army values, and leadership attributes. Through their leaders' programs, soldiers and DA civilians develop character through education, experience, and reflection. By educating their subordinates and setting the example, Army leaders enable their subordinates to make ethical decisions that in turn contribute to excellence. Subordinates gain deeper understanding from experiencing, observing, and reflecting on the aspects of Army leadership under the guidance of their leaders.

E-6. Inculcating Army values doesn't end with basic training. All Army leaders should seek to deepen subordinates' understanding of the ethical aspects of character through classes, informal discussions, one-on-one coaching, and formal developmental counseling. Army leaders can also improve their own understanding through study, reflection, and discussions with peers and superiors.

LEADERS REINFORCE VALUES; SUBORDINATES COMPLY

E-7. Leaders reinforce and discipline behavior to guide subordinates' development. To help subordinates live according to Army values, leaders enforce rules, policies, and regulations. Still, soldiers and DA civilians of character do more than merely comply with established institutional rules. Acting correctly but without complete understanding or sound motivation is not good enough in America's values-based Army. People of character behave correctly through correct understanding and personal desire. Understanding comes from training and self-development. Personal desire comes from a person's realization that Army values are worth adopting and living by and from that person's decision to do just that.

E-8. Character stems from a thorough understanding of Army values; however, this understanding must go beyond knowing the one-line definitions. Individuals must also know why Army values are important and how to apply them to everyday Army life. Leaders can promote Army values by setting the example themselves and pointing out other examples of Army values in both normal and exceptional activities. Army leaders can use unit histories and traditions, prominent individuals, and recent events to bring Army values to life and explain why adhering to them is important.

LEADERS SHAPE THE ETHICAL CLIMATE; SUBORDINATES INTERNALIZE ARMY VALUES

E-9. Doing the right thing is good. Doing the right thing for the right reason and with the right intention is better. People of character must possess the desire to act ethically in all situations. One of the Army leader's primary responsibilities is to maintain an ethical climate that supports development of such a character. When an organization's ethical climate nurtures ethical behavior, over time, people think, feel, and act ethically—they internalize the aspects of sound character.

E-10. Leaders should influence others' character development and foster correct actions through role modeling, teaching, and coaching. Army leaders seek to build a climate in which subordinates and organizations can reach their full potential. Together, these actions promote organizational excellence.

E-11. Army leaders can use the ECAS to assess ethical aspects of their own character and actions, the workplace, and the external environment. Once they have done their assessment, leaders prepare and carry out a plan of action. The plan of action focuses on solving ethical problems within the leaders' span of influence; leaders pass ethical problems they cannot change to higher headquarters. E-12. Becoming a person of character and a leader of character is a career-long process involving both self-development and developmental counseling. While individuals are responsible for their own character development, leaders are responsible for encouraging, supporting, and assessing the efforts of their subordinates. Leaders of character can develop only through continual study, reflection, experience, and feedback.

Appendix F

The Constitution of the United States

As a member of the Army, you have taken an oath to "support and defend the Constitution of the United States against all enemies, foreign and domestic [and to] bear true faith and allegiance to the same." But what is this document that you have sworn to protect? In essence, the Constitution is a blueprint establishing the powers and responsibilities of the three branches of the United States government as well as the rights of American citizens. Especially important for the Army are those provisions in the Constitution that place fundamental military authority in Congress and the president. As part of that authority, Congress has the power to "provide for the common Defense," which includes the power to "raise and support Armies," and the president is the commander in chief of the armed forces. So the Constitution establishes the critical principle that America's military leaders are subordinate to the nation's civilian authorities. Given the importance of that concept to our system of government, this appendix contains a copy of the US Constitution. As you read it, you will see that, although the Constitution was written over 200 years ago, it remains relevant today for Army leaders and all Americans.

THE PREAMBLE

We the People of the United States, in Order to form a more perfect Union, establish Justice, insure domestic Tranquility, provide for the common defence, promote the general Welfare, and secure the Blessings of Liberty to ourselves and our Posterity, do ordain and establish this Constitution for the United States of America.

THE CONSTITUTION

ARTICLE I

Section 1

All legislative Powers herein granted shall be vested in a Congress of the United States, which shall consist of a Senate and House of Representatives.

Section 2

The House of Representatives shall be composed of Members chosen every second Year by the People of the several States, and the Electors in each State shall have the Qualifications requisite for Electors of the most numerous Branch of the State Legislature.

No Person shall be a Representative who shall not have attained to the Age of twenty five Years, and been seven Years a Citizen of the United States, and who shall not, when elected, be an Inhabitant of that State in which he shall be chosen.

Representatives and direct Taxes shall be apportioned among the several States which may be included within this Union, according to their respective Numbers, which shall be determined by adding to the whole Number of free Persons, including those bound to Service for a Term of Years, and excluding Indians not taxed, three fifths of all other Persons. The actual Enumeration shall be made within three Years after the first Meeting of the Congress of the United States, and within every subsequent Term of ten Years, in such Manner as they shall by Law direct. The Number of Representatives shall not exceed one for every thirty Thousand, but each State shall have at Least one Representative; and until such enumeration shall be made, the State of New Hampshire shall be entitled to chuse three, Massachusetts eight, Rhode-Island and Providence Plantations one, Connecticut five, New-York six, New Jersey four, Pennsylvania eight, Delaware one, Maryland six, Virginia ten, North Carolina five, South Carolina five, and Georgia three.

When vacancies happen in the Representation from any State, the Executive Authority thereof shall issue Writs of Election to fill such Vacancies.

The House of Representatives shall chuse their Speaker and other Officers; and shall have the sole Power of Impeachment.

Section 3

The Senate of the United States shall be composed of two Senators from each State, chosen by the Legislature thereof, for six Years; and each Senator shall have one Vote.

Immediately after they shall be assembled in Consequence of the first Election, they shall be divided as equally as may be into three Classes. The Seats of the Senators of the first Class shall be vacated at the Expiration of the second Year, of the second Class at the Expiration of the fourth Year, and of the third Class at the Expiration of the sixth Year, so that one third may be chosen every second Year; and if Vacancies happen by Resignation, or otherwise, during the Recess of the Legislature of any State, the Executive thereof may make temporary Appointments until the next Meeting of the Legislature, which shall then fill such Vacancies.

No Person shall be a Senator who shall not have attained to the Age of thirty Years, and been nine Years a Citizen of the United States, and who shall not, when elected, be an Inhabitant of that State for which he shall be chosen.

The Vice President of the United States shall be President of the Senate, but shall have no Vote, unless they be equally divided.

The Senate shall chuse their other Officers, and also a President pro tempore, in the Absence of the Vice President, or when he shall exercise the Office of President of the United States.

The Senate shall have the sole Power to try all Impeachments. When sitting for that Purpose, they shall be on Oath or Affirmation. When the President of the United States is tried, the Chief Justice shall preside: And no Person shall be convicted without the Concurrence of two thirds of the Members present.

Judgment in Cases of Impeachment shall not extend further than to removal from Office, and disqualification to hold and enjoy any Office of honor, Trust or Profit under the United States: but the Party convicted shall nevertheless be liable and subject to Indictment, Trial, Judgment and Punishment, according to Law.

Section 4

The Times, Places and Manner of holding Elections for Senators and Representatives, shall be prescribed in each State by the Legislature thereof; but the Congress may at any time by Law make or alter such Regulations, except as to the Places of chusing Senators.

The Congress shall assemble at least once in every Year, and such Meeting shall be on the first Monday in December, unless they shall by Law appoint a different Day.

Section 5

Each House shall be the Judge of the Elections, Returns and Qualifications of its own Members, and a Majority of each shall constitute a Quorum to do Business; but a smaller Number may adjourn from day to day, and may be authorized to compel the Attendance of absent Members, in such Manner, and under such Penalties as each House may provide.

Each House may determine the Rules of its Proceedings, punish its Members for disorderly Behaviour, and, with the Concurrence of two thirds, expel a Member.

Each House shall keep a Journal of its Proceedings, and from time to time publish the same, excepting such Parts as may in their Judgment require Secrecy; and the Yeas and Nays of the Members of either House on any question shall, at the Desire of one fifth of those Present, be entered on the Journal.

Neither House, during the Session of Congress, shall, without the Consent of the other, adjourn for more than three days, nor to any other Place than that in which the two Houses shall be sitting.

Section 6

The Senators and Representatives shall receive a Compensation for their Services, to be ascertained by Law, and paid out of the Treasury of the United States. They shall in all Cases, except Treason, Felony and Breach of the Peace, be privileged from Arrest during their Attendance at the Session of their respective Houses, and in going to and returning from the same; and for any Speech or Debate in either House, they shall not be questioned in any other Place.

No Senator or Representative shall, during the Time for which he was elected, be appointed to any civil Office under the Authority of the United States, which shall have been created, or the Emoluments whereof shall have been encreased during such time; and no Person holding any Office under the United States, shall be a Member of either House during his Continuance in Office.

Section 7

All Bills for raising Revenue shall originate in the House of Representatives; but the Senate may propose or concur with Amendments as on other Bills.

Every Bill which shall have passed the House of Representatives and the Senate, shall, before it become a Law, be presented to the President of the United States; If he approve he shall sign it, but if not he shall return it, with his Objections to that House in which it shall have originated, who shall enter the Objections at large on their Journal, and proceed to reconsider it. If after such Reconsideration two thirds of that House shall agree to pass the Bill, it shall be sent, together with the Objections, to the other House, by which it shall likewise be reconsidered, and if approved by two thirds of that House, it shall become a Law. But in all such Cases the Votes of both Houses shall be determined by Yeas and Nays, and the Names of the Persons voting for and against the Bill shall be entered on the Journal of each House respectively. If any Bill shall not be returned by the President within ten Days (Sundays excepted) after it shall have been presented to him, the Same shall be a Law, in like Manner as if he had signed it, unless the Congress by their Adjournment prevent its Return, in which Case it shall not be a Law.

Every Order, Resolution, or Vote to which the Concurrence of the Senate and House of Representatives may be necessary (except on a question of Adjournment) shall be presented to the President of the United States; and before the Same shall take Effect, shall be approved by him, or being disapproved by him, shall be repassed by two thirds of the Senate and House of Representatives, according to the Rules and Limitations prescribed in the Case of a Bill.

Section 8

The Congress shall have Power To lay and collect Taxes, Duties, Imposts and Excises, to pay the Debts and provide for the common Defence and general Welfare of the United States; but all Duties, Imposts and Excises shall be uniform throughout the United States;

To borrow Money on the credit of the United States;

To regulate Commerce with foreign Nations, and among the several States, and with the Indian Tribes;

To establish an uniform Rule of Naturalization, and uniform Laws on the subject of Bankruptcies throughout the United States;

To coin Money, regulate the Value thereof, and of foreign Coin, and fix the Standard of Weights and Measures;

To provide for the Punishment of counterfeiting the Securities and current Coin of the United States;

To establish Post Offices and post Roads;

To promote the Progress of Science and useful Arts, by securing for limited Times to Authors and Inventors the exclusive Right to their respective Writings and Discoveries;

To constitute Tribunals inferior to the supreme Court;

To define and punish Piracies and Felonies committed on the high Seas, and Offences against the Law of Nations;

To declare War, grant Letters of Marque and Reprisal, and make Rules concerning Captures on Land and Water;

To raise and support Armies, but no Appropriation of Money to that Use shall be for a longer Term than two Years;

To provide and maintain a Navy;

To make Rules for the Government and Regulation of the land and naval Forces;

To provide for calling forth the Militia to execute the Laws of the Union, suppress Insurrections and repel Invasions;

To provide for organizing, arming, and disciplining, the Militia, and for governing such Part of them as may be employed in the Service of the United States, reserving to the States respectively, the Appointment of the Officers, and the Authority of training the Militia according to the discipline prescribed by Congress;

To exercise exclusive Legislation in all Cases whatsoever, over such District (not exceeding ten Miles square) as may, by Cession of particular States, and the Acceptance of Congress, become the Seat of the Government of the United States, and to exercise like Authority over all Places purchased by the Consent of the Legislature of the State in which the Same shall be, for the Erection of Forts, Magazines, Arsenals, dock-Yards, and other needful Buildings;—And

To make all Laws which shall be necessary and proper for carrying into Execution the foregoing Powers, and all other Powers vested by this Constitution in the Government of the United States, or in any Department or Officer thereof.

Section 9

The Migration or Importation of such Persons as any of the States now existing shall think proper to admit, shall not be prohibited by the Congress prior to the Year one thousand eight hundred and eight, but a Tax or duty may be imposed on such Importation, not exceeding ten dollars for each Person.

The Privilege of the Writ of Habeas Corpus shall not be suspended, unless when in Cases of Rebellion or Invasion the public Safety may require it.

No Bill of Attainder or ex post facto Law shall be passed.

No Capitation, or other direct, Tax shall be laid, unless in Proportion to the Census or Enumeration herein before directed to be taken.

No Tax or Duty shall be laid on Articles exported from any State.

No Preference shall be given by any Regulation of Commerce or Revenue to the Ports of one State over those of another: nor shall Vessels bound to, or from, one State, be obliged to enter, clear, or pay Duties in another.

No Money shall be drawn from the Treasury, but in Consequence of Appropriations made by Law; and a regular Statement and Account of the Receipts and Expenditures of all public Money shall be published from time to time.

No Title of Nobility shall be granted by the United States: And no Person holding any Office of Profit or Trust under them, shall, without the Consent of the Congress, accept of any present, Emolument, Office, or Title, of any kind whatever, from any King, Prince, or foreign State.

Section 10

No State shall enter into any Treaty, Alliance, or Confederation; grant Letters of Marque and Reprisal; coin Money; emit Bills of Credit; make any Thing but gold and silver Coin a Tender in Payment of Debts; pass any Bill of Attainder, ex post facto Law, or Law impairing the Obligation of Contracts, or grant any Title of Nobility.

No State shall, without the Consent of the Congress, lay any Imposts or Duties on Imports or Exports, except what may be absolutely necessary for executing its inspection Laws: and the net Produce of all Duties and Imposts, laid by any State on Imports or Exports, shall be for the Use of the Treasury of the United States; and all such Laws shall be subject to the Revision and Controul of the Congress.

No State shall, without the Consent of Congress, lay any Duty of Tonnage, keep Troops, or Ships of War in time of Peace, enter into any Agreement or Compact with another State, or with a foreign Power, or engage in War, unless actually invaded, or in such imminent Danger as will not admit of delay.

ARTICLE II
Section 1

The executive Power shall be vested in a President of the United States of America. He shall hold his Office during the Term of four Years, and, together with the Vice President, chosen for the same Term, be elected, as follows:

Each State shall appoint, in such Manner as the Legislature thereof may direct, a Number of Electors, equal to the whole Number of Senators and Representatives to which the State may be entitled in the Congress: but no Senator or Representative, or Person holding an Office of Trust or Profit under the United States, shall be appointed an Elector.

The Electors shall meet in their respective States, and vote by Ballot for two Persons, of whom one at least shall not be an Inhabitant of the same State with themselves. And they shall make a List of all the Persons voted for, and of the Number of Votes for each; which List they shall sign and certify, and transmit sealed to the Seat of the Government of the United States, directed to the President of the Senate. The President of the Senate shall, in the Presence of the Senate and House of Representatives, open all the Certificates, and the Votes shall then be counted. The Person having the greatest Number of Votes shall be the President, if such Number be a Majority of the whole Number of Electors appointed; and if there be more than one who have such Majority, and have an equal Number of Votes, then the House of Representatives shall immediately chuse by Ballot one of them for President; and if no Person have a Majority, then from the five highest on the List the said House shall in like Manner chuse the President. But in chusing the President, the Votes shall be taken by States, the Representation from each State having one Vote; A quorum for this Purpose shall consist of a Member or Members from two thirds of the States, and a Majority of all the States shall be necessary to a Choice. In every Case, after the Choice of the President, the Person having the greatest Number of Votes of the Electors shall be the Vice President. But if there should remain two or more who have equal Votes, the Senate shall chuse from them by Ballot the Vice President.

The Congress may determine the Time of chusing the Electors, and the Day on which they shall give their Votes; which Day shall be the same throughout the United States.

No Person except a natural born Citizen, or a Citizen of the United States, at the time of the Adoption of this Constitution, shall be eligible to the Office of President; neither shall any Person be eligible to that Office who shall not have attained to the Age of thirty five Years, and been fourteen Years a Resident within the United States.

In Case of the Removal of the President from Office, or of his Death, Resignation, or Inability to discharge the Powers and Duties of the said Office, the Same shall devolve on the Vice President, and the Congress may by Law provide for the Case of Removal, Death, Resignation or Inability, both of the President and Vice President, declaring what Officer shall then act as President, and such Officer shall act accordingly, until the Disability be removed, or a President shall be elected.

The President shall, at stated Times, receive for his Services, a Compensation, which shall neither be encreased nor diminished during the Period for which he shall have been elected, and he shall not receive within that Period any other Emolument from the United States, or any of them.

Before he enter on the Execution of his Office, he shall take the following Oath or Affirmation:—"I do solemnly swear (or affirm) that I will faithfully execute the Office of President of the United States, and will to the best of my Ability, preserve, protect and defend the Constitution of the United States."

Section 2

The President shall be Commander in Chief of the Army and Navy of the United States, and of the Militia of the several States, when called into the actual Service of the United States; he may require the Opinion, in writing, of the principal Officer in each of the executive Departments, upon any Subject relating to the Duties of their respective Offices, and he shall have Power to grant Reprieves and Pardons for Offences against the United States, except in Cases of Impeachment.

He shall have Power, by and with the Advice and Consent of the Senate, to make Treaties, provided two thirds of the Senators present concur; and he shall nominate, and by and with the Advice and Consent of the Senate, shall appoint Ambassadors, other public Ministers and Consuls, Judges of the supreme Court, and all other Officers of the United States, whose Appointments are not herein otherwise provided for, and which shall be established by Law: but the Congress may by Law vest the Appointment of such inferior Officers, as they think proper, in the President alone, in the Courts of Law, or in the Heads of Departments.

The President shall have Power to fill up all Vacancies that may happen during the Recess of the Senate, by granting Commissions which shall expire at the End of their next Session.

Section 3

He shall from time to time give to the Congress Information of the State of the Union, and recommend to their Consideration such Measures as he shall judge necessary and expedient; he may, on extraordinary Occasions, convene both Houses, or either of them, and in Case of Disagreement between them, with Respect to the Time of Adjournment, he may adjourn them to such Time as he shall think proper; he shall receive Ambassadors and other public Ministers; he shall take Care that the Laws be faithfully executed, and shall Commission all the Officers of the United States.

Section 4

The President, Vice President and all civil Officers of the United States, shall be removed from Office on Impeachment for, and Conviction of, Treason, Bribery, or other high Crimes and Misdemeanors.

ARTICLE III

Section 1

The judicial Power of the United States, shall be vested in one supreme Court, and in such inferior Courts as the Congress may from time to time ordain and establish. The Judges, both of the supreme and inferior Courts, shall hold their Offices during good Behaviour, and shall, at stated Times, receive for their Services, a Compensation, which shall not be diminished during their Continuance in Office.

Section 2

The judicial Power shall extend to all Cases, in Law and Equity, arising under this Constitution, the Laws of the United States, and Treaties made, or which shall be made, under their Authority;—to all Cases affecting Ambassadors, other public Ministers and Consuls;—to all Cases of admiralty and maritime Jurisdiction;—to Controversies to which the United States shall be a Party;—to Controversies between two or more States;—between a State and Citizens of another State;—between Citizens of different States, —between Citizens of the same State claiming Lands under Grants of different States, and between a State, or the Citizens thereof, and foreign States, Citizens or Subjects.

In all Cases affecting Ambassadors, other public Ministers and Consuls, and those in which a State shall be Party, the supreme Court shall have original Jurisdiction. In all the other Cases before mentioned, the supreme Court shall have appellate Jurisdiction, both as to Law and Fact, with such Exceptions, and under such Regulations as the Congress shall make.

The Trial of all Crimes, except in Cases of Impeachment, shall be by Jury; and such Trial shall be held in the State where the said Crimes shall have been committed; but when not committed within any State, the Trial shall be at such Place or Places as the Congress may by Law have directed.

Section 3

Treason against the United States, shall consist only in levying War against them, or in adhering to their Enemies, giving them Aid and Comfort. No Person shall be convicted of Treason unless on the Testimony of two Witnesses to the same overt Act, or on Confession in open Court.

The Congress shall have Power to declare the Punishment of Treason, but no Attainder of Treason shall work Corruption of Blood, or Forfeiture except during the Life of the Person attainted.

ARTICLE IV
Section 1

Full Faith and Credit shall be given in each State to the public Acts, Records, and judicial Proceedings of every other State. And the Congress may by general Laws prescribe the Manner in which such Acts, Records and Proceedings shall be proved, and the Effect thereof.

Section 2

The Citizens of each State shall be entitled to all Privileges and Immunities of Citizens in the several States.

A Person charged in any State with Treason, Felony, or other Crime, who shall flee from Justice, and be found in another State, shall on Demand of the executive Authority of the State from which he fled, be delivered up, to be removed to the State having Jurisdiction of the Crime.

No Person held to Service or Labour in one State, under the Laws thereof, escaping into another, shall, in Consequence of any Law or Regulation therein, be discharged from such Service or Labour, but shall be delivered up on Claim of the Party to whom such Service or Labour may be due.

Section 3

New States may be admitted by the Congress into this Union; but no new State shall be formed or erected within the Jurisdiction of any other State; nor any State be formed by the Junction of two or more States, or Parts of States, without the Consent of the Legislatures of the States concerned as well as of the Congress.

The Congress shall have Power to dispose of and make all needful Rules and Regulations respecting the Territory or other Property belonging to the United States; and nothing in this Constitution shall be so construed as to Prejudice any Claims of the United States, or of any particular State.

Section 4

The United States shall guarantee to every State in this Union a Republican Form of Government, and shall protect each of them against Invasion; and on Application of the Legislature, or of the Executive (when the Legislature cannot be convened) against domestic Violence.

ARTICLE V

The Congress, whenever two thirds of both Houses shall deem it necessary, shall propose Amendments to this Constitution, or, on the Application of the Legislatures of two thirds of the several States, shall call a Convention for proposing Amendments, which, in either Case, shall be valid to all Intents and Purposes, as Part of this Constitution, when ratified by the Legislatures of three fourths of the several States, or by Conventions in three fourths thereof, as the one or the other Mode of Ratification may be proposed by the Congress; Provided that no Amendment which may be made prior to the Year One thousand eight hundred and eight shall in any Manner affect the first and fourth Clauses in the Ninth Section of the first Article; and that no State, without its Consent, shall be deprived of its equal Suffrage in the Senate.

ARTICLE VI

All Debts contracted and Engagements entered into, before the Adoption of this Constitution, shall be as valid against the United States under this Constitution, as under the Confederation.

This Constitution, and the Laws of the United States which shall be made in Pursuance thereof; and all Treaties made, or which shall be made, under the Authority of the United States, shall be the supreme Law of the Land; and the Judges in every State shall be bound thereby, any Thing in the Constitution or Laws of any State to the Contrary notwithstanding.

The Senators and Representatives before mentioned, and the Members of the several State Legislatures, and all executive and judicial

Officers, both of the United States and of the several States, shall be bound by Oath or Affirmation, to support this Constitution; but no religious Test shall ever be required as a Qualification to any Office or public Trust under the United States.

ARTICLE VII

The Ratification of the Conventions of nine States, shall be sufficient for the Establishment of this Constitution between the States so ratifying the same.

Done in Convention by the Unanimous Consent of the States present the Seventeenth Day of September in the Year of our Lord one thousand seven hundred and Eighty seven and of the Independence of the United States of America the Twelfth. In witness whereof We have hereunto subscribed our Names,

G. WASHINGTON—President. And deputy from Virginia

Delaware
Geo: Read
Gunning Bedford jun
John Dickinson
Richard Bassett
Jaco: Broom
South Carolina
J. Rutledge
Charles Cotesworth
Pinckney
Charles Pinckney
Pierce Butler
Georgia
William Few
Abr Baldwin
New Hampshire
John Langdon
Nicholas
Massachusetts
Nathaniel Gorham
Rufus King
Connecticut
Wm: Saml. Johnson
Roger Sherman

Maryland
James McHenry
Dan of St Thos. Jenifer
Danl Carroll.
Virginia
John Blair---
James Madison Jr.
New York
Alexander Hamilton
New Jersy
Wil: Livingston
David Brearley
Wm. Paterson
Jona: Dayton
Pennsylvania
B Franklin
Thomas Mifflin
Robt Morris
Geo. Clymer
Thos. FitzSimons
Jared Ingersoll
James Wilson
Gouv Morris

Attest William Jackson
Secretary

North Carolina
Wm. Blount
Richd. Dobbs Spaight
Hu Williamson

THE PREAMBLE TO THE BILL OF RIGHTS

Congress of the United States begun and held at the City of New York, on Wednesday the fourth of March, one thousand seven hundred and eighty-nine.

The Conventions of a number of the States, having at the time of their adopting the Constitution expressed a desire in order to prevent misconstruction or abuse of its powers, that further declaratory and restrictive clauses should be added: And as extending the ground of public confidence in the Government will best ensure the beneficent ends of its institution.

Resolved by the Senate and House of Representatives of the United States of America in Congress assembled, two thirds of both Houses concurring that the following Articles be proposed to the Legislatures of the several states as Amendments to the Constitution of the United States, all or any of which articles, when ratified by three fourths of the said Legislatures to be valid to all intents and purposes as part of the said Constitution. viz.

Articles in addition to, and Amendment of the Constitution of the United States of America, proposed by Congress and Ratified by the Legislatures of the several States, pursuant to the fifth Article of the original Constitution.

THE BILL OF RIGHTS

AMENDMENT I

Congress shall make no law respecting an establishment of religion, or prohibiting the free exercise thereof; or abridging the freedom of speech, or of the press; or the right of the people peaceably to assemble, and to petition the Government for a redress of grievances.

AMENDMENT II

A well regulated Militia being necessary to the security of a free State, the right of the people to keep and bear Arms, shall not be infringed.

AMENDMENT III

No Soldier shall, in time of peace be quartered in any house, without the consent of the Owner, nor in time of war, but in a manner to be prescribed by law.

AMENDMENT IV

The right of the people to be secure in their persons, houses, papers, and effects, against unreasonable searches and seizures, shall not be violated, and no Warrants shall issue, but upon probable cause, supported by Oath or affirmation, and particularly describing the place to be searched, and the persons or things to be seized.

AMENDMENT V

No person shall be held to answer for a capital, or otherwise infamous crime, unless on a presentment or indictment of a Grand Jury, except in cases arising in the land or naval forces, or in the Militia, when in actual service in time of War or public danger; nor shall any person be subject for the same offence to be twice put in jeopardy of life or limb; nor shall be compelled in any criminal case to be a witness against himself, nor be deprived of life, liberty, or property, without due process of law; nor shall private property be taken for public use, without just compensation.

AMENDMENT VI

In all criminal prosecutions, the accused shall enjoy the right to a speedy and public trial, by an impartial jury of the State and district wherein the crime shall have been committed, which district shall have been previously ascertained by law, and to be informed of the nature and cause of the accusation; to be confronted with the witnesses against him; to have compulsory process for obtaining witnesses in his favor, and to have the Assistance of Counsel for his defence.

AMENDMENT VII

In suits at common law, where the value in controversy shall exceed twenty dollars, the right of trial by jury shall be preserved, and no fact tried by a jury, shall be otherwise reexamined in any Court of the United States, than according to the rules of the common law.

AMENDMENT VIII

Excessive bail shall not be required, nor excessive fines imposed, nor cruel and unusual punishments inflicted.

AMENDMENT IX

The enumeration in the Constitution, of certain rights, shall not be construed to deny or disparage others retained by the people.

AMENDMENT X

The powers not delegated to the United States by the Constitution, nor prohibited by it to the States, are reserved to the States respectively, or to the people.

AMENDMENTS 11 THROUGH 27

AMENDMENT XI

Passed by Congress March 4, 1794. Ratified February 7, 1795.

The Judicial power of the United States shall not be construed to extend to any suit in law or equity, commenced or prosecuted against one of the United States by Citizens of another State, or by Citizens or Subjects of any Foreign State.

Note: *Article III, section 2, of the Constitution was modified by amendment 11.*

AMENDMENT XII

Passed by Congress December 9, 1803. Ratified June 15, 1804.

The Electors shall meet in their respective states and vote by ballot for President and Vice-President, one of whom, at least, shall not be an inhabitant of the same state with themselves; they shall name in their ballots the person voted for as President, and in distinct ballots the person voted for as Vice-President, and they shall make distinct lists of all persons voted for as President, and of all persons voted for as Vice-President, and of the number of votes for each, which lists they shall sign and certify, and transmit sealed to the seat of the government of the United States, directed to the President of the Senate; the President of the Senate shall, in the presence of the Senate and House of Representatives, open all the certificates and the votes shall then be counted;

The person having the greatest number of votes for President, shall be the President, if such number be a majority of the whole number of Electors appointed; and if no person have such majority, then from the persons having the highest numbers not exceeding three on the list of those voted for as President, the House of Representatives shall choose immediately, by ballot, the President. But in choosing the President, the votes shall be taken by states, the representation from each state having one vote; a quorum for this purpose shall consist of a member or members from two-thirds of the states, and a majority of all the states shall be necessary to a choice. [And if the House of Representatives shall not choose a President whenever the right of choice shall devolve upon them, before the fourth day of March next following, then the Vice-President shall act as President, as in case of the death or other constitutional disability of the President.]* The person having the greatest number of votes as Vice-President, shall be the Vice-President, if such number be a majority of the whole number of Electors appointed, and if no person have a majority, then from the two highest numbers on the list, the Senate shall choose the Vice-President; a quorum for the purpose shall consist of two-thirds of the whole number of Senators, and a majority of the whole number shall be necessary to a choice. But no person constitutionally ineligible to the office of President shall be eligible to that of Vice-President of the United States.

* Superseded by section 3 of the 20th amendment.

***Note**: A portion of Article II, section 1 of the Constitution was superseded by the 12th amendment.*

AMENDMENT XIII

Passed by Congress January 31, 1865. Ratified December 6, 1865.

Section 1

Neither slavery nor involuntary servitude, except as a punishment for crime whereof the party shall have been duly convicted, shall exist within the United States, or any place subject to their jurisdiction.

Section 2

Congress shall have power to enforce this article by appropriate legislation.

***Note**: A portion of Article IV, section 2, of the Constitution was superseded by the 13th amendment.*

AMENDMENT XIV

Passed by Congress June 13, 1866. Ratified July 9, 1868.

Section 1

All persons born or naturalized in the United States, and subject to the jurisdiction thereof, are citizens of the United States and of the State wherein they reside. No State shall make or enforce any law which shall abridge the privileges or immunities of citizens of the United States; nor shall any State deprive any person of life, liberty, or property, without due process of law; nor deny to any person within its jurisdiction the equal protection of the laws.

Section 2

Representatives shall be apportioned among the several States according to their respective numbers, counting the whole number of persons in each State, excluding Indians not taxed. But when the right to vote at any election for the choice of electors for President and Vice-President of the United States, Representatives in Congress, the Executive and Judicial officers of a State, or the members of the Legislature thereof, is denied to any of the male inhabitants of such State, being twenty-one years of age,* and citizens of the United States, or in any way abridged, except for participation in rebellion, or other crime, the basis of representation therein shall be reduced in the proportion which the number of such male citizens shall bear to the whole number of male citizens twenty-one years of age in such State.

Section 3

No person shall be a Senator or Representative in Congress, or elector of President and Vice-President, or hold any office, civil or military, under the United States, or under any State, who, having previously taken an oath, as a member of Congress, or as an officer of the United States, or as a member of any State legislature, or as an executive or judicial officer of any State, to support the Constitution of the United States, shall have engaged in insurrection or rebellion against the same, or given aid or comfort to the enemies thereof. But Congress may by a vote of two-thirds of each House, remove such disability.

Section 4

The validity of the public debt of the United States, authorized by law, including debts incurred for payment of pensions and bounties for services in suppressing insurrection or rebellion, shall not be questioned. But neither the United States nor any State shall assume or pay any debt or obligation incurred in aid of insurrection or rebellion against the United States, or any claim for the loss or emancipation of any slave; but all such debts, obligations and claims shall be held illegal and void.

Section 5

The Congress shall have the power to enforce, by appropriate legislation, the provisions of this article.

* Changed by section 1 of the 26th amendment.

***Note**: Article I, section 2, of the Constitution was modified by section 2 of the 14th amendment.*

AMENDMENT XV

Passed by Congress February 26, 1869. Ratified February 3, 1870.

Section 1

The right of citizens of the United States to vote shall not be denied or abridged by the United States or by any State on account of race, color, or previous condition of servitude.

Section 2

The Congress shall have the power to enforce this article by appropriate legislation.

AMENDMENT XVI

Passed by Congress July 2, 1909. Ratified February 3, 1913.

The Congress shall have power to lay and collect taxes on incomes, from whatever source derived, without apportionment among the several States, and without regard to any census or enumeration.

Note: Article I, section 9, of the Constitution was modified by amendment 16.

AMENDMENT XVII

Passed by Congress May 13, 1912. Ratified April 8, 1913.

The Senate of the United States shall be composed of two Senators from each State, elected by the people thereof, for six years; and each Senator shall have one vote. The electors in each State shall have the qualifications requisite for electors of the most numerous branch of the State legislatures.

When vacancies happen in the representation of any State in the Senate, the executive authority of such State shall issue writs of election to fill such vacancies: Provided, That the legislature of any State may empower the executive thereof to make temporary appointments until the people fill the vacancies by election as the legislature may direct.

This amendment shall not be so construed as to affect the election or term of any Senator chosen before it becomes valid as part of the Constitution.

Note: Article I, section 3, of the Constitution was modified by the 17th amendment.

AMENDMENT XVIII

Passed by Congress December 18, 1917. Ratified January 16, 1919. Repealed by amendment 21.

Section 1

After one year from the ratification of this article the manufacture, sale, or transportation of intoxicating liquors within, the importation thereof into, or the exportation thereof from the United States and all territory subject to the jurisdiction thereof for beverage purposes is hereby prohibited.

Section 2

The Congress and the several States shall have concurrent power to enforce this article by appropriate legislation.

Section 3

This article shall be inoperative unless it shall have been ratified as an amendment to the Constitution by the legislatures of the several States, as provided in the Constitution, within seven years from the date of the submission hereof to the States by the Congress.

AMENDMENT XIX

Passed by Congress June 4, 1919. Ratified August 18, 1920.

The right of citizens of the United States to vote shall not be denied or abridged by the United States or by any State on account of sex.

Congress shall have power to enforce this article by appropriate legislation.

AMENDMENT XX

Passed by Congress March 2, 1932. Ratified January 23, 1933.

Section 1

The terms of the President and the Vice President shall end at noon on the 20th day of January, and the terms of Senators and Representatives at noon on the 3d day of January, of the years in which such terms would have ended if this article had not been ratified; and the terms of their successors shall then begin.

Section 2

The Congress shall assemble at least once in every year, and such meeting shall begin at noon on the 3d day of January, unless they shall by law appoint a different day.

Section 3

If, at the time fixed for the beginning of the term of the President, the President elect shall have died, the Vice President elect shall become President. If a President shall not have been chosen before the time fixed for the beginning of his term, or if the President elect shall have failed to qualify, then the Vice President elect shall act as President until a President shall have qualified; and the Congress may by law provide for the case wherein neither a President elect nor a Vice President shall have qualified, declaring who shall then act as President, or the manner in which one who is to act shall be selected, and such person shall act accordingly until a President or Vice President shall have qualified.

Section 4

The Congress may by law provide for the case of the death of any of the persons from whom the House of Representatives may choose a President whenever the right of choice shall have devolved upon them, and for the case of the death of any of the persons from whom the Senate may choose a Vice President whenever the right of choice shall have devolved upon them.

Section 5

Sections 1 and 2 shall take effect on the 15th day of October following the ratification of this article.

Section 6

This article shall be inoperative unless it shall have been ratified as an amendment to the Constitution by the legislatures of three-fourths of the several States within seven years from the date of its submission.

Note: *Article I, section 4, of the Constitution was modified by section 2 of this amendment. In addition, a portion of the 12th amendment was superseded by section 3.*

AMENDMENT XXI

Passed by Congress February 20, 1933. Ratified December 5, 1933.

Section 1

The eighteenth article of amendment to the Constitution of the United States is hereby repealed.

Section 2

The transportation or importation into any State, Territory, or Possession of the United States for delivery or use therein of intoxicating liquors, in violation of the laws thereof, is hereby prohibited.

Section 3

This article shall be inoperative unless it shall have been ratified as an amendment to the Constitution by conventions in the several States, as provided in the Constitution, within seven years from the date of the submission hereof to the States by the Congress.

AMENDMENT XXII

Passed by Congress March 21, 1947. Ratified February 27, 1951.

Section 1

No person shall be elected to the office of the President more than twice, and no person who has held the office of President, or acted as President, for more than two years of a term to which some other person was elected President shall be elected to the office of President more than once. But this Article shall not apply to any person holding the office of President when this Article was proposed by Congress, and shall not prevent any person who may be holding the office of President, or acting as President, during the term within which this Article becomes operative from holding the office of President or acting as President during the remainder of such term.

Section 2

This article shall be inoperative unless it shall have been ratified as an amendment to the Constitution by the legislatures of three-fourths of the several States within seven years from the date of its submission to the States by the Congress.

AMENDMENT XXIII

Passed by Congress June 16, 1960. Ratified March 29, 1961.

Section 1

The District constituting the seat of Government of the United States shall appoint in such manner as Congress may direct:

A number of electors of President and Vice President equal to the whole number of Senators and Representatives in Congress to which the District would be entitled if it were a State, but in no event more than the least populous State; they shall be in addition to those appointed by the States, but they shall be considered, for the purposes of the election of President and Vice President, to be electors appointed by a State; and they shall meet in the District and perform such duties as provided by the twelfth article of amendment.

Section 2

The Congress shall have power to enforce this article by appropriate legislation.

AMENDMENT XXIV

Passed by Congress August 27, 1962. Ratified January 23, 1964.

Section 1

The right of citizens of the United States to vote in any primary or other election for President or Vice President, for electors for President or Vice President, or for Senator or Representative in Congress, shall not be denied or abridged by the United States or any State by reason of failure to pay poll tax or other tax.

Section 2

The Congress shall have power to enforce this article by appropriate legislation.

AMENDMENT XXV

Passed by Congress July 6, 1965. Ratified February 10, 1967.

Section 1

In case of the removal of the President from office or of his death or resignation, the Vice President shall become President.

Section 2

Whenever there is a vacancy in the office of the Vice President, the President shall nominate a Vice President who shall take office upon confirmation by a majority vote of both Houses of Congress.

Section 3

Whenever the President transmits to the President pro tempore of the Senate and the Speaker of the House of Representatives his written declaration that he is unable to discharge the powers and duties of his office, and until he transmits to them a written declaration to the contrary, such powers and duties shall be discharged by the Vice President as Acting President.

Section 4

Whenever the Vice President and a majority of either the principal officers of the executive departments or of such other body as Congress may by law provide, transmit to the President pro tempore of the Senate and the Speaker of the House of Representatives their written declaration that the President is unable to discharge the powers and duties of his office, the Vice President shall immediately assume the powers and duties of the office as Acting President.

Thereafter, when the President transmits to the President pro tempore of the Senate and the Speaker of the House of Representatives his written declaration that no inability exists, he shall resume the powers and duties of his office unless the Vice President and a majority of either the principal officers of the executive department or of such other body as Congress may by law provide, transmit within four days to the President pro tempore of the Senate and the Speaker of the House of Representatives their written declaration that the President is unable to discharge the powers and duties of his office. Thereupon Congress shall decide the issue, assembling within forty-eight hours for that purpose if not in session. If the Congress, within twenty-one days after receipt of the latter written declaration, or, if Congress is not in session, within twenty-one days after Congress is required to assemble, determines by two-thirds vote of both Houses that the President is unable to discharge the powers and duties of his office, the Vice President shall continue to discharge the same as Acting President; otherwise, the President shall resume the powers and duties of his office.

Note: Article II, section 1, of the Constitution was affected by the 25th amendment.

AMENDMENT XXVI

Passed by Congress March 23, 1971. Ratified July 1, 1971.

Section 1

The right of citizens of the United States, who are eighteen years of age or older, to vote shall not be denied or abridged by the United States or by any State on account of age.

Section 2

The Congress shall have power to enforce this article by appropriate legislation.

Note: Amendment 14, section 2, of the Constitution was modified by section 1 of the 26th amendment.

AMENDMENT XXVII

Originally proposed Sept. 25, 1789. Ratified May 7, 1992.

No law, varying the compensation for the services of the Senators and Representatives, shall take effect, until an election of representatives shall have intervened.

FULL INDEXES AVAILABLE ON ACCOMPANYING DVD

ATTP 3-34.39 (FM 20-3)/MCRP 3-17.6A

CAMOUFLAGE, CONCEALMENT, AND DECOYS

NOVEMBER 2010

DISTRIBUTION RESTRICTION: Approved for public release; distribution is unlimited.

HEADQUARTERS, DEPARTMENT OF THE ARMY

**This publication is available
at Army Knowledge Online** www.us.army.mil
**and the General Dennis J. Reimer Training and
Doctrine Digital Library at** www.train.army.mil.

*ATTP 3-34.39 (FM 20-3)/MCRP 3-17.6A

Army Tactics, Techniques, and Procedures
No. 3-34.39/Marine Corps Reference Publication 3-17.6A

Headquarters
Department of the Army
Washington, DC, 26 November 2010

Camouflage, Concealment, and Decoys

Contents

DISTRIBUTION RESTRICTION: Approved for public release; distribution is unlimited.

*This publication supersedes FM 20-3, dated 30 August 1999.

i

Figures

Tables

ESSENTIAL ARMY MANUAL
(479) CAMOUFLAGE, CONCEALMENT AND DECOYS

Preface

This Army Tactics, Techniques, and Procedures (ATTP) is intended to help company-level leaders understand the principles and techniques of camouflage, concealment, and decoys (CCD). To remain viable, all units must apply CCD to personnel and equipment. Ignoring a threat's ability to detect friendly operations on the battlefield is shortsighted and dangerous. Friendly units enhance their survivability capabilities if they are well versed in CCD principles and techniques.

CCD is equal in importance to marksmanship, maneuver, and mission. It is an integral part of a soldier's duty. CCD encompasses individual and unit efforts such as movement, light, and noise discipline; letter control; dispersal; and deception operations. Each soldier's actions must contribute to the unit's overall CCD posture to maximize effectiveness.

Increased survivability is the goal of a CCD plan. A unit commander must encourage each soldier to think of survivability and CCD as synonymous terms. Training soldiers to recognize this correlation instills a greater appreciation of CCD values.

A metric conversion chart is provided in appendix A.

This publication applies to the Active Army, Army National Guard (ARNG)/Army National Guard of the United States (ARNGUS), and the United States Army Reserve (USAR) unless otherwise stated.

The proponent of this publication is United States Army Training and Doctrine Command (TRADOC). Send comments and recommendations of Department of the Army (DA) Form 2028 (Recommended Changes to Publications and Blank Forms) directly to Commandant, United States Army Engineer School (USAES), ATTN: ATSE-DOT-DD, Fort Leonard Wood, Missouri 65473-6650.

This publication implements Standardization Agreement (STANAG) 2931, *Orders for the Camouflage of the Red Cross and Red Crescent on Land in Tactical Operations*.

Unless otherwise stated, masculine nouns and pronouns do not refer exclusively to men.

Chapter 1

Basics

CCD is the use of materials and techniques to hide, blend, disguise, decoy, or disrupt the appearance of military targets and/or their backgrounds. CCD helps prevent an enemy from detecting or identifying friendly troops, equipment, activities, or installations. Properly designed CCD techniques take advantage of the immediate environment and natural and artificial materials. One of the imperatives of current military doctrine is to conserve friendly strength for decisive action. Such conservation is aided through sound operations security (OPSEC) and protection from attack. Protection includes all actions that make soldiers, equipment, and units difficult to locate.

DOCTRINAL CONSIDERATIONS

1-1. CCD degrades the effectiveness of enemy reconnaissance, surveillance, and target-acquisition (RSTA) capabilities. Skilled observers and sophisticated sensors can be defeated by obscuring telltale signs (signatures) of units on the battlefield. Preventing detection impairs enemy efforts to assess friendly operational patterns, functions, and capabilities.

1-2. CCD enhances friendly survivability by reducing an enemy's ability to detect, identify, and engage friendly elements. Survivability encompasses all actions taken to conserve personnel, facilities, and supplies from the effects of enemy weapons and actions. Survivability techniques include using physical measures such as fighting and protective positions; nuclear, biological, chemical (NBC) equipment; and armor. These actions include interrelated tactical countermeasures such as dispersion, movement techniques, OPSEC, communications security (COMSEC), CCD, and smoke operations (a form of CCD). Improved survivability from CCD is not restricted to combat operations. Benefits are also derived by denying an enemy the collection of information about friendly forces during peacetime.

1-3. Deception helps mask the real intent of primary combat operations and aids in achieving surprise. Deception countermeasures can delay effective enemy reaction by disguising information about friendly intentions, capabilities, objectives, and locations of vulnerable units and facilities. Conversely, intentionally poor CCD can project misleading information about friendly operations. Successful tactical deception depends on stringent OPSEC.

1-4. Smoke and obscurants are effective CCD tools and greatly enhance the effectiveness of other traditionally passive CCD techniques. Smoke and obscurants can change battlefield dynamics by blocking or degrading the spectral bands used by an enemy's target-acquisition and weapons systems. More recently developed obscurants are now able to degrade nonvisual detection systems such as thermal infrared (IR) imaging systems, selected radar systems, and laser systems.

RESPONSIBILITIES

1-5. Each soldier is responsible for camouflaging and concealing himself and his equipment. Practicing good CCD techniques lessens a soldier's probability of becoming a target. Additionally, a thorough knowledge of CCD and its guiding principles allows a soldier to easily recognize CCD as employed by an enemy.

1-6. A commander is responsible for CCD of his unit, and noncommissioned officers (NCOs) supervise well-disciplined soldiers in executing CCD. They use established standing operating procedures (SOPs) and battle drills to guide their efforts. CCD is a combat multiplier that should be exploited to the fullest extent.

1-7. An engineer is a battlefield expert on CCD. He integrates CCD into higher unit operations and advises commanders on all aspects of CCD employment as it relates to a unit's current mission.

PRIORITIES

1-8. Every soldier and military unit has an inherent mission of self-protection, and they should use all CCD means available. However, CCD countermeasures have become more complicated due to advancing technology. Commanders must recognize that advanced technologies have—

- Enhanced the performance of enemy recon and surveillance equipment.
- Increased an enemy's ability to use electromagnetic (EM) signature analysis for detecting friendly units.
- Reduced the time available to apply CCD because units must perform nearly all aspects of battlefield operations at an increased speed.

1-9. When time, camouflage materials, or other resources are insufficient to provide adequate support to units, commanders must prioritize CCD operations. Considerations for establishing these priorities involve analyzing the mission, enemy, terrain, weather, troops, time available, and civilian considerations (METT-TC). The following sets forth a METT-TC methodology to help determine CCD priorities:

- **Mission.** The mission is always the first and most important consideration. CCD efforts must enhance the mission but not be so elaborate that they hinder a unit's ability to accomplish the mission.
- **Enemy.** An enemy's RSTA capabilities often influence the camouflage materials and CCD techniques needed to support a unit's mission. Before beginning a mission, conduct an intelligence analysis to identify the enemy's RSTA capabilities.
- **Terrain and weather.** The battlefield terrain generally dictates what CCD techniques and materials are necessary. Different terrain types or background environments (urban, mountain, forest, plains, desert, arctic) require specific CCD techniques. (See chapter 7 for more information.)
- **Troops.** Friendly troops must be well trained in CCD techniques that apply to their mission, unit, and equipment. A change in the environment or the mission often requires additional training on effective techniques. Leaders must also consider the alertness of troops. Careless CCD efforts are ineffective and may disclose a unit's location, degrade its survivability, and hamper its mission accomplishment. Intelligence analysis should address the relative detectability of friendly equipment and the target signatures that unit elements normally project.
- **Time.** Time is often a critical consideration. Elaborate CCD may not be practical in all tactical situations. The type and amount of CCD needed are impacted by the time a unit occupies a given area, the time available to employ CCD countermeasures, and the time necessary to remove and reemploy camouflage during unit relocation. Units should continue to improve and perfect CCD measures as time allows.
- **Civilian considerations.** From conflict to war and from tactical to strategic, civilians in the area of operation (AO) may be active or passive collectors of information. Commanders and their staffs should manage this collection capability to benefit the command and the mission.

TRAINING

1-10. CCD training must be included in every field exercise. Soldiers must be aware that an enemy can detect, identify, and acquire targets by using resources outside the visual portion of the EM spectrum.

INDIVIDUAL

1-11. Each member of the unit must acquire and maintain critical CCD skills. These include the ability to analyze and use terrain effectively; to select an individual site properly; and to hide, blend, disguise, disrupt, and decoy key signatures using natural and artificial materials.

CAUTION

Ensure that local environmental considerations are addressed before cutting live vegetation or foliage in training areas.

UNIT

1-12. Unit CCD training refines individual and leader skills, introduces the element of team coordination, and contributes to tactical realism. If CCD is to conserve friendly strength, it must be practiced with the highest degree of discipline. The deployment and teardown of camouflage; light, noise, and communications discipline; and signal security must be practiced and evaluated in an integrated mission-training environment. CCD proficiency is developed through practicing and incorporating lessons learned from exercises and operations. A unit must incorporate CCD (who, what, where, when, and how) into its tactical standing operating procedure (TACSOP). (Appendix B provides additional guidance on integrating CCD into a unit's field TACSOP.) Generally, CCD is additive and synergistic with other defensive measures. CCD enhances unit survivability and increases the likelihood of mission success. A unit that is well trained in CCD operations more easily recognizes CCD as employed by an enemy, and this recognition enhances a unit's lethality.

EVALUATION

1-13. CCD training should be realistic and integrated with a unit's training evaluations. Employ the following techniques to enhance training evaluations:

- Have small-unit leaders evaluate their unit's CCD efforts from an enemy's viewpoint. How a position looks from a few meters away is probably of little importance. Evaluators should consider the following:
  - Could an approaching enemy detect and place aimed fire on the position?
  - From what distance can an enemy detect the position?
  - Which CCD principle was ignored that allowed detection?
  - Which CCD technique increased the possibility of detection?
- Use binoculars or night-vision or thermal devices, when possible, to show a unit how it would appear to an enemy.
- Use photographs and videotapes, if available, of a unit's deployments and positions as a method of self-evaluation.
- Incorporate ground-surveillance-radar (GSR) teams in training when possible. Let the troops know how GSR works and have them try to defeat it.
- Request aerial multispectral (visual, IR, radar) imagery of friendly unit positions. This imagery shows how positions appear to enemy aerial recon. Unit leaders should try to obtain copies of opposing forces (OPFOR) cockpit heads-up display (HUD) or videotapes, which are excellent assessment tools for determining a unit's detectability from an enemy's perspective. Another valuable assessment tool is the overhead imagery of a unit's actions and positions. Overhead imagery is often difficult to obtain; but if a unit is participating in a large-scale exercise or deployment, the imagery probably exists and can be accessed through the unit's intelligence channels.
- Use OPFOR to make training more realistic. Supporting aviation in an OPFOR role also helps. When possible, allow the OPFOR to participate in the after-action review (AAR) following each

mission. The unit should determine what factors enabled the OPFOR to locate, identify, and engage the unit and what the unit could have done to reduce its detectability.

OTHER CONSIDERATIONS

1-14. Warfare often results in personnel losses from fratricide. Fratricide compels commanders to consider CCD's effect on unit recognition by friendly troops.

1-15. Army policy prescribes that camouflage aids be built into equipment and supplies as much as possible. Battle-dress uniforms (BDUs), paint, Lightweight Camouflage Screen Systems (LCSSs), and decoys help achieve effective camouflage. These aids are effective only if properly integrated into an overall CCD plan that uses natural materials and terrain. During training exercises, ensure that cutting vegetation or foliage does not adversely effect the natural environment (coordinate with local authorities). CCD aids should not interfere with the battlefield performance of soldiers or equipment or the installations that they are designed to protect. (See *appendix C* for more information on LCSSs.)

1-16. When employed correctly, expedient CCD countermeasures are often the most effective means of confusing an enemy. Along with the standard items and materials listed above, soldiers can use battlefield by-products, construction materials, and indigenous or locally procurable items to enhance unit CCD posture. For example, a simple building decoy can be constructed with two-by-fours and plywood. With the addition of a heat source, such as a small charcoal pit, the decoy becomes an apparently functional building. However, as with all CCD countermeasures, ensure that expedient treatments project the desired signatures to the enemy and do not actually increase the unit's vulnerability to detection. Expedient CCD countermeasures are also beneficial because the enemy has less time to study and become familiar with the selected countermeasures.

Chapter 2

Threat

The enemy employs a variety of sensors to detect and identify US soldiers, equipment, and supporting installations. These sensors may be visual, near infrared (NIR), IR, ultraviolet (UV), acoustic, or multispectral/hyperspectral. They may be employed by dismounted soldiers or ground-, air-, or space-mounted platforms. Such platforms are often capable of supporting multiple sensors. Friendly troops rarely know the specific sensor systems or combination of systems that an enemy employs. When possible, friendly troops should protect against all known threat surveillance systems.

DOCTRINE

2-1. Many threat forces were trained and equipped by the former Soviet Union. Its long-standing battlefield doctrine of *maskirovka* is a living legacy in many former Soviet-client states. *Maskirovka* incorporates all elements of CCD and tactical battlefield deception into a cohesive and effective philosophy. During the Gulf War, Iraq used *maskirovka* to effectively maintain its capability of surface-to-surface missiles (Scuds) in the face of persistent coalition-force attacks. Enemy forces that are trained in *maskirovka* possess a strong fundamental knowledge of CCD principles and techniques. Friendly forces must be very careful to conduct CCD operations so that a well-trained enemy will not easily recognize them.

2-2. Typical threat doctrine states that each battalion will continuously maintain two observation posts when in close contact with its enemy. An additional observation post is established when the battalion is in the defense or is preparing for an offense.

2-3. Patrolling is used extensively, but particularly during offensive operations. Patrols are used to detect the location of enemy indirect- and direct-fire weapons, gaps in formations, obstacles, and bypasses.

2-4. Enemy forces use raids to capture prisoners, documents, weapons, and equipment. A recon-in-force (usually by a reinforced company or battalion) is the most likely tactic when other methods of tactical recon have failed. A recon-in-force is often a deceptive tactic designed to simulate an offensive and cause friendly forces to reveal defensive positions.

ORGANIZATION

2-5. A typical enemy force conducts recon activities at all echelons. A troop recon is usually conducted by specially trained units. The following types of enemy units might have specific intelligence-collection missions:

- **Troops.** An enemy uses ordinary combat troops to perform recon. One company per battalion trains to conduct recon operations behind enemy lines.
- **Motorized rifle and tank regiments.** Each regiment has a recon company and a chemical recon platoon.
- **Maneuver divisions.** Divisions have a recon battalion, an engineer recon platoon, a chemical recon platoon, and a target-acquisition battery.

DATA COLLECTION

2-6. An enemy collects information about United States (US) forces for two basic reasons—target acquisition and intelligence production. Enemy weapons systems often have sensors that locate and identify targets at long ranges in precise detail. Soldiers and units should take actions to hinder the enemy's target-acquisition process. These actions include all practical CCD operations expected to reduce the identification of soldiers, units, and facilities.

2-7. An enemy uses sensor systems to locate and identify large Army formations and headquarters (HQ) and to predict their future activities. Enemy detection of rear-area activities, such as logistics centers and communications nodes, may also reveal friendly intentions.

2-8. An enemy uses tactical recon to provide additional information on US forces' dispositions and the terrain in which they are going to operate. The enemy's tactical recon also attempts to identify targets for later attack by long-range artillery, rockets, aircraft, and ground forces.

SENSOR SYSTEMS

2-9. An enemy uses many different types of electronic surveillance equipment. Sensor systems are classified according to the part of the EM spectrum in which they operate. Figure 2-1 shows the EM spectrum and some typical enemy sensors operating within specific regions of the spectrum. An enemy uses detection sensors that operate in the active or passive mode:

- **Active.** Active sensors emit energy that reflects from targets and is recaptured by the emitting or other nearby sensor, indicating the presence of a target. Examples of active sensors are searchlights and radar.
- **Passive.** Passive sensors do not emit energy; they collect energy, which may indicate the presence of a target. Examples of passive sensors are the human eye, night-vision devices (NVDs), IR imaging devices, acoustic sensors, and photographic devices.

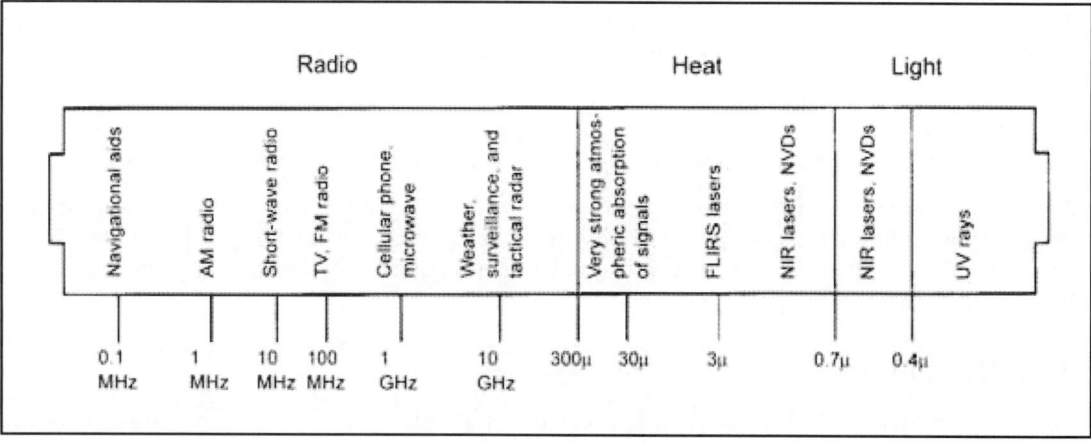

Figure 2-1. EM spectrum

VISUAL

2-10. Visual sensors work in the parts of the EM spectrum that are visible to the human eye. Enemy soldiers' eyes are the principle sensors on a battlefield. They may be aided by binoculars, telescopic sights, and image intensifiers. Civilian populations, enemy agents, recon teams, and patrols are visual-sensor systems from the enemy's intelligence viewpoint. Three types of enemy visual sensors are—

- **Image intensifiers.** Image intensifiers are passive night-observation devices. They amplify the low-level light that is present on even the darkest nights. These devices are used for surveillance and as weapon sights on small arms and vehicles. Airborne platforms are also capable of supporting image intensifiers.

- **Low-light television (LLTV).** LLTV combines image intensification with television technology, and it is usually mounted on airborne platforms.
- **Aerial recon, remote sensing, and imagery.** Aerial photography, satellite imagery, and video imagery allow image analysts to record and study visual information. These analysts then produce target nomination lists that are, in effect, priority lists of targets in a given target scene. Since analysts often have to make subjective determinations of the identity and/or importance of a given target, the ranking of targets provides the defender with an opportunity to use CCD to impact an enemy's target-prioritization process. Video systems allow transmission of visual images to the ground while the manned aircraft, satellite, or unmanned aerial vehicle (UAV) is still in flight.

NEAR INFRARED

2-11. NIR sensors operate at a wavelength immediately above the visible light wavelength of the EM spectrum (figure 2-1). NIR energy reflects well from live vegetation but reflects better from dead vegetation and most man-made materials. NIR sensors, such as sights and periscopes, allow the human eye to detect targets based on differences in their reflection of NIR energy. NIR sensors are partially blocked by fog, mist, and smoke operations, although not as completely as visual sensors. An enemy's combat vehicles use active NIR sensors that employ searchlights, scopes, and sights; but these sensors are rapidly being replaced with image intensifiers and thermal gun sights.

INFRARED

2-12. IR sensors detect the contrasts in heat energy that targets radiate on the battlefield and display the contrasts as different colors or shades. Because longer wavelength IR radiation is more susceptible to atmospheric absorption than NIR radiation, IR sensors are less affected by typical concentrations of fog or conventional smoke.

2-13. Differences in thermal mass and surface properties (reflectivity) of man-made and natural materials result in target-to-background contrasts. These contrast levels change dramatically over a daily cycle. For example, operating vehicles and generators, heated buildings and tents, and soldiers are usually hotter than their background. Also, equipment exposed to direct sunlight appears hotter than most natural backgrounds. At night, however, equipment might appear cooler than its background if it is treated with special emissivity coatings. In other words, military equipment, particularly metallic equipment, generally heats up and cools off more quickly than its background.

2-14. Sophisticated, passive IR sensors (such as the Forward-Looking Infrared System [FLIRS]) can be mounted on aircraft. FLIRS sensors provide aircrews and enemy ground forces with real-time IR imagery that is displayed on video monitors.

2-15. Recon aircraft often employ special IR films to record temperature differences. Due to film processing, however, these systems are subject to time delays in obtaining the data. Newer versions of this sensor produce non-film-based images.

ULTRAVIOLET

2-16. The UV area is the part of the EM spectrum immediately below visible light. UV sensors are more important in snow-covered areas, because snow reflects UV energy well and most white paints and man-made objects do not reflect UV energy very well. Photographic intelligence systems with simple UV filters highlight military targets as dark areas against snow-covered backgrounds. These backgrounds require specially designed camouflage that provides a high UV reflectance.

RADAR

2-17. Radar uses high-frequency radio waves to penetrate atmospheric impediments such as fog, mist, and smoke. Radar works by transmitting a very strong burst of radio waves and then receiving and processing the reflected waves. In general, metal objects reflect radar waves well, while radar waves are either weakly reflected by or pass through most other objects. The shape and size of a metal object determine the strength of the reflected signal. A large, metal object generally reflects more signal than a small object. Therefore, large, metal objects can be detected from greater distances. The method by which the received radio wave is processed determines the type of radar. Radar systems commonly used against ground forces on the battlefield include—

- **Moving-target indicators (MTIs).** When an EM wave hits a moving target, the wave is reflected and changes frequency. The faster the target moves, the larger the changes in frequency. The simplest and most common battlefield radar detects this frequency change. Threat forces use MTIs for target acquisition. More sophisticated developmental radar systems, such as the Joint Surveillance Target Attack Radar System (JSTARS), use airborne surveillance platforms that downlink captured data to ground-station modules in near real time. Ground-based operators are then able to manipulate the data and gain heightened situational information, which is forwarded to command-and-control (C^2) nodes to enhance tactical decision-making.
- **Imaging radar.** An imaging radar's receiver and processor are so sensitive that an image of the detected target is displayed on a scope. Imaging radar, such as side-looking airborne radar (SLAR), is generally used on airborne or space-borne platforms. Imaging radar typically does not provide the same resolution as the FLIRS and is less likely to be used for terminal target acquisition.
- **Countermortar (CM) and counterbattery (CB) radar.** CM and CB radar usually transmit two beams of energy that sweep above the horizon. An artillery or mortar round or a rocket passing through the beams reflects two signals that are received and plotted to determine the origin of the round.

ACOUSTIC

2-18. The three predominant types of acoustical detection systems are—

- **Human ear.** Every soldier, whether engaged in normal operations or at a listening post, is an acoustic sensor. However, visual confirmation is usually preferred.
- **Flash-sound ranging.** Flash-sound ranging is used against artillery. Light travels faster than sound, so enemy sound-ranging teams can determine the distance to a gun tube by accurately measuring the time between seeing a muzzle flash and hearing the sound. If the sound is detected by two or more teams, analysts plot the ranges using automated data-processing computers. The target is located where the plots intersect.
- **Ground-based microphone array.** Ground-based microphone-array systems allow listeners to record acoustic signatures and accurately triangulate their positions.

RADIO

2-19. Threat forces make a great effort to search for, detect, and locate the sources of US radio communications. They use various direction-finding techniques to locate opposing emitters. Once an emitter is detected, an enemy can take a number of actions, ranging from simply intercepting the transmissions to jamming or targeting the emitter for destruction.

MULTISPECTRAL AND HYPERSPECTRAL

2-20. Recent advancements in sensor acquisition and information-processing technologies have fostered the advent of multispectral and hyperspectral sensors:

- **Multispectral.** Multispectral sensors typically scan a few broad-band channels within the EM spectrum. An example of a multispectral sensor might be one which coincidentally scans the visual and thermal IR portions of the EM spectrum. Such sensors allow an enemy to assess a cross section of EM wavelengths and acquire a target in one wavelength even though it might be effectively concealed in another.

- **Hyperspectral.** Hyperspectral sensors collect data across a continuous portion of the EM spectrum. These sensors scan many channels across a relatively narrow bandwidth and provide detailed information about target spatial and spectral patterns. Absorption and emission bands of given substances often occur within very narrow bandwidths. They allow high-resolution, hyperspectral sensors to distinguish the properties of the substances to a finer degree than an ordinary broadband sensor.

CCD VERSUS THREAT SENSORS

2-21. Target acquisition can be accomplished by a variety of sensors that operate throughout the EM spectrum. This poses a challenge in CCD planning and employment—determining which enemy sensor(s) that CCD operations should be designed to defeat. Unfortunately, no single answer is correct for all situations. Unit commanders without specific guidance from higher echelons assess their tactical situation and plan CCD operations accordingly. If intelligence data indicate that an enemy will use visual sensors for recon and target acquisition, then visual countermeasures must be employed. For IR or radar sensors, countermeasures that are effective in those spectra must be employed. If a multispectral or hyperspectral threat is anticipated, CCD operations are conducted to protect a unit in its most vulnerable EM bandwidths. Very few available camouflage materials or techniques provide complete broadband protection.

Chapter 3

Fundamentals

To remain a viable force on the battlefield, units must understand CCD fundamentals because they are essential to survivability. To design and place effective CCD, soldiers must constantly consider an enemy's point of view. (What will it see? What characteristics will its sensors detect?) Placing a low priority on CCD because of time constraints, minimal resources, or inconvenience could result in mission failure and unnecessary loss of life. (Appendix D contains more information on individual CCD.)

SECTION I - PRINCIPLES

AVOIDING DETECTION

3-1. The primary goal of CCD is to avoid enemy detection; however, this is not always feasible. In some cases, CCD may succeed by merely preventing an enemy from identifying a target. Simply avoiding identification is often sufficient to increase survivability. The following seven rules are critical when considering how to avoid detection or identification:

- Identify the enemy's detection capabilities.
- Avoid detection by the enemy's routine surveillance.
- Take countermeasures against the enemy's sensors.
- Employ realistic, CCD countermeasures.
- Minimize movement.
- Use decoys properly.
- Avoid predictable operational patterns.

IDENTIFYING THE THREAT

3-2. Obtain as much information as possible about an enemy's surveillance capability. Intelligence preparation of the battlefield (IPB) should—

- Include the sensors that an enemy may use in a particular AO.
- Include information on the enemy's tactical employment of the sensors, if possible.
- Assess the impact of the enemy's surveillance potential on the target under consideration. This assessment varies with the relative positions of the sensor and the target on the battlefield, the role of the target, and the physical characteristics of the sensor and the target.

AVOIDING DETECTION BY ROUTINE SURVEILLANCE

3-3. Sophisticated sensors often have narrow fields of view. Furthermore, sensors can be very expensive and are unlikely to be deployed in such numbers as to enable coverage of the entire battlefield at all times. Sophisticated sensors are most likely to be deployed in those areas where an enemy suspects that friendly targets are deployed. The enemy may suspect that an area contains targets because of detection by less sophisticated, wider-coverage sensors or because of tactical analysis. Therefore, an important aspect of remaining undetected is to avoid detection by routine enemy surveillance.

3-4. Many sensors operate as well at night as they do during the day. Therefore, darkness does not provide effective protection from surveillance. Passive sensors are very difficult to detect, so assume that

ESSENTIAL ARMY MANUAL
(491) CAMOUFLAGE, CONCEALMENT AND DECOYS

they are being used at night. Do not allow antidetection efforts to lapse during the hours of darkness. For example, conceal spoil while excavating a fighting position, even at night. Certain types of smoke will also defeat NVDs.

TAKING COUNTERMEASURES

3-5. In some cases, it might be appropriate to take action against identified enemy sensors. The ability to deploy countermeasures depends on a number of factors—the effective range of friendly weapons, the distance to enemy sensors, and the relative cost in resources versus the benefits of preventing the enemy's use of the sensor. An additional factor to consider is that the countermeasure itself may provide an enemy with an indication of friendly intentions.

EMPLOYING REALISTIC CCD

3-6. The more closely a target resembles its background, the more difficult it is for an enemy to distinguish between the two. Adhering to this fundamental CCD principle requires awareness of the surroundings, proper CCD skills, and the ability to identify target EM signatures that enemy sensors will detect.

VISUAL SENSORS

3-7. The most plentiful, reliable, and timely enemy sensors are visual. Therefore, CCD techniques effective in the visual portion of the EM spectrum are extremely important. Something that cannot be seen is often difficult to detect, identify, and target. BDUs, standard camouflage screening paint patterns (SCSPPs), LCSS, and battlefield obscurants are effective CCD techniques against visual sensors. Full-coverage CCD helps avoid visual detection by the enemy. When time is short, apply CCD first to protect the target from the most likely direction of attack and then treat the remainder of the target as time allows.

NEAR INFRARED SENSORS

3-8. NIR sights are effective at shorter ranges (typically 900 meters) than enemy main guns. While red filters help preserve night vision, they cannot prevent NIR from detecting light from long distances. Therefore, careful light discipline is an important countermeasure to NIR sensors and visual sensors (such as image intensifiers). BDUs, LCSS, battlefield obscurants, and SCSPPs are designed to help defeat NIR sensors.

INFRARED SENSORS

3-9. Natural materials and terrain shield heat sources from IR sensors and break up the shape of cold and warm military targets viewed on IR sensors. Do not raise vehicle hoods to break windshield glare because this exposes a hot spot for IR detection. Even if the IR system is capable of locating a target, the target's actual identity can still be disguised. Avoid building unnecessary fires. Use vehicle heaters only when necessary. BDU dyes, LCSSs, IR-defeating obscurants, and chemical-resistant paints help break up IR signatures; but they will not defeat IR sensors.

ULTRAVIOLET SENSORS

3-10. UV sensors are a significant threat in snow-covered areas. Winter paint patterns, the arctic LCSS, and terrain masking are critical means for defending against these sensors. Any kind of smoke will defeat UV sensors. Field-expedient countermeasures, such as constructing snow walls, also provide a means of defeating UV sensors.

RADAR

3-11. An enemy uses MTI, imaging, CM, and CB radars. Mission dictates the appropriate defense, while techniques depend on the equipment available.

Moving-Target Indicator

3-12. MTI radar is a threat to ground forces near a battle area. Radar-reflecting metal on uniforms has been reduced, and Kevlar helmets and body armor are now radar-transparent. Plastic canteens are standard issue, and buttons and other nonmetal fasteners have replaced metal snaps on most field uniforms. A soldier wearing only the BDU cannot be detected until he is very close to MTI radar.

3-13. Soldiers still carry metal objects (ammunition, magazines, weapons) to accomplish their mission, and most radar can detect these items. Therefore, movement discipline is very important. Moving by covered routes (terrain masking) prevents radar detection. Slow, deliberate movements across areas exposed to radar coverage helps avoid detection by MTI radar.

3-14. Vehicles are large radar-reflecting targets, and a skilled MTI operator can even identify the type of vehicle. Moving vehicles can be detected by MTI radar from 20 kilometers, but travelling by covered routes helps protect against surveillance.

Imaging

3-15. Imaging radar is not a threat to individual soldiers. Concealing vehicles behind earth, masonry walls, or dense foliage effectively screens them from imaging radar. Light foliage may provide complete visual concealment; however, it is sometimes totally transparent to imaging radar. When properly deployed, the LCSS effectively scatters the beam of imaging radar. (See appendix C for more information.)

Countermortar and Counterbattery

3-16. Radar is subject to overload. It is very effective and accurate when tracking single rounds; however, it cannot accurately process data on multiple rounds (four or more) that are fired simultaneously. Chaff is also effective against CM and CB radar if it is placed near the radar.

ACOUSTIC SENSORS

3-17. Noise discipline defeats detection by the human ear. Pyrotechnics or loudspeakers can screen noise, cover inherently noisy activities, and confuse sound interpretation.

3-18. It is possible to confuse an enemy by screening flashes or sounds. Explosives or pyrotechnics, fired a few hundred meters from a battery's position within a second of firing artillery, will effectively confuse sound-ranging teams. Coordinating fire with adjacent batteries (within two seconds) can also confuse enemy sound-ranging teams.

RADIO SENSORS

3-19. The best way to prevent an enemy from locating radio transmitters is to minimize transmissions, protect transmissions from enemy interception, and practice good radiotelephone-operator (RATELO) procedures. Preplanning message traffic, transmitting as quickly as possible, and using alternate communication means whenever possible ensure that transmissions are minimized. To prevent the enemy from intercepting radio communications, change the radio frequencies and use low-power transmissions, terrain masking, or directional or short-range antennas.

MINIMIZING MOVEMENT

3-20. Movement attracts the enemy's attention and produces a number of signatures (tracks, noise, hot spots, dust). In operations that inherently involve movement (such as offensive operations), plan, discipline, and manage movement so that signatures are reduced as much as possible. (See chapter 4 for information on disciplined movement techniques.)

USING DECOYS

3-21. Use decoys to confuse an enemy. The goal is to divert enemy resources into reporting or engaging false targets. An enemy who has mistakenly identified decoys as real targets is less inclined to search

ESSENTIAL ARMY MANUAL
(493) CAMOUFLAGE, CONCEALMENT AND DECOYS

harder for the actual, well-hidden targets. The keys to convincing an enemy that it has found the real target are—

- Decoy fidelity (realism), which refers to how closely the multispectral decoy signature represents the target signature.
- Deployment location, which refers to whether or not a decoy is deployed so that the enemy will recognize it as typical for that target type. For example, a decoy tank is not properly located if it is placed in the middle of a lake.

3-22. A high-fidelity decoy in a plausible location often fools an enemy into believing that it has acquired the real target. Deploying low-fidelity decoys, however, carries an associated risk. If an enemy observes a decoy and immediately recognizes it as such, it will search harder for the real target since decoys are generally deployed in the same vicinity as the real targets. Plausible, high-fidelity decoys specifically designed to draw enemy fire away from real targets should be deployed to closely represent the multispectral signatures of the real targets. Properly deployed decoys have been proven in operational employment and experimental field tests to be among the most effective of all CCD techniques.

AVOIDING OPERATIONAL PATTERNS

3-23. An enemy can often detect and identify different types of units or operations by analyzing the signature patterns that accompany their activities. For example, an offensive operation is usually preceded by the forward movement of engineer obstacle-reduction assets; petroleum, oils, and lubricants (POL); and ammunition. Such movements are very difficult to conceal; therefore, an alternative is to modify the pattern of resupply. An enemy will recognize repetitive use of the same CCD techniques.

APPLYING RECOGNITION FACTORS

3-24. To camouflage effectively, continually consider the threat's viewpoint. Prevent patterns in antidetection countermeasures by applying the following recognition factors to tactical situations. These factors describe a target's contrast with its background. If possible, collect multispectral imagery to determine which friendly target signatures are detectable to enemy sensors.

REFLECTANCE

3-25. Reflectance is the amount of energy returned from a target's surface as compared to the energy striking the surface. Reflectance is generally described in terms of the part of the EM spectrum in which the reflection occurs:

- *Visual reflectance* is characterized by the color of a target. Color contrast can be important, particularly at close ranges and in homogeneous background environments such as snow or desert terrain. The longer the range, the less important color becomes. At very long ranges, all colors tend to merge into a uniform tone. Also, the human eye cannot discriminate color in poor light.
- *Temperature reflectance* is the thermal energy reflected by a target (except when the thermal energy of a target is self-generated, as in the case of a hot engine). IR imaging sensors measure and detect differences in temperature-reflectance levels (known as thermal contrast).
- *Radar-signal reflectance* is the part of the incoming radio waves that is reflected by a target. Radar sensors detect differences in a target's reflected radar return and that of the background. Since metal is an efficient radio-wave reflector and metals are still an integral part of military equipment, radar return is an important reflectance factor.

SHAPE

3-26. Natural background is random, and most military equipment has regular features with hard, angular lines. Even an erected camouflage net takes on a shape with straight-line edges or smooth curves between support points. An enemy can easily see silhouetted targets, and its sensors can detect targets against any background unless their shape is disguised or disrupted. Size, which is implicitly related to shape, can also distinguish a target from its background.

SHADOW

3-27. Shadow can be divided into two types:

- *A cast shadow* is a silhouette of an object projected against its background. It is the more familiar type and can be highly conspicuous. In desert environments, a shadow cast by a target can be more conspicuous than the target itself.
- *A contained shadow* is the dark pool that forms in a permanently shaded area. Examples are the shadows under the track guards of an armored fighting vehicle (AFV), inside a slit trench, inside an open cupola, or under a vehicle. Contained shadows show up much darker than their surroundings and are easily detected by an enemy.

MOVEMENT

3-28. Movement always attracts attention against a stationary background. Slow, regular movement is usually less obvious than fast, erratic movement.

NOISE

3-29. Noise and acoustic signatures produced by military activities and equipment are recognizable to the enemy.

TEXTURE

3-30. A rough surface appears darker than a smooth surface, even if both surfaces are the same color. For example, vehicle tracks change the texture of the ground by leaving clearly visible track marks. This is particularly true in undisturbed or homogeneous environments, such as a desert or virgin snow, where vehicle tracks are highly detectable. In extreme cases, the texture of glass or other very smooth surfaces causes a shine that acts as a beacon. Under normal conditions, very smooth surfaces stand out from the background. Therefore, eliminating shine must be a high priority in CCD.

PATTERNS

3-31. Rows of vehicles and stacks of war materiel create equipment patterns that are easier to detect than random patterns of dispersed equipment. Equipment patterns should be managed to use the surroundings for vehicle and equipment dispersal. Equipment dispersal should not be implemented in such a way that it reduces a unit's ability to accomplish its mission.

3-32. Equipment paint patterns often differ considerably from background patterns. The critical relationships that determine the contrast between a piece of equipment and its background are the distance between the observer and the equipment and the distance between the equipment and its background. Since these distances usually vary, it is difficult to paint equipment with a pattern that always allows it to blend with its background. As such, no single pattern is prescribed for all situations. Field observations provide the best match between equipment and background.

3-33. The overall terrain pattern and the signatures produced by military activity on the terrain are important recognition factors. If a unit's presence is to remain unnoticed, it must match the signatures produced by stationary equipment, trucks, and other activities with the terrain pattern. Careful attention must also be given to vehicle tracks and their affect on the local terrain during unit ingress, occupation, and egress.

SITE SELECTION

3-34. Site selection is extremely important because the location of personnel and equipment can eliminate or reduce recognition factors. If a tank is positioned so that it faces probable enemy sensor locations, the thermal signature from its hot engine compartment is minimized. If a vehicle is positioned under foliage, the exhaust will disperse and cool as it rises, reducing its thermal signature and blending it more closely with the background. Placing equipment in defilade (dug-in) positions prevents detection by ground-mounted radar. The following factors govern site selection:

MISSION

3-35. The mission is the most important factor in site selection. A particular site may be excellent from a CCD standpoint, but the site is useful only if the mission is accomplished. If a site is so obvious that the enemy will acquire and engage a target before mission accomplishment, the site was poorly selected to begin with. Survivability is usually a part of most missions, so commanders must first evaluate the worthiness of a site with respect to mission accomplishment and then consider CCD.

DISPERSION

3-36. Dispersion requirements dictate the size of a site. A site has limited usefulness if it will not permit enough dispersal for survivability and effective operations.

TERRAIN PATTERNS

3-37. Every type of terrain, even a flat desert, has a discernible pattern. Terrain features can blur or conceal the signatures of military activity. By using terrain features, CCD effectiveness can be enhanced without relying on additional materials. The primary factor to consider is whether using the site will disturb the terrain pattern enough to attract an enemy's attention. The goal is not to disturb the terrain pattern at all. Any change in an existing terrain pattern will indicate the presence of activity. Terrain patterns have distinctive characteristics that are necessary to preserve. The five general terrain patterns are—

- **Agricultural.** Agricultural terrain has a checkerboard pattern when viewed from aircraft. This is a result of the different types of crops and vegetation found on most farms.
- **Urban.** Urban terrain is characterized by uniform rows of housing with interwoven streets and interspersed trees and shrubs.
- **Wooded.** Woodlands are characterized by natural, irregular features, unlike the geometric patterns of agricultural and urban terrains.
- **Barren.** Barren terrain presents an uneven, irregular work of nature without the defined patterns of agricultural and urban areas. Desert environments are examples of barren terrain.
- **Arctic.** Arctic terrain is characterized by snow and ice coverage.

CCD DISCIPLINE

3-38. CCD discipline is avoiding an activity that changes the appearance of an area or reveals the presence of military equipment. CCD discipline is a continuous necessity that applies to every soldier. If the prescribed visual and audio routines of CCD discipline are not observed, the entire CCD effort may fail. Vehicle tracks, spoil, and debris are the most common signs of military activity. Their presence can negate all efforts of proper placement and concealment.

3-39. CCD discipline denies an enemy the indications of a unit's location or activities by minimizing disturbances to a target area. To help maintain unit viability, a unit must integrate all available CCD means into a cohesive plan. CCD discipline involves regulating light, heat, noise, spoil, trash, and movement. Successful CCD discipline depends largely on the actions of individual soldiers. Some of these actions may not be easy on a soldier, but his failure to observe CCD discipline could defeat an entire unit's CCD efforts and possibly impact the unit's survivability and mission success.

3-40. TACSOPs prescribing CCD procedures aid in enforcing CCD discipline, and they should—

- List specific responsibilities for enforcing established CCD countermeasures and discipline.
- Detail procedures for individual and unit conduct in assembly areas (AAs) or other situations that may apply to the specific unit.

3-41. Units should have frequent CCD battle drills. CCD discipline is a continuous requirement that calls for strong leadership, which produces a disciplined CCD consciousness throughout the entire unit. Appendix B contains additional guidance for incorporating CCD into a unit TACSOP.

LIGHT AND HEAT

3-42. Light and heat discipline, though important at all times, is crucial at night. As long as visual observation remains a primary recon method, concealing light signatures remains an important CCD countermeasure. Lights that are not blacked out at night can be observed at great distances. For example, the human eye can detect camp fires from 8 kilometers and vehicle lights from 20 kilometers. Threat surveillance can also detect heat from engines, stoves, and heaters from great distances. When moving at night, vehicles in the forward combat area should use ground guides and blackout lights. When using heat sources is unavoidable, use terrain masking, exhaust baffling, and other techniques to minimize thermal signatures of fires and stoves.

NOISE

3-43. Individuals should avoid or minimize actions that produce noise. For example, muffle generators by using shields or terrain masking or place them in defilade positions. Communications personnel should operate their equipment at the lowest possible level that allows them to be heard and understood. Depending on the terrain and atmospheric conditions, noise can travel great distances and reveal a unit's position to an enemy.

SPOIL

3-44. The prompt and complete policing of debris and spoil is an essential CCD consideration. Proper spoil discipline removes a key signature of a unit's current or past presence in an area.

TRACK

3-45. Vehicle tracks are clearly visible from the air, particularly in selected terrain. Therefore, track and movement discipline is essential. Use existing roads and tracks as much as possible. When using new paths, ensure that they fit into the existing terrain's pattern. Minimize, plan, and coordinate all movement; and take full advantage of cover and dead space.

SECTION II – TECHNIQUES AND MATERIALS

TECHNIQUES

3-46. CCD is an essential part of tactical operations. It must be integrated into METT-TC analyses and the IPB process at all echelons. CCD is a primary consideration when planning OPSEC. The skillful use of CCD techniques is necessary if a unit is to conceal itself and survive. A general knowledge of CCD methods and techniques also allows friendly troops to recognize CCD better when the enemy uses it. Table 3-1, page 3-8, lists the five general techniques of employing CCD—hiding, blending, disguising, disrupting, and decoying.

HIDING

3-47. Hiding is screening a target from an enemy's sensors. The target is undetected because a barrier hides it from a sensor's view. Every effort should be made to hide all operations; this includes using conditions of limited visibility for movement and terrain masking. Examples of hiding include—

- Burying mines.
- Placing vehicles beneath tree canopies.
- Placing equipment in defilade positions.
- Covering vehicles and equipment with nets.

- Hiding roads and obstacles with linear screens.
- Using battlefield obscurants, such as smoke.

Table 3-1. CCD techniques

CCD Techniques	Sensor Systems		
	Optical	Thermal	Radar
Hiding	Earth cover Earth embankments Vegetation LCSS Screens Smoke	Earth cover Earth embankments Vegetation LCSS Screens Smoke	Chaff Earth cover Earth embankments Vegetation Nets RAM LCSS
Blending	Paint Foam Lights Vegetation LCSS Textured mats	Thermal paint Foam Air conditioning/heating Vegetation LCSS Textured mats Water Insulation	Vegetation LCSS RAM Reshaping Textured mats
Disguising	Reshaping Paint LCSS	Reshaping Paint	Corner reflectors
Disrupting	Camouflage sails FOS Pyrotechnics Smudge pots Balloons Strobe lights Tracer simulators Smoke	Flares Smoke	Chaff Corner reflectors
Decoying	Decoy target (pneumatic or rigid structures) Lights Smoke	Decoy target Flares Air conditioning/heating Smoke	Decoy target Corner reflectors Signal generators

BLENDING

3-48. Blending is trying to alter a target's appearance so that it becomes a part of the background. Generally, it is arranging or applying camouflage material on, over, and/or around a target to reduce its contrast with the background. Characteristics to consider when blending include the terrain patterns in the vicinity and the target's size, shape, texture, color, EM signature, and background.

DISGUISING

3-49. Disguising is applying materials on a target to mislead the enemy as to its true identity. Disguising changes a target's appearance so that it resembles something of lesser or greater significance. For example, a missile launcher might be disguised to resemble a cargo truck or a large building might be disguised to resemble two small buildings.

DISRUPTING

3-50. Disrupting is altering or eliminating regular patterns and target characteristics. Disrupting techniques include pattern painting, deploying camouflage nets over selected portions of a target, and using shape disrupters (such as camouflage sails) to eliminate regular target patterns.

DECOYING

3-51. Decoying is deploying a false or simulated target(s) within a target's scene or in a position where the enemy might conclude that it has found the correct target(s). Decoys generally draw fire away from real targets. Depending on their fidelity and deployment, decoys will greatly enhance survivability.

TESTS AND EVALUATIONS

3-52. Until recently, the effectiveness of CCD techniques had not been scientifically quantified. As such, CCD was not widely accepted in the US military as an effective means of increasing survivability. However, the Joint Camouflage, Concealment, and Deception (JCCD) Joint Test and Evaluation (JT&E) completed in 1995 measured the effectiveness of CCD against manned aerial attacks. It provided military services the basis for guidance on CCD-related issues. JCCD field tests were conducted in multiple target environments using a broad cross section of US attack aircraft flying against different classes of military targets. In controlled attack sorties, targets were attacked before and after employing CCD techniques.

3-53. The presence of CCD greatly reduced correct target attacks, particularly when decoys were employed as part of the CCD plan. Other JCCD findings included the following:

- CCD significantly increased aircrew aim-point error.
- CCD increased the target's probability of survival.
- Each CCD technique (hiding, blending, disguising, disrupting, and decoying) was effective to some degree in increasing the probability of survival.
- CCD was effective in all tested environments (desert, temperate, and subarctic).

NATURAL CONDITIONS

3-54. Properly using terrain and weather is a first priority when employing CCD. Cover provided by the terrain and by conditions of limited visibility is often enough to conceal units. The effective use of natural conditions minimizes the resources and the time devoted to CCD. The terrain's concealment properties are determined by the number and quality of natural screens, terrain patterns, and the type and size of targets.

FORESTS

3-55. Forests generally provide the best type of natural screen against optical recon, especially if the crowns of the trees are wide enough to prevent aerial observation of the ground. Forests with undergrowth also hinder ground observation. Deciduous (leafing) forests are not as effective during the months when trees are bare, while coniferous (evergreen) forests preserve their concealment properties all year. When possible, unit movements should be made along roads and gaps that are covered by tree crowns. Shade should be used to conceal vehicles, equipment, and personnel from aerial observation.

OPEN TERRAIN

3-56. Limited visibility is an especially important concealment tool when conducting operations in open terrain. The threat, however, will conduct recon with a combination of night-surveillance devices, radar, IR sensors, and terrain illumination. When crossing open terrain during limited visibility, supplement concealment with smoke.

DEAD SPACE

3-57. Units should not locate or move along the topographic crests of hills or other locations where they are silhouetted against the sky. They should use reverse slopes of hills, ravines, embankments, and other terrain features as screens to avoid detection by ground-mounted sensors. IPB concealment and terrain overlays should identify areas of dead space. If overlays are not available, use the line-of-sight (LOS) method to identify areas of dead space.

WEATHER

3-58. Conditions of limited visibility (fog, rain, snowfall) hamper recon by optical sensors. Dense fog is impervious to visible sensors and some thermal sensors, making many threat night-surveillance devices unusable. Dense fog and clouds are impenetrable to thermal sensors (IR). Rain, snow, and other types of precipitation hinder optical, thermal, and radar sensors.

SMOKE

3-59. Smoke is an effective CCD tool when used by itself or with other CCD techniques. It can change the dynamics of a battle by blocking or degrading the spectral bands that an enemy's target-acquisition and weapons systems use, including optical and thermal bands.

DATA SOURCES

3-60. Commanders must be able to evaluate natural conditions in their area to effectively direct unit concealment. They must know the terrain and weather conditions before mission execution. In addition to IPB terrain overlays, weather reports, and topographic maps, commanders should use aerial photographs, recon, and information gathered from local inhabitants to determine the terrain's natural concealment properties.

MATERIALS

3-61. Using natural conditions and materials is the first CCD priority, but using man-made materials can greatly enhance CCD efforts. Available materials include pattern-painted equipment, camouflage nets (LCSS), radar-absorbing paint (RAP), radar-absorbing material (RAM), false operating surfaces (FOSs), vegetation, expedient paint, decoys, and battlefield by-products (construction materials, dirt). (Appendix E lists man-made CCD materials that are available through the supply system.)

PATTERN PAINT

3-62. Pattern-painted vehicles blend well with the background and can hide from optical sensors better than those painted a solid, subdued color. Pattern-painted equipment enhances antidetection by reducing shape, shadow, and color signatures. Improved paints also help avoid detection by reducing a target's reflectance levels in the visible and IR portions of the EM spectrum. The result is a vehicle or an item of equipment that blends better with its background when viewed by threat sensors. While a patterned paint scheme is most effective in static positions, it also tends to disrupt aim points on a moving target. (See appendix E for a list of available paints.)

CAMOUFLAGE NETS

3-63. The LCSS is the standard Army camouflage net currently available, and it can be ordered through normal unit supply channels (see appendix E). The LCSS reduces a vehicle's visual and radar signatures. Stainless steel fibers in the LCSS material absorb some of the radar signal and reflect most of the remaining signal in all directions. The result is a small percentage of signal return to the radar for detection. The radar-scattering capabilities of the LCSS are effective only if there is at least 2 feet of space between the LCSS and the camouflaged equipment and if the LCSS completely covers the equipment. Do not place a radar-scattering net over a radar antenna because it interferes with transmission. The LCSS is also available in a radar-transparent model.

3-64. The three different LCSS color patterns are desert, woodland, and arctic. Each side of each LCSS has a slightly different pattern to allow for seasonal variations. The LCSS uses modular construction that allows the coverage of various sizes of equipment. (Appendix C discusses the required components and the instructions for assembling LCSS structures for different sizes of equipment.)

VEGETATION

3-65. Use branches and vines to temporarily conceal vehicles, equipment, and personnel. Attach vegetation to equipment with camouflage foliage brackets, spring clips, or expedient means (such as plastic tie-wraps). Use other foliage to complete the camouflage or to supplement natural-growing vegetation. Also use cut foliage to augment other artificial CCD materials, such as branches placed on an LCSS to break up its outline. Be careful when placing green vegetation since the underside of leaves presents a lighter tone in photographs. Replace cut foliage often because it wilts and changes color rapidly. During training exercises, ensure that cutting vegetation and foliage does not adversely effect the natural environment (coordinate with local authorities).

Living Vegetation

3-66. Living vegetation can be obtained in most environments, and its color and texture make it a good blending agent. However, foliage requires careful maintenance to keep the material fresh and in good condition. If branches are not placed in their proper growing positions, they may reveal friendly positions to enemy observers. Cutting large amounts of branches can also reveal friendly positions, so cut all vegetation away from target areas.

3-67. Living vegetation presents a chlorophyll response at certain NIR wavelengths. As cut vegetation wilts, it loses color and its NIR-blending properties, which are related to the chlorophyll response. Replace cut vegetation regularly because over time it becomes a detection cue rather than an effective concealment technique.

Dead Vegetation

3-68. Use dead vegetation (dried grass, hay, straw, branches) for texturing. It provides good blending qualities if the surrounding background vegetation is also dead. Dead vegetation is usually readily available and requires little maintenance; however, it is flammable. Due to the absence of chlorophyll response, dead vegetation offers little CCD against NIR sensors and hyperspectral sensors operating in the IR regions.

Foliage Selection

3-69. When selecting foliage for CCD, consider the following:

- Coniferous vegetation is preferred to deciduous vegetation since it maintains a valid chlorophyll response longer after being cut.
- Foliage cut during periods of high humidity (at night, during a rainstorm, or when there is fog or heavy dew) will wilt more slowly.
- Foliage with leaves that feel tough to the fingers and branches with large leaves are preferred because they stay fresher longer.
- Branches that grow in direct sunlight are tougher and will stay fresher longer.
- Branches that are free of disease and insects will not wilt as rapidly.

CHLOROPHYLL RESPONSE

3-70. Standard-issue camouflage materials (LCSS) are designed to exhibit an artificial chlorophyll response at selected NIR wavelengths. Nonstandard materials (sheets, tarps) are not likely to exhibit a chlorophyll response and will not blend well with standard CCD material or natural vegetation. Use nonstandard materials only as CCD treatments against visual threat sensors, not against NIR or hyperspectral threat sensors.

EXPEDIENT PAINT

CAUTION

Expedient paint containing motor oil should be used with extreme caution.

3-71. Use earth, sand, and gravel to change or add color, provide a coarse texture, simulate cleared spots or blast marks, and create shapes and shadows. Mud makes an excellent field expedient for toning down bright, shiny objects (glass, bayonets, watches). Add clay (in mud form) of various colors to crankcase oil to produce a field-expedient paint. Table 3-2 provides instructions on how to mix soil-based expedient paints. Use surface soils to mimic natural surface color and reflectivity.

Table 3-2. Expedient paints

Paint Materials	Mixing	Color	Finish
Earth, GI soap, water, soot, paraffin	Mix soot with paraffin, add to solution of 8 gal water and 2 bars soap, and stir in earth.	Dark gray	Flat, lusterless
Oil, clay, water, gasoline, earth	Mix 2 gal water with 1 gal oil and ¼ to ½ gal clay, add earth, and thin with gasoline or water.	Depends on earth colors	Glossy on metal, otherwise dull
Oil, clay, GI soap, water, earth	Mix 1½ bars soap with 3 gal water, add 1 gal oil, stir in 1 gal clay, and add earth for color.	Depends on earth colors	Glossy on metal, otherwise dull
Note. Use canned milk or powdered eggs to increase the binding properties of field-expedient paints.			

RADAR-ABSORBING MATERIAL

3-72. RAM was designed for placement on valuable military equipment. It absorbs radar signals that are transmitted in selected threat wave bands and reduces the perceived radar cross section (RCS) of the treated equipment. RAM is expensive relative to other CCD equipment and is not yet widely available. RAP offers the same RCS reduction benefits as RAM, and it is also expensive.

BATTLEFIELD BY-PRODUCTS

3-73. Battlefield by-products (construction materials, dirt) can be used to formulate expedient CCD countermeasures. For example, use plywood and two-by-fours to erect expedient target decoys or use dirt to construct concealment berms.

DECOYS

3-74. Decoys are among the most effective of all CCD tools. The proper use of decoys provides alternate targets against which an enemy will expend ammunition, possibly revealing its position in the process. Decoys also enhance friendly survivability and deceive an enemy about the number and location of friendly weapons, troops, and equipment.

Employment Rationale

3-75. Decoys are used to attract an enemy's attention for a variety of tactical purposes. Their main use is to draw enemy fire away from high-value targets (HVTs). Decoys are generally expendable, and they—

- Can be elaborate or simple. Their design depends on several factors, such as the target to be decoyed, a unit's tactical situation, available resources, and the time available to a unit for CCD employment.
- Can be preconstructed or made from field-expedient materials. Except for selected types, preconstructed decoys are not widely available (see appendix E). A typical Army unit can construct effective, realistic decoys to replicate its key equipment and features through imaginative planning and a working knowledge of the EM signatures emitted by the unit.

3-76. Proper decoy employment serves a number of tactical purposes, to include—

- Increasing the survivability of key unit equipment and personnel.
- Deceiving the enemy about the strength, disposition, and intentions of friendly forces.
- Replacing friendly equipment removed from the forward line of own troops (FLOT).
- Drawing enemy fire, which reveals its positions.
- Encouraging the enemy to expend munitions on relatively low-value targets (decoys).

Employment Considerations

3-77. The two most important factors regarding decoy employment are location and fidelity (realism):

- **Location.** Logically placing decoys will greatly enhance their plausibility. Decoys are usually placed near enough to the real target to convince an enemy that it has found the target. However, a decoy must be far enough away to prevent collateral damage to the real target when the decoy draws enemy fire. Proper spacing between a decoy and a target depends on the size of the target, the expected enemy target-acquisition sensors, and the type of munitions directed against the target.
- **Fidelity.** Decoys must be constructed according to a friendly unit's SOP and must include target features that an enemy recognizes. The most effective decoys are those that closely resemble the real target in terms of EM signatures. Completely replicating the signatures of some targets, particularly large and complex targets, can be very difficult. Therefore, decoy construction should address the EM spectral region in which the real target is most vulnerable. The seven recognition factors that allow enemy sensors to detect a target are conversely important for decoys. When evaluating a decoy's fidelity, it should be recognizable in the same ways as the real target, only more so. Try to make the decoy slightly more conspicuous than the real target.

Chapter 4
Offensive Operations

CCD countermeasures implemented during an offensive operation deceive the enemy or prevent it from discovering friendly locations, actions, and intentions. Successful CCD contributes to achieving surprise and reduces subsequent personnel and equipment losses.

PREPARATIONS

4-1. The main CCD concern in preparing for offensive operations is to mask tactical unit deployment. While CCD is the primary means of masking these activities, deceptive operations frequently achieve the same goals.

SIGNATURES

4-2. Offensive operations create signatures that are detectable to an enemy. Analyzing these signatures may alert an enemy to the nature of an offensive operation (such as planning and location). Commanders at all levels should monitor operation signatures and strive to conceal them from enemy surveillance. These signatures include—

- Increasing scouting and recon activity.
- Preparing traffic routes.
- Moving supplies and ammunition forward.
- Breaching obstacles.
- Preparing and occupying AAs (engineer function).
- Preparing and occupying forward artillery positions.
- Increasing radio communications.

ASSEMBLY AREAS

4-3. Prepare AAs during limited visibility. They should then suppress the signatures that their preparations produced and remove any indications of their activities upon mission completion.

4-4. Designate AAs on terrain with natural screens and a developed network of roads and paths. Thick forests and small towns and villages often provide the best locations. If natural screens are unavailable, use spotty sectors of the terrain or previously occupied locations. Place equipment on spots of matching color, and take maximum advantage of artificial CCD materials.

4-5. Designate concealed routes for movement into and out of an area. Mask noise by practicing good noise discipline. For instance, armor movements can be muffled by the thunder of artillery fire, the noise of low-flying aircraft, or the transmission of sounds from broadcast sets.

4-6. Position vehicles to take full advantage of the terrain's natural concealment properties, and cover the vehicles with camouflage nets. Apply paint and cut vegetation to vehicles to enhance CCD at AAs and during battle. (When using vegetation for this type of CCD treatment, do not cut it from areas close to vehicles.) AAs are particularly vulnerable to aerial detection. Strictly enforce track, movement, and radio discipline. Remove tracks by covering or sweeping them with branches.

4-7. While at an AA, personnel should apply individual CCD. Applying stick paint and cut vegetation enhances CCD during all phases of an operation.

DECOYS

4-8. An enemy may interpret decoy construction as an effort to reinforce a defensive position. Laying false minefields and building bunkers and positions can conceal actual offensive preparations and give the enemy the impression that defenses are being improved. If necessary, conduct engineer preparation activities on a wide front so that the area and direction of the main attack are not revealed.

MOVEMENT

4-9. Move troops, ammunition, supplies, and engineer breaching equipment forward at night or during limited visibility. Although an enemy's use of radar and IR aerial recon hinders operations at night, darkness remains a significant concealment tool. Select routes that take full advantage of the terrain's screening properties. Commanders must understand how to combine darkness and the terrain's concealing properties to conceal troop and supply movements.

4-10. When conducting a march, convoy commanders must strictly enforce blackout requirements and the order of march. Guidelines concerning lighting, march orders, and other requirements are usually published in SOPs or operation orders (OPORDs). Required lighting conditions vary depending on the type of movement (convoy versus single vehicle) and a unit's location (forward edge of the battle area [FEBA], division area, corps rear area). Inspect each vehicle's blackout devices for proper operation.

4-11. Enemy aerial recon usually focuses on open and barely passable route sectors. When on a march, vehicles should pass these types of sectors at the highest possible speeds. If prolonged delays result from encountering an unexpected obstacle, halt the column and disperse into the nearest natural screens. If a vehicle breaks down during a movement, push it off the road and conceal it.

4-12. When conducting a march during good visibility, consider movement by infiltration (single or small groups of vehicles released at different intervals). Movement in stages, from one natural screen to the next, will further minimize possible detection. Use smoke screens at critical crossings and choke points.

4-13. During brief stops, quickly disperse vehicles under tree crowns or other concealment along the sides of the road. Strictly enforce CCD discipline. Watch for glare from vehicle windshields, headlights, or reflectors; and remedy the situation if it does occur. Try to control troop movement on the road or in other open areas. Conduct recon to select areas for long halts. The recon party should select areas that are large enough to allow sufficient CCD and dispersion. The quartering party should predetermine vehicle placement, develop a vehicle circulation plan, and guide vehicles into suitable and concealed locations. The first priority, however, is to move vehicles off the road as quickly as possible, even at the expense of initial dispersion. Use camouflage nets and natural vegetation to enhance concealment, and carefully conceal dug-in positions.

4-14. Traffic controllers have a crucial role in enforcing convoy CCD. Commanders should issue precise instructions for traffic controllers to stop passing vehicles and have the drivers correct the slightest violation of CCD discipline. Convoy commanders are responsible for the convoy's CCD discipline.

4-15. Pass through friendly obstacles at night, in fog, or under other conditions of poor visibility. Also use smoke screens because these conditions will not protect against many types of threat sensors. Lay smoke on a wide front, several times before actually executing the passage of lines. Doing this helps deceive an enemy about the time and place of an attack. Conceal lanes through obstacles from the enemy's view.

DECEPTIVE OPERATIONS

4-16. Conduct demonstrations and feints to confuse an enemy about the actual location of the main attack. Such deceptive operations are effective only if prior recon activities were conducted on a wide front, thereby preventing the enemy from pinpointing the likely main-attack area.

BATTLE

4-17. Units should adapt to the terrain during a battle. Deploying behind natural vegetation, terrain features, or man-made structures maximizes concealment from enemy observation. Make optimum use of concealed routes, hollows, gullies, and other terrain features that are dead-space areas to enemy observation and firing positions. A trade-off, however, usually exists in terms of a slower rate of movement when using these types of routes.

4-18. Movement techniques emphasizing fire and maneuver help prevent enemy observation and targeting. Avoid dusty terrain because clouds of dust will alert an enemy to the presence of friendly units. However, if the enemy is aware of a unit's presence, dust can be an effective means of obscuring the unit's intentions in the same way as smoke. When natural cover and concealment are unavailable or impractical, the coordinated employment of smoke, suppressive fires, speed, and natural limited-visibility conditions minimize exposure and avoid enemy fire sacks. However, offensive operations under these conditions present unique training and C^2 challenges.

4-19. Breaching operations require concealing the unit that is conducting the breach. Use conditions of poor visibility, and plan the use of smoke and suppressive fires to screen breaching operations.

4-20. Deliberate river crossings are uniquely difficult and potentially hazardous. Plan the coordinated use of terrain masking, smoke, decoys, and deceptive operations to ensure successful crossings.

Chapter 5

Defensive Operations

Successful defensive operations require strong emphasis on OPSEC. Proper OPSEC denies an enemy information about a friendly force's defensive preparations. Particularly important is the counterrecon battle, where defensive forces seek to blind an enemy by eliminating its recon forces. The winner of this preliminary battle is often the winner of the main battle. CCD, by virtue of its inherent role in counterefforts, plays an important role in both battles.

PREPARATIONS

5-1. The purpose of CCD during defensive preparations is to mask key or sensitive activities. Successful CCD of these activities leads to an enemy force that is blinded or deceived and therefore more easily influenced to attack where the defender wants (at the strengths of the defense). These key activities include—

- Preparing reserve and counterattack forces' locations.
- Preparing survivability positions and constructing obstacles (minefields, tank ditches).
- Establishing critical C^2 nodes.

SIGNATURES

5-2. A number of signatures may indicate the intentions of friendly defensive preparations, and an enemy analyzes these signatures to determine the defensive plan. Specific signatures that could reveal defensive plans include—

- Working on survivability positions.
- Emplacing minefields and other obstacles.
- Moving different types of combat materiel into prepared positions.
- Preparing routes and facilities.
- Constructing strongpoints or hardened artillery positions.

COUNTERATTACK AND RESERVE FORCES

5-3. Due to the similarity of missions, the concerns for concealing counterattack and reserve forces are similar to those of maneuver forces engaged in offensive operations. Chapter 4 discusses considerations about AAs, troop and supply movements, passages of lines, and deception operations. This information is also useful as a guide when planning CCD for a counterattack.

Planning

5-4. Proper planning is essential to avoid threat detection and prevent successful enemy analysis of the engineer efforts that are integral to defensive preparations. Engineer equipment creates significant signatures, so minimize its use to a level that is commensurate with available time and manpower. Disperse engineer equipment that is not required at the job site. Complete as much work as possible without using heavy equipment, and allow heavy equipment on site only when necessary. Engineers should minimize their time on site by conducting thorough, extensive planning and preparation. Additional signatures include—

- Supplies, personnel, and vehicles arriving to and departing from the unit area.

- Survivability positions being constructed.
- Smoke and heat emitting from kitchens, fires, or stoves.
- Communications facilities being operated.
- Educational and training exercises being conducted.

Movement

5-5.   Reserve forces should move along preplanned, concealed routes. They should also move and occupy selected locations at night or during other conditions of limited visibility. Quartering parties should preselect individual positions and guide vehicles and personnel to assigned locations. Light, noise, and track discipline are essential; but they are difficult to control during this phase. The quartering party should also develop a traffic-flow plan that minimizes vehicle and troop movement to and from the unit area.

5-6.   Arriving units should immediately begin to conceal their positions. Commanders should detail the priorities for CCD in the OPORD, based on their assessment of which signatures present the greatest opportunity for threat detection.

Assembly Areas

5-7.   While AA CCD actions are similar to those of counterattack and reserve positions, the latter positions are more likely to be occupied longer. Therefore, CCD needs are more extensive and extended for counterattack and reserve forces. In fact, their CCD operations are often indistinguishable from those of support units.

5-8.   Counterattack and reserve forces awaiting employment should capitalize on the time available to conduct rehearsals. While essential, these activities are prone to detection by an enemy's sensors so observe CCD discipline at all times and locations.

Placement And Dispersal

5-9.   Site selection is crucial when concealing engineer effort. Proper placement and dispersal of equipment and operations are essential. Use natural screens (terrain masking); however, urban areas often provide the best concealment for counterattack and reserve forces. (Chapter 7 discusses placement and dispersal in more detail.) When using forests as natural screens, carefully consider factors such as the height and density of vegetation, the amount and darkness of shadows cast by the screen, and the appropriateness of the particular screen for the season. The condition and quality of natural screens have a decisive effect on the capability to conceal units. Commanders should evaluate natural screens during engineer recon missions and conduct the missions on a timely, extensive basis.

5-10. The probability of detection increases considerably when survivability positions are prepared. Detection is easier due to the increased size of the targets to be concealed, the contrasting upturned soil, and the difficulty of concealing survivability effort. Despite these considerations, the enhanced protection afforded by survivability positions usually dictates their use. To minimize the probability of detection, employ a combination of natural screens and overhead nets to conceal construction sites.

CAMOUFLAGE NETS

5-11. Use camouflage nets (LCSS) to conceal vehicles, tents, shelters, and equipment. Use vegetation to further disrupt the outline of the target rather than completely hide it. Ensure that vegetation is not removed from a single location, because it could leave a signature for threat detection. Gather vegetation sparingly from as many remote areas as possible. This technique allows the immediate area to remain relatively undisturbed.

STOVES AND FIRES

5-12. Strictly control the use of stoves and fires because they produce visual and thermal signatures detectable to threat sensors. If fires are necessary, permit them only during daylight hours and place them in dead ground or under dense foliage. Use nets and other expedient thermal screens to dissipate rising heat and reduce the fire's thermal signature.

COMMUNICATIONS

5-13. Monitor communications to prevent enemy intelligence teams from identifying unit locations.

CCD DISCIPLINE

5-14. Strict CCD discipline allows the continued concealment of a unit's position. The longer a unit stays in one location, the harder it is for it to maintain CCD discipline. Extended encampments require constant command attention to CCD discipline. The evacuation of an area also requires CCD discipline to ensure that evidence (trash, vehicle tracks) is not left for enemy detection.

SURVIVABILITY POSITIONS AND OBSTACLES

5-15. Survivability positions include fighting positions, protective positions (shelters), and trench-work connections. Such positions are usually constructed of earth and logs but may also be composed of man-made building materials such as concrete.

PLACEMENT

5-16. Properly occupying positions and placing obstacles are critical CCD considerations. When possible, place obstacles and occupy positions out of the direct view of threat forces (such as a reverse-slope defense), at night, or under conditions of limited visibility.

BACKGROUNDS

5-17. Select backgrounds that do not silhouette positions and obstacles or provide color contrast. Use shadows to hinder an enemy's detection efforts. If possible, place positions and obstacles under overhead cover, trees, or bushes or in any other dark area of the terrain. This technique prevents the disruption of terrain lines and hinders aerial detection. CCD efforts, however, should not hinder the integration of obstacles with fires.

5-18. When using the terrain's natural concealment properties, avoid isolated features that draw the enemy's attention. Do not construct positions directly on or near other clearly defined terrain features (tree lines, hedge rows, hill crests). Offsetting positions into tree lines or below hill crests avoids silhouetting against the background and also counters enemy fire.

NATURAL MATERIALS

5-19. Use natural materials to supplement artificial materials. Before constructing positions and obstacles, remove and save natural materials (turf, leaves, humus) for use in restoring the terrain's natural appearance for deception purposes. During excavation, collect spoil in carrying devices for careful disposal. When preparing survivability positions and obstacles—

- Avoid disturbing the natural look of surroundings. Use camouflage nets and natural vegetation to further distort the outline of a position, to hide the bottom of an open position or trench, and to mask spoil used as a parapet. To further avoid detection, replace natural materials regularly or when they wilt or change color.
- Consider the effect of backblasts from rocket launchers, missile systems, and antitank weapons. Construct a concealed open space to the position's rear to accommodate backblasts. A backblast area should not contain material that will readily burn or generate large dust signatures.
- Use natural materials to help conceal machine-gun emplacements. Machine guns are priority targets, and concealing them is an essential combat task. Although CCD is important, placement is the primary factor in concealing machine guns.

- Place mortars in defilade positions. Proper placement, coupled with the use of artificial and natural CCD materials, provides the maximum possible concealment. Also consider removable overhead concealment.
- Use decoy positions and phony obstacles to draw enemy attention away from actual survivability positions and traces of obstacle preparation. Decoys serve the additional function of drawing enemy fire, allowing easier targeting of an enemy's weapons systems.

BATTLE

5-20. CCD during the defensive battle is essentially the same as for the offensive battle. While a majority of the battle is normally fought from prepared, concealed positions, defensive forces still maneuver to prevent enemy breakthroughs or to counterattack. When maneuvering, units should—

- Adapt to the terrain.
- Make optimum use of concealed routes.
- Preselect and improve concealed routes to provide defensive forces with a maneuver advantage.
- Plan smoke operations to provide additional concealment for maneuvering forces.

Chapter 6

High-Value Targets

The purpose of threat doctrine is for enemy forces to locate, target, and destroy deep targets, thereby degrading friendly capabilities while adding offensive momentum to attacking enemy forces. Enemy commanders focus their most sophisticated sensors in search of HVTs. By attacking these targets, enemy forces hope to deny adequate C^2, combat support, or resupply operations to forward friendly forces throughout the battlespace. Therefore, properly employing CCD at key fixed installations, such as command posts (CPs) and Army aviation sites (AASs) is essential to survival on a battlefield. HVTs fall into two general classifications—fixed installations (section II) and relocatable units (section III). For information on camouflaging medical facilities, see appendix F.

SECTION I – CCD PLANNING

PLANS

6-1.   No single solution exists for enhancing the survivability of HVTs with CCD (except for large-area smoke screens). The characteristics of many such targets are unique and require the creative application of CCD principles and techniques. Therefore, the CCD planning process presented in this section is not intended to impose a regimen that must be followed at all costs. Rather, it suggests a logical sequence that has proven successful over time. In fact, the steps outlined below often lead to creative CCD solutions simply because they allow designers to consider the many options, benefits, and pitfalls of CCD employment. No CCD plan is wrong if it achieves the intended signature-management goals and does not impair mission accomplishment.

6-2.   Each commander should develop his unit's CCD plan based on an awareness, if not a comprehensive assessment, of the detectable EM signatures emitted by HVTs under his command. He should evaluate these signatures by considering the enemy's expected RSTA capabilities (airborne and ground-based), knowledge of the target area, and weapons-on-target capability.

OBJECTIVE

6-3.   A CCD plan increases target survivability within the limits of available resources. The design procedure must systematically determine which features of a given target are conspicuous, why those features are conspicuous, and how CCD principles and techniques can best eliminate or reduce target signatures. CCD should decrease the effectiveness of enemy attacks by interfering with its target-acquisition process, which in turn increases target survivability.

PLANNING PROCESS

6-4.   The steps outlined below provide guidance for designing CCD plans for HVTs. The detailed planning approach is applicable in any situation where CCD employment is necessary, but more so when the plans include HVTs.

   Step 1.   Identify the threat. Identify the principal threat sensors, weapon-delivery platforms, and likely directions of attack.

Step 2. Identify critical facilities. Identify critical HVTs. Include those that are critical from an operational standpoint and those that may provide reference points (cues) for an attack on more lucrative targets.

Step 3. Evaluate facilities. Once the critical HVTs are identified, focus efforts on identifying the target features that might be conspicuous to an enemy RSTA. Consider multispectral (visual, thermal, NIR, radar) signatures in this assessment. The seven recognition factors (chapter 3) are an excellent framework for conducting this assessment. Include a review of area maps, site plans, photographs, and aerial images of the target area.

Step 4. Quantify signatures. Quantify the multispectral signatures that are emitted by high-value facilities. Base the quantification on actual surveys of critical facilities, using facsimiles of threat sensors when possible. Specify the EM wavelengths in which targets are most vulnerable, and develop signature-management priorities.

Step 5. Establish CCD goals. Establish specific CCD goals for HVTs. These goals should indicate the signature reduction (or increase) desired and the resources available for CCD implementation. Base these goals on the results of steps 1 through 4. Change the CCD goals as the planning process develops and reiterate them accordingly.

Step 6. Select materials and techniques. Select CCD materials and techniques that best accomplish signature-management goals within logistical, maintenance, and resource constraints. Expedient, off-the-shelf materials and battlefield by-products are not identified in this manual, but they are always optional CCD materials.

Step 7. Organize the plan. Develop a CCD plan that matches goals with available materials, time and manpower constraints, and operational considerations. If the goals are unobtainable, repeat steps 5 and 6 until a manageable plan is developed.

Step 8. Execute the plan. Once a feasible CCD plan is developed, execute it. Store temporary or expedient materials inconspicuously. Conduct deployment training on a schedule that denies enemy intelligence teams the opportunity to identify the countermeasures or develop methods to defeat the CCD.

Step 9. Evaluate the CCD. The final step in the CCD planning process is to evaluate the deployed CCD materials and techniques. Important questions to ask in this evaluation include the following:

- Does CCD increase the survivability of HVTs?
- Does deployed CCD meet the signature-management goals outlined in the plan?
- Is deployed CCD operationally compatible with the treated target(s)?
- Are CCD materials and techniques maintainable within manpower and resource constraints?

SECTION II – FIXES INSTALLATIONS

CONCEPT

6-5. Fixed installations (base camps, AASs, CPs, warehouses, roadways, pipelines, railways, and other lines-of-communication [LOC] facilities) provide scarce, nearly irreplaceable functional support to ground maneuver forces. The threat to these facilities is both ground-based and aerial. The CCD techniques for the two attack types do not necessarily change, but the defender must be aware of the overall implications of his CCD plan.

GROUND ATTACKS

6-6. Ground attacks against fixed installations (enemy offensives, terrorist attacks, and enemy special-force incursions) require constant operational awareness by the defenders. While most CCD techniques are conceptually designed to defend against an aerial attack, these same techniques can affect the target-acquisition capabilities of an enemy's ground forces to the benefit of the defender. SCSPP, LCSS, and natural vegetation provide CCD against a ground attack.

6-7.   CCD discipline (light, noise, spoil) involves prudent operational procedures that friendly troops should observe in any tactical situation, particularly in the presence of hostile ground forces. (See chapter 5 for more information.)

AERIAL ATTACKS

6-8.   Fixed installations are susceptible to aerial attacks because of their long residence time and immobility. However, fighter-bomber and helicopter aircrews face unique target-acquisition problems due to the relatively short time available to locate, identify, and lock onto targets. Fighter-bombers typically travel at high speeds, even during weapons delivery. This means attacking aircrews have limited search time once they reach the target area. Helicopters travel at slower speeds but generally encounter similar time-on-target limitations. Because of lower flying altitudes and slower speeds, helicopters are more vulnerable to ground defenses. In either case, proper CCD can increase aircrew search time, thereby reducing available time to identify, designate, and attack an HVT. The longer an aircrew is forced to search for a target in a defended area, the more vulnerable the aircraft becomes to counterattack.

ENEMY INTELLIGENCE

6-9.   The location and configuration of most fixed installations are usually well known. CCD techniques that protect against sophisticated surveillance sensor systems, particularly satellite-based systems, can be costly in terms of manpower, materials, and time. Steps can be taken to reduce an enemy's detection of relocatable targets. Fixed installations are difficult to conceal from RSTA sensors due to the relatively long residence time of fixed installations versus relocatable targets. Unless the construction process for a given fixed installation was conducted secretly, defenders can safely assume that enemy RSTA sensors have previously detected and catalogued its location. Defenders can further assume that attacking forces have intelligence data leading them to the general area of the fixed installation. CCD design efforts, therefore, should focus on the multispectral defeat or impairment of the enemy's local target-acquisition process.

CCD TECHNIQUES

6-10.   Selected CCD techniques should capitalize on terrain features that are favorable to the defender and on the short time available to attacking aircrews for target acquisition. Use artificial and natural means to camouflage the installation. Where time and resources allow, deploy alternative targets (decoys) to draw the attention of the attacking aircrews away from the fixed installation.

6-11.   Comprehensive CCD designs and techniques for fixed installations can be costly, yet field tests have shown that simple, expedient techniques can be effective. HVTs are usually supplied with artificial CCD materials. If they are not, soldiers increase the survivability of an installation by using CCD principles.

OTHER CONSIDERATIONS

6-12.   While standard CCD materials are designed to enhance fixed-installation survivability, they have practical limitations that are not easily overcome. Materials applied directly to a fixed installation may achieve the signature-management goals stated in the CCD plan. However, if other features of the target scene are not treated accordingly, the target may be well hidden but remain completely vulnerable.

6-13.   For example, three weapons-storage-area (WSA) igloos are in a row. The middle igloo is treated with CCD materials while the other two are not. The middle igloo will still be vulnerable. The enemy knows that three igloos exist and will probably locate the middle one no matter how well the CCD plan is designed. However, if all three igloos are treated with CCD materials and three decoy igloos are placed away from them, the treated igloos' survivability will increase.

6-14.   Furthermore, if a man-made object (traffic surface) or a natural feature (tree line) is close to the igloos, attacking forces will use these cues to proceed to the target area even if all three igloos are treated with CCD materials. Remember, an HVT is part of an overall target scene and an attacker must interpret the scene. Do not make his task easy. CCD plans that treat only the target and ignore other cues (man-made or natural) within the target scene are insufficient.

COMMAND POSTS

6-15. C^2 systems provide military leaders with the capability to make timely decisions, communicate the decisions to subordinate units, and monitor the execution of the decisions. CPs contain vital C^2 systems.

SIGNATURES

6-16. Since World War II, the size and complexity of CPs have increased dramatically. Their signatures have correspondingly increased from a physical and communications perspective (more types of antennas and transmission modes at a wider range of frequencies). As a result, the enemy can use several conspicuous signatures to detect and target CPs for attack. Therefore, CPs require excellent CCD to survive on the battlefield.

Lines Of Communication

6-17. CPs are usually located near converging LOC, such as road or rail junctions, and often require new access and egress routes. Consider the following regarding CCD and CPs:

- **Vehicle traffic.** When evaluating EM signatures that CPs emit, consider concentrations of vehicles, signs of heavy traffic (characteristic wear and track marks), and air traffic. Park vehicles and aircraft a significant distance from CPs.
- **Antennas.** Antennas and their electronic emissions and numerous support towers are common to most CPs. Paint antennas and support equipment with nonconductive green, black, or brown paint if the surfaces are shiny. If tactically feasible, use remote antennas to reduce the vulnerability of the radio system to collateral damage.
- **Security emplacements.** Security measures (barbwire, barriers, security and dismount points, and other types of emplacements) can indicate CP operations. Barbwire exhibits a measurable RCS at radar frequencies. Ensure that barbwire and concertina wire follow natural terrain lines and are concealed as much as possible.

Equipment

6-18. Power generators and other heat sources produce signatures that an enemy's surveillance and target-acquisition sensors can detect. Place heat-producing equipment and other thermal sources in defilade positions, within structures, or under natural cover. Heat diffusers, which tone down and vent vehicle exhaust away from threat direction, are an expedient means of thermal-signature reduction.

Defensive Positions

6-19. Defensive positions (berms, revetments, fighting positions) for protection against direct- and indirect-fire attackers typically create scarred earth signatures and detectable patterns due to earth excavation.

CCD

6-20. CCD improves OPSEC and increases survivability by minimizing the observable size and EM signatures of CPs. CP CCD requires recon, planning, discipline, security, and maintenance. Carefully controlled traffic plans decrease the possibility of disturbing natural cover and creating new, observable paths. Decoys are a highly effective means of confusing the enemy's target-acquisition process, particularly against airborne sensors. Against ground threats, the same general rules of CCD discipline apply; however, recon and heightened security patrols enhance CCD efforts against ground attack.

SITES

6-21. CP sites, which could move every 24 hours, are still occupied for a longer period than AAs. CP site selection is crucial, therefore units should—

- Consider the needs of supporting an extended occupation while minimizing changes to natural terrain patterns. When constructing defensive positions, minimize earth scarring as much as possible. If scarred earth is unavoidable, cut vegetation, toned-down agents (paint), and camouflage nets help conceal scarred areas.
- Use existing LOC (roads, trails, streams). If a site requires construction of roads or paths, make maximum use of natural concealment and existing terrain. The fewer new lines required, the better the CP blends, leaving natural features relatively unchanged.
- Never locate a CP at a road junction. Road junctions are high-priority targets for enemy forces and are easily detectable.
- Locate a CP in an existing civilian structure, if possible, which simplifies hiding military activity. However, choose a structure in an area where a sufficient number of buildings with similar EM signatures can mask its location.

TELECOMMUNICATIONS PROCEDURES

6-22. By strictly complying with proper radio, telephone, and digital communications procedures, the opportunities for an enemy to detect friendly telecommunications activities are minimized. Consider the following:

- Place antennas in locations using natural supports when possible (trees for dipoles). As a rule of thumb, place antennas a minimum of one wavelength away from surrounding structures or other antennas.

Note. One wavelength is 40 meters (typically) for low frequencies and 1 meter for very high frequencies (VHFs).

- Move antennas as often as possible within operational constraints.
- Use directional antennas when possible. If using nondirectional antennas, employ proper terrain-masking techniques to defeat the threat's radio direction-finding efforts.
- Use existing telephone lines as much as possible. Newly laid wire is a readily observable signature that can reveal a CP's location. Communications wire and cable should follow natural terrain lines and be concealed in the best way possible.

CCD DISCIPLINE

6-23. Maintain CCD discipline after occupying a site. Establish and use designated foot paths to, from, and within a CP's area. If a unit occupies a site for more than 24 hours, consider periodically rerouting foot paths to avoid detectable patterns. Conceal security and dismount points and other individual emplacements, and make paths to the CP inconspicuous. Enforce proper disposal procedures for trash and spoil. Rigidly enforce light and noise discipline. Enhance the realism of a decoy CP by making it appear operational. Allow CCD discipline to be lax in the decoy CP, thus making it a more conspicuous target than the real CP.

SUPPLY AND WATER POINTS

6-24. Supply and water points provide logistical support—the backbone of sustained combat operations. As these targets are relatively immobile and the object of an enemy's most sophisticated sensors, using CCD is one of the most effective means to improve their survivability.

OPERATIONS

6-25. Many CCD methods associated with AAs and CPs also apply to supply and water points, but with additional requirements. Large amounts of equipment and supplies are quickly brought into tactical areas and delivered to supply points located as close to the FLOT as possible. Supplies must be unloaded and concealed quickly, while supply points remain open and accessible for distribution. Under these conditions, multiple supply points are generally easier to camouflage than single, large ones. Decoy supply and water points can also confuse a threat's targeting efforts.

CCD

6-26. Take maximum advantage of natural cover and concealment. Configure logistics layouts to conform with the local ground pattern. Creativity can play a role in this effort. The following guidance enhances concealment of these operations:

- Avoid establishing regular (square or rectangular) perimeter shapes for an area.
- Select locations where concealed access and egress routes are already established and easily controlled.
- Use roads with existing overhead concealment if you need new access roads. Conceal access over short, open areas with overhead nets.
- Control movement into and out of the supply area.
- Mix and disperse supply-point stocks to the maximum extent possible. This not only avoids a pattern of stockpile shapes but also avoids easy destruction of one entire commodity.
- Space stocks irregularly (in length and depth) to avoid recognizable patterns. Stack supplies as low as possible to avoid shadows. Dig supplies in if resources allow. In digging operations, disperse the spoil so as not to produce large piles of earth.
- Cover stocks with nets and other materials that blend with background patterns and signatures. Flattops (large, horizontal CCD nets) are effective for concealing supply-point activities when resources allow their construction and when supply points are not too large. Dunnage from supply points provides excellent material for expedient decoys.

TRAFFIC CONTROL

6-27. Ensure that vehicles cause minimal changes to the natural terrain as a result of movement into, within, and out of the area. Provide concealment and control of vehicles waiting to draw supplies. Rigidly practice and enforce CCD discipline and OPSEC. Debris control could be a problem and requires constant attention.

WATER POINTS

6-28. CCD for water points include the following additional considerations:

- **Spillage.** Water spillage can have positive and negative effects on a unit's CCD posture. Standing pools of water reflect light that is visible to observers. Pools can also act as forward scatterers of radar waves, resulting in conspicuous black-hole returns on radar screens. Therefore, minimize water spillage and provide adequate drainage for runoff. On the other hand, dispersed water can be used to reduce the thermal signatures of large, horizontal surfaces. However, use this technique sparingly and in such a way that pools do not form.
- **Equipment.** Use adequate natural and artificial concealment for personnel, storage tanks, and specialized pumping and purification equipment. Conceal water-point equipment to eliminate shine from damp surfaces. Conceal shine by placing canvas covers on bladders, using camouflage nets, and placing foliage on and around bladders. This also distorts the normal shape of the bladders.
- **Scheduling.** Enhance CCD discipline at water points by establishing and strictly enforcing a supply schedule for units. The lack of or violation of a supply schedule produces a concentration of waiting vehicles that is difficult to conceal.

ARMY AVIATION SITES

6-29. AASs are among the most important of all battlefield HVTs. AASs are typically comprised of several parts that make up the whole, including tactical assembly areas (TAAs), aviation maintenance areas (AMAs), forward operating bases (FOBs), and forward arming and refueling points (FARPs). The positioning of AAS elements with respect to each other is dynamic and often depends on the existing tactical situation. In the following discussion, an AAS will be defined as a TAA, an AMA, and a FARP collocated in the same area. While these elements are not always collocated, the CCD techniques for individual elements will not greatly differ based on positioning. Untreated AASs are detectable in most threat sensor wavelengths.

- **TAA.** A TAA is typically a parking area for helicopters. Helicopters are highly conspicuous targets because of their awkward shape, distinctive thermal signatures, and large RCS. An enemy expends a lot of time and energy attempting to locate TAAs. Once it finds them, the enemy aggressively directs offensive operations against them.

- **AMA.** The most conspicuous features of an AMA are the large transportable maintenance shelters. These shelters are highly visible and indicate the presence of helicopters to an enemy. AMAs occupy large areas to allow for ground handling of aircraft. Traffic patterns around AMAs are also strong visual cues to the enemy. Maintenance assets, including aviation shop sets, have characteristically distinct multispectral cues.

- **FARP.** A FARP provides POL and ammunition support to AASs and other tactical units. A FARP consists of fuel bladders, heavy expanded mobility tactical trucks (HEMTTs), fueling apparatus, and bulk ammunition. Due to safety requirements, FARP elements are dispersed as much as possible within terrain and operational constraints. Each element is detectable with multispectral radar. In a FARP—

  - Fuel bladders contain petroleum liquids whose thermal mass is a strong IR cue relative to the background. Bladders are often bermed, which means that visible earth scarring is necessary to construct the berm.

  - Large HEMTTs are conspicuous in all wavelengths.

  - Fueling areas are generally arranged in such a way that the fueling apparatus (hoses, pumps) are arranged linearly in an open area for safe and easy access. The linear deployment of these hoses is a strong visual cue, and their dark color usually contrasts with the background. The dark hoses experience solar loading, and the POL liquids within the hoses can provide a thermal cue.

- **Equipment.** Palletized ammunition and support equipment accompany AASs. Such equipment is often stacked in regular, detectable patterns.

- **Aircraft.** Aircraft create large dust plumes when deployed to unpaved areas. Such plumes are distinct visual cues and indicate the presence of rotary aircraft to an enemy.

  - **Parked aircraft.** Camouflage nets, berms, stacked equipment, and revetments can effectively conceal parked aircraft. Vertical screens constructed from camouflage nets help conceal parked aircraft, particularly against ground-based threats. However, CCD techniques for rapid-response aircraft must not impair operational requirements, meaning that obtrusive, permanent CCD techniques are generally not an option. Also, foreign object damage (FOD) is a critical concern for all aviation assets. CCD for parked aircraft depends on the expected ground time between flights. The commanding officer must approve all aircraft CCD techniques before implementation.

  - **Aircraft refueling.** Aircraft refueling positions, particularly fuel hoses, should be dispersed and arrayed in a nonlinear configuration. The hoses can be concealed at periodic locations with cut vegetation or a light earth/sod covering to reduce visual and thermal signatures.

- **Defensive positions.** Constructing defensive positions can create detectable areas of scarred earth.

- **CCD.** AASs are extremely valuable targets; therefore, try to prevent their initial detection by an enemy.

- **Vehicles.** Large vehicles can be effectively concealed with camouflage nets. Also, properly placing these vehicles to use terrain features and indigenous vegetation increases their survivability. Expedient vehicle decoys provide an enemy with alternate targets, and proper CCD discipline is essential.
- **Dunnage.** Quickly conceal all dunnage (packing materials) to minimize the evidence of AASs.
- **Dust.** To avoid dust, park aircraft in grassy areas or where the earth is hard-packed. If such areas are unavailable, disperse water on the area to minimize dust plumes. However, water-soaked earth can also be an IR detection cue so use this option sparingly and, if possible, at night. Several chemical dust palliatives are available that provide excellent dust control for aviation areas.
- **Construction.** When constructing defensive positions, minimize disturbances to the surrounding area. Cover scarred earth with cut vegetation, camouflage nets, or toned-down agents.

SECTION III – RELOCATABLE UNITS

MOBILITY AND CCD

6-30. Examples of valuable relocatable units include TOCs, tactical-missile-defense (TMD) units (Patriot batteries), refuel-on-the-move (ROM) sites, and FARPs. These units are critical to offensive and defensive operations, and their protection should receive a high priority.

6-31. Mobility and CCD enhance the survivability of relocatable units. A CCD plan must include the techniques for units to deploy rapidly and conduct mobile operations continuously. The CCD techniques available to mobile units are basically the same as for fixed installations, and the principles of CCD still apply. However, the mission of relocatable units differs from that of fixed installations so CCD execution also differs.

6-32. Relocatable units spend from a few hours to several weeks in the same location, depending on their tactical situation. CCD techniques must be planned accordingly. If a unit is at a location for a few hours, it should employ expedient CCD techniques. If a unit is at a location for several days, it should employ robust CCD plans. The resources a unit expends on CCD execution must be weighed against the length of time that it remains in the same location. As CCD plans increase in complexity, subsequent assembly and teardown times also increase. Commanders must ensure that the unit's manpower and resources dedicated to CCD execution are equal to the tactical mobility requirements.

BUILT-IN CAPABILITIES

6-33. CCD should be built into systems to the maximum extent possible. Supplemental CCD is usually necessary and should be designed to enhance the built-in CCD. Apply the same rules for avoiding detection and the same considerations regarding the seven recognition factors that are discussed in chapter 3. The CCD planning process outlined at the beginning of this chapter also applies.

Chapter 7

Special Environments

The fundamentals of CCD do not change between environments. The seven rules for avoiding detection and the seven recognition factors that are listed in chapter 3 and the three CCD principles—preventing detection, improving survivability, and improving deception capabilities—still apply. However, the guidelines for their application change. Different environments require thoughtful, creative, and unique CCD techniques. This chapter discusses different CCD techniques that have proven effective in three special environments—desert, snow-covered areas, and urban terrain.

DESERT

7-1. The color of desert terrain varies from pink to blue, depending on the minerals in the soil and the time of the day. No color or combination of colors matches all deserts. Patches of uniform color in the desert are usually 10 times larger than those in wooded areas. These conditions have led to the development of a neutral, monotone tan as the best desert CCD paint color.

TOPOGRAPHY

7-2. Although desert terrain may appear featureless, it is not completely flat. In some ways, desert terrain resembles unplowed fields; barren, rocky areas; grasslands; and steppes.

SHADOWS

7-3. The closer a target is to the ground, the smaller its shadow; and a small shadow is easier to conceal from aerial observation. The proper draping of CCD nets will alter or disrupt the regular, sharp-edged shadows of military targets and allow target shadows to appear more like natural shadows. When supplemented by artificial materials, natural shadows cast by folds of the ground can be used for CCD purposes. The best solution to the shadow problem in desert terrain is to dig in and use overhead concealment or cover. Otherwise, park vehicles in a way that minimizes their broadside exposure to the sun.

PLACEMENT

7-4. Proper placement and shadow disruption remain effective techniques. Place assets in gullies, washes, wadis, and ravines to reduce their shadows and silhouettes and to take advantage of terrain masking. More dispersion is necessary in desert terrain than in wooded areas. Move assets as the sun changes position to keep equipment in shadows.

TERRAIN MOTTLING

7-5. Use terrain mottling when the ground offers little opportunity for concealment. This technique involves scarring the earth with bulldozers, which creates darker areas on which to place equipment for better blending with the background. Ensure that the mottled areas are irregularly shaped and at least twice the size of the target you are concealing. Place the target off center in the mottled area and drape it with camouflage nets. When employing the scarring technique, dig two to three times as many scars as pieces of equipment being concealed. Doing this prevents the mere presence of mottled areas from giving away a unit's location.

MOVEMENT DISCIPLINE

7-6.   Movement discipline is especially important in the desert. Desert terrain is uniform and fragile, making it easily disturbed by vehicle tracks. Vehicle movement also produces dust and diesel plumes that are easily detectable in the desert. When movement is necessary, move along the shortest route and on the hardest ground. Shine is a particularly acute desert problem due to the long, uninterrupted hours of sunlight. To deal with this problem, remove all reflective surfaces or cover them with burlap. Use matte CCD paint or expedient paints (see table 3-2, page 3-12) to dull the gloss of a vehicle's finish. Shade optical devices (binoculars, gun sights) when using them.

NOISE AND LIGHT DISCIPLINE

7-7.   Noise and light discipline is particularly important in desert terrain since sound and light can be detected at greater distances on clear desert nights. The techniques for reducing these signatures remain the same as for other environments. Be aware that thermal sensors, while not as effective during the day, have an ideal operating environment during cold desert nights. Starting all vehicle and equipment engines simultaneously is a technique that can be used to confuse enemy acoustical surveillance efforts.

SNOW-COVERED AREAS

7-8.   When the main background is white, apply white paint or whitewash over the permanent CCD paint pattern. The amount of painting should be based on the percentage of snow coverage on the ground:

- If the snow covers less than 15 percent of the background, do not change the CCD paint pattern.
- If the snow cover is 15 to 85 percent, substitute white for green in the CCD paint pattern.
- If the snow cover is more than 85 percent, paint the vehicles and equipment completely white.

PLACEMENT

7-9.   A blanket of snow often eliminates much of the ground pattern, causing natural textures and colors to disappear. Blending under these conditions is difficult. However, snow-covered terrain is rarely completely white so use the dark features of the landscape. Place equipment in roadways, in streambeds, under trees, under bushes, in shadows, and in ground folds. Standard BDUs and personal equipment contrast with the snow background, so use CCD to reduce these easily recognized signatures.

MOVEMENT

7-10. Concealing tracks is a major problem in snow-covered environments. Movement should follow wind-swept drift lines, which cast shadows, as much as possible. Vehicle drivers should avoid sharp turns and follow existing track marks. Wipe out short lengths of track marks by trampling them with snowshoes or by brushing them out.

THERMAL SIGNATURES

7-11. Snow-covered environments provide excellent conditions for a threat's thermal and UV sensors. Terrain masking is the best solution to counter both types of sensors. Use arctic LCSS and winter camouflage paint to provide UV blending, and use smoke to create near-whiteout conditions.

URBAN TERRAIN

7-12. Urbanization is reducing the amount of open, natural terrain throughout the world. Therefore, modern military units must be able to apply effective urban CCD. Many of the CCD techniques used in natural terrain are effective in urban areas.

PLANNING

7-13. Planning for operations in urban areas presents unique difficulties. Tactical maps do not show man-made features in enough detail to support tactical operations. Therefore, they must be supplemented with aerial photographs and local city maps. Local government and military organizations are key sources of information that can support tactical and CCD operations. They can provide diagrams of underground facilities, large-scale city maps, and/or civil-defense or air-raid shelter locations.

SELECTING A SITE

7-14. The physical characteristics of urban areas enhance CCD efforts. The dense physical structure of these areas generates clutter (an abundance of EM signatures in a given area) that increases the difficulty of identifying specific targets. Urban clutter greatly reduces the effectiveness of a threat's surveillance sensors, particularly in the IR and radar wavelengths. Urban terrain, therefore, provides an excellent background for concealing CPs, reserves, combat-service-support (CSS) complexes, or combat forces. The inherent clutter in urban terrain generally makes visual cues the most important consideration in an urban CCD plan.

7-15. The regular pattern of urban terrain; the diverse colors and contrast; and the large, enclosed structures offer enhanced concealment opportunities. Established, hardened road surfaces effectively mask vehicle tracks. Depending on the nature of the operation, numerous civilian personnel and vehicles may be present and may serve as clutter. This confuses an enemy's ability to distinguish between military targets and the civilian population. Underground structures (sewers, subways) are excellent means of concealing movement and HVTs.

7-16. When augmented by artificial means, man-made structures provide symmetrical shapes that provide ready-made CCD. The CCD for fighting positions is especially important because of the reduced identification and engagement ranges (100 meters or less) typical of urban fighting. Limit or conceal movement and shine. These signatures provide the best opportunity for successful threat surveillance in urban terrain. Careful placement of equipment and fighting positions remains important to provide visual CCD and avoid detection by contrast (thermal sensors detecting personnel and equipment silhouetted against colder buildings or other large, flat surfaces).

ESTABLISHING FIGHTING POSITIONS

7-17. The fundamental CCD rule is to maintain the natural look of an area as much as possible. Buildings with large, thick walls and few narrow windows provide the best concealment. When selecting a position inside a building, soldiers should—

- Avoid lighted areas around windows.
- Stand in shadows when observing or firing weapons through windows.
- Select positions with covered and concealed access and egress routes (breaches in buildings, underground systems, trenches).
- Develop decoy positions to enhance CCD operations.

PLACING VEHICLES

7-18. Hide vehicles in large structures, if possible, and use local materials to help blend vehicles with the background environment. Paint vehicles and equipment a solid, dull, dark color. If you cannot do this, use expedient paints to subdue the lighter, sand-colored portions of the SCSPP. When placing vehicles outdoors, use shadows for concealment. Move vehicles during limited visibility or screen them with smoke.

Appendix A
Metric Conversion Chart

This appendix complies with current Army directives which state that the metric system will be incorporated into all new publications. Table A-1 is a conversion chart.

Table A-1. Metric conversion chart

US Units	Multiplied By	Metric Units
Cubic feet	0.0283	Cubic meters
Feet	0.3048	Meters
Gallons	3.7854	Liters
Inches	2.54	Centimeters
Inches	0.0254	Meters
Inches	25.4001	Millimeters
Miles, statute	1.6093	Kilometers
Miles, statute	0.9144	Yards
Ounces	28.349	Grams
Pounds	0.454	Kilograms
Tons, short	0.9072	Tons, metric
Square feet	0.093	Square meters
Metric Units	**Multiplied By**	**US Units**
Centimeters	0.3937	Inches
Cubic meters	35.3144	Cubic feet
Cubic meters	1.3079	Cubic yards
Grams	0.035	Ounces
Kilograms	2.205	Pounds
Kilometers	0.62137	Miles, statute
Kilometers	1,093.6	Yards
Liters	0.264	Gallons
Meters	3.2808	Feet
Meters	39.37	Inches
Meters	1.0936	Yards
Millimeters	0.03937	Inches
Square meters	10.764	Square feet
Tons, metric	2,204.6	Pounds

Appendix B

Guidelines for Tactical Standing Operating Procedures

TACSOPs are critical to battlefield success. All commanders should establish camouflage guidelines in their TACSOPs and ensure that their soldiers are familiar with them. TACSOPs provide guidelines that help reduce the time required to perform routine tasks. Commanders can achieve these ends by defining the responsibilities, identifying the expected tasks, and providing supervisors with a memory aid when planning or inspecting. TACSOPs, coupled with battle drills (appendix C), provide units with guidance on how to execute anticipated battlefield tasks. CCD employment is a task that should be routine for all units.

CONTENT

B-1. The following CCD considerations may be included in a unit TACSOP:

- A review of CCD fundamentals.
- Rules of unit CCD discipline.
- Memory aids for supervisors, which should include an inspection checklist (figure B-1, pages B-2 through B-4) and a chart of an enemy's sensor systems with possible countermeasures.
- Guidelines on CCD discipline to provide uniformity among all subunits.
- The different CCD postures.
- Procedures for blackout, the quartering party, unit movement, and the deployment area.
- Appropriate CCD postures in OPORDs for different missions.

COMMANDERS' RESPONSIBILITIES

B-2. Commanders must ensure that each soldier has the required quantities of serviceable BDUs and that these uniforms are properly maintained to protect their IR screening properties. Based on unit requirements, supply personnel forecast, request, and store adequate quantities of expendable CCD supplies (paint, makeup, repair kits). Commanders ensure that authorized quantities of CCD screens (LCSS) and support systems (to include repair kits and spare parts) are on hand and continually maintained in a clean, serviceable condition.

FRATRICIDE

B-3. Since warfare often results in the loss of life from fratricide, the unit TACSOP should include a way to reduce fratricide. Commanders should consider ways for friendly and allied units to identify each other on the battlefield. Fratricide compels commanders to consider the effect CCD and deception operations have on the necessity of being recognized by friendly troops.

CCD Inspection Checklist

1. Command Emphasis.

   a. The commander—

      (1) Establishes CCD goals.

      (2) Executes CCD plans.

      (3) Inspects frequently for CCD deficiencies.

      (4) Conducts follow-up inspection of CCD deficiencies.

      (5) Integrates CCD into training exercises.

   b. The unit—

      (1) Integrates CCD into its TACSOP.

      (2) Follows the TACSOP.

2. Discipline.

   a. The unit—

      (1) Observes noise discipline.

      (2) Observes light discipline with respect to smoking, fires, and lights.

      (3) Conceals highly visible equipment.

      (4) Covers shiny surfaces.

      (5) Keeps exposed activity to a minimum.

      (6) Uses cut vegetation properly.

      (7) Uses and conceals dismount points properly.

   b. Soldiers—

      (1) Wear the correct uniform.

      (2) Control litter and spoil.

3. Techniques. The unit—

   a. Places and disperses vehicles and equipment.

   b. Disperses the CP.

   c. Employs camouflage nets (LCSS).

   d. Uses (or minimizes) shadows.

   e. Minimizes movement.

   f. Hides operations and equipment.

   g. Blends operations and equipment with backgrounds.

   h. Employs pattern-painting techniques.

   i. Employs decoys.

   j. Integrates smoke operations with unit movement.

Figure B-1. Sample CCD checklist

k. Practices individual CCD on—

    (1) Helmet.

    (2) Face.

    (3) Weapon.

    (4) Other equipment.

l. Employs CCD on fighting positions by—

    (1) Eliminating or minimizing target silhouettes.

    (2) Practicing spoil control.

    (3) Eliminating or minimizing regular or geometric shapes and layouts.

    (4) Maintaining overhead concealment.

    (5) Practicing dust control.

m. Employs CCD on tactical vehicles by—

    (1) Minimizing and concealing track marks.

    (2) Minimizing or eliminating the shine on vehicles and equipment.

    (3) Reducing or using shadows to the unit's advantage.

    (4) Employing camouflage nets (LCSS).

    (5) Painting vehicles to match their surroundings.

    (6) Dispersing vehicles and equipment.

    (7) Concealing vehicles and supply routes.

    (8) Controlling litter and spoil.

    (9) Storing and concealing ammunition.

n. Employs CCD on AAs by—

    (1) Facilitating mission planning for access and egress concealment.

    (2) Marking guideposts for route junctions.

    (3) Ensuring that turn-ins are not widened by improper use.

    (4) Dispersing dismount, mess, and maintenance areas.

    (5) Dispersing the CP.

    (6) Maintaining CCD by—

        (a) Inspecting CCD frequently.

        (b) Controlling litter and garbage.

        (c) Observing blackout procedures.

Figure B-1. Sample CCD checklist (continued)

(7) Observing evacuation procedures by—

    (a) Policing the area.

    (b) Covering or eliminating tracks.

    (c) Preventing traffic congestion.

    (d) Concealing spoil.

o. Employs CCD on the CP by—

(1) Ensuring that LOC are not converged.

(2) Dispersing vehicles.

(3) Ensuring that turn-ins are not widened through improper use.

(4) Ensuring that protective barriers follow terrain features.

(5) Concealing defensive weapons.

(6) Ensuring that existing poles are used for LOC.

(7) Digging in the CP (when in open areas).

(8) Maintaining camouflage nets (LCSS).

(9) Using civilian buildings properly by—

    (a) Controlling access and egress.

    (b) Observing blackout procedures.

    (c) Avoiding obvious locations.

p. Employs CCD on supply points by—

(1) Dispersing operations.

(2) Concealing access and egress routes.

(3) Using the track plan.

(4) Providing concealed loading areas.

(5) Developing and implementing a schedule for the units being serviced.

q. Employs CCD on water points by—

(1) Concealing access and egress routes.

(2) Ensuring that the track plan is used.

(3) Controlling spillage.

(4) Controlling shine and reflections.

(5) Developing and implementing a schedule for the units being serviced.

Figure B-1. Sample CCD checklist (continued)

Appendix C

Camouflage Requirements and Procedures

This appendix provides information on the LCSS and describes how to erect it. Also included is a figure for determining the amount of modules needed to camouflage the various vehicles in the Army's inventory. This appendix also includes a sample battle drill that can be used to train soldiers.

LIGHTWEIGHT CAMOUFLAGE SCREEN SYSTEM

C-1. The LCSS is a modular system consisting of a hexagon screen, a diamond-shaped screen, a support system, and a repair kit. You can join any number of screens to cover a designated target or area (figure C-1, page C-2). Use figure C-2, page C-3, to determine the number of modules needed for camouflaging a given area. Measure the vehicle or use table C-1, page C-4, to determine the vehicle's dimensions.

Notes.

1. See appendix E for a list of LCSS national stock numbers (NSNs) and ordering information.

2. See TM 5-1080-200-13&P for more information on maintenance, erection, and characteristics of the LCSS.

CAPABILITIES

C-2. The LCSS protects targets in four different ways. It—

- Casts patterned shadows that break up the characteristic outlines of a target.
- Scatters radar returns (except when radar-transparent nets are used).
- Traps target heat and allows it to disperse.
- Simulates color and shadow patterns that are commonly found in a particular region.

ERECTING PROCEDURES

C-3. To erect camouflage nets effectively—

- Keep the net structure as small as possible.
- Maintain the net a minimum of 2 feet from the camouflaged target's surface. This prevents the net from assuming the same shape and thermal signature as the target it is meant to conceal.
- Ensure that the lines between support poles are gently sloped so that the net blends into its background. Sloping the net over the target also minimizes sharp edges, which are more easily detectable to the human eye.
- Extend the net completely to the ground to prevent creating unnatural shadows that are easily detected. This ensures that the net effectively disrupts the target's shape and actually absorbs and scatters radar energy.
- Extend the net all the way around the target to ensure complete protection from enemy sensors.

ESSENTIAL ARMY MANUAL
(527) CAMOUFLAGE, CONCEALMENT AND DECOYS

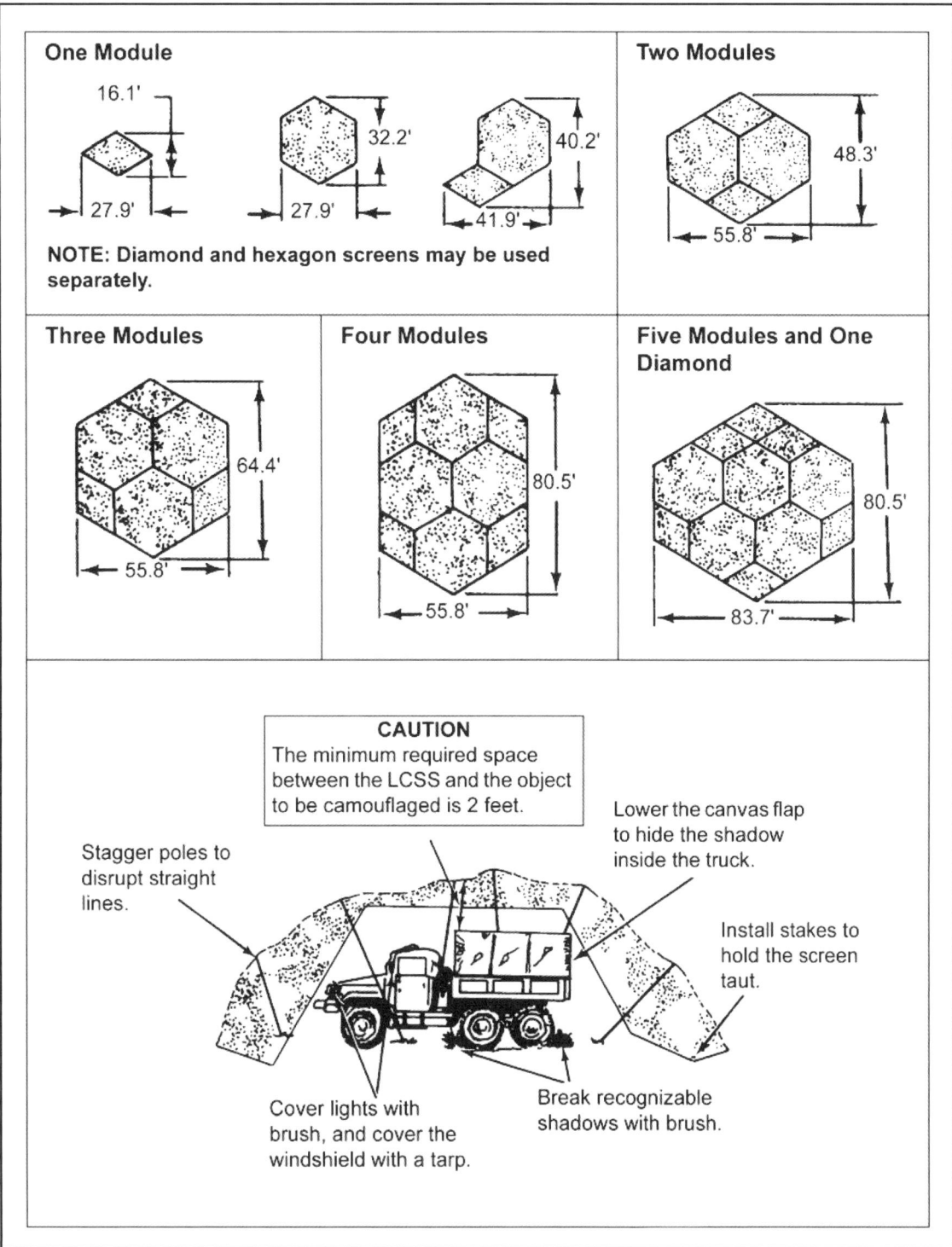

Figure C-1. LCSS modular system

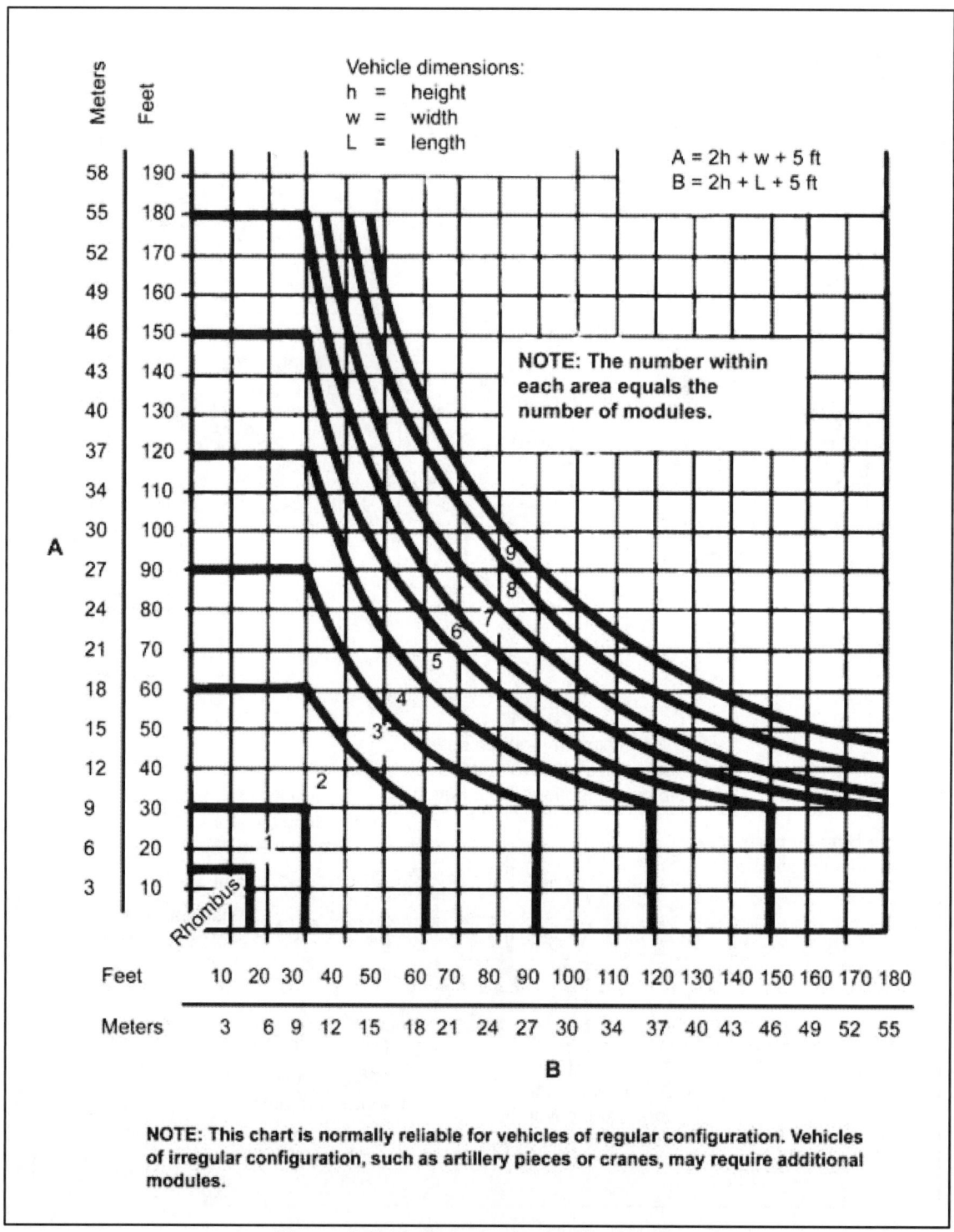

Figure C-2. Module determination chart

Table C-1. Vehicle dimensions

Nomenclature	Height (feet)	Width (feet)	Length (feet)	No. of Modules
AVLB	17	13	37	5
C7 loader, scoop, 2½-ton, w/o cage	9	9	25	2
D7 dozer, with blade	10	12	19	2
M106A1 carrier, mortar, 107-mm	7	10	16	2
M109A3 howitzer, 155-mm (SP)	11	12	30	3
M113A2 carrier, personnel	7	9	16	2
M113A3 carrier, personnel	7	9	19	2
M125A1 carrier, mortar, 81-mm	7	9	16	2
M149 trailer, water, 400-gal	6	7	7	1
M172 trailer, low-bed, 25-ton	6	10	35	2
M1A1 tank, with mine roller	10	12	40	3
M1A1(2) tank, combat, 105- and 120-mm	10	12	28	3
M2 fighting vehicle, infantry	10	11	22	2
M2 TOW vehicle, improved	11	9	15	2
M3 fighting vehicle, cavalry	10	11	22	2
M35A2 truck, cargo, 2½-ton	9	8	23	2
M520 truck, cargo, 8-ton	11	9	32	3
M548 carrier, cargo, 6-ton	10	9	21	2
M54A2 truck, cargo, 5-ton	10	8	26	2
M553 truck, wrecker, 10-ton	11	9	33	3
M559 truck, fuel, 2,500-gal	11	9	33	3
M577A1 carrier, CP	9	9	19	2
M578 vehicle, recovery, light	11	10	21	2
M60A3 tank, combat, 105-mm	11	12	27	3
M713 truck, ambulance, ¼-ton	7	6	12	2
M728 vehicle, combat engineer	11	12	29	3
M792 truck, ambulance, 1½-ton	8	7	19	2
M816 truck, wrecker, 5-ton	10	8	30	3
M880 truck, cargo, 1¼-ton	8	7	19	2
M88A1 vehicle, recovery, medium	10	11	27	3
M9 vehicle, ACE	9	13	21	2
M920 truck, tractor, 20-ton	12	11	27	3
M930 truck, dump, 5-ton	9	8	24	2
M977 truck, cargo, HEMTT	9	8	34	3
M978 truck, tanker, HEMTT	9	8	34	3
M992 ammo carrier (FAAS-V)	11	11	23	3
M998, HMMWV, carrier, personnel	6	7	15	2
MLRS	9	10	23	2
MT250 crane, hydraulic, 25-ton	10	8	45	3
RT crane, boom, 20-ton	14	11	44	4

SUPPLEMENTAL CAMOUFLAGE

C-4. Camouflage nets are often employed in conjunction with supplemental camouflage because nets alone do not make a target invisible to a threat's multispectral sensors. Use other CCD techniques to achieve effective concealment. Cover or remove all of the target's reflective surfaces (mirrors, windshields, lights). Also ensure that the target's shadow is disrupted or disguised. Use native vegetation, because placing a target in dense foliage provides natural concealment and a smoother transition between the edges of the camouflage net and the target's background. Cover exposed edges of the net with dirt or cut vegetation to enhance the transition.

VEHICLE CAMOUFLAGE

C-5. Measure the vehicle or determine its dimensions from table C-1. Use the following equations and figure C-2, page C-3, to determine the number of modules needed to camouflage a vehicle.

Equation 1: $A = 2h + w + 5\,feet$

Equation 2: $B = 2h + L + 5\,feet$

where—

$h = height, in\,feet$

$w = width, in\,feet$

$L = length, in\,feet$

Step 1. Determine the vehicle's dimensions (measure or use table C-1). For the M2 fighting vehicle, the height is 10 feet, the width is 11 feet, and the length is 22 feet.

Step 2. Use the above equations and the measurements from Step 1 to determine the total dimensions.

$A = 2(10) + 11 + 5 = 36\,feet$

$B = 2(10) + 22 + 5 = 47\,feet$

Step 3. Determine the number of modules needed (use figure C-2). Since A equals 36 and B equals 47, two modules of camouflage are required to cover the M2 fighting vehicle.

TRAINING

C-6. Units should develop and practice battle drills that cover the requirements and procedures for erecting nets over assigned equipment. Table C-2, page C-6, shows a sample battle drill.

Table C-2. Sample battle drill

Standards: • Complete camouflage net setup drills within 20 minutes. • Complete camouflage net teardown drills within 15 minutes.
Personnel Required: Three crew members.
Equipment Required: Two modules or the following items: • Nets, hexagonal, 2 each. • Nets, diamond, 2 each. • Pole sections, 24 each. • Stakes, 36 each. • Lanyards, 6 each. • Spreaders, 12 each.
Stowage Location: The camouflage net is strapped to the right side of the trim vane.
Setup Drill: • The gunner and the assistant gunner remove the camouflage net from the trim vane and place it on top of the M2. • The driver removes poles and stakes from the bag and places them around the vehicle. • The gunner and the assistant gunner remove the vehicle's antenna, position the net on top of the vehicle, and roll the net off the sides of the vehicle. • The driver stakes the net around the vehicle. • The driver and the assistant gunner assemble plies and spreaders and then erect the net. • The gunner inspects the camouflage from a distance. • The crew adjusts the camouflage as necessary.
Teardown Drill: • The driver and the assistant gunner take down and disassemble plies and spreaders. • The gunner and the assistant gunner unstake the net and roll it to the top of the M2. • The gunner and the assistant gunner complete rolling the net on top of the vehicle and replace the vehicle's antenna. • The driver stores the net on the trim vane. • The gunner and the assistant gunner store poles, spreaders, and stakes on the trim vane.
Notes. 1. Preassemble the nets before placing them on the M2. 2. Supplement camouflage nets by properly placing vehicles and using natural vegetation.

Appendix D

Individual Camouflage, Concealment, and Decoys

Each soldier is responsible for camouflaging himself, his equipment, and his position. CCD reduces the probability of an enemy placing aimed fire on a soldier.

MATERIALS

D-1.   Use natural and artificial materials for CCD. Natural CCD includes defilade, grass, bushes, trees, and shadows. Artificial CCD for soldiers includes BDUs, camouflage nets, skin paint, and natural materials removed from their original positions. To be effective, artificial CCD must blend with the natural background.

DISCIPLINE

D-2.   Noise, movement, and light discipline contribute to individual CCD:

- Noise discipline muffles and eliminates sounds made by soldiers and their equipment.
- Movement discipline minimizes movement within and between positions and limits movement to routes that cannot be readily observed by an enemy.
- Light discipline controls the use of lights at night. Avoid open fires, do not smoke tobacco in the open, and do not walk around with a lit flashlight.

DISPERSAL

D-3.   Dispersal is the deliberate deployment of soldiers and equipment over a wide area. It is a key individual survival technique. Dispersal creates a smaller target mass for enemy sensors and weapons systems. Therefore, it reduces casualties and losses in the event of an attack and also makes enemy detection efforts more difficult.

CONSIDERATIONS

D-4.   Every soldier should have a detailed understanding of the recognition factors described in chapter 3. While all of these factors remain important when applying individual CCD, the following factors are critical:

- **Movement.** Movement draws attention, whether it involves vehicles on the road or individuals walking around positions. The naked eye, IR, and radar sensors can detect movement. Minimize movement while in the open and remember that darkness does not prevent observation by an enemy equipped with modern sensors. When movement is necessary, slow, smooth movement attracts less attention than quick, irregular movement.
- **Shape.** Use CCD materials to break up the shapes and shadows of positions and equipment. Stay in the shadows whenever possible, especially when moving, because shadows can visually mask objects. When conducting operations close to an enemy, disguise or distort helmet and body shapes with artificial CCD materials because an enemy can easily recognize them at close range.
- **Shine and light.** Shine can also attract attention. Pay particular attention to light reflecting from smooth or polished surfaces (mess kits, mirrors, eyeglasses, watches, windshields, starched uniforms). Plastic map cases, dust goggles worn on top of a helmet, and clear plastic garbage bags also reflect light. Cover these items or remove them from exposed areas. Vehicle headlights, taillights, and safety reflectors not only reflect light but also reflect laser energy used in weapon systems. Cover this equipment when the vehicle is not in operation.

Red filters on vehicle dome lights and flashlights, while designed to protect a soldier's night vision, are extremely sensitive to detection by NVDs. A tank's red dome light, reflecting off the walls and out through the sight and vision blocks, can be seen with a starlight scope from 4 kilometers. Red-lensed flashlights and lit cigarettes and pipes are equally observable. To reduce the chances of detection, replace red filters with blue-green filters and practice strict light discipline. Use measures to prevent shine at night because moonlight and starlight can be reflected as easily as sunlight.

- **Color.** The contrast of skin, uniforms, and equipment with the background helps an enemy detect OPFOR. Individual CCD should blend with the surroundings; or at a minimum, objects must not contrast with the background. Ideally, blend colors with the background or hide objects with contrasting colors.

EMPLOYMENT

D-5.   Study nearby terrain and vegetation before applying CCD to soldiers, equipment, or the fighting position. During recon, analyze the terrain in lieu of the CCD considerations listed above and then choose CCD materials that best blend with the area. Change CCD as required when moving from one area to another.

SKIN

D-6.   Exposed skin reflects light and may draw attention. Even very dark skin, because of natural oils, will reflect light. CCD paint sticks cover these oils and help blend skin with the background. Avoid using oils or insect repellent to soften the paint stick because doing so makes skin shiny and defeats the purpose of CCD paint. Soldiers applying CCD paint should work in pairs and help each other. Self-application may leave gaps, such as behind ears. Use the following technique:

- Paint high, shiny areas (forehead, cheekbones, nose, ears, chin) with a dark color.
- Paint low, shadow areas with a light color.
- Paint exposed skin (back of neck, arms, hands) with an irregular pattern.

D-7.   When CCD paint sticks are unavailable, use field expedients such as burnt cork, bark, charcoal, lampblack, or mud. Mud contains bacteria, some of which is harmful and may cause disease or infection, so consider mud as the last resource for individual CCD field-expedient paint.

UNIFORMS

D-8.   BDUs have a CCD pattern but often require additional CCD, especially in operations occurring very close to the enemy. Attach leaves, grass, small branches, or pieces of LCSS to uniforms and helmets. These items help distort the shape of a soldier, and they blend with the natural background. BDUs provide visual and NIR CCD. Do not starch BDUs because starching counters the IR properties of the dyes. Replace excessively faded and worn BDUs because they lose their CCD effectiveness as they wear.

EQUIPMENT

D-9.   Inspect personal equipment to ensure that shiny items are covered or removed. Take corrective action on items that rattle or make other noises when moved or worn. Soldiers assigned equipment, such as vehicles or generators, should be knowledgeable of their appropriate camouflage techniques (see chapters 3, 4, and 5).

INDIVIDUAL FIGHTING POSITIONS

> *Note.* Review the procedures for camouflaging positions in chapter 5, which include considerations for camouflaging individual positions.

D-10. While building a fighting position, camouflage it and carefully dispose of earth spoil. Remember that too much CCD material applied to a position can actually have a reverse effect and disclose the position to the enemy. Obtain CCD materials from a dispersed area to avoid drawing attention to the position by the stripped area around it.

D-11. Camouflage a position as it is being built. To avoid disclosing a fighting position, never—

- Leave shiny or light-colored objects exposed.
- Remove shirts while in the open.
- Use fires.
- Leave tracks or other signs of movement.
- Look up when aircraft fly overhead. (One of the most obvious features on aerial photographs is the upturned faces of soldiers.)

D-12. When CCD is complete, inspect the position from an enemy's viewpoint. Check CCD periodically to see that it stays natural-looking and conceals the position. When CCD materials become ineffective, change or improve them.

Appendix E

Standard Camouflage Materials

Table E-1, lists standard camouflage items available to the soldier. Items on this list are ordered through normal unit-procurement channels:

- A complete list of Department of Defense (DOD) stock materials is available from the Defense Logistics Service Center (DLSC), Battle Creek, Michigan, leon.cdidcodddengdoc@conus.army.mil .
- A complete list of Army materials is available from the Army Materiel Command (AMC), Logistics Support Activity, Redstone Arsenal, Alabama, leon.cdidcodddengdoc@conus.army.mil.

Table E-1. Camouflage items

Item	NSN	Mil No.	Remarks
Camo enamel, black	8010-00-111-8356	NA	5 gal
Camo enamel, black	8010-00-111-8005	NA	1 gal
Camo enamel, sand	8010-00-111-8336	NA	5 gal
Camo enamel, sand	8010-00-111-7988	NA	1 gal
Camo screen, ultralite, asphalt/ concrete	1080-01-338-4468	PN88116169	CVU-165/G
Camo screen, ultralite, green/tan	1080-01-338-4471	PN88116003	CVU-166/G
Camo screen, ultralite, snow/partial snow	1080-01-338-4469	PN88116170	CVU-164/G
Camo support set, ultralite (A-frame)	1080-01-338-4472	PN88116154	MTU-96/G
Connector plug, w/o gen-test	5935-01-050-6586	MS3456W16S-1P	Use 5935-00-431-4935
Connector, receptacle, electrical CCK-77/E	1370-01-171-1336	293E663P404	1.4G class/div, 49 ea
Control, remote smoke gen, MXK-856/E32	1080-01-338-7051	PN88115510	For SG-18-02
Decoy target, bailey bridge	1080-00-650-1098	MIL-D-52165	None
Decoy target, how, 105-mm	1080-00-570-6519	MIL-D-52165B	PN EB 306D4904-IT08
Decoy units, inflating, radar, AN/SLQ-49	5865-01-266-3840	MRIIRVIN820/821	Passive radar freq respondent
Decoy, aircraft, ground (F-16)	1080-01-301-8273	PN160002	Only 25 produced
Decoy, close combat, M1A1 tank	1080-01-242-7251	PN13277E9830	None
Decoy, close combat, M60A3 tank	1080-01-242-7250	PN3228E1979	None
Decoy, runway (FOS)	1080-01-338-5201	PN88116100	50 x 1,000 ft
Diesel fuel, DF-1	9140-00-286-5288	VV-F-800D	Smoke/obsc-alt
Diesel fuel, DF-2	9140-00-286-5296	VV-F-800D	Smoke/obsc-alt
Diesel fuel, DF-2	9140-00-286-5297	VV-F-800D	Smoke/obsc-alt
Drum, S&S, 55-gal	8110-00-292-9783	NA	18-gauge steel, painted
Drum, S&S, 55-gal	8110-00-597-2353	NA	16-gauge steel, painted

Table E-1. Camouflage items

Item	NSN	Mil No.	Remarks
Explosive, airburst projectile launch atk	1055-01-175-4002	PN102575	Smoky flak, LMK-25
Federal standard colors 595-B	7690-01-162-2210	NA	2-ft x 10-in fan deck of color
Gen set, smoke, mech, M157	1040-01-206-0147	PN31-15-255	None
Gen, signal radio freq	6625-00-937-4029	NA	SM-422/GRC
Gen, smoke, mech, A/E32U-13	1040-01-338-8839	PN88115460	SG-18-02
Gen, smoke, mech, M3A	1040-00-587-3618	MILSTD604	None
Gen, smoke, mech, M3A4	1040-01-143-9506	MILSTD604	PN E31-15-2000
Indiv camo cover, 3-color woodland	8415-01-280-3098	MIL-C-44358	8 oz, 5- x 8-ft coverage
Indiv camo cover, 6-color desert	8415-01-280-5234	MIL-C-44358	8 oz, 5- x 8-ft coverage
Indiv camo cover, snow	8415-01-282-3160	MIL-C-44358	8 oz, 5- x 8-ft coverage
Launcher rckt, 1-bay launcher, LMU-23E	1055-01-131-7857	PN1335AS380	Smoky SAM
Launcher rckt, 4-bay launcher, OMU-24E	1055-01-144-0864	PN1335AS700	Smoky SAM
LCSS support set, desert	1080-00-623-7295	MIL-C-52765	Can use 1080-01-253-0522
LCSS support set, snow	1080-00-556-4954	MIL-C-52765	Same as 1080-01-179-6024
LCSS support set, woodland	1080-00-108-1173	MIL-C-52765	Same as 1080-01-179-6025
LCSS support set, woodland	1080-00-108-1173	MIL-C-52765	Plastic poles
LCSS, desert, radar-scattering	1080-00-103-1211	MIL-C-52771	Can use 1080-01-266-1828
LCSS, desert, radar-scattering	1080-01-266-1825	PN13228E5930	Can use 1080-01-266-1828
LCSS, desert, radar-scattering	1080-01-266-1828	PN13228E5933	Use 1080-01-266-1825 first
LCSS, desert, radar-transparent	1080-00-103-1217	MIL-C52765	PN13226E1357
LCSS, snow, radar-scattering	1080-00-103-1233	MIL-C-52765	Can use 1080-01-266-1826
LCSS, snow, radar-scattering	1080-00-103-1234	MIL-C-52765	PN13226E1355
LCSS, snow, radar-scattering	1080-01-266-1823	PN13228E5928	Can use 1080-01-266-1826
LCSS, snow, radar-scattering	1080-01-266-1826	PN13228E5931	Can use 1080-00-103-1233
LCSS, woodland, radar-scattering	1080-00-103-1246	MIL-C-53004	Can use 1080-01-266-1827
LCSS, woodland, radar-scattering	1080-00-103-1322	MIL-C-53004	PN13226E1356
LCSS, woodland, radar-scattering	1080-01-266-1824	PN13228E5929	Can use 1080-01-266-1827
LCSS, woodland, radar-scattering	1080-01-266-1827	PN13228E5932	Use 1080-01-266-1824 first
Lead acid btry, 24V, BB-297U	6140-00-059-3528	MS75047-1	For SG 18-02 w/o gen
Mounting kit, smoke gen, M284	1040-01-249-0272	PN31-14-2680	For M157 gen
Net, multipurpose, olive-green mesh	8465-00-889-3771	MIL-N-43181	108- x 60-in coverage
Paint, temp, tan	8010-01-326-8078	MIL-P-52905	Fed-std-595B 33446
Paint, temp, tan	8010-01-326-8079	MIL-P-52905	Fed-std-595B 33446
Paint, temp, white	8010-01-129-5444	MIL-P-52905	None

ESSENTIAL ARMY MANUAL

(537) CAMOUFLAGE, CONCEALMENT AND DECOYS

Table E-1. Camouflage items

Item	NSN	Mil No.	Remarks
Pump inflating, manual, smoky flak	4320-00-822-9036	XX-P-746	Need 1 ea TO 11A-1-46
Reflector, radar, Coast Guard buoy marker	2050-01-225-2779	120768	1 cu ft, 10-lb, aluminum
Simulator, atomic explosion, M142	1370-00-474-0270	MIL-S-46528(1)	PM8864243
Simulator, projectile airburst, PJU-7/E	1370-01-180-5856	PN102549	1.1G class/div, 48 ea
Simulator, projectile airburst, PJU-7A/E	1370-01-279-9505	PN8387310	1.3G class/div, 48 ea
Smoke pot, 30-lb, HC, M5	1365-00-598-5207	MIL-S-13183	PH E36-1-18, ±17 min
Smoke pot, floating, HC	1365-00-939-6599	MIL-S-51235	w/M208/M209 fuse
Smoke pot, floating, HC, M4A2	1365-00-598-5220	MIL-S-51235B	w/M207a fuse, ±12 min
Smokey SAM rocket, GTR-18A	1340-01-130-6282	DL1335AS100	Firing cartridge and rocket
Support poles, camo net, ultralite	1080-01-338-4470	PN88116153	MTU-99/G, 2 poles/battens
Tool, special purpose, smoky flak	5120-01-176-2188	PN103320	Need 1 ea
Trailer, ground-handling, MHU-141/M	1740-01-031-5868	MIL-BK-300	5,500-lb cap, for SG-18-02
Valve adapter assy, smoky flak	1055-01-216-4803	PN8523971-10	Need 1 each
Valve, pneumatic tank, smoky flak	4820-00-427-5047	GV500RK2	Need 1 ea
Wrench, bung	5120-00-045-5055	Cage #07227	2- x 3/4-in plugs

Appendix F

The Geneva Emblem and Camouflage of Medical Facilities

> This appendix implements STANAG 2931.

STANAG 2931 covers procedures for using the Geneva emblem and camouflaging medical facilities. This STANAG requires signatories to display the Geneva emblem (red cross) on medical facilities to help identify and protect the sick and wounded. All signatories, however, are allowed to display the Geneva emblem according to their national regulations and procedures. STANAG 2931 also defines medical facilities as medical units, medical vehicles, and medical aircraft on the ground. A tactical commander may order the camouflage of medical facilities, including the Geneva emblem, when the failure to do so will endanger or compromise tactical operations. Such an order is considered temporary and must be rescinded as soon as the tactical situation permits. The camouflage of large, fixed medical facilities is not envisaged under the guidelines of STANAG 2931.

Glossary

Acronym/Term	Definition
AA	assembly area
AAR	after-action review
AAS	Army aviation site
ACE	armored combat earthmover, M9
AFJPAM	Air Force joint pamphlet
AFV	armored fighting vehicle
alt	alternate
AM	amplitude modulation
AMA	aviation maintenance area
AMC	Army Materiel Command
ammo	ammunition
AO	area of operation
assy	assembly
atk	attack
ATTN	attention
AVLB	armored vehicle-launched bridge
background	The features in a target area that surround the target.
BDU	battle-dress uniform
blending	A CCD technique that causes a target to appear as part of the background. Many target characteristics must be considered when attempting a blending treatment, including target size and shape, regular patterns in the target scene, and rough or smooth target contours.
btry	battery
C^2	command and control
C^2W	command and control warfare. The integrated use of PSYOP, military deception, OPSEC, EW, and physical destruction supported by intelligence to deny information to, influence, degrade, or destroy adversary C^2 capabilities while protecting friendly C^2 capabilities against such actions.
C^3	command, control, and communications
C^3CM	command, control, and communications countermeasure. The integrated use of OPSEC, military deception, jamming, and physical destruction supported by intelligence to deny information to, influence, degrade, or destroy adversary C^3 capabilities while protecting friendly C^3 capabilities against such actions.
camo	camouflage. The use of natural or artificial materials on personnel, objects, and tactical positions to confuse, mislead, or evade the enemy.

camouflage net	Part of a system designed to blend a target with its surroundings and conceal the identity of critical assets (aircraft, fixed targets, vehicles, personnel) where natural cover and/or concealment might be absent or inadequate.
camouflage net set	Standard DOD set consisting of a hexagon-shaped net (673.6 sq ft), a diamond-shaped net (224.5 sq ft), and a net repair kit.
camouflage net spreader	A plastic or aluminum disc or paddle that is supported by a lightweight pole and used to support camouflage nets above the ground, buildings, or vehicles.
cap	capacity
CB	counterbattery
CCD	camouflage, concealment, and decoys. Methods and resources to prevent adversary observation or surveillance; confuse, mislead, or evade the adversary; or induce the adversary to act in a manner prejudicial to his interests.
CCD treatment	A combination of CCD equipment and techniques applied to a selected target and/or its background to reduce or delay target acquisition.
chaff	Material consisting of thin, narrow, metallic strips of various lengths and frequency responses used as artificial clouds to scatter radar signals.
clutter	EM radiation from sources around the target that tend to hinder target detection.
CM	countermortar
countermeasure	Any technique intended to confuse or mislead hostile sensors.
COMSEC	communications security
concealment	The protection from observation or surveillance.
conspicuity	A term peculiar to the CCD community that denotes the perceived difference of one feature in a scene as compared to other features in the scene.
corner reflector	An object that reflects multiple signals from smooth surfaces mounted mutually perpendicular and produces a radar return of greater magnitude than expected from the size of the object the reflector conceals.
counterreconnaissance	All measures taken to prevent hostile observation of a force, an area, or a place.
countersurveillance	All measures, active or passive, taken to counteract hostile surveillance.
cover	Any natural or artificial protection from enemy observation and fire.
covered approach	Any route that offers protection against enemy observation or fire.
CP	command post
CSS	combat service support
cu	cubic
DA	Department of the Army

DC	District of Columbia
deceive	Any action that causes the enemy to believe the false or purposely causes the enemy to make incorrect conclusions based on false evidence.
deception	Those measures designed to mislead the enemy by manipulation, distortion, or falsification of evidence, inducing him to react in a manner prejudicial to his interests.
decoy	An imitation in any sense of a person, an object, or a phenomenon that is intended to deceive enemy surveillance devices or mislead enemy evaluation.
detection	The discovery of an existence or presence.
disguise	Any alteration of identity cues for items, signals, or systems sufficient to cause misidentification by the enemy.
dispersal	Relocation of forces for the purpose of increasing survivability.
disrupt	Any action intended to interrupt the shape or outline of an object or an individual, making it less recognizable.
div	division
DLSC	Defense Logistics Service Center
DOD	Department of Defense
DSN	Defense Switched Network
ea	each
ECCM	electronic counter-countermeasure. Any action involving effective use of the EM spectrum by friendly forces, despite the enemy's use of EW.
ECM	electronic countermeasure. Any action involving prevention or reduction of an enemy's effective use of the EM spectrum. ECMs include electronic jamming and electronic deception.
electronics security	The protection resulting from all measures designed to deny unauthorized persons information of value that, when analyzed, might alert the enemy to the intentions of friendly forces (for example, a signal security provided by encryption equipment).
EM	electromagnetic
EM spectrum	electromagnetic spectrum. The range of frequencies from zero to infinity where energy is transferred by electric and magnetic waves. EM waves at the lower end of this spectrum (low-frequency navigation aids and AM and shortwave radio services) are refracted back to earth by the ionosphere to frequencies as high as 50 MHz. At frequencies above 50 MHz, propagation is generally limited to LOS. These frequencies are used by TV, FM radio, and land-mobile and point-to-point communication services. They extend on to parts of the EM spectrum generally termed as radar, IR, visible light, UV light, and cosmic rays.
EW	electronic warfare. Any military action involving the use of EM energy to determine, exploit, reduce, or prevent hostile use of the EM spectrum; action which retains friendly use of the EM spectrum.
FAAS-V	field artillery ammunition support vehicle
FARP	forward arming and refueling point

FEBA	forward edge of the battle area
fed	federal
FLIRS	Forward-Looking Infrared System. An imaging IR sensor used to acquire a target's heat signature.
FLOT	forward line of own troops
FM	field manual
FM	frequency modulation
FOB	forward operating base
FOD	foreign object damage
fog oil	Petroleum compounds of selected molecular weight and composition to facilitate the formation of smoke by atomization, vaporization, and subsequent recondensation.
FOS	false operating surface. A simulated horizontal construction placed to represent operating surfaces such as runways, taxiways, parking pads, and access roads.
freq	frequency
ft	foot, feet
gal	gallon(s)
gen	generator
GHz	gigahertz
GI	government issue
GSR	ground-surveillance radar
hardening	The construction of a facility to provide protection against the effects of conventional or nuclear explosions. The facility may also be equipped to provide protection against chemical or biological attacks. Construction usually involves reinforced concrete placement and/or burying the structure.
HC	hydrogen chloride
HEMTT	heavy expanded mobility tactical truck
hiding	The choice of a position or materials to obstruct direct observation.
HMMWV	high-mobility multipurpose wheeled vehicle
how	howitzer
HQ	headquarters
HTF	how to fight
HUD	heads-up display
HVT	high-value target
hyperspectral	Refers to a sensor or data with many bands extending over a range of the EM spectrum.
imaging radar	An electronic or optical process for recording or displaying a scene generated by a radar sensor.
in	inch(es)
indiv	individual

intervisibility	The condition of the atmosphere that allows soldiers the ability to see from one point to another. This condition may be altered or interrupted by weather, smoke, dust, or debris.
IPB	intelligence preparation of the battlefield. A systematic approach to analyzing the enemy, weather, and terrain in a specific geographic area. It integrates enemy doctrine with the weather and terrain conditions as they relate to the mission and the specific battlefield environment. IPB provides the framework for determining and evaluating enemy capabilities, vulnerabilities, and probable courses of action.
IR	infrared
IR smoke screen	It produces obscuration in one or more of the transparent IR spectral bands between 0.7 and 14 microns. In most cases, an effective IR smoke screen is also an effective visual smoke screen. However, effective visual smoke screens are not necessarily effective IR smoke screens.
JCCD	Joint Camouflage, Concealment, and Deception
JSTARS	Joint Surveillance Target Attack Radar System
JT&E	Joint Test and Evaluation
lb	pound(s)
LCSS	Lightweight Camouflage Screen System
LLTV	low-light television
LOC	lines of communication
LOS	line of sight
low emissivity paint	Paint used to lower the apparent temperature of a target (or nearby scene features), thus making the *hot* target less conspicuous to a thermal target-acquisition sensor. Using a paint that has too low an emissivity (less than 0.6) causes the target to become more visually conspicuous (or shiny).
maskirovka	The battlefield doctrine of the former Soviet Union.
MCRP	Marine Corps reference publication
MCWP	Marine Corps warfighting publication
mech	mechanized
METT-TC	mission, enemy, terrain, weather, troops, time available, and civilian considerations
MHz	megahertz
mil	military
min	minute(s)
MLRS	Multiple Launch Rocket System
mm	millimeter(s)
MOUT	military operations on urbanized terrain
movement techniques	The methods used by a unit to travel from one point to another (traveling, traveling overwatch, and bounding overwatch) are considered movement techniques. The likelihood of enemy contact determines which technique is used.
MTI	moving-target indicator

multispectral	Refers to a sensor or data in two or more regions of the EM spectrum.
NA	not applicable
NBC	nuclear, biological, chemical
NCO	noncommissioned officer
NIR	near infrared
No.	number
NSN	national stock number. A 13-digit number assigned to each item of supply purchased, stocked, or distributed within the federal government.
NVD	night-vision device
NWP	Navy warfighting publication
obsc	obscurant. Suspended particulates or entrained liquid droplets that can absorb and/or scatter EM radiation in various parts of the EM spectrum (visual, IR, radar).
obscuration	The effects of weather, battlefield dust, and debris; the use of smoke munitions to hamper observation and target acquisition; and the concealment of activities or movement.
OPFOR	opposing forces
OPORD	operation order
OPSEC	operations security. The process of denying adversaries information about friendly capabilities and intentions by identifying, controlling, and protecting signatures associated with planning for and conducting military operations and other activities. It includes countersurveillance and physical, signal, and information security.
oz	ounce(s)
POL	petroleum, oils, and lubricants
PSYOP	psychological operations
pub	publication
radar	A device that uses EM waves to provide information on the range, the azimuth, or the elevation of objects.
radar camouflage	Any radar-absorbing or -reflecting material that changes the radar-echoing properties of an object's surface.
radar clutter	Unwanted signals, echoes, or images displayed by a radar unit that interfere with the observation of desired signals.
radar imagery	The picture produced on a radar screen by recording the EM waves reflected from a given target surface.
radio detection	The detection of a radio's presence by intercepting its signals without precise determination of its position.
radio direction-finding	The act of determining the azimuth to a radio transmitter, from a specific location, using signal-detecting equipment.
radio fix	The location of a radio transmitter determined by simultaneously using two direction-finding devices stationed at different locations and plotting the results on a map. The intersection of the two azimuths indicates the transmitter's location.

radio range-finding	The act of determining the distance to a radio transmitter. This technique involves using electronic equipment to intercept and measure a transmitter's emissions and then translating this information into a distance.
RAM	radar-absorbing material. Material that absorbs and dissipates incident radar energy as contrasted to radar-scattering material, which reflects the incident energy in a different direction.
RAP	radar-absorbing paint. A coating that can absorb incident radar energy.
RATELO	radiotelephone operator
rckt	rocket
RCS	radar cross section. The size of a conducting square, metal plate that would return the same signal to a radar sensor as a target, provided that the radar energy received at the target is reradiated equally in all directions.
recon	reconnaissance. An exploratory survey of a particular area or airspace by visual, aural, electronic, photographic, IR, or other means. It may imply a physical visit to the area.
redundancy	The use of multiple systems with similar perceived functional capabilities to provide higher system survivability.
relocatable asset	A military asset that normally stays in place for a short period of time relative to a fixed asset.
reverse-slope position	A position on the ground that is not exposed to direct fire or observation; for example, a slope that descends away from the enemy.
revetment	A barrier used to protect assets against attack.
ROM	refuel on the move
RSTA	reconnaissance, surveillance, and target acquisition
RT	rough terrain
S&S	supply and service
SAM	surface-to-air missile
SCSPP	standard camouflage screening paint pattern
Scud	A surface-to-surface missile.
signature	Detectable indications that forces are occupying or operating in an area. Signatures can be EM (visible, IR, NIR, radar) or mechanical (acoustic, seismic). Common detectable EM signatures include visible vehicle tracks, thermal flames, and radar signal returns. Common mechanical signatures include radio noise, humans conversing, and seismic ground waves produced by tanks and heavy vehicles.
SLAR	side-looking airborne radar
smk	smoke. An artificially produced aerosol of solid, liquid, or vapor deposited in the atmosphere that inhibits the passage of visible light or other forms of EM radiation.
smky	smoky
smoke generator	A machine that produces large volumes of smoke to support hasty or deliberate operations for screening, protecting, and/or sustaining

airfields, ports, staging areas, and bridge crossings. Present smoke generators vaporize liquid aerosol materials such as fog oil, diesel fuel, and polyethylene glycol. These generators consist of a heat source to vaporize the liquid aerosol material and an apparatus for the production of airflow to efficiently disseminate the smoke vapor into the atmosphere where it disperses and condenses.

smoke pot	An expendable bucket- or pot-like munition that produces dense smoke by burning combustible material.
smoke screen	Smoke generated to deceive or confuse an enemy as to the activities of tactical elements.
SOP	standing operating procedure
SP	self-propelled
sq	square
STANAG	standardization agreement
std	standard
surveillance	A systematic observation of airspace or surface areas by visual, aural, electronic, photographic, IR, or other means.
survivability operations	Activities involving the development and construction of fighting and protective positions (earth berms, defilade positions, overhead protection, camouflage) that reduce the effectiveness of enemy detection systems.
TAA	tactical assembly area
TACSOP	tactical standing operating procedure
target acquisition	The process involving the detection and identification of hostile operations and equipment for subsequent engagement.
target scene	The view of a target area that includes both the target and its surroundings.
temp	temporary
terrain analysis	The process of examining a geographic area to determine what effects its natural and man-made features may have on military operations.
terrain mottling	A camouflage technique normally used in desert terrain. It involves scarring the earth with heavy equipment to expose patches of bare ground. Equipment and supplies are placed on the bare patches to avoid detection by aerial reconnaissance.
thermal contrast	The difference in radiance (as usually measured in the 8-to-14 micron band) between two features of a scene; for example, a target and its background.
thermal crossover	A temporary situation, in the morning or evening, when the target and background temperatures become equal.
thermal emissivity	The ratio of the emissive power of a surface to that of a black body. The emissivity is 1 for a black body and 0.9 for most natural and man-made materials. The apparent temperature of a target can be reduced by reducing its real temperature and/or lowering its emissivity. Unfortunately, as the thermal emissivity is lowered, its reflectivity in the visual portion of the spectrum increases, thus making the target more conspicuous to a visual sensor. A typical compromise is 0.7, which lowers the apparent target temperature but does not make it too shiny in the visible spectrum.

TM	technical manual
TMD	tactical missile defense
tone down	The process of blending a target or other high-value asset with the background by reducing its brightness characteristics using nets or coatings. The recommended reflectance of a target as compared with the surrounding scene is 10 percent or less.
TOW	tube-launched, optically tracked, wire-guided
TRADOC	United States Army Training and Doctrine Command
TV	television
UAV	unmanned aerial vehicle
US	United States
USAES	United States Army Engineer School
UV	ultraviolet
V	volt
w/	with
w/o	without
WSA	weapons storage area
μ	micron(s)

References

SOURCES USED

These are the sources quoted or paraphrased in this publication. DA Forms are available on the APD website (www.apd.army.mil).

DA Form 2028, *Recommended Changes to Publications and Blank Forms.*

FM 2-0, *Intelligence*, 23 March 2010.

FM 3-06. *Urban Operations.* 26 October 2006.

FM 3-11, *Multiservice Tactics, Techniques, and Procedures for Nuclear, Biological, and Chemical Defense Operations*, 10 March 2003.

FM 3-25.26, *Map Reading and Land Navigation*, 18 January 2005.

FM 5-34, *Engineer Field Data*, 19 July 2005.

FM 5-103, *Survivability*, 10 June 1985.

FM 21-10, *Field Hygiene and Sanitation*, 21 June 2000.

JP 1-02, *Department of Defense Dictionary of Military and Associated Terms*, 23 April 2001.

JP 3-13.1, *Electronic Warfare*, 23 January 2007.

STANAG 2931 (Edition 2), *Orders for the Camouflage of the Red Cross and Red Crescent on Land in Tactical Operations*, 19 January 1998.

TM 5-1080-200-13&P, *Operator's, Unit, and Direct Support Maintenance Manual (Including Repair Parts and Special Tools List) for Lightweight Camouflage Screen Systems and Support Systems*, 29 January 1987.

DOCUMENTS NEEDED

None.

READINGS RECOMMENDED

None.

FULL INDEXES AVAILABLE ON ACCOMPANYING DVD

PHYSICAL FITNESS TRAINING

FM 21 - 20

HEADQUARTERS,
DEPARTMENT OF THE ARMY

Field Manual
No. 21-20

*FM 21-20
Headquarters
Department of the Army
Washington, DC, 30 September 1992

TABLE OF CONTENTS

This publication supersedes FM 21-20, 28 August 1985.

TABLE OF CONTENTS (CONT.)

PAGE

PAGE

Preface

On 5 July 1950, U.S. troops, who were unprepared for the physical demands of war, were sent to battle. The early days of the Korean war were nothing short of disastrous, as U.S. soldiers were routed by a poorly equipped, but well-trained, North Korean People's Army. As American soldiers withdrew, they left behind wounded comrades and valuable equipment their training had not adequately prepared them to carry heavy loads.

The costly lessons learned by Task Force Smith in Korea are as important today as ever. If we fail to prepare our soldiers for their physically demanding wartime tasks, we are guilty of paying lip service to the principle of "Train as you fight." Our physical training programs must do more for our soldiers than just get them ready for the semiannual Army Physical Fitness Test (APFT').

FM 21 -20 is directed at leaders who plan and conduct physical fitness training. It provides guidelines for developing programs which will improve and maintain physical fitness levels for all Army personnel. These programs will help leaders prepare their soldiers to meet the physical demands of war. This manual can also be used as a source book by all soldiers. FM 21-20 was written to conform to the principles outlined in FM 25-100, Training the Force.

The benefits to be derived from a good physical fitness program are many. It can reduce the number of soldiers on profile and sick call, invigorate training, and enhance productivity and mental alertness. A good physical fitness program also promotes team cohesion and combat survivability. It will improve soldiers' combat readiness.

The proponent of this publication is HQ TRADOC. Send comments and recommendations on DA Form 2028 (Recommended Changes to Publications and Blank Forms) directly to Headquarters, US Army Infantry Center, US Army Physical Fitness School (ATZB-PF), Fort Benning, GA31905-5000.

Unless this publication states otherwise, masculine nouns and pronouns do not refer exclusively to men.

CHAPTER 1

Introduction

A soldier's level of physical fitness' has a direct impact on his combat readiness. The many battles in which American troops have fought underscore the important role physical fitness plays on the battlefield. The renewed nationwide interest in fitness has been accompanied by many research studies on the effects of regular participation in sound physical fitness programs. The overwhelming conclusion is that such programs enhance a person's quality of life, improve productivity, and bring about positive physical and mental changes. Not only are physically fit soldiers essential to the Army, they are also more likely to have enjoyable, productive lives.

This chapter provides an overview of fitness. It defines physical fitness, outlines the phases of fitness, and discusses various types of fitness programs and fitness evaluation. Commanders and leaders can use this information to develop intelligent, combat-related, physical fitness programs.

Physical fitness, the emphasis of this manual, is but one component of total fitness. Some of the "others are weight control, diet and nutrition, stress management, dental health, and spiritual and ethical fitness, as well as the avoidance of hypertension, substance abuse, and tobacco use. This manual is primarily concerned with issues relating directly to the development and maintenance of the five components of physical fitness.

The Army's physical fitness training program extends to all branches of the total Army. This includes the USAR and ARNG and encompasses all ages and ranks and both sexes. Its purpose is to physically condition all soldiers throughout their careers beginning with initial entry training (IET). It also includes soldiers with limiting physical profiles who must also participate in physical fitness training.

Commanders and leaders must ensure that all soldiers in their units maintain the highest level of physical fitness in accordance with this manual and with AR 350-15 which prescribes policies, procedures, and responsibilities for the Army physical fitness program.

Leadership Responsibilities

Effective leadership is critical to the success of a good physical training program. Leaders, especially senior leaders, must understand and practice the new Army doctrine of physical fitness. They must be visible and active participants in physical training programs. In short, leaders must lead PT! Their example will emphasize the importance of physical fitness training and will highlight it as a key element of the unit's training mission.

Leaders must emphasize the value of physical training and clearly explain the objectives and benefits of the program. Master Fitness Trainers (MFTs), graduates of a special course taught by the U.S. Army Physical Fitness School, can help commanders do this. However, regardless of the level of technical experience MFTs have, the sole responsibility for good programs rests with leaders at every level.

A poorly designed and executed physical fitness program hurts morale. A good program is well planned and organized, has reasonable yet challenging requirements, and is competitive and progressive. It also has command presence at every level with leaders setting the example for their soldiers.

Leaders should also continually assess their units to determine which specific components of fitness they lack. Once they identify the shortcomings, they should modify their programs to correct the weaknesses.

Leaders should not punish soldiers who fail to perform to standard. Punishment, especially excessive repetitions or additional PT, often does more harm than good. Leaders must

Components of physical fitness include weight control, diet, nutrition, stress management, and spiritual and ethical fitness.

plan special training to help soldiers who need it. The application of sound leadership techniques is especially important in bringing physically deficient soldiers up to standard.

'COMMAND FUNCTIONS

Commanders must evaluate the effectiveness of physical fitness training and ensure that it is focused on the unit's missions. They can evaluate its effectiveness by participating in and observing training, relating their fitness programs to the unit's missions, and analyzing individual and unit APFT performance.

Leaders should regularly measure the physical fitness level of every soldier to evaluate his progress and determine the success of the unit's program.

Commanders should assure that qualified leaders supervise and conduct fitness training and use their MFTs, for they have received comprehensive training in this area.

Leaders can learn about fitness training in the following ways:
• Attend the four-week MFT course or one-week Exercise Leaders Course.
• Request a fitness workshop from the Army Physical Fitness School.
• Become familiar with the Army's fitness publications. Important examples include this manual, AR 350-15, and DA Pamphlets 350-15, 350-18, and 350-22.

Commanders must provide adequate facilities and funds to support a program which will improve each soldier's level of physical fitness. They must also be sure that everyone participates, since all individuals, regardless of rank, age, or sex, benefit from regular exercise. In some instances, leaders will need to make special efforts to overcome recurring problems which interfere with regular training.

Leaders must also make special efforts to provide the correct fitness training for soldiers who are physically substandard. "Positive profiling" (DA Form 3349) permits and encourages profiled soldiers to do as much as they can within the limits of their profiles. Those who have been away from the conditioning process because of leave, sickness, injury, or travel may also need special consideration.

Commanders must ensure that the time allotted for physical fitness training is used effectively.

Training times is wasted by the following:
• Unprepared or unorganized leaders.
• Assignment fo a group which us too large for one leader.
• Insufficient training intensity: it will result in no improvement.
• Rates of progression that are too slow or too fast.
• Extreme faomality that usually emphasizes form over substance. An example would be too many units runs at slow paces or "daily dozen" activities that look impressive but do not result in impovement.
• Inadequate facilities which cause long waiting periods between exercises during a workout and/or between workouts.
• Long rest periods which interfere with progress.

To foster a positive attitude, unit leaders and instructors must be knowledgeable, understanding, and fair, but demanding. They must recognize individual differences and motivate soldiers to put forth their best efforts. However, they must also emphasize training to standard. Attaining a high level of physical fitness cannot be done simply by going through the motions. Hard training is essential.

Commanders must ensure that leaders are familiar with approved

Commanders must ensure that the time alloted for physical fitness training is used effectively.

techniques, directives, and publications and that they use them. The objective of every commander should be to incorporate the most effective methods of physical training into a balanced program. This program should result in the improved physical fitness of their soldiers and an enhanced ability to perform mission-related tasks.

MFTs can help commanders formulate sound programs that will attain their physical training goals, but commanders must know and apply the doctrine. However, since the responsibility for physical training is the commander's, programs must be based on his own training objectives. These he must develop from his evaluation of the unit's mission-essential task list (METL). Chapter 10 describes the development of the unit's program.

MASTER FITNESS TRAINERS

A Master Fitness Trainer (MFT) is a soldier who has completed either the four-week active-component, two-week reserve-component, or U.S. Military Academy's MFT course work. Although called "masters," MFTs are simply soldiers who know about all aspects of physical fitness training and how soldiers' bodies function. Most importantly, since MFTs are taught to design individual and unit programs, they should be used by commanders as special staff assistants for this purpose.

MFTs can do the following:
• Assess the physical fitness levels of individuals and units.
• Analyze the unit's mission-related tasks and develop sound fitness training programs to support those tasks.
• Train other trainers to conduct sound, safe physical training.
• Understand the structure and function of the human body, especially as it relates to exercise.

Components of Fitness

Physical fitness is the ability to function effectively in physical work, training, and other activities and still have enough energy left over to handle any emergencies which may arise.

The components of physical fitness are as follows:
• Cardiorespiratory (CR) endurance-the efficiency with which the body delivers oxygen and nutrients needed for muscular activity and transports waste products from the cells.
• Muscular strength - the greatest amount of force a muscle or muscle group can exert in a single effort.
• Muscular endurance - the ability of a muscle or muscle group to perform repeated movements with a sub-maximal force for extended periods of times.
• Flexibility-the ability to move the joints (for example, elbow, knee) or any group of joints through an entire, normal range of motion.
• Body composition-the amount of body fat a soldier has in comparison to his total body mass.

Improving the first three components of fitness listed above will have a positive impact on body composition and will result in less fat. Excessive body fat detracts from the other fitness components, reduces performance, detracts from appearance, and negatively affects one's health.

Factors such as speed, agility, muscle power, eye-hand coordination, and eye-foot coordination are classified as components of "motor" fitness. These factors affect a soldier's survivability on the battlefield. Appropriate training can improve these factors within the limits of each soldier's potential. The Army's fitness program seeks to improve or maintain all the components of physical and motor fitness through sound, progressive, mission-specific physical training for individuals and units.

Principles of Exercise

Adherence to certain basic exercise principles is important for developing an effective program. The principles of exercise apply to everyone at all levels of physical training, from the Olympic-caliber athlete to the weekend jogger. They also apply to fitness training for military personnel.

These basic principles of exercise must be followed:

● Regularity. To achieve a training effect, a person must exercise often. One should strive to exercise each of the first four fitness components at least three times a week. Infrequent exercise can do more harm than good. Regularity is also important in resting, sleeping, and following a good diet.

● Progression. The intensity (how hard) and/or duration (how long) of exercise must gradually increase to improve the level of fitness.

e Balance. To be effective, a program should include activities that address all the fitness components, since overemphasizing any one of them may hurt the others.

● Variety. Providing a variety of activities reduces boredom and increases motivation and progress.

● Specificity. Training must be geared toward specific goals. For example, soldiers become better runners if their training emphasizes running. Although swimming is great exercise, it does not improve a 2-mile-run time as much as a running program does.

● Recovery. A hard day of training for a given component of fitness should be followed by an easier training day or rest day for that component and/or muscle group(s) to help permit recovery. Another

way to allow recovery is to alternate the muscle groups exercised every other day, especially when training for strength and/or muscle endurance.

● Overload. The work load of each exercise session must exceed the normal demands placed on the body in order to bring about a training effect.

FITT Factors

Certain factors must be part of any fitness training program for it to be successful. These factors are Frequency, Intensity, Time, and Type. The acronym FITT makes it easier to remember them. (See Figure 1- 1.)

FREQUENCY

Army Regulation 350-15 specifies that vigorous physical fitness training will be conducted 3 to 5 times per week. For optimal results, commanders must strive to conduct 5 days of physical training per week. Ideally, at least three exercise sessions for CR fitness, muscle endurance, muscle strength, and flexibility should be performed each week to improve fitness levels. Thus, for example, to obtain maximum gains in muscular strength, soldiers should have at least three strength-training sessions per week. Three physical activity periods a week, however, with only one session each of cardiorespiratory, strength, and flexibility training will not improve any of these three components.

With some planning, a training program for the average soldier can be developed which provides fairly equal emphasis on all the components of physical fitness. The following training program serves as an example.

In the first week, Monday, Wednesday, and Friday are devoted to CR fitness, and Tuesday and Thursday are devoted to muscle endurance and strength. During the second week, the

*Factors for a succe[...]
training program [...]
Frequency, Intensi[...]
Time, and Typ[...]
"FITT".*

FITT Factors Applied to Physical Conditioning Program

	Cardiorespiratory Endurance	Muscular Strength	Muscular Endurance	Muscular Strength and Muscular Endurance	Flexibility
F Frequency	3-5 times/week	3 times/week	3-5 times/week	3 times/week	Warm-up and Cool-down: Stretch before and after each exercise session Developmental Stretching: To improve flexibility, stretch 2-3 times/week
I Intensity	60-90% HRR*	3-7 RM*	12+ RM	8-12 RM	Tension and slight discomfort, NOT PAIN
T Time	20 minutes or more	The time required to do 3-7 repetitions of each exercise	The time required to do 12+ repetitions of each exercise	The time required to do 8-12 repetitions of each exercise	Warm-up and Cool-down Stretches: 10-15 seconds/stretch Developmental Stretches: 30-60 seconds/stretch
T Type	Running Swimming Cross-Country Skiing Rowing Bicycling Jumping Rope Walking/Hiking Stair Climbing	Free Weights Resistance Machines Partner-Resisted Exercises Body-Weight Exercises (Pushups/Situps/Pullups/Dips, etc.)			Stretching: Static Passive P.N.F.

* HRR = Heart Rate Reserve   * RM = Repetition Maximum

Figure 1-1

training days are flip-flopped: muscle endurance and strength are trained on Monday, Wednesday, and Friday, and CR fitness is trained on Tuesday and Thursday. Stretching exercises are done in every training session to enhance flexibility. By training continuously in this manner, equal emphasis can be given to developing muscular endurance and strength and to CR fitness while training five days per week.

If the unit's mission requires it, some muscular and some CR training can be done during each daily training session as long as a "hard day/recovery day" approach is used. For example, if a unit has a hard run on Monday, Wednesday, and Friday, it may also choose to run on Tuesday and Thursday. However, on Tuesday and Thursday the intensity and/or distance/time should be reduced to allow recovery. Depending on the time available for each session and the way training sessions are conducted, all components of fitness can be developed using a three-day-per-week schedule. However, a five-day-per-week program is much better than three per week. (See Training Program in Chapter 10.)

Numerous other approaches can be taken when tailoring a fitness program to meet a unit's mission as long as the principles of exercise are not violated. Such programs, when coupled with good nutrition, will help keep soldiers fit to win.

INTENSITY

Training at the right intensity is the biggest problem in unit programs. The intensity should vary with the type of exercise being done. Exercise for CR development must be strenuous enough to elevate the heart rate to between 60 and 90 percent of the heart rate reserve (HRR). (The calculation of percent HRR is explained in Chapter 2.) Those with low fitness levels should start exercising at a lower training heart rate (THR) of about 60 percent of HRR.

For muscular strength and endurance, intensity refers to the percentage of the maximum resistance that is used for a given exercise. When determining intensity in a strength-training program, it is easier to refer to a "repetition maximum" or "RM." For example, a 10-RM is the maximum weight that can be correctly lifted 10 times. An 8-12 RM is the weight that can be lifted 8 to 12 times correctly. Doing an exercise "correctly" means moving the weight steadily and with proper form without getting help from

other muscle groups by jerking, bending, or twisting the body. For the average person who wants to improve both muscular strength and endurance, an 8-12 RM is best.

The person who wants to concentrate on muscular strength should use weights which let him do three to seven repetitions before his muscles fatigue. Thus, for strength development, the weight used should be a 3-7 RM. On the other hand, the person who wants to concentrate on muscular endurance should use a 12+ RM. When using a 12+ RM as the training intensity, the more repetitions performed per set, over time, the greater will be the improvement in muscular endurance. Conversely, the greater the number of repetitions performed, the smaller will be the gains in strength. For example, a person who regularly trains with a weight which lets him do 100 repetitions per exercise (a 100-RM) greatly increases his muscular endurance but minimally improves his muscular strength. (See Chapter 3 for information on resistance training.)

All exercise sessions should include stretching during the warm-up and cool-down. One should stretch so there is slight discomfort, but no pain, when the movement is taken beyond the normal range of motion. (See Chapter 4 for information on stretching.)

TIME

Like intensity, the time spent exercising depends on the type of exercise being done. At least 20 to 30 continuous minutes of intense exercise must be used in order to improve cardiorespiratory endurance.

For muscular endurance and strength, exercise time equates to the number of repetitions done. For the average soldier, 8 to 12 repetitions with enough resistance to cause muscle failure improves both muscular endurance and strength. As soldiers progress, they

All exercises sess should inclu stretching du the warm-up and down.

will make better strength gains by doing two or three sets of each resistance exercise.

Flexibility exercises or stretches should be held for varying times depending on the objective of the session. For warming-up, such as before a run, each stretch should be held for 10 to 15 seconds. To improve flexibility, it is best to do stretching during the cool-down, with each stretch held for 30 to 60 seconds. If flexibility improvement is a major goal, at least one session per week should be devoted to developing it.

TYPE

Type refers to the kind of exercise performed. When choosing the type, the commander should consider the principle of specificity. For example, to improve his soldiers' levels of CR fitness (the major fitness component in the 2-mile run), he should have them do CR types of exercises. These are discussed in Chapter 2.

Ways to train for muscular strength and endurance are addressed in Chapter 3, while Chapter 4 discusses flexibility. These chapters will help commanders design programs which are tailor-made to their soldiers' needs. The basic rule is that to improve performance, one must practice the particular exercise, activity, or skill he wants to improve. For example, to be good at push-ups, one must do push-ups. No other exercise will improve push-up performance as effectively.

Warm-up and Cool-Down

One must prepare the body before taking part in organized PT, unit sports competition, or vigorous physical activity. A warm-up may help prevent injuries and maximize performance. The warm-up increases the body's internal temperature and the heart rate. The chance of getting injured decreases when the heart, muscles,

ligaments, and tendons are properly prepared for exertion. A warm-up should include some running-in-place or slow jogging, stretching, and calisthenics. It should last five to seven minutes and should occur just before the CR or muscular endurance and strength part of the workout. After a proper warm-up, soldiers are ready for a more intense conditioning activity.

Soldiers should cool down properly after each exercise period, regardless of the type of workout. The cool-down serves to gradually slow the heart rate and helps prevent pooling of the blood in the legs and feet. During exercise, the muscles squeeze the blood through the veins. This helps return the blood to the heart. After exercise, however, the muscles relax and no longer do this, and the blood can accumulate in the legs and feet. This can cause a person to faint. A good cool-down will help avoid this possibility.

Soldiers should walk and stretch until their heart rates return to less than 100 beats per minute (BPM) and heavy sweating stops. This usually happens five to seven minutes after the conditioning session.

Phases of Fitness Conditioning

The physical fitness training program is divided into three phases: preparatory, conditioning, and maintenance. The starting phases for different units or individuals vary depending on their age, fitness levels, and previous physical activity.

Young, healthy persons may be able to start with the conditioning phase, while those who have been exercising regularly may already be in the maintenance phase. Factors such as extended field training, leave time, and illness can cause soldiers to drop from a maintenance to a conditioning phase.

Persons who have not been active, especially if they are age 40 or older, should start with the preparatory phase. Many soldiers who fall into this category may be recovering from illness or injury, or they may be just out of high school. Most units will have soldiers in all three phases of training at the same time.

PREPARATORY PHASE

The preparatory phase helps both the cardiorespiratory and muscular systems get used to exercise, preparing the body to handle the conditioning phase. The work load in the beginning must be moderate. Progression from a lower to a higher level of fitness should be achieved by gradual, planned increases in frequency, intensity, and time.

Initially, poorly conditioned soldiers should run, or walk if need be, three times a week at a comfortable pace that elevates their heart rate to about 60 percent HRR for 10 to 15 minutes. Recovery days should be evenly distributed throughout the week, and training should progress slowly. Soldiers should continue at this or an appropriate level until they have no undue fatigue or muscle soreness the day following the exercise. They should then lengthen their exercise session to 16 to 20 minutes and/or elevate their heart rate to about 70 percent HRR by increasing their pace. To be sure their pace is faster, they should run a known distance and try to cover it in less time. Those who feel breathless or whose heart rate rises beyond their training heart rate (THR) while running should resume walking until the heart rate returns to the correct training level. When they can handle an intensity of 70 percent HRR for 20 to 25 minutes, they should be ready for the next phase. Chapter 2 shows how to determine the THR, that is, the right training level during aerobic training.

The preparatory phase for improving muscular endurance and strength through weight training should start easily and progress gradually. Beginning weight trainers should select about 8 to 12 exercises that work all the body's major muscle groups. They should use only very light weights the first week (that is, the first two to three workouts). This is very important, as they must first learn the proper form for each exercise. Light weights will also help minimize muscle soreness and decrease the likelihood of injury to the muscles, joints, and ligaments. During the second week, they should use progressively heavier weights on each resistance exercise. By the end of the second week (four to six workouts), they should know how much weight will let them do 8 to 12 repetitions to muscle failure for each exercise. At this point the conditioning phase begins.

CONDITIONING PHASE

To reach the desired level of fitness, soldiers must increase the amount of exercise and/or the workout intensity as their strength and/or endurance increases.

To improve cardiorespiratory endurance, for example, they must increase the length of time they run. They should start with the preparatory phase and gradually increase the running time by one or two minutes each week until they can run continuously for 20 to 30 minutes. At this point, they can increase the intensity until they reach the desired level of fitness. They should train at least three times a week and take no more than two days between workouts.

For weight trainers, the conditioning phase normally begins during the third week. They should do one set of 8 to 12 repetitions for each of the selected resistance exercises. When they can do more than 12 repetitions of any exercise, they should increase the

Soldiers and units should be encouraged progress beyond minimum requiremen

weight used on that exercise by about five percent so they can again do only 8 to 12 repetitions. This process continues throughout the conditioning phase. As long as they continue to progress and get stronger while doing only one set of each exercise, it is not necessary for them to do more than one set per exercise. When they stop making progress with one set, they should add another set on those exercises in which progress has slowed. As training progresses, they may want to increase the sets to three to help promote further increases in strength and/ or muscle mass.

For maximum benefit, soldiers should do strength training three times a week with 48 hours of rest between workouts for any given muscle group. It helps to periodically do a different type of exercise for a given muscle or muscle group. This adds variety and ensures better strength development.

The conditioning phase ends when a soldier is physically mission-capable and all personal, strength-related goals and unit-fitness goals have been met.

MAINTENANCE PHASE

The maintenance phase sustains the high level of fitness achieved in the conditioning phase. The emphasis here is no longer on progression. A well-designed, 45- to 60-minute workout (including warm-up and cool-down) at the right intensity three times a week is enough to maintain almost any appropriate level of physical fitness. These workouts give soldiers time to stabalize their flexibility, CR endurance, and muscular endurance and strength. However, more frequent training may be needed to reach and maintain peak fitness levels.

Soldiers and units should always be encouraged to progress beyond minimum requirements. Maintaining an optimal level of fitness should become part of every soldier's life-style and should be continued throughout his life.

An effective program uses a variety of activities to develop muscular endurance and strength, CR endurance, and flexibility, and to achieve good body composition. It should also promote the development of coordination as well as basic physical skills. (See Chapter 10 for guidance in constructing a unit program.)

Types of Fitness Programs

The Army has too many types of units with different missions to have one single fitness program for everyone. Therefore, only broad categories of programs and general considerations are covered here. They are classified as unit, individual, and special programs.

UNIT PROGRAMS

Unit programs must support unit missions. A single unit may require several types of programs. Some units, such as infantry companies, have generally the same types of soldiers and MOSS. On the other hand, certain combat--service-support units have many different types of soldiers, each with unique needs. Commanders can develop programs for their own unit by following the principles in this chapter. MFTs know how to help commanders develop programs for their units/soldiers.

Commanders of units composed of both men and women must also understand the physiological differences between the sexes. These are summarized in Appendix A. Although women are able to participate in the same fitness programs as men, they must work harder to perform at the same absolute level of work or exercise. The same holds true for poorly-conditioned soldiers running with well-conditioned soldiers.

To overcome this problem in the case of running, for example, the unit should use ability group runs rather than unit runs. Soldiers in a given ability group will run at a set pace, with groups based on each soldier's most recent 2-mile-run time. Three to six groups per company-sized unit are usually enough. Within each group, each soldier's heart rate while running should be at his own THR. When the run is not intense enough to bring one or more of the soldiers to THR, it is time for those soldiers to move up to the next ability group.

Ability group running does two things more effectively than unit runs: 1) it lets soldiers improve to their highest attainable fitness level; and, 2) it more quickly brings subpar performers up to minimum standards.

It also allows soldiers to train to excel on the APFT which, in turn, helps promotion opportunities. Holding a fit soldier back by making him run at a slow, unit-run pace (normally less than his minimum pace for the 2-mile run on the APFT) hurts his morale and violates the principle of training to challenge.

initial Entry Training (IET)

The training program in basic training (BT) brings soldiers up to the level of physical fitness they need to do their jobs as soldiers. However, the program requires good cadre leadership to ensure that it is appropriate, demanding, and challenging.

Trainees report to active duty at various levels of physical fitness and ability. During basic training they pass through the preparatory into the conditioning phase. During "fill" periods and the first week of training, the focus is on learning and developing the basics of physical fitness.

Training emphasizes progressive conditioning of the whole body. To minimize the risk of injury, exercises must be done properly, and the intensity must progress at an appropriate rate. Special training should be considered for soldiers who fail to maintain the unit's or group's rate of progression. Commanders should evaluate each basic trainee who falls below standard and give him individualized, special assistance to improve his deficiencies.

Additional training should not be used as punishment for a soldier's inability to perform well.

More PT is not necessarily better. Chapter 11 describes how to develop physical training programs in IET units.

Advanced Individual Training (AIT)

Although AIT focuses on technical and MOS-oriented subjects, physical fitness must be emphasized throughout. Most soldiers arriving from basic training are already well into the conditioning phase. Therefore, AIT unit training should focus on preparing soldiers to meet the physical requirements of their initial duty assignments. (See TRADOC Reg. 350-6, Chapter 4.)

Walking, running, and climbing during unit training contribute to physical fitness, but they are not enough. Physical training in AIT requires continued, regular, vigorous exercise which stresses the whole body and addresses all the components of fitness.

By the end of AIT, soldiers must meet APFT standards. With good programs and special training, all healthy AIT graduates should easily be able to demonstrate that they, possess the required level of physical fitness.

By the end of AIT, soldiers must meet APFT standards.

TOE and TDA Units–Active Component

There are many types of units in the Army, and their missions often require different levels of fitness. TOE and TDA units must emphasize attaining and maintaining the fitness level required for the mission.

The unit's standards may exceed the Army's minimums. By regulation (AR 350- 15), the unit's standards can be established by the unit's commander, based on mission requirements.

TOE and TDA Units--Reserve Components

The considerations for the active component also apply to reserve components (RCS). However, since members of RC units cannot participate together in collective physical training on a regular basis, RC unit programs must focus on the individual's fitness responsibilities and efforts. Commanders, however, must still ensure that the unit's fitness level and individual PT programs are maintained. MFTs can give valuable assistance to RC commanders and soldiers.

INDIVIDUAL PROGRAMS

There must be a ositive approach to all special fitness t r a i n i n g .

Many soldiers are assigned to duty positions that offer little opportunity to participate in collective unit PT programs. Examples are HQDA, MACOM staffs, hospitals, service school staff and faculty, recruiting, and ROTC. In such organizations, commanders must develop leadership environments that encourage and motivate soldiers to accept individual responsibility for their own physical fitness. Fitness requirements are the same for these personnel as for others. Section chiefs and individual soldiers need to use the fundamental principles and techniques outlined in this manual to help them attain and maintain a high level of physical fitness. MFTs can help develop individual fitness programs.

SPECIAL PROGRAMS

The day-to-day unit PT program conducted for most soldiers may not be appropriate for all unit members. Some of them may not be able to exercise at the intensity or duration best suited to their needs.

At least three groups of soldiers may need special PT programs. They are as follows:

• Those who fail the APFT and do not have medical profiles.

• Those who are overweight/overfat according to AR 600-9

• Those who have either permanent or temporary medical profiles.

Leaders must also give special consideration to soldiers who are age 40 or older and to recent arrivals who cannot meet the standards of their new unit.

Special programs must be tailored to each soldier's needs, and trained, knowledgeable leaders should develop and conduct them. This training should be conducted with the unit, If this is impossible, it should at least occur at the same time.

There must be a positive approach to all special fitness training. Soldiers who lack enough upper body strength to do a given number of push-ups or enough stamina to pass the 2-mile run should not be ridiculed. Instead, their shortcomings should be assessed and the information used to develop individualized programs to help them remedy their specific shortcomings. A company-sized unit may have as many as 20 soldiers who need special attention. Only smart planning will produce good programs for all of them.

Commanders must counsel soldiers, explaining that special programs are being developed in their best interests. They must make it clear that standards will be enforced. Next, they should coordinate closely with medical personnel to develop programs that fit the capabilities of soldiers with medical limitations. Each soldier should then begin an individualized program based on his needs.

MFTs know how to assess CR endurance, muscular strength and endurance, flexibility, and body composition. They can also develop thorough, tailor-made programs for all of a unit's special population.

APFT Failures

Although it is not the heart of the Army's physical fitness program, the APFT is the primary instrument for evaluating the fitness level of each soldier. It is structured to assess the muscular endurance of specific muscle groups and the functional capacity of the CR system.

Soldiers with reasonable levels of overall physical fitness should easily pass the APFT. Those whose fitness levels are substandard will fail. Soldiers who fail the APFT must receive special attention. Leaders should analyze their weaknesses and design programs to overcome them. For example, if the soldier is overweight, nutrition and dietary counseling may be needed along with a special exercise program. DA Pam 350-22 outlines several ways to improve a soldier's performance on each of the APFT events.

When trying to improve APFT performances, leaders must ensure that soldiers are not overloaded to the point where the fitness training becomes counterproductive. They should use ability groups for their running program and, in addition to a total-body strength-training program, should include exercises designed for push-up and sit-up improvement. When dealing with special populations, two very important principles are overload and recovery. The quality, not just the

quantity, of the workout should be emphasized. Two-a-day sessions, unless designed extremely well, can be counter-productive. More PT is not always better.

Overweight Soldiers

Designers of weight loss and physical training programs for overweight soldiers should remember this: even though exercise is the key to sensible weight loss, reducing the number of calories consumed is equally important. A combination of both actions is best.

The type of exercise the soldier does affects the amount and nature of the weight loss. Both running and walking burn about 100 calories per mile. One pound of fat contains 3,500 calories. Thus, burning one pound of fat through exercise alone requires a great deal of running or walking. On the other hand, weight lost through dieting alone includes the loss of useful muscle tissue. Those who participate in an exercise program that emphasizes the development of strength and muscular endurance, however, can actually increase their muscle mass while losing body fat. These facts help explain why exercise and good dietary practices must be combined.

Unit MFTs can help a soldier determine the specific caloric requirement he needs to safely and successfully lose excess fat. They can devise a sound, individualized plan to arrive at that reduced caloric intake. Likewise, unit MFTs can also develop training programs which will lead to fat loss without the loss of useful muscle tissue.

Generally, overweight soldiers should strive to reduce their fat weight by two pounds per week. When a soldier loses weight, either by diet or exercise or both, a large initial weight loss is not unusual. This may be due to water loss associated with the using up of the body's carbohydrate stores. Although these losses may be encouraging to the soldier, little of this initial weight loss is due to the loss of fat.

Soldiers should be weighed under similar circumstances and at the same time each day. This helps avoid false measurements due to normal fluctuations in their body weight during the day. As a soldier develops muscular endurance and strength, lean muscle mass generally increases. Because muscle weighs more per unit of volume than fat. caution is advised in assessing his progress. Just because a soldier is not losing weight rapidly does not necessarily mean he is not losing fat. In fact, a good fitness program often results in gaining muscle mass while simultaneously losing fat weight. If there is reasonable doubt, his percentage of body fat should be determined.

Soldiers with Profiles

This manual stresses what soldiers can do while on medical profile rather than what they cannot do.

DOD Directive 1308.1 requires that, "Those personnel identified with medically limiting defects shall be placed in a physical fitness program consistent with their limitations as advised by medical authorities."

AR 350-15 states, "For individuals with limiting profiles, commanders will develop physical fitness programs in cooperation with health care personnel."

The Office of the Surgeon General has developed DA Form 3349 to ease the exchange of information between health care personnel and the units. On this form, health care personnel list, along with limitations, those activities that the profiled soldier can do to maintain his fitness level. With this information, the unit should direct profiled soldiers to participate in the activities they can do. (An example of DA Form 3349 is in Appendix B.)

All profiled soldiers should take part in as much of the regular fitness

program as they can. Appropriate activities should be substituted to replace those regular activities in which they cannot participate.

Chapter 2 describes some aerobic activities the soldier can do to maintain cardiorespiratory fitness when he cannot run. Chapter 3 shows how to strengthen each body part. Applying this information should allow some strength training to continue even when body parts are injured. The same principle applies to flexibility (Chapter 4).

Medical treatment and rehabilitation should be aimed at restoring the soldier to a suitable level of physical fitness. Such treatment should use appropriate, progressive physical activities with medical or unit supervision.

MFTs can help profiled soldiers by explaining alternative exercises and how to do them safely under the limitations of their profile. MFTs are not, however, trained to diagnose injuries or prescribe rehabilitative exercise programs. This is the domain of qualified medical personnel.

The activity levels of soldiers usually decrease while they are recovering from sickness or injury. As a result, they should pay special attention to their diets to avoid gaining body fat. This guidance becomes more important as soldiers grow older. With medical supervision, proper diet, and the right PT programs, soldiers should be able to overcome their physical profiles and quickly return to their normal routines and fitness levels.

Age as a Factor in Physical Fitness

Soldiers who are age 40 and older represent the Army's senior leadership. On the battlefield, they must lead other soldiers under conditions of severe stress. To meet this challenge

All profiled soldiers should do as much of the regular fitness program as they can, along with substitute activities.

and set a good example, these leaders must maintain and demonstrate a high level of physical fitness. Since their normal duties may be stressful but nonphysical, they must take part regularly in a physical fitness program. The need to be physically fit does not decrease with increased age.

People undergo many changes as they grow older. For example, the amount of blood the heart can pump per beat and per minute decreases during maximal exercise, as does the maximum heart rate. This lowers a person's physical ability, and performance suffers. Also, the percent of body weight composed of fat generally increases, while total muscle mass decreases. The result is that muscular strength and endurance, CR endurance, and body composition suffer. A decrease in flexibility also occurs.

Men tend to maintain their peak levels of muscular strength and endurance and CR fitness until age 30. After 30 there is a gradual decline throughout their lives. Women tend to reach their peak in physical capability shortly after puberty and then undergo a progressive decline.

Although a decline in performance normally occurs with aging, those who stay physically active do not have the same rate of decline as those who do not. Decreases in muscular strength and endurance, CR endurance, and flexibility occur to a lesser extent in those who regularly train these fitness components.

Soldiers who are fit at age 40 and continue to exercise show a lesser decrease in many of the physiological functions related to fitness than do those who seldom exercise. A trained 60-year-old, for example, may have the same level of CR fitness as a sedentary 20-year-old. In short, regular exercise can help add life to your years and years to your life.

The assessment phase of a program is especially important for those age 40 and over. However, it is not necessary or desirable to develop special fitness programs for these soldiers. Those who have been exercising regularly may continue to exercise at the same level as they did before reaching age 40. A program based on the principles of exercise and the training concepts in this manual will result in a safe, long-term conditioning program for all soldiers. Only those age 40 and over who have not been exercising regularly may need to start their exercise program at a lower level and progress more slowly than younger soldiers. Years of inactivity and possible abuse of the body cannot be corrected in a few weeks or months.

As of 1 January 1989, soldiers reaching age 40 are no longer required to get clearance from a cardiovascular screening program before taking the APFT. Only a medical profile will exempt them from taking the biannual record APFT. They must, however, have periodic physical examinations in accordance with AR 40-501 and NGR 40-501. These include screening for cardiovascular risk factors.

Evaluation

To evaluate their physical fitness and the effectiveness of their physical fitness training programs, all military personnel are tested biannually using the APFT in accordance with AR 350-15. (Refer to Chapter 14.) However, commanders may evaluate their physical fitness programs more frequently than biannually.

SCORING CATEGORIES

There are two APFT categories of testing for all military personnel Initial Entry Training (IET) and the Army Standard.

IET **Standard**

The APFT standard for basic training is a minimum of 50 points per event and no less than 150 points overall by the end of basic training. Graduation requirements for AIT and One Station Unit Training (OSUT) require 60 points per event.

Army Standard

All other Army personnel (active and reserve) who are non-IET soldiers must attain the minimum Army standard of at least 60 points per event. To get credit for a record APFT, a medically profiled soldier must, as a minimum, complete the 2-mile run or one of the alternate aerobic events.

Safety is a major consideration when planning and evaluating physical training programs

SAFETY

Safety is a major consideration when planning and evaluating physical training programs. Commanders must ensure that the programs do not place their soldiers at undue risk of injury or accident. They should address the following items:

•Environmental conditions (heat/cold/traction).

• Soldiers' levels of conditioning (low/high/age/sex).

•Facilities (availability/instruction/repair).

•Traffic (routes/procedures/formations).

•Emergency procedures (medical/communication/transport).

The objective of physical training in the Army is to enhance soldiers' abilities to meet the physical demands of war. Any physical training which results in numerous injuries or accidents is detrimental to this goal. As in most training, common sense must prevail. Good, sound physical training should challenge soldiers but should not place them at undue risk nor lead to situations where accidents or injuries are likely to occur.

Cardiorespiratory Fitness

Cardiorespiratory (CR) fitness, sometimes called CR endurance, aerobic fitness, or aerobic capacity, is one of the five basic components of physical fitness. CR fitness is a condition in which the body's cardiovascular (circulatory) and respiratory systems function together, especially during exercise or work, to ensure that adequate oxygen is supplied to the working muscles to produce energy. CR fitness is needed for prolonged, rhythmic use of the body's large muscle groups. A high level of CR fitness permits continuous physical activity without a decline in performance and allows for rapid recovery following fatiguing physical activity.

Activities such as running, road marching, bicycling, swimming, cross-country skiing, rowing, stair climbing, and jumping rope place an extra demand on the cardiovascular and respiratory systems. During exercise, these systems attempt to supply oxygen to the working muscles. Most of this oxygen is used to produce energy for muscular contraction. Any activity that continuously uses large muscle groups for 20 minutes or longer taxes these systems. Because of this, a wide variety of training methods is used to improve cardiorespiratory endurance.

Physiology of Aerobic Training

Aerobic exercise uses oxygen to produce most of the body's energy needs. It also brings into play a fairly complex set of physiological events.

To provide enough energy-producing oxygen to the muscles, the following events occur:
- Greater movement of air through the lungs.
- Increased movement of oxygen from the lungs into the blood stream.
- Increased delivery of oxygen-laden blood to the working muscles by the heart's accelerated pumping action.
- Regulation of the blood vessel's size to distribute blood away from inactive tissue to working muscle.
- Greater movement of oxygen from the blood into the muscle tissue.
- Accelerated return of veinous blood to the heart.

Correctly performed aerobic exercise, over time, causes positive changes in the body's CR system. These changes allow the heart and vascular systems to deliver more oxygen-rich blood to the working muscles during exercise. Also, those muscles regularly used during aerobic exercise undergo positive changes. By using more oxygen, these changes let the muscles make and use more energy during exercise and, as a result, the muscles can work longer and harder.

During maximum aerobic exercise, the trained person has an increased maximum oxygen consumption ($\dot{V}O_2max$). He is better able to process oxygen and fuel and can therefore provide more energy to the working muscles.

$\dot{V}O_2max$, also called aerobic capacity, is the most widely accepted single indicator of one's CR fitness level.

CR fitness is needed for prolonged, rhythmic use of the body's large muscle groups.

erobic exercise is the t type of activity for ttaining and aintaining a low rcentage of body fat.

The best way to determine aerobic capacity is to measure it in the laboratory. It is much easier, however, to estimate maximum oxygen uptake by using other methods.

It is possible to determine a soldier's CR fitness level and get an accurate estimate of his aerobic capacity by using his APFT 2-mile-run time. (Appendix F explains how to do this.) Other tests - the bicycle, walk, and step tests - may also be used to estimate one's aerobic capacity and evaluate one's CR fitness level.

In the presence of oxygen, muscle cells produce energy by breaking down carbohydrates and fats. In fact, fats are only used as an energy source when oxygen is present. Hence, aerobic exercise is the best type of activity for attaining and maintaining a low percentage of body fat.

A person's maximum aerobic capacity can be modified through physical training. To reach very high levels of aerobic fitness, one must train hard. The best way to improve CR fitness is to participate regularly in a demanding aerobic exercise program.

Many factors can negateively affect one's ability to perform well aerobically. These include the following:

• Age.
• Anemia.
• Carbon monoxide from tobacco smoke or pollution.
• High altitude (reduced oxygen pressure).
• Illness (heart disease).
• Obesity.
• Sedentary life-style.

Any condition that reduces the body's ability to bring in, transport, or use oxygen reduces a person's ability to perform aerobically. Inactivity causes much of the decrease in physical fitness that occurs with increasing age. Some of this decrease in aerobic fitness

can be slowed by taking part in a regular exercise program.

Certain medical conditions also impair the transport of oxygen. They include diseases of the lungs, which interfere with breathing, and disabling heart conditions. Another is severe blocking of the arteries which inhibits blood flow to the heart and skeletal muscles.

Smoking can lead to any or all of the above problems and can, in the long and short term, adversely affect one's ability to do aerobic exercise.

FITT Factors

As mentioned in Chapter 1, a person must integrate several factors into any successful fitness training program to improve his fitness level. These factors are summarized by the following words which form the acronym FITT. Frequency, Intensity, Time, and Type. They are described below as they pertain to cardiorespiratory fitness. A warm-up and cool-down should also be part of each workout. Information on warming up and cooling down is given in Chapters 1 and 4.

FREQUENCY

Frequency refers to how often one exercises. It is related to the intensity and duration of the exercise session. Conditioning the CR system can best be accomplished by three adequately intense workouts per week. Soldiers should do these on alternate days. By building up gradually, soldiers can get even greater benefits from working out five times a week. However, leaders should recognize the need for recovery between hard exercise periods and should adjust the training intensity accordingly. They must also be aware of the danger of overtraining and recognize that the risk of injury increases as the intensity and duration of training increases.

INTENSITY

Intensity is related to how hard one exercises. It represents the degree of effort with which one trains and is probably the single most important factor for improving performance. Unfortunately, it is the factor many units ignore.

Changes in CR fitness are directly related to how hard an aerobic exercise is performed. The more energy expended per unit of time, the greater the intensity of the exercise. Significant changes in CR fitness are brought about by sustaining training heart rates in the range of 60 to 90 percent of the heart rate reserve (HRR). Intensities of less than 60 percent HRR are generally inadequate to produce a training effect, and those that exceed 90 percent HRR can be dangerous.

Soldiers should gauge the intensity of their workouts for CR fitness by determining and exercising at their training heart rate (THR). Using the THR method lets them find and prescribe the correct level of intensity during CR exercise. By determining one's maximum heart rate, resting heart rate, and relative conditioning level, an appropriate THR or intensity can be prescribed.

One's ability to monitor the heart rate is the key to success in CR training. (Note: Ability-group running is better than unit running because unit running does not accommodate the individual soldier's THR. For example, some soldiers in a formation may be training at 50 percent HRR and others at 95 percent HRR. As a result, the unit run will be too intense for some and not intense enough for others.)

The heart rate during work or exercise is an excellent indicator of how much effort a person is exerting. Keeping track of the heart rate lets one gauge the intensity of the CR exercise being done. With this information, one can be sure that the intensity is enough to improve his CR fitness level.

Following are two methods for determining training heart rate (THR). The first method, percent maximum heart rate (% MHR), is simpler to use, while the second method, percent heart rate reserve (% HRR), is more accurate. Percent HRR is the recommended technique for determining THR.

Percent MHR **Method**

With this method, the THR is figured using the estimated maximal heart rate. A soldier determines his estimated maximum heart rate by subtracting his age from 220. Thus, a 20-year-old would have an estimated maximum heart rate (MHR) of 200 beats per minute (220 -20 = 200).

To figure a THR that is 80 percent of the estimated MHR for a 20-year-old soldier in good physical condition, multiply 0.80 times the MHR of 200 beats per minute (BPM). This example is shown below.

FORMULA

% x MHR = THR

CALCULATION

0.80 x 200 BPM = 160 BPM

When using the MHR method, one must compensate for its built-in weakness. A person using this method may exercise at an intensity which is not high enough to cause a training effect. To compensate for this, a person who is in poor shape should exercise at 70 percent of his MHR; if he is in relatively good shape, at 80 percent MHR; and, if he is in excellent shape, at 90 percent MHR.

Intensity is probably the single most important factor for improving performance.

By determining one's maximum heart rate, resting rate, and conditioning level, an appropriate THR can be prescribed.

Percent HRR Method

A more accurate way to calculate THR is the percent HRR method. The range from 60 to 90 percent HRR is the THR range in which people should exercise to improve their CR fitness levels. If a soldier knows his general level of CR fitness, he can determine which percentage of HRR is a good starting point for him. For example, if he is in excellent physical condition, he could start at 85 percent of his HRR; if he is in reasonably good shape, at 70 percent HRR; and, if he is in poor shape, at 60 percent HRR.

Most CR workouts should be conducted with the heart rate between 70 to 75 percent HRR to attain, or maintain, an adequate level of fitness. Soldiers who have reached a high level of fitness may derive more benefit from working at a higher percentage of HRR, particularly if they cannot find more than 20 minutes for CR exercise. Exercising at any lower percentage of HRR does not give the heart, muscles, and lungs an adequate training stimulus.

Before anyone begins aerobic training, he should know his THR (the heart rate at which he needs to exercise to get a training effect).

The example below shows how to figure the THR by using the resting heart rate (RHR) and age to estimate heart rate reserve (HRR). A 20-year-old male soldier in reasonably good physical shape is the example.

STEP 1: Determine the MHR by subtracting the soldier's age from 220.

FORMULA
220 - age = MHR
(GIVEN)

CALCULATION
220 - 20 = 200 BPM

STEP 2: Determine the RHR in beats per minute (BPM) by counting the resting pulse for 30 seconds, and multiply the count by two. A shorter period can be used, but a 30-second count is more accurate. This count should be taken while the soldier is completely relaxed and rested. How to determine heart rate is described below. Next, determine the heart rate reserve (HRR) by subtracting the RHR from the estimated MHR. If the soldier's RHR is 69 BPM, the HRR is calculated as shown here.

FORMULA
MHR - RHR = HRR

CALCULATION
200 BPM - 69 BPM = 131 BPM

STEP 3: Calculate the THR based on 70 percent of HRR (a percentage based on a good level of CR fitness).

FORMULA
(% x HRR) + RHR = THR

CALCULATION
(0.70x131 BPM)+69 BPM=160.7 BPM

As shown, the percentage (70 percent in this example) is converted to the decimal form (0.70) before it is multiplied by the HRR. The result is then added to the resting heart rate (RHR) to get the THR. Thus, the product obtained by multiplying 0.70 and 131 is 91.7. When 91.7 is added to the RHR of 69, a THR of 160.7 results. When the calculations produce a fraction of a heart beat, as in the example, the value is rounded off to the nearest whole number. In this case, 160.7 BPM is rounded off to give a THR of 161 BPM. In summary, a reasonably fit 20-year-old soldier with a resting heart rate of 69 BPM has a training heart rate goal of 161 BPM. To determine the RHR, or to see if one is within the THR during and right after exercise, place the tip of the third finger lightly over one of the carotid arteries in the neck. These arteries are located to the left and right of the Adam's apple. (See Figure 2-1A.) Another convenient spot from which to monitor the pulse is on the radial artery on the wrist just above the base of the thumb. (See Figure 2-1B.) Yet another way is to place the hand over the heart and count the number of heart beats. (See Figure 2-1 C.)

During aerobic exercise, the body will usually have reached a "Steady State" after five minutes of exercise, and the heart rate will have leveled off. At this time, and immediately after exercising, the soldier should monitor his heart rate.

He should count his pulse for 10 seconds, then multiply this by six to get his heart rate for one minute. This will let him determine if his training intensity is high enough to improve his CR fitness level.

For example, use the THR of 161 BPM figured above. During the 10-second period, the soldier should get a count of 27 beats (161/6= 26.83 or 27) if he is exercising at the right intensity. If his pulse rate is below the THR, he must exercise harder to increase his pulse to the THR. If his pulse is above the THR, he should normally exercise at a lower intensity to reduce the pulse rate to the prescribed THR. He should count as accurately as possible, since one missed beat during the 10-second count, multiplied by six, gives an error of six BPM.

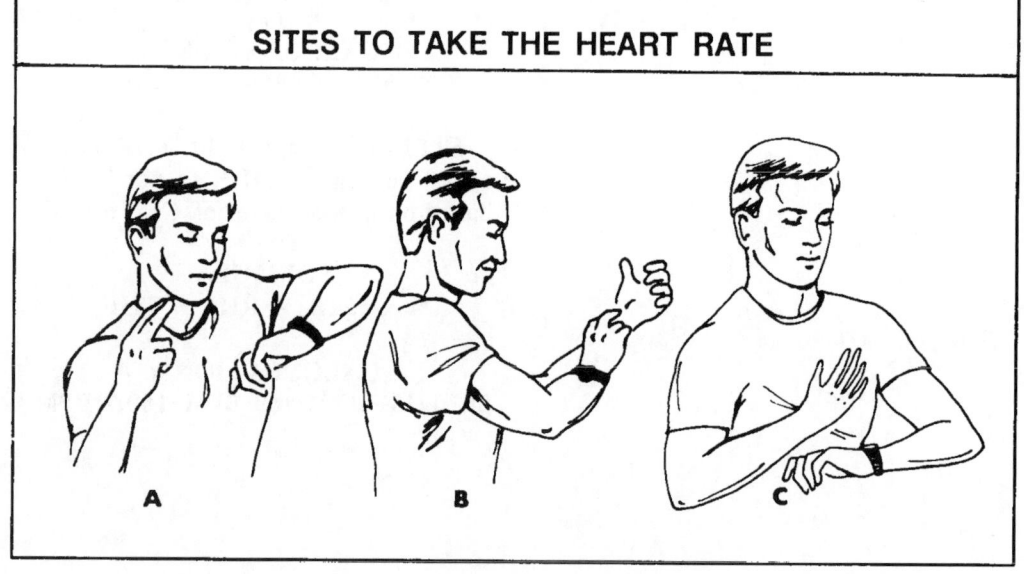

SITES TO TAKE THE HEART RATE

A   B   C

Figure 2-1

A soldier who maintains his THR throughout a 20-30-minute exercise period is doing well and can expect improvement in his CR fitness level.

A soldier who maintains his THR throughout a 20- to 30-minute exercise period is doing well and can expect improvement in his CR fitness level. He should check his exercise and post-exercise pulse rate at least once each workout. If he takes only one pulse check, he should do it five minutes into the workout.

Figure 2-2 is a chart that makes it easy to determine what a soldier's THR should be during a 10-second count. Using this figure, a soldier can easily find his own THR just by knowing his age and general fitness level. For example, a 40-year-old soldier with a low fitness level should, during aerobic exercise. have a THR of 23 beats in 10 seconds. He can determine this from the table by locating his age and then tracking upward until he reaches the percent HRR for his fitness level. Again, those with a low fitness level should work at about 60 percent HRR and those with a good fitness level at **70** percent HRR. Those with a high level of fitness may benefit most by training at 80 to 90 percent HRR.

Another way to gauge exercise intensity is "perceived exertion." This method relies on how difficult the exercise seems to be and is described in Appendix G.

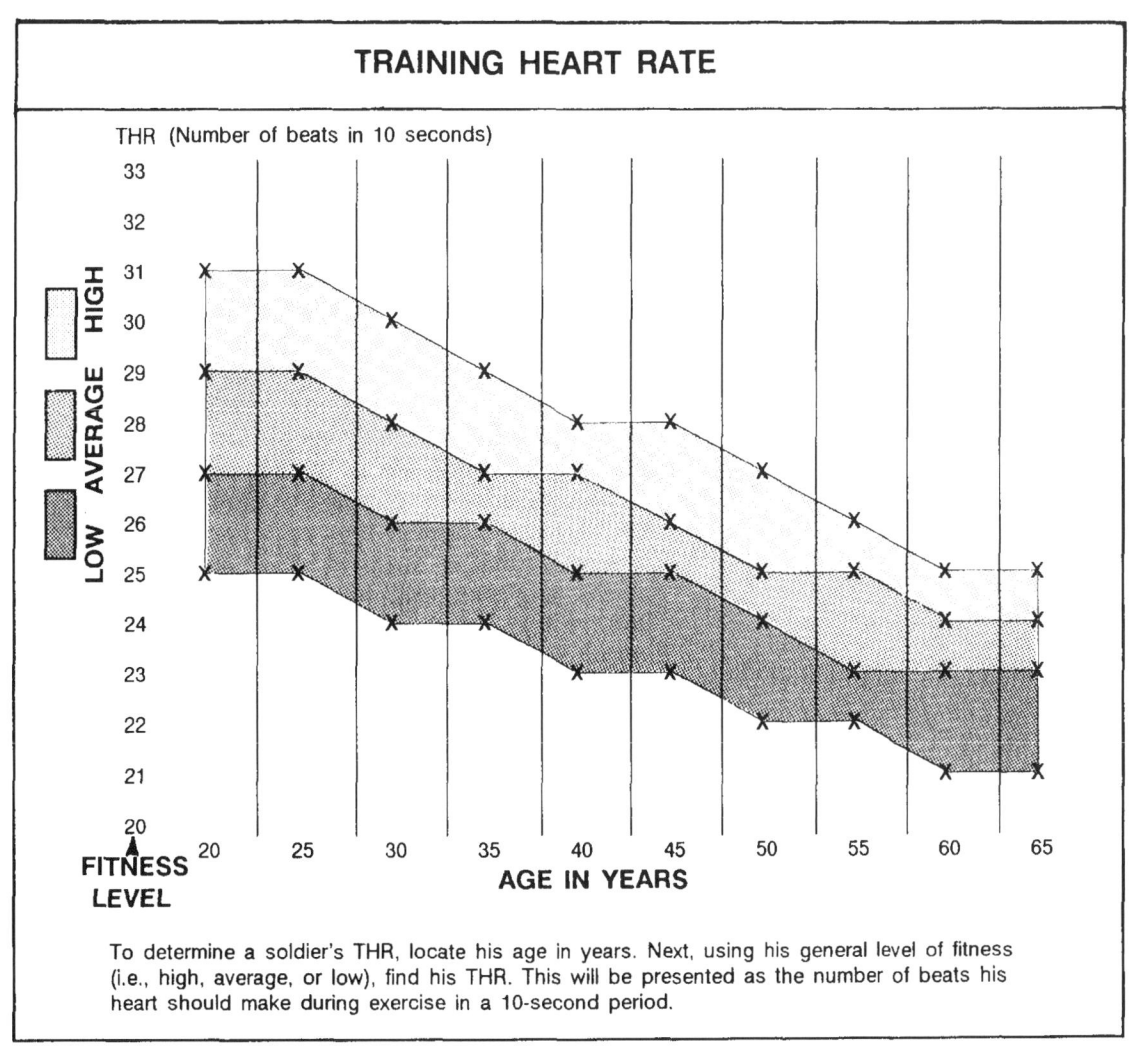

To determine a soldier's THR, locate his age in years. Next, using his general level of fitness (i.e., high, average, or low), find his THR. This will be presented as the number of beats his heart should make during exercise in a 10-second period.

Figure 2-2

TIME

Time, or duration, refers to how long one exercises. It is inversely related to intensity. The more intense the activity, the shorter the time needed to produce or maintain a training effect; the less intense the activity, the longer the required duration. To improve CR fitness, the soldier must train for at least 20 to 30 minutes at his THR.

TYPE

Only aerobic exercises that require breathing in large volumes of air improve CR fitness. Worthwhile aerobic activities must involve the use of large muscle groups and must be rhythmic. They must also be of sufficient duration and intensity (60 to 90 percent HRR). Examples of primary and secondary exercises for improving CR fitness are as follows:

PRIMARY
- Running.
- Rowing.
- Jogging.
- Skiing (cross-country).
- Walking (vigorous).
- Exercising to music.
- Road marching.
- Rope skipping.
- Bicycling (stationary).
- Swimming.
- Bicycling (road/street).
- Stair climbing.

SECONDARY (Done with partners or opponents of equal or greater ability.)
- Racquetball (singles).
- Basketball (full court).
- Handball (singles).
- Tennis (singles).

The primary exercises are more effective than the secondary exercises in producing positive changes in CR fitness.

The secondary activities may briefly elevate the heart rate but may not keep it elevated to the THR throughout the entire workout.

Every activity has its advantages and disadvantages. Trainers must weigh these and design programs that fit the unit's needs.

Running

Running enables the body to improve the transport of blood and oxygen to the working muscles and brings about positive changes in the muscles' ability to produce energy. Running fits well into any physical training program 'because a training effect can be attained with only three 20-minute workouts per week.

Some soldiers may need instruction to improve their running ability. The following style of running is desired. The head is erect with the body in a straight line or slightly bent forward at the waist. The elbows are bent so the forearms are relaxed and held loosely at waist level. The arms swing naturally from front to rear in straight lines. (Cross-body arm movements waste energy. The faster the run, the faster the arm action.) The toes point straight ahead, and the feet strike on the heel and push off at the big toe.

Besides learning running techniques, soldiers need information on ways to prevent running injuries. The most common injuries associated with PT in the Army result from running and occur to the feet, ankles, knees, and legs. Proper warm-up and cool-down, along with stretching exercises and wearing appropriate clothing and well-fitting running shoes, help prevent injuries. Important information on safety factors and common running injuries is presented in Chapter 13 and Appendix E.

Failure to allow recovery between hard bouts of running cannot only lead to overtraining, but can also be a major

Every activity has its advantages and disadvantages. Trainers must design programs that fit the unit's needs.

Important information on safety factors and common running injuries is presented in Chapter 13 and Appendix E.

cause of injuries. A well-conditioned soldier can run five to six times a week. However, to do this safely, he should do two things: 1) gradually buildup to running that frequently; and, 2) vary the intensity and/or duration of the running sessions to allow recovery between them.

ABILITY GROUP RUNNING

Traditionally, soldiers have run in unit formations at a pace prescribed by the PT leader. Commanders have used unit runs to improve unit cohesion and fitness levels. Unfortunately, too many soldiers are not challenged enough by the intensity or duration of the unit run, and they do not receive a training benefit. For example, take a company that runs at a nine-minute-per-mile pace for two miles. Only soldiers who cannot run two miles in a time faster than 18 minutes will receive a significant training effect. Therefore, in terms of conditioning, most soldiers who can pass the 2-mile-run test are wasting their time and losing the chance to train hard to excel. Ability group running (AGR) is the best way to provide enough intensity so each soldier can improve his own level of CR fitness.

AGR lets soldiers train in groups of near-equal ability. Each group runs at a pace intense enough to produce a training effect for that group and each soldier in it. Leaders should program these runs for specific lengths of time, not miles to be run. This procedure lets more-fit groups run a greater distance than the less-fit groups in the same time period thus enabling every soldier to improve.

The best way to assign soldiers to ability groups is to make a list, in order, of the unit's most recent APFT 2-mile-run times. The number of groups depends on the unit size, number of leaders available to conduct the runs, and range of 2-mile-run times. A company-sized unit broken down

into four to six ability groups, each with a leader, is best for aerobic training, For activities like circuits, strength training, and competitive events, smaller groups are easier to work with than one large group.

Because people progress at different rates, soldiers should move to faster groups when they are ready. To help them train at their THR and enhance their confidence, those who have a hard time keeping up with a group should be placed in a slower group. As the unit's fitness level progresses, so should the intensity at which each group exercises. Good leadership will prevent a constant shifting of soldiers between groups due to lack of effort.

AGR is best conducted at the right intensity at least three times a week. As explained, the CR system should not be exercised "hard" on consecutive days. If AGR is used on hard CR-training days, unit runs at lower intensities are good for recovery days. Using this rotation, soldiers can gain the desired benefits of both unit and ability-group runs. The problem comes when units have a limited number of days for PT and there is not enough time for both. In this case, unit runs should seldom, if ever, be used and should be recognized for what they are -- runs to build unit cohesion.

Leaders can use additional methods to achieve both goals. The unit can begin in formation and divide into ability groups at a predetermined release point. The run can also begin with soldiers divided into ability groups which join at a link-up point. Alternately, ability groups can be started over the same route in a stagger, with the slowest group first. Link-ups occur as each faster group overtakes slower groups.

With imagination and planning, AGR will result in more effective training workouts for each soldier. The argument that ability-group running detracts from unit cohesion is invalid. Good leadership and training in all

The best way to assign soldiers to ability groups is to, make a list, in order, of the unit's most recent APFT 2-mile-run times.

areas promote unit cohesion and team spirit; training that emphasizes form over substance does not.

INTERVAL TRAINING

Interval training also works the cardiorespiratory system. It is an advanced form of exercise training which helps a person significantly improve his fitness level in a relatively short time and increase his running speed.

In interval training, a soldier exercises by running at a pace that is slightly faster than his race pace for short periods of time. This may be faster than the pace he wants to maintain during the next APFT 2-mile run. He does this repeatedly with periods of recovery placed between periods of fast running. In this way, the energy systems used are allowed to recover, and the exerciser can do more fast-paced running in a given workout than if he ran continuously without resting. This type of intermittent training can also be used with activities such as cycling, swimming, bicycling, rowing, and road marching.

The following example illustrates how the proper work-interval times and recovery times can be calculated for interval training so that it can be used to improve a soldier's 2-mile-run performance.

The work-interval time (the speed at which a soldier should run each 440-yard lap) depends on his actual race pace for one mile. If a soldier's actual 1-mile-race time is not known, it can be estimated from his last APFT by taking one half of his 2-mile-run time. Using a 2-mile-run time of 1600 minutes as an example, the pace for an interval training workout is calculated as follows:

Step 1. Determine (or estimate) the actual 1-mile-race pace. The soldier's 2-mile-run time is 16:00 minutes, and his estimated pace for 1 mile is one half of this or 8:00 minutes.

Step 2. Using the time from Step 1, determine the time it took to run 440 yards by dividing the 1-mile-race pace by four. (8:00 minutes/4 = 2:00 minutes per 440 yards.)

Step 3. Subtract one to four seconds from the 440-yard time in Step 2 to find the time each 440-yard lap should be run during an interval training session. (2:00 minutes - 1 to 4 seconds = 1:59 to 1:56.)

Thus, each 440-yard lap should be run in 1 munute, 56 seconds to 1 minute, 59 seconds during interval training based on the soldier's 16:00, 2-mile run time. Recovery periods, twice the length of the work-interval periods. These recovery periods, therefore, will be 3 minutes, 52 seconds long (1:56 + 1:56 = 3:52).

Using the work-interval time for each 440-yard lap from Step 3, the soldier can run six to eight repetitions of 440 yards at a pace of 1 minute, 56 seconds (1:56) for each 440-yard run. This can be done on a 440-yard track (about 400 meters) as follows:

1. Run six to eight 440-yard repetitions with each interval run at a 1:56 pace.

2. Follow each 440-yard run done in 1 minute, 56 secons by an easy jog of 440 yards for recovery. Each 440-yard jog should take twice as much time as the work interval (that is, 3:52). For each second of work, there are two seconds of recovery. Thus, the work-to-rest ratio is 1:2.

440-YARD TIMES FOR INTERVAL TRAINING

1-Mile Time	440- Yard Time
4:45 - 5:00*	1:05 - 1:09*
5:01 - 5:59	1:14 - 1:25
6:00 - 6:59	1:25 - 1:40
7:00 - 7:59	1:41 - 1:55
8:00 - 8:59	1:55 - 2:10
9:00 - 9:59	2:10 - 2.25
10:00 - 10:59	2:25 +

* The slower 1-mile run times correspond to the slower 440-yard times, as do the faster 1-mile times with the faster 440-yard times.

Table 2-1

Fartlek training, the ldier varies the 'ensity (speed) of the nning throughout the orkout.

To help determine the correct time intervals for a wide range of fitness levels, refer to Table 2-1. It shows common 1 -mile times and the corresponding 440-yard times.

Monitoring the heart-rate response during interval training is not as important as making sure that the work intervals are run at the proper speed. Because of the intense nature of interval training, during the work interval the heart rate will generally climb to 85 or 90 percent of HRR. During the recovery interval, the heart rate usually falls to around 120 to 140 beats per minute. Because the heart rate is not the major concern during interval training, monitoring THR and using it as a training guide is not necessary.

As the soldier becomes more conditioned, his recovery is quicker. As a result, he should either shorten the recovery interval (jogging time) or run the work interval a few seconds faster.

After a soldier has reached a good CR fitness level using the THR method, he should be ready for interval training. As with any other new training method, interval training should be introduced into his training program gradually and progressively. At first, he should do it once a week. If he responds well, he may do it twice a week at the most, with at least one recovery day in between. He may also do recovery workouts of easy jogging on off days. It is recommended that interval training be done two times a week only during the last several weeks before an APFT. Also, he should rest the few days before the test by doing no, or very easy, running.

As with any workout, soldiers should start interval workouts with a warm-up and end them with a cool-down.

FARTLEK TRAINING

In Fartlek training, another type of CR training sometimes called speed play, the soldier varies the intensity (speed) of the running during the workout. Instead of running at a constant speed, he starts with veryslow jogging. When ready, he runs hard for a few minutes until he feels the need to slow down. At this time he recovers by jogging at an easy pace. This process of alternating fast and recovery running (both of varying distances) gives the same results as interval training. However, neither the running *nor* recovery interval is timed, and the running is not done on a track. For these reasons, many runners prefer Fartlek training to interval training.

LAST-MAN-UP RUNNING

This type of running, which includes both sprinting and paced running, improves CR endurance and conditions the legs. It consists of 40- to 50-yard sprints at near-maximum effort. This type of running is best done by squads and sections. Each squad leader places the squad in an evenly-spaced, single-file line on a track or a smooth, flat course. During a continuous 2- to 3-mile run of moderate intensity, the squad leader, running in the last position, sprints to the front of the line and becomes the leader. When he reaches the front, he resumes the moderate pace of the whole squad. After he reaches the front, the next soldier, who is now at the rear, immediately sprints to the front. The rest of the soldiers continue to run at a moderate pace. This pattern of sprinting by the last person continues until each soldier has resumed his original position in line. This pattern of sprinting and running is repeated several times during the run. The distance run and number of sprints performed should increase as the soldiers' conditioning improves.

CROSS-COUNTRY RUNNING

Cross-country running conditions the leg muscles and develops CR endurance. It consists of running a certain distance on a course laid out across fields, over hills, through woods, or on any other irregular terrain. It can be used as both a physical conditioning activity and a competitive event. The object is to cover the distance in the shortest time.

The unit is divided into ability groups using 2-mile-run times. Each group starts its run at the same time. This lets the better-conditioned groups run farther and helps ensure that they receive an adequate training stimulus.

The speed and distance can be increased gradually as the soldiers' conditioning improves. At first, the distance should be one mile or less, depending on the terrain and fitness level. It should then be gradually increased to four miles. Cross-country runs have several advantages: they provide variety in physical fitness training, and they can accommodate large numbers of soldiers. Interest can be stimulated by competitive runs after soldiers attain a reasonable level of fitness. These runs may also be combined with other activities such as compass work (orienteering).

Road Marches

The road or foot march is one of the best ways to improve and maintain fitness. Road marches are classified as either administrative or tactical, and they can be conducted in garrison or in the field. Soldiers must be able to move quickly, carry a load (rucksack) of equipment, and be physically able to perform their missions after extended marching.

BENEFITS OF ROAD MARCHES

Road marches are an excellent aerobic activity. They also help develop endurance in the muscles of the lower body when soldiers carry a heavy load. Road marches offer several benefits when used as part of a fitness program. They are easy to organize, and large numbers of soldiers can participate. In addition, when done in an intelligent, systematic, and progressive manner, they produce relatively few injuries. Many soldier-related skills can be integrated into road marches. They can also help troops acclimatize to new environments. They help train leaders to develop skills in planning, preparation, and supervision and let leaders make first-hand observations of the soldiers' physical stamina. Because road marches are excellent fitness-training activities, commanders should make them a regular part of their unit's PT program.

Cross-country runs can accommodate large numbers of soldiers.

Road marches help troops acclimatize to new environments,

TYPES OF MARCHES

The four types of road marches - day, limited visibility, forced, and shuttle - are described below. For more information on marches, see FM 21-18.

Day Marches

Day marches, which fit easily into the daily training plan, are most conducive to developing physical fitness. They are characterized by dispersed formations and ease of control and reconnaissance.

Limited Visibility Marches

Limited visibility marches require more detailed planning and supervision and are harder to control than day marches. Because they move more slowly and are in tighter formations, soldiers may not exercise hard enough to obtain a conditioning effect. Limited visibility marches do have some advantages, however. They protect soldiers from the heat of the day, challenge the ability of NCOS and officers to control their soldiers, and provide secrecy and surprise in tactical situations.

Forced Marches

Forced marches require more than the normal effort in speed and exertion. Although they are excellent conditioners, they may leave soldiers too fatigued to do other required training tasks.

Shuttle Marches

Shuttle marches alternate riding and marching, usually because there are not enough vehicles to carry the entire unit. These marches may be modified and used as fitness activities. A shuttle march can be planned to move troops of various fitness levels from one point to another, with all soldiers arriving at about the same time. Soldiers who have high fitness levels can generally march for longer stretches than those who are less fit.

PLANNING A ROAD MARCH

Any plan to conduct a road march to improve physical fitness should consider the following:
• Load to be carried.
• Discipline and supervision.
• Distance to be marched.
• Route reconnaissance.
• Time allotted for movement.
• Water stops.
• Present level of fitness.
• Rest stops.
• Intensity of the march.
• Provisions for injuries.
• Terrain an weather conditions.
• Safety precautions.

Soldiers should usually receive advance notice before going on a march. This helps morale and gives them time to prepare. The leader should choose an experienced soldier as a pacesetter to lead the march. The pacesetter should carry the same load as the other soldiers and should be of medium height to ensure normal strides. The normal stride for a foot march, according to FM 21-18, is 30 inches. This stride, and a cadence of 106 steps per minute, results in a speed of 4.8 kilometers per hour (kph). When a 10-minute rest is taken each hour, a net speed of 4 kph results.

The pacesetter should keep in mind that ground slope and footing affect stride length. For example, the length decreases when soldiers march up hills or down steep slopes. Normal stride and cadence are maintained easily on moderate, gently rolling terrain unless the footing is muddy, slippery, or rough.

Personal hygiene is important in preventing unnecessary injuries. Before the march, soldiers should cut their toenails short and square them

ldiers should receive
vance notcie before
ing on a march, to
lp morale and give
m time to prepare.

off, wash and dry their feet, and lightly apply foot powder. They should wear clean, dry socks that fit well and have no holes. Each soldier should take one or more extra pair of socks depending on the length of the march. Soldiers who have had problems with blisters should apply a thin coating of petroleum jelly over susceptible areas. Leaders should check soldiers' boots before the march to make sure that they fit well, are broken in and in good repair, with heels that are even and not worn down.

During halts soldiers should lie down and elevate their feet. If time permits, they should massage their feet, apply powder, and change socks. Stretching for a few minutes before resuming the march may relieve cramps and soreness and help prepare the muscles to continue exercising. To help prevent lower back strain, soldiers should help each other reposition the rucksacks and other loads following rest stops. Soldiers can relieve swollen feet by slightly loosening the laces across their arches.

After marches, soldiers should again care for their feet, wash and dry their socks, and dry their boots.

PROGRAMS TO IMPROVE LOAD-CARRYING ABILITY

The four generalized programs described below can be used to improve the soldiers' load-carrying ability. Each program is based on a different number of days per week available for a PT program.

If only two days are available for PT, both should include exercises for improving CR fitness and muscular endurance and strength. Roughly equal emphasis should be given to each of these fitness components.

If there are only three days available for PT, they should be evenly dispersed throughout the week. Two of the days should stress the development of muscular endurance and strength for the whole body. Although all of the major muscle groups of the body should be trained, emphasis should be placed on the leg (hamstrings and quadriceps), hip (gluteal and hip flexors), low back (spinal erector), and abdominal (rectus abdominis) muscles. These two days should also include brief (2-mile) CR workouts of light to moderate intensity (65 to 75 percent HRR). On the one CR fitness day left, soldiers should take a long distance run (4 to 6 miles) at a moderate pace (70 percent HRR), an interval workout, or an aerobic circuit. They should also do some strength work of light volume and intensity. If four days are available, a road march should be added to the three-day program at least twice monthly. The speed, load, distance, and type of terrain should be varied.

If there are five days, leaders should devote two of them to muscular strength and endurance and two of them to CR fitness. One CR fitness day will use long distance runs; the other can stress more intense workouts including interval work, Fartlek running, or last-man-up running. At least two times per month, the remaining day should include a road march.

Soldiers can usually begin road-march training by carrying a total load equal to 20 percent of their body weight. This includes all clothing and equipment. However, the gender make-up and/or physical condition of a unit may require using a different starting load. Beginning distances should be between five and six miles, and the pace should be at 20 minutes per mile over flat terrain with a hard surface. Gradual increases should be made in speed, load, and distance until soldiers can do the anticipated, worst-case, mission-related scenarios without excessive difficulty or exhaustion. Units should take maintenance marches at least twice a month. Distances should vary from six to eight miles, with loads of 30 to 40 percent of body weight. The pace should be 15 to 20 minutes per mile.

Leaders must train and march with their units as much as possible.

Units should do maintenance marches at least twice a month.

A recent Army study showed that road-march training two times a month and four times a month produced similar improvements in road-marching performance. Thus, twice-monthly road marches appear to produce a favorable improvement in soldiers' abilities to road march if they are supported by a sound PT program (five days per week)

Commanders must establish realistic goals for road marching based on assigned missions. They should also allow newly assigned soldiers and those coming off extended profiles to gradually build up to the unit's fitness level before making them carry maximum loads. This can be done with ability groups.

Road marching should be integrated into all other training. Perhaps the best single way to improve load-earring capacity is to have a regular training program which systematically increases the load and distance. It must also let the soldier regularly practice carrying heavy loads over long distances.

As much as possible, leaders at all levels must train and march with their units. This participation enhances leaders' fitness levels and improves team spirit and confidence, both vital elements in accomplishing difficult and demanding road marches.

Alternate Forms of Aerobic Exercise

Some soldiers cannot run. In such cases, they may use other activities as supplements *or* alternatives. Swimming, bicycling, and cross-country skiing are all excellent endurance exercises and are good substitutes for running. Their drawback is that they require special equipment and facilities that are not always available. As with all exercise, soldiers should start slowly and progress gradually. Those who use non-running activities to such training may not improve running ability. To prepare a soldier for the APFT 2-mile run, there is no substitute for running.

SWIMMING

Swimming is a good alternative to running. Some advantages of swimming include the following:
o Involvement of all the major muscle groups.
o Body position that enhances the blood's return to the heart.
o Partial support of body weight by the water, which minimizes lower body stress in overweight soldiers.

Swimming may be used to improve one's CR fitness level and to maintain and improve CR fitness during recovery from an injury. It is used to supplement running and develop upper body endurance and limited strength. The swimmer should start slowly with a restful stroke. After five minutes, he should stop to check his pulse, compare it with his THR and, if needed, adjust the intensity.

Compared with all the other modes of aerobic exercise presented in this manual (e.g., running, walking, cycling, cross-country skiing, rope jumping, etc.) in swimming alone, one's THR should be lower than while doing the other forms of aerobic exercise. This is because, in swimming, the heart does not beat as fast as when doing the other types of exercise at the same work rate. Thus, in order to effectively train the CR system during swimming, a soldier should set his THR about 10 bpm lower than while running. For example, a soldier whose THR while running is 150 bpm should have a THR of about 140 bpm while swimming. By modifying their THRs in this manner while swimming, soldiers will help to ensure that they are working at the proper intensity.

Non-swimmers can run in waist-to chest-deep water, tread water, and do pool-side kicking for an excellent aerobic workout. They can also do calisthenics in the water. Together these activities combine walking and running with moderate resistance work for the upper body.

For injured soldiers, swimming and aerobic water-training are excellent for improving CR fitness without placing undue stress on injured weight-bearing parts of the body.

CYCLING

Cycling is an excellent exercise for developing CR fitness. Soldiers can bicycle outdoors or on a stationary cycling machine indoors. Road cycling should be intense enough to allow the soldier to reach and maintain THR at least 30 minutes.

Soldiers can alter the cycling intensity by changing gears, adding hill work, and increasing velocity. Distance can also be increased to enhance CR fitness, but the distance covered is not as important as the amount of time spent training at THR. The intensity of a workout can be increased by increasing the resistance against the wheel or increasing the pedaling cadence (number of RPM), For interval training, the soldier can vary the speed and resistance and use periods of active recovery at low speed and/or low resistance.

WALKING

Walking is another way to develop cardiorespiratory fitness. It is enjoyable, requires no equipment, and causes few injuries. However, unless walking is done for a long time at the correct intensity, it will not produce any significant CR conditioning.

Sedentary soldiers with a low degree of fitness should begin slowly with 12 minutes of walking at a comfortable pace. The heart rate should be monitored to determine the intensity. The soldier should walk at least four times a week and add two minutes each week

Cycling should be intense enough to let the soldier reach and maintain THR at least 30 minutes.

For swimming, a soldier should set his THR at about 10 beats per minute lower then when running.

to every workout until the duration reaches 45 to 60 minutes per workout. He can increase the intensity by adding hills or stairs.

As the walker's fitness increases, he should walk 45 to 60 minutes at a faster pace. A simple way to increase walking speed is to carry the arms the same way as in running. With this technique the soldier has a shorter arm swing and takes steps at a faster rate. Swinging the arms faster to increase the pace is a modified form of race walking (power walking) which allows for more upper-body work. This method may also be used during speed marches. After about three months, even the most unfit soldiers should reach a level of conditioning that lets them move into a running program.

CROSS-COUNTRY SKIING

Cross-country or Nordic skiing is another excellent alternative to the usual CR activities. It requires vigorous movement of the arms and legs which develops muscular and CR endurance and coordination. Some of the highest levels of aerobic fitness ever measured have been found in cross-country skiers.

Although some regions lack snow, one form or another of cross-country skiing can be done almost anywhere-- on country roads, golf courses, open fields, and in parks and forests.

Cross-country skiing is easy to learn. The action is similar to that used in brisk walking, and the intensity may be varied as in running. The work load is determined by the difficulty of terrain, the pace, and the frequency and duration of rest periods. Equipment is reasonably priced, with skis, boots, and poles often obtainable from the outdoor recreation services.

Cross-country skiing requires vigorous movement of the arms and legs, developing muscular and CR endurance.

ROPE SKIPPING

Rope skipping is also a good exercise for developing CR fitness. It requires little equipment, is easily learned, may be done almost anywhere, and is not affected by weather. Some runners use it as a substitute for running during bad weather.

A beginner should select a jump rope that, when doubled and stood on, reaches to the armpits. Weighted handles or ropes may be used by better-conditioned soldiers to improve upper body strength. Rope skippers should begin with five minutes of jumping rope and then monitor their heart rate. They should attain and maintain their THR to ensure a training effect, and the time spent jumping should be increased as the fitness level improves.

Rope jumping, however, may be stressful to the lower extremities and therefore should be limited to no more than three times a week. Soldiers should skip rope on a cushioned surface such as a mat or carpet and should wear cushioned shoes.

HANDBALL AND RACQUET SPORTS

Handball and the racquet sports (tennis, squash, and racquetball) involve bursts of intense activity for short periods. They do not provide the same degree of aerobic training as exercises of longer duration done at lower intensities. However, these sports are good supplements and can provide excellent aerobic benefits depending on the skill of the players. If played vigorously each day, they may be an adequate substitute for low-level aerobic training. Because running increases endurance, it helps improve performance in racket sports, but the reverse is not necessarily true.

EXERCISE TO MUSIC

Aerobic exercise done to music is another excellent alternative to running. It is a motivating, challenging activity that combines exercise and rhythmic movements. There is no prerequisite skill, and it can be totally individualized to every fitness level by varying the frequency, intensity, and duration. One can move to various tempos while jogging or doing jumping jacks, hops, jumps, or many other calisthenics.

Workouts can be done in a small space by diverse groups of varying fitness levels. Heart rates should be taken during the conditioning phase to be sure the workout is sufficiently intense. If strengthening exercises are included, the workout addresses every component of fitness. Holding relatively light dumbbells during the workout is one way to increase the intensity for the upper body and improve muscular endurance. Warm-up and cooldown stretches should be included in the aerobic workout.

On today's battlefield, in addition to cardiorespiratory fitness, soldiers need a high level of muscular endurance and strength. In a single day they may carry injured comrades, move equipment, lift heavy tank or artillery rounds, push stalled vehicles, or do many other strength-related tasks. For example, based on computer-generated scenarios of an invasion of Western Europe, artillerymen may have to load from 300 to 500, 155mm-howitzer rounds (95-lb rounds) while moving from 6 to 10 times each day over 8 to 12 days. Infantrymen may need to carry loads exceeding 100 pounds over great distances, while supporting units will deploy and displace many times. Indeed, survival on the battlefield may, in large part, depend on the muscular endurance and strength of the individual soldier.

Muscular Fitness

Muscular fitness has two components: muscular strength and muscular endurance.

Muscular strength is the greatest amount of force a muscle or muscle group can exert in a single effort.

Muscular endurance is the ability of a muscle or muscle group to do repeated contractions against a less-than-maximum resistance for a given time.

Although muscular endurance and strength are separate fitness components, they are closely related. Progressively working against resistance will produce gains in both of these components.

Muscular Contractions

Isometric, isotonic, and isokinetic muscular endurance and strength are best produced by regularly doing each specific kind of contraction. They are described here.

Isometric contraction produces contraction but no movement, as when pushing against a wall. Force is produced with no change in the angle of the joint.

Isotonic contraction causes a joint to move through a range of motion against a constant resistance. Common examples are push-ups, sit-ups, and the lifting of weights.

Isokinetic contraction causes the angle at the joint to change at a constant rate, for example, at 180 degrees per second. To achieve a constant speed of movement, the load or resistance must change at different joint angles to counter the varying forces produced by the muscle(s) at different angles. This requires the use of isokinetic machines. There are other resistance-training machines which, while not precisely controlling the speed of movement, affect it by varying the resistance throughout the range of motion. Some of these devices are classified as pseudo-isokinetic and some as variable-resistance machines.

Isotonic and isokinetic contractions have two specific phases - the concentric or "positive" phase and the eccentric or "negative" phase. In the concentric phase (shortening) the muscle contracts, while in the eccentric phase (elongation) the muscle returns to its normal length. For example, on the upward phase of the biceps curl, the biceps are shortening. This is a concentric (positive) contraction. During the lowering phase of the curl the biceps are lengthening. This is an eccentric (negative) contraction.

A muscle can control more weight in the eccentric phase of contraction than it can lift concentrically. As a result, the muscle may be able to handle more of an overload eccentrically. This greater overload, in return, may produce greater strength gains.

The nature of the eccentric contraction, however, makes the muscle and connective tissue more susceptible to damage, so there is more muscle soreness following eccentric work.

When a muscle is overloaded, whether by isometric, isotonic, or isokinetic contractions, it adapts by becoming stronger. Each type of contraction has advantages and disadvantages, and each will result in strength gains if done properly.

The above descriptions are more important to those who assess strength than to average people trying to develop strength and endurance. Actually, a properly designed weight training program with free weights or resistance machines will result in improvements in all three of these categories.

Principles of Muscular Training

To have a good exercise program, the seven principles of exercise, described in Chapter 1, must be applied to all muscular endurance and strength training. These principles are overload, progression, specificity, regularity, recovery, balance, and variety.

OVERLOAD

The overload principle is the basis for all exercise training programs. For a muscle to increase in strength, the workload to which it is subjected during exercise must be increased beyond what it normally experiences. In other words, the muscle must be overloaded. Muscles adapt to increased workloads by becoming larger and stronger and by developing greater endurance.

To understand the principle of overload, it is important to know the following strength-training terms:
- Full range of motion. To obtain optimal gains, the overload must be applied thoughout the full range of motion. Exercise a joint and its associated muscles through its complete range starting from the pre-stretched position (stretched past the relaxed position) and ending in a fully contratcted position. This is crucial to strength development.
- Repetition. When an exercise has progressed through one complete range of motion and back to the beginning, one repetition has been completed.
- One-repetition maximum (1-RM). This is a repetition performed against the greatest possible resistance (the maximum weight a person can lift one time). A 10-RM is the maximum weight one can lift correctly 10 times. Similarly, an 8-12 RM is that weight which allows a person to do from 8 to 12 correct repetitions. The intensity for muscular endurance and strength training is often expressed as a percentage of. the 1-RM.
- Set. This is a series of repetitions done without rest.
- Muscle Failure. This is the inability of a person to do another correct repetition in a set.

The minimum resistance needed to obtain strength gains is 50 percent of the 1 -RM. However, to achieve enough overload, programs are designed to require sets with 70 to 80 percent of one's 1 -RM. (For example, if a soldier's 1 -RM is 200 pounds, multiply 200 pounds by 70 percent [200 X 0.70 = 140 pounds] to get 70 percent of the 1 -RM.)

When a muscle is overloaded by isometric, isotonic, or isokinetic contractions, it adapts by becoming stronger.

A better and easier method is the repetition maximum (RM) method. The exerciser finds and uses that weight which lets him do the correct number of repetitions. For example, to develop both muscle endurance and strength, a soldier should choose a weight for each exercise which lets him do 8 to 12 repetitions to muscle failure. (See Figure 3-1.) The weight should be heavy enough so that, after doing from 8 to 12 repetitions, he momentarily cannot correctly do another repetition. This weight is the 8-12 RM for that exercise.

MUSCULAR ENDURANCE/ STRENGTH DEVELOPMENT

To develop muscle strength, the weight selected should be heavier and the RM will also be different. For example, the soldier should find that weight for each exercise which lets him do 3 to 7 repetitions correctly. This weight is the 3-7 RM for that exercise. Although the greatest improvements seem to come from resistances of about 6-RM, an effective range is a 3-7 RM. The weight should be heavy enough so that an eighth repetition would be impossible because of muscle fatigue.

The weight should also not be too heavy. If one cannot do at least three repetitions of an exercise, the resistance is too great and should be reduced. Soldiers who are just beginning a resistance-training program should not start with heavy weights. They should first build an adequate foundation by training with an 8-12 RM or a 12+ RM.

To develop muscular endurance, the soldier should choose a resistance that lets him do more than 12 repetitions of a given exercise. This is his 12+ repetition maximum (12+ RM). With continued training, the greater the number of repetitions per set, the greater will be the improvement in muscle endurance and the smaller the gains in strength. For example, when a soldier trains with a 25-RM weight, gains in muscular endurance will be greater than when using a 15-RM weight, but the gain in strength will not be as great. To optimize a soldier's performance, his RM should be determined from an analysis of the critical tasks of his mission. However, most soldiers will benefit most from a resistance-training program with an 8-12 RM.

FITT Factors Applied to Conditioning Programs for Muscular Endurance and/or Strength		
Muscular Strength	Muscular Endurance	Muscular Strength and Muscular Endurance
3 times/week	3-5 times/week	3 times/week
3-7 RM*	12+ RM	8-12 RM
The time required to do 3-7 repetitions of each resistance exercise	The time required to do 12+ repetitions of each resistance exercise	The time required to do 8-12 repetitions of each resistance exercise
Free Weights Resistance Machines Partner-Resisted Exercises Body-Weight Exercises (Push-ups/Sit-ups/Pull-ups/Dips, etc.) * RM = Repetition Maximum		

Figure 3-1

Whichever RM range is selected, the soldier must always strive to overload his muscles. The key to overloading a muscle is to make that muscle exercise harder than it normally does.

An overload may be achieved by any of the following methods:

- Increasing the resistance.
- Increasing the number of repetitions per set.
- Increasing the number of sets.
- Reducing the rest time between sets.
- Increasing the speed of movement in the concentric phase.
  (Good form is more important than the speed of movement.)
- Using any combination of the above.

PROGRESSION

When an overload is applied to a muscle, it adapts by becoming stronger and/or by improving its endurance. Usually significant increases in strength can be made in three to four weeks of proper training depending on the individual. If the workload is not progressively increased to keep pace with newly won strength, there will be no further gains. When a soldier can correctly do the upper limit of repetitions for the set without reaching muscle failure, it is usually time to increase the resistance. For most soldiers, this upper limit should be 12 repetitions.

For example, if his plan is to do 12 repetitions in the bench press, the soldier starts with a weight that causes muscle failure at between 8 and 12 repetitions (8- 12 RM). He should continue with that weight until he can do **12** repetitions correctly. He then should increase the weight by about 5 percent but no more than 10 percent. In a multi-set routine, if his goal is to do three sets of eight repetitions of an exercise, he starts with a weight that causes muscle failure before he com -

pletes the eighth repetition in one or more of the sets. He continues to work with that weight until he can complete all eight repetitions in each set, then increases the resistance by no more than 10 percent.

SPECIFICITY

A resistance-training program should provide resistance to the specific muscle groups that need to be strengthened. These groups can be identified by doing a simple assessment. The soldier slowly does work-related movements he wants to improve and, at the same time, he feels the muscles on each side of the joints where motion occurs. Those muscles that are contracting or becoming tense during the movement are the muscle groups involved. If the soldier's performance of a task is not adequate or if he wishes to improve, strength training for the identified muscle(s) will be beneficial. To improve his muscular endurance and strength. in a given task, the soldier must do resistance movements that are as similar as possible to those of doing the task. In this way, he ensures maximum carryover value to his soldiering tasks.

REGULARITY

Exercise must be done regularly to produce a training effect. Sporadic exercise may do more harm than good. Soldiers can maintain a moderate level of strength by doing proper strength workouts only once a week, but three workouts per week are best for optimal gains. The principle of regularity also applies to the exercises for individual muscle groups. A soldier can work out three times a week, but when different muscle groups are exercised at each workout, the principle of regularity is violated and gains in strength are minimal.

Exercise must be done regularly to produce a training effect.

RECOVERY

Consecutive days of hard resistance training for the same muscle group can be detrimental. The muscles must be allowed sufficient recovery time to adapt. Strength training can be done every day only if the exercised muscle groups are rotated, so that the same muscle or muscle group is not exercised on consecutive days. There should be at least a 48-hour recovery period between workouts for the same muscle groups. For example, the legs can be trained with weights on Monday, Wednesday, and Friday and the upper body muscles on Tuesday, Thursday, and Saturday.

Recovery is also important within a workout. The recovery time between different exercises and sets depends, in part, on the intensity of the workout. Normally, the recovery time between sets should be 30 to 180 seconds.

There should be at least a 48-hour recovery period between workouts for the same muscle group.

BALANCE

When developing a strength training program, it is important to include exercises that work all the major muscle groups in both the upper and lower body. One should not work just the upper body, thinking that running will strengthen the legs.

Most muscles are organized into opposing pairs. Activating one muscle results in a pulling motion, while activating the opposing muscle results in the opposite, or pushing, movement. When planning a training session, it is best to follow a pushing exercise with a pulling exercise which results in movement at the same joint(s). For example, follow an overhead press with a lat pull-down exercise. This technique helps ensure good strength balance between opposing muscle groups which may, in turn, reduce the risk of injury. Sequence the program to exercise the larger muscle groups first, then

It is important to include exercises that work all the major *muscle groups in both* the upper and lower *body.*

the smaller muscles. For example, the lat pull-down stresses both the larger latissimus dorsi muscle of the back and the smaller biceps muscles of the arm. If curls are done first, the smaller muscle group will be exhausted and too weak to handle the resistance needed for the lat pull-down. As a result, the soldier cannot do as many repetitions with as much weight as he normally could in the lat pull-down. The latissimus dorsi muscles will not be overloaded and, as a result, they may not benefit very much from the workout.

The best sequence to follow for a total-body strength workout is to first exercise the muscles of the hips and legs, followed by the muscles of the upper back and chest, then the arms, abdominal, low back, and neck. As long as all muscle groups are exercised at the proper intensity, improvement will occur.

VARIETY

A major challenge for all fitness training programs is maintaining enthusiasm and interest. A poorly designed strength-training program can be very boring. Using different equipment, changing the exercises, and altering the volume and intensity are good ways to add variety, and they may also produce better results. The soldier should periodically substitute different exercises for a given muscle group(s). For example, he can do squats with a barbell instead of leg presses on a weight machine. Also, for variety or due to necessity (for example, when in the field), he can switch to partner-resisted exercises or another form of resistance training. However, frequent wholesale changes should be avoided as soldiers may become frustrated if they do not have enough time to adapt or to see improvements in strength.

Workout Techniques

Workouts for improving muscular endurance or strength must follow the principles just described. There are also other factors to consider, namely, safety, exercise selection, and phases of conditioning.

SAFETY FACTORS

Major causes of injury when strength training are improper lifting techniques combined with lifting weights that are too heavy. Each soldier must understand how to do each lift correctly before he starts his strength training program.

The soldier should always do weight training with a partner, or spotter, who can observe his performance as he exercises. To ensure safety and the best results, both should know how to use the equipment and the proper spotting technique for each exercise.

A natural tendency in strength training is to see how much weight one can lift. Lifting too much weight forces a compromise in form and may lead to injury. All weights should be selected so that proper form can be maintained for the appropriate number of repetitions.

Correct breathing is another safety factor in strength training. Breathing should be constant during exercise. The soldier should never hold his breath, as this can cause dizziness and even loss of consciousness. As a general rule, one should exhale during the positive (concentric) phase of contraction as the weight or weight stack moves away from the floor, and inhale during the negative (eccentric) phase as the weight returns toward the floor.

EXERCISE SELECTION

When beginning a resistance-training program, the soldier should choose about 8 to 16 exercises that work all of the body's major muscle groups. Usually eight well-chosen exercises will serve as a good starting point. They should include those for the muscles of the leg, low back, shoulders, and so forth. The soldier should choose exercises that work several muscle groups and try to avoid those that isolate single muscle groups. This will help him train a greater number of muscles in a given time. For example, doing lat pull-downs on the "lat machine" works the latissimus dorsi of the back and the biceps muscles of the upper arm. On the other hand, an exercise like concentration curls for the biceps muscles of the upper arm, although an effective exercise, only works the arm flexor muscles. Also, the concentration curl requires twice as much time as lat pull-downs because only one arm is worked at a time.

Perhaps a simpler way to select an exercise is to determine the number of joints in the body where movement occurs during a repetition. For most people, especially beginners, most of the exercises in the program should be "multi-joint" exercises. The exercise should provide movement at more than one joint. For example, the pull-down exercise produces motion at both the shoulder and elbow joints. The concentration curl, however, only involves the elbow joint.

PHASES OF CONDITIONING

There are three phases of conditioning: preparatory, conditioning, and maintenance. These are also described in Chapter 1.

Preparatory Phase

The soldier should use very light weights during the first week (the preparatory phase) which includes the first two to three workouts. This is very important, because the beginner must concentrate at first on learning

The three phases of conditioning are preparatory, conditioning, and maintenance.

the proper form for each exercise. Using light weights also helps minimize muscle soreness and decreases the likelihood of injury to the muscles, joints, and ligaments. During the second week, he should use progressively heavier weights. By the end of the second week (4 to 6 workouts), he should know how much weight on each exercise will allow him to do 8 to 12 repetitions to muscle failure. If he can do only seven repetitions of an exercise, the weight must be reduced; if he can do more than 12, the weight should be increased.

Conditioning Phase

The third week is normally the start of the conditioning phase for the beginning weight trainer. During this phase, the soldier should increase the amount of weight used and/or the intensity of the workout as his muscular strength and/or endurance increases. He should do one set of 8 to 12 repetitions for each of the heavy-resistance exercises. When he can do more than 12 repetitions of any exercise, he should increase the weight until he can again do only 8 to 12 repetitions. This usually involves an increase in weight of about five percent. This process continues indefinitely. As long as he continues to progress and get stronger, he does not need to do more than one set per exercise. If he stops making progress with one set of 8 to 12 repetitions per exercise, he may benefit from adding another set of 8 to 12 repetitions on those exercises in which progress has slowed. As time goes on and he progresses, he may increase the number to three sets of an exercise to get even further gains in strength and/or muscle mass. Three sets per exercise is the maximum most soldiers will ever need to do.

Maintenance Phase

Once the soldier reaches a high level of fitness, the maintenance phase is used to maintain that level. The emphasis in this phase is no longer on progression but on retention. Although training three times a week for muscle endurance and strength gives the best results, one can maintain them by training the major muscle groups properly one or two times a week. More frequent training, however, is required to reach and maintain peak fitness levels. Maintaining the optimal level of fitness should become part of each soldier's life-style and training routine. The maintenance phase should be continued throughout his career and, ideally, throughout his life.

As with aerobic training, the soldier should do strength training three times a week and should allow at least 48 hours of rest from resistance training between workouts for any given muscle group.

TIMED SETS

Timed sets refers to a method of physical training in which as many repetitions as possible of a given exercise are performed in a specified period of time. After an appropriate period of rest, a second, third, and so on, set of that exercise is done in an equal or lesser time period. The exercise period, recovery period, and the number of sets done should be selected to make sure that an overload of the involved muscle groups occurs.

The use of timed sets, unlike exercises performed in cadence or for a specific number of repetitions, helps to ensure that each soldier does as many repetitions of an exercise as possible within a period of time. It does not hold back the more capable performer by restricting the number of repetitions he may do. Instead, soldiers at all levels of fitness can individually do the number of repetitions they are capable of and thereby be sure they obtain an adequate training stimulus.

In this FM, timed sets will be applied to improving soldier's sit-up and push-up performance. (See Figures 3-2 and 3-3.) Many different but equally valid approaches can be taken when using timed sets to improve push-up and sit-up performance. Below, several of these will be given.

It should first be stated that improving sit-up and push-up performance, although important for the APFT, should not be the main goal of an Army physical training program. It must be to develop an optimal level of physical fitness which will help soldiers carry out their mission during combat. Thus, when a soldier performs a workout geared to develop muscle endurance and strength, the goal should be to develop sufficient strength and/or muscle endurance in all the muscle groups he will be called upon to use as he performs his mission. To meet this goal, and to be assured that all emergencies can be met, a training regimen which exercises all the body's major muscle groups must

be developed and followed. Thus, as a general rule, a muscle endurance or strength training workout should not be designed to work exclusively, or give priority to, those muscle groups worked by the sit-up or push-up event.

For this reason, the best procedure to follow when doing a resistance exercise is as follows. First, perform a workout to strengthen all of the body's major muscles. Then, do timed sets to improve push-up and sit-up performance. Following this sequence ensures that all major muscles are worked. At the same time, it reduces the amount of time and work that must be devoted to push-ups and sit-ups. This is because the muscles worked by those two exercises will already be pre-exhausted.

The manner in which timed sets for push-ups and sit-ups are conducted should occasionally be varied. This ensures continued gains and minimizes boredom. This having been said, here is a very time-efficient way of conducting push-up/sit-up improvement. Alternate timed sets of push-ups and timed sets of sit-ups with little or no time between sets allowed for recovery. In this way, the muscle groups used by the push-up can recover while the muscles used in the sit-up are exercised, and vice versa. The following is an example of this type of approach:

TIMED SETS

SET NO.	ACTIVITY	TIME PERIOD	REST INTERVAL
1	Push-ups	45 seconds	0
2	Sit-ups	45 seconds	0
3	Push-ups	30 seconds	0
4	Sit-ups	30 seconds	0
5	Push-ups	30 seconds	0
6	Sit-ups	30 seconds	0

Figure 3-2

If all soldiers exercise at the same time, the above activity can be finished in about 3.5 minutes. As the soldiers' levels of fitness improve, the difficulty of the activity can be increased. This is done by lengthening the time period of any or all timed sets, by decreasing any rest period between timed sets, by increasing the number of timed sets performed, or by any combination of these.

To add variety and increase the overall effectiveness of the activity, different types of push-ups (regular, feet-elevated, wide-hand, close-hand, and so forth) and sit-ups (regular, abdominal twists, abdominal curls, and so forth) can be done. When performing this type of workout, pay attention to how the soldiers are responding, and make adjustments accordingly. For example, the times listed in the chart above may prove to be too long or too short for some soldiers. In the same way, because of the nature of the sit-up, it may become apparent that some soldiers can benefit by taking slightly more time for timed sets of sit-ups than for push-ups.

When using timed sets for push-up and sit-up improvement, soldiers can also perform all sets of one exercise before doing the other. For example, several timed sets of push-ups can be done followed by several sets of sit-ups, or vice versa. With this approach, rest intervals must be placed between timed sets. The following example can be done after the regular strength workout and is reasonable starting routine for most soldiers.

During a timed set of push-ups, a soldier may reach temporary muscle failure at any time before the set is over. If this happens, he should immediately drop to his knees and continue doing modified push-ups on his knees.

Finally, as in any endeavor, soldiers must set goals for themselves. This applies when doing each timed set and when planning for their next and future APFTs.

Major Muscle Groups

In designing a workout it is important to know the major muscle groups, where they are located, and their primary action. (See Figure 3-4.)

To ensure a good, balanced workout, one must do at least one set of exercises for each of the major muscle groups.

TIMED SETS

SET NO.	ACTIVITY	TIME PERIOD	REST INTERVAL
1	Regular Push-ups	30 seconds	30 seconds
2	Wide-hand Push-ups	30 seconds	30 seconds
3	Close-hand Push-ups	30 seconds	30 seconds
4	Regular Push-ups	20 seconds	30 seconds
5	Regular Push-ups done on knees	30 seconds	30 seconds
6	Regular Sit-ups	60 seconds	30 seconds
7	Abdominal Twists	40 seconds	30 seconds
8	Curl-ups	30 seconds	30 seconds
9	Abdominal Crunches	30 seconds	End

Figure 3-3

MAJOR MUSCLE GROUPS

The Major Skeletal Muscles of the Human Body

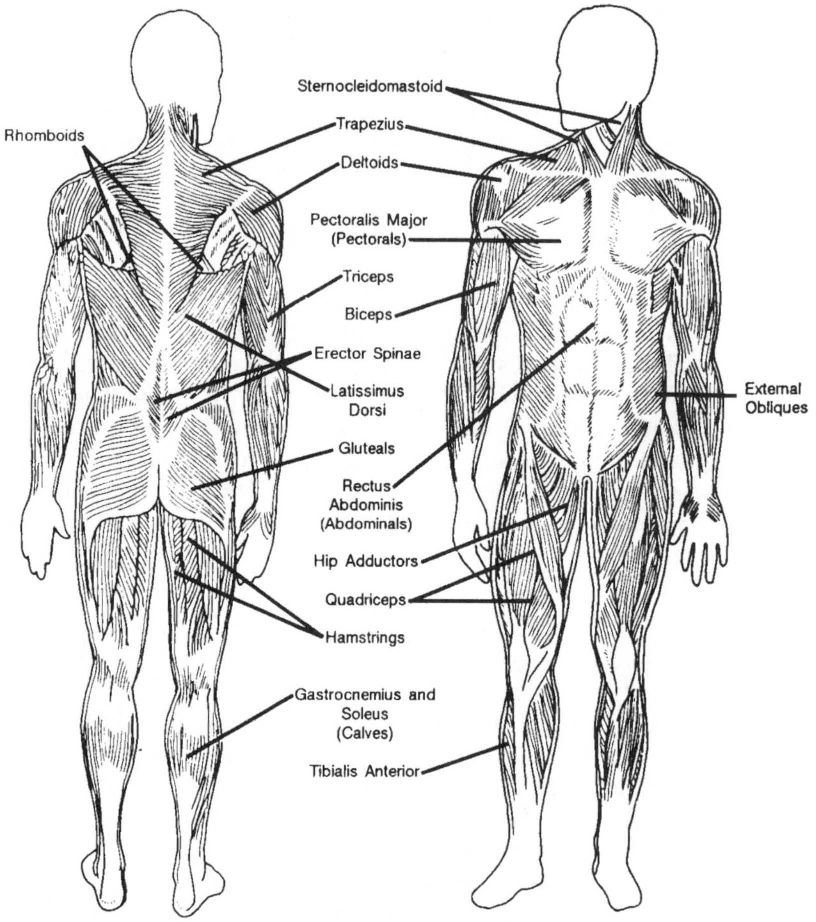

Rhomboids

Sternocleidomastoid

Trapezius

Deltoids

Pectoralis Major (Pectorals)

Triceps

Biceps

Erector Spinae

Latissimus Dorsi

Gluteals

Rectus Abdominis (Abdominals)

Hip Adductors

Quadriceps

Hamstrings

Gastrocnemius and Soleus (Calves)

Tibialis Anterior

External Obliques

The iliopsoas muscle (a hip flexor) cannot be seen as it lies beneath other muscles. It attaches to the lumbar, the pelvis, the vertebrae and the femur.

Figure 3-4

BEGINNING EXERCISE PROGRAM

NAME OF EXERCISE	MAJOR MUSCLE GROUP(S) WORKED*
1. Leg press or squat	---Quadriceps, Gluteals
2. Leg curl	---Hamstrings
3. Heel raise	---Gastrocnemius
4. Bench press	---Pectorals, Triceps, Deltoids
5. Lat pull-down or pull-up	---Latissimus Dorsi, Biceps
6. Overhead press	---Deltoids, Triceps
7. Sit-up	---Rectus Abdominus, Iliopsoas, oblique muscles
8. Bent-leg dead-lift	---Erector Spinae, Quadriceps, Gluteals

Figure 3-5

The beginning weight-training program shown at Figure 3-5 will work most of the important, major muscle groups. It is a good program for beginners and for those whose time is limited. The exercises should be done in the order presented.

The weight-training program shown at Figure 3-6 is a more comprehensive program that works the major muscle groups even more thoroughly. It has some duplication with respect to the muscles that are worked. For example, the quadriceps are worked by the leg press/squat and leg extensions, and the biceps are worked by the seated row,

lat pull-down, and biceps curl. Thus, for the beginner, this program may overwork some muscle groups. However, for the more advanced lifter, it will make the muscles work in different ways and from different angles thereby providing a better over-all development of muscle strength. This program also includes exercises to strengthen the neck muscles.

When doing one set of each exercise to muscle failure, the average soldier should be able to complete this routine and do a warm-up and cool-down within the regular PT time.

MORE ADVANCED EXERCISE PROGRAM

NAME OF EXERCISE	MAJOR MUSCLE GROUP(S) WORKED
1. Leg press or squat	---Quadriceps, Gluteals
2. Leg raises	---Iliopsoas (hip flexors)
3. Leg extension	---Quadriceps
4. Leg curl	---Hamstrings
5. Heel raise	---Gastrocnemius, Soleus
6. Bench press	---Pectorals, Triceps, Deltoids
7. Seated row	---Rhomboids, Latissimus dorsi, Biceps
8. Overhead press	---Deltoids, Triceps
9. Lat pull-down or pull-up	---Latissimus dorsi, Biceps
10. Shoulder shrug	---Upper trapezius
11. Triceps extension	---Triceps
12. Biceps curl	---Biceps
13. Sit-up	---Rectus abdominus, iliopsoas
14. Bent-leg dead lift	---Erector spinae, Quadriceps, Gluteals
15. Neck flexion	---Sternocleidomastoid
16. Neck extension	---Upper trapezius

Figure 3-6

Key Points to Emphasize

Some key points to emphasize when doing resistance training tire as follows

- Train with a partner if possible, This helps to increase motivation, the intensity of the workout, and safety,
- Always breathe when lifting. Exhale during the concentric (positive] phase of contraction, and inhale during the eccentric (negative) phase,
- Accelerate the weight through the concentric phase of contraction, and return the weight to the starting position in a controlled manner during the eccentric phase,
- Exercise the large muscle groups first, then the smaller ones.
- Perform all exercises through their full range of motion. Begin from a fully extended, relaxed position (pre-stretched), and end the concentric phase in a fully contracted position,
- Always use strict form. Do not twist, lurch, lunge, or arch the body, This can cause serious injury. These motions also detract from the effectiveness of the exercise because they take much of the stress off the targeted muscle groups and place it on other muscles.
- Rest from 30 to 180 seconds between different exercises and sets of a given exercise.
- Allow at least 48 hours of recovery between workouts, but not more than 96 hours, to let the body recover and help prevent over training and injury.
- Progress slowly, Never increase the resistance used by more than 10 percent at a time.
- Alternate pulling and pushing exercises. For example, follow triceps extensions with biceps curls.
- Ensure that every training program is balanced. Train the whole body, not just specific areas. Concentrating on weak areas is all right, but the rest of the body must also be trained.

Exercise Programs

When developing strength programs for units, there are limits to the type of training that can be done. The availability of facilities is always a major concern. Although many installations have excellent strength-training facilities, it is unreasonable to expect that all units can use them on a regular basis. However, the development of strength does not require expensive equipment. All that is required is for the soldier, three times a week, to progressively overload his muscles.

TRAINING WITHOUT SPECIAL EQUIPMENT

Muscles do not care what is supplying the resistance. Any regular resistance exercise that makes the muscle work harder than it is used to causes it to adapt and become stronger. Whether the training uses expensive machines, sandbags, or partners, the result is largely the same.

Sandbags are convenient for training large numbers of soldiers, as they are available in all military units. The weight of the bags can be varied depending on the amount of fill. Sandbag exercises are very effective in strength-training circuits. Logs, ammo boxes, dummy rounds, or other equipment that is unique to a unit can also be used to provide resistance for strength training. Using a soldier's own body weight as the resistive force is another excellent alternative method of strength training. Pull-ups, push-ups, dips, sit-ups, and single-leg squats are examples of exercises which use a person's body weight. They can improve an untrained soldier's level of strength.

Partner-resisted exercises (PREs) are another good way to develop muscular strength without equipment, especially when training large numbers of soldiers at one time. As with all training, safety is a critical factor. Soldiers should warm up, cool down, and follow the principles of exercise previously outlined.

PARTNER-RESISTED EXERCISE

In partner-resisted exercises (PREs) a person exercises against a partner's opposing resistance. The longer the partners work together, the more effective they should become in providing the proper resistance for each exercise. They must communicate with each other to ensure that neither too much nor too little resistance is applied. The resister must apply enough resistance to bring the exerciser to muscle failure in 8 to 12 repetitions. More resistance usual] y can and should be applied during the eccentric (negative) phase of contraction (in other words, the second half of each repetition as the exerciser returns to the starting position). The speed of movement for PREs should always be slow and controlled. As a general rule, the negative part of each exercise should take at least as long to complete as the positive part. Proper exercise form and regularity in performance are key ingredients when using PREs for improving strength.

Following are descriptions and illustrations of several PREs. They should be done in the order given to ensure that the exercising soldier is working his muscle groups from the largest to the smallest. More than one exercise per muscle group may be used. The PT leader can select exercises which meet the unit's specific goals while considering individual limitations:

A 36-to 48-inch stick or bar one inch in diameter may be used for some of the exercises. This gives the resister a better grip and/or leverage and also provides a feel similar to that of free weights and exercise machines.

SPLIT-SQUAT
This exercise is for beginning trainees' quadriceps and gluteal muscles.

Exerciser
Position: Stand erect with both feet pointed straight ahead, the left foot placed in a forward position and the right foot placed about 2.5 feet behind the left foot.
Action: Keeping the back straight and the head up, bend both legs at the same time, and lower yourself slowly until the right knee barely touches the floor. Return to the starting position. This is one repetition. After 8 to 12 repetitions to muscle failure, repeat the action with the opposite leg forward.

Resister
Position: Stand directly behind the exerciser with the fleshy portion of your forearms resting squarely on the exerciser's shoulders. You may clasp your hands to gain extra leverage as long as you do not squeeze the exerciser's neck. Be sure to place the same foot forward as the exerciser.
Action: As the exerciser lowers himself, apply a steady, forceful pressure downward against his shoulders. A slightly lesser pressure should be applied as the exerciser returns to the starting position.

SINGLE-LEG SQUAT
This exercise is for advanced trainees' quadriceps and gluteal muscles.

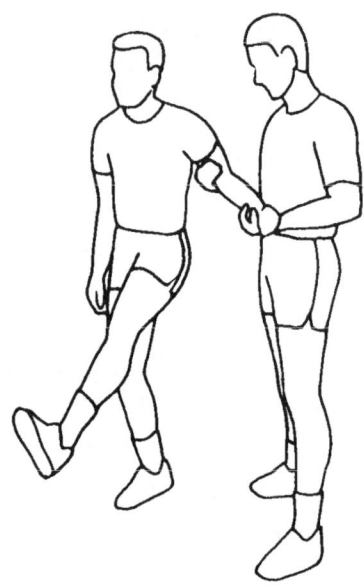 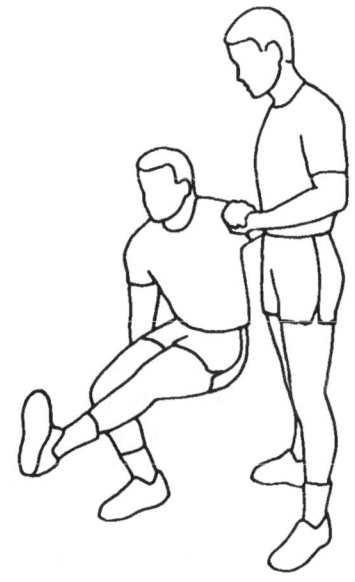

Exerciser
Position: Face your partner and grasp his wrists. Extend your right leg in front; keep it straight but do not let it contact your partner.
Action: Lower yourself in a controlled manner. Next, return to the upright position. After 8-12 repetitions to muscle failure, repeat this exercise with the other leg.

Resister
Position: Face the exerciser with your arms extended obliquely forward.
Action: Provide stability to the exerciser along with resistance or assistance as needed. When the exerciser can do more than 12 repetitions, apply an appropriate resistance that results in muscle failure in 8-12 repetitions.

LEG EXTENSION
This exercise is for the quadriceps muscles.

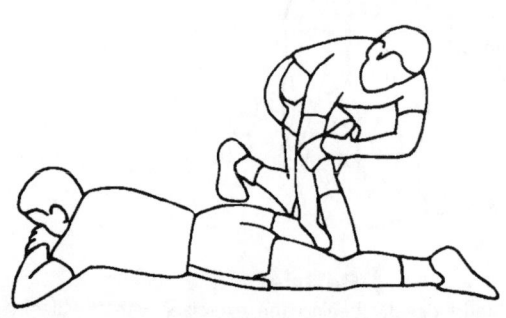 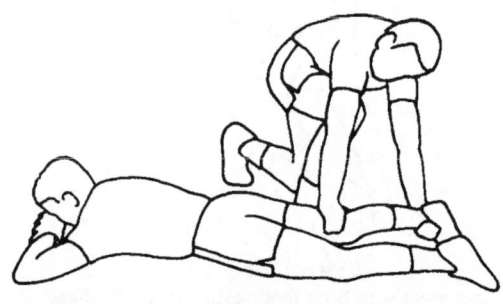

Exerciser
Position: Lie face down with one leg straight and the other flexed at the knee. Move your heel as close to your buttocks as possible.
Action: Extend your knee against the partner's resistance. Next, resist as your partner returns you to the starting position. Do 8 to 12 repetitions to muscle failure. Repeat this exercise with the other leg.

Resister
Position: Support the leg being exercised by placing your foot under the exerciser's thigh just above his knee.
Action: Resist while exerciser extends his leg. Next, apply upward pressure to return the exerciser to the starting position.

LEG CURL
This exercise is for the hamstring muscles.

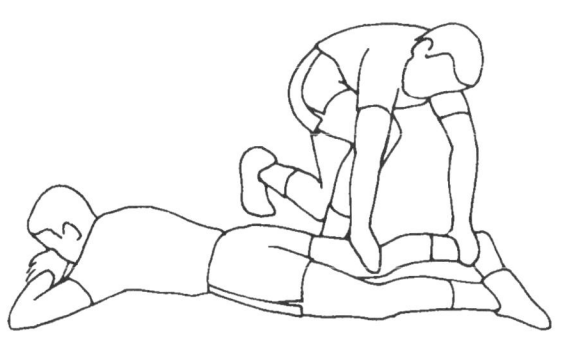 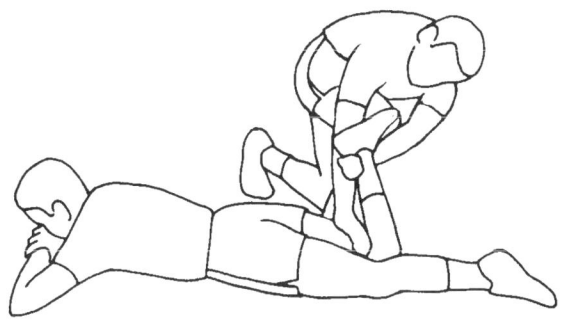

Exerciser
Position: Lie face down with your legs extended.
Action: Flex one leg against your partner's resistance until your heel is as close to your buttocks as possible. Next, resist your partner's efforts as he returns you to the starting position. Do 8 to 12 repetitions to muscle failure, Repeat this exercise with the other leg.

Resister
Position: Support the exerciser's leg as in the Leg Extension exercise.
Action: Resist the exerciser's movement with your hand(s) placed on his heel. Next, apply downward pressure to return the exerciser to the starting position.

HEEL RAISE (BENT OVER)
This exercise is for the gastrocnemius and soleus muscles.

 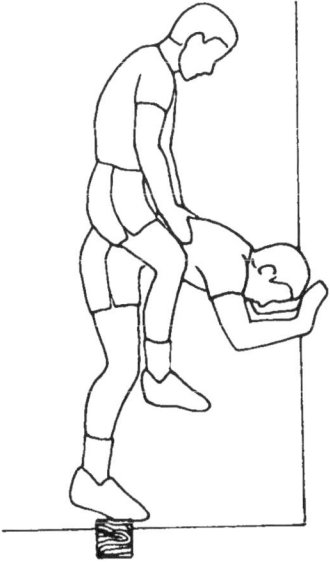

Exerciser
Position: Form a 90-degree angle between your upper body and legs by bending over at the hips. Use an additional partner or a fixed object for support.
Action: Keep your legs straight and rise up on the balls of your feet. Do 8 to 12 repetitions to muscle failure. If possible, perform the exercise by placing the balls of your feet firmly on a 4" x 4" board or the edge of a curb. Be sure to lower and raise your heels as far as possible.

Resister
Position: Sit on the upper part of the exerciser's buttocks; DO NOT SIT ON THE EXERCISER'S LOW BACK. (Properly positioning your body places less pressure on the exerciser's back and helps him better work his gastrocnemius and soleus muscles.)
Action: Provide resistance to the exerciser with your body weight.

TOE RAISE
This exercise is for the tibialis anterior muscle.

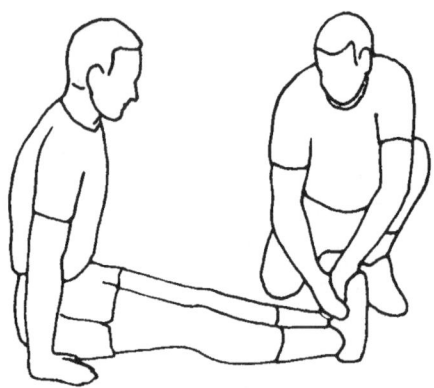 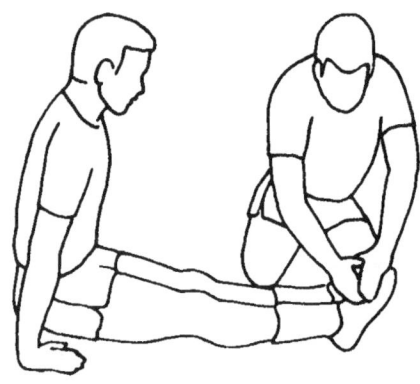

Exerciser
Position: Sit on the floor with your legs together, knees straight, and feet fully extended.

Action: Against the resister's efforts, move your toes toward the knees; then have the resister pull your toes back to the starting position while you resist. Do 8 to 12 repetitions to muscle failure.

Resister
Position: Place your hand(s) on the exerciser's shoelaces near the toes. Press your palms against the exerciser's insteps to resist his foot and ankle movements.

Action: Resist the exerciser's effort to pull his toes toward his knees. Next, pull the exerciser's toes back to the starting position against his resistance.

PUSH-UP
This exercise is for the pectoral and triceps muscles.

 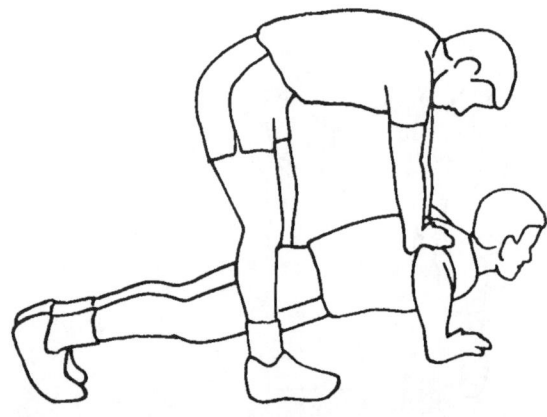

Exerciser
Position: Assume a front-leaning-rest position.

Action: Perform a push-up against your partner's resistance. Do 8 to 12 repetitions to muscle failure.

Resister
Position: Straddle the exerciser's hips. Place your hands on top of his shoulders. Be careful to place your left hand on the upper left part and your right hand on the upper right part of his shoulder.

Action: Apply pressure against the exerciser's push-up movements. As stated earlier, slightly more resistance should be applied during the eccentric phase of contraction (in this case, as the exerciser moves closer to the floor)

SEATED ROW
This exercise is for the biceps, latissimus dorsi, and rhomboid muscles.

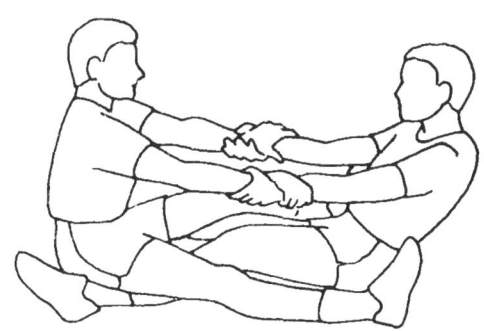 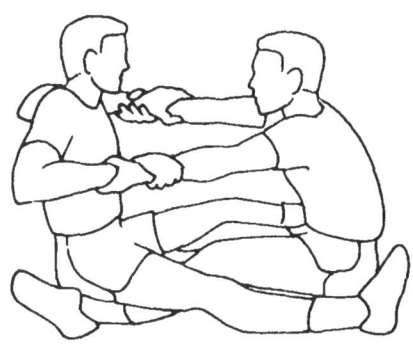

Exerciser

Position: Sit facing the resister with your back straight. Overlap your legs with the resister's, being sure to place your legs on top. Establish a good grip by interlocking your hands with the resister's or by firmly grasping his wrists. The exerciser's palms should be facing downward.

Action: Pull the resister toward you with a rowing motion while keeping your elbows elevated to shoulder height. Be sure to keep your back straight, and move only the arms.Next, slowly return to the starting position as the resister pulls your arms forward. Do 8 to 12 repetitions to muscle failure.

Resister

Position: Face the exerciser and sit with your back straight. Place your legs under the exerciser's legs; establish a good grip by interlocking hands with the resister or by firmly grasping his wrists.

Action: As the exerciser pulls, resist his pulling motion. Next, slowly pull the exerciser back to the starting position by pulling with the muscles of the lower back.

OVERHEAD PRESS
This exercise is for the deltoid and triceps muscles.

 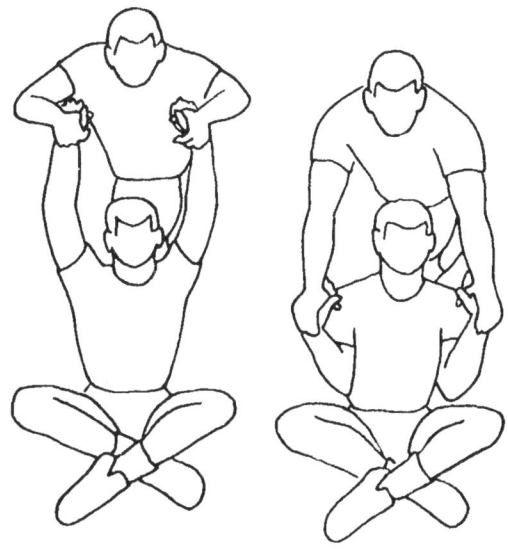

Exerciser

Position: Sit with your legs crossed and your back straight. Raise your hands to shoulder height with your palms flat and facing upward.

Action: Move your arms slowly upward to full extension against your partner's resistance. Next, slowly return to the starting position as the resister applys downward pressure. Do 8 to 12 repetitions to muscle failure.

Resister

Position: Stand behind the exerciser; interlock your thumbs with the exerciser's, and place your hands with the palms down on his hands. Support the exerciser's back with the side of your lower leg.

Action: Resist the exerciser's upward movement; then push his arms back to the starting position. A bar or stick may be used for a better grip and improved leverage.

PULL-DOWN
This exercise is for the latissimus dorsi muscles.

Exerciser

Position: Sit with your legs crossed and back straight. Raise and cross your arms behind your head with your elbows bent.
Action: Pull out and down with your elbows against the partner's resistance until your elbows touch your ribcage. Next, resist as your partner pulls your elbows back to the starting position. Do 8 to 12 repetitions to muscle failure.

Resister

Position: Stand behind the exerciser, and support his back with the side of your lower leg. Place your palms underneath the exerciser's elbows.
Action: Resist the exerciser's downward movements; then pull his elbows back to the up or starting position. VARIATION: A bar or stick may be used for a better grip and leverage and to exercise the biceps and forearm muscles.

SHRUG
This exercise is for the upper trapezius muscle.

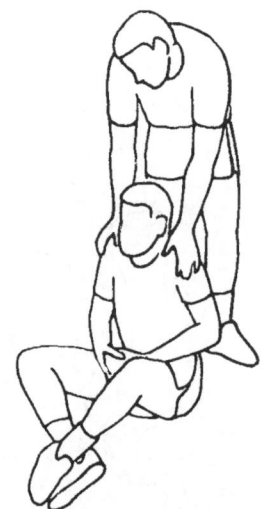

Exerciser

Position: Sit with your legs crossed, back straight, and hands resting in your lap.
Action: Shrug your shoulders as high as possible against your partner's resistance, then resist your partner's pushing motion as you return to the starting position. Do 8 to 12 repetitions to muscle failure.

Resister

Positon: Stand behind the exerciser, and support his back with the side of your lower leg. Place your hands on each of the exerciser's shoulders.
Action: Apply pressure downward with your hands to resist the upward, shrugging movements of the exerciser and, during the second phase of the exercise, push downward as the exerciser resists your pushing movements.

TRICEPS EXTENSION
This exercise is for the triceps muscles.

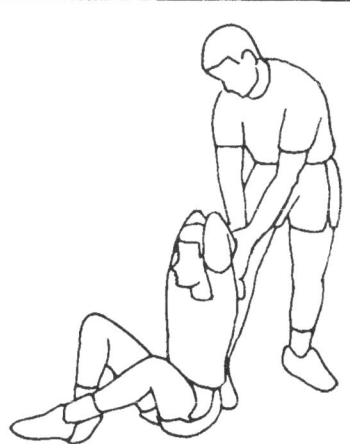

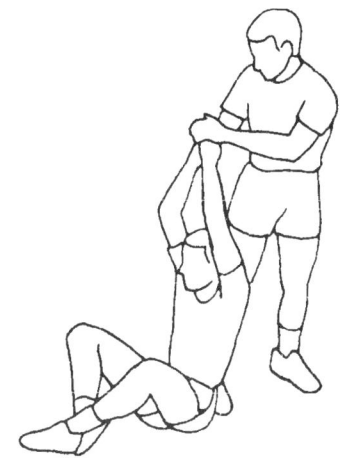

Exerciser

Position: Sit with your legs crossed and back straight. Clasp your hands and place them behind your head while bending your elbows.

Action: Extend your arms upward against the partner's resistance. Next, return to the starting position while resisting your partner's force. Always keep your elbows stationary and pointing straight ahead. Do 8 to 12 repetitions to muscle failure.

Resister

Position: Stand behind the exerciser and support his back with the side of your lower leg. Place your hands, palms down, over the exerciser's hands.

Action: Apply pressure to resist the upward movement of the exerciser, and then push his hands back to the starting position. A bar or a stick may be used for a better grip and/ or improved leverage. This exercise may also be done in the prone position with the resister applying a force against the exerciser's movements.

BICEPS CURL
This exercise is for the biceps muscles.

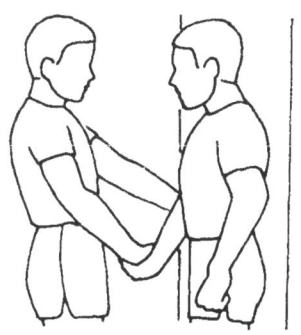 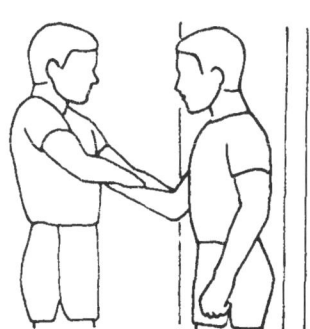  

Exerciser

Position: Stand straight with your back supported. Hold the arm to be exercised close to your side.

Action: Bend the elbow, bringing your hand up to your shoulder against your partner's resistance. Return to the starting position by resisting the pushing efforts of your partner. Do 8 to 12 repetitions to muscle failure; repeat with the other arm.

Resister

Position: Face the exerciser with your feet staggered. Use one of your hands to grasp the exerciser's wrist; place the other hand behind his elbow to stabilize it during the exercise movement.

Action: Resist the exerciser's upward movement and provide a downward, pushing force during the lowering movement. A bar may also be used for a better grip and leverage.

VARIATION: A variation may be used if the resister is unable to provide enough resistance to the exerciser with the first exercise. Using this variation, the exerciser places the back of his hand on the non-exercising arm behind the elbow of his exercising arm for support; the resister places both hands on the hand, wrist, or lower part of the exerciser's forearm to apply resistance to the exerciser's movements. The action is the same as before.

ABDOMINAL CURL

This exercise is for the rectus abdominus, iliopsoas, and external and internal oblique muscles.

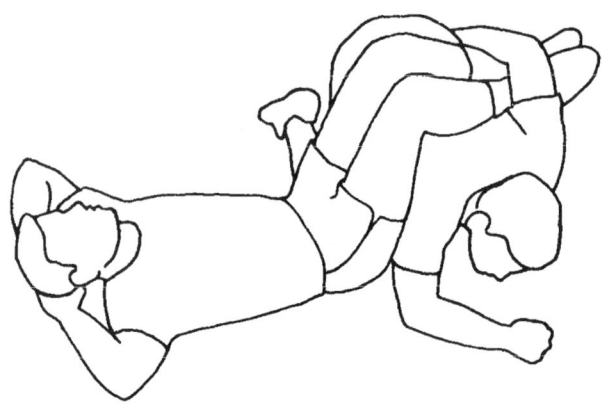 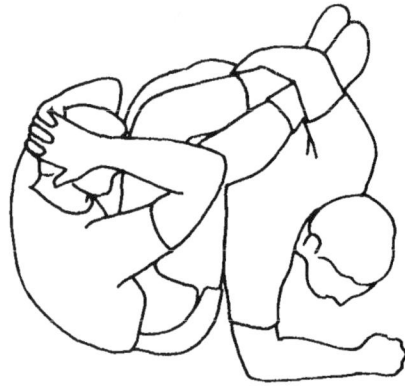

Exerciser

Position: Lie on your back with both legs bent at the knee to about a 90-degree angle. Place your bent legs over the resister's back. Interlace your fingers behind your neck.

Action: Do regular sit-ups, bringing both elbows to your knees.

Do 20 to 50 repetitions to muscle failure.

NOTE: A variation to this exercise is the ROCKY SIT-UP where the exerciser moves the left elbow up to the right knee and then reverses the action, right elbow to left knee.

Resister

Position: Kneel with your inside elbow resting on the ground. With your outside arm, reach back and hold the exerciser's ankles.

Action: Provide a firm foundation upon which the exerciser can place his legs, and keep them tightly anchored during the exercise.

ABDOMINAL CRUNCH

This exercise is for the rectus abdominis and external and internal oblique muscles.

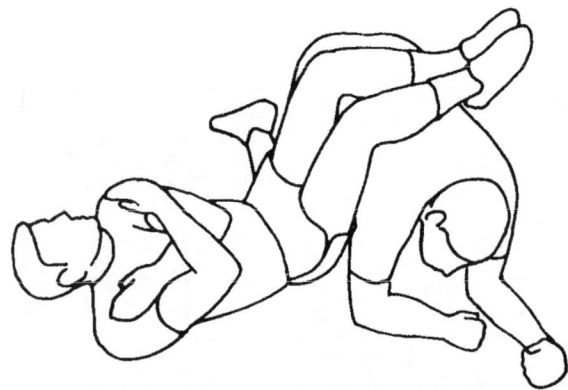 

Exerciser

Position: Lie down with your arms crossed over your chest, the backs of your lower legs resting over your partner's back, and your upper leg placed at right angles to the floor.

Action: Curl your neck off the ground, and curl your upper body up toward your knees. (Progressively lift your shoulders, upper back, and finally, lower back off the ground.) Hold this position briefly while forcefully tensing your abdominal muscles. Return slowly to the starting position and repeat. Do 20 to 50 repetitions to muscle failure.

Resister

Position: Kneel with both forearms on the ground.

Action: Allow the exerciser to place the back of his lower legs on your back. DO NOT HOLD HIS LEGS DOWN. (This eliminates the iliopsoas muscle from the exercise and instead isolates the rectus abdominis and external and internal oblique muscles.)

TRAINING WITH EQUIPMENT

Units in garrison usually have access to weight rooms with basic equipment for resistance-training exercises. The exercises described here require free weights and supporting equipment. Although not shown below for the sake of simplicity, all exercises done with free weights require a partner, or spotter, to ensure proper form and the safety of the lifter.

Free-Weight Exercises

SQUAT
This exercise is for the quadriceps and gluteal muscles.

Position: Stand with the feet about shoulder width apart. Hold the weight on your shoulders.
Action: Bend the knees until the tops of your thighs are parallel to the ground. Keep your head and shoulders upright and back straight. In the lowest position, the top of your thighs should not go lower than parallel to the ground. Do 8 to 12 repetitions to muscle failure. A 2" x 4" block may be placed under the heels to increase stability.

HEEL RAISE
This exercise is for the gastrocnemius and soleus muscles.

Position: Place a bar on your shoulders behind your neck. Stand with the toes and the balls of the feet on a platform or a 4" x 4" board.

Action: Rise upward on the toes and balls of the feet to full extension, then slowly lower the heels as far as possible. Do not bend the knees or jerk the hips. Do 8 to 12 repetitions to muscle failure.

BENCH PRESS
This exercise is for the pectoralis major, triceps, and deltoid muscles.

 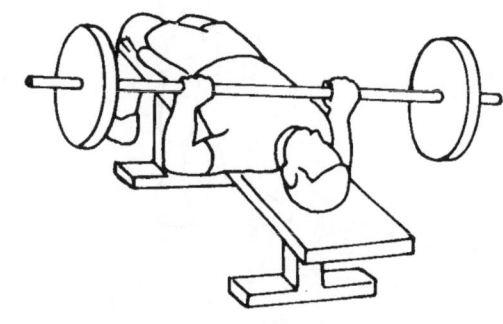

Position: Hold a weight with an overhand grip (palms facing away) slightly wider than shoulder width. Hold the bar directly above your chest at arm's length.

Action: Lower the bar to your chest, keeping the feet flat on the floor. Push the bar up to arm's length. The elbows should be kept wide and away from the body. Keep the buttocks in contact with the bench at all times. Do 8 to l2 repetitions to muscle failure.

BENT-OVER ROW
This exercise is for the latissimus dorsi and biceps muscles.

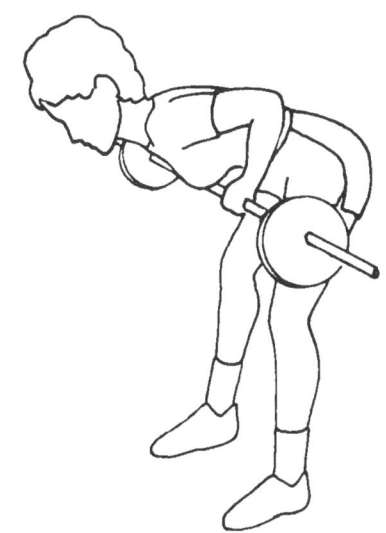

Position:. Lean forward at the hips with the back flat; let your arms hang straight down from the shoulders. Keep your knees slightly flexed.

Action: Use an overhand grip with the hands 12 to 24 inches apart. Bend the elbows, bringing the bar up in a straight motion up to the lower portion of the chest. Slowly lower the weight back to the starting position. Do 8 to 12 repetitions to muscle failure.

OVERHEAD PRESS
This exercise is for the deltoids and triceps muscles.

Position: With a barbell, use the overhand grip with the hands spaced slightly greater than shoulder width apart.

Action: Push the bar overhead, moving it upward in a straight line until the elbows are straight. Lower the bar until it touches the chest. Do not bounce the bar off the chest. Dumbbells may also be used. Do 8 to 12 repetitions to muscle failure.

SHRUG
This exercise is for the trapezius muscles.

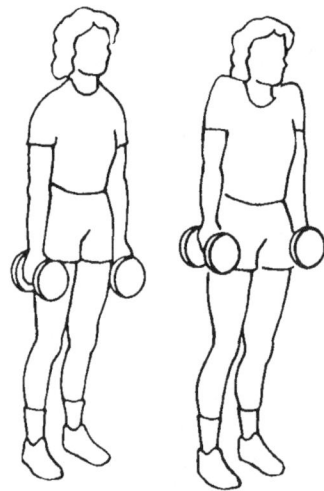

Position: Using a barbell, start with the bar at thigh-rest position. Use an overhand or reverse grip.

Action: Elevate the bar by contracting the trapezius and raising the shoulders upward toward the ears. In the top position, roll your shoulders backward. Then, slowly lower the shoulders until the bar returns to the starting, thigh-rest position. Keep the arms straight throughout the entire repetition. Dumbbells may also be used. Do 8 to 12 repetitions to muscle failure.

TRICEPS EXTENSION
This exercise is for the triceps muscles.

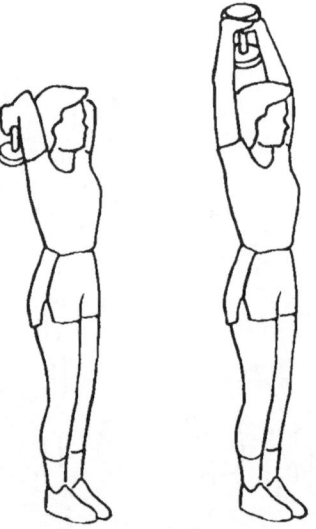

Position: Using a barbell, hold the bar directly overhead with an overhand grip. Keep the elbows high, close to the head, and stationary.

Action: Lower the bar slowly without bouncing it when it reaches the lower neck area. Extend the bar back to the overhead position while keeping the heels flat and the knees and elbows stationary. A dumbbell may also be used. Do 8 to 12 repetitions to muscle failure.

BICEPS CURL
This exercise is for the biceps muscles.

Position: Start with an underhand grip. Hold the bar at thigh-rest position.

Action: Keep the elbows stationary and close to your sides as you curl the bar to your chest. Do not use your legs or bend your back for assistance. A cambered (bent) bar or dumbbells may also be used. Do 8 to 12 repetitions to muscle failure.

WRIST CURL
This exercise is for the development of the forearm muscles.

Position: Holding your hand with the palms facing upward, grasp a barbell using only the fingers.

Action: Curl the fingers, then the wrist up as far as possible and then down, keeping the elbows stationary. For the best results, do not grip the barbell; keep it placed on the last few digits of the fingers. Do 8 to 12 repetitions to muscle failure.

BENT-LEG DEAD-LIFTS
This exercise is for the quadriceps, the erector spinae, the gluteals, and the trapezius muscles.

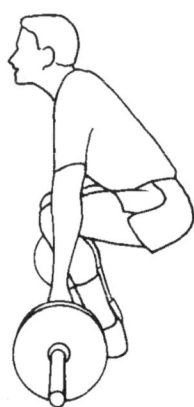 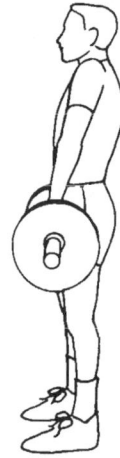

Position: Bend and grasp the bar with the hands shoulder width apart. The legs should be bent, the back flat but inclined forward at a 45 degree angle, the arms straight, and the head up.

Action: Keeping the head erect, gradually straighten the legs and the back together at the same time. Make sure that the back remains flat and the arms remain straight. When the entire body is straight, shrug the shoulders upward as high as possible. In a controlled manner, return to the starting position by first lowering the shoulders. Then, bend at the knees and at the waist simultaneously until the beginning position is attained. Keep the back flat, head up, and the arms straight at all times. Do 8 to 12 repetitions to muscle failure.

Exercises Performed with an Exercise Machine

If exercise machines are available, the exercises described below are also good for strength training. All movements, particularly during the eccentric (negative) phase of contraction, should be done in a deliberate, controlled manner.

LEG PRESS
This exercise is for the gluteal and quadriceps muscles.

 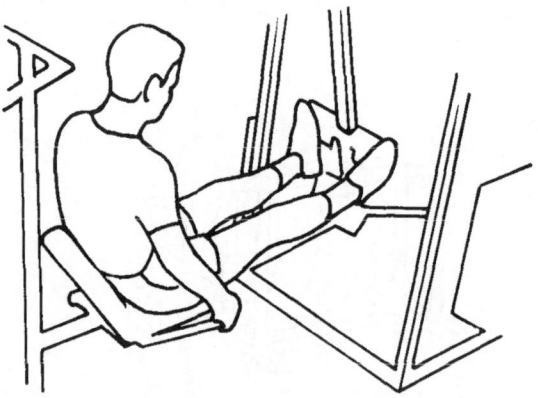

Position: Sit at the leg-press station with the legs bent no more than 90 degrees. Ensure that the balls of both feet are very securely placed on the pedals.

Action: Push the weight with the legs until your knees are straight but not locked. In a controlled manner, return to the starting position. This is one repetition. Do 8 to 12 repetitions to muscle failure.

LEG EXTENSION
This exercise is for the quadriceps muscles.

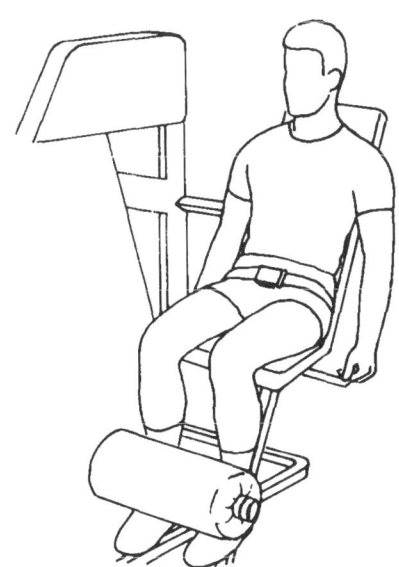 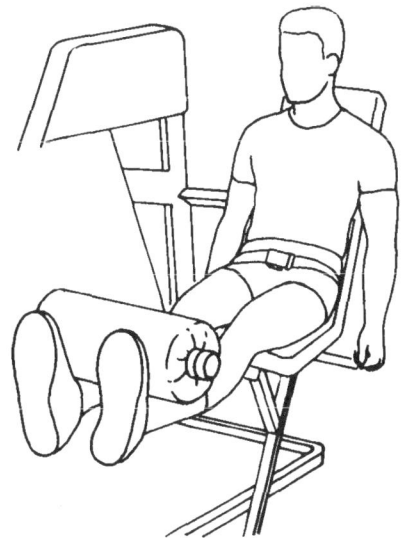

Position: Sit on a bench with your lower legs behind the padded lever. Hold on to the bench or provided handles with your hands to keep the upper body in the correct position.

Action: Straighten the legs as much as possible. In a controlled manner, return to the starting position. This is one repetition. Do 8 to 12 repetitions to muscle failure.

LEG CURL
This exercise is for the hamstring muscles.

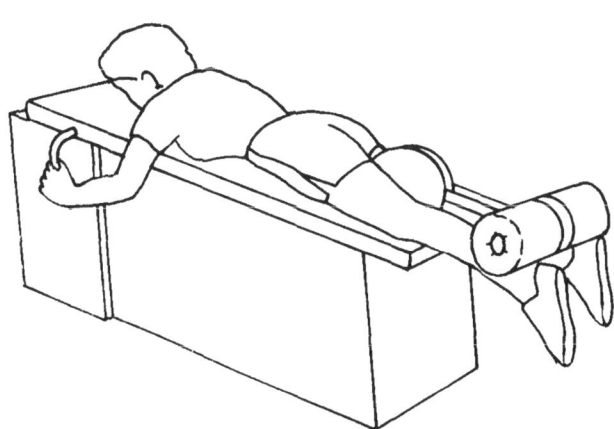 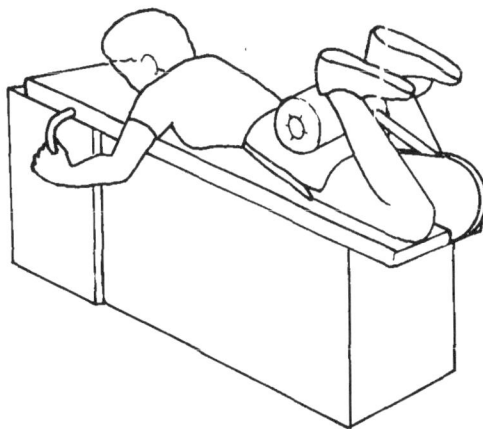

Position: Lie on your stomach with the legs straight and the ankles under the padded lever. Maintain correct upper body position by loosely grasping the sides of the bench or provided handles.

Action: Bend your legs at the knee until the lower legs pass well beyond the perpendicular position and the heels are as close to your buttocks as possible. Return to the starting position. Do 8 to 12 repetitions to muscle failure.

HEEL RAISE
This exercise is for the gastrocnemius and soleus muscles.

Position: Stand with a weight on your shoulders and the balls of your feet placed firmly on a 4-inch raised surface.

Action: Raise your heels off the floor as far as possible while maintaining your balance. Then, lower them as far as possible. This is one repetition. Do 8 to 12 repetitions to muscle failure.

TOE RAISE
This exercise is for the tibialis anterior muscle.

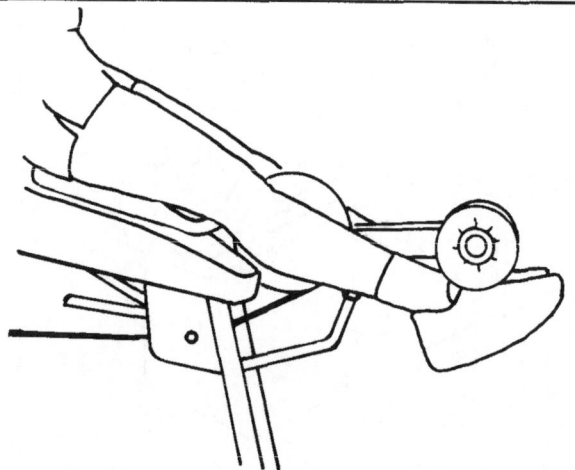 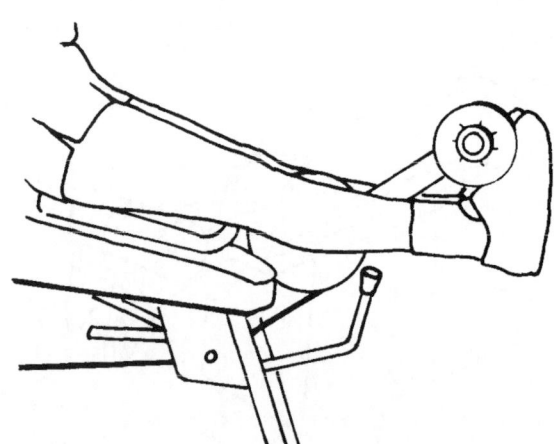

Position: Sit on the leg curl machine with your legs together, knees straight, and toes pointed. Place the top of your feet under the roller pad.

Action: Move your toes toward the knees as far as possible. Then lower the weight to the starting position in a controlled manner. Do 8 to 12 repetitions to muscle failure.

BENCH PRESS

This exercise is for the pectoralis major, triceps, and deltoid muscles.

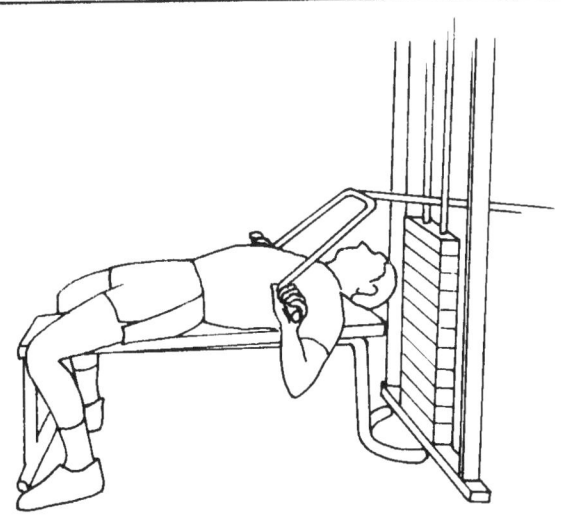 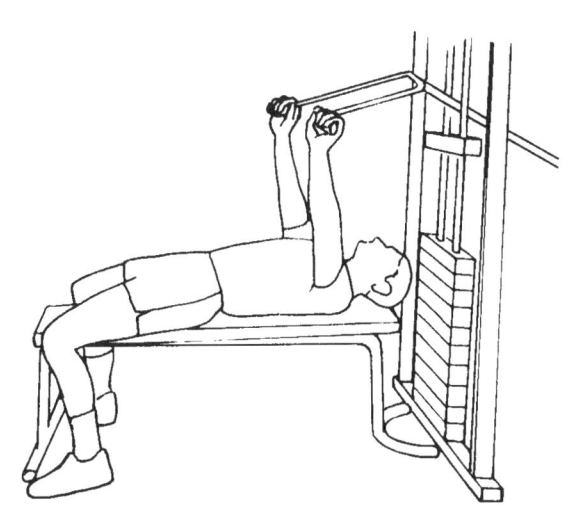

Position: Lie on your back with your hands placed about shoulder width apart on the bar. Generally, the bar or handles should be located at the lower half of the chest.

Action: Push the bar up until your arms are straight. Then, lower the bar to the starting position. This is one repetition. Do 8 to 12 repetitions to muscle failure.

SEATED ROW

This exercise is for the latissimus dorsi and biceps muscles.

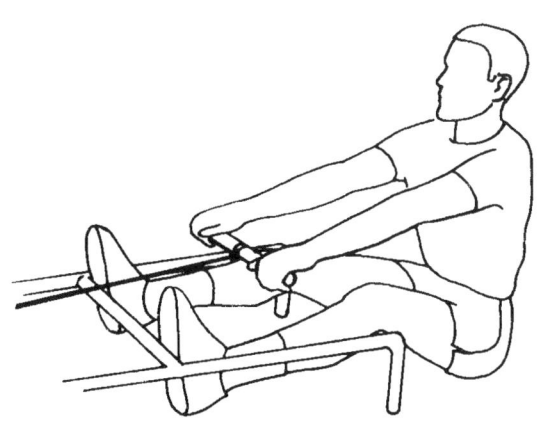 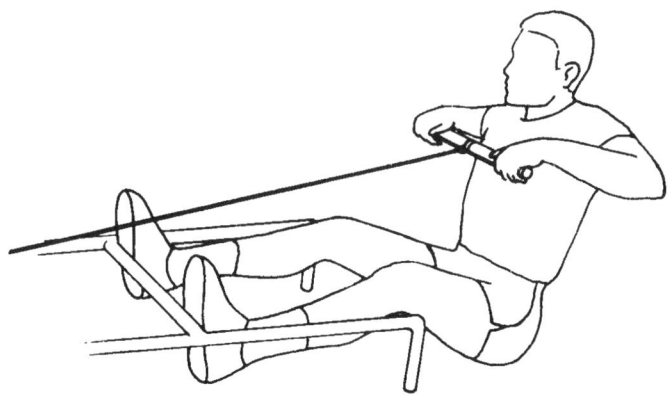

Position: Sit and assume the straight arm position shown above. Use the overhand grip with your hands spaced 6 to 8 inches apart.

Action: Pull the bar to the lower part of your chest, while keeping your elbows elevated to shoulder height, then slowly extend the arms and lower the weight to the beginning position. Be sure to keep the back straight, and move only the arms. Do 8 to 12 repetitions to muscle failure.

LAT PULL-DOWN
This exercise is for the latissimus dorsi and biceps muscles. (Pull-ups or chin-ups may be substituted for this exercise.)

 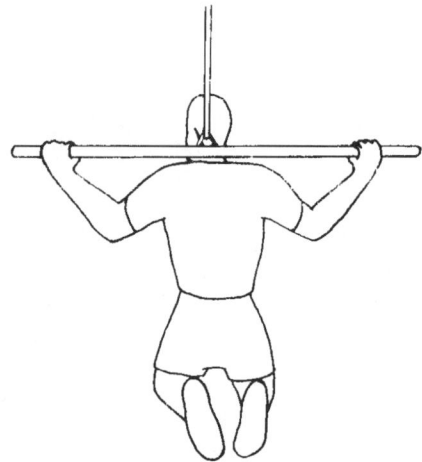

Position: Sit or kneel and grasp the bar with your palms facing away from the body.

Action: Pull the bar down until it touches the back of your neck; return the bar in a controlled manner to that starting position. This is one repetition. Do 8 to 12 repetitions to muscle failure.

SHRUG
This exercise is for the trapezius muscles of the upper back.

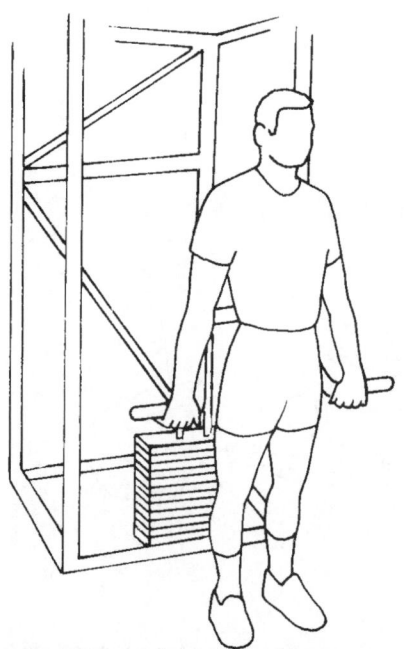 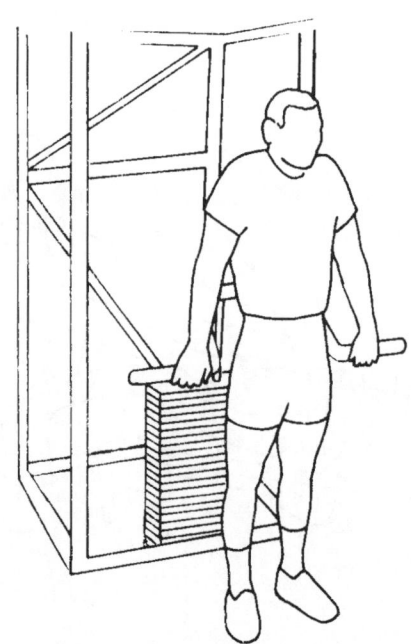

Position: Stand with the feet shoulder width apart. Hold a weight in your hands with the arms locked in a straight position.

Action: Pull the shoulders up toward your ears as far as possible and then backward. Always keep your arms completely straight. Next, lower your shoulders to the starting position. This is one repetition. Do 8 to 12 repetitions to muscle failure.

3-30

PARALLEL BAR DIP
This exercise is for the pectoralis major and triceps muscles.

 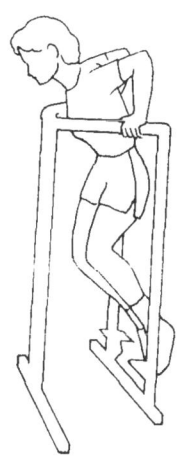

Position: Keep your feet off the floor and support the body's weight on straight arms.

Action: Bend the arms and lower your body until the upper arms are at least parallel to the floor. If necessary, bend your legs at the knees to keep the feet from touching the floor. Straighten your arms to return to the starting position. This is one repetition. Do 8 to 12 repetitions to muscle failure. A weight belt may be worn if additional resistance is needed.

CHIN-UP
This exercise is for the latissimus dorsi and biceps muscles. (Lat pull-downs or pull-ups may be substituted for this exercise.)

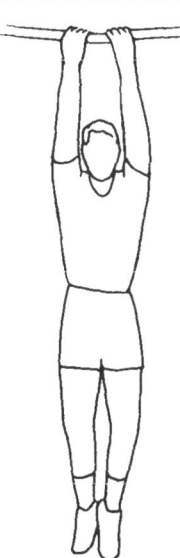 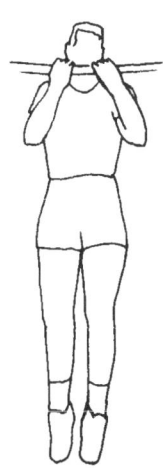

Position: From a standing position, grasp the bar with your palms facing the body.

Action: Bending both arms, pull your body up until your chin clears the bar. Return to the starting position in a controlled manner. If necessary, bend your knees to keep the feet from touching the floor. Do 8 to 12 repetitions to muscle failure. A weight belt may be worn if additional resistance is needed.

TRICEPS EXTENSION
This exercise is for the triceps muscles.

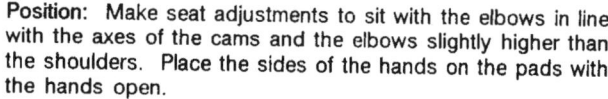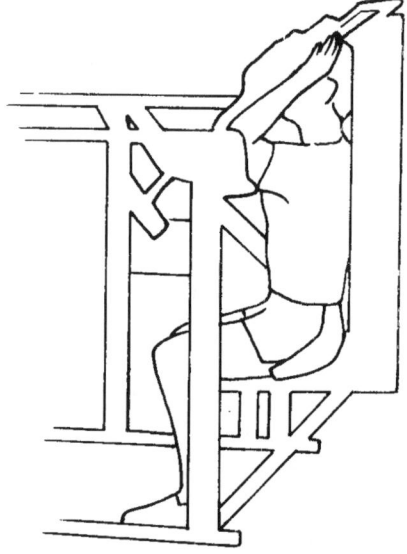

Position: Make seat adjustments to sit with the elbows in line with the axes of the cams and the elbows slightly higher than the shoulders. Place the sides of the hands on the pads with the hands open.

Action: Straighten the arms against the resistance. After doing this, bend the elbows, and return to the starting position in a controlled manner. Do 8 to 12 repetitions to muscle failure.

BICEPS CURL
This exercise is for the biceps muscles.

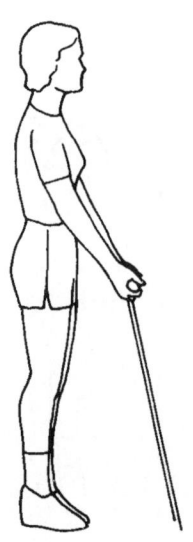

Position: Stand with the bar in front of your body, arms straight and elbows at the sides. Your hands should be spaced about shoulder width apart and the palms should face away from the body.

Action: Without moving your elbows, bend the arms, bringing the bar to shoulder level. In a controlled manner, lower the weight to the starting position. This is one repetition. Do 8 to 12 repetitions to muscle failure.

The following exercises can be performed to condition the muscles of the mid-section (erector spinae, rectus abdominus and external and internal obliques). As the soldier becomes more conditioned on these exercises, resistance can be added.

BACK EXTENSION
This exercise is for the erector spinae muscle group.

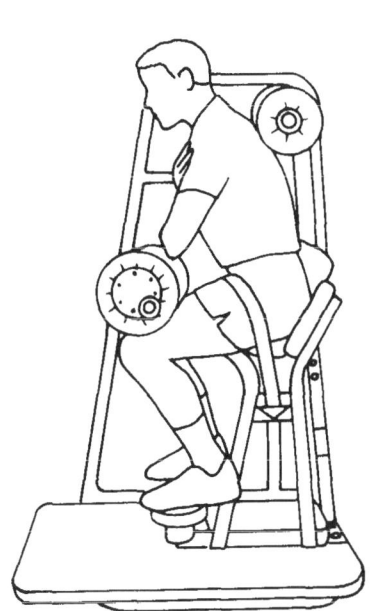  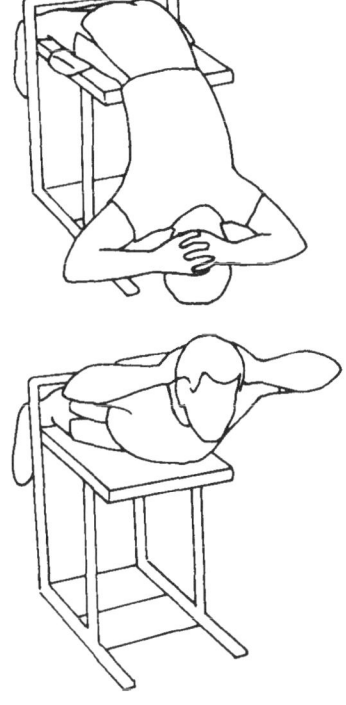

Position: Sit in the machine with your back underneath the highest roller pad. Stabilize your lower body by moving your thighs under the lower roller pads. Place the feet firmly on the platform and fasten the seat belt. Interlace your fingers across your waist, or fold your arms across your chest.

Action: Move the torso backward smoothly until the upper body forms a straight line with the lower body. Do not arch the back excessively by moving past this point. Return to the starting position in a controlled manner. Do 8 to 12 repetitions to muscle failure.

Alternative: If a low-back machine is not available, the following exercise may be performed.

Position: While face-down, anchor your feet securely and position the supporting pad under the upper part of the front thighs. Position your upper body as close to vertical as possible. The hands may be placed behind the head with fingers interlocked; or, the arms may be folded across the chest, provided they do not restrict the downward range of motion.

Action: Straighten the back and raise the upper body until it forms a straight line with the legs. Do not allow your upper body to come any higher than parallel to the floor. Lower your upper body to the starting position in a controlled manner. Do 8 to 12 repetitions to muscle failure.

SIT-UP
This exercise is for the rectus abdominis and iliopsoas (hip flexor) muscles.

Position: Lie on your back with your knees bent at approximately a 90 degree angle and feet anchored. Place your hands behind your head.

Action: Sit up until your trunk is in a vertical position relative to the floor while keeping the knees bent. Lower yourself in a controlled manner to the starting position. The number of repetitions you should do depends on the maximum number of sit-ups you perform in two minutes. Do three sets of 50 percent of your maximum number. For example, if you can do 60 sit-ups in two minutes, do three sets of 30 or more repetitions per set.

INCLINE SIT-UP
This exercise is for the rectus abdominis and iliopsoas muscles.

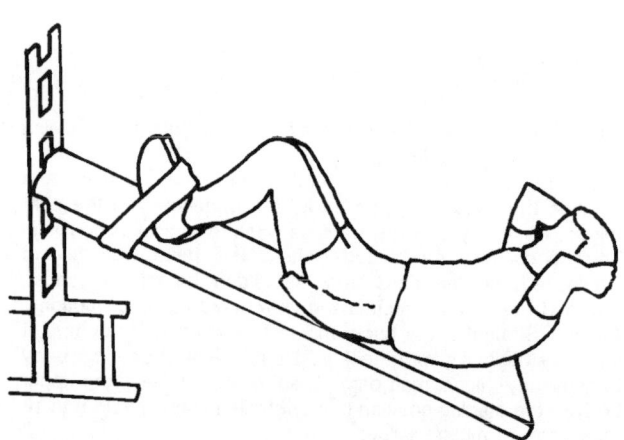 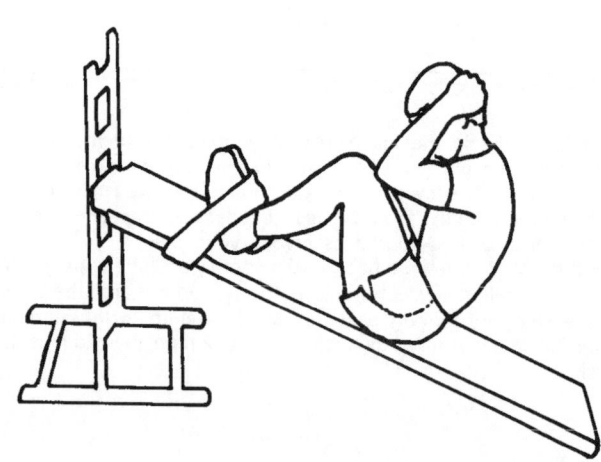

Position: Lie on an incline board with your knees bent at approximately a 90 degree angle and your feet anchored. The steeper the incline of the board, the more difficult the sit-up will be. Interlace the fingers behind your head.

Action: Curl your torso up as far as comfortably possible. Return to the starting position. This is one repetition. Do 20 to 50 repetitions to muscle failure.

ABDOMINAL CRUNCH
This exercise is for the rectus abdominis muscle.

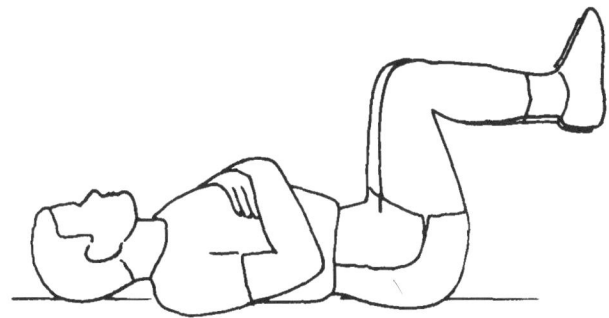 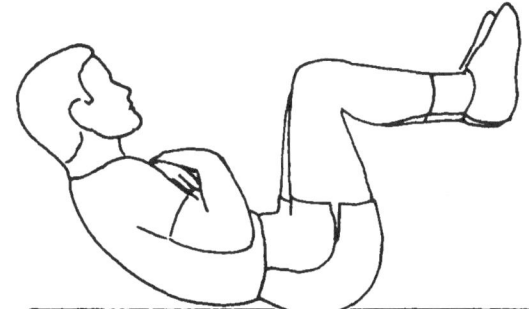

Position: Lie on your back with both legs bent at the knees and the upper legs at right angles to the floor. Your arms should be crossed at chest level with the palms of the hands on their opposite shoulders. Your ankles may be crossed but, in all cases, the feet should not touch the floor.

Action: Roll up your upper body by first lifting your head and tucking the chin. Next, curl your spine by rolling the upper back and then the lower back off the floor. Pause briefly in the up position while tensing the abdominal muscles. Return in a slow, controlled manner to the starting position by "unrolling" the upper body. Do 20 to 50 repetitions to muscle failure.

Exercise Chart

The chart labeled Figure 3-5 will help the soldier select appropriate exercises for use in developing a good muscular endurance and strength workout. For example, if the soldier wants to develop his upper leg muscles, he has several options. He may choose from the following: 1) PREs, concentrating on the split- or single-leg squat; 2) exercises with equipment, doing free weight squats; or, 3) exercises with a machine, doing leg presses, leg curls, and leg extensions.

EXERCISE CHART FOR MUSCULAR STRENGTH AND ENDURANCE

	EXERCISES	LOWER LEGS	UPPER LEGS	WAIST	CHEST	UPPER ARMS	LOWER ARMS	SHOULDERS	BACK
Partner-Resisted Exercises	Split-Squat		x						
	Single-Leg Squat		x						
	Leg Extension		x						
	Leg Curl		x						
	Heel Raise	x							
	Toe Raise	x							
	Push-Up				x	x			
	Seated Row					x			x
	Overhead Press					x		x	
	Pull-Down					x			x
	Shrug							x	
	Triceps Extension					x			
	Biceps Curl					x			
	Abdominal Twist			x					
	Abdominal Curl			x					
	Abdominal Crunch			x					
Exercises with Equipment (Barbell/Dumbbell)	Squat		x						
	Heel Raise	x							
	Bench Press				x	x			
	Bent-Over Row					x			x
	Overhead Press					x		x	
	Shrug							x	
	Triceps Extension					x			
	Biceps Curl					x			
	Wrist Curl						x		
	Bent-Leg Dead Lift		x					x	x
Exercises with an Exercise Machine	Leg Press		x						
	Leg Extension		x						
	Leg Curl		x						
	Heel Raise	x							
	Toe Raise	x							
	Bench Press				x	x			
	Seated Row					x			x
	Lat Pull-Down					x			x
	Shrug							x	
	Parallel Bar Dip				x	x			
	Chin-up					x			x
	Triceps Extension					x			
	Biceps Curl					x			
	Back Extension								x
	Sit-Up			x					
	Incline Sit-Up			x					
	Abdominal Twist			x					
	Abdominal Crunch			x					

Figure 3-5

CHAPTER 4

Flexibility

...ty refers to the

...f movement of a

joint.

Flexibility is a component of physical fitness. Developing and maintaining it are important parts of a fitness program. Good flexibility can help a soldier accomplish such physical tasks as lifting, loading, climbing, parachuting, running, and rappelling with greater efficiency and less risk of injury.

Flexibility is the range of movement of a joint or series of joints and their associated muscles. It involves the ability to move a part of the body through the full range of motion allowed by normal, disease-free joints.

No one test can measure total-body flexibility. However, field tests can be used to assess flexibility in the hamstring and low-back areas. These areas are commonly susceptible to injury due, in part, to loss of flexibility. A simple toe-touch test can be used. Soldiers should stand with their legs straight and feet together and bend forward slowly at the waist. A soldier who cannot touch his toes without bouncing or bobbing needs work to improve his flexibility in the muscle groups stretched by this test. The unit's Master Fitness Trainer can help him design a stretching program to improve his flexibility.

Stretching during the warm-up and cool-down helps soldiers maintain overall flexibility. Stretching should not be painful, but it should cause some discomfort because the muscles are being stretched beyond their normal length. Because people differ somewhat anatomically, comparing one person's flexibility with another's should not be done. People with poor flexibility who try to stretch as far as others may injure themselves.

Stretching Techniques

Using good stretching techniques can improve flexibility. There are four commonly recognized categories of stretching techniques: static, passive, proprioceptive neuromuscular facilitation (PNF), and ballistic. These are described here and shown later in this chapter.

STATIC STRETCHING

Static stretching involves the gradual lengthening of muscles and tendons as a body part moves around a joint. It is a safe and effective method for improving flexibility. The soldier assumes each stretching position slowly until he feels tension or tightness. This lengthens the muscles without causing a reflex contraction in the stretched muscles. He should hold each stretch for ten seconds or longer. This lets the lengthened muscles adjust to the stretch without causing injury.

The longer a stretch is held, the easier it is for the muscle to adapt to that length. Static stretching should not be painful. The soldier should feel slight discomfort, but no pain. When pain results from stretching, it is a signal that he is stretching a muscle or tendon too much and may be causing damage.

PASSIVE STRETCHING

Passive stretching involves the soldier's use of a partner or equipment, such as a towel, pole, or rubber tubing, to help him stretch. This produces a safe stretch through a range of motion he could not achieve without help. He should talk with his partner to ensure that each muscle is stretched safely through the entire range of motion.

PNF STRETCHING

PNF stretching uses the neuromuscular patterns of each muscle group to help improve flexibility. The soldier performs a series of intense contractions and relaxations using a partner or equipment to help him stretch. The PNF technique allows for greater muscle relaxation following each contraction and increases the soldier's ability to stretch through a greater range of motion.

...categories of

...techniques are

...ssive,

...ceptive

...uscular

(PNF), and

...stic.

BALLISTIC STRETCHING

Ballistic, or dynamic, stretching involves movements such as bouncing or bobbing to attain a greater range of motion and stretch. Although this method may improve flexibility, it often forces a muscle to stretch too far and may result in an injury. Individuals and units should not use ballistic stretching.

FITT Factors

Commanders should include stretching exercises in all physical fitness programs.

The following FITT factors apply when developing a flexibility program.

Frequency: Do flexibility exercises daily. Do them during the warm-up to help prepare the muscles for vigorous activity and to help reduce injury. Do them during the cool-down to help maintain flexibility.

Intensity: Stretch a muscle beyond its normal length to the point of tension or slight discomfort, not pain.

Time: Hold stretches for 10 to 15 seconds for warming up and cooling down and for 30 seconds or longer to improve flexibility.

Type: Use static stretches, assumed slowly and gradually, as well as passive stretching and/or PNF stretching.

Warm-Up and Cool-Down

The warm-up and cool-down are very important parts of a physical training session, and stretching exercises should be a major part of both.

THE WARM-UP

Before beginning any vigorous physical activity, one should prepare the body for exercise. The warm-up increases the flow of blood to the muscles and tendons, thus helping reduce the risk of injury. It also increases the joint's range of motion and positively affects the speed of muscular contraction.

A recommended sequence of warm-up activities follows. Soldiers should do these for five to seven minutes before vigorous exercise.

• Slow joggin-in-place or walking for one to two minutes. This causes a gradual increase in the heart rate, blood pressure, circulation, and increases the temperature of the active muscles.

• Slow joint rotation exercises (for example, arm circles, knee/ankle rotations) to gradually increase the joint's range of motion. Work each major joint for 5 to 10 seconds.

• Slow, static stretching of the muscles to be used during the upcoming activity. This will "loosen up" muscles and tendons so they can achieve greater ranges of motion with less risk of injury. Hold each stretch position for 10 to 15 seconds, and do not bounce or bob.

• Calisthenic exerciese, as described in Chapter 7, to increase the intensity level before the activity or conditioning period.

• Slowly mimic the activities to be performed. For example, lift a lighter weight to warm-up before lifting a heavier one. This helps prepare the neuromuscular pathways.

The warm-up warms the muscles, increasing the flow of blood and reducing the risk of injury.

THE COOL-DOWN

The following information explains the importance of cooling down and how to do it correctly.

- Do not stop suddenly after vigorous exercise, as this can be very dangerous. Gradually bring the body back to its resting state by slowly decreasing the intensity of the activity. After running, for example, one should walk for one to two minutes. Stopping exercise suddenly can cause blood to pool in the muscles, thereby reducing blood flow to the heart and brain. This may cause fainting or abnormal rhythms in the heart which could lead to serious complications.
- Repeat the stretches done in the warm-up to help ease muscle tension and any immediate feeling of muscle soreness. Be careful not to overstretch. The muscles are warm from activity and can possibly be overstretched to the point of injury.
- Hold stretches 30 seconds or more during the cool-down to improve flexiblity. Use partner-assisted or PNF techniques, if possible.

The soldier should not limit flexibility training to just the warm-up and cool-down periods. He should sometimes use an entire PT session on a "recovery" or "easy"training day to work on flexibility improvement. He may also work on it at home. Stretching is one form of exercise that takes very little time relative to the benefits gained.

Rotation Exercises

Rotation exercises are used to gently stretch the tendons, ligments, and muscles associated with a joint and to stimulate lubrication of the joint with synovial fluid. This may provide better movement and less friction in the joint.

The following exercises should be performed slowly.

NECK

Position: Stand with the back straight and feet shoulder width apart. Place the hands on the hips.

Action: Roll the head slowly to the left, making a complete circle with the path of the head. Do this three times in each direction.

ARMS AND SHOULDERS

Position: Stand with the back straight and feet shoulder width apart. Extend the arms outward to shoulder height.

Action: Rotate the shoulders forward, and make a large circular motion with the arms. Repeat the action in the opposite direction. Do this three times in each direction.

HIPS

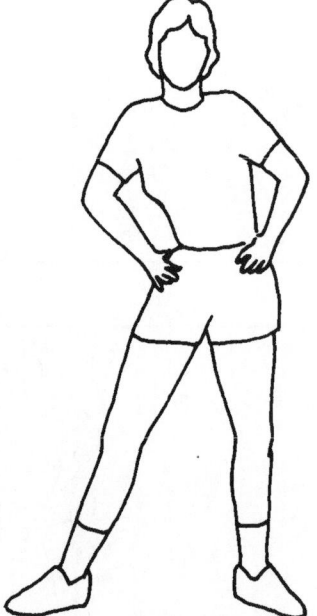 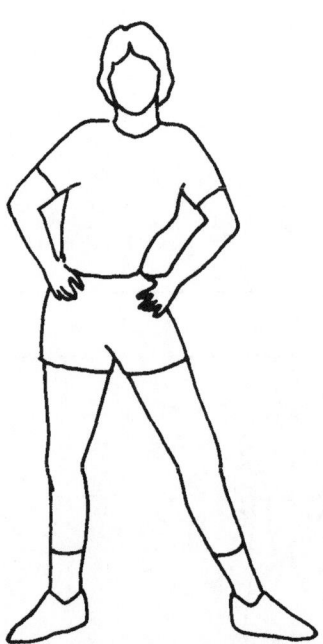

Position: Stand in the same manner as for the neck rotation.

Action: Rotate the hips clockwise while keeping the back straight. Repeat the action in a counterclockwise direction. Do this three times in each direction.

KNEES AND ANKLES

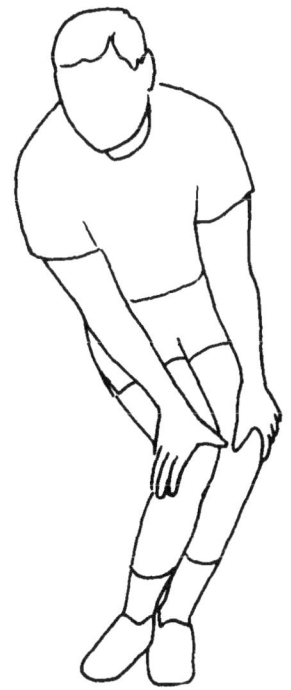

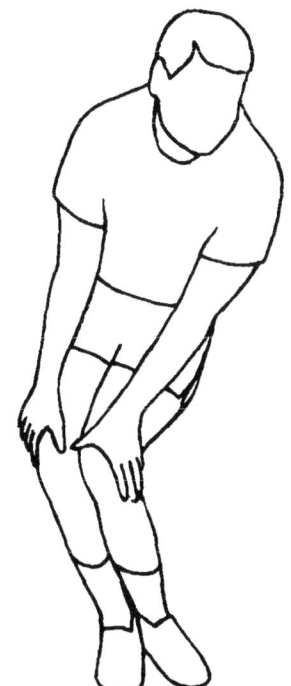

Position: Stand with the feet together, and bend at the waist with the knees slightly bent.

Actio⁻ Place the hands above the knees, and rotate the legs in a clockwise direction. Repeat the action in a counterclockwise direction. Do this three times in each direction.

Common Stretching Exercises

The following exercises improve flexibility when performed slowly, regularly, and with gradual progression. Static, passive and PNF stretches are shown.
CAUTION Some of these exercises may be difficult or too strenuous for unfit or medically limited soldiers. Common sense should be used ;n selecting stretching exercises.

STATIC STRETCHES

Assume all stretching positions slowly until you feel tension or slight discomfort. Hold each position for at least 10 to 15 seconds during the warm-up and cool-down. Developmental stretching to improve flexibility requires holding each stretch for 30 seconds or longer.
Choose the appropriate stretch for the muscle groups which you will be working.

NECK AND SHOULDER STRETCH
This stretches the sternocleidomastoid, pectoralis major, and deltoid muscles.

Position: Stand with the feet shoulder width apart and the arms behind the body.

Action: Grasp the left wrist with the right hand. Pull the left arm down and to the right. Tilt the head to the right. Hold this position for 10 to 15 seconds. Repeat the action with the right wrist, pulling the right arm down and to the left. Tilt the head to the left.

ABDOMINAL STRETCH
This stretches the abdominals, obliques, latissimus dorsi, and biceps.

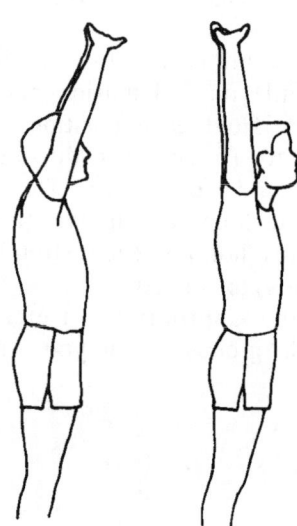 

Position: Stand and extend the arms upward and over the head. Interlace the fingers with the palms turned upward.

Action: Stretch the arms up and slightly back. Hold this position for 10 to 15 seconds.
Variation: This stretches the rectus abdominis muscles. Stretch to one side, then the other. Return to the starting position.

CHEST STRETCH

This stretches the pectoralis major, deltoids, and biceps muscle groups.

Position: Stand and interlace the fingers behind the back.

Action: Lift the arms behind the back so that they move outward and away from the body. Lean forward from the waist. Hold this position for 10 to 15 seconds. Bend the knees before moving to the upright position. Return to the starting position.

UPPER-BACK STRETCH

This stretches the lower trapezius and posterior deltoid muscles of the upper back.

Position: Stand with the arms extended to the front at shoulder height with the fingers interlaced and palms facing outward.

Action: Extend the arms and shoulders forward. Hold this position for 10 to 15 seconds. Return to the starting position.

OVERHEAD ARM PULL
This stretches the external and internal obliques, latissimus dorsi, and triceps.

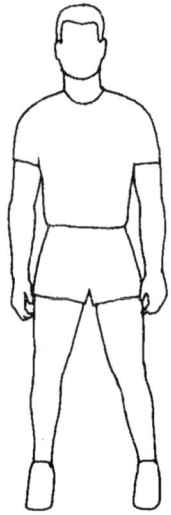 

Position: Stand with the feet shoulder width apart. Raise the right arm, bending the right elbow and touching the right hand to the back of the neck.

Action: Grab the right elbow with the left hand, and pull to the left. Hold this position for 10 to 15 seconds. Return to the starting position. Do the same stretch, and pull the left elbow with the right hand for 10 to 15 seconds.

THIGH STRETCH
This stretches the quadriceps and anterior tibialis.

 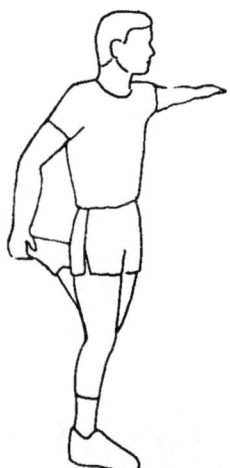 

Position: Stand. (For variation, lie on the stomach.)

Action: Bend the left leg up toward the buttocks. Grasp the toes of the left foot with the right hand, and pull the heel to the left buttock. Extend the left arm to the side for balance. Hold this position for 10 to 15 seconds. Return to the starting position. Bend the right leg, grasp the toes of the right foot with the left hand, and pull the heel to the right buttock. Extend the right arm for balance. Hold this position for 10 to 15 seconds. Return to the starting position.

HAMSTRING STRETCH (STANDING)
This stretches the hamstrings, erector spinae, and gluteal muscles.

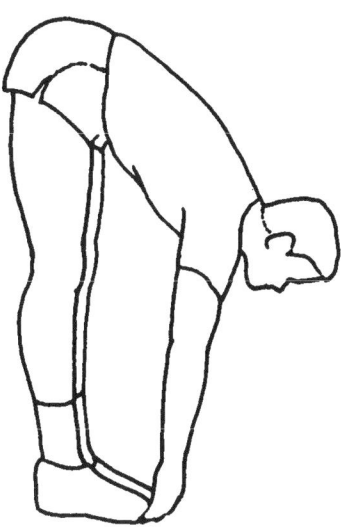

Position: Stand with the knees slightly bent.

Action: Bend forward keeping the head up, and reach toward the toes. Straighten the legs, and hold this position for 10 to 15 seconds.

HAMSTRING STRETCH (SEATED)
In addition to the muscles mentioned in the standing hamstring stretch, this stretches the calf (gastrocnemius and soleus) muscles.

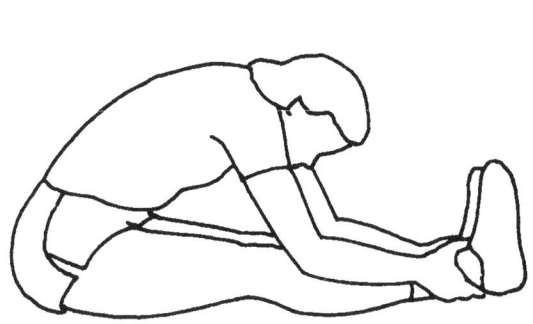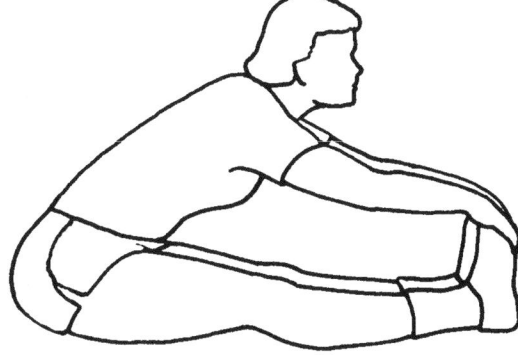

Position: Sit on the ground with both legs straight and extended forward with the feet upright about six inches apart. Put the hands on the ankles or toes.

Action: Bend from the hips, keeping the back and head in a comfortable, straight line. Hold this position for 10 to 15 seconds. (Variation for greater stretch: Stretch and pull back on the toes.)

GROIN STRETCH (STANDING)
This stretches the hip adductor muscles.

Position: Lunge slowly to the left while keeping the right leg straight, the right foot facing straight ahead and entirely on the floor.

Action: Lean over the left leg while stretching the right groin muscles. Hold this position for 10 to 15 seconds. Repeat with the opposite leg.

GROIN STRETCH (SEATED)
This stretches the hip adductor and erector spinae muscles.

Position: Sit on the ground with the soles together. Place the hands on or near the feet.

Action: Bend forward from the hips, keeping the head up. Hold this position for 10 to 15 seconds.

GROIN STRETCH (SEATED STRADDLE)
This stretches the hip adductor (on the inside of the upper leg), gluteals, erector spinae, and hamstring muscles.

Position: Sit on the ground with the legs straight and spread as far apart as possible.

Action: Bend forward at the hips, keep the head up, and reach toward the feet. Hold this position for 10 to 15 seconds.
Variation: Stretch to one side while trying to touch the toes. Next, stretch to the other side.

CALF STRETCH
This stretches the calf (gastrocnemius and soleus) muscles.

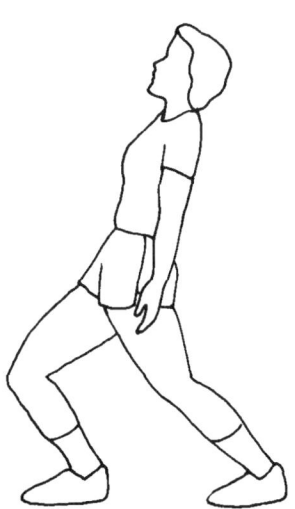

Position: Stand straight with the feet together, arms extended downward, elbows locked, palms facing backward, fingers extended and joined, and head and eyes facing front.

Action: Move the right foot to the rear about two feet, and place the ball of the foot on the ground. Slowly press the right heel to the ground. Slowly bend the left knee while pushing the hips forward and arching the back slightly. Hold this position for 10 to 15 seconds. Return to the starting position. Repeat with the left foot. Return to the starting position.

CALF STRETCH (VARIATION: TOE PULL)

This stretches the calf (gastrocnemius) and to a lesser extent the hamstrings, gluteus maximus, and erector spinae muscles.

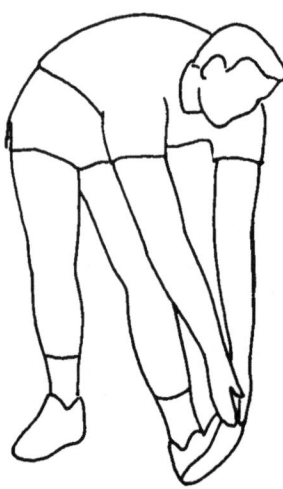

Position: Stand with the feet shoulder width apart and the left foot slightly forward.

Action: Bend forward at the waist. Slightly bend the right knee, and fully extend the left leg. Reach down and pull the toes of the left foot toward the left shin. Hold this position for 10 to 15 seconds. Return to the starting position. In a similar manner, pull the toes of the right foot toward the right shin, and hold for 10 to 15 seconds.

HIP AND BACK STRETCH (SEATED)

This stretches the hip abductors, erector spinae, latissimus dorsi, and oblique muscle groups.

Position: Sit on the ground with the right leg forward and straight. Cross the left leg over the right while sitting erect. Keep the heels of both feet in contact with the ground.

Action: Slowly rotate the upper body to the left and look over the left shoulder. Reach across the left leg with the right arm, and push the left leg to your right. Use the left hand for support by placing it on the ground. Hold this position for 10 to 15 seconds. Repeat this stretch for the other side by crossing and turning in the opposite direction.

HIP AND BACK STRETCH (LYING DOWN)
This stretches the gluteal and erector spinae muscles.

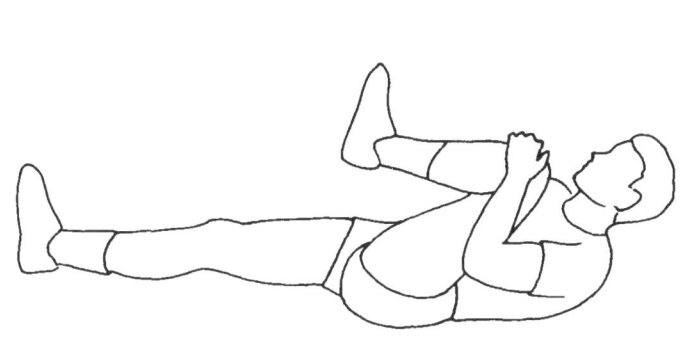
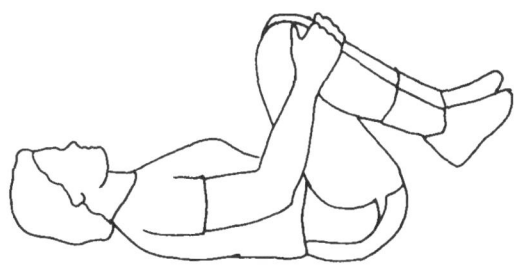
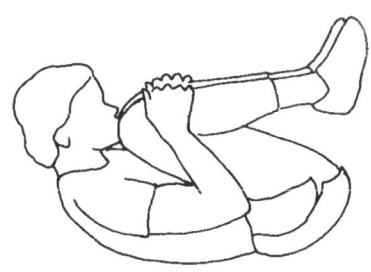

Action 1

Action 2

Position: Lie on the back with the arms straight beside the body. Keep the legs straight and the knees and feet together.
Action 1: Bring the left leg straight back toward the head, leaving the right leg in the starting position. Bring the head and arms up. Grab the bent left leg below the knee, and pull it gradually to the chest. Hold this position for 10 to 15 seconds. Gradually return to the starting position. Repeat these motions with the *opposite leg.*

Action 2: *Pull both knees to the chest. Pull the head up to the knees. Hold for 10 to 15 seconds. Return to the starting position.*

PASSIVE STRETCHES

Passive stretching is done with the help of a partner or equipment. The examples in this chapter show passive stretching done with a towel or with a partner. When stretching alone, using a towel may help the exerciser achieve a greater range of motion.

TOWEL STRETCHES

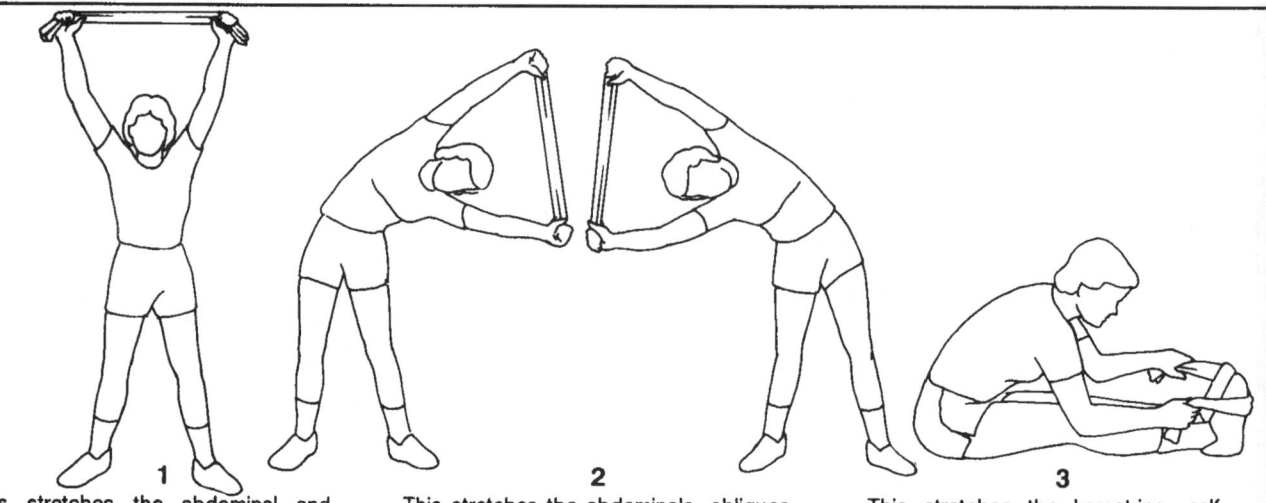

1
This stretches the abdominal and pectoral muscles.
Position: Stand erect with the hands overhead and grasping a towel.
Action: Pull tightly on the towel while reaching up and slightly arching the back. Hold for 10 to 15 seconds.

2
This stretches the abdominals, obliques, and latissimus dorsi.
Position: Stand erect with the hands overhead and grasping a towel.
Action: Slowly bend sideways to the left as far as possible. Hold for 10 to 15 seconds. Repeat for the opposite side. While doing this stretch, pulling on the towel with the bottom arm will enhance the stretch.

3
This stretches the hamstring, calf, and low back muscles.
Position: Sit with the legs straight and together. Grasping each end of a short towel, place the middle of the towel over the balls of the feet.
Action: Pulling on the towel, come forward as far as possible keeping the legs straight and the toes pulled back.

PARTNER-ASSISTED CHEST STRETCH
This exercise stretches the pectoralis major, deltoids, and biceps muscles.

Position: Sit erect with the arms straight, elevated to shoulder height, and the palms facing forward. The partner stands behind the exerciser grasping the arms between the wrists and the elbows.

Action: The partner gradually pulls both of the exerciser's arms toward the rear until the stretch causes the exerciser mild discomfort. Hold this position for 10 to 15 seconds.

PARTNER-ASSISTED HAMSTRING STRETCH
This exercise stretches the hamstrings and erector spinae muscle groups.

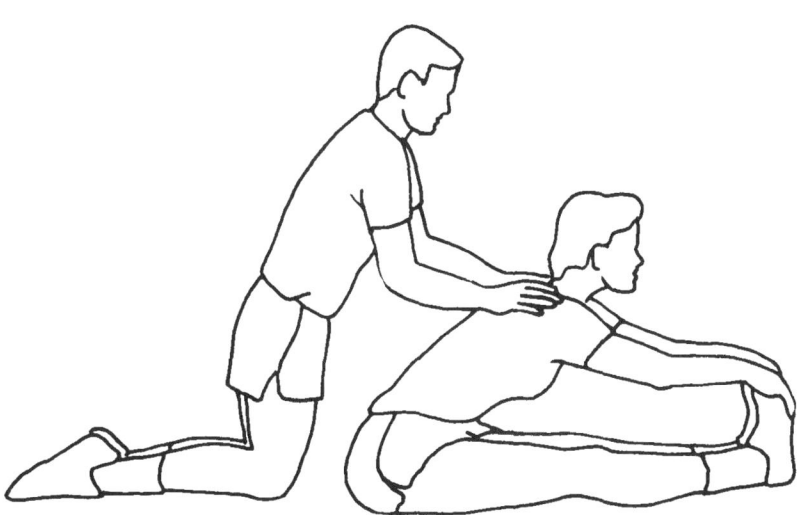

Position: Sit erect on the ground with the legs together. The partner kneels behind the exerciser. If the partner stands, he may apply too much pressure.

Action: The partner places light pressure on the exerciser's upper back until the exerciser's forward motion results in mild discomfort. This position is held for 10 to 15 seconds.

PARTNER-ASSISTED GROIN STRETCH
This exercise stretches the hip adductor and erector spinae muscle groups.

Position: Sit on the ground with knees bent and soles together. The partner kneels behind the exerciser. If the partner stands, he may apply too much pressure.

Action: The partner places light pressure on the exerciser's knees with his hands and leans gently on the exerciser's back with his chest until the stretch causes the exerciser mild discomfort. This position is held for 10 to 15 seconds.

Soldiers can do PNF (Propriocep-tive Neuromuscular Facilitation) stretches for most major muscle groups. PNF stretches use a series of contrac-tions, done against a partner's resis-tance, and relaxations.

Obtaining a safe stretch beyond the muscle's normal length requires a part-ner's assistance. The following four steps provide general guidance as to how PNF stretches are done. Both the exerciser and partner should follow these instructions:

1. Assume the stretch position slowly with the partner's help.
2. Isometrically contract the muscles to be stretched. Hold the contraction for 5 to 10 seconds against the partner's unyielding resistance.
3. Relax. Next, contract the antago-nistic muscles for 5 to 10 seconds while the partner helps the exerciser obtain a greater stretch.
4. Repeat this sequence three times, and try to stretch a little further each time. (Caution: The exerciser should not hold his breath. He should breathe out during each contraction.)

Several examples of PNF stretches are provided below in a stepwise fash-ion. The numbers given above for each step correspond to the general descrip-tion listed below.

PNF HAMSTRING AND GLUTEAL STRETCH

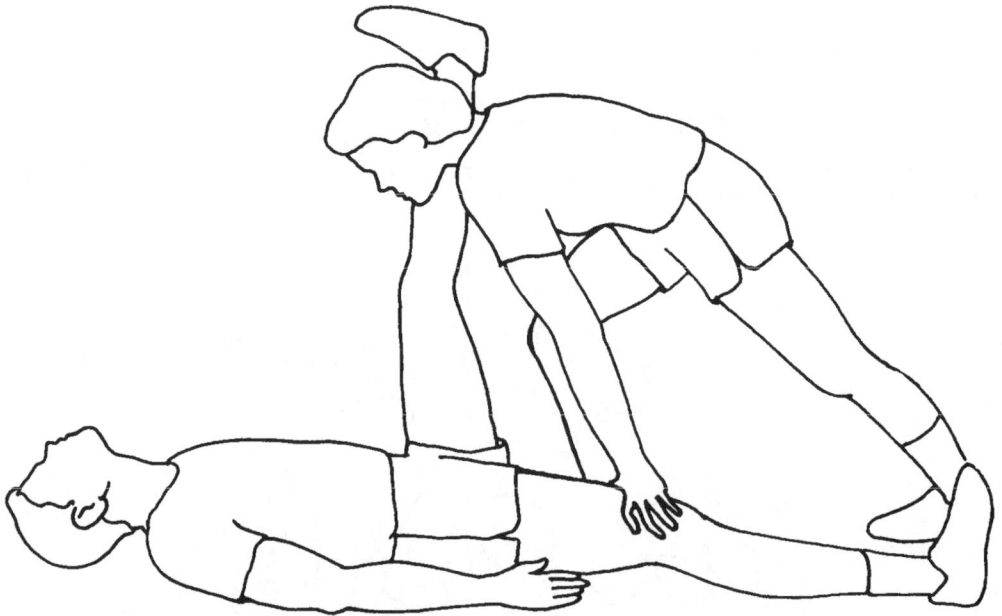

1. The exerciser lies on his back and places the lower part of his left leg on the partner's right shoulder. The exerciser slowly stretches the hamstring and gluteal muscles by gradually bringing the straightened leg toward his head until he feels tension in the stretched muscles. The partner then applies light pressure on the exerciser's lower leg to help maintain or further the stretch.
2. The exerciser isometrically contracts his hamstring and gluteal muscles for 5 to 10 seconds by trying to move his leg downward and away from his head. The partner steadily resists the exerciser's efforts and does not allow any movement to occur.
3. The exerciser relaxes the hamstring and gluteal muscles. He then tries to stretch them farther by using the partner's help and by contracting the antagonistic, hip flexor muscles (the iliopsoas and quadriceps) and the tibialis anterior muscle for 5 to 10 seconds.
4. Perform these movements three times for each leg. Try to stretch a little further each time.

PNF CHEST STRETCH

1. The exerciser and partner assume the positions used in the partner-assisted chest stretch. The exerciser slowly stretches until he feels tension in the stretched muscles. The partner then applies light pressure on the forearm to help maintain or further the stretch.

2. The exerciser contracts the pectorals for 5 to 10 seconds, trying to pull his arms and hands together against the partner's unyielding resistance.

3. The exerciser relaxes his pectorals and attempts to stretch further with the partner's help while contracting the antagonistic muscles of the upper back. He does this for 5 to 10 seconds.
4. Repeat this sequence three times, trying to stretch a little further each time.

PNF GROIN STRETCH

1. The exerciser assumes the position used for the partner-assisted groin stretch. The partner kneels behind him. The exerciser slowly lowers his legs and leans forward until tension is felt in the muscles of the groin (his abductors) and lower back (erector spinae muscles). Next, the partner applies light pressure on the exerciser's thighs and back to help maintain or further increase the stretch.

2. The exerciser then attempts to push upward for 5 to 10 seconds by contracting the groin and lower back muscles while the partner resists and allows no movement to occur.
3. The exerciser relaxes the groin and lower back muscles and tries to stretch further with the partner's help and by contracting the antagonistic muscles (hip abductor and abdominal muscles) for 5 to 10 seconds.
4. Perform these movements three times. Try to stretch a little further each time.

CHAPTER 5

Body Composition

Body composition, which refers to the body's relative amounts of fat and lean body mass (organs, bones, muscles), is one of the five components of physical fitness. Good body composition is best gained through proper diet and exercise. Examples of poor body composition are underdeveloped musculature or excessive body fat. Being overweight (that is, overly fat) is the more common problem.

Poor body composition causes problems for the Army. Soldiers with inadequate muscle development cannot perform as well as soldiers with good body composition. As a soldier gets fat, his ability to perform physically declines, and his risk of developing disease increases. Soldiers with high percentages of body fat often have lower APFT scores than those with lower percentages. Poor body composition, especially obesity, has a negative effect on appearance, self--esteem, and negatively influences attitude and morale.

The Army's weight control program is described in AR 600-9. It addresses body composition standards, programs for the overly fat, and related administrative actions.

The amount of fat on the body, when expressed as a percentage of total body weight, is referred to as the percent body fat. The Army's maximum allowable percentages of body fat, by age and sex, are listed in Figure 5-1.

Evaluation Methods

The Army determines body fat percentage using the girth method. (This is described in AR 600-9, pages 12 to 21.)

Body composition is influenced by age, diet, fitness level, and genetic factors (gender and body type). The Army's screening charts for height and weight (shown in AR 600-9) make allowances for these differences. A soldier whose weight exceeds the standard weight shown on the charts may not necessarily be overfat. For example, some well-muscled athletes have body weights that far exceed the values for weight listed on the charts for their age, gender, and height. Yet, only a small percentage of their total body mass may be fat. In such cases, the lean body mass accounts for a large share of their total body composition, while only a small percentage of the total body mass is composed of fat.

Soldiers who do not meet the weight standards for their height and/or soldiers whose appearance suggests that they have excessive fat are to be evaluated using the circumference (girth measurement) method described in AR 600-9.

Body composition is influenced by age, fitness level, and genetic factors.

BODY FAT STANDARDS				
AGES:	17-20	21-27	28-39	40+
MALES 20%	20%	22%	24%	26%
FEMALES	30%	32%	34%	36%

Figure 5-1

A more accurate way to determine body composition is by hydrostatic or underwater weighing. However, this method is very time-consuming and expensive and usually done only at hospitals and universities.

Soldiers who do not meet Army body fat standards are placed on formal, supervised weight (fat) loss programs as stipulated in AR 600-9. Such programs include sensible diet and exercise regimens.

Diet and Exercise

A combination of exercise and diet is the best way to lose excessive body fat. Losing one to two pounds a week is a realistic goal which is best accomplished by reducing caloric intake and increasing energy expenditure. In other words, one should eat less and exercise more. Dieting alone can cause the body to believe it is being starved. In response, it tries to conserve its fat reserves by slowing down its metabolic rate and, as a result, it loses fat at a slower rate.

Soldiers must consume a minimum number of calories from all the major food groups, with the calories distributed over all the daily meals including snacks. This ensures an adequate consumption of necessary vitamins and minerals. A male soldier who is not under medical supervision when dieting requires a caloric intake of at least 1,500; women require at least 1,200 calories. Soldiers should avoid diets that fail to meet these criteria.

Trying to lose weight with fad diets and devices or by skipping meals does not work for long-term fat loss, since weight lost through these practices is mostly water and lean muscle tissue, not fat. Losing fat safely takes time

A combination of exercise and diet is the best way to lose unwanted body fat.

Aerobic exercise is best for burning fat. examples include jogging, walking, swimming, bicycling, cross-country skiing, and rowing.

and patience. There is no quick and easy way to improve body composition.

The soldier who diets and does not exercise loses not only fat but muscle tissue as well. This can negatively affect his physical readiness. Not only does exercise burn calories, it helps the body maintain its useful muscle mass, and it may also help keep the body's metabolic rate high during dieting.

Fat can only be burned during exercise if oxygen is used. Aerobic exercise, which uses lots of oxygen, is the best type of activity for burning fat. Aerobic exercises include jogging, walking, swimming, bicycling, cross-country skiing, rowing, stair climbing, exercise to music, and jumping rope. Anaerobic activities, such as sprinting or lifting heavy weights, burn little, if any, fat.

Exercise alone is not the best way to lose body fat, especially in large amounts. For an average-sized person, running or walking one mile burns about 100 calories. Because there are 3,500 calories in one pound of fat, he needs to run or walk 35 miles if pure fat were being burned. In reality, fat is seldom the only source of energy used during aerobic exercise. Instead, a mixture of both fats and carbohydrates is used. As a result, most people would need to run or walk over 50 miles to burn one pound of fat.

A combination of proper diet and aerobic exercise is the proven way to lose excessive body fat. Local dietitians and nutritionists can help soldiers who want to lose weight by suggesting safe and sensible diet programs. In addition, the unit's MFT can design tailored exercise programs which will help soldiers increase their caloric expenditure and maintain their lean body mass.

CHAPTER 6

Nutrition and Fitness

In addition to exercise, proper nutrition plays a major role in attaining and maintaining total fitness. Good dietary habits (see Figure 6-1) greatly enhance the ability of soldiers to perform at their maximum potential. **A** good diet alone, however, will not make up for poor health and exercise habits. This chapter gives basic nutritional guidance for enhancing physical performance. Soldiers must know and follow the basic nutrition principles if they hope to maintain weight control as well as achieve maximum physical fitness, good health, and mental alertness.

Guidelines for Healthy Eating

Eating a variety of foods and maintaining an energy balance are basic guidelines for a healthy diet. Good nutrition is not complicated for those who understand these dietary guidelines.

To be properly nourished, soldiers should regularly eat a wide variety of foods fro-m the major food groups, selecting a variety of foods from within each group. (See Figure 6-2.) A well-balanced diet provides all the nutrients needed to keep one healthy.

Most healthy adults do not need vitamin or mineral supplements if they eat a proper variety of foods. There are no known advantages in consuming excessive amounts of any nutrient, and there may be risks in doing so.

For soldiers to get enough fuel from the food they eat and to obtain the variety of foods needed for nutrient balance, they should eat three meals a day. Even snacking between meals can contribute to good nutrition if the right foods are eaten.

Another dietary guideline is to consume enough calories to meet one's energy needs. Weight is maintained as long as the body is in energy balance,

---

DIETARY GUIDELINES

- **Eat a Variety of Foods**
- **Maintain a Healthy Body Weight**
- **Choose a Diet Low in Fat, Saturated Fat, and Cholesterol**
- **Choose a Diet with Plenty of Vegetables, Fruits, and Grain Products**
- **Use Sugars Only in Moderation**
- **Use Salt and Sodium Only in Moderation**
- **If you Drink Alcoholic Beverages, Do So in Moderation**

Figure 6-1

DAILY FOOD GUIDE

Eat a variety of foods from each food group. Most people should have the minimum number of servings; others need more due to their body size and activity level.

FOOD GROUP	SUGGESTED NUMBER OF SERVINGS	SUGGESTED SIZE OF SERVINGS
Vegetables (Include dark green, leafy, or deep yellow ones)	3 to 5	1 cup of raw, leafy greens or 1/2 cup of cooked vegetables
Fruits (Include citrus fruits or juices, melons, or berries)	2 to 4	1 medium fruit or 1/2 cup of diced or small fruit or 3/4 cup of juice
Breads, Cereals, Rice, and Pasta (Include whole grain varieties)	6 to 11	1 slice of bread, 1/2 bun or roll, 1/2 cup of cooked cereal, rice or pasta, 1 oz. of ready-to- eat cereal
Milk, Yogurt, and Cheese (Include skim or lowfat varieties)	2 to 3	1 cup of milk or yogurt, 1-1/2 oz. of hard cheese
Meats, Poultry, Fish, Dry Beans or Peas, Eggs, Nuts (Use lean meats and remove skin from poultry)	2 to 3	2 or 3 oz. of cooked meat, fish, or poultry (TOTAL 6 oz/day) 2 eggs, or 1 cup of cooked beans or peas

Figure 6-2

that is, when the number of calories used equals the number of calories consumed.

The most accurate way to control caloric intake is to control the size of food portions and thus the total amount of food ingested. One can use standard household measuring utensils and a small kitchen scale to measure portions of foods and beverages. Keeping a daily record of all foods eaten and physical activity done is also helpful.

Figure 6-3 shows the number of calories burned during exercise periods of different types, intensities, and durations. For example, while participating in archery, a person will burn 0.034 calories per pound per minute. Thus, a 150-pound person would burn 5.1 calories per minute (150 lbs. x 0.034 calories/minute/lb. = 5.1 calories/minute) or about 305 calories/hour, as shown in Figure 6-4. Similarly, a person running at 6 miles per hour (MPH) will burn 0.079 cal./min./lb. and a typical, 150-pound male will burn 11.85 calories/minute (150 lbs. x 0.079 cal./lb./min. = 11.85) or about 710 calories in one hour, as shown in Figure 6-3.

To estimate the number of calories you use in normal daily activity, multiply your body weight by 13 if you are sedentary, 14 if somewhat active, and 15 if moderately active. The result is a rough estimate of the number of calories you need to maintain your present body weight. You will need still more calories if you are more than moderately active. By comparing caloric intake with caloric expenditure, the state of energy balance (positive, balanced, or negative) can be determined.

CALORIC EXPENDITURE CHART

ACTIVITY	CAL/MIN/LB	CAL/HR/150 LB	ACTIVITY	CAL/MIN/LB	CAL/HR/150 LB *
Archery	.034	305	Judo, Karate	.087	785
Badminton:			Motor Boating	.016	145
Moderate	.039	350	Mountain Climbing	.086	775
Vigorous	.065	585	Rowing		
Basketball:			(Rec 2.5 MPH)	.036	325
Moderate	.047	420	Vigorous	.118	1000
Vigorous	.066	595	Running:		
Baseball:			6 MPH (10 min/mi)	.079	710
Infield-outfield	.031	280	10 MPH (6 min/mi)	.1	900
Pitching	.039	350	12 MPH (5 min/mi)	.13	1170
Bicycling:			Sailing	.02	180
Slow (5 MPH)	.025	225	Skating:		
Moderate (10 MPH)	.05	450	Moderate (Rec)	.036	325
Fast (13 MPH)	.072	650	Vigorous	.064	575
Bowling	.028	255	Skiing (Snow):		
Calisthenics:			Downhill	.059	530
General	.045	405	Level (5 MPH)	.078	700
Canoeing:			Soccer	.06	570
2.5 MPH	.023	210	Squash	.07	630
4.0 MPH	.047	420	Stationary Run:		
Dancing:			70-80 cts/min	.078	705
Slow	.029	260	Strength Training		
Moderate	.045	405	(10 rep circuit)		
Fast	.064	575	60% 1RM	.022	198
Fencing:			80% 1RM	.048	432
Moderate	.033	300	Swimming (crawl):		
Vigorous	.057	515	20 yds/min	.032	290
Fishing	.016	145	45 yds/min	.058	520
Football (tag)	.04	360	50 yds/min	.071	640
Gardening	.024	220	Table Tennis:		
Gardening-Weeding	.039	260	Moderate	.026	235
Golf	.029	260	Vigorous	.06	540
Gymnastics:			Tennis:		
Light	.022	200	Moderate	.046	415
Heavy	.056	505	Vigorous	.04	540
Handball	.063	570	Volleyball:		
Hiking	.042	375	Moderate	.036	325
Hill Climbing	.06	540	Vigorous	.065	585
Hoeing, Raking,			Walking:		
Planting	.031	280	2.0 MPH	.022	200
Horseback Riding:			3.0 MPH	.03	270
Walk	.019	175	4.0 MPH	.039	350
Trot	.046	415	5.0 MPH	.064	575
Gallop	.067	600	Water Skiing	.053	480
Jogging:			Wrestling	.091	820
4.5 MPH (13:330 mi.)	.063	565			

* A 150-pound person will expend the number of calories indicated in one hour for any given activity.

Figure 6-3

Avoiding an excessive intake of fats is an important fundamental of nutrition.

Avoiding an excessive intake of fats is another fundamental dietary guideline. A high intake of fats, especially saturated fats and cholesterol, has been associated with high levels of blood cholesterol.

The blood cholesterol level in most Americans is too high. Blood cholesterol levels can be lowered by reducing both body fat and the amount of fat in the diet. Lowering elevated blood cholesterol levels reduces the risk of developing coronary artery disease (CAD) and of having a heart attack. CAD, a slow, progressive disease, results from the clogging of blood vessels in the heart. Good dietary habits help reduce the likelihood of developing CAD.

It is recommended that all persons over the age of two should reduce their fat intake to 30 percent or less of their total caloric intake. The current national average is 38 percent. In addition, we should reduce our intake of saturated fat to less than 10 percent of the total calories consumed. We should increase our intake of polyunsaturated fat, but to no more than 10 percent of our total calories. Finally, we should reduce our daily cholesterol intake to 300 milligrams or less. Figure 6-4 suggests actions commanders can take to support sound dietary guidelines. Most of these actions concern dining-facility management.

Carbohydrates are the primary fuel source for muscles during short-term, high-intensity activities.

Concerns for Optimal Physical Performance

Carbohydrates, in the form of glycogen (a complex sugar), are the primary fuel source for muscles during short-term, high-intensity activities. Repetitive, vigorous activity can use up most of the carbohydrate stores in the exercised muscles.

The body uses fat to help provide energy for extended activities such as a one-hour run. Initially, the chief fuel burned is carbohydrates, 'but as the duration increases, the contribution from fat gradually increases.

The intensity of the exercise also influences whether fats or carbohydrates are used to provide energy. Very intense activities use more carbohydrates. Examples include weight training and the APFT sit-up and push-up events.

Eating foods rich in carbohydrates helps maintain adequate muscle-glycogen reserves while sparing amino acids (critical building-blocks needed for building proteins). At least 50 percent of the calories in the diet should come from carbohydrates. Individual caloric requirements vary, depending on body size, sex, age, and training mission. Foods rich in complex carbohydrates (for example, pasta, rice, whole wheat bread, potatoes) are the best sources of energy for active soldiers.

COMMANDER'S CHECKLIST FOR NUTRITION

PRINCIPLES OF NUTRITION

1. Eat a variety of foods.
   No single food item provides all essential nutrients.

2. Maintain a desirable body weight.
   Excess body fat detracts from fitness. Weight loss is achieved by increasing physical activity and decreasing total food intake, especially fats, refined sugars, and alcohol.

3. Avoid excess dietary fat.
   Too much fat (especially cholesterol and saturated fat) can lead to heart disease and weight problems. Fats contain twice as many calories as equal amounts of carbohydrates or protein.

4. Avoid too much sugar.
   Sweets are empty calories and may lead to dental cavities and weight problems.

5. Eat foods with adequate starch and fiber.
   Eating complex carbohydrates adds to the diet and reduces symptoms of constipation.

6. Avoid too much sodium.
   Eating highly-salted foods may lead to excessive sodium intake. This may be a problem for those "at risk" for high blood pressure.

7. If you drink alcoholic beverages, do so in moderation.
   Alcoholic beverages are high in calories and and low in nutrients. One or two standard-size drinks daily appears to cause no harm in normal, healthy, nonpregnant adults.

8. Know the nutrition principles.
   Educating soldiers maximizes efforts to improve nutritional fitness.

Reference: AR 30-1, Appendix J.

SUPPORTING ACTIONS

In the dining facility:
- Ensure menus provide foods from the basic 4 food groups: fruits and vegetables, meats, dairy products, and breads and cereals.
- Establish serving lines in the following order, if possible:
  (1) salads, (2) fruits, (3) entrees, (4) hot vegetables, (5) breads, (6) beverages, (7) desserts.

In the dining facility, provide:
- Low-calorie menu, including short-order items at each meal. Use the Master Menu (SB 10-260) menu patterns.
- Reduced-portion sizes.
- No-calorie beverages.
- Low-calorie salad dressings.
- Posted list of caloric values of menu items, before or on the serving line.

In the dining facility, provide:
- Non-fried eggs as an alternative.
- Margarine as a butter alternative.
- Two percent milk as the primary milk in bulk dispensers.
- Skim milk in 1/2-pint cartons.
- Sauces, gravies, and margarine separately from the entree or vegetable.
- Avoid animal fats, palm oil, and hydrogenated vegetable oil.

In the dining facility, provide:
- Fruit as a dessert alternative.
- Unsweetened juices.
- No-calorie, unsweetened beverages.
- Non-nutritive, sugar substitute as a granulated sugar alternative.
- Unsweetened cereal.

In the dining facility, provide:
- Whole-grain breads, cereals and legumes.
- Fresh fruit.
- Salad bars at lunch and dinner.

- Reduce salt in recipes by 25 percent.

- Avoid alcohol; it is detrimental to good health and weight management.

- Display educational materials on nutrition; (posters, table tents, bulletin boards, and handouts).
- Provide food-service personnel with training programs on nutrition standards.
- Provide unit-training programs on nutrition for soldiers. (Use installation dietitian).

Figure 6-4

Because foods eaten one to three days before an activity provide part of the fuel for that activity, it is important to eat foods every day that are rich in complex carbohydrates. It is also important to avoid simple sugars, such as candy, up to 60 minutes before exercising, because they can lead to low blood sugar levels during exercise.

Soldiers often fail to drink enough water, especially when training in the heat. Water is an essential nutrient that is critical to optimal physical performance. It plays an important role in maintaining normal body temperature. The evaporation of sweat helps cool the body during exercise. As a result, water lost through sweating must be replaced or poor performance, and possibly injury, can result. Sweat consists primarily of water with small quantities of minerals like sodium. Cool, plain water is the best drink to use to replace the fluid lost as sweat. Soldiers should drink water before, during, and after exercise to prevent dehydration and help enhance performance. Figure 6-5 shows recommendations for fluid intake when exercising.

Sports drinks, which are usually simple carbohydrates (sugars) and electrolytes dissolved in water, are helpful under certain circumstances. There is evidence that solutions containing up to 10 percent carbohydrate will enter the blood fast enough to deliver additional glucose to the active muscles. This can improve endurance.

During prolonged periods of exercise (1.5+ hours) at intensities over **50** percent of heart rate reserve, one can benefit from periodically drinking sports drinks with a concentration of 5 to 10 percent carbohydrate. Soldiers on extended road marches can also benefit from drinking these types of glucose-containing beverages. During intense training, these beverages can provide a source of carbohydrate for working muscles. On the other hand, drinks that exceed levels of 10 percent carbohydrate, as do regular soda pops and most fruit juices, can lead to abdominal cramps, nausea, and diarrhea. Therefore, these drinks should be used with caution during intense endurance training and other similar activities.

Many people believe that body builders need large quantities of

---

RECOMMENDATIONS FOR FLUID INTAKE

- Drink cool (40 degrees F) water. This is the best drink to sustain performance. Fluid also comes from juice, milk, soup, and other beverages.

- Do not drink coffee, tea, and soft drinks even though they provide fluids. The caffeine in them acts as a diuretic which can increase urine production and fluid loss. Avoid alcohol for the same reason.

- Drink large quantities (20 oz.) of water one or two hours before exercise to promote hyperhydration. This allows time for adequate hydration and urination.

- Drink three to six ounces of fluid every 15 to 30 minutes during exercise.

- Replace fluid sweat losses by monitoring pre-and post-exercise body weights. Drink two cups of fluid for every pound of weight lost.

Figure 6-5

protein to promote better muscle growth. The primary functions of protein are to build and repair body tissue and to form enzymes. Protein is believed to contribute little, if any, to the total energy requirement of heavy-resistance exercises. The recommended dietary allowance of protein for adults is 0.8 grams per kilogram of body weight. Most people meet this level when about 15 percent of their daily caloric intake comes from protein. During periods of intense aerobic training, one's need for protein might be somewhat higher (for example, 1.0 to 1.5 grams per kilogram of body weight per day). Weight lifters, who have a high proportion of lean body mass, can easily meet their protein requirement with a well-balanced diet which has 15 to 20 percent of its calories provided by protein. Recent research suggests that weight trainers may need no more protein per kilogram of body weight than average, nonathletic people. Most Americans routinely consume these levels of protein, or more. The body converts protein consumed in excess of caloric needs to fat and stores it in the body.

Nutrition in the Field

Soldiers in the field must eat enough food to provide them with the energy they need. They must also drink plenty of water or other non-alcoholic beverages. The "meal, ready to eat" (MRE) supplies the needed amount of carbohydrates, protein, fat, vitamins, and minerals. It is a nutritionally adequate ration when all of its components are eaten and adequate amounts of water are consumed. Because the foods are enriched and fortified with vitamins and minerals, each component is a major source of nutrients. Soldiers must eat all the components in order to get the daily military recommended dietary allowances (MRDA) and have an adequate diet in the field. Soldiers who are in weight control programs or who are trying to lose weight can eat part of each MRE item, as recommended by dietitians.

CHAPTER 7

Circuit Training and Exercise Drills

This chapter gives commanders and trainers guidance in designing and using exercise circuits. It describes calisthenic exercises for developing strength, endurance, coordination, and flexibility. It also describes grass drills and guerilla exercises which are closely related to soldiering skills and should be regularly included in the unit's physical fitness program.

Circuit training is a term associated with specific training routines. Commanders with a good understanding of the principles of circuit training may apply them to a wide variety of training situations and environments.

Circuits

A circuit is a group of stations or areas where specific tasks or exercises are performed.

A circuit is a group of stations or areas where specific tasks or exercises are performed. The task or exercise selected for each station and the arrangement of the stations is determined by the objective of the circuit.

Circuits are designed to provide exercise to groups of soldiers at intensities which suit each person's fitness level. Circuits can promote fitness in a broad range of physical and motor fitness areas. These include CR endurance, muscular endurance, strength, flexibility, and speed. Circuits can also be designed to concentrate on sports skills, soldiers' common tasks, or any combination of these. In addition, circuits can be organized to exercise all **the** fitness components in a short period of time. A little imagination can make circuit training an excellent addition to a unit's total physical fitness program. At the same time, it can provide both fun and a challenge to soldiers' physical and mental abilities. Almost any area can be used, and any number of soldiers can exercise for various lengths of time.

TYPES OF CIRCUITS

The two basic types of circuits are the free circuit and the fixed circuit. Each has distinct advantages.

Free Circuit

In a free circuit, there is no set time for staying at each station, and no signal is given to move from one station to the next. Soldiers work at their own pace, doing a fixed number of repetitions at each station. Progress is measured by the time needed to complete a circuit. Because soldiers may do incomplete or fewer repetitions than called for to reduce this time, the quality and number of the repetitions done should be monitored. Aside from this, the free circuit requires little supervision.

Fixed Circuit

In a fixed circuit, a specific length of time is set for each station. The time is monitored with a stopwatch, and soldiers rotate through the stations on command.

There are three basic ways to increase the intensity or difficulty of a fixed circuit:

- Keep the time for completion the same, but increase the number of repetitions.
- Increase the time per station along with the number of repetitions.
- Increase the number of times soldiers go through the circuit.

VARIABLES IN CIRCUIT TRAINING

Several variables in circuit training must be considered. These include the time, number of stations, number of time, number of stations, number of soldiers, number of times the circuit is completed, and sequence of stations. These are discussed below.

Time

One of the first things to consider is how long it should take to complete the circuit. When a fixed circuit is run, the time at each station should always be the same to avoid confusion and help maintain control. Consider also the time it takes to move from one station to the next. Further, allow from five to seven minutes both before and after running a circuit for warming up and cooling down, respectively.

Number of Stations

The objective of the circuit and time and equipment available strongly influence the number of stations. A circuit geared for a limited objective (for example, developing lower-body strength) needs as few as six to eight stations. On the other hand, circuits to develop both strength and CR fitness may have as many as 20 stations.

Number of Soldiers

If there are 10 stations and 40 soldiers to be trained, the soldiers should be divided into 10 groups of four each. Each station must then be equipped to handle four soldiers. For example, in this instance a rope jumping station must have at least four jump ropes. It is vital in a free circuit that no soldier stand around waiting for equipment. Having enough equipment reduces bottlenecks, slowdowns, and poor results.

Number of Times a Circuit is Completed

To achieve the desired training effect, soldiers may have to repeat the same circuit several times. For example, a circuit may have ten stations. Soldiers may run through the circuit three times, exercising for 30 seconds at each station, and taking 15 seconds to move between stations. The exercise time at each station may be reduced to 20 seconds the second and third time through. The whole workout takes less than 45 minutes including warm-up and cool-down. As soldiers become better conditioned, exercise periods may be increased to 30 seconds or longer for all three rotations. Another option is to have four rotations of the circuit.

Sequence of Stations

Stations should be arranged in a sequence that allows soldiers some recovery time after exercising at strenuous stations. Difficult exercises can be alternated with less difficult ones. After the warm-up, soldiers can start a circuit at any station and still achieve the objective by completing the full circuit.

DESIGNING A CIRCUIT

The designer of a circuit must consider many factors. The six steps below cover the most important aspects of circuit development.

Determine Objectives

The designer must consider the specific parts of the body and the components of fitness on which soldiers need to concentrate. For example, increasing muscular strength may be the primary objective, while muscular endurance work may be secondary. On the other hand, improving cardiorespiratory endurance may be the top priority. The designer must first identify the training objective in order to choose the appropriate exercises.

The designer must consider the specific parts of the body and the components of fitness on which soldiers need to concentrate.

Select the Activities

The circuit designer should list all the exercises or activities that can help meet the objectives. Then he should look at each item on the list and ask the following questions:

● Will equipment be needed? Is it available?

● Will supervision be needed? Is it available?

● Are there safety factors to consider?

Answering these questions helps the designer decide which exercises to use. He can choose from the exercises, calisthenics, conditioning drills, grass drills, and guerrilla drills described in this chapter. However, he should not limit the circuit to only these activities. Imagination and field expediency are important elements in developing circuits that hold the interest of soldiers. (See Figures 7-1 through 7-3.)

Arrange the Stations

A circuit usually has 8 to 12 stations, but it may have as many as 20. After deciding how many stations to include, the designer must decide how to arrange them. For example, in a circuit for strength training, the same muscle group should not be exercised at consecutive stations.

One approach is to alternate "pushing" exercises with "pulling" exercises which involve movement at the same joint(s). For example, in a strength training circuit, exercisers may follow the pushing motion of a bench press with the pulling motion of the seated row. This could be followed by the pushing motion of the overhead press which could be followed by the pulling motion of the lat pull-down. Another approach might be to alternate between upper and lower body exercises.

By not exercising the same muscle group twice in a row, each muscle has

The choice of exercises for circuit training depends on the objectives of the circuit.

a chance to recover before it is used in another exercise. If some exercises are harder than others, soldiers can alternate hard exercises with easier ones. The choice of exercises depends on the objectives of the circuit.

Select the Training Sites

Circuits may be conducted outdoors or indoors. If the designer wants to include running or jogging a certain distance between stations, he may do this in several ways. In the gymnasium, soldiers may run five laps or for 20 to 40 seconds between stations. Outdoors, they may run laps or run between spread-out stations if space is available. However, spreading the stations too far apart may cause problems with control and supervision.

Prepare a Sketch

The designer should draw a simple sketch that shows the location of each station in the training area. The sketch should include the activity and length of time at each station, the number of stations, and all other useful information.

Lay Out the Stations

The final step is to lay out the stations which should be numbered and clearly marked by signs or cards. In some cases, instructions for the stations are written on the signs. The necessary equipment is placed at each station.

Sample Conditioning Circuits

Figures 7-1, 7-2, and 7-3 show different types of conditioning circuits. Soldiers should work at each station 45 seconds and have 15 seconds to rotate to the next station.

SAMPLE CIRCUIT FOR STRENGTH DEVELOPMENT

STATION #1

Leg Press
8-12 reps

STATION #13

Incline Sit-Up
8-12 reps

STATION #2

Leg Raise
8-12 reps

Do 1-2 complete rotations.
Lift weight with slow,
controlled movements.

STATION #12

Biceps Curl
8-12 reps

STATION #3

Leg Extension
8-12 reps

Try to achieve muscle
failure within 8-12 reps.

STATION #11

Triceps Extension
8-12 reps

STATION #4

Leg Curl
8-12 reps

STATION #10

Shrug
8-12 reps

STATION #5

Heel Raise
8-12 reps

STATION #9

Lat Pull-Down
8-12 reps

STATION #6

Bench Press
8-12 reps

STATION #8

Military Press
8-12 reps

STATION #7

Seated Row
8-12 reps

Figure 7-1

SAMPLE CIRCUIT FOR CARDIORESPIRATORY ENDURANCE

STATION #1

Stationary Run
30 seconds

STATION #14

All-Fours Run
30 seconds

STATION #2

Push-Up
30 seconds

Do 2-3 complete rotations.

STATION #13

Mule Kicks
30 seconds

STATION #3

Side-Straddle Hop
30 seconds

Stations may be 25-30 meters
apart to allow more running.

STATION #12

Twisting Sit-up
30 seconds

STATION #4

Sit-Up
30 seconds

STATION #11

Steam Engine
30 seconds

STATION #5

Ski Jumps
30 seconds

STATION #10

Knee Bender
30 seconds

STATION #6

Flutter Kicks
30 seconds

STATION #9

Bicycle
30 seconds

STATION #8

Wide-Hand Push-Ups
30 seconds

STATION #7

Bend and Reach
(done slowly)
30 seconds

Figure 7-2

SAMPLE CIRCUIT FOR
PUSH-UP AND SIT-UP IMPROVEMENT

STATION #1

Elevated Push-Up
30 seconds

STATION #8

Bicycle
30 seconds

STATION #2

Twisting Sit-Up
30 seconds

Do 1-2 complete rotations.

STATION #7

Close-Hand Push-Up
30 seconds

STATION #3

Parallel Dips
30 seconds

Time may decrease to 20 sec
on the second rotation.

STATION #6

Flutter Kick
30 seconds

STATION #4

Sit-Up
30 seconds

Move immediately from
station to station. If too
fatigued, push-ups may be
done on the knees.

STATION #5

Wide-Hand Push-Up
30 seconds

Figure 7-3

Calisthenics

Calisthenics can be used to help develop coordination. CR and muscular encurance, flexibility, and strength.

Calisthenics can be used to exercise most of the major muscle groups of the body. They can help develop coordination, CR and muscular endurance, flexibility, and strength. Poorly-coordinated soldiers, however, will derive the greatest benefit from many of these exercises

Although calisthenics have some value when included in a CR circuit or when exercising to music, for the average soldier, calisthenics such as the bend and reach, squat bender, lunger, knee bender, and side-straddle hop can best be used in the warm-up and cooldown periods. Exercises such as the push-up, sit-up, parallel bar dip, and chin-up/pull-up, on the other hand, can effectively be used in the conditioning period to develop muscular endurance or muscular strength.

Please note that exercises such as the bend and reach, lunger, and leg spreader, which were once deleted from FM 21-20 because of their potential risk to the exerciser, have been modified and re-introduced in this edition. All modifications should be strictly adhered to.

Few exercises are inherently unsafe. Nonetheless, some people, because of predisposing conditions or injuries, may find certain exercises less safe than others. Leaders must consider each of their soldier's physical limitations and use good judgment before letting a soldier perform these exercises. However, for the average soldier who is of sound body, following the directions written below will produce satisfactory results with a minimum risk of injury.

Finally, some of the calisthenics listed below may be done in cadence. These calisthenics are noted, and directions are provided below with respect to the actions and cadence. When doing exercises at a moderate cadence, use 80 counts per minute. With a slow cadence, use 50 counts per minute unless otherwise directed.

SAFETY FACTORS

While injury is always possible in any vigorous physical activity, few calisthenic exercises are really unsafe or dangerous. The keys to avoiding injury while gaining training benefits are using correct form and intensity. Also, soldiers with low fitness levels, such as trainees, should not do the advanced exercises highly fit soldiers can do. For example, with the lower back properly supported, flutter kicks are an excellent way to condition the hip flexor muscles. However, without support, the possibility of straining the lower back increases. It is not sensible to have recruits do multiple sets of flutter kicks because they probably are not conditioned for them. On the other hand, a conditioned Ranger company may use multiple sets of flutter' kicks with good results.

The key to doing calisthenic exercises safely is to use common sense. Also, ballistic (that is, quick-moving) exercises that combine rotation and bending of the spine increase the risk of back injury and should be avoided. This is especially true if someone has had a previous injury to the back. If this type of action is performed, slow stretching exercises, not conditioning drills done to cadence, should be used.

Some soldiers complain of shoulder problems resulting from rope climbing, horizontal ladder, wheelbarrow, and crab-walk exercises. These exercises are beneficial when the soldier is fit and he does them in a regular, progressive manner. However, a certain level of muscular strength is needed to do them safely. Therefore, soldiers should progressively train to build up to these exercises. Using such exercises for unconditioned soldiers increases the risk of injury and accident.

Progression and Recovery

Other important principles for avoiding injury are progression and recovery. Programs that try to do too much too soon invite problems. The day after a "hard" training day, if soldiers are working the same muscle groups and/or fitness components, they should work them at a reduced intensity to minimize stress and permit recovery.

The best technique is to train alternate muscle groups and/or fitness components on different days. For example, if the Monday-Wednesday-Friday (M-W-F) training objective is CR fitness, soldiers can do ability group running at THR with some light calisthenics and stretching. If the Tuesday-Thursday (T-Th) objective is muscular endurance and strength, soldiers can benefit from doing partner-resisted exercises followed by a slow run. To ensure balance and regularity in the program, the next week should have muscle endurance and strength development on M-W-F and training for CR endurance on T-Th. Such a program has variety, develops all the fitness components, and follows the seven principles of exercise while, at the same time, it minimizes injuries caused by overuse.

Leaders should plan PT sessions to get a positive training effect, not to conduct "gut checks." They should know how to correctly do all the exercises in their program and teach their soldiers to train using good form to help avoid injuries.

Key Points for Safety

Doing safe exercises correctly improves a soldier's fitness with a minimum risk of injury.

The following are key points for ensuring safety during stretching and calisthenic exercises:

- Stretch slowly and without pain and unnatural stress to a joint. Use static (slow and sustained) stretching for warming up, cooling down, ballistic (bouncy or jerky) stretching movements.
- Do not allow the angle formed by the upper and lower legs to become less than 90 degrees when the legs are bearing weight.
- A combination of spinal rotation and bending should generally be avoided. However, if done, use only slow, controlled movements with little or no extra weight.

Leaders must be aware of the variety of methods they may use to attain their physical training goals. The unit's Master Fitness Trainer is schooled to provide safe, effective training methods and answer questions about training techniques.

CALISTHENIC EXERCISES

The following are some common calisthenic exercises.

SIDE-STRADDLE HOP

Position: Assume the position of attention.

Action: (1) Jump slightly into the air while moving the legs more than shoulder-width apart, swinging the arms overhead, and clapping the palms together. (2) Jump slightly into the air while swinging the arms sideward and downward and returning to the position of attention. (3) Repeat action 1. (4) Repeat action 2. Use a moderate cadence.

Variation: (1) Jump slightly into the air while moving the left leg forward and the right leg backward, swinging the arms overhead, and clapping the palms together. (2) Jump slightly into the air while swinging the arms sideward and downward and returning to the position of attention. (3) Repeat the jumping and arm movements of action 1 while moving the right leg forward and the left leg backward. (4) Repeat action 2. Use a moderate cadence.

MULE KICK

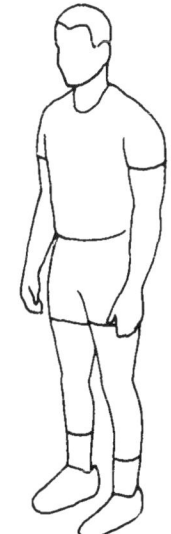

Position: Stand with the feet shoulder-width apart.

Action: Jump up repeatedly while kicking the heels to the buttocks. To do the Mule Kick to cadence, do one repetition per count. Use a moderate cadence.

SKI JUMP

Position: Stand with the feet together, the hands placed behind the head with the fingers interlaced.

Action: (1) Keeping the feet together, jump sideways to the left. (2) Keeping the feet together, jump sideways to the right. (3) Repeat action 1. (4) Repeat action 2. Use a moderate cadence.

FLUTTER KICK

Position: Lie on your back with the hands beneath the buttocks, the head raised, and the knees slightly bent.

Action: Alternately raise and lower the legs, keeping the knees slightly bent and the feet elevated 6 to 18 inches above the floor. To do the flutter kick to cadence, do one repetition per count. Use a moderate cadence.

BEND AND REACH

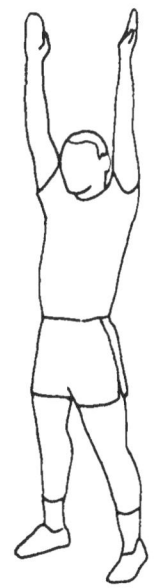

Position: Stand in a wide, side-straddle position with the palms facing each other and the arms overhead and straight.

Action: (1) Bend at the knees and waist. Slowly bring the arms down, and reach between the legs as far as possible. **Make sure the angle formed by the upper and lower leg is never less than 90 degrees.** (2) Recover slowly to the start position. (3) Repeat action 1. (4) Repeat action 2. Use a slow cadence.

HIGH JUMPER

Position: Place the feet about shoulder-width apart with the knees flexed. Bend forward at the waist, aligning the arms with the trunk and hips. Keep the arms straight at all times during the exercise. Keep the palms facing each other with the head and eyes initially to the front.

Action: (1) Take a slight jump into the air while swinging the arms forward and up to shoulder level. (2) Take a slight jump while swinging the arms backward, returning to the start position. (3) Jump strongly upward while swinging the arms forward and up to the overhead position; at the same time, briefly look skyward. While descending, return the head and eyes to the front, and flex the knees. (4) Repeat action 2. Use a moderate cadence.

SQUAT BENDER

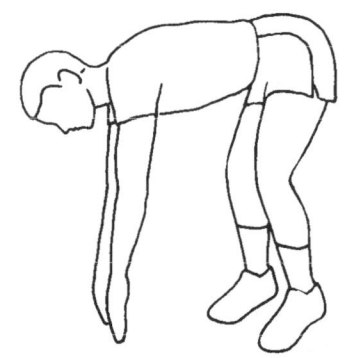

Position: Stand with the feet shoulder-width apart, hands on hips, thumbs in the small of the back, and the elbows back.

Action: (1) Bending the knees, lower yourself to a half-squat position while maintaining balance on the balls of the feet. With the trunk inclined slightly forward, thrust the arms forward to shoulder level with the elbows locked and the palms down. (2) Recover to the start position. (3) Keeping the knees slightly bent, bend forward at the waist, touching the ground in front of the toes. (4) Recover to the start position. Use a moderate cadence.

LUNGER

Position: Start from the position of attention.
Action: (1) Lunge diagonally forward to the left by stepping in that direction with the left foot, placing the left knee over the left foot. At the same time, place the arms sideward at shoulder level, the palms up, and the head and shoulders squarely to the front.

(2) Bend **slowly** forward and downward over the left thigh, and wrap the arms around the thigh, hands grasping the opposite arms above the elbows. (3) Recover **slowly** to the second position by releasing the arms, straightening the trunk, and extending the arms sideward, palms up. (4) Resume the position of attention by dropping the arms and returning the left foot to the side of the right. Repeat the exercise to the right side. Use a moderate cadence.

KNEE BENDER

Position: Stand with the feet shoulder-width apart, hands on the hips, the thumbs in the small of the back, and the elbows back.

Action: Bend at the knees, lean slightly forward at the waist with the head up, and slide the hands along the outside of the legs until the extended fingers reach the top of the boots or the middle of the lower leg. (2) Recover to the start position. (3) Repeat action 1. (4) Repeat action 2. Use a moderate cadence.

THE SWIMMER

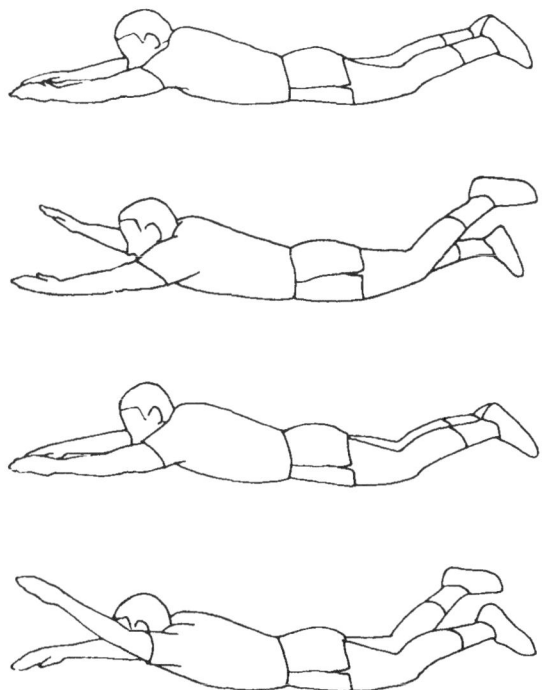

Position: Lie prone with the feet together and with the arms together and extended forward in front of the body. Keep the arms and legs straight at all times during this exercise.

Action: (1) Move the right arm and left leg up. (2) Return to the start position. (3) Move the left arm and right leg up. (4) Return to the start position. Continue in an alternating manner. Use a moderate cadence.

SUPINE BICYCLE

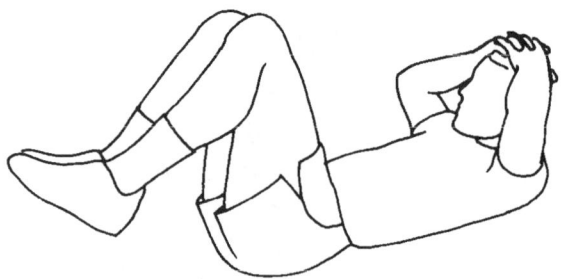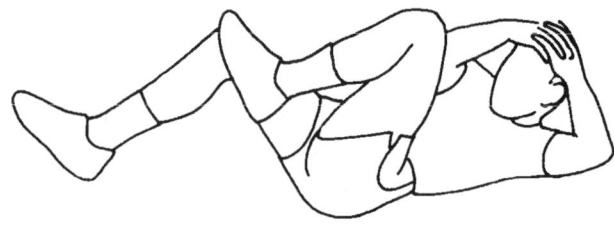

Position: Assume a supine position with the hips and knees flexed. Place the palms directly on top of the head with the fingers interlaced.

Action: (1) Bring the left knee upward while curling the trunk upward, and touch the right elbow to the left knee. (2) Repeat action 1 with the other leg and elbow. (3) Repeat action 1. (4) Repeat action 2. Use a slow cadence.

THE ENGINE

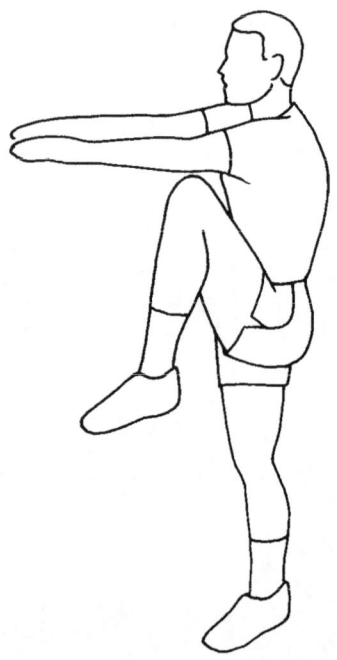

Position: Stand with the arms straight and in front of the body. The arms should be parallel to the ground with the palms facing downward.

Action: (1) Bring the left knee upward to the left elbow. (2) Return to the start position. (3) Touch the right knee to the right elbow. (4) Recover to the start position. Be sure to keep the arms parallel to the ground throughout the entire exercise. Use a moderate cadence.

CROSS-COUNTRY SKIER

Position: Assume a position of attention.

Action: Jump slightly into the air, and move the left foot forward and the right foot backward, landing with both knees slightly bent. At the same time, move the right arm upward and forward to shoulder height and the left arm back as far as possible, always keeping the arms straight and the palms facing each other.

(2) Jump slightly into the air, and move the right foot forward and the left foot backward. At the same time, move the left arm upward and forward to shoulder height and the right arm back as far as possible. (3) Repeat action 1. (4) Repeat action 2. Use a moderate cadence.

PUSH-UP

SHOULDER - WIDTH
HAND POSITION

CLOSE - HAND
POSITION

WIDE - HAND
POSITION

FEET - ELEVATED
POSITION

PUSH - UP ON KNEES

Position: Assume the front-leaning rest position with the hands placed comfortably apart, the feet together or up to 12 inches apart, and the body forming a generally straight line from the shoulders to the ankles.

Action: Keeping the body straight throughout the exercise, lower the body until the upper arms are at least parallel to the ground. Then, push yourself up to the initial position by completely straightening the arms.

Push-Up Variations: To train the muscles more completely, place the hands at varying widths. They may be wider apart or closer together than shoulder width. Elevating the feet to different heights makes push-ups more difficult. The higher the feet, the more difficult the exercise. Push-ups are also more difficult when the hands and feet are placed on boxes or chairs. This helps the soldier exercise through a fuller range of motion. To do extra repetitions when fatigued, drop to the knees while keeping the knees, hips, and shoulders in a straight line.

SIT-UP

Position: Lie on the back with the feet together or up to 12 inches apart, the knees bent so that an angle of 90 degrees is formed by the upper and lower legs, and the fingers interlocked behind the head.

Action: Raise your upper body forward to the vertical position so that the base of the neck is above the base of the spine, then lower yourself in a controlled manner until the bottom of the shoulder blades touch the ground.

Sit-Up Variations: Variations include keeping the feet elevated and crossing the hands on the chest.

CHIN-UP (PULL-UP)

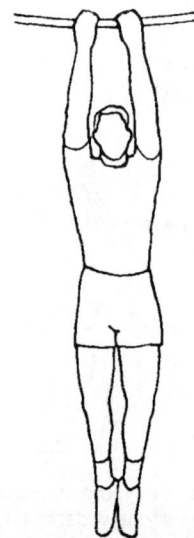

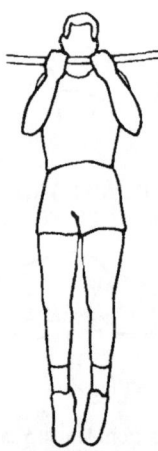

Position: Hang from a horizontal bar with the arms fully extended.
Action: Bend your elbows and pull yourself upward until your chin is above the bar; do not swing or kick your legs. Return to the starting position in a controlled manner.

Variations: Use overhand (pull-up), underhand (chin-up), or alternating grips, with the hands close together, far apart, or at shoulder-width. If unable to complete a chin-up using proper form, elevate yourself to the up position with help and hang there, or slowly lower yourself to the starting position. Repeat this several times, gradually adding more repetitions from workout to workout.

PARALLEL BAR DIP

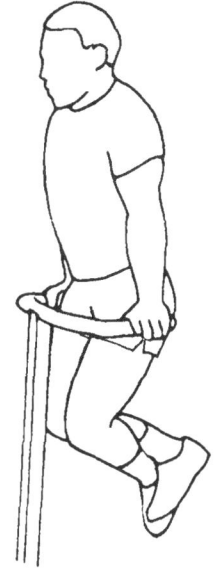

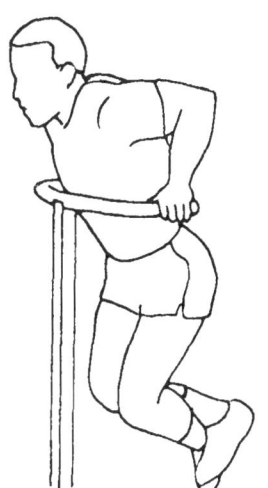

Position: Keep the feet off the floor and support the body's weight on straight arms.

Action: Bend the arms and lower the body in a controlled manner until the upper arms are at least parallel to the floor. If necessary, bend the legs at the knees to keep the feet from touching the floor. Straighten the arms to return to the starting position.

CONDITIONING DRILLS

Conditioning drills are intended to supplement muscular strength and endurance training sessions.

Some large units prefer to use sets of calisthenic exercises as part of their PT sessions. Figure 7-4 shows three calisthenic conditioning drills for both the poorly conditioned and physically fit soldiers. The drills are designed to be done progressively and are intended to supplement muscular strength and endurance training sessions.

Leaders can mix the exercises to provide greater intensity, based on the fitness level of the soldiers being trained. However, they should choose and sequence them to alternate the muscle groups being worked. Soldiers should do each exercise progressively from 15 to 40 or more repetitions (20 to 60 seconds for timed sets) based on their level of conditioning. They may also do each exercise in cadence unless timed sets are specified. For timed sets, soldiers do as many repetitions of an exercise as possible in the allowed time. Using timed sets, both the well-conditioned and less-fit soldiers can work themselves to their limits.

The following conditioning drills (Figure 7-4) are arranged according to the phase of training.

Grass Drills

Grass drills are exercise movements that feature rapid changes in body position. These are vigorous drills which, when properly done, exercise all the major muscle groups. Soldiers should respond to commands as fast as possible and do all movements at top speed. They continue to do multiple repetitions of each exercise until the next command is given. No cadence is counted.

TRAINING-PHASE CONDITIONING DRILLS

#1 PREPARATORY TRAINING

High Jumper
Push-Up (TS* 20-45** seconds)
Sit-Up (TS 20-45** seconds)
Side-Straddle Hop
Side Bender
Knee Bender
Stationary Run

#2 CONDITIONING TRAINING

Push-UP (varied hand positions)
   (TS 30-60 seconds)
Supine Bicycle
High Jumper
Sit-Up (all types)
   (TS 30-60 seconds)
The Engine or Cross-Country Skiier
All-Fours Run (stationary)

#3 MAINTENANCE TRAINING

Ski Jump
Sit-Ups (all types) (TS 30-60 seconds)
Push-Up (varied hand positions) (TS 30-60 seconds)
Mule Kick
Flutter Kick
The Engine
The Swimmer

*TS = timed set

** Because of a lower level of fitness, 45 seconds will usually be the upper limit.

Figure 7-4

Performing grass drills can improve CR endurance, help develop muscular endurance and strength, and speed up reaction time. Since these drills are extremely strenuous, they should last for short periods (30 to 45 seconds per exercise). The two drills described here each have four exercises. Leaders can develop additional drills locally.

The soldiers should do a warm-up before performing the drills and do a cool-down afterward. The instructor does all the activities so that he can gauge the intensity of the session. The commands for grass drills are given in rapid succession without the usual preparatory commands. To prevent confusion, commands are given sharply to distinguish them from comments or words of encouragement.

As soon as the soldiers are familiar with the drill, they do all the exercises as vigorously and rapidly as possible, and they do each exercise until the

Grass drills are exercis[e]
movements that featur[e]
rapid changes in
body position.

Soldiers should do a warm-up before performing grass drills and do a cool-down afterward.

next command is given. Anything less than a top-speed performance decreases the effectiveness of the drills.

Once the drills start, soldiers do not have to resume the position of attention. The instructor uses the command "Up" to halt the drill for instructions or rest. At this command, soldiers assume a relaxed, standing position.

Grass drills can be done in a short time. For example, they may be used when only a few minutes are available for exercise or when combined with another activity. Sometimes, if time is limited, they are a good substitute for running.

Most movements are done in place. The extended-rectangular formation is best for a platoon- or company-sized unit. The circle formation is more suitable for squad- or section-sized groups.

When soldiers are starting an exercise program, a 10- to 15-minute workout may be appropriate. Progression is made by a gradual increase in the time devoted to the drills. As the fitness of the soldiers improves, the times should be gradually lengthened to 20 minutes. The second drill is harder than the first. Therefore, as soldiers progress in the first drill, the instructor should introduce the second. If he sees that the drill needs to be longer, he can repeat the exercises or combine the two drills.

STARTING POSITIONS

After the warm-up, bring the soldiers to a position of ATTENTION. The drills begin with the command *GO*. Other basic commands are FRONT, BACK, and STOP. (See Figure 7-5 for the positions and actions associated with these commands.)

- ATTENTION: The position of at tention is described in FM 22-5, Drill and Ceremonies.

Progression with grass drills is made by a gradual increase in the time devoted to the drills.

- GO This involves running in place at top speed on the balls of the feet. The soldier raises his knees high, pumps his arms, and bends forward slightly at the waist.
- FRONT The soldier lies prone with elbows bent and palms directly under the shoulders as in the down position of the push up. The legs are straight and together with the head toward the instructor. BACK: The soldier lies flat on his back with his arms extended along his sides and his palms facing down ward. His legs are straight and to gether; his feet face the instructor.
- STOP The soldier assumes the stance of a football lineman with feet spread and staggered. His left arm is across his left thigh; his right arm is straight. His knuckles are on the ground; his head is up, and his back is roughly parallel to the ground.

To assume the FRONT or BACK position from the standing GO or STOP positions, the soldier changes positions vigorously and rapidly. (See Figure 7-5.)

To change from the FRONT to the BACK position (Figure 7-5), the soldier does the following:

- Takes several short steps to the right or left.
  Lifts his arm on the side toward which his feet move.
- Thrusts his legs vigorously to the front.

To change from the BACK to the FRONT position, the soldier sits up quickly. He places both hands on the ground to the right or left of his legs. He takes several short steps to the rear on the side opposite his hands. When his feet are opposite his hands, he thrusts his legs vigorously to the rear and lowers his body to the ground. (See Figure 7-5.)

STARTING POSITIONS FOR GRASS DRILLS

GO   FRONT   BACK   STOP

CHANGING FROM FRONT TO BACK

CHANGING FROM BACK TO FRONT

Figure 7-5

GRASS DRILL ONE

Exercises for grass drill one are described below and shown in Figure 7-6.

Bouncing Ball

From the FRONT position, push up and support the body on the hands (shoulder-width apart) and feet. Keep the back and legs generally in line and the knees straight. Bounce up and down in a series of short, simultaneous, upward springs from the hands, hips, and feet.

Supine Bicycle

From the BACK position, flex the hips and knees. Place the palms directly on top of the head, and interlace the fingers. Bring the knee of one leg upward toward the chest. At the same time, curl the trunk and head upward while touching the opposite elbow to the elevated knee. Repeat with the other leg and elbow. Continue these movements as opposite legs and arms take turns.

Knee Bender

From the position of ATTENTION, do half-knee bends with the feet in line and the hands at the sides. Make sure the knees do not bend to an angle less than 90 degrees.

Roll Left and Right

From the FRONT position, continue to roll in the direction commanded until another command is given. Then, return to the FRONT position.

GRASS DRILL TWO

Exercises for grass drill two are described below and shown in Figure 7-6.

The Swimmer

From the FRONT position, extend the arms forward. Move the right arm and left leg up and down; then, move the left arm and right leg up and down. Continue in an alternating manner.

Bounce and Clap Hands

The procedure is almost the same as for the bouncing ball in grass drill one. However, while in the air, clap the hands. This action requires a more vigorous bounce or spring. The push-up may be substituted for this exercise.

Leg Spreader

From the BACK position, raise the legs until the heels are no higher than six inches off the ground. Spread the legs apart as far as possible, then put them back together. Keep the head off the ground. Throughout, place the hands under the upper part of the buttocks, and slightly bend the knees to ease pressure on the lower back. Open and close the legs as fast as possible. The curl-up may be substituted for this exercise.

Forward Roll

From the STOP position, place both hands on the ground, tuck the head, and roll forward. Keep the head tucked while rolling.

Stationary Run

From the position of ATTENTION, start running in place at the GO command by lifting the left foot first. Follow the instructor as he counts two repetitions of cadence. For example, "One, two, three, four; one, two, three, four." The instructor then gives informal commands such as the following: "Follow me," "Run on the toes and balls of your feet," "Speed it up," "Increase to a sprint, raise your knees high, lean forward at your waist, and pump your arms vigorously," and "Slow it down."

To halt the exercise, the instructor counts two repetitions of cadence as the left foot strikes the ground: "One, two, three, four, one, two, three, HALT."

Figure 7-6

Guerilla Exercises

Guerrilla exercises, which can be used to improve agility, CR endurance, muscular endurance, and to some degree muscular strength, combine individual and partner exercises. These drills require soldiers to change their positions quickly and do various basic skills while moving forward. Figures 7-7 and 7-8 show these exercises.

The instructor decides the duration for each exercise by observing its effect on the soldiers. Depending on how vigorously it is done, each exercise should be continued for 20 to 40 seconds.

The group moves in circle formation while doing the exercises. If the platoon exceeds 30 soldiers, concentric circles may be used. A warm-up activity should precede these exercises, and a cool-down should follow them. After the circle is formed, the

Soldiers progress with guerilla exercises by shortening the quick-time marching periods between exercises and by doing all the exercises a second time.

instructor steps into the center and issues commands.

EXERCISE AND PROGRESSION

Soldiers progress by shortening the quick-time marching periods between exercises and by doing all exercises a second time. This produces an *over-*load that improves fitness.

Many soldiers have not had a chance **to** do the simple skills involved in guerrilla exercises. However, they can do these exercises easily and quickly in almost any situation.

The preparatory command is always the name of the exercise, and the command of execution is always "March." The command "Quick time, march" ends each exercise.

For the double guerrilla exercises (in circle formation) involving two soldiers, the commands for pairing are as follows:

GUERILLA EXERCISES

ALL-FOURS RUN

BOTTOMS-UP WALK

CRAB WALK

THE ENGINE

Figure 7-7

7-23

- "Platoon halt."
- "From (soldier is designated), by twos, count off." (For example: 1-2, 1-2, 1-2.)
- "Even numbers, move up behind odd numbers." (Pairs are adjusted according to height and weight.)
- "You are now paired up for double guerrillas." The command "Change" is given to change the soldiers' positions.

After the exercises are completed, the instructor halts the soldiers and positions the base soldier or platoon guide by commanding, "Base man (or platoon guide), post." He then commands "Fall out and fall in on the base man (or platoon guide)."

EXERCISE DESCRIPTIONS

Brief explanations of guerrilla exercises follow.

All-Fours Run

Face downward, supporting the body on the hands and feet. Advance forward as fast as possible by moving the arms and legs forward in a coordinated way.

Bottoms-Up Walk

Take the front-leaning rest position, and move the feet toward the hands in short steps while keeping the knees locked. When the feet are as close to the hands as possible, walk forward on the hands to the front-leaning-rest position.

Crab Walk

Assume a sitting position with the hips off the ground and hands and feet supporting the body's weight. Walk forward, feet first.

The Engine

Stand with the arms straight and in front of the body. The arms should be parallel to the ground with the palms facing downward. While walking forward, bring the left knee upward to the left elbow. Return to the start position. Continuing to walk forward, touch the right knee to the right elbow. Recover to the start position. Be sure to keep the arms parallel to the ground throughout the entire exercise.

Double Time

Do a double-time run while maintaining the circle formation.

Broad Jump

Jump forward on both feet in a series of broad jumps. Swing the arms vigorously to help with the jumps.

Straddle Run

Run forward, leaping to the right with the left foot and to the left with the right foot.

Hobble Hopping

Hold one foot behind the back with the opposite hand and hop forward. On the command "Change," grasp the opposite foot with the opposite hand and hop forward.

Two-Man Carry

For two-man carries, soldiers are designated as number one (odd-numbered) and number two (even-numbered). A number-one and number-two soldier work as partners.

Fireman's Carry

Two soldiers do the carry. On command, number-two soldier bends at the waist, with feet apart in a balanced stance. Number-one soldier moves toward his partner. He places himself by his partner's left shoulder and bends himself over his partner's shoulders and back. When in position, number-two soldier, with his left hand, reaches between his partner's legs and grasps his left wrist. On command, they move forward until the command for changeover. They then change positions. The fireman's carry can also be done from the other side.

Single-Shoulder Carry

Two soldiers do the carry. On command, number-two soldier bends at the waist with feet apart in a balanced stance. At the same time, number-one soldier moves toward his partner. He places his abdominal area onto his partner's right or left shoulder and leans over. Number-two soldier puts his arms around the back of his partner's knees and stands up. On command, they move forward until the command for changeover. They then change positions.

Cross Carry

On command, number-two soldier bends over at the waist. He twists slightly to the left with feet spread apart in a balanced position. At the same time, number-one soldier moves toward his partner's left side and leans over his partner's back. Number two soldier, with his left arm, reaches around his partner's legs. At the same time, he reaches around his partner's back with his right arm, being careful not to grab his partner's neck or head. He then stands up straight, holding his partner on his back. On command, they move forward until the command for changeover. They then change positions.

Saddle-Back (Piggyback) Carry

On command, number-two soldier bends at the waist and knees with his hand on his knees and his head up. To assume the piggyback position, number-one soldier moves behind his partner, places his hands on his partner's shoulders, and climbs carefully onto his partner's hips. As number-one soldier climbs on, number-two soldier grasps his partner's legs to help support him. Number-one soldier places his arms over his partner's shoulders and crosses his hands over his partner's upper chest. They move forward until the command for changeover is given. They then change positions.

ADDITIONAL GUERILLA EXERCISES

DOUBLE TIME

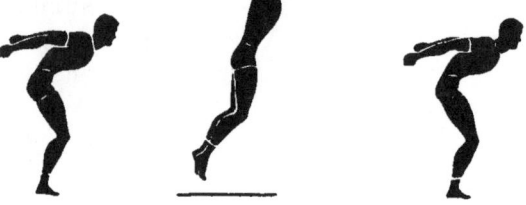

BROAD JUMP

STRADDLE RUN

HOBBLE HOPPING

FIREMAN'S CARRY

SINGLE-SHOULDER
CARRY

CROSS CARRY

SADDLE-BACK CARRY

Figure 7-8

CHAPTER 8

ESSENTIAL ARMY MANUAL

(673) PHYSICAL FITNESS TRAINING

Obstacle Courses and Additional Drills

This chapter describes obstacle courses as well as rifle drills, log drills, and aquatic exercises. These are not designed to develop specific components of physical fitness. Commanders should use them to add variety to their PT programs and to help soldiers develop motor fitness including speed, agility, coordination, and related skills and abilities. Many of these activities also give soldiers the chance to plan strategy, make split-second decisions, learn teamwork, and demonstrate leadership.

Obstacle Courses

There are two types of obstacle courses-- conditioning and confidence.

Physical performance and success in combat may depend on a soldier's ability to perform skills like those required on the obstacle course. For this reason, and because they help develop and test basic motor skills, obstacle courses are valuable for physical training.

There are two types of obstacle courses--conditioning and confidence. The conditioning course has low obstacles that must be negotiated quickly. Running the course can be a test of the soldier's basic motor skills and physical condition. After soldiers *receive* instruction and practice the skills, they run the course against time.

A confidence course has higher, more difficult obstacles than a conditioning course. It gives soldiers confidence in their mental and physical abilities and cultivates their spirit of daring. Soldiers are encouraged, but not forced, to go through it. Unlike conditioning courses, confidence courses are not run against time.

NONSTANDARD COURSES AND OBSTACLES

Commanders may build obstacles and courses that are nonstandard (that is, not covered in this manual) in order to create training situations based on their unit's METL.

When planning and building such facilities, designers should, at a minimum, consider the following guidance:

- Secure approval from the local installation's commander.
- Prepare a safety and health-risk assessment to support construction of each obstacle.
- Coordinate approval for each obstacle with the local or supporting safety office. Keep a copy of the approval in the permanent records.
- Monitor and analyze all injuries.
- Inspect all existing safety precautions on-site to verify their effectiveness.
- Review each obstacle to determine the need for renewing its approval.

SAFETY PRECAUTIONS

Instructors must always be alert to safety. They must take every precaution to minimize injuries as soldiers go through obstacle courses. Soldiers must do warm-up exercises before they begin. This prepares them for the physically demanding tasks ahead and helps minimize the chance of injury. A cool-down after the obstacle course is also necessary, as it helps the body recover from strenuous exercise.

Commanders should use ingenuity in building courses, making good use of streams, hills, trees, rocks, and other natural obstacles. They must inspect courses for badly built obstacles, protruding nails, rotten logs, unsafe landing pits, and other safety hazards.

There are steps which designers can take to reduce injuries. For example, at the approach to each obstacle, they should post an instruction board or sign with text and pictures showing how to negotiate it. Landing pits for jumps or vaults, and areas under or around obstacles where soldiers may fall from a height, should be filled with loose sand or sawdust, All landing areas should be raked and refilled before each use. Puddles of water under obstacles can cause a false sense of security. These could result in improper landing techniques and serious injuries. Leaders should postpone training on obstacle courses when wet weather makes them slippery.

Units should prepare their soldiers to negotiate obstacle courses by doing conditioning exercises beforehand. Soldiers should attain an adequate level of conditioning before they run the confidence course, Soldiers who have not practiced the basic skills or run the conditioning course should not be allowed to use the confidence course.

Instructors must explain and demonstrate the correct ways to negotiate all obstacles before allowing soldiers to run them. Assistant instructors should supervise the negotiation of higher, more dangerous obstacles. The emphasis is on avoiding injury. Soldiers should practice each obstacle until they are able to negotiate it. Before they run the course against time, they should make several slow runs while the instructor watches and makes needed corrections. Soldiers should never be allowed to run the course against time until they have practiced on all the obstacles.

CONDITIONING OBSTACLE COURSES

If possible, an obstacle course should be shaped like a horseshoe or figure eight so that the finish is close to the start. Also, signs should be placed to show the route.

A course usually ranges from 300 to 450 yards and has 15 to 25 obstacles that are 20 to 30 yards apart. The obstacles are arranged so that those which exercise the same groups of muscles are separated from one another.

The obstacles must be solidly built. Peeled logs that are six to eight inches

wide are ideal for most of them. Sharp points and corners should be eliminated, and landing pits for jumps or vaults must be filled with sand or sawdust. Courses should be built and marked so that soldiers cannot sidestep obstacles or detour around them. Sometimes, however, courses can provide alternate obstacles that vary in difficulty.

Each course should be wide enough for six to eight soldiers to use at the same time, thus encouraging competition. The lanes for the first few obstacles should be wider and the obstacles easier than those that follow. In this way, congestion is avoided and soldiers can spread out on the course. To minimize the possibility of falls and injuries due to fatigue, the last two or three obstacles should not be too difficult or involve high climbing.

Trainers must always be aware that falls from the high obstacles could cause serious injury. Soldiers must be in proper physical condition, closely supervised, and adequately instructed.

The best way for the timer to time the runners is to stand at the finish and call out the minutes and seconds as each soldier finishes. If several watches are available, each wave of soldiers is timed separately. If only one watch is available, the waves are started at regular intervals such as every 30 seconds. If a soldier fails to negotiate an obstacle, a previously determined penalty is imposed.

When the course is run against time, stopwatches, pens, and a unit roster are needed. Soldiers may run the course with or without individual equipment.

Obstacles for Jumping

These obstacles are ditches to clear with one leap, trenches to jump into, heights to jump from, or hurdles. (See Figure 8-1.)

Instructors must explain and demonstrate the correct ways to negotiate all obstacles before allowing soldiers to run them.

Obstacles for Dodging

These obstacles are usually mazes of posts set in the ground at irregular intervals. (See Figure 8-2.) The spaces between the posts are narrow so that soldiers must pick their way carefully through and around them. Lane guides are built to guide soldiers in dodging and changing direction.

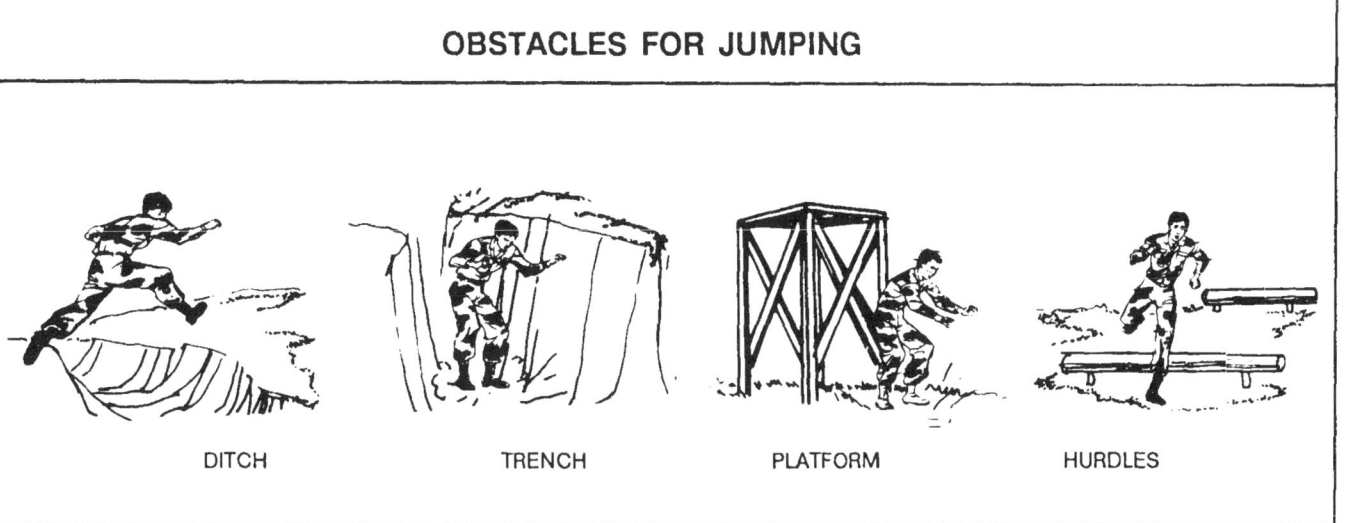

OBSTACLES FOR JUMPING

DITCH   TRENCH   PLATFORM   HURDLES

Figure 8-1

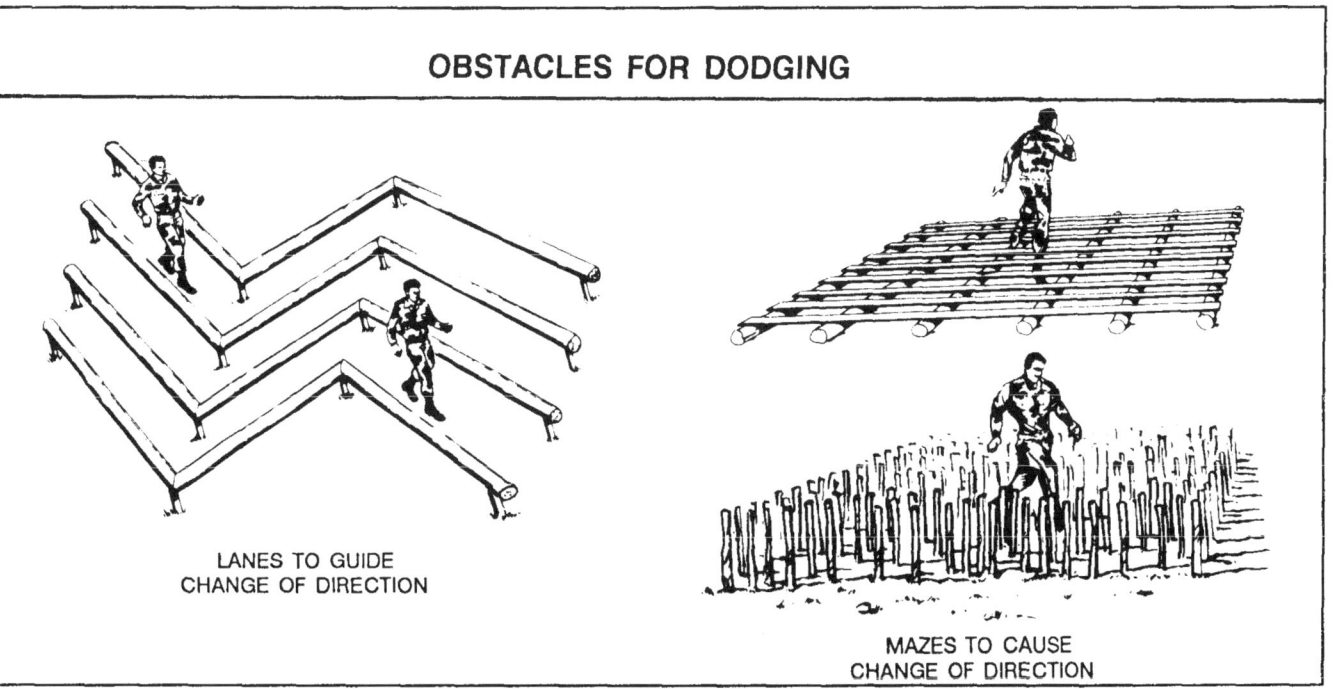

OBSTACLES FOR DODGING

LANES TO GUIDE
CHANGE OF DIRECTION

MAZES TO CAUSE
CHANGE OF DIRECTION

Figure 8-2

Obstacles for Vertical Climbing and Surmounting

These obstacles are shown at Figure 8-3 and include the following:
- Climbing ropes that are 1 1/2 inches wide and either straight or knotted.
- Cargo nets.
- Walls 7 or 8 feet high.
- Vertical poles 15 feet high and 6 to **8** inches wide.

Obstacles for Horizontal Traversing

Horizontal obstacles may be ropes, pipes, or beams. (See Figure 8-4.)

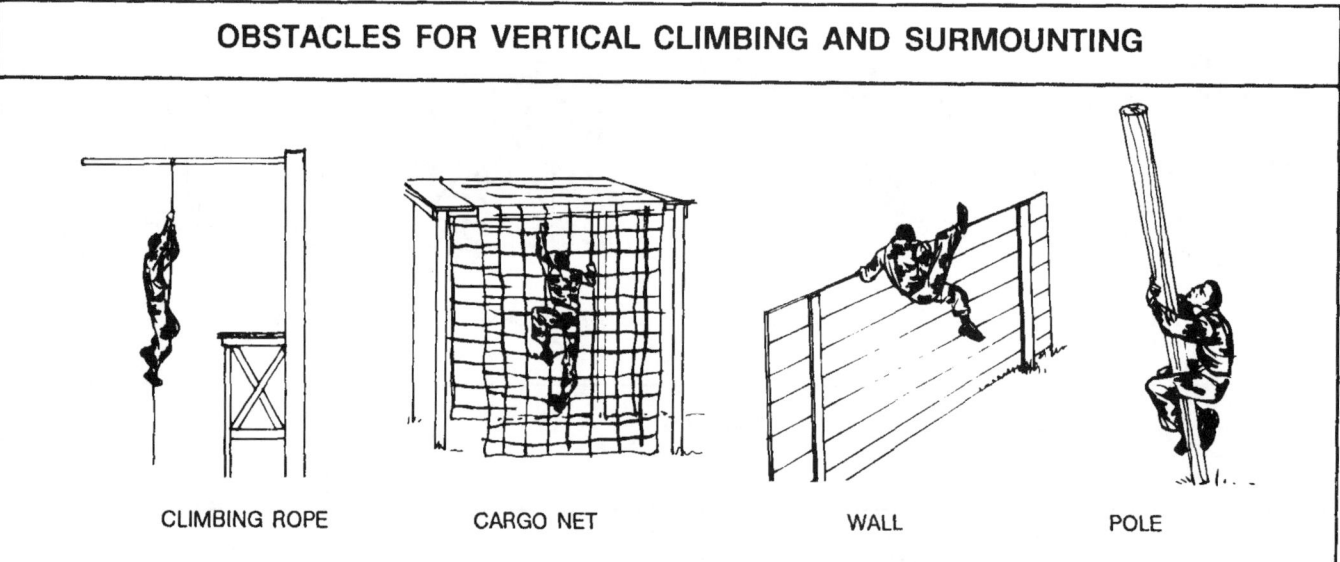

OBSTACLES FOR VERTICAL CLIMBING AND SURMOUNTING

CLIMBING ROPE   CARGO NET   WALL   POLE

Figure 8-3

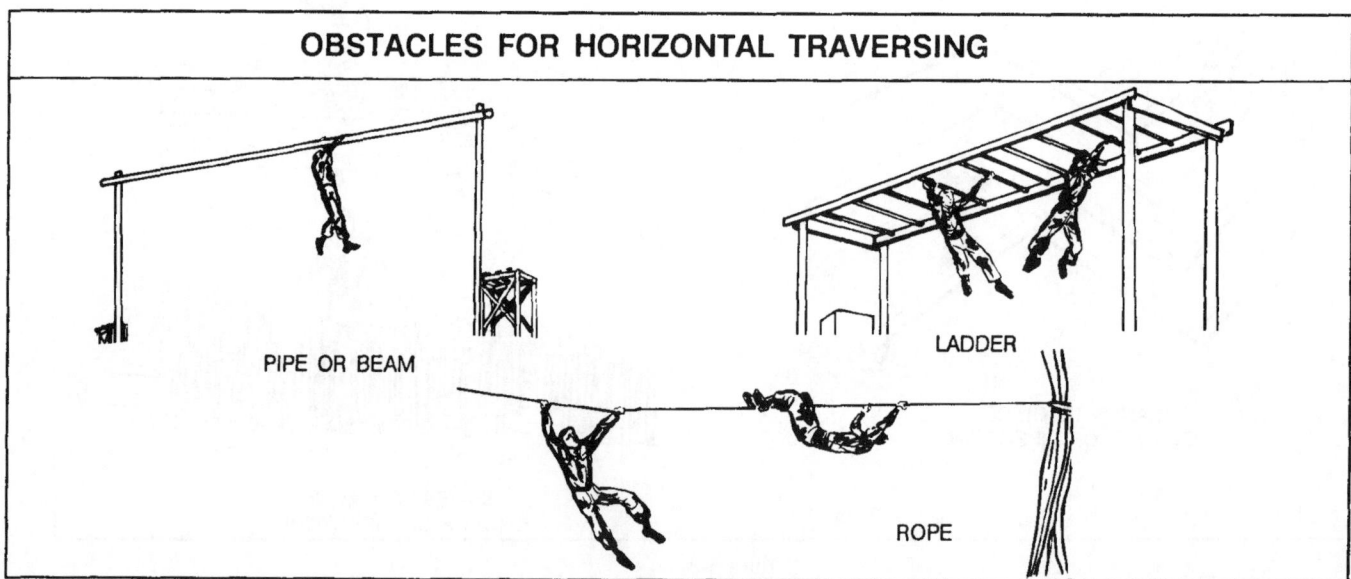

OBSTACLES FOR HORIZONTAL TRAVERSING

PIPE OR BEAM   LADDER   ROPE

Figure 8-4

Obstacles for Crawling

These obstacles may be built of large pipe sections, low rails, or wire. (See Figure 8-5.)

Obstacles for Vaulting

These obstacles should be 3 to 3 1/2 feet high. Examples are fences and low walls. (See Figure 8-6.)

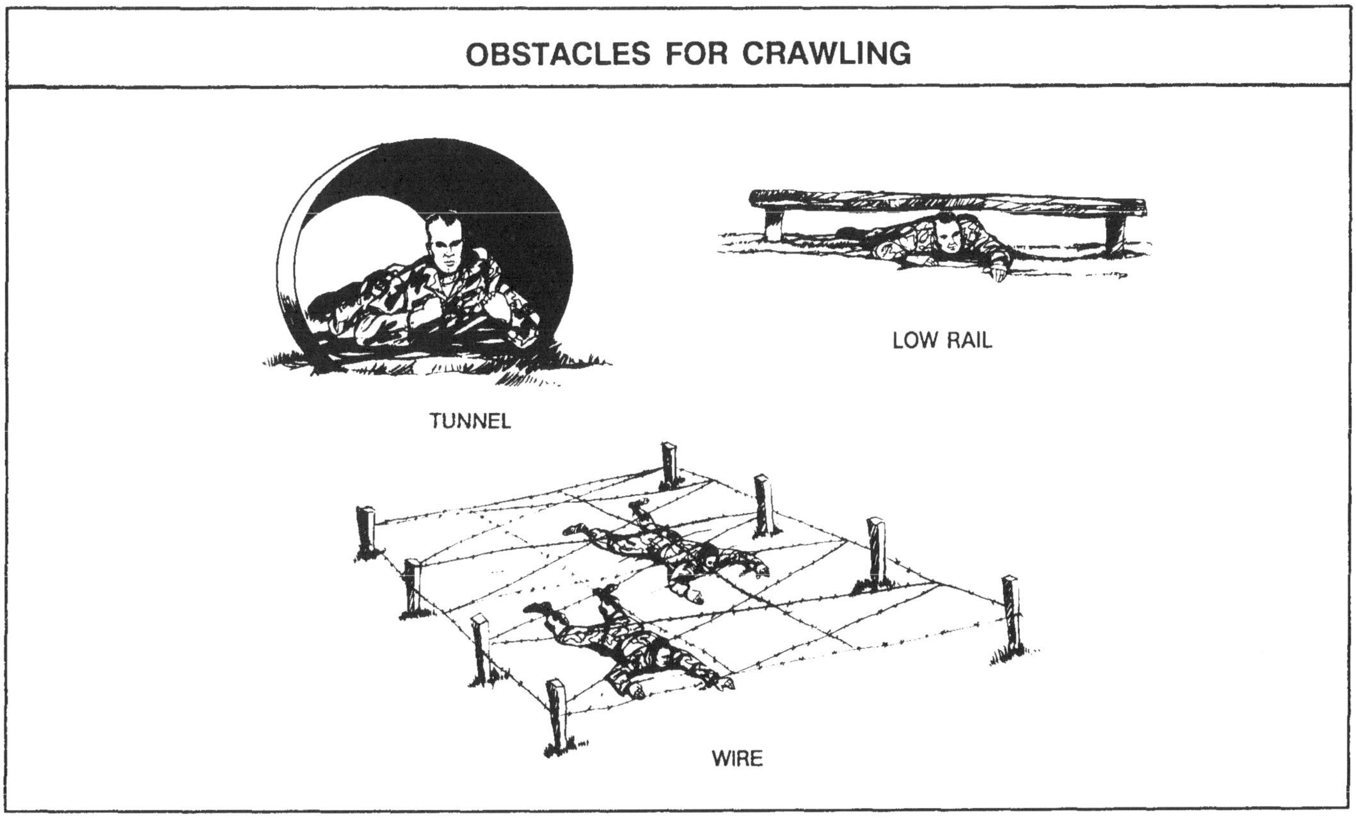

OBSTACLES FOR CRAWLING

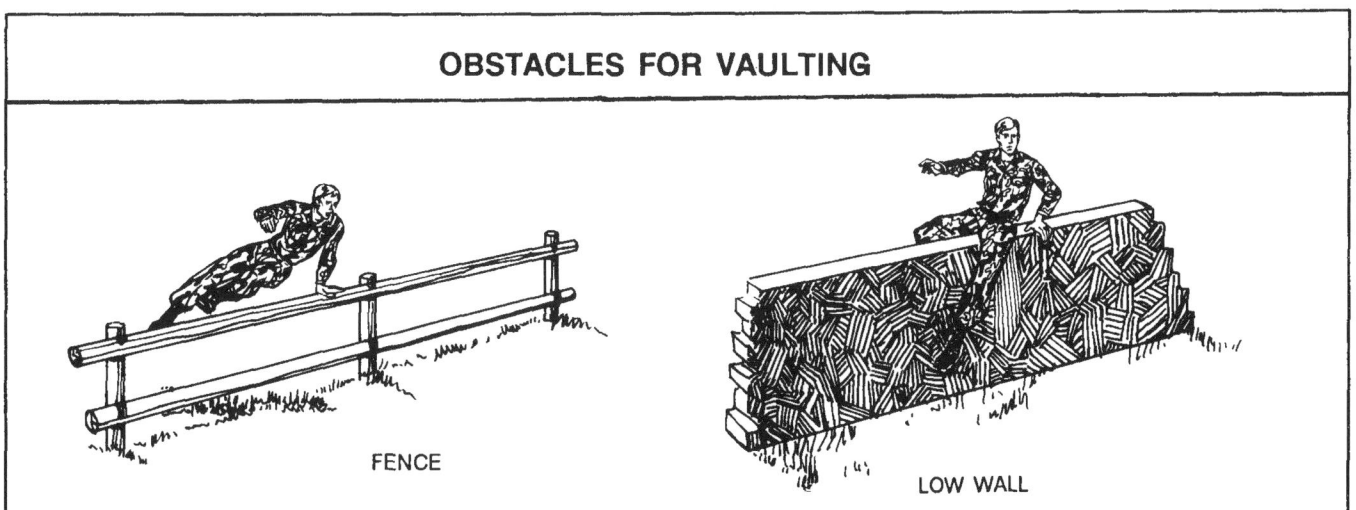

TUNNEL

LOW RAIL

WIRE

Figure 8-5

OBSTACLES FOR VAULTING

FENCE

LOW WALL

Figure 8-6

Obstacles for Balancing

Beams, logs, and planks may be used. These may span water obstacles and dry ditches, or they may be raised off the ground to simulate natural depressions. (See Figure 8-7.)

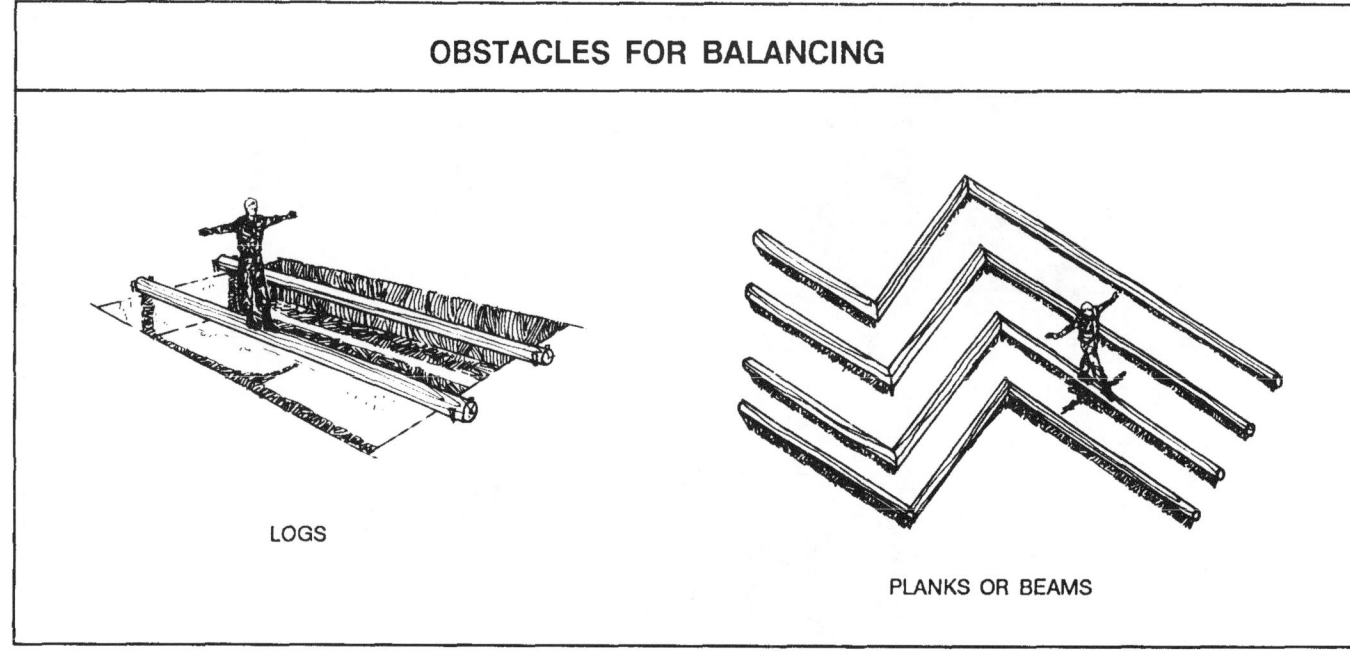

OBSTACLES FOR BALANCING

LOGS

PLANKS OR BEAMS

Figure 8-7

CONFIDENCE OBSTACLE COURSES

Confidence obstacle courses must be built in accordance with Folio No. 1, "Training Facilities," Corps of Engineers Drawing Number 28-13-95. You can obtain this publication from the Directorate of Facilities Engineering at most Army installations.

Confidence courses can develop confidence and strength by using obstacles that train and test balance and muscular strength. Soldiers do not negotiate these obstacles at high speed or against time. The obstacles vary from fairly easy to difficult, and some are high. For these, safety nets are provided. Soldiers progress through the course without individual equipment. Only one soldier at a time negotiates an obstacle unless it is designed for use by more than one.

Confidence courses should accommodate four platoons, one at each group of six obstacles. Each platoon begins at a different starting point. In the example below, colors are used to group the obstacles. Any similar method may be used to spread a group over the course. Soldiers are separated into groups of 8 to 12 at each obstacle. At the starting signal, they proceed through the course.

Soldiers may skip any obstacle they are unwilling to try. Instructors should encourage fearful soldiers to try the easier obstacles first. Gradually, as their confidence improves, they can take their places in the normal rotation. Soldiers proceed from one obstacle to the next until time is called. They then assemble and move to the next group of obstacles.

Rules for the Course

Supervisors should encourage, but not force, soldiers to try every obstacle. Soldiers who have not run the course before should receive a brief orientation at each obstacle, including an explanation and demonstration of the best way to negotiate it. Instructors should help those who have problems. Trainers and soldiers should not try to make obstacles more difficult by shaking ropes, rolling logs, and so forth. Close supervision and common sense must be constantly used to enhance safety and prevent injuries.

Soldiers need not conform to any one method of negotiating obstacles, but there is a uniformity in the general approach. Recommended ways to negotiate obstacles are described below.

Red Group

This group contains the first six obstacles. These are described below and numbered 1 through 6 in Figure 8-8.
Belly Buster. Soldiers vault, jump, or climb over the log. They must be warned that it is not stationary. Therefore, they should not roll or rock the log while others are negotiating it.
Reverse Climb. Soldiers climb the reverse incline and go down the other side to the ground.
Weaver. Soldiers move from one end of the obstacle to the other by weaving their bodies under one bar and over the next.
Hip-Hip. Soldiers step over each bar; they either alternate legs or use the same lead leg each time.
Balancing Logs. Soldiers step up on a log and walk or run along it while keeping their balance.
Island Hopper. Soldiers jump from one log to another until the obstacle is negotiated.

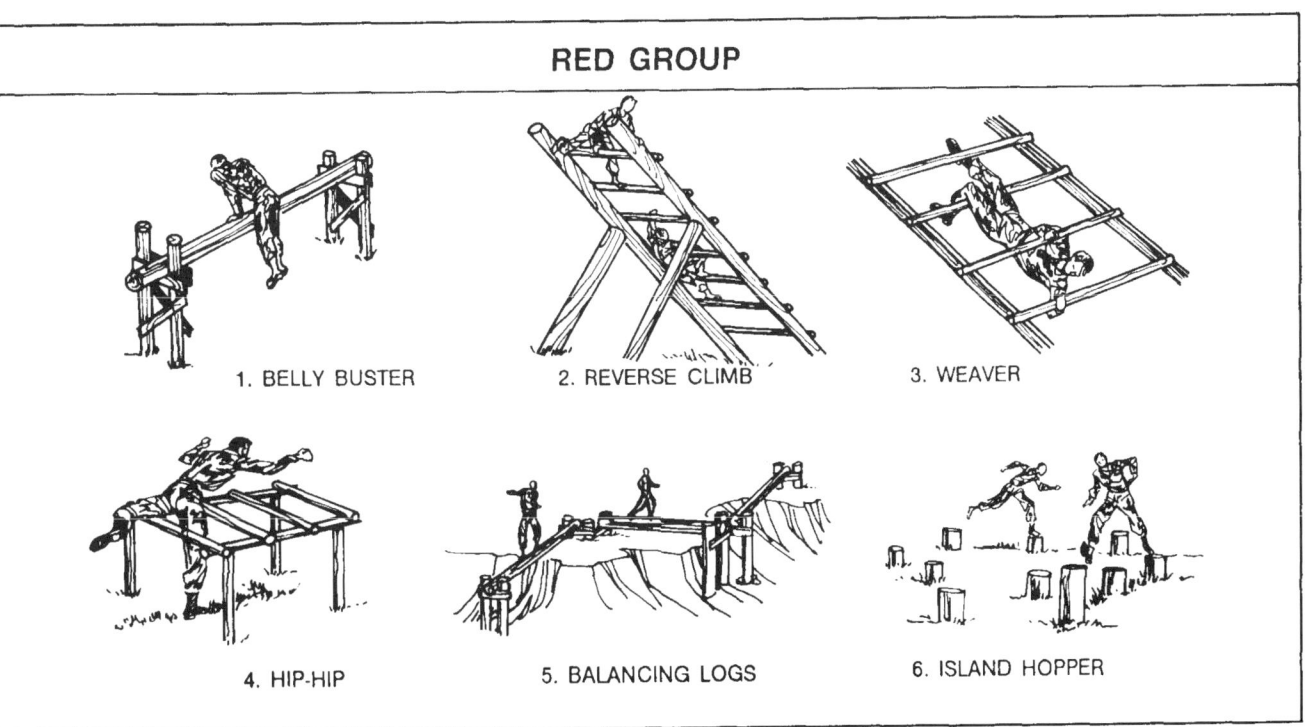

RED GROUP

1. BELLY BUSTER
2. REVERSE CLIMB
3. WEAVER
4. HIP-HIP
5. BALANCING LOGS
6. ISLAND HOPPER

Figure 8-8

White Group

This group contains the second six obstacles. These are described below and numbered 7 through 12 in Figure 8-9.

Tough Nut. Soldiers step over each X in the lane.

Inverted Rope Descent. Soldiers climb the tower, grasp the rope firmly, and swing their legs upward. They hold the rope with their legs to distribute the weight between their legs and arms. Braking the slide with their feet and legs, they proceed down the rope. Soldiers must be warned that they may get rope burns on their hands. This obstacle can be dangerous when the rope is slippery. Soldiers leave the rope at a clearly marked point of release. Only one soldier at a time is allowed on the rope. Soldiers should not shake or bounce the ropes. This obstacle requires two instructors--one on the platform and the other at the base.

Low Belly-Over. Soldiers mount the low log and jump onto the high log.

They grasp over the top of the log with both arms, keeping the belly area in contact with it. They swing their legs over the log and lower themselves to the ground.

Belly Crawl. Soldiers move forward under the wire on their bellies to the end of the obstacle. To reduce the tendency to push the crawling surface, it is filled with sand or sawdust to the far end of the obstacle. The direction of negotiating the crawl is reversed from time to time.

Easy Balancer. Soldiers walk up one inclined log and down the one on the other side to the ground.

Tarzan. Soldiers mount the lowest log, walk the length of it, then each higher log until they reach the horizontal ladder. They grasp two rungs of the ladder and swing themselves into the air. They negotiate the length of the ladder by releasing one hand at a time and swinging forward, grasping a more distant rung each time.

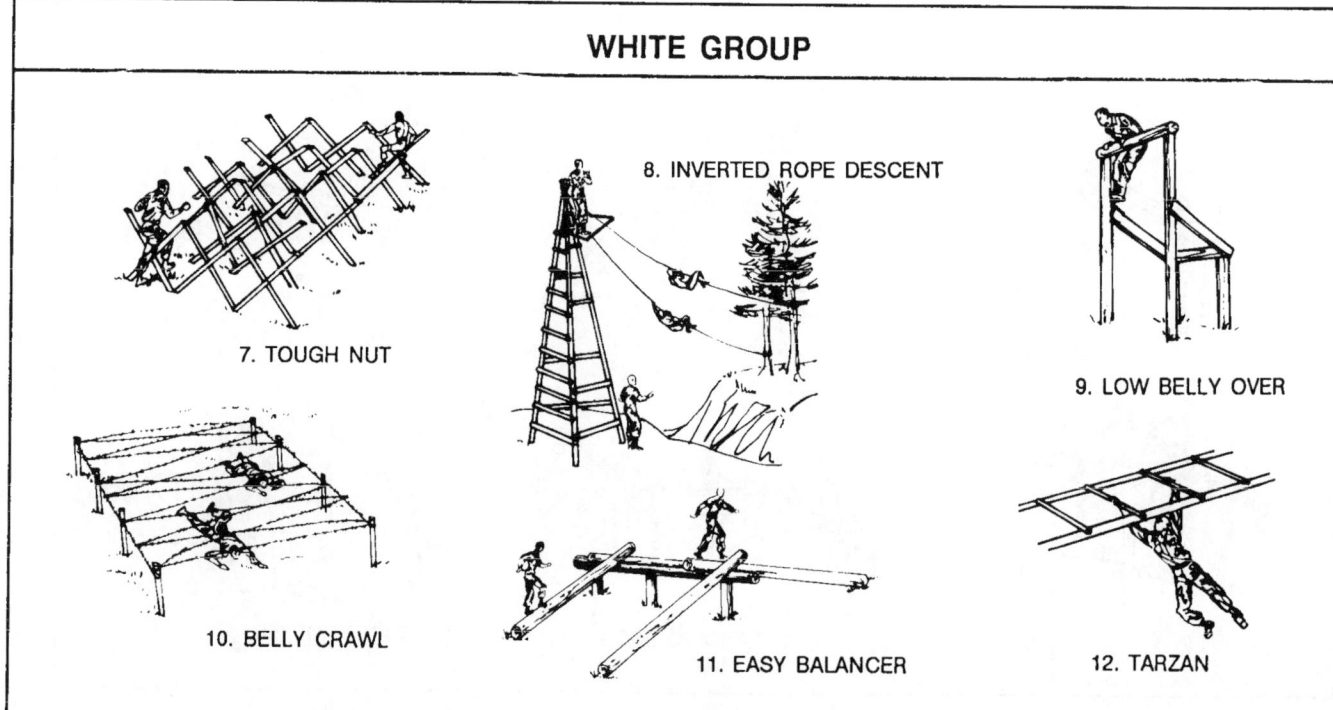

WHITE GROUP

7. TOUGH NUT

8. INVERTED ROPE DESCENT

9. LOW BELLY OVER

10. BELLY CRAWL

11. EASY BALANCER

12. TARZAN

Figure 8-9

Blue Group

This group contains the third six obstacles. These are described below and numbered 13 through 18 in Figure 8-10.

High **Step-over.** **Soldiers step over** each log while alternating their lead foot or using the same one.

Swinger. Soldiers climb over the swing log to the ground on the opposite side.

Low Wire. Soldiers move under the wire on their backs while raising the wire with their hands to clear their bodies. To reduce the tendency to push the crawling surface, it is filled with sand or sawdust to the far end of the obstacle. The direction of negotiating the obstacle is alternated.

Swing, Stop, and Jump. Soldiers gain momentum with a short run, grasp the rope, and swing their bodies forward to the top of the wall. They release the rope while standing on the wall and jump to the ground.

Six Vaults. Soldiers vault over the logs using one or both hands.

Wall Hanger. Soldiers walk up the wall using the rope. From the top of the wall, they grasp the bar and go hand-over-hand to the rope on the opposite end. They use the rope to descend,

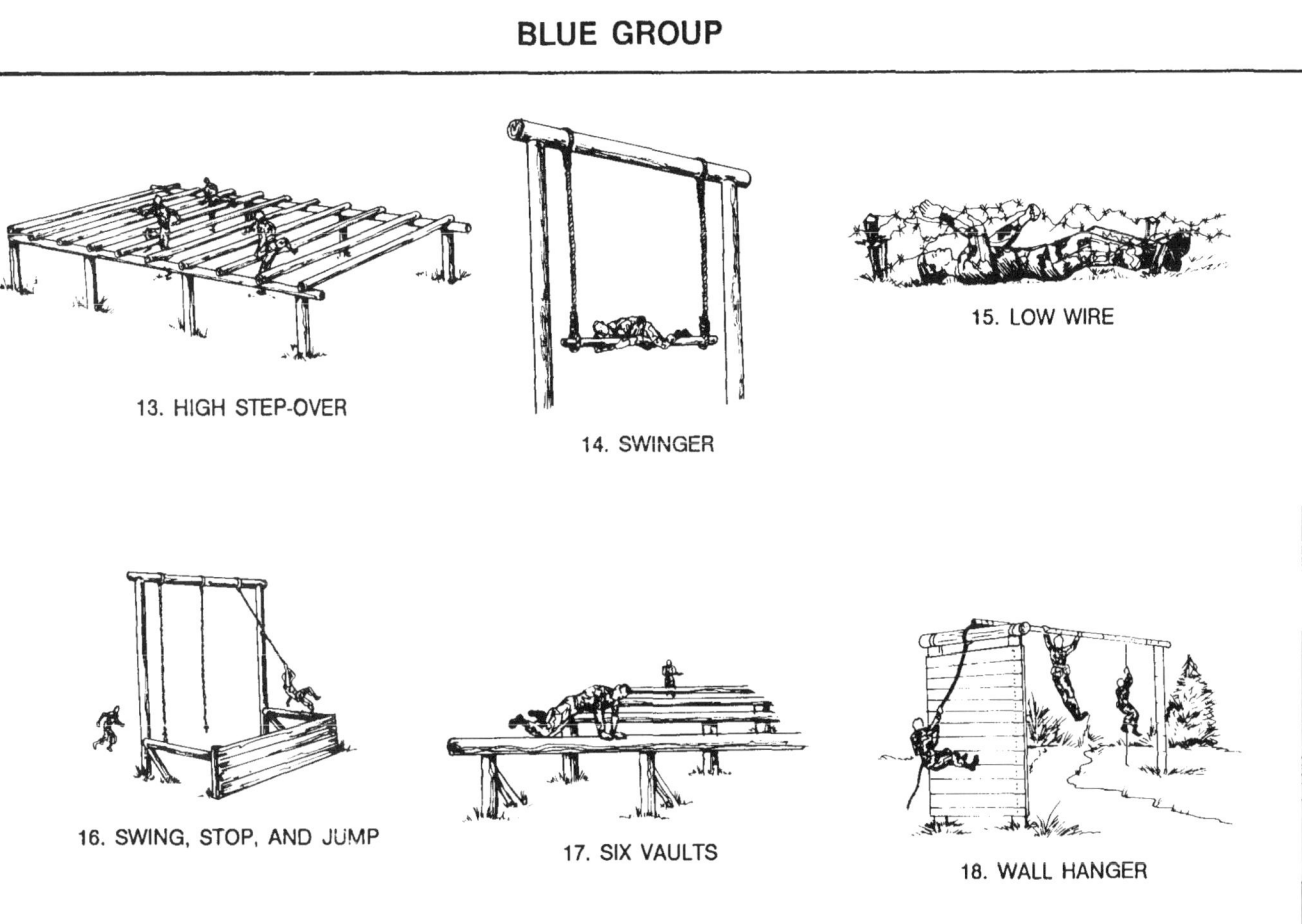

BLUE GROUP

13. HIGH STEP-OVER

14. SWINGER

15. LOW WIRE

16. SWING, STOP, AND JUMP

17. SIX VAULTS

18. WALL HANGER

Figure 8-10

Black Group

This group contains the last six obstacles. These are described below and numbered 19 through 24 in Figure 8-11.

Inclining Wall. Soldiers approach the underside of the wall, jump up and grasp the top, and pull themselves up and over. They slide or jump down the incline to the ground.

Skyscraper. Soldiers jump or climb to the first floor and either climb the corner posts or help one another to the higher floors. They descend to the ground individually or help one another down. The top level or roof is off limits, and the obstacle should not be overloaded. A floor must not become so crowded that soldiers are bumped off. Soldiers should not jump to the ground from above the first level.

Jump and Land. Soldiers climb the ladder to the platform and jump to the ground.

Confidence Climb. Soldiers climb the inclined ladder to the vertical ladder. they go to the top of the vertical ladder, then down the other side to the ground.

Belly Robber. Soldiers step on the lower log and take a prone position on the horizontal logs. They crawl over the logs to the opposite end of the obstacle. Rope gaskets must be tied to the ends of each log to keep the hands from being pinched and the logs from falling.

The Tough One. Soldiers climb the rope or pole on the lowest end of the obstacle. They go over or between the logs at the top of the rope. They move across the log walkway, climb the ladder to the high end, then climb down the cargo net to the ground.

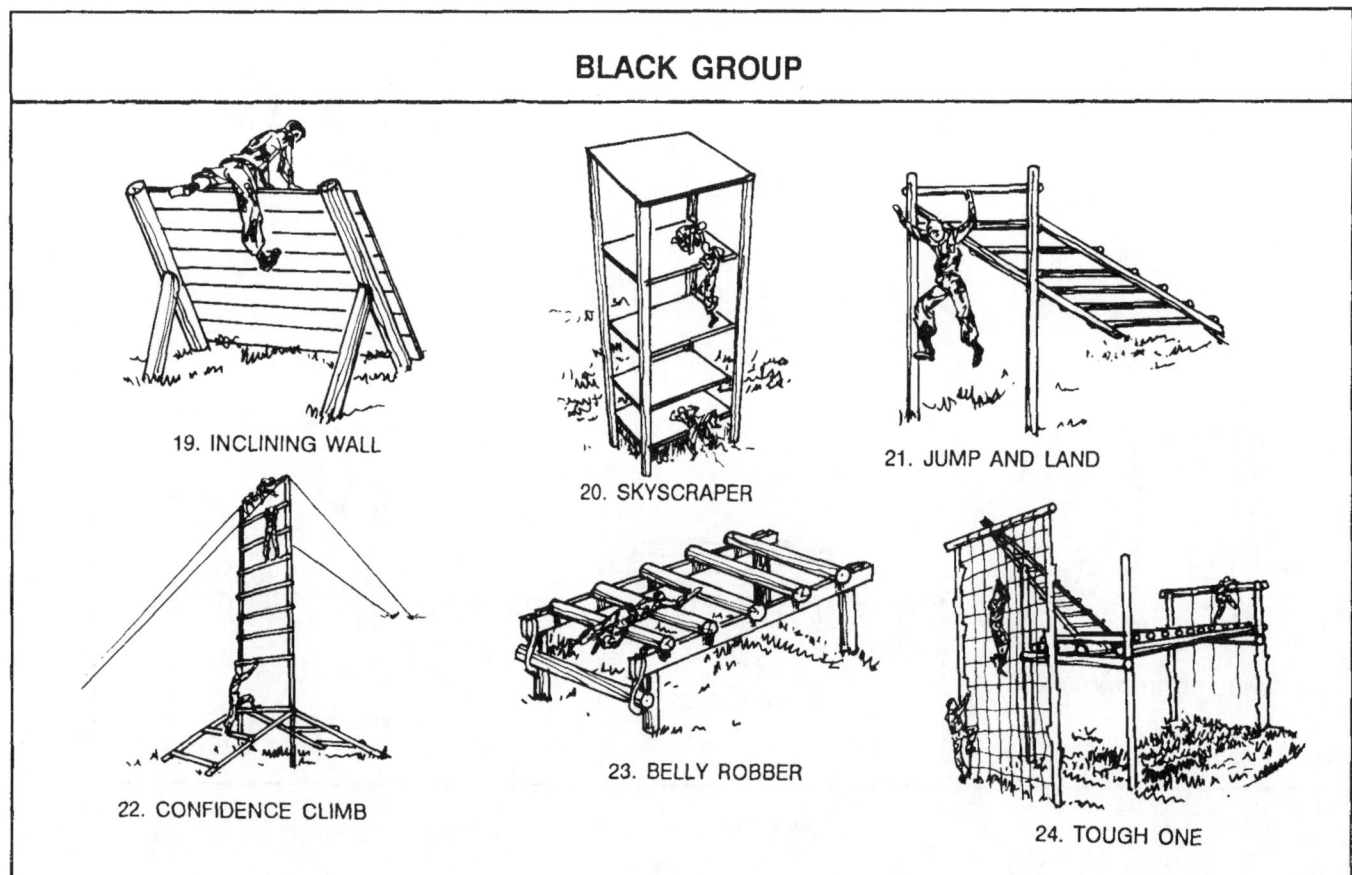

BLACK GROUP

19. INCLINING WALL

20. SKYSCRAPER

21. JUMP AND LAND

22. CONFIDENCE CLIMB

23. BELLY ROBBER

24. TOUGH ONE

Figure 8-11

Rifle Drills

Rifle drills are suitable activities for fitness training while bivouacking or during extended time in the field. In most situations, the time consumed in drawing weapons makes this activity cumbersome for garrison use. However, it is a good conditioning activity, and the use of individual weapons in training fosters a warrior's spirit.

There are four rifle-drill exercises that develop the upper body. They are numbered in a set pattern. The main muscle groups strengthened by rifle drills are those of the arms, shoulders, and back.

Rifle drill is a fast-moving method of exercising that soldiers can do in as little as 15 minutes. With imagination, the number of steps and/or rifle exercises can be expanded beyond those described here.

EXERCISE PROGRESSION

The rifle-drill exercise normally begins with six repetitions and increases by one repetition for each three periods of exercise. This rate continues until soldiers can do 12 repetitions. However, the number of repetitions can be adjusted as the soldiers improve.

In exercises that start from the rifle-downward position, on the command "Move," soldiers execute port arms and assume the starting position. At the end of the exercise, the command to return soldiers to attention is "Position of attention, move."

In exercises that end in other than the rifle-downward position, soldiers assume that position before executing port arms and order arms.

These movements are done without command and need not be precise. Effective rifle exercises are strenuous enough to tire the arms. When the arms are tired, moving them with precision is difficult.

RIFLE DRILL EXERCISES

The following exercises are for use in rifle drills.

Up and Forward

This is a four-count exercise done at a fast cadence. (See Figure 8-12.)

Fore-Up, Squat

This is a four-count exercise done at a moderate cadence. (See Figure 8-13.)

Fore-Up, Behind Back

This is a four-count exercise done at a moderate cadence. (See Figure 8-14.)

Fore-Up, Back Bend

This is a four-count exercise done at moderate cadence. (See Figure 8-15.)

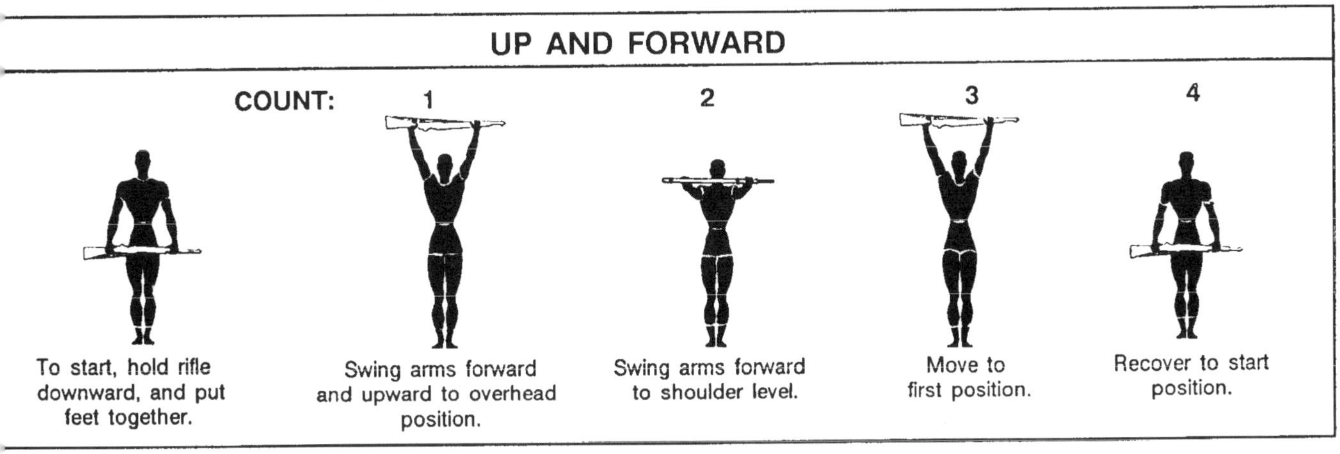

UP AND FORWARD

COUNT: 1 2 3 4

To start, hold rifle downward, and put feet together.

Swing arms forward and upward to overhead position.

Swing arms forward to shoulder level.

Move to first position.

Recover to start position.

Figure 8-12

FORE-UP, SQUAT

COUNT:	1	2	3	4
To start, hold rifle downward, and put feet about shoulder width apart.	Swing arms forward and upward to overhead position.	Swing arms down to shoulder level, and assume half-knee-bend position.	Move to first position.	Recover to start position.

Figure 8-13

FORE UP, BEHIND BACK

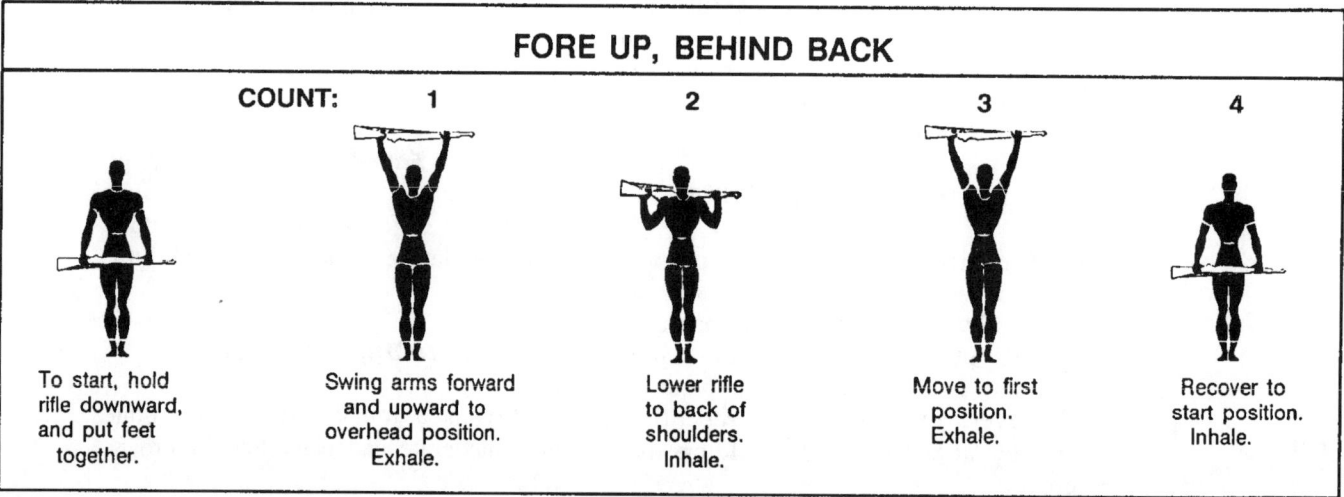

COUNT:	1	2	3	4
To start, hold rifle downward, and put feet together.	Swing arms forward and upward to overhead position. Exhale.	Lower rifle to back of shoulders. Inhale.	Move to first position. Exhale.	Recover to start position. Inhale.

Figure 8-14

FORE-UP, BACK BEND

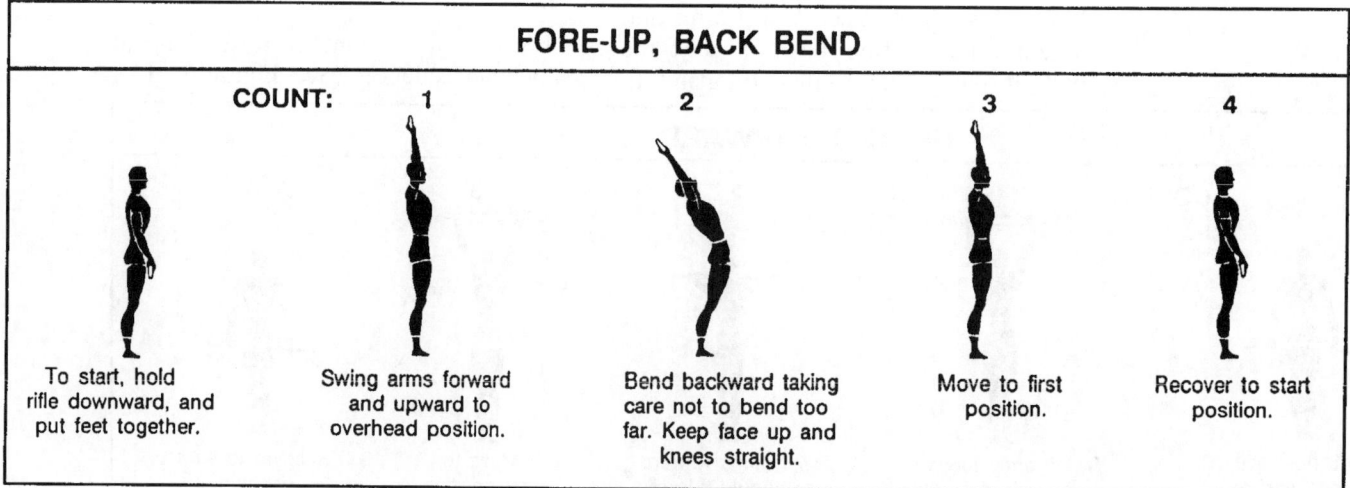

COUNT:	1	2	3	4
To start, hold rifle downward, and put feet together.	Swing arms forward and upward to overhead position.	Bend backward taking care not to bend too far. Keep face up and knees straight.	Move to first position.	Recover to start position.

Figure 8-15

Log Drills

Log drills are excellent for developing strength and muscular endurance, because they require the muscles to contract under heavy loads.

Log drills are team-conditioning exercises. They are excellent for developing strength and muscular endurance because they require the muscles to contract under heavy loads. They also develop teamwork and add variety to the PT program.

Log drills consist of six different exercises numbered in a set pattern. The drills are intense, and teams should complete them in 15 minutes. The teams have six to eight soldiers per team. A principal instructor is required to teach, demonstrate, and lead the drill. He must be familiar with leadership techniques for conditioning exercises and techniques peculiar to log drills.

AREA AND EQUIPMENT

Any level area is good for doing log drills. All exercises are done from a standing position. If the group is larger than a platoon, an instructor's stand may be needed.

The logs should be from six to eight inches thick, and they may vary from 14 to 18 feet long for six and eight soldiers, respectively. The logs should be stripped, smoothed, and dried. The 14-foot logs weigh about 300 pounds, the 18-foot logs about 400 pounds. Rings should be painted on the logs to show each soldier's position. When not in use, the logs are stored on a rack above the ground.

FORMATION

All soldiers assigned to a log team should be about the same height at the shoulders. The best way to divide a platoon is to have them form a single file or column with short soldiers in front and tall soldiers at the rear. They take their positions in the column according to shoulder height, not head height. When they are in position, they are divided into teams of six or eight.

The command is "Count off by sixes (or eights), count off." Each team, in turn, goes to the log rack, shoulders a log, and carries it to the exercise area.

The teams form columns in front of the instructor. Holding the logs in chest position, they face the instructor and ground the log. Ten yards should separate log teams within the columns. If more than one column is used, 10 yards should separate columns.

STARTING DOSAGE AND PROGRESSION

The starting session is six repetitions of each exercise. The progression rate is an increase of one repetition for each three periods of exercise. Soldiers continue this rate until they do 12 repetitions with no rest between exercises. This level is maintained until another drill is used.

START POSITIONS

The soldiers fall in facing their log, with toes about four inches away. Figure 8-16 shows the basic starting positions and commands.

Right-Hand Start Position, Move

On the command "Move," move the left foot 12 inches to the left, and lower the body into a flatfooted squat. Keep the back straight, head up, and arms between the legs. Encircle the far side of the log with the left hand. Place the right hand under the log. (See 1, Figure 8-16.)

Left-Hand Start Position, Move

This command is done the same way as the preceding command. However, the left hand is under the log, and the right hand encircles its far side. (See 2, Figure 8-16.)

Right-Shoulder Position, Move

This command is given from the right-hand-start position. On the command "Move," pull the log upward in one continuous motion to the right shoulder. At the same time, move the left foot to the rear and stand up, facing left. Balance the log on the right shoulder with both hands. (See 3, Figure 8-16.) This movement cannot be done from the left-hand-start position because of the position of the hands.

1. RIGHT-HAND-START POSITION

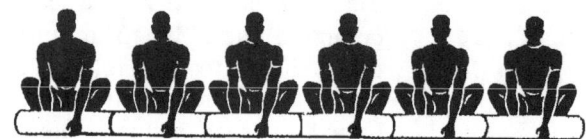

2. LEFT-HAND-START POSITION

3. RIGHT-SHOULDER POSITION

Figure 8-16

Left-Shoulder Position, Move

This command is given from the left-hand-start position. On the command "Move, " pull the log upward to the left shoulder in one continuous motion. At the same time, move the right foot to the rear, and stand up facing right. Balance the log on the left shoulder with both hands. (See 4, Figure 8-17.) This movement cannot be done from the right-hand-start position.

Waist Position, Move

From the right-hand-start position, pull the log waist high. Keep the arms straight and fingers laced under the log. The body is inclined slightly to the rear, and the chest is lifted and arched. (See 5, Figure 8-17.)

Chest Position, Move

This command is given after taking the waist position. On the command "Move," shift the log to a position high on the chest, bring the left arm under the log, and hold the log in the bend of the arms. (See 6, figure 8-17.) Keep the upper arms parallel to the ground.

To move the log from the right to the left shoulder, the command is "Left-shoulder position, move." Push the log overhead, and lower it to the opposite shoulder.

To return the log to the ground from any of the above positions, the command is "Start position, move." At the command "Move," slowly lower the log to the ground. Position the hands and fingers so they are not under the log.

4. LEFT-SHOULDER POSITION

5. WAIST POSITION

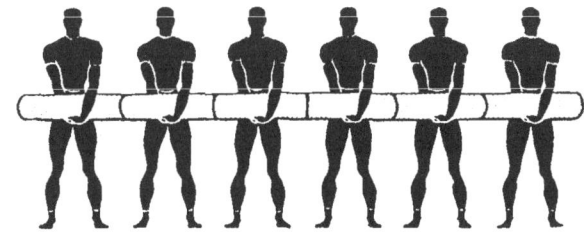

6. CHEST POSITION

Figure 8-17

LOG-DRILL EXERCISES

The following are log-drill exercises.

Exercise 1. Two-Arm Push-Up

Start Position: Right- or left-shoulder position, with feet about shoulder-width apart. (See 1, Figure 8-18.)
Cadence: Moderate.
Movement: A four-count exercise; at the count of --
"One"-Push the log overhead until the elbows lock.
"Two"-Lower the log to the opposite shoulder.
"Three"-Repeat the action of count one.
"Four"-Recover to the start position.

Exercise 2. Forward Bender

Start Position: Chest position, with feet about shoulder-width apart. (See 2, Figure 8-18.)
Cadence: Moderate.
Movement A four-count exercise; at the count of --
"One"-Bend forward at the waist while keeping the back straight and the knees slightly bent.
"Two"-Recover to the start position.
'Three"-Repeat the action of count one.
"Four"-Recover to the start position.

EXERCISE 1: TWO-ARM PUSH-UP

Start Position: 1   2   3   4

EXERCISE 2: FORWARD BENDER

Start Position: 1   2   3   4

Figure 8-18

Exercise 3. Straddle Jump

Start Position Right- or left-shoulder position, with feet together, and fingers locked on top of the log. Pull the log down with both hands to keep it from bouncing on the shoulder. (See 3, Figure 8-19.)

Cadence: Moderate.

Movement A four-count exercise; at the count of--

"One"-Jump to a side straddle.

"Two"-Recover to the start position.

'Three"-Repeat the action of count one.

"Four"-Recover to the start position.

Exercise 4. Side Bender

Start Position: Right-shoulder position with the feet about shoulder-width apart. (See 4, Figure 8- 19.)

Cadence Moderate.

Movement: A four-count exercise; at the count of--

"One"-Bend sideward to the left as far as possible, bending the left knee.

"Two"-Recover to the start position.

"Three"-Repeat the action of count one.

"Four"-Recover to the start position.

NOTE: After doing the required number of repetitions, change shoulders and do an equal number to the right side.

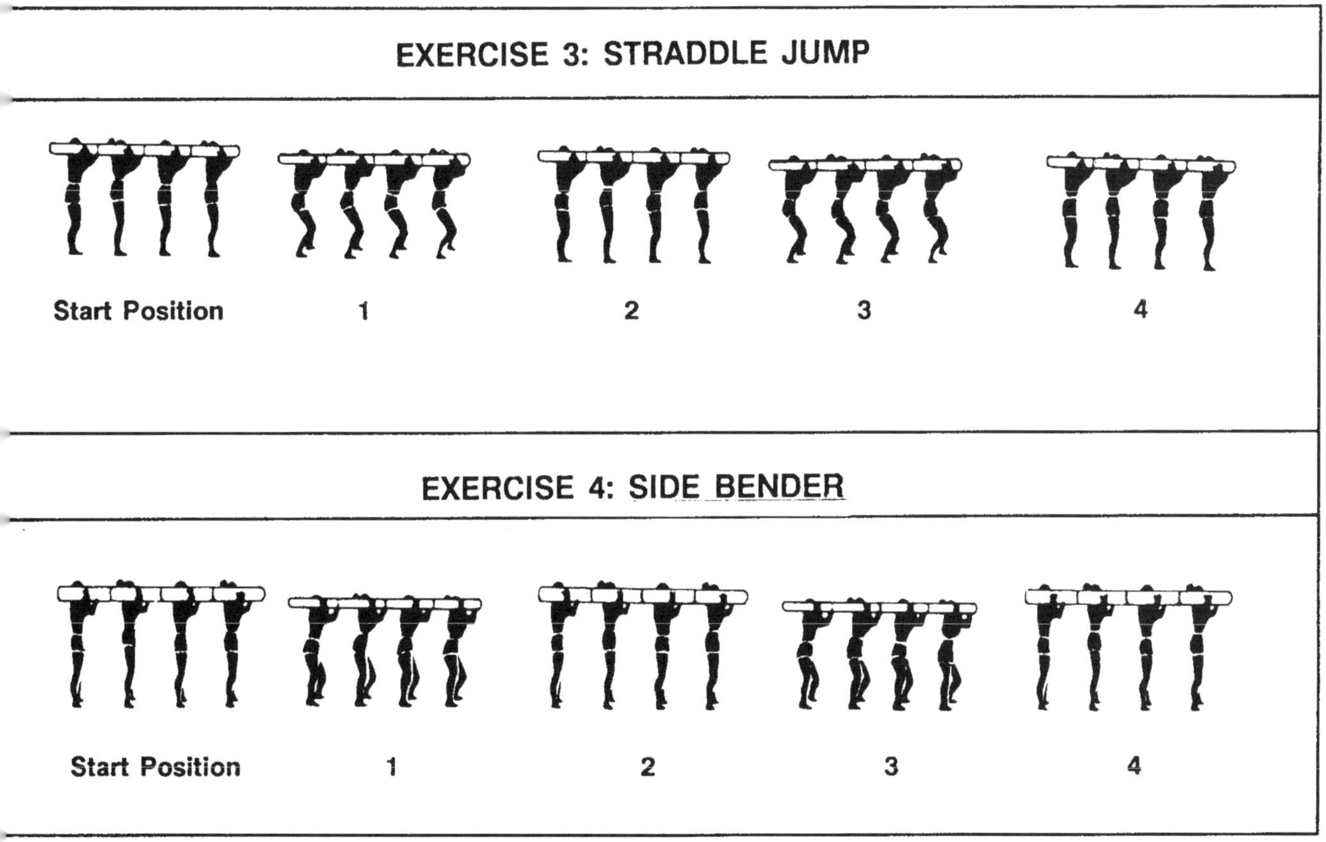

EXERCISE 3: STRADDLE JUMP

Start Position   1   2   3   4

EXERCISE 4: SIDE BENDER

Start Position   1   2   3   4

Figure 8-19

Exercise 5. Half-Knee Bend
Start Position: Right- or left-shoulder position, with feet about shoulder-width apart, and fingers locked on top of the log. (See 5, Figure 8-20.)
Cadence: Slow.
Movement: A four-count exercise; at the count of --
"One"-Flex the knees to a half-knee bend.
"Two"-Recover to the start position.
"Three"-Repeat the action of count one.
"Four"-Recover to the start position.
(NOTE: Pull forward and downward on the log throughout the exercise.)

Exercise 6. Overhead Toss (NOTE: Introduce this exercise only after soldiers have gained experience and strength by doing the other exercises for several sessions.)
Start Position: Right-shoulder position with the feet about shoulder-width part. The knees are at a quarter bend. (See 6, Figure 8-20.)
Cadence: Moderate.
Movement: A four-count exercise; at the count of --
"One"-Straighten the knees and toss the log about 12 inches overhead. Catch the log with both hands, and lower it toward the opposite shoulder. As the log is caught, lower the body into a quarter bend.
"Two"-Again, toss the log into the air and, when caught, return it to the original shoulder.
"Three"-Repeat the action of count one.
"Four"-Recover to the start position.

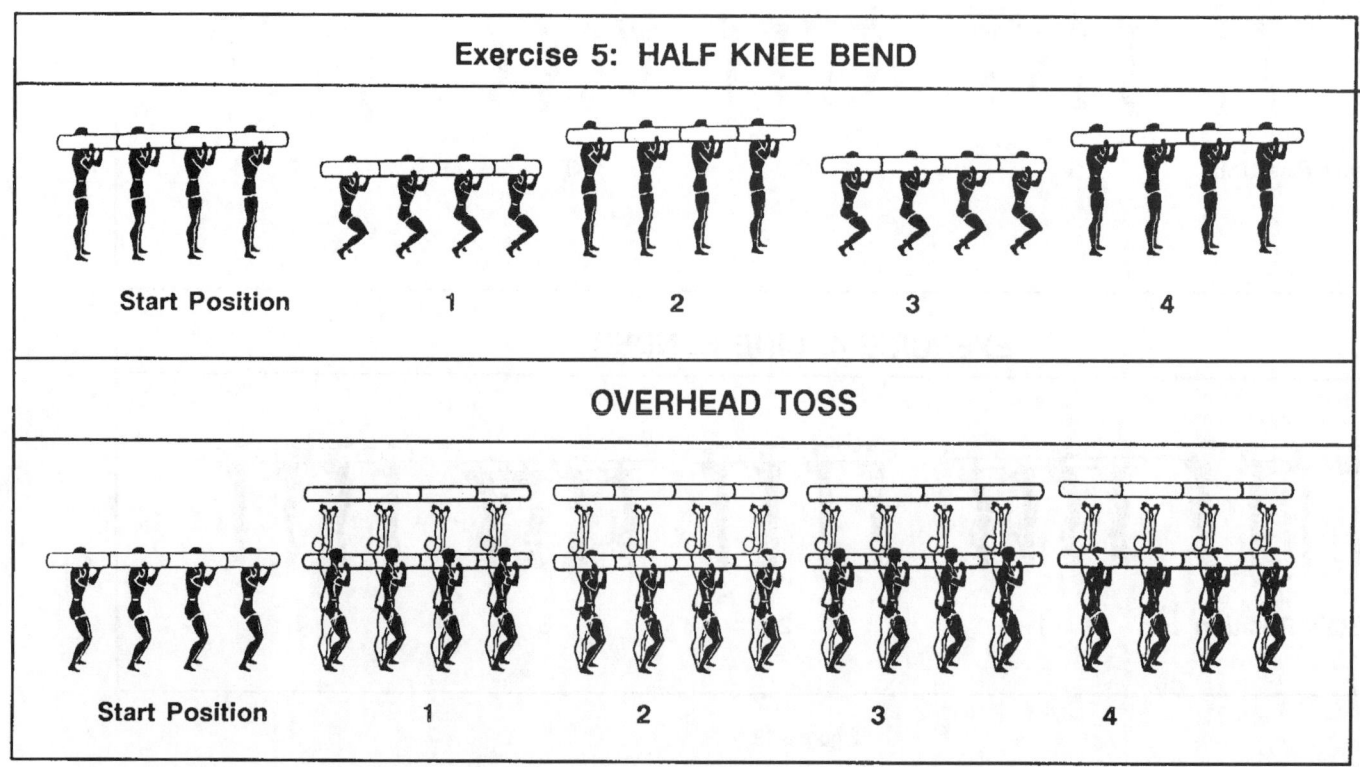

Figure 8-20

Aquatic Exercise

Aquatics is a mode of physical training which helps one attain and maintain physical fitness through exercises in the water. It is sometimes called slimnastics. Aquatic training can improve muscular endurance, CR endurance, flexibility, coordination, and muscular strength.

Because of its very low impact to the body, an aquatic exercise program is ideal for soldiers who are overweight and those who are limited due to painful joints, weak muscles, or profiles. The body's buoyancy helps minimize injuries to the joints of the lower legs and feet. It exercises the whole body without jarring the bones and muscles. Leaders can tailor the variety and intensity of the exercises to the needs of all the soldiers in the unit.

Aquatic training is a good supplement to a unit's PT program. Not only is it fun, it exposes soldiers to water and can make them more comfortable around it. Most Army installations have swimming pools for conducting aquatic, physical training sessions.

SAFETY CONSIDERATIONS

One qualified lifeguard is needed for every 40 soldiers at all aquatic training sessions. Nonswimmers must remain in the shallow end of the pool. They should never exercise in the deep end with or without flotation devices.

EQUIPMENT

Soldiers normally wear swim suits for aquatics, but they can wear boots and fatigues to increase the intensity of the activities. The following equipment is optional for training:
- Goggles.
- Kickboard.
- Pull buoy.
- Ear/nose plugs.
- Fins.
- Hand paddles.

SAMPLE TRAINING PROGRAM

'Warm-Up

As in any PT session, a warm-up is required. It can be done in the water or on the deck. Allow five to seven minutes for the warm-up.

Conditioning Phase

Soldiers should exercise vigorously to get a training effect. Energetic music may be used to keep up the tempo of the workout. The following are some exercises that can be used in an aquatic workout. (See Figure 8-21.)

Side Leg-Raises. Stand in chest to shoulder-deep water with either side of the body at arm's length to the wall of the pool, and grasp the edge with the nearest hand. Raise the outside leg sideward and upward from the hip. Next, pull the leg down to the starting position. Repeat these actions. Then, turn the other side of the body to the wall, and perform the exercise with the other leg. DURATION: 30 seconds (15 seconds per leg).

Leg-Over. Stand in chest-to shoulder-deep water, back facing the wall of the pool. Reach backward with the arms extended, and grasp the pool's edge. Next, raise one leg in front of the body away from the wall, and move it sideward toward the other leg as far as it can go. Then, return the leg to the front-extended position, and lower it to the starting position. Repeat these actions with the other leg, and continue to alternate legs. DURATION: 30 seconds (15 seconds per leg).

Rear Leg Lift. Stand in chest-to shoulder-deep water with hands on the pool's edge, chest to the wall. Raise one leg back and up from the hip, extend it, and point the foot. Then, pull the leg back to the starting position. Alternate these actions back and forth with each leg. DURATION: 20 seconds (10 seconds each leg).

AN AQUATIC EXERCISE WORKOUT CENTER

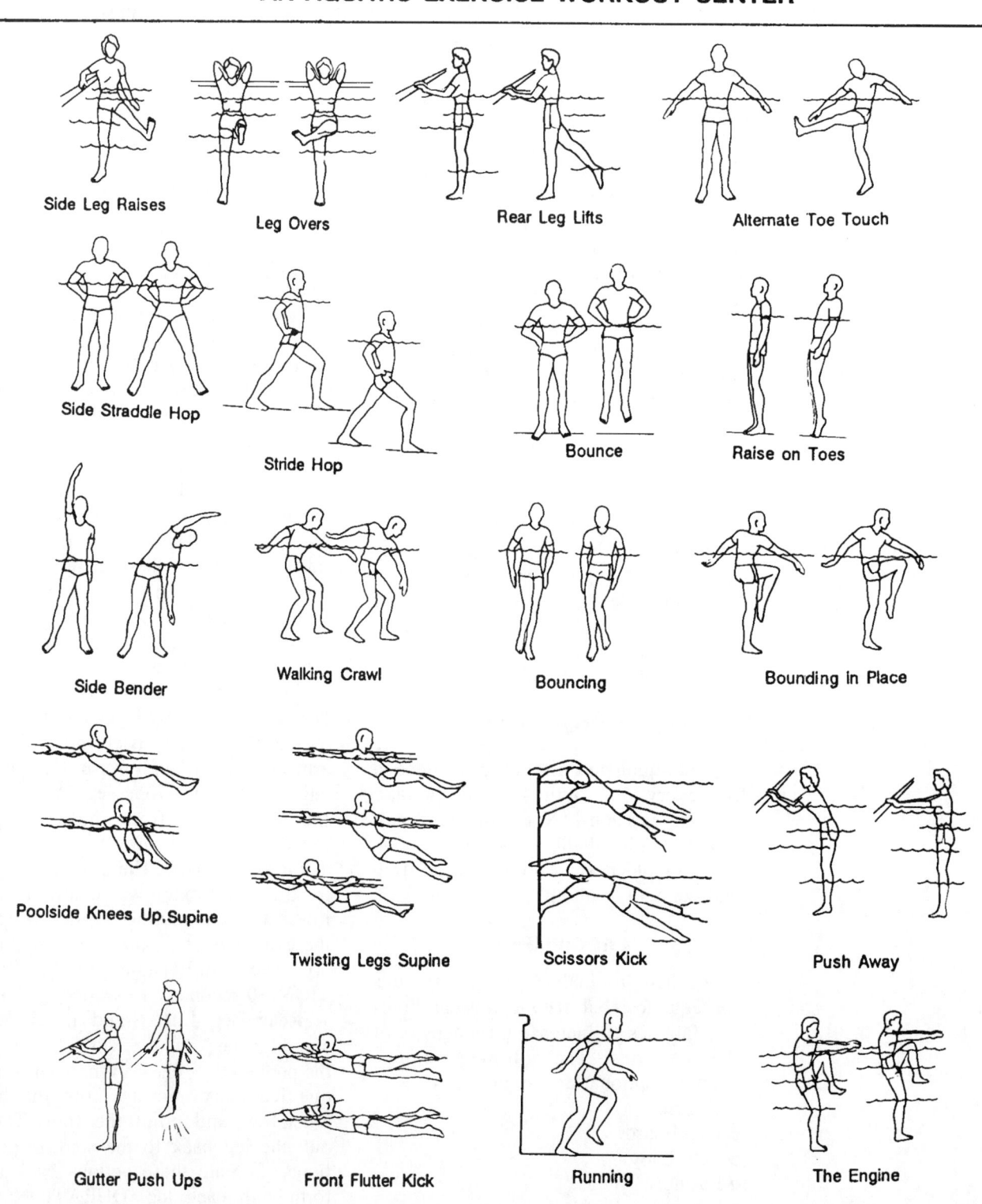

Side Leg Raises

Leg Overs

Rear Leg Lifts

Alternate Toe Touch

Side Straddle Hop

Stride Hop

Bounce

Raise on Toes

Side Bender

Walking Crawl

Bouncing

Bounding in Place

Poolside Knees Up, Supine

Twisting Legs Supine

Scissors Kick

Push Away

Gutter Push Ups

Front Flutter Kick

Running

The Engine

Figure 8-21

Alternate Toe Touch. Stand in waist-deep water. Raise the left leg as in kicking while touching the elevated toe with the right hand. At the same time, rotate the head toward the left shoulder, and push the left arm backward through the water. Alternate these actions back and forth with each leg and opposite hand. DURATION 2 minutes.

Side Straddle Hop. Stand in waist-deep water with hands on hips and feet together. Jump sideward and land with feet about two feet apart. Then, return to the starting position, and repeat the jumping action. DURATION 2 minutes.

Stride Hop. Stand in waist-deep water with hands on hips and feet together. Jump, moving the left leg forward and right leg backward. Then, jump again moving the right leg forward and left leg backward. Repeat these actions. DURATION 2 minutes.

The Bounce. Stand in waist-deep water with hands on hips and feet together. Jump high with feet together. Upon landing, use a bouncing motion, and repeat the action. DURATION: 1 minute.

Rise on Toes. Stand in chest-to shoulder-deep water with arms at sides and feet together. Rise up using the toes. Then, lower the body to the starting position. Repeat the action. DURATION: 1 minute.

Side Bender. Stand in waist-deep water with the left arm at the side and the right arm extended straight overhead. Stretch slowly, bending to the left. Recover to the starting position, and repeat the action. Next, reverse to the right arm at the side and the left arm extended straight overhead. Repeat the stretching action to the right side. DURATION: 1 minute.

Walking Crawl. Walk in waist- to chest-deep water. Simulate the overhand crawl stroke by reaching out with the left hand cupped and pressing the water downward to the thigh. Repeat the action with the right hand. Alter-

nate left and right arm action. DURATION: 2 minutes.

Bouncing. Stand in chest-deep water, arms at sides. Bounce on the left foot while pushing down vigorously with both hands. Repeat the action with the right foot. Alternate bouncing on the left and right foot. DURATION: 2 minutes.

Bounding in Place with Alternate Arm Stretch, Forward. Bound in place in waist-deep water using high knee action. Stretch the right arm far forward when the left knee is high and the left arm is stretched backward. When the position of the arm is reversed, simulate the action of the crawl stroke by pulling down and through the water with the hand. DURATION 1 minute.

Poolside Knees Up, Supine. Stand in chest-to shoulder-deep water, back against the wall of the pool. Extend the arms backward, and grasp the pool's edge. With feet together, extend the legs in front of the torso, and assume a supine position. Then with the legs together, raise the knees to the chin. Return to the starting position, and repeat the action. DURATION: 2 minutes (maximum effort).

Twisting Legs, Supine. Stand in chest-to shoulder-deep water, back against the wall of the pool. Extend the arms backward, and grasp the pool's edge. With feet together, extend the legs in front of the torso, and assume a supine position. Then, twist the legs slowly to the left, return to the starting position, and twist the legs slowly to the right. Repeat this twisting action. DURATION: 1 minute (2 sets, 30 seconds each).

Scissor Kick. Float in chest- to shoulder- deep water on either side of the body with the top arm extended, hand holding the pool's edge. Brace the bottom hand against the pool's wall with feet below the water's surface. Next, assume a crouching position by gringing the heels toward the hips by bending the knees. Then, straighten and spread the legs with the top leg extending backward. When the legs are extended and spread, squeeze them back together (scissoring). Pull with the top hand, and push with the bottom hand. The propulsive force of the kick will tend to cause the body to rise to the water's surface. DURATION 1 minute (2 sets, 30 seconds each, maximum effort).

Push Away. Stand in chest-to shoulder-deep water facing the pool's wall and at arm's length from it. Grasp the pool's edge, and bend the arms so that the body is leaning toward the wall of the pool. Vigorously push the chest back from the wall by straightening the arms. Then, with equal vigor, pull the upper body back to the wall. Repeat these actions. DURATION: 2 minutes (maximum effort).

Gutter Push-Ups. Stand in chest-to shoulder- deep water facing the pool's wall. Place the hands on the edge or gutter of the pool. Then, raise the body up and out of the water while extending the arms. repeat this action. DURATION: 2 minutes (4 sets, 30 seconds each with 5-second rests between sets).

Front Flutter Kick. Stand in chest-to shoulder-deep water facing the pool's wall. Grasp the pool's edge or gutter and assume a prone position with legs extended just below the water's surface. Then, kick flutter style, toes pointed, ankles flexible, knee joint loose but straight. The legs should simulate a whip's action. DURATION 1 minute (2 sets, 30 seconds each).

Running. Move in a running gait in chest-to shoulder-deep water with arms and hands under the water's surface. This activity can be stationary, or the exerciser may run from poolside to poolside. Runners must concentrate on high knee action and good arm movement. DURATION 10 to 20 minutes.

The Engine. Stand in chest-to shoulder-deep water, arms straight and in front of the body and parallel to the water with the palms facing downward. While walking forward, raise the left knee to the left elbow, then return to the starting position. Continuing to walk forward, touch the right knee to the right elbow, and return to the starting position. Be sure to keep the arms parallel to the water throughout the exercise. DURATION 1 to 2 minutes (2 sets).

Cool-Down

This is required to gradually bring the body back to its pre-exercise state. It should last from five to seven minutes.

CHAPTER 9

Competitive Fitness Activities

Physical fitness is one of the foundations of combat readiness, and maintaining it must be an integral part of every soldier's life. This chapter discusses competitive fitness activities and athletic events that commanders can use to add variety to a unit's physical fitness program. There is also a section on developing a unit intramural program. Athletic and competitive fitness activities are sports events which should only be used to supplement the unit's PT program. They should never replace physical training and conditioning sessions but, rather, should exist to give soldiers a chance for healthy competition. Only through consistent, systematic physical conditioning can the fitness components be developed and maintained.

Crucial to the success of any program is the presence and enthusiasm of the leaders who direct and participate in it. The creativity of the physical training planners also plays a large role. Competitive fitness and athletic activities must be challenging. They must be presented in the spirit of fair play and good competition.

It is generally accepted that competitive sports have a tremendous positive influence on the physical and emotional development of the participants. Sports competition can enhance a soldier's combat readiness by promoting the development of coordination, agility, balance, and speed. Competitive fitness activities also help develop assets that are vital to combat effectiveness. These include team spirit, the will to win, confidence, toughness, aggressiveness, and teamwork.

Competitive fitness activities help in the development of assets that are vital to combat effectiveness.

Intramural

The Army's sports mission is to give all soldiers a chance to participate in sports activities. A unit-level intramural program can help achieve this important goal. DA Pam 28-6 describes how to organize various unit-level intramural programs.

Factors that affect the content of the sports program differ at every Army installation and unit. Initiative and ingenuity in planning are the most vital assets. They are encouraged in the conduct of every program.

OBJECTIVES

A well-organized and executed intramural program yields the following:

• Team spirit, the will to win, confidence, aggressiveness, and teamwork. All are vital to combat effectiveness.

• A change from the routine PT program.

• The chance for all soldiers to take part in organized athletics.

ORGANIZATION

The command level best suited to organize and administer a broad intramural program varies according to a unit's situation. If the objective of maximum participation is to be achieved, organization should start at company level and then provide competition up through higher unit levels. Each command level should have its own program and support the next higher program level.

To successfully organize and conduct an intramural program, developers should consider the following factors and elements.

Authority

The unit commander should publish and endorse a directive giving authorization and guidance for a sports program. A detailed SOP should also be published.

Personnel

Leaders at all levels of the intramural program should plan, organize, and supervise it. Appointments at all echelons should be made for at least one year to provide continuity. The commander must appoint a qualified person to be the director, regardless of the local situation, type, and size of the unit. The director must be a good organizer and administrator and must have time to do the job correctly. He should also have a sense of impartiality and some athletic experience.

Commanders should form an intramural sports council in units of battalion size or larger and should appoint members or require designated unit representatives. The council should meet at least once a month or as often as the situation requires. The council serves as an advisory body to the unit commander and intramural director. It gives guidance about the organization and conduct of the program.

Facilities and Equipment

Adequate facilities and equipment must be available. When facilities are limited, leaders must plan activities to ensure their maximum use. In all cases, activities must be planned to ensure the safety of participants and spectators.

Funds and Budget

Adequate funds are essential to successfully organize and operate a sports program. Therefore, beforehand, organizers must determine how much money is available to support it. To justify requests for funds they must prepare a budget in which they justify each sports activity separately. The budget must include special equipment, supplies, awards, pay for officials, and other items and services. Units can reduce many of their costs by being resourceful.

AWARD SYSTEM

Commanders can stimulate units and soldiers to participate in competitive athletics by using an award system. One type is a point-award system where teams get points based on their win/loss records and/or final league standings. This reflects the unit's standings in the overall intramural sports program. The recognition will help make units and individuals participate throughout the year. Trophies can then be given for overall performance and individual activities.

PROGRAM PLANNING

A successful program depends on sound plans and close coordination between the units involved. The intramural director should meet with subordinate commanders or a sports representative to determine what program of activities is compatible with the mission and training activities of each unit. Unless they resolve this issue, they may not get command support which, in turn, could result in forfeitures or lack of participation. The less-popular activities may not be supported because of a lack of interest.

Commanders can stimulate soldiers to participate in competitive athletics b using an award system

Evaluations

Before the program is developed, leaders must study the training and availability situation at each unit level. They should include the following items in a survey to help them determine the scope of the program and to develop plans:

- *General.* Evaluate the commander's attitude, philosophy, and policy about the sports program. Understand the types of units to be served, their location, the climate, and military responsibilities.
- *Troops.* Determine the following: 1) number and types of personnel; 2) training status and general duty assignment; 3) special needs, interests, and attitudes.
- *Time available.* Coordinate the time available for the sports program with the military mission. Determine both the on-duty and off-duty time soldiers have for taking part in sports activities.
- *Equipment.* Consider the equipment that will be needed for each sport.
- *Facilities.* Determine the number, type, and location of recreational facilities both within the unit and in those controlled by units at higher levels.

- *Funds.* Determine how much each unit can spend on the intramural program.
- *Personnel.* Assess how many people are needed to run the program. The list should include a director and assistants, sports council, officials, and team captains, as well as volunteers for such tasks as setting up a playing field.
- *Coordination.* Coordinate with the units' operations sections to avoid conflict with military training schedules.
- *Activities.* The intramural director should plan a tentative program of activities based on the season, local situation, and needs and interests of the units. Both team and individual sports should be included. Some team sports are popular at all levels and need little promotional effort for success. Among these are volleyball, touch football, basketball, and softball. Some individual competitive sports have direct military value. They include boxing, wrestling, track and field, cross country, triathlon, biathlon, and swimming. While very popular, these sports are harder to organize than team sports. See Figures 9-1 and 9-2 for a list of sports activities.

SPORTS ACTIVITIES			
Team Sports			
Baseball Flag Football Softball	Basketball Water Polo Speedball	Field Hockey Pushball Tug-of-War	Football Soccer Volleyball
Field-Type-Meets			
Athletic Carnivals Physical Fitness Meet Track and Field Urban Orienteering	Cross Country Relay Carnival Water Carnival	Military Field Meets Swimming and Diving Unit Olympics	

Figure 9-1

SPORTS ACTIVITIES			
Individual Sports			
Archery Boxing Handball Marathon Track & Field Triathlon	Badminton Canoeing Judo Squash Rowing Skating	Tennis Table Tennis Horseshoes Skating Sky Diving Weightlifting	Bowling Gymnastics Modern Biathlon Mountain Climbing Skeet Shooting Swimming and Diving

Figure 9-2

ESSENTIAL ELEMENTS

Intramural Handbook

- Commander's foreward.
- Personnel directory.
- Title page.
- Purpose.
- References.
- Objectives.
- Duties of the personnel.
- Eligibility rules.
- Intramural sports council.
- Protest and sportsmanship board.
- Budgets and funding.
- Officials association.

- Master calendar of activities.
- Organization of leagues and units of competition.
- Command- points award system.
- Facilities and their hours of operation.
- Equipment regulations.
- Rules and regulations of each sport.
- Reporting time for competition.
- Postponement of contests.
- Protest procedures.
- Awards.
- Records and results.
- Bulletin boards and publicity.

Table 9-1

Functions

Once the evaluations have been made, the following functions should be performed:

- *Make a handbook.* An intramural handbook should be published at each level of command from installation to company to serve as a standing operating procedure (SOP). This handbook should include the essential elements listed in Table 9-1 above.
- *Plan the calendar.* Local situations and normal obstacles may conflict with the intramural program. How ever, a way can be found to provide a scheduled program for every season of the year.
- *Choose the type of competition.* Intramural directors should be able to choose the type of competition best suited for the sport and local circumstances. They should also know how to draw up tournaments. Unless the competition must take place in a short time, elimination tournaments should not be used. The round-robin tournament has the greatest advantage because individuals and teams are never eliminated. This type of competition is adaptable to both team and individual play. It is appropriate for small numbers of entries and league play in any sport.

- *Make a printed schedule.* Using scheduling forms makes this job easier. The form should include game number, time, date, court or field, and home or visiting team. Space for scores and officials is also helpful. Championship games or matches should be scheduled to take place at the best facility.

Unit Activities

The following games and activities may be included in the unit's PT program, They are large-scale activities which can combine many components of physical and motor fitness. In addition, they require quick thinking and the use of strategy. When played vigorously, they are excellent activities for adding variety to the program.

NINE-BALL SOCCER

The object of this game is for each of a team's five goalies to have one ball.

Players

There are 25 to 50 players on each team, five of whom are goalies. The other players are divided into four equal groups. The goalies play between the goal line and 5-yard line of a standard football field. The other four groups start the game between the designated 10-yard segments of the field. (See Figure 9-3.) The goalies and all other players must stay in their assigned areas throughout the game. The only exceptions are midfielder who stand between the 35- and 45-yard lines. These players may occupy both their assigned areas and the 10-yard free space at the center of the field.

The Game

The game starts with all players inside their own areas and midfielder on their own 40-yard line. The nine balls are placed as follows. Four are on each 45-yard line with at least five yards between balls. One is centered on the 50-yard line. The signal to start play is one long whistle blast. Players must pass the balls through the opposing team's defenses into the goal area using only their feet or heads. The first team whose goalies have five balls wins a point. The game then stops, and the balls are placed for the start of a new set. The first team to score five points wins.

There are no time-outs except in case of injury, which is signaled by two sharp whistle blasts. The teams change positions on the field after each set. Team members move to different zones after the set.

Rules

A ball is played along the ground or over any group or groups of players. The ball may travel any distance if it is played legally.

Goalies may use their hands in playing the ball and may give a ball to other goalies on their team. For a set to officially end, each goalie must have a ball.

If players engage in unnecessary roughness or dangerous play, the referee removes them from the game for the rest of the set and one additional set. He also removes players for the rest of the set if they step on or over a boundary or sideline or use their hands outside the goal area.

If a goalie steps on or over a boundary or sideline, the referee takes the ball being played plus another ball from the goalie's team and gives these balls to the nearest opposing player. If

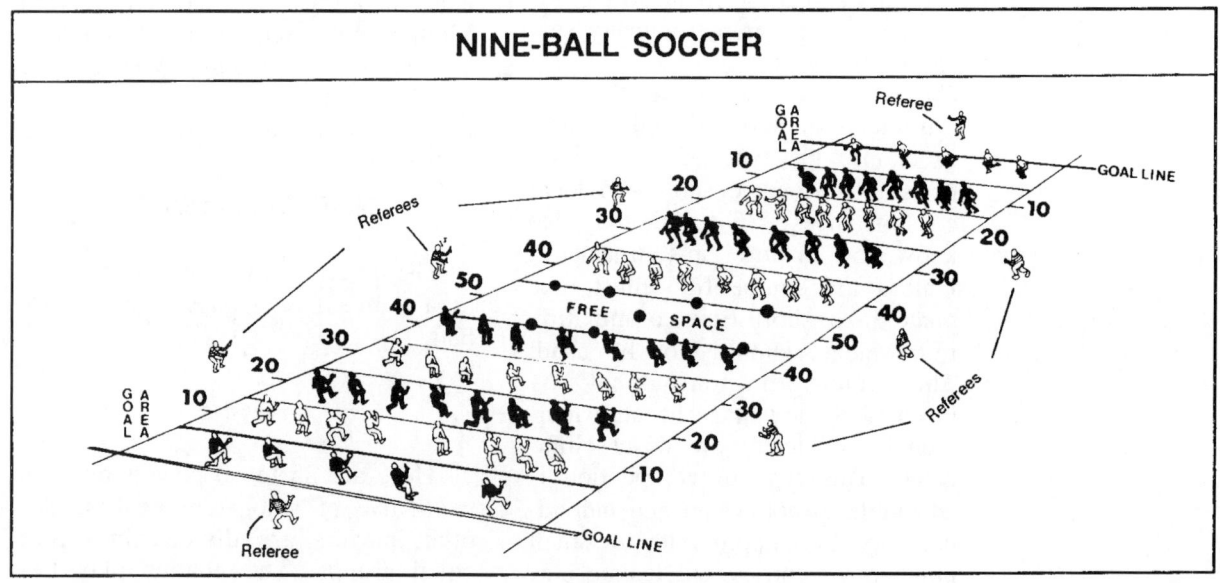

Figure 9-3

the team has no other ball in the goal area, the referee limits the penalty to the ball that is being played.

If a ball goes out of bounds, the referee retrieves it. The team that caused it to go out of bounds or over the goal line loses possession. The referee puts the ball back into play by rolling it to the nearest opposing player.

PUSHBALL

This game requires a large pushball that is five to six feet in diameter. It also requires a level playing surface that is 240 to 300 feet long and 120 to 150 feet wide. The length of the field is divided equally by a center line. Two more lines are marked 15 feet from and parallel to the end lines and extending across the entire field. (See Figure 9-4.)

Players

There are 10 to 50 soldiers on each of two teams.

The Game

The object of the game is to send the ball over the opponent's goal line by pushing, rolling, passing, carrying, or using any method other than kicking the ball.

The game begins when the ball is placed on the centerline with the opposing captains three feet away from it. The other players line up 45 feet from the ball on their half of the field. At the referee's starting whistle, the captains immediately play the ball, and their teams come to their aid.

At quarter time, the ball stays dead for two minutes where it was when the quarter ended. At halftime, the teams exchange goals, and play resumes as if the game were beginning.

A team scores a goal when it sends the ball across the opposing team's end line. A goal counts five points. The team that scores a goal may then try for an extra point. For the extra point, the ball is placed on the opposing team's 5-yard line, and the teams line

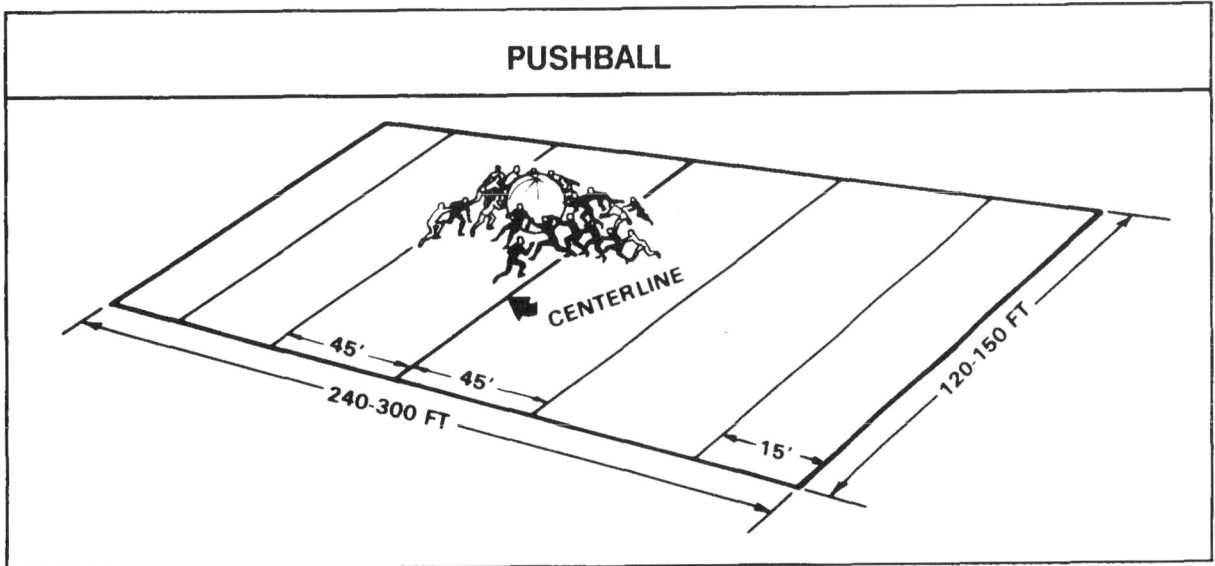

PUSHBALL

Figure 9-4

up across the field separated by the width of the ball. Only one player may place his hands on the ball. The player who just scored is directly in front of the ball. At the referee's signal, the ball is put into play for one minute. If any part of the ball is driven across the goal line in this period, the offense scores one point. The defense may not score during the extra point attempt.

The game continues until four 10-minute quarters have been played. Rest periods are allowed for two minutes between quarters and five minutes at halftime.

Rules

Players may use any means of interfering with the opponents' progress except striking and clipping. Clipping is throwing one's body across the back of an opponent's legs as he is running or standing. Force may legally be applied to all opponents whether they are playing the ball or not. A player who strikes or clips an opponent is removed from the game, and his team is penalized half the distance to its goal.

When any part of the ball goes out of bounds, it is dead. The teams line up at right angles to the sidelines. They should be six feet apart at the point where the ball went out. The referee tosses the ball between the teams.

When, for any reason, the ball is tied up in one spot for more than 10 seconds, the referee declares it dead. He returns the ball into play the same way he does after it goes out of bounds.

STRATEGY PUSHBALL

Strategy pushball is similar to pushball except that it is played on two adjacent fields, and opposing teams supply soldiers to the games on both fields. Team commanders assess the situation on the fields and distribute their soldiers accordingly. The commander decides the number of soldiers used, within limits imposed by the rules. This number may be adjusted throughout the game. Play on both fields occurs at the same time, but each game progresses independently. At the end of play, a team's points from both fields are added together to determine the overall winner.

This game requires two pushballs that are five to six feet in diameter. Pull-over vests or jerseys of two different colors are used by each team for a total of four different colors. Starters and reserves should be easily distinguishable. Starters and substitutes should wear vests of one color, while the team commander and reserves wear vests of the second color.

Players may wear any type of athletic shoes except those with metal cleats. Combat boots may be worn, but extra caution must be used to prevent injuries caused by kicking or stepping on other players. Soldiers wearing illegal equipment may not play until the problem has been corrected.

The playing area is two lined-off fields. These are 240 to 300 feet long by 120 to 150 feet wide. They are separated lengthwise by a 20-foot-wide divider strip. The length of each field is divided equally by a centerline that is parallel to the goal lines. Lines are also marked 45 feet from each side of the centerline and parallel to it. The lines extend across both fields. Dimensions may be determined locally based on available space and the number of players. The space between the fields is the team area. Each team occupies the third of the team space that immediately adjoins its initial playing field.

Time periods should be adjusted to suit weather conditions and soldiers' fitness levels.

Players

There are 25 to 40 soldiers on each team. A typical, 25-member team has the following:

- One team commander. He is responsible for overall game strategy and for determining the number and positions of players on the field.
- Sixteen starting members. Eight are on each field at all times; one is appointed field captain.
- Four reserve members. These are players the team commander designates as reinforcements.
- Three substitutes. These are replacements for starters or reserves.
- One runner. He is designated to convey messages from the team commander to field captains.

The proportion of soldiers in each category stays constant regardless of the total number on a team. Before the event, game organizers must coordinate with participating units and agree on the number on each team.

Runners serve at least one period; they may not play during that period. They are allowed on the field only during breaks in play after a dead ball or goal.

Reserves are used at any point in the game on either field and are committed as individuals or groups. They may enter or leave the playing field at any time whether the ball is in play or not. Team commanders may enter the game as reserves if they see the need for such action.

Reserves, substitutes, and starting members may be redesignated into any of the other components on a one-for-one basis only during dead balls, injury time-outs, or quarter- and half-time breaks. A reserve may become a starter by switching vests with an original starter, who then becomes a reserve.

When possible, senior NCOS and officers from higher headquarters or other units should be used as officials. Players must not question an official's

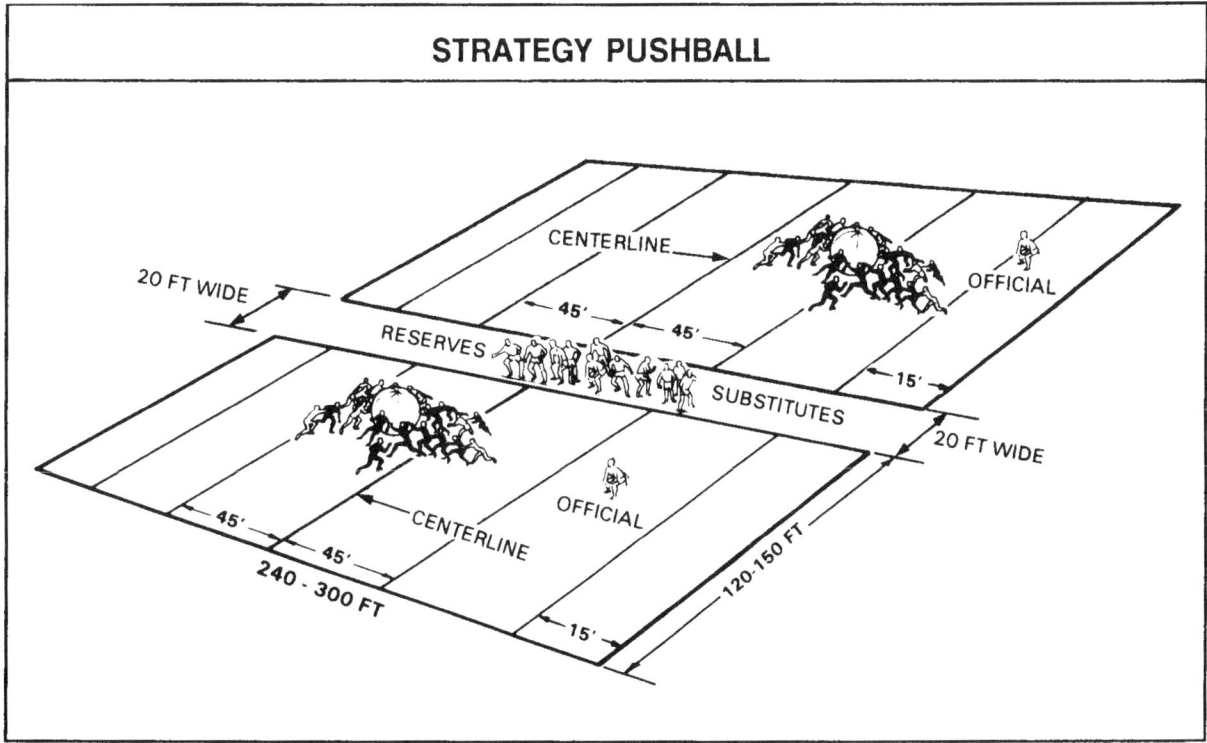

STRATEGY PUSHBALL

Figure 9-5

authority during play. Otherwise, the game can quickly get out of control.

Chain-of-command personnel should act as team commanders and field captains whenever possible.

The Game

The object is to propel the ball over the opponent's goal line by pushing, rolling, passing, carrying, or using any means other than kicking.

The game is officiated by two referees on each field, a chief umpire, and a scorekeeper. Referees concentrate on player actions so that they can quickly detect fouls and assess penalties. The chief umpire and scorekeeper occupy any area where they can best officiate the games. The chief umpire monitors the use of substitutes and reserves and ensures smooth progress of the games on both fields. The number of officials may be increased if teams have more than 25 players. Referees use their whistles to stop and start play except at the start and end of each quarter. The scorekeeper, who times the game with a stopwatch, starts and ends each quarter and stops play for injuries with some noisemaker other than a whistle. He may use such devices as a starter's pistol, klaxon, or air horn.

The game begins after the ball is placed on each field's center mark. Opposing field captains are three feet from the ball (six feet from the centerline). The rest of the starters are lined up 45 feet from the ball on their half of the field. (See Figure 9-5.) At the scorekeeper's signal, field captains immediately play the ball, and their teams come to their aid.

Starters may be exchanged between the fields if the minimum number of starters or substitutes per field is maintained.

Substitutes may enter the game only during breaks in play after a dead ball, goal, or time-out for injury.

A substitute may not start to play until the player being replaced leaves the field.

When any part of the ball goes out of bounds, it is dead. The teams line up at right angles to the sidelines; they are 10 feet apart at the point where the ball went out of bounds. The referee places the ball between the teams at a point 15 feet inside the sideline. Play resumes when the referee blows the whistle.

When the ball gets tied up in one spot for more than 10 seconds for any reason, the referee declares it dead. He restarts play as with an out-of-bounds dead ball, except that he puts the ball on the spot where it was stopped.

Time does not stop for dead balls or goals. Play continues on one field while dead balls are restarted on the other.

At each quarter break, the ball stays on the spot where it was when the quarter ended. The next quarter, signaled by the scorekeeper, starts as it does after a ball goes out of bounds. At halftime the teams exchange goals, and play resumes as if the game were beginning.

A goal is scored when any part of the ball breaks the plane of the goal line between the sidelines. A goal counts one point. At the end of the fourth quarter, the points of each team from both fields are added together to determine the winner.

If there is a tie, a three-minute overtime is played. It is played the same as in regulation play, but only one field is used, with starting squads from both teams opposing each other. For control purposes, no more than 15 players per team are allowed on the field at once. The team with more points at the end of the overtime wins the game. If the game is still tied when time expires, the winner is the team that has gained more territory.

The game continues until four 10-minute quarters have been played. There is a 10-minute halftime between the second and third quarters. The clock stops at quarter breaks and halftime. Time-out is allowed only for serious injury. Play is then stopped on both fields.

Rules

Players may use any means of interfering with their opponents' progress, but they are penalized for striking or clipping opponents or throwing them to the ground. These penalties are enforced by the referees. Force maybe legally applied to any opponent whether or not they are playing the ball. Blocking is allowed if blockers stay on their feet and limit contact to the space between waist and shoulders. Blockers may not swing, throw, or flip their elbows or forearms. Tackling opposing soldiers who are playing the ball is allowed. The chief umpire or any referee may call infractions and impose penalties for unsportsmanlike conduct or personal fouls on either field. Penalties may also be called for infractions committed on the field or sidelines during playing time, quarter- and halftime breaks, and time-outs. Personal fouls are called for the following:

- Illegal blocking (below an opponent's waist).
- Clipping (throwing the body across the back of the opponent's legs as he is running or standing).
- Throwing an opponent to the ground (that is, lifting and dropping or slamming a player to the ground in stead of tackling cleanly).
- Spearing, tackling, or piling on an opponent who is already on the ground.
- Striking or punching with closed fist(s).
- Grasping an opponent's neck or head.
- Kicking.
- Butting heads.

Unsportsmanlike conduct is called for abusive or insulting language that the referee judges to be excessive and blatant. It is also called against a player on the sidelines who interferes with the ball or with his opponents on the field. A player who violates these rules should be removed from the game and made to run one lap around both playing fields. A penalized player leaves the team shorthanded until he completes the penalty lap and the next break in play occurs on the field from which he was removed. The penalized player or a substitute then enters the game. Referees and the chief umpire may, at their discretion, eject any player who is a chronic violator or who is judged to be dangerous to other players, Once ejected, the player must leave both the field of play and team area. Substitutes for ejected players may enter during the next break in play that follows a goal scored by either team. They enter on the field from which the players were ejected.

BROOM-BALL HOCKEY

This game is played on ice or a frozen field using hockey rules. Players wear boots with normal soles and carry broom-shaped sticks with which they hit the ball into the goals.

The object of this game is for teams to score goals through the opponent's defenses. Using only brooms, players pass the ball through the opposing team to reach its goal. The first team to score five points wins. Broom ball provides a good cardiorespiratory workout.

Players

There are 15 to 20 players on each team. One is a goalie and the others are divided into three equal groups. The goalie plays in the goal area of a standard soccer or hockey field or along the goal line if the two opposing goals are the same size. One soccer ball, or some other type of inflated ball, is used. The players need no padding.

The three groups begin the game in center field. All players must stay in their designated space throughout the game. A diagram of the field is shown at Figure 9-6.

The Game

The face-off marks the start of the game, the second half, and the restart of play after goals. Each half lasts 15 minutes. For the face-off, each player is on his own half of the field. All players, except the two centers, are outside the center circle. The referee places the ball in the center of the circle between the two centers. The signal to begin play is one long blast on the whistle. The ball must travel forward and cross the center circle before being played by another player. There are no time-outs except for injury. The time-out signal is two sharp whistle blasts.

Rules

All players, including goalies, must stay inside their legal boundaries at all times. Only goalies may use their hands to play the ball, but they must always keep control of their sticks. Other players must stay in their respective zones of play (Attack, Defense, Centerfield). The ball is played along the ground or over one or more groups of players. It may travel any distance as long as it is legally played.

The referee calls infractions and imposes penalties. Basic penalties are those called for the following:
● Unnecessary roughness or dangerous play. (The player is removed from the game; he stays in the penalty box for two minutes.)
● Ball out-of-bounds. (The team that caused it to go out loses possession, and the opposing team puts the ball back into play by hitting it to the nearest player.)
● Use of hands by a player other than a goalie. (The player must stay in the penalty box one minute.)
● Improper crossing of boundaries. (When a member of the team in possession of the ball crosses the boundary line of his zone of play, possession will be awarded to the other team.)

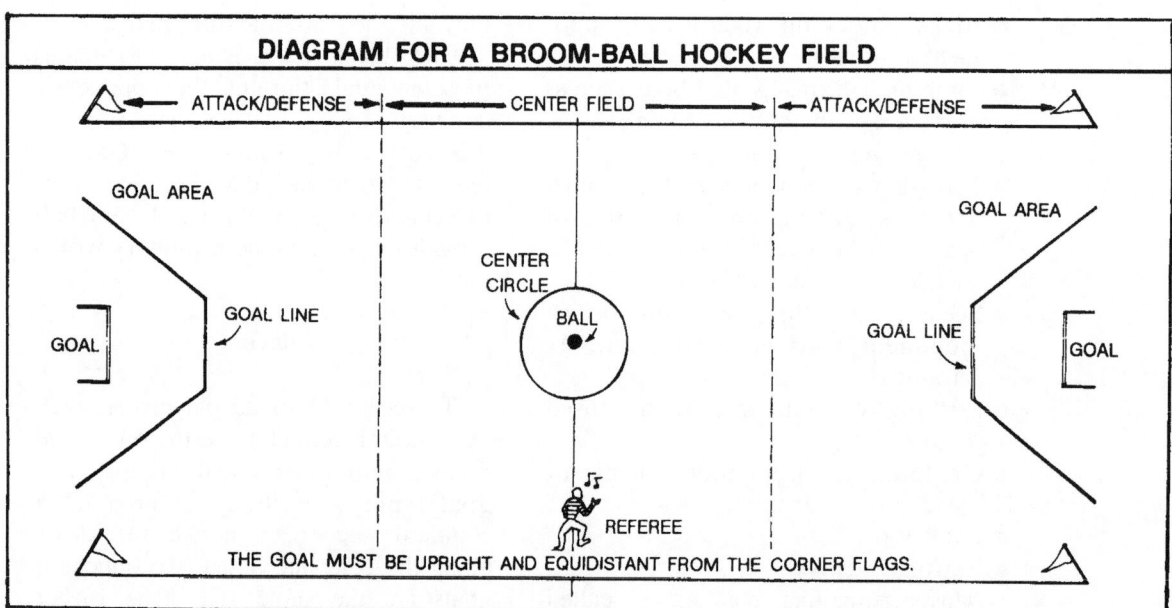

Figure 9-6

Orienteering

Orienteering combines map reading, compass use, and terrain study with strategy, competition, and exercise.

Orienteering is a competitive form of land navigation. It combines map reading, compass use, and terrain study with strategy, competition, and exercise. This makes it an excellent activity for any training schedule.

An orienteering course is set up by placing control points or marker signs over a variety of terrain. The orienteer or navigator uses a detailed topographical map and a compass to negotiate the course. The map should be 1:25,000 scale or larger. A liquid-filled orienteering compass works best. The base of the compass is transparent plastic, and it gives accurate readings on the run. The standard military, lensatic compass will work even though it is not specifically designed for the sport.

The best terrain for an orienteering course is woodland that offers varied terrain. Several different courses can be setup in an area 2,000 to 4,000 yards square. Courses can be short and simple for training beginners or longer and more difficult to challenge the advanced competitors.

The various types of orienteering are described below.

CROSS-COUNTRY ORIENTEERING

This popular type of orienteering is used in all international and championship events. Participants navigate to a set number of check or control points in a designated order. Speed is important since the winner is the one who reaches all the control points in the right order and returns to the finish area in the least time.

SCORE ORIENTEERING

Quick thinking and strategy are major factors in score orienteering. A competitor selects the check-points to find based on point value and location. Point values throughout the course are high or low depending on how hard the

markers are to reach. Whoever collects the most points within a designated time is the winner. Points are deducted for returning late to the finish area.

LINE ORIENTEERING

Line orienteering is excellent for training new orienteers. The route is premarked on the map, but checkpoints are not shown. The navagator tries to walk or run the exact map route. While negotiating the course, he looks for checkpoints or control-marker signs. The winner is determined by the time taken to run the course and the accuracy of marking the control points when they are found.

ROUTE ORIENTEERING

This variation is also excellent for beginners. The navigator follows a route that is clearly marked with signs or streamers. While negotiating the course, he records on the map the route being taken. Speed and accuracy of marking the route determine the winner.

NIGHT ORIENTEERING

Competitors in this event carry flashlights and navigate with map and compass. The night course for cross-country orienteering is usually shorter than the day course. Control points are marked with reflective material or dim lights. Open, rolling terrain, which is poor for day courses, is much more challenging at night.

URBAN ORIENTEERING

Urban orienteering is very similar to traditional types, but a compass, topographical map, and navigation skills are not needed. A course can be set up on any installation by using a map of the main post or cantonment area. Soldiers run within this area looking for coded location markers, which are numbered and marked on the map before the start. This eliminates the need for a compass. Soldiers only need a combination map-scorecard, a watch, and a pencil. (Figure 9-7 shows a sample scorecard.)

Urban orienteering adds variety and competition to a unit's PT program and is well suited for an intramural program. It also provides a good cardiovascular workout.

Participants and Rules

Urban orienteering is conducted during daylight hours to ensure safety and make the identification of checkpoint markers easy. Soldiers form two-man teams based on their APFT 2-mile-run times. Team members should have similar running ability. A handicap is given to slower teams. (See Figure 9-8.) At the assembly area, each team gets identical maps that show the

URBAN ORIENTEERING

LOCATION MARKER	POINT VALUE	LOCATION MARKER CODE	LOCATION MARKER	POINT VALUE	LOCATION MARKER CODE
1	10		26	10	
2	10		27	15	
3	15		28	5	
4	10		29	15	
5	15		30	15	
6	10		31	15	
7	25		32	25	
8	15		33	15	
9	25		34	15	
10	15		35	25	
11	15		36	15	
12	25		37	15	
13	15		38	25	
14	15		39	15	
15	25		40	25	
16	15		41	25	
17	25		42	15	
18	10		43	10	
19	10		44	15	
20	15		45	10	
21	10		46	25	
22	5		47	10	
23	15		48	15	
24	10		49	15	
25	10		50	10	

Figure 9-7

location of markers on the course. Location markers are color-coded on the map based on their point value. The markers farthest from the assembly area have the highest point values. The maps are labeled with a location number corresponding to the location marker on the course. A time limit is given, and teams finishing late are penalized. Five points are deducted for each minute a team is late. While on the course, team members must stay together and not separate to get two markers at once. A team that separates is disqualified. Any number of soldiers may participate, the limiting factors being space and the number of points on the course.

Playing the Game

Once the soldiers have been assigned a partner, the orienteering marshal briefs them on the rules and objectives of the game. He gives them their time limitations and a reminder about the overtime penalty. He also gives each team a combination map/scorecard with a two-digit number on it to identify their team. When a team reaches a location marker, it records on the scorecard the letters that correspond to its two-digit number.

Point values of each location marker are also annotated on the scorecard. When the orienteering marshal signals the start of the event, all competitors

HANDICAPS FOR URBAN ORIENTEERING

2-MILE RUN TIME	POINTS	2-MILE RUN TIME	POINTS
12:00 or faster	0	14:31-15:00	60
12:01-12:30	10	15:01-15:30	70
12:31-13:00	20	15:31-16:00	80
13:01-13:30	30	16:01-16:30	90
13:31-14:00	40	16:31-17:00	100
14:01-14:30	50	17:01+	100

Figure 9-8

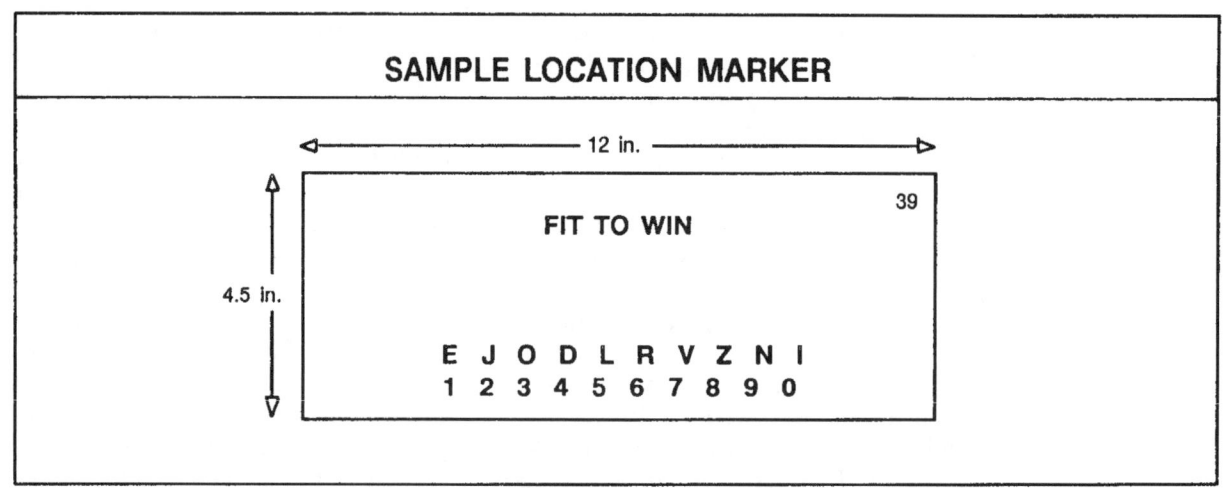

SAMPLE LOCATION MARKER

12 in.

4.5 in.

FIT TO WIN   39

E J O D L R V Z N I
1 2 3 4 5 6 7 8 9 0

Figure 9-9

leave the assembly area at the same time. One to two hours is the optimal time for conducting the activity. A sample location marker is shown at Figure 9-9.

For this example, team number 54 found the marker. The letters corresponding to 54 are LD, so they place "LD" on line 39 of their scorecard. This line number corresponds to the location's marker number. When the location marker code is deciphered, the team moves on to the next marker of its choice. Each team goes to as many markers as possible within the allotted time. After all teams have found as many location markers as possible and have turned in their map/scorecards, the points are computed by the orienteering marshal to determine the teams' standings. He has the key to all the points and can determine each team's accuracy. Handicap points are then added. Each soldier gets points if his 2-mile-run time is slower than 12 minutes. (See Figure 9-8.) The teams' standings are displayed shortly after the activity ends.

Safety Briefing

The orienteering marshal gives a safety briefing before the event starts. He reminds soldiers to be cautious while running across streets and to emphasize that team members should always stay together.

Set Up and Materials

The course must be well thought out and set up in advance. Setting up requires some man-hours, but the course can be used many times. The major tasks are making and installing location markers and preparing map/scorecard combinations. Once the location marker numbers are marked and color coded on the maps, they are covered with combat acetate to keep them useful for a long time. Combat acetate (also called plastic sheet) can be purchased in the self-service supply center store under stock number 9330-00-618-7214.

The course organizer must decide how many location markers to make and where to put them. He should use creativity to add excitement to the course. Suggestions for locations to put point markers are as follows: at intersections, along roads in the tree line, on building corners, and along creek beds and trails. They should not be too hard to find. To help teams negotiate the course, all maps must be precisely marked to correspond with the placement of the course-location markers.

Unit Olympics

Unit olympics incorporate athletic events that represent all five fitness components.

The unit olympics is a multifaceted event that can be tailored to any unit to provide athletic participation for all soldiers. The objective is to incorporate into a team-level competition athletic. events that represent all five fitness components. The competition can be within a unit or between competing units. When conducted with enthusiasm, it promotes team spirit and provides a good workout. It is a good diversion from the regular PT session.

A unit olympics, if well promoted from the top and well staged by the project NCO or officer, can be a good precursor to an SDT or the EIB test.

TYPES OF EVENTS

The olympics should include events that challenge the soldiers' muscular strength and endurance, aerobic endurance, flexibility, agility, speed, and related sports skills.

Events can be held for both individuals and teams, and they should be designed so that both male and female soldiers can take part. Each soldier should be required to do a minimum number of events. Teams should wear a distinctively marked item such as a T-shirt or arm band. This adds character to the event and sets teams apart from each other. A warm-up should precede and a cool-down should follow the events.

The following are examples of athletic events that could be included in a unit olympics:

Push-Up Derby

This is a timed event using four-member teams. The objective is for the team to do as many correct push-ups as possible within a four-minute time limit. Only one team member does push-ups at a time. The four team members may rotate as often as desired,

Sandbag Relay

This event uses four-man teams for a running relay around a quarter-mile track carrying sandbags. One player from each team lines up at the starting line with a full sandbag *in* each hand. He hands the sandbags off to a teammate when he finishes his part of the race. This continues until the last team player crosses the finish line. Placings are determined by the teams' order of finish.

Team Flexibility

In this event, if teams are numerically equal, all members of each team should participate. If not, as many team members should participate as possible. Each team's anchor person places his foot against a wall or a curb. He stretches his other foot as far away as possible as in doing a split. The next team member puts one foot against the anchor man's extended foot and does a split-stretch. This goes on until all team members are stretched. They cover as much distance as possible keeping in contact with each other. The team that stretches farthest from the start point without a break in their chain is the winner.

Medicine-Ball Throw

This event uses four-member teams. The teams begin by throwing the ball from the same starting line. When it lands, the ball is marked for each team thrower, and the next team player throws from this spot. This is repeated until all the team's players have thrown. The team whose combined throws cover the most distance is the winner.

Job-Related Events

The organizer should use his imagination when planning activities. He may incorporate soldier skills required of an MOS. For instance, he could devise a timed land-navigation event geared toward soldiers with an MOS of 11 C. The team would carry an 81 -mm mortar (tube, tripod, and baseplate) to three different locations, each a mile apart, and set it up in a firing configuration. This type of event is excellent for fine-tuning job skills and is also physically challenging.

OPENING CEREMONY

The commander, ranking person, or ceremony host gives an inspirational speech before the opening ceremonies, welcoming competitors and wishing them good luck. The olympics is officially opened with a torch lighting. This is followed by a short symbolic parade of all the teams. The teams are then put back into formation, and team captains lead motivating chants. The master of ceremonies (MC) announces the sequence of events and rules for each event. The games then begin.

JUDGING AND SCORING

The MC should have one assistant per team who will judge that one team during each event. Assistants give input on events that need a numerical count. The MC monitors the point accumulation of each team. Points are awarded for each event as follows:
• First = 4 points.
• Second = 3 points.
• Third = 2 points.
• Fourth = 1 point.

When two teams tie an event, the points are added together and split equally between them. After the competition ends, the totaled point scores for each team are figured. The first- through fourth-place teams are then recognized.

Developing the Program

Commanders must develop prgrams that train soldiers to maximize their physical performance.

The goal of the Army's physical fitness program is to improve each soldier's physical ability so he can survive and win on the battlefield. Physical fitness includes all aspects of physical performance, not just performance on the APFT. Leaders must understand the principles of exercise, the FITT factors, and know how to apply them in order to develop a sound PT program that will improve all the fitness components. To plan PT successfully, the commander and MFT must know the training management system. (See FM 25-100.)

Commanders should not be satisfied with merely meeting the minimum requirements for physical training which is having all of their soldiers pass the APFT. They must develop programs that train soldiers to maximize their physical performance. Leaders should use incentives. More importantly, they must set the example through their own participation.

The unit PT program is the commander's program. It must reflect his goals and be based on sound, scientific principles. The wise commander also uses his PT program as a basis for building team spirit and for enhancing other training activities. Tough, realistic training is good. However, leaders must be aware of the risks involved with physical training and related activities. They should, therefore, plan wisely to minimize injuries and accidents.

Steps in Planning

STEP 1: ANALYZE THE MISSION

When planning a physical fitness program, the commander must consider the type of unit and its mission. Missions vary as do the physical requirements necessary to complete them. As stated in FM 25-100, "The wartime mission drives training." A careful analysis of the mission, coupled with the commander's intent, yields the mission-essential task list (METL) a unit must perform.

Regardless of the unit's size or mission, reasonable goals are essential. According to FM 25-100, the goals should provide a common direction for all the commander's programs and systems. An example of a goal is as follows because the exceptional physical fitness of the soldier is a critical combat-multiplier in the division, it must be our goal to ensure that our soldiers are capable of roadmarching 12 miles with a 50-pound load in less than three hours.

STEP 2: DEVELOP FITNESS OBJECTIVES

Objectives direct the unit's efforts by prescribing specific actions. The commander, as tactician, and the MFT, as physical fitness advisor, must analyze the METL and equate this to specific fitness objectives. Examples of fitness objectives are the following:

- Improve the unit's overall level of strength by ensuring that all soldiers in the unit can correctly perform at least one repetition with 50 percent of their bodyweight on the overhead press using a barbell.
- Improve the unit's average APFT score through each soldier obtaining a minimum score of 80 points on the push-up and sit-up events and 70 points on the 2-mile run.
- Decrease the number of physical training injuries by 25 percent through properly conducted training.

The commander and MFT identify and prioritize the objectives.

STEP 3: ASSESS THE UNIT

With the training objectives established, the commander and MFT are ready to find the unit's current fitness level and measure it against the desired level.

Giving a diagnostic APFT is one way to find the current level. Another way is to have the soldiers road march a certain distance within a set time while carrying a specified load. Any quantifiable, physically demanding, mission-essential task can be used as an assessment tool. Training records and reports, as well as any previous ARTEP, EDREs, and so forth, can also provide invaluable information.

STEP 4: DETERMINE TRAINING REQUIREMENTS

By possessing the unit's fitness capabilities and comparing them to the standards defined in training objectives, leaders can determine fitness training requirements. When, after extensive training, soldiers cannot reach the desired levels of fitness, training requirements may be too idealistic. Once training requirements are determined, the commander reviews higher headquarters' long- and short-range training plans to identify training events and allocations of resources which will affect near-term planning.

STEP 5: DEVELOP FITNESS TASKS

Fitness tasks provide the framework for accomplishing all training requirements. They identify what has to be done to correct all deficiencies and sustain all proficiencies. Fitness tasks establish priorities, frequencies, and the sequence for training requirements. They must be adjusted for real world constraints before they become a part of the training plan. The essential elements of fitness tasks can be cataloged into four groups:
(1) Collective tasks
(2) Individual tasks
(3) Leader tasks
(4) Resources required for training

Collective tasks. Collective tasks are the training activities performed by the unit. They are keyed to the unit's specific fitness objectives. An example would be to conduct training to develop strength and muscular endurance utilizing a sandbag circuit.

Individual tasks. Individual tasks are activities that an individual soldier must do to accomplish the collective training task. For example, to improve CR endurance the individual soldier must do ability-group running, road marching, Fartlek training, interval training, and calculate/monitor his THR when appropriate.

Leader tasks. Leader tasks are the specific tasks leaders must do in order for collective and individual training to take place. These will involve procuring resources, the setting up of training, education of individual soldiers, and the supervision of the actual training.

Resources. Identifying the necessary equipment, facilities, and training aids during the planning phase gives the trainer ample time to prepare for the training. The early identification and acquisition of resources is necessary to fully implement the training program. The bottom line is that training programs must be developed using resources which are available.

STEP 6: DEVELOP A TRAINING SCHEDULE

The fitness training schedule results from leaders' near-term planning. Leaders must emphasize the development of all the fitness components and follow the principles of exercise and the FITT factors. The training schedule shows the order, intensity, and duration of activities for PT. Figure 10-1 illustrates a typical PT session and its component parts.

There are three distinct steps in planning a unit's daily physical training activities. They are as follows:
1. Determine the minimum frequency of training. Ideally, it should include three cardiorespiratory and three muscular conditioning sessions each weeks. (See the FITT factors in Chapter 1.)
2. Determine the type of activity. This depends on the specific purpose of the training session. (See Figure 10-2.) For more information on this topic, see Chapters 1, 2, and 3.
3. Determine the intensity and time of the selected activity. (See the FITT factors in Chapter 1.)

Each activity period should include a warm-up, a workout that develops cardiorespiratory fitness and/or muscular endurance and strength, and a cool-down. (See Figure 10-1).

At the end of a well-planned and executed PT session, all soldiers should feel that they have been physically stressed. They should also understand the objective of the training session and how it will help them improve their fitness levels.

STEP 7: CONDUCT AND EVALUATE TRAINING

The commander and MFT now begin managing and supervising the day-to-day training. They evaluate how the training is performed by monitoring its intensity, using THR or muscle failure, along with the duration of the daily workout.

The key to evaluating training is to determine if the training being conducted will result in improvements in physical conditioning. If not, the training needs revision. Leaders should

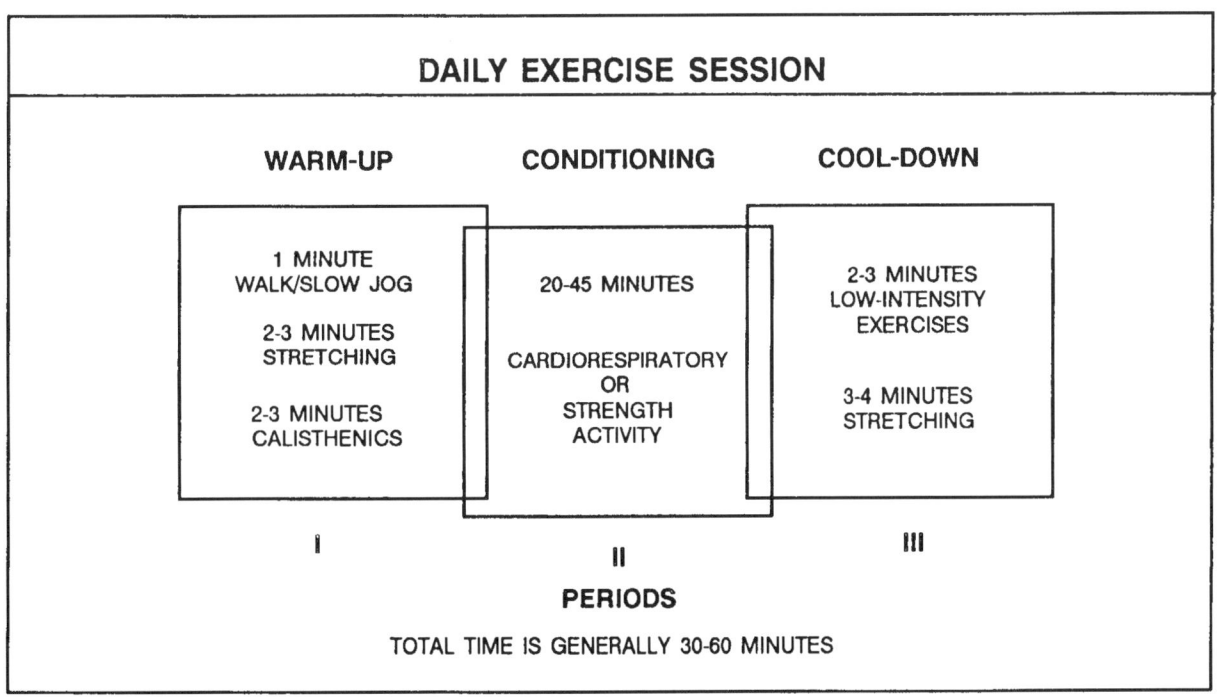

Figure 10-1

<table>
<tr><td colspan="10">ACTIVITY SELECTION GUIDE</td></tr>
</table>

PURPOSE	MUSCULAR STRENGTH	MUSCULAR ENDURANCE	CARDIO-RESPIRATORY ENDURANCE	FLEXIBILITY	BODY COMPOSITION	SPEED/AGILITY	COORDINATION	TEAMWORK	SOLDIER SKILLS
Aerobics		x	x	x	x		x		
Bicycling		x	x		x				
Circuits	x	x	x	x	x	x	x	x	x
Competitive Activities						x	x	x	x
Calisthenics		x		x		x	x		
Cross Country Skiing	x	x	x	x	x		x		
Grass/Guerrilla Drills	x	x	x		x		x		
Obstacle Courses	x	x	x		x	x	x	x	x
Partner-Resisted Exercises	x	x					x	x	
Relays		x	x		x	x	x	x	
Rifle Drills	x	x					x		x
Road Marching	x	x	x		x				x
Running		x	x		x				
Stretching				x					
Weight Training	x	x				x	x		

Figure 10-2

not be sidetracked by PT that is all form and little substance. Such training defeats the concept of objective-based training and results in little benefit to soldiers.

Education

Teaching soldiers about physical fitness is vital. It must be an ongoing effort that uses trained experts like MFTs. Soldiers must understand why the program is organized the way it is and what the basic fitness principles are. When they know why they are training in a certain way, they are more likely to wholeheartedly take part. This makes the training more effective.

Education also helps the Army develop its total fitness concept. Total fitness should be reinforced throughout each soldier's career. Classroom instruction in subjects such as principles of exercise, diet and nutrition, tobacco cessation, and stress management should be held at regular intervals. Local "Fit to Win" coordinators (AR 600-63) can help develop classes on such subjects.

Common Errors

There are some common errors in unit programs. The most common error concerns the use of unit runs. When all soldiers must run at the same pace as with a unit run, many do not receive a training effect because they do not reach their training heart rate (THR). The least-fit soldiers of the unit may be at risk because they may be training at heart rates above their THR. Another error is exclusively using activities such as the "daily dozen." These exercises emphasize form over substance and do little to improve fitness.

Yet another error is failing to strike a balance in a PT program between CR endurance training and muscular endurance and strength training. In addition, imbalances often stem from a lack of variety in the program which

Total fitness should be reinforced throughout each soldier's career by classroom instruction.

leads to boredom. The principles of exercise are described in Chapter 1, and their application is shown in the sample program below.

A Sample Program

The following sample program shows a commander's thought processes as he develops a 12-week fitness training program for his unit.

Captain Frank Jones's company has just returned from the field where it completed an ARTEP. Several injuries occurred including a broken foot, resulting from a dropped container, and three low back strains. After evaluating his unit during this ARTEP, CPT Jones concluded that its level of physical fitness was inadequate. He thought this contributed to the injuries and poor performance. The soldiers' flexibility was poor, and there was an apparent lack of prior emphasis on, and training in, good lifting techniques. This, combined with poor flexibility in the low back and hamstrings, may have contributed to the unacceptably high number of low back strains. Captain Jones decided to ask the battalion's MFT to help him develop a good unit program for the company. They went through the following steps.

7-STEP PLANNING PROCESS

ANALYZE THE MISSION

DEVELOP FITNESS OBJECTIVES

ASSESS THE UNIT

DETERMINE TRAINING REQUIREMENTS

DESIGN FITNESS TASKS

DEVELOP A TRAINING SCHEDULE

CONDUCT AND EVALUATE TRAINING

ANALYZE THE MISSION

First, they analyzed the recently completed ARTEP and reviewed the ARTEP manual to find the most physically demanding, mission-oriented tasks the unit performs. The analysis showed that, typically, the company does a tactical road march and then occupies a position. It establishes a perimeter, improves its positions, and selects and prepares alternate positions. One of the most demanding missions while in position requires soldiers to move by hand, for 15 to 30 minutes, equipment weighing up to 95 pounds. If his unit received artillery fire, it would need to be able to move to alternate positions as quickly as possible. This requires much lifting, digging, loading, unloading, and moving of heavy equipment. All of these tasks require good muscular endurance and strength and a reasonable level of cardiorespiratory endurance.

DEVELOP FITNESS OBJECTIVES

Next, CPT Jones reviewed his battalion commander's physical training guidance. It showed that the commander was aware that the unit's tasks require muscular endurance and strength and cardiorespiratory fitness. The guidance and objectives issued are as follows:

a. Units will do PT five days a week (0600-0700) when in garrison. In the field, organized PT will beat the commander's discretion.

Captain Jones determined that the major PT emphasis should be to improve muscular endurance and strength. He based this on his unit's mission, training schedule, available resources, and on his commander's guidance and objectives. With this information and the MIT's recommendations, CPT Jones developed the following fitness objectives.

- Improve the unit's overall level of muscular endurance and strength.
- Improve the unit's overall level of flexibility.
- Improve the unit's average APFT score. Each soldier will score at least 80 points on the push-up and

sit-up events and 70 points on the 2-mile run.

- Improve the unit's road marching capability so that 100 percent of the unit can complete a 12-mile road march with a 35-pound load in at least 3.5 hours.
- Decrease the number of profiles.
- Reduce tobacco use.

ASSESS THE UNIT

The next step CPT Jones accomplished was to assess his unit.

The MFT studied the results of the unit's latest APFT and came up with the following information:

- The average push-up score was 68 points.
- The average sit-up score was 72 points.
- The average number of points scored on the 2-mile run was 74.
- There were six failures, two on the 2-mile run and four on the push-up.

The MFT also recommended that the unit be assessed in the following areas: road march performance, strength, flexibility, substance abuse, and profiled soldiers.

Following the MFT's recommendations, subordinate leaders made the following assessments/determinations:

- Eighty-eight percent of the company finished the 12-mile road march with a 35-pound load in under 3 hours 30 minutes.
- A formation toe-touch test revealed that over half the company could not touch their toes while their knees were extended.
- Thirty percent of the unit uses tobacco.
- Two soldiers are in the overweight program.
- Eight percent of the unit is now on temporary profile, most from back problems.

DETERMINE TRAINING REQUIREMENTS

The next step CPT Jones accomplished was to determine the training requirements.

Training requirements are determined by analyzing the training results and the data obtained from the unit assessment. The next step is to compare this data to the standards identified in the training objectives. When performance is less than the established standard, the problem must be addressed and corrected.

Captain Jones established the following training requirements.

Units will do flexibility exercises during the warm-up and cool-down phase of every PT session. During the cool-down, emphasis on will be placed on developing flexibility in the low back, hamstrings, and hip extensor muscle groups.

Each soldier will do 8 to 12 repetitions of bent-leg, sandbag dead-lifts at least two times a week to develop strength. The section leader will supervise lifts.

Each soldier will do heavy resistance/weight training for all the muscle groups of the body two to three times a week.

Each soldier will perform timed sets of push-ups and sit-ups.

Each soldier will train at least 20 to 30 minutes at THR two to three times a week.

Road marches will be conducted at least once every other week.

Tobacco cessation classes will be established to reduce the number of tobacco users.

DESIGN FITNESS TASKS

Once all training requirements are identified, the next step is to use them to design fitness tasks which relate to the fitness objectives. In developing the fitness tasks, CPT Jones must address collective, individual, and leader tasks as well as resources required.

Fitness tasks provide the framework for accomplishing the training requirements. By accurately listing the fitness tasks that must be done and the resources required to do them, the subsequent step of developing a training schedule is greatly facilitated.

An example of designing fitness tasks is provided in Figure 10-3 by using the activities which might occur during one week of physical training.

The collective tasks for the unit are to perform the following: develop muscular endurance and strength, improve CR endurance, and improve flexibility.

The individual tasks all soldiers must perform during the week are as follows. For developing strength and muscular endurance, they must perform appropriate strength circuit exercises, PREs, sandbag circuits, to include performing bent-leg dead lifts exercises, and training for push-up/sit-up improvement. To improve cardiorespiratory endurance, they must

do ability-group runs, interval training, road marching, and they must calculate their THR and monitor THR when appropriate. To improve their flexibility, they must do stretching exercises during their daily warm-up and cool-down.

The leader's tasks are to organize and supervise all strength- and muscle endurance-training sessions and CR training sessions so as to best meet all related fitness objectives. Similarly, the leader must organize and supervise all warm-up and cool-down sessions to best meet the fitness objectives for the development and maintenance of flexibility.

To provide specific examples of leaders tasks in the area of training for strength and muscle endurance, the leader will ensure the following:

● Each strength- and/or muscle endurance-training session works all the major muscle groups of the body.
● High priority is given to training those muscles and muscle groups used in mission-essential tasks.
● Areas where weaknesses exist, with respect to strength/muscle

FITNESS TASKS FOR ONE WEEK OF PHYSICAL TRAINING

COLLECTIVE	INDIVIDUAL	LEADER	RESOURCES
Improve strength and muscle endurance	Do STR CIR EX*, PRE, SNDBG CIR, SU-PU IMP	Organize & supervise STR CIR EX, PRE, & SNDBG CIR	STR RM, Gym, Sandbags, PT Field
Improve CR endurance	Do AGR, CAL/MON THR, road march, do intervals (4x440) IND AB	Organize & supervise CR workouts, CAL/MON THR, MON work/relief ratio for intervals	Track, Running Trails
Improve flexibility	Do stretching exercises	Organize & supervise activity	Gym

* A list of abbreviations appears at the end of Figure 10-4.

Figure 10-3

endurance, are targeted in all workouts.

- Problem areas related to APFT performance are addressed in appropriate workouts.
- The duration of each strength training session is 20-40 minutes.
- Soldiers train to muscle failure.
- All the principles of exercise, to include regularity, overload, recovery, progression, specificity, balance are used.

In a similar manner, the leader would ensure that the guidelines and principles outlined in this and earlier chapters are used to organize training sessions for improving CR endurance and flexibility.

The resources needed for the one-week period are as follows: a strength room, a gym, a PT field, a running track and/or running trails, and sandbags.

DEVELOP A TRAINING SCHEDULE

The next step was to develop a fitness training schedule (shown at Figure 10-4). It lists the daily activities and their intensity and duration.

12-WEEK TRAINING PLAN				
JULY				
MONDAY	TUESDAY	WEDNESDAY	THURSDAY	FRIDAY
START ASSESSMENT* ACT: AGR** INT: 70% HRR*** DUR: 20 MIN	FINISH ASSESSMENT ACT: PLT 1 & 2 STR CIR; PLT 3 & 4 SNDBG CIR/PU-SU IMP INT: MF/MF DUR: 30/4 MIN	ACT: AGR/LINE SOCCER INT: 70% HRR/NA DUR: 20/30 MIN	ACT: PRE/PU-SU IMP INT: MF DUR: 35/4 MIN	ACT: ROAD MARCH, 5 MLE W 35 LBS IN 90 MIN
ACT: PLT 1 & 2 WT STR CIR; PLT 3 & 4 SNDBG CIR/PU-SU IMP INT: MF DUR : 30-35/4 MIN	ACT: AGR/ PAR COURSE INT: 70% HRR DUR: 20/15-20 MIN	FLIP - FLOP MONDAY's WORKOUT	ACT: AGR/GDR INT: 70% HRR DUR: 20/15-20 MIN	ACT: PRE/PU-SU IMP INT: MF DUR: 35/4 MIN
ACT: FIXED CIR I INT: 70% HRR DUR: 30-40 MIN	ACT: PRE/PU-SU IMP INT: MF DUR: 40/5 MIN	ACT: AGR/GDR INT: 70% HRR DUR: 22/20 MIN	ACT: SNDBG CIR/PU-SU IMP INT: MF DUR: 35-40/5 MIN	ACT: ROAD MARCH, 5 MLE W 35 LBS IN 90 MIN
ACT: PLT 1 & 2 STR CIR; PLT 3 & 4 SNDBG CIR/PU-SU IMP INT: MF DUR: 30-40/5 MIN	ACT: AGR INT: 70% HRR DUR: 25 MIN	FLIP - FLOP MONDAY'S WORKOUT	ACT: AGR INT: 70% HRR DUR: 25 MIN	ACT: OBS CRS/ PRE/PU-SU IMP INT: MF DUR: 25/20/5 MIN

* Initially, assessments must be made of each soldier's level of physical fitness. Particularly important is assessing a soldier's strength and muscular endurance by determining his 8-12 RM for each resistance exercise he will be doing. As mentioned in the Phases of Conditioning section in Chapter 3, this will take two weeks and should be planned for accordingly. The other components of fitness should also be addressed as the need arises.
** A list of abbreviations and acronyms appears at the end of this training plan.
*** Those soldiers with a fairly good level of CR fitness (that is, the average soldier) should exercise at about 70 percent HRR. Those with very high levels of CR fitness may benefit most from training at around 80 to 85 percent HRR during a CR training workout.

Figure 10-4

12-WEEK TRAINING PLAN

AUGUST

MONDAY	TUESDAY	WEDNESDAY	THURSDAY	FRIDAY
ACT: AGR INT: 70% HRR DUR: 27 MIN	ACT: PLT 1 & 2 SNDBG CIR; PLT 3 & 4 PRE/PU-SU IMP INT: MF DUR: 40/6	ACT: INTERVALS INT: 8 x 440 IND AB DUR: 45 MIN	FLIP - FLOP TUESDAY'S WORKOUT	ACT: ROAD MARCH, 7.5 MLE W 35 LBS IN 2.5 HOURS
ACT: PLT 1 & 2 SNDBG CIR; PLT 3 & 4 STR CIR/PU SU IMP INT: MF DUR: 40/6 MIN	ACT: LAST MAN-UP RUN IN AG/ PAR CRS INT: 70-80% HRR*/70%HRR DUR: 30/20 MIN	FLIP - FLOP MONDAY'S WORKOUT	ACT: AGR/FIT-NESS RELAYS INT: 70% HRR/NA DUR: 30/20 MIN	ACT: PLT 1 & 2 SNDBG CIR; PLT 3 & 4 PRE/ PU-SU IMP INT: MF DUR: 40/6 MIN
IN FIELD: PLAN FOR	IN FIELD PLAN FOR	IN FIELD: PLAN FOR	IN FIELD: PLAN FOR	IN FIELD: PLAN FOR
ACT: LAST MAN-UP RUN IN AG INT: 70-90% HRR* DUR: 32 MIN	ACT: PRE/PU-SU IMP INT: MF DUR: 40/7 MIN	ACT: FARTLEK IN AG INT: 60-90% HRR* DUR: 32 MIN	ACT: PRE/PU-SU IMP INT: MF DUR: 40/8 MIN	ACT: ROAD MARCH, 10 MLE W 35 LBS IN 3.5 HOURS
ACT: PRE/PU-SU IMP INT: MF DUR: 35/10 MIN	ACT: INTERVALS INT: 8 x 440 IND AB DUR: 45 MIN	ACT: PU-SU, PULL-UP IMP INT: MF DUR: 45 MIN	ACT: FARTLEK IN AG INT: 60-90% HRR* DUR: 35 MIN	ACT: PU-SU, PULL-UP IMP INT: MF DUR: 45 MIN LC: APFT FOR GRADERS

* During the Last-Man-Up and Fartlek running, the heart rate will vary depending on whether it is taken during the slower or the faster portion of the run. The smaller and larger numbers provided for percent HRR should set the lower and upper limits, respectively, for a soldier's heart rate during this type of training. During interval running, the soldier should concern himself with running at the appropriate pace; he should not monitor THR during interval work.

Figure 10-4 (continued)

12-WEEK TRAINING PLAN

SEPTEMBER

MONDAY	TUESDAY	WEDNESDAY	THURSDAY	FRIDAY
ACT: DEVEL-OPMENTAL STRETCHING INT: SLIGHT TENSION, NOT PAIN DUR: 20-30 MIN	ACT: APFT DUR: NA	ACT: UNIT OLYMPICS, PART I DUR: NA	ACT: UNIT OLYMPICS, PART II DUR: NA	ACT: APFT & OLYMPIC AWARDS CEREMONY/ UNIT RUN INT: NA/CD DUR: NA/30-40 MIN
ACT: PLT 1 & 2 STR CIR; PLT 3, & 4 PRE INT: MF DUR: 40 MIN	ACT: PLT 1 & 2 AGR; PLT 3 & 4 EX TO MUSIC INT: 70% HRR DUR: 35/45 MIN	FLIP - FLOP MONDAY'S WORKOUT	ACT: ROAD MARCH, 10 MLE W 35 LBS IN 3 HOURS	ACT: PLT 1 & 2 SNDBG CIR; PLT 3 & 4 PRE INT; MF DUR: 40 MIN
ACT: AGR INT: 70% HRR DUR: 35 MIN	ACT: PRE/PU-SU IMP INT: MF DUR: 40/8 MIN	ACT: FIXED CIRCUIT/RELAYS INT: 70% HRR/NA DUR: 20/20 MIN	ACT: UPPER BODY PRE/PU-SU IMP INT: MF DUR: 30/8 MIN	ACT: ROAD MARCH, 12 MLE W 35 LBS IN UNDER 4 HOURS
ACT: PLT 1 & 2 LOG DRILLS; PLT 3 & 4 SNDBG CIR/PU-SU IMP INT: MF DUR: 30/8 MIN	ACT: PLT 1 & 2 FIXED CIR; PLT 3 & 4 AQUATICS INT: 70% HRR/NA DUR: 30 MIN	FLIP - FLOP MONDAY'S WORKOUT	FLIP - FLOP TUESDAY'S WORKOUT	ACT: PRE/PU-SU IMP INT: MF DUR: 35/8 MIN

NOTES

1. Push-ups and sit-ups are done as part of each strength workout. In the above sessions, they have been placed near the end of the workout. However, they can occasionally be done before the strength workout for variety. An example of a beginning PU-SU improvement workout lasting about three minutes follows:

   a. Perform one timed set of push-ups for 50 seconds. Follow this immediately with one 50-second, timed set of sit-ups. For all timed sets, each soldier must perform as many repetitions of the exercise as he can during the alloted time period.

   b. Perform a second set of push-ups for 40 seconds. Follow this immediately with a timed set of sit-ups of equal duration.

   As the soldier adapts to this, the difficulty of the session can be increased by adding more timed sets and/or by decreasing the rest interval between like or unlike sets of exercises. For example, the rest period between timed sets of push-ups and sit-ups can be decreased. Also, all of the timed sets for push-ups may be done back-to-back (as can the sit-ups), the rest interval between these timed sets of push-ups can be progressively reduced to make the workout more demanding. Many more options exist for increasing the difficulty of, and adding variety to, these sessions.

2. Activities are planned for the FTX; duration is determined on site.

3. The unit's olympic events include the following:

   a. Ammo-box shuttle-run (fastest time by section).

   b. Biceps, barbell curl (most reps with 60 lbs., total by section).

   c. Leg press (most weight lifted by section).

   d. Standing-toe touch (most soldiers touching toes by section/must hold five seconds).

   e. Highest APFT score by section.

Figure 10-4 (continued)

ABBREVIATIONS AND ACRONYMS

ACT	activity
AG	ability groups
AGR	ability group run
ANA ACT	anaerobic activity
AOTR	assessment of training requirements
CAL/MON	calculate/monitor
CD	commander's decision
CFA	competitive fitness activities
CIR	circuit
CR	cardiorespiratory training
DL	dead-lift (bent-leg)
EX	exercise
GDR	grass drills
GUD	guerilla drills
HRR	heart rate reserve
IMP	improvement
IND AB	at the individual's ability
INT	intensity
LBS	pounds
LC	leader's class
MIN	minute(s)
MF	muscle failure (due to fatigue)
MLE	mile(s)
MS/E	muscle strength/endurance
NA	not applicable
OBS CRS	obstacle course
PLT	platoon
PRE	partner-resisted exercise(s)
PT FLD	physical training field
PU-SU IMP	push-up, sit-up improvement
R	run
SNDBG	sandbag
STR CIR EX	strength circuit exercise
STR RM	strength room
THR	training heart rate
TNG	training
W	with
WT STR CIR	strength circuit with weights
% HRR	percent of heart rate reserve
2-MR	2-mile run

Figure 10-4 (continued)

CONDUCT AND EVALUATE TRAINING

Conducting and evaluating training is the final phase of the training process. This phase includes the evaluation of performance, assessment of capabilities, and feedback portions of the training management cycle. These portions of the cycle must be simultaneous and continuous. To be effective, the evaluation process must address why weaknesses exist, and it must identify corrective actions to be taken. Evaluations should address the following:

● Assessment of proficiency in mission-essential tasks.
● Status of training goals and objectives.
● Status of training in critical individual and collective tasks.
● Shortfalls in training.
● Recommendations for next training cycle (key in on correcting weaknesses).
● Results of educational programs.

Using the Principles of Exercise

As CPT Jones developed his program, he made sure he used the seven principles of exercise. He justified his program as follows:

● Balance. This program is balanced because all the fitness components are addressed. The emphasis is on building muscular endurance and strength in the skeletal muscular system because of the many lifting tasks the unit must do. The program also trains cardiorespiratory endurance and flexibility, and warm-up and cool-down periods are included in every workout.
● Specificity. The unit's fitness goals are met. The sand-bag lifting and weight training programs help develop muscular endurance and strength. The movements should, when possible, stress muscle groups

used in their job-related lifting tasks. Developmental stretching should help reduce work-related back injuries. The different types of training in running will help ensure that soldiers reach a satisfactory level of CR fitness and help each soldier score at least 70 points on the APFT's 2-mile run. Soldiers do push-ups and sit-ups at least two or three times a week to improve the unit's performance in these events. The competitive fitness activities will help foster teamwork and cohesion, both of which are essential to each section's functions.

● Overload. Soldiers reach overload in the weight circuit by doing each exercise with an 8- to 12-RM lift for a set time and/or until they reach temporary muscle failure. For the cardiorespiratory workout, THR is calculated initially using 70 percent of the HRR. They do push-ups and sit-ups in multiple, timed sets with short recovery periods to ensure that muscle failure is reached. They also do PREs to muscle failure.

● Progression. To help soldiers reach adequate overload as they improve, the program is made gradually more difficult. Soldiers progress in their CR workout by increasing the time they spend at THR up to 30 to 45 minutes per session and by maintaining THR. They progress on the weight training circuit individually. When a soldier can do an exercise for a set time without reaching muscle failure, the weight is increased so that the soldier reaches muscle failure between the 8th and 12th repetition again. Progression in push-ups and sit-ups involves slowly increasing the duration of the work intervals.

● Variety. There are many different activities for variety. For strength and muscular endurance training the soldiers use weight circuits, sandbag circuits, and PREs. Ability group runs, intervals, Par courses,

Fartlek running, and guerrilla drills are all used for CR training. Varied stretching techniques, including static, partner-assisted, and contract-relax, are used for developmental stretching.

- Regularity. Each component of fitness is worked regularly. Soldiers will spend at least two to three days a week working each of the major fitness components. They will also do push-ups and sit-ups regularly to help reach their peak performance on the APFT.
- Recovery. The muscular and cardiorespiratory systems are stressed in alternate workouts. This allows one system to recover on the day the other is working hard.

Conclusion

CPT Jones's step-by-step process of developing a sound PT program for his unit is an example of what each commander should do in developing his own unit program.

Good physical training takes no more time to plan and execute than does poor training. When commanders use a systematic approach to develop training, the planning process bears sound results and the training will succeed.

CHAPTER 11

Physical Training During Initial Entry Training

Soldiers report to initial entry training (IET) ranging widely in their levels of physical fitness. Because of this, there are special considerations when designing a physical training program for IET soldiers. Physical training involves safely training and challenging all soldiers while improving their fitness level to meet required standards. The regulations which govern the conduct of physical training in IET and explain the graduation requirements are TRADOC Reg. 350-6 and AR 350-15.

The mission of physical training in IET is twofold: to safely train soldiers to meet the graduation requirements of each course and to prepare soldiers to meet the physical demands of their future assignments.

Program Development

All physical training programs in IET must do the following: 1) progressively condition and toughen soldiers for military duties; 2) develop soldiers' self-confidence, discipline, and team spirit; 3) develop healthy life-styles through education; and, 4) improve physical fitness to the highest levels possible in all five components of physical fitness (cardiorespiratory endurance, muscular strength, muscular endurance, flexibility, and body composition).

Because each IET school is somewhat different, commanders must examine the graduation requirements for the course and establish appropriate fitness objectives. They can then design a program that attains these objectives. The seven principles of exercise outlined in Chapter 1 are universal, and they apply to all PT programs including those in IET. Commanders of initial entry training should look beyond the graduation requirements of their own training course to ensure that their soldiers are prepared for the physical challenges of their future assignments. This means developing safe training programs which will produce the maximum physical improvement possible.

MFTs are skilled at assessing soldiers' capabilities. They use the five components of physical fitness in designing programs to reach the training objectives established by the commander. They also know how to conduct exercise programs that are effective and safe. MFTs are not, however, trained to diagnose or treat injuries.

The commander's latitude in program development varies with the length and type of the IET course. For example, commanders of basic combat training (BCT) may do a standard PT program at one installation, while AIT commanders may design their own programs. Regardless of the type of course, all leaders must strive to train their soldiers to attain the highest level of physical fitness possible. This means using the established principles of exercise to develop a safe physical training program.

Safety Considerations

Overuse injuries are common in IET. However, they can be avoided by carefully following the exercise principles of "recovery" and "progression."

Research suggests that soldiers are more prone to injuries of the lower extremities after the third week of IET. High-impact activities, such as road marching and running on hard surfaces, should be carefully monitored during at this time. During this period, fixed circuits and other activities that develop CR fitness are good, low-impact alternatives.

Properly fitted, high-quality running shoes are important, especially when PT sessions require running on hard surfaces. Court shoes, like basketball or tennis shoes, are not designed to absorb the repetitive shock of running. Activities such as running obstacle courses and road marching require combat boots to protect and support the feet and ankles. Naturally, common sense dictates a reasonable break-in period for new combat boots, especially before long marches.

Examples of recommended PT sessions and low-risk exercises are in Chapter 7. Specific health and safety considerations are in TRADOC Reg. 350-6, paragraph 4-2.

Road Marching

One road march should be conducted weekly with the difficulty of the marches progressing gradually throughout IET.

In the first two weeks of IET, soldiers can be expected to road march up to 5 kilometers with light loads. Loads should be restricted to the standard LCE, kevlar helmet, and weapon. Bones, ligaments, and tendons respond slowly to training and may be injured if the load and/or duration are increased too quickly.

After the initial adaptations in the early weeks of IET, soldiers can be expected to carry progressively heavier loads including a rucksack. By he start of the fourth week, they should be accustomed to marching in boots, and their feet should be less prone to blistering. By the sixth week, the load may be increased to 40 pounds including personal clothing and equipment. At no time during IET or one-station unit training (OSUT) should loads exceed 40 pounds.

A sample regimen for road marches during IET is at Figure 11-1.

SAMPLE ROAD MARCH PROGRAM

WEEK	DISTANCE	EQUIPMENT	REMARKS
1	5 km	LCE, kevlar helmet, weapon	
2	5 km	SAME AS WEEK 1	
3	7 km	LCE, kevlar helmet, weapon, and 10-pound rucksack	With all equipment, the total load is 30 pounds.
4	7 km	SAME AS WEEK 3	SAME TOTAL LOAD AS WEEK 3
5	7 km	SAME AS WEEK 3	SAME TOTAL LOAD AS WEEK 3
6	10 km	LCE, hevlar helmet, weapon, and 20-pound rucksack	With all equipment, the total load is 40 pounds.
7	10 km	SAME AS WEEK 6	SAME TOTAL LOAD AS WEEK 6
8	10 km	SAME AS WEEK 6	SAME TOTAL LOAD AS WEEK 6

Note: The total load carried (to include the LCE, kevlar helmet, weapon, and ruck load) should not exceed that shown in the remarks column. If the road marches are to or from non-tactical training, they need not be tactical road marches.

Figure 11-1

CHAPTER 12

Environmental Considerations

In today's Army, soldiers may deploy anywhere in the world. They may go into the tropical heat of Central America, the deserts of the Middle East, the frozen tundra of Alaska, or the rolling hills of Western Europe. Each environment presents unique problems concerning soldiers' physical performance. Furthermore, physical exertion in extreme environments can be life-threatening. While recognizing such problems is important, preventing them is even more important. This requires an understanding of the environmental factors which affect physical performance and how the body responds to those factors.

Temperature Regulation

The body constantly produces heat, especially during exercise. To maintain a constant normal temperature, it must pass this heat on to the environment. Life-threatening circumstances can develop if the body becomes too hot or too cold. Body temperature must be maintained within fairly narrow limits, usually between 74 and 110 degrees Fahrenheit. However, hypothermia and heat injuries can occur within much narrower limits. Therefore, extreme temperatures can have a devastating effect on the body's ability to control its temperature.

Overheating is a serious threat to health and physical performance. During exercise, the body can produce heat at a rate 10 to 20 times greater than during rest. To survive, it must get rid of the excess heat.

The four ways in which the body can gain or lose heat are the following:

- Conduction-the transfre of heat from a warm object to a cool one that is touching it. (Warming boots by putting them on is an example.)
- Convection-the transfer of heat by circulation or movement of air. (Using a fan on a hot day is an example.)
- Radiation-the transfer of heat by electromagnetic waves. (Sitting under a heat lamp is an example.)
- Evaporation- the transfer of heat by changing a liquid into a gas. (Evaporating sweat cooling the skin is an example.)

Heat moves from warm to cool areas. During exercise, when the body is extremely warm, heat can be lost by a combination of the four methods. Sweating, however, is the body's most important means for heat loss, especially during exercise. Any condition that slows or blocks the transfer of heat from the body by evaporation causes heat storage which results in an increase in body temperature.

The degree to which evaporative cooling occurs is also directly related to the air's relative humidity (a measure of the amount of water vapor in the air). When the relative humidity is 100 percent, the air is completely saturated at its temperature. No more water can evaporate into the surrounding air. As a result, sweat does not evaporate, no cooling effect takes place, and the body temperature increases. This causes even more sweating. During exercise in the heat, sweat rates of up to two quarts per hour are not uncommon.

If the lost fluids are not replaced, dehydration can occur. This condition, in turn, can result in severe heat injuries.

Thus, in hot, humid conditions when a soldier's sweat cannot evaporate, there is no cooling effect through the process of evaporation. High relative humidities combined with high temperatures can cause serious problems. Weather of this type occurs in the tropics and equatorial regions such as Central America and southern Asia. These are places where soldiers have been or could be deployed.

Heat Injuries and Symptoms

The following are common types of heat injuries and their symptoms.
● Heat cramps-muscles cramps of the abdomen, legs, or arms.
● Heat exhaustion-headache, excessive sweating, dizziness, nausea, clammy skin.
● Heat stroke-hot, dry skin, cessation of sweating, rapid pulse, mental confusion, unconsciousness.

Adapting to differing environmental conditions is called acclimatization.

To prevent heat injuries while exercising, trainers must adjust the intensity to fit the temperature and humidity. They must ensure that soldiers drink enough water before and during the exercise session. Body weight is a good gauge of hydration. If rapid weight loss occurs, dehydration should be suspected. Plain water is the best replacement fluid to use. Highly concentrated liquids such as soft drinks and those with a high sugar content may hurt the soldier's performance because they slow the absorption of water from the stomach.

To prevent heat injuries, the following hydration guidelines should be used:
● Type of drink: cool water (45 to 55 degrees F).
● Before the activity: drink 13 to 20 ounces at least 30 minutes before.
● During the activity: drink 3 to 6 ounces at 15 to 30 minute intervals.
● After the activity: drink to satisfy thirst, then drink a little more.

Acclimatization to Hot, Humid Environments

Adapting to differing environmental conditions is called acclimatization. Soldiers who are newly introduced to a hot, humid climate and are moderately active in it can acclimatize in 8 to 14 days. Soldiers who are sedentary take much longer. Until they are acclimatized, soldiers are much more likely to develop heat injuries.

A soldier's ability to perform effectively in hot, humid conditions depends on both his acclimatization and level of fitness. The degree of heat stress directly depends on the relative workload. When two soldiers do the same task, the heat stress is less for the soldier who is in better physical condition, and his performance is likely to be better. Therefore, it is important to maintain high levels of fitness.

Increased temperatures and humidity cause increased heart rates. Consequently, it takes much less effort to elevate the heart rate into the training zone, but the training effect is the same. These facts underscore the need to use combat-development running and to monitor heart rates when running, especially in hot, humid conditions.

Some important changes occur as a result of acclimatization to a hot climate. The following physical adaptations help the body cope with a hot environment

- Sweating occurs at a lower body temperature.
- Sweat production is increased.
- Blood volume is increased.
- Heart rate is less at any given work rate.

Exercising in Cold Environments

Contrary to popular belief, there are few real dangers in exercising at temperatures well below freezing. Since the body produces large amounts of heat during exercise, it has little trouble maintaining a normal temperature. There is no danger of freezing the lungs. However, without proper precautions, hypothermia, frostbite, and dehydration can occur.

HYPOTHERMIA

If the body's core temperature drops below normal, its ability to regulate its temperature can become impaired or lost. This condition is called hypothermia. It develops because the body cannot produce heat as fast as it is losing it. This can lead to death. The chance of a soldier becoming hypothermic is a major threat any time he is exposed to the cold.

Some symptoms of hypothermia are shivering, loss of judgment, slurred speech, drowsiness, and muscle weakness.

During exercise in the cold, people usually produce enough heat to maintain normal body temperature. As they get fatigued, however, they slow down and their bodies produce less heat. Also, people often overdress for exercise in the cold. This makes the body sweat. The sweat dampens the clothing next to the skin making it a good conductor of heat. The combination of decreased heat production and increased heat loss can cause a rapid onset of hypothermia.

Some guidelines for dressing for cold weather exercise are shown in Figure 12-1.

Hypothermia develops when the body cannot produce heat as fast as it is losing it.

GUIDELINES FOR DRESSING FOR EXERCISE IN THE COLD

Clothing for cold weather should protect, insulate, and ventilate.

- Protect by covering as large an area of the body as possible.
- Insulation will occur by trapping air which has been warmed by the body and holding it near the skin.
- Ventilate by allowing a two-way exchange of air through the various layers of clothing.

Clothing should leave your body slightly cool rather than hot.

Clothing should also be loose enough to allow movement.

Clothing soaked with perspiration should be removed if reasonably possible.

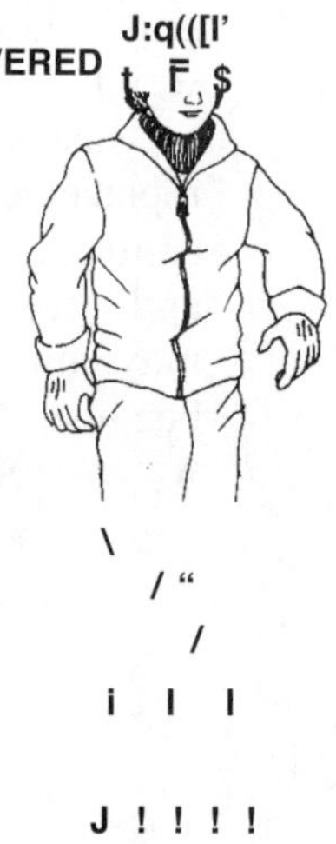

40% HEAT LOSS THROUGH HEAD AND NECK WHEN UNCOVERED

LIGHTWEIGHT WARM-UPS (NOT WATERPROOF)

FEET SHOULD BE KEPT DRY

Figure 12-1

FROSTBITE

Frostbite is the freezing of body tissue. It commonly occurs in body parts located away from the core and exposed to the cold such as the nose, ears, feet, hands, and skin. Severe cases of frostbite may require amputation.

Factors which lead to frostbite are cold temperatures combined with windy conditions. The wind has a great cooling effect because it causes rapid convective heat transfer from the body. For a given temperature, the higher the wind speed, the greater the cooling effect. Figure 12-2 shows how the wind can affect cooling by providing information on windchill factors.

A person's movement through the air creates an effect similar to that caused by wind. Riding a bicycle at 15 mph is the same as standing in a 15-mph wind. If, in addition, there is a 5-mph headwind, the overall effect is equivalent to a 20-mph wind. Therefore, an exercising soldier must be very cautious to avoid getting frostbite. Covering exposed parts of the body will substantially reduce the risks.

DEHYDRATION

Dehydration can result from losing body fluids faster than they are replaced. Cold environments are often dry, and water may be limited. As a result, soldiers may in time become dehydrated. While operating in extremely cold climates, trainers should check the body weights of the soldiers regularly and encourage them to drink liquids whenever possible.

WINDCHILL FACTOR

Cooling Power of Wind on Exposed Flesh Expressed as an Equivalent Temperature (under calm conditions)

Estimated wind speed (in mph)	Actual Thermometer Reading (°F)											
	50	40	30	20	10	0	−10	−20	−30	−40	−50	−60
	EQUIVALENT CHILL TEMPERATURE (°F)											
Calm	50	40	30	20	10	0	−10	−20	−30	−40	−50	−60
5	48	37	27	16	6	−5	−15	−26	−36	−47	−57	−68
10	40	28	16	4	−9	−24	−33	−46	−58	−70	−83	−95
15	36	22	9	−5	−18	−32	−45	−58	−72	−85	−99	−112
20	32	18	4	−10	−25	−39	−53	−67	−82	−96	−110	−124
25	30	16	0	−15	−29	−44	−59	−74	−88	−104	−118	−133
30	28	13	−2	−18	−33	−48	−63	−79	−94	−109	−125	−140
35	27	11	−4	−21	−35	−51	−67	−82	−98	−113	−129	−145
40	26	10	−6	−21	−37	−53	−69	−85	−100	−116	−132	−148

(wind speed greater than 40mph have little additional effect)	LITTLE DANGER In 0.5 hr with dry skin Maximum danger of false sense of security		INCREASING DANGER Danger from freezing of exposed flesh within one minute	GREAT DANGER Flesh may freeze within 30 seconds
	Trenchfoot and immersion foot may occur at any point on this chart.			

INSTRUCTIONS

MEASURE local temperature and wind speed if possible. If not, ESTIMATE. Enter table at closest 5°F interval along the top and with appropriate wind speed along left side. Intersection gives approximate equivalent chill temperature. That is, the temperature that would cause the same rate of cooling under calm conditions. Note that regardless of cooling rate, you do not cool below the actual air temperature unless wet.

Figure 12-2

Acclimatization to High Altitudes

Elevations below 5,000 feet have little noticeable effect on healthy people. However, at higher elevations the atmospheric pressure is reduced, and the body tissues get less oxygen. This means that soldiers cannot work or exercise as well at high altitudes. The limiting effects of high elevation are often most pronounced in older soldiers and persons with low levels of fitness.

Due to acclimatization, the longer a soldier remains at high altitude, the better his performance becomes. Generally, however, he will not perform as well as at sea level and should not be expected to. For normal activities, the time required to acclimatize depends largely on the altitude. In order to insure that soldiers who are newly assigned to altitudes above 5,000 feet are not at a disadvantage, it is recommended that 30 days of acclimatization, including regular physical activity, be permitted before they are administered a record APFT.

Before acclimatization is complete, people at high altitudes may suffer acute mountain sickness. This includes such symptoms as headache, rapid pulse, nausea, loss of appetite, and an inability to sleep. The primary treatment is further acclimatization or returning to a lower altitude.

Once soldiers are acclimatized to altitudes above 5,000 feet, deacclimatization will occur if they spend 14 or more days at lower altitudes. For this reason, soldiers should be permitted twice the length of their absence, not to exceed 30 days, to reacclimatize before being required to take a record APFT. A period of 30 days is adequate for any given reacclimatization.

Air Pollution and Exercise

Pollutants are substances in the environment which lower the environment's quality. Originally, air pollutants were thought to be only by-products of the industrial revolution. However, many pollutants are produced naturally. For example, volcanoes emit sulfur oxides and ash, and lightning produces ozone.

There are two classifications of air pollutants - primary and secondary. Primary pollutants are produced directly by industrial sources. These include carbon monoxide (CO), sulfur oxides (SO), hydrocarbons, and particulate (ash). Secondary pollutants are created by the primary pollutant's interaction with the environment. Examples of these include ozone **(O3),** aldehydes, and sulfates. Smog is a combination of primary and secondary pollutants.

Some pollutants have negative effects on the body. For example, carbon monoxide binds to hemoglobin in the red blood cells and reduces the amount of oxygen carried in the blood. Ozone and the oxides irritate the air passageways in the lungs, while other pollutants irritate the eyes.

When exercisers in high-pollution areas breathe through the mouth, the nasal mucosa's ability to remove impurities is bypassed, and many pollutants can be inhaled. This irritates the respiratory tract and makes the person less able to perform aerobically.

Pollutants ◆
the respira
and make t
less able t
aerob◆

The following are some ways to deal with air pollution while exercising:
- Avoid exposure to pollutants before and during exercise, if possible.
- In areas of high ozone concentration, train early in the day and after dark.
- Avoid exercising near heavily traveled streets and highways during rush hours.
- Consult your supporting preventive-medicine activity for advice in identifying or defining training restrictions during periods of heavy air pollution.

CHAPTER 13

Injuries

Injuries are not an uncommon occurrence during intense physical training. It is, nonetheless, a primary responsibility of all leaders to minimize the risk of injury to soldiers. Safety is always a major concern.

Most injuries can be prevented by designing a well-balanced PT program that does not overstress any body parts, allows enough time for recovery, and includes a warm-up and cool-down. Using strengthening exercises and soft, level surfaces for stretching and running also helps prevent injuries. If, however, injuries do occur, they should be recognized and properly treated in a timely fashion. If a soldier suspects that he is injured, he should stop what he is doing, report the injury, and seek medical help.

Many common injuries are caused by overuse, that is, soldiers often exercise too much and too often and with too rapid an increase in the workload. Most overuse injuries can be treated with rest, ice, compression, and elevation (RICE). Following any required first aid, health-care personnel should evaluate the injured soldier.

Most injuries can be prevented by designing a well-balanced PT program.

Typical Injuries Associated with Physical Training

Common injuries associated with exercise are the following:

- Abrasion (strawberry) - the rubbing off of skin by friction.
- Dislocation - "the displacement of one or more bones of a joint from their natural positions.
- Hot spot - a hot or irritated feeling of the skin which occurs just before a blister forms. These can be prevented by using petroleum jelly over friction-prone areas.
- Blister - a raised spot on the skin filled with liquid. These can generally be avoided by applying lubricants such as petroleum jelly to areas of friction, keeping footwear (socks, shoes, boots) in good repair, and wearing the proper size of boot or shoe.
- Shinsplints - a painful injury to the soft tissues and bone in the shin area. These are generally caused by wearing shoes with inflexible soles or inadequate shock absorption, running on the toes or on hard surfaces, and/or having calf muscles with a limited range of motion.
- Sprain - a stretching or tearing of the ligament(s) at a joint.
- Muscle spasm (muscle cramp) - a sudden, involuntary contraction of one or more muscles.
- Contusion - a bruise with bleeding into the muscle tissue.
- Strain - a stretching or tearing of the muscles.
- Bursitis - an inflammation of the bursa (a sack-like structure where tendons pass over bones). This occurs at a joint and produces pain when the joint is moved or touched. Sometimes swelling occurs.
- Tendinitis - an inflammation of a tendon that produces pain when the attached muscle contracts. Swelling may not occur.
- Stress fractures of the feet.
- Tibial stress fractures - overuse injuries which seem like shinsplints except that the pain is in a specific area.
- Knee injuries - caused by running on uneven surfaces or with worn out shoes, overuse, and improper body alignment. Soldiers who have problems with their knees can benefit from doing leg exercises that strengthen the front (quadriceps) and rear (hamstrings) thigh muscles.
- Low back problems - caused by poor running, sitting, or lifting techniques, and by failing to stretch the back and hip-flexor muscles and to strengthen the abdominal muscles.

The most common running injuries occur in the feet, ankles, knees, and legs. Although they are hard to eliminate, much can be done to keep them to a minimum. Preventive measures include proper warm-up and cool-down along with stretching exercises. Failure to allow recovery between hard bouts of running can lead to overtraining and can also be a major cause of injuries. A well-conditioned soldier can run five to six times a week. However, to do this safely, he should do two things: gradually build up to running that frequently and vary the intensity of the running sessions to allow recovery between them.

Many running injuries can be prevented by wearing proper footwear. Soldiers should train in running shoes. These are available in a wide range of prices and styles. They should fit properly and have flexible, multi-layered soles with good arch and heel support. Shoes made with leather and nylon uppers are usually the most comfortable. See Appendix E for more information on running shoes.

Since injuries can also be caused by running on hard surfaces, soldiers should, if possible, avoid running on concrete. Soft, even surfaces are best for injury prevention. Whenever possible, soldiers should run on grass paths, dirt paths, or park trails. However, with adequate footwear and recovery periods, running on roads and other hard surfaces should pose no problem.

Common running injuries include the following:
- Black toenails.
- Ingrown toenails.
- Stress fractures of the feet.
- Ankle sprains and fractures.
- Achilles tendinitis (caused by improper stretching and shoes that do not fit.

- Upper leg and groin injuries (which can usually be prevented by using good technique in stretching and doing strengthening exercises).

Tibial stress fractures, knee injuries, low back problems, shinsplints, and blisters, which were mentioned earlier, are also injuries which commonly occur in runners.

Other Factors

Proper clothing can also help prevent injuries. Clothes used for physical activity should be comfortable and fit loosely. A T-shirt or sleeveless undershirt and gym shorts are best in warm weather. In cold weather, clothing may be layered according to personal preference. For example, soldiers can wear a BDU, sweat suit, jogging suit, or even Army-issued long underwear. In very cold weather, soldiers may need gloves or mittens and ear-protecting caps. Rubberized or plastic suits should never be worn during exercise. They cause excessive sweating which can lead to dehydration and a dangerous increase in body temperature.

Army Regulation 385-55 (paragraph B-12, C) prohibits the use of headphones or earphones while walking, jogging, skating, or bicycling on the roads and streets of military installations. However, they may be worn on tracks and running trails.

Road safety equipment is required on administative-type walks, marches, or runs which cross highways, roads, or tank trails or which are conducted on traffic ways. If there is reduced visibility, control personnel must use added caution to ensure the safety of their soldiers.

Many running injuries can be prevented by wearing proper footwear.

The APFT is a three-event physical performance test used to assess muscular endurance and cardiorespiratory (CR) fitness .

Performance on the APFT is strongly linked to the soldier's fitness level and his ability to do fitness-related tasks.

All soldiers in the Active Army, Army National Guard, and Army Reserve must take the Army Physical Fitness Test (APFT) regardless of their age. The APFT is a three-event physical performance test used to assess muscular endurance and cardiorespiratory (CR) fitness. It is a simple way to measure a soldier's ability to effectively move his body by using his major muscle groups and CR system. Performance on the APFT is strongly linked to the soldier's fitness level and his ability to do fitness-related tasks. An APFT with alternate test events is given to soldiers with permanent profiles and with temporary profiles greater than three months' duration.

While the APFT testing is an important tool in determining the physical readiness of individual soldiers and units, it should not be the sole basis for the unit's physical fitness training. Commanders at every level must ensure that fitness training is designed to develop physical abilities in a balanced way, not just to help soldiers do well on the APFT.

Commanders should use their unit's APFT results to evaluate its physical fitness level. APFT results may indicate a need to modify the fitness programs to attain higher fitness levels. However, mission-essential tasks, not the APFT, should drive physical training.

Additional physical performance tests and standards which serve as prerequisites for Airborne/Ranger/Special Forces/SCUBA qualification are provided in DA Pam 351-4.

Methods of Evaluation

Commanders are responsible for ensuring that their soldiers are physically fit (AR 350-15). There are several ways they can assess fitness including the following

• Testing. This is an efficient way to evaluate both the individual's and the unit's physical performance levels.

• Inspection. This evaluates training procedures and indicates the soundness of the unit's physical fitness program.

• Observation. This is an ongoing way to review training but is not as reliable as testing as an indicator of the unit's level of fitness.

• Medical examination. This detects individual disabilities, health-related problems, and physical problems.

Over-Forty Cardiovascular Screening Program

The Army's over-40 cardiovascular screening program (CVSP) does the following:

• Identifies soldiers with a risk of coronary heart disease.

• Provides guidelines for safe, regular CR exercise.

• Gives advice and help in controlling heart-disease risk factors.

• Uses treadmill testing only for high-risk soldiers who need it.

All soldiers, both active and reserve component, must take the APFT for record regardless of age unless prohibited by a medical profile. For soldiers who reached age 40 on or after 1 January 1989, there is no requirement for clearance in the cardiovascular screening program before taking a record APFT. Soldiers who reached age 40 before 1 January 1989 must be cleared through the cardiovascular screening program before taking a record APFT. Prior to their CVSP evaluation, however, they may still take part in physical training to include diagnostic APFTs unless profiled or contraindications to exercise exist. All soldiers must undergo periodic physical examinations in accordance with AR 40-501 and NGR 40-501. These include screening for cardiovascular risk factors.

Overview

As stated, APFT events assess muscular endurance and CR fitness. The lowest passing APFT standards reflect the minimum acceptable fitness level for all soldiers, regardless of MOS or component. When applied to a com - mand, APFT results show a unit's overall level of physical fitness. However, they are not all-inclusive, overall measures of physical-combat readiness. To assess this, other physical capabilities must be measured. The APFT does, however, give a commander a sound measurement of the general fitness level of his unit.

Service schools, agencies, and units may set performance goals which are above the minimum APFT standards in accordance with their missions (AR 350- 15). Individual soldiers are also encouraged to set for themselves a series of successively higher APFT performance goals. They should always strive to improve themselves physically and never be content with meeting minimum standards. Competition on the APFT among soldiers or units can also be used to motivate them to improve their fitness levels.

Testing is not a substitute for a regular, balanced exercise program. Diagnostic testing is important in monitoring training progress but, when done too often, may decrease motivation and waste training time.

The test period is defined as *the* period of time which elapses from starting to finishing the three events. It must not take more than two hours. Soldiers must do all three events in the same test period.

Test Administration

The APFT must be administered properly and to standard in order to accurately evaluate a soldier's physical fitness and to be fair to all soldiers. (Test results are used for personnel actions.)

Individual soldiers are not authorized to administer the APFT to themselves for the purpose of satisfying a unit's diagnostic or record APFT requirement.

REQUIRED EQUIPMENT

The OIC or NCOIC at the test site must have a copy of FM 21-20 on hand. The supervisor of each event must have the event instructions and standards. Scorers should have a clipboard and an ink pen to record the results on the soldiers' scorecards.

Two stopwatches are needed. They must be able to measure time in both minutes and seconds.

Runners must wear numbers or some other form of identification for the 2-mile run. The numbers may be stenciled or pinned onto pullover vests or sleeveless, mesh pullovers or attached to the runners themselves.

Soldiers should wear clothing that is appropriate for PT such as shorts, T-shirts, socks, and running shoes (not tennis shoes). They should not wear basketball shoes or other types of court shoes. BDUs may be worn but may be a hindrance on some events.

Anything that gives a soldier an unfair advantage is not permitted during the APFT. Wearing devices such as weight belts or elastic bandages may or may not provide an advantage. However, for standardization, such additional equipment is not authorized unless prescribed by medical personnel. The only exception is gloves. They may be worn in cold weather when approved by the local commander.

Each soldier needs a DA Form 705, Army Physical Fitness Test Scorecard. The soldier fills in his name, social security number, grade, age, and sex.

(See Figure 14-1.) The unit will complete the height and weight data.

Scorers record the raw score for each event and initial the results. If a soldier fails an event or finds it difficult to perform, the scorer should write down the reasons and other pertinent information in the comment block. After the entire APFT has been completed, the event scorer will convert raw scores to point scores using the scoring standards on the back of the scorecards. (See Figure 14-1.)

See page 14-8.1 for instructions on completing DA Form 705.

ARMY PHYSICAL FITNESS TEST SCORECARD

Army Physical Fitness Test Scorecard

NAME (LAST, FIRST, MIDDLE)

SSN | GENDER

UNIT

TEST FIVE	TEST SIX	TEST SEVEN	TEST EIGHT
DATE / GRADE / AGE	DATE / GRADE / AGE	DATE / GRADE / AGE	DATE / GRADE / AGE
HEIGHT (IN INCHES)	HEIGHT (IN INCHES)	HEIGHT (IN INCHES)	HEIGHT (IN INCHES)
WEIGHT: ____ lbs GO / NO-GO; BODY FAT: ____ % GO / NO-GO	WEIGHT: ____ lbs GO / NO-GO; BODY FAT: ____ % GO / NO-GO	WEIGHT: ____ lbs GO / NO-GO; BODY FAT: ____ % GO / NO-GO	WEIGHT: ____ lbs GO / NO-GO; BODY FAT: ____ % GO / NO-GO
PU RAW SCORE / INITIALS / POINTS	PU RAW SCORE / INITIALS / POINTS	PU RAW SCORE / INITIALS / POINTS	PU RAW SCORE / INITIALS / POINTS
SU RAW SCORE / INITIALS / POINTS	SU RAW SCORE / INITIALS / POINTS	SU RAW SCORE / INITIALS / POINTS	SU RAW SCORE / INITIALS / POINTS
2MR RAW SCORE / INITIALS / POINTS	2MR RAW SCORE / INITIALS / POINTS	2MR RAW SCORE / INITIALS / POINTS	2MR RAW SCORE / INITIALS / POINTS
ALTERNATE AEROBIC EVENT: EVENT ____ TIME ____ GO NO-GO; TOTAL POINTS	ALTERNATE AEROBIC EVENT: EVENT ____ TIME ____ GO NO-GO; TOTAL POINTS	ALTERNATE AEROBIC EVENT: EVENT ____ TIME ____ GO NO-GO; TOTAL POINTS	ALTERNATE AEROBIC EVENT: EVENT ____ TIME ____ GO NO-GO; TOTAL POINTS
NCOIC/OIC Signature	NCOIC/OIC Signature	NCOIC/OIC Signature	NCOIC/OIC Signature
COMMENTS	COMMENTS	COMMENTS	COMMENTS

SPECIAL INSTRUCTION: USE INK

LEGEND: PU - PUSH UPS; SU - SIT UPS; 2MR - 2 MILE RUN; APFT - ARMY PHYSICAL FITNESS TEST

Data Required by the Privacy Act of 1974

Title DA Form 705
Authority 10 USC 301 2(g)
Principal purpose record of individual scores on physical fitness events
Mandatory or voluntary disclosure and effect on individual not providing information: mandatory - individuals not providing information cannot be rated/scored

DA FORM 705, JUN 98

SAMPLE

***Figure 14-1**

PUSH-UP STANDARDS

Repetitions	17-21 M	17-21 F	22-26 M	22-26 F	27-31 M	27-31 F	32-36 M	32-36 F	37-41 M	37-41 F	Repetitions	42-46 M	42-46 F	47-51 M	47-51 F	52-56 M	52-56 F	57-61 M	57-61 F	62+ M	62+ F	Repetitions
77					100						77											77
76					99						76											76
75			100		98		100				75											75
74			99		97		99				74											74
73			98		96		98		100		73											73
72			97		95		97		99		72											72
71	100		95		94		96		98		71											71
70	99		94		93		95		97		70											70
69	97		93		92		94		96		69											69
68	96		92		91		93		95		68											68
67	94		91		89		92		94		67											67
66	93		90		88		91		93		66	100										66
65	92		89		87		90		92		65	99										65
64	90		87		86		89		91		64	98										64
63	89		86		85		88		90		63	97										63
62	88		85		84		87		89		62	96										62
61	86		84		83		86		88		61	94										61
60	85		83		82		85		87		60	93										60
59	83		82		81		84		86		59	92		100								59
58	82		81		80		83		85		58	91		99								58
57	81		79		79		82		84		57	90		98								57
56	79		78		78		81		83		56	89		96		100						56
55	78		77		77		79		82		55	88		95		99						55
54	77		76		76		78		81		54	87		94		98						54
53	75		75		75		77		79		53	86		93		97		100				53
52	74		74		74		76		78		52	84		92		96		99				52
51	72		73		73		75		77		51	83		91		94		98				51
50	71		71		72	100	74		76		50	82		89		93		97		100		50
49	70		70		71	99	73		75		49	81		88		92		95		99		49
48	68		69		69	98	72		74		48	80		87		91		94		98		48
47	67		68		68	96	71		73		47	79		86		90		93		96		47
46	66		67	100	67	95	70		72		46	78		85		89		92		95		46
45	64		66	99	68	94	69	100	71		45	77		84		88		91		94		45
44	63		65	97	65	93	68	99	70		44	76		87		87		90		93		44
43	61		63	96	64	92	67	97	69		43	74		81		86		89		92		43
42	60	100	62	94	63	90	66	96	68		42	73		80		85		87		91		42
41	59	98	61	93	62	89	65	95	67		41	72		79		83		86		89		41
40	57	97	60	92	61	88	64	93	66	100	40	71		78		82		85		88		40
39	56	95	59	90	60	87	63	92	65	99	39	70		76		81		84		87		39
38	54	93	58	89	58	85	62	91	64	97	38	69		75		80		83		86		38
37	53	91	57	88	58	84	61	89	63		37	68	100	74		79		82		85		37
36	52	90	55	86	57	83	60	88	62	94	36	67	98	73		78		81		84		36
35	50	88	54	85	55	82	59	87	61	93	35	66	97	72		77		79		82		35
34	49	86	52	83	55	81	58	85	60	91	34	64	95	71	100	76		78		81		34
33	48	84	52	82	54	79	57	84	59	90	33	63	94	69	98	74		77		80		33
32	46	83	51	81	53	78	56	83	58	88	32	62	92	68	97	73		76		79		32
31	45	81	50	79	55	77	55	81	57	87	31	61	90	67	95	72	100	75		78		31
30	43	79	49	78	50	75	54	80	56	85	30	60	89	66	93	71	98	74		76		30
29	42	77	47	77	49	75	53	79	55	84	29	59	87	65	92	70	95	73		75		29
28	41	76	46	75	48	73	52	77	54	82	28	58	86	64	90	69	95	71	100	74		28
27	39	74	45	74	47	72	51	76	53	81	27	57	84	62	88	68	93	70	98	73		27
26	38	72	44	72	46	71	50	75	52	79	26	56	82	61	87	67	91	69	96	72		26
25	37	70	43	71	45	70	49	73	51	78	25	54	81	60	85	66	89	68	94	71	100	25
24	35	69	42	70	44	68	48	72	50	76	24	53	79	59	83	64	87	67	92	69	98	24
23	34	67	39	67	43	67	47	71	49	75	23	52	78	58	82	63	85	66	90	68	96	23
22	32	65	39	67	42	66	46	69	48	73	22	51	76	56	80	62	84	65	88	67	93	22
21	31	63	38	66	41	65	45	68	47	72	21	50	74	55	78	61	82	63	86	66	91	21
20	30	62	37	64	40	64	44	67	46	70	20	49	73	54	77	60	80	62	84	65	89	20
19	28	60	36	63	38	61	43	65	45	69	19	48	71	53	75	59	78	61	82	64	87	19
18	27	58	35	61	38	61	42	64	44	67	18	47	70	52	73	58	76	60	80	62	84	18
17	26	57	34	60	37	60	41	63	43	66	17	46	68	51	72	57	75	59	78	61	82	17
16	24	55	33	59	36	58	39	61	41	64	16	45	65	49	70	56	73	58	76	60	80	16
15	23	53	31	57	35	58	38	60	41	63	15	43	63	48	68	54	71	57	74	59	78	15
14	21	51	30	56	34	56	37	59	39	61	14	42	63	47	67	53	69	55	72	58	76	14
13	20	50	29	54	33	55	36	58	38	60	13	41	62	46	65	52	67	54	70	56	73	13
12	19	48	28	52	32	54	35	56	37	59	12	40	60	45	63	51	66	53	68	55	71	12
11	17	46	27	50	31	52	34	54	36	57	11	39	58	44	62	50	64	52	66	54	69	11
10	16	44	26	49	29	50	33	52	35	56	10	38	57	42	60	49	62	51	64	53	67	10
9	14	43	25	49	28	49	32	50	34	54	9	37	55	41	58	48	60	50	62	52	64	9
8	13	41	23	48	27	49	31	49	33	53	8	36	54	40	57	47	58	49	60	51	62	8
7	12	39	22	46	26	48	30	49	32	51	7	34	52	39	55	46	56	47	58	49	60	7
6	10	37	21	45	25	47	29	48	31	50	6	33	50	38	53	44	55	46	56	48	58	6
5	9	36	20	43	24	45	28	47	30	48	5	32	49	36	52	43	53	45	54	47	56	5
4	8	34	19	42	23	44	27	45	29	47												4
3	6	32	18	41	22	43	26	43	28	45												3
2	5	30	17	39	21	42	25	43	27	44												2
1	3	29	15	38	20	41	24	41	26	42												1
Repetitions	M	F	M	F	M	F	M	F	M	F	Repetitions	M	F	M	F	M	F	M	F	M	F	Repetitions
Age Group	17-21		22-26		27-31		32-36		37-41		Age Group	42-46		47-51		52-56		57-61		62+		Age Group

Scoring standards are used to convert raw scores to point scores after test events are completed. Male point scores are indicated by the M at the top and bottom of the shaded column. Female point scores are indicated by the F at the top of the unshaded column. To convert raw scores to point scores, find the number of repetitions performed in the left-hand column. Next, move right along that row and locate the intersection of the soldier's appropriate age column. Record that number in the Push-Up points block on the front of the scorecard.

*Figure 14-1 (continued)

SIT-UP STANDARDS

Repetitions	17-21 M/F	22-26 M/F	27-31 M/F	32-36 M/F	37-41 M/F	Repetitions	42-46 M/F	47-51 M/F	52-56 M/F	57-61 M/F	62+ M/F	Repetitions
82			100			82						82
81			99			81						81
80		100	98			80						80
79		99	97			79						79
78	100	97	96			78						78
77	98	96	95			77						77
76	97	95	94	100	100	76						76
75	95	93	92	99	99	75						75
74	94	92	91	98	98	74						74
73	92	91	90	96	97	73						73
72	90	89	89	95	96	72	100					72
71	89	88	88	94	95	71	99					71
70	87	87	87	93	94	70	98					70
69	86	85	86	92	93	69	97					69
68	84	84	85	91	92	68	96					68
67	82	83	84	89	91	67	95					67
66	81	81	83	88	89	66	94	100	100			66
65	79	80	82	87	88	65	93	99	99			65
64	78	79	81	86	87	64	92	98	98	100		64
63	76	77	79	85	86	63	91	97	97	99	100	63
62	74	76	78	84	85	62	90	96	96	98	99	62
61	73	75	77	82	84	61	89	95	95	97	98	61
60	71	73	76	81	83	60	88	93	94	96	97	60
59	70	72	75	80	82	59	87	92	93	95	96	59
58	68	71	74	79	81	58	86	91	92	94	95	58
57	66	69	73	78	80	57	85	90	91	92	94	57
56	65	68	72	76	79	56	84	89	89	91	92	56
55	63	67	71	75	78	55	83	88	88	90	91	55
54	62	65	70	74	77	54	82	87	87	89	90	54
53	60	64	69	73	76	53	81	86	86	88	89	53
52	58	63	68	72	75	52	80	84	85	87	88	52
51	57	61	66	71	74	51	79	83	84	86	87	51
50	55	60	65	69	73	50	78	82	83	85	86	50
49	54	59	64	68	72	49	77	81	82	84	85	49
48	52	57	63	67	71	48	76	80	81	83	84	48
47	50	56	62	66	69	47	75	79	80	82	83	47
46	49	55	61	65	68	46	74	78	79	81	82	46
45	47	53	60	64	67	45	73	77	78	79	81	45
44	46	52	59	62	66	44	72	76	77	78	79	44
43	44	50	58	61	65	43	71	74	76	77	78	43
42	42	49	57	60	64	42	70	73	75	76	77	42
41	41	48	56	59	63	41	69	72	74	75	76	41
40	39	47	55	58	62	40	68	71	73	74	75	40
39	38	45	54	56	61	39	67	70	72	73	74	39
38	36	44	52	55	60	38	66	69	71	72	73	38
37	34	43	51	54	59	37	65	68	69	71	72	37
36	33	41	50	53	58	36	64	67	68	70	71	36
35	31	40	49	52	57	35	63	66	67	69	70	35
34	30	39	48	50	56	34	62	64	66	68	69	34
33	28	37	47	49	55	33	61	63	65	66	68	33
32	27	36	46	48	54	32	60	62	64	65	66	32
31	25	35	45	47	53	31	59	61	63	64	65	31
30	23	33	44	46	52	30	58	60	62	63	64	30
29	22	32	43	45	50	29	57	59	61	62	63	29
28	20	31	42	44	49	28	56	58	60	61	62	28
27	18	29	41	42	48	27	55	57	59	60	61	27
26	17	28	39	41	47	26	54	56	58	59	60	26
25	15	27	38	40	46	25	53	54	57	58	59	25
24	14	25	37	39	45	24	52	53	56	57	58	24
23	12	24	36	38	44	23	51	52	55	56	57	23
22	10	23	35	36	43	22	50	51	54	55	56	22
21	9	21	34	35	42	21	49	50	53	54	55	21
AGE GROUP	17-21 M/F	22-26 M/F	27-31 M/F	32-36 M/F	37-41 M/F	AGE GROUP	42-46 M/F	47-51 M/F	52-56 M/F	57-61 M/F	62+ M/F	AGE GROUP

Scoring standards are used to convert raw scores to point scores after test events are completed. To convert raw scores to point scores, find the number of repetitions performed in the left-hand column. Next, move right along that row and locate the intersection of the soldier's appropriate age column. Record that number in the Sit-Up points block on the front of the scorecard.

*Figure 14-1 (continued)

2-MILE RUN STANDARDS

Time	17-21 M	17-21 F	22-26 M	22-26 F	27-31 M	27-31 F	32-36 M	32-36 F	37-41 M	37-41 F	Time	42-46 M	42-46 F	47-51 M	47-51 F	52-56 M	52-56 F	57-61 M	57-61 F	62+ M	62+ F	Time
12:54											12:54											12:54
13:00	100		100								13:00											13:00
13:06	99		99								13:06											13:06
13:12	97		98								13:12											13:12
13:18	96		97		100		100				13:18											13:18
13:24	94		96		99		99				13:24											13:24
13:30	93		94		98		98				13:30											13:30
13:36	92		93		97		97		100		13:36											13:36
13:42	90		92		96		96		99		13:42											13:42
13:48	89		91		95		95		98		13:48											13:48
13:54	88		90		94		95		97		13:54											13:54
14:00	86		89		92		94		97		14:00											14:00
14:06	85		88		91		93		96		14:06	100										14:06
14:12	83		87		90		92		95		14:12	99										14:12
14:18	82		86		89		91		94		14:18	98										14:18
14:24	81		84		88		90		93		14:24	97		100								14:24
14:30	79		83		87		89		92		14:30	97		99								14:30
14:36	78		82		86		88		91		14:36	96		98								14:36
14:42	77		81		85		87		91		14:42	95		98		100						14:42
14:48	75		80		84		86		90		14:48	94		97		99						14:48
14:54	74		79		83		85		89		14:54	93		96		98						14:54
15:00	72		78		82		85		88		15:00	92		95		98						15:00
15:06	71		77		81		84		87		15:06	91		95		97						15:06
15:12	70		76		79		83		86		15:12	90		94		96						15:12
15:18	68		74		78		82		86		15:18	90		93		95		100				15:18
15:24	67		73		77		81		85		15:24	89		92		95		99				15:24
15:30	66		72		76		80		84		15:30	88		91		94		93				15:30
15:36	64	100	71	100	75		79		83		15:36	87		91		93		97				15:36
15:42	63	99	70	99	74		78		82		15:42	86		90		92		97		100		15:42
15:48	61	98	69	98	73	100	77		81		15:48	85		89		91		96		99		15:48
15:54	60	96	68	97	72	99	76	100	80		15:54	84		88		91		95		98		15:54
16:00	59	95	67	98	71	98	75	99	80		16:00	83		87		90		94		97		16:00
16:06	57	94	66	95	70	97	75	99	79		16:06	83		87		89		93		96		16:06
16:12	56	93	64	94	69	97	74	98	78		16:12	82		86		88		92		95		16:12
16:18	54	92	63	93	68	96	73	97	77		16:18	81		85		87		91		94		16:18
16:24	53	90	62	92	66	95	72	97	76		16:24	80		84		87		90		93		16:24
16:30	52	89	61	91	65	94	71	96	75		16:30	79		84		86		90		93		16:30
16:36	50	88	60	90	64	93	70	95	74		16:36	78		83		85		89		92		16:36
16:42	49	87	59	89	63	92	69	94	74		16:42	77		82		84		88		91		16:42
16:48	48	85	58	88	62	91	68	94	73		16:48	77		81		84		87		90		16:48
16:54	46	84	57	87	61	91	67	93	72		16:54	76		80		83		86		89		16:54
17:00	45	83	56	86	60	90	66	92	71	100	17:00	75		80		82		85		88		17:00
17:06	43	82	54	85	59	89	65	92	70	99	17:06	74		79		81		84		87		17:06
17:12	42	81	53	84	58	88	65	91	69	99	17:12	73		78		80		83		86		17:12
17:18	41	79	52	83	57	87	64	90	69	98	17:18	72		77		80		83		85		17:18
17:24	39	78	51	82	56	86	63	90	68	97	17:24	71	100	76		79		82		84		17:24
17:30	38	77	50	81	55	86	62	89	67	96	17:30	70	99	76		78		81		83		17:30
17:36	37	76	49	80	54	85	61	88	66	96	17:36	70	99	75	100	77		80		82		17:36
17:42	35	75	48	79	52	84	60	88	65	95	17:42	69	98	74	99	76		79		81		17:42
17:48	34	73	47	78	51	83	59	87	64	94	17:48	68	97	73	99	76		78		80		17:48
17:54	32	72	46	77	50	82	58	86	63	94	17:54	67	97	73	98	75		77		80		17:54
18:00	31	71	44	76	49	81	57	86	62	93	18:00	66	96	72	97	74		77		79		18:00
18:06	30	70	43	75	48	80	56	85	62	92	18:06	65	96	71	97	73		76		78		18:06
18:12	28	68	42	74	47	80	55	84	61	92	18:12	64	95	70	96	73		75		77		18:12
18:18	27	67	41	73	46	79	55	83	60	91	18:18	63	94	69	96	72		74		76		18:18
18:24	26	66	40	72	45	78	54	83	59	90	18:24	62	94	69	95	71		73		75		18:24
18:30	24	65	39	71	44	77	53	82	58	89	18:30	62	93	68	94	70		72		74		18:30
18:36	23	64	38	70	43	76	52	81	57	89	18:36	61	92	67	94	69		71		73		18:36
18:42	21	62	37	69	42	75	51	81	57	88	18:42	60	92	66	93	69		70		72		18:42
18:48	20	61	36	68	41	74	50	80	56	87	18:48	59	91	65	92	68		70		71		18:48
18:54	19	60	34	67	39	74	49	79	55	87	18:54	58	90	65	92	67		69		70		18:54
19:00	17	59	33	66	38	73	48	79	54	86	19:00	57	90	64	91	66	100	68		69		19:00
19:06	16	58	32	65	37	72	47	78	53	85	19:06	57	89	63	91	65	99	67		68		19:06
19:12	14	56	31	64	36	71	46	77	52	85	19:12	56	89	62	90	65	99	66		67		19:12
19:18	13	55	30	63	35	70	45	77	51	84	19:18	55	88	62	89	64	98	65		67		19:18
19:24	12	54	29	62	34	69	45	76	51	83	19:24	54	87	61	89	63	97	64		66		19:24
19:30	10	53	28	61	33	69	44	75	50	82	19:30	53	87	60	88	62	96	63		65		19:30
19:36	9	52	27	60	32	68	43	74	49	82	19:36	52	86	59	87	62	96	63		64		19:36
19:42	8	50	26	59	31	67	42	74	48	81	19:42	51	85	58	87	61	95	62	100	63		19:42
19:48	6	49	24	58	30	66	41	73	47	80	19:48	50	85	58	86	60	94	61	99	62		19:48
19:54	5	48	23	57	29	65	40	72	46	80	19:54	50	84	57	86	59	93	60	98	61		19:54
20:00	3	47	22	56	28	64	39	72	46	79	20:00	49	83	56	85	58	93	59	98	60	100	20:00
20:06	2	45	21	55		63	38	71		78	20:06	48		55	84	58		58	97	59		20:06

*Figure 14-1 (continued)

14-6

Time	M	F	M	F	M	F	M	F	M	F	Time	M	F	M	F	M	F	M	F	M	F	Time
20:00	3	41		28	64		46	79			20:00	49	83	5?		93	5?					20:00
20:06	2	45	21	55	26	63	38	71	45	78	20:06	48	83	55	84	58	92	58	97	59	99	20:06
20:12	1	44	20	54	25	63	37	70	44	78	20:12	47	82	55	84	57	91	57	96	58	98	20:12
20:18	0	43	19	53	24	62	36	70	43	77	20:18	46	82	54	83	56	90	57	95	57	98	20:18
20:24		42	18	52	23	61	35	69	42	76	20:24	45	81	53	82	55	90	56	95	56	97	20:24
20:30		41	17	51	22	60	35	68	41	75	20:30	44	80	52	82	55	89	55	94	55	96	20:30
20:36		39	16	50	21	59	34	68	40	75	20:36	43	80	51	81	54	88	54	93	54	95	20:36
20:42		38	14	49	20	58	33	67	40	74	20:42	43	79	51	81	53	87	53	92	53	94	20:42
20:48		37	13	48	19	57	32	66	39	73	20:48	42	78	50	80	52	87	52	91	53	94	20:48
20:54		36	12	47	18	57	31	66	38	73	20:54	41	78	49	79	51	86	51	91	52	93	20:54
21:00		35	11	46	17	56	30	65	37	72	21:00	40	77	48	79	51	85	50	90	51	92	21:00
21:06		33	10	45	16	55	29	64	36	71	21:06	39	77	47	78	50	84	50	89	50	91	21:06
21:12		32	9	44	15	54	28	63	35	71	21:12	38	76	47	77	49	84	49	88	49	90	21:12
21:18		31	8	43	14	53	27	63	34	70	21:18	37	75	46	77	48	83	48	87	48	90	21:18
21:24		30	7	42	12	52	26	62	34	69	21:24	37	75	45	76	47	82	47	87	47	89	21:24
21:30		28	6	41	11	51	25	61	33	68	21:30	36	74	44	76	47	81	46	86	46	88	21:30
21:36		27	4	40	10	51	25	61	32	68	21:36	35	73	44	75	46	81	45	85	45	87	21:36
21:42		26	3	39	9	50	24	60	31	67	21:42	34	73	43	74	45	80	44	84	44	86	21:42
21:48		25	2	38	8	49	23	59	30	66	21:48	33	72	42	74	44	79	44	84	43	86	21:48
21:54		24	1	37	7	48	22	59	29	66	21:54	32	71	41	73	44	79	43	83	42	85	21:54
22:00		22	0	36	6	47	21	58	29	65	22:00	31	71	40	72	43	78	42	82	41	84	22:00
22:06		21		35	5	46	20	57	28	64	22:06	30	70	40	72	42	77	41	81	40	83	22:06
22:12		20		34	4	46	19	57	27	64	22:12	30	70	39	71	41	76	40	80	40	82	22:12
22:18		19		33	3	45	18	56	26	63	22:18	29	69	38	71	40	76	39	80	39	82	22:18
22:24		18		32	2	44	17	55	25	62	22:24	28	68	37	70	40	75	38	79	38	81	22:24
22:30		16		31	1	43	16	54	24	61	22:30	27	68	36	69	39	74	37	78	37	80	22:30
22:36		15		30	0	42	15	54	23	61	22:36	26	67	36	68	38	73	37	77	36	79	22:36
22:42		14		29		41	15	53	23	60	22:42	25	66	35	68	37	73	36	76	35	78	22:42
22:48		13		28		40	14	52	22	59	22:48	24	66	34	67	36	72	35	76	34	78	22:48
22:54		12		27		40	13	52	21	59	22:54	23	65	33	67	36	71	34	75	33	77	22:54
23:00		10		26		39	12	51	20	58	23:00	23	64	33	66	35	70	33	74	32	76	23:00
23:06		9		25		38	11	50	19	57	23:06	22	64	32	66	34	70	32	73	31	75	23:06
23:12		8		24		37	10	49	18	56	23:12	21	63	31	65	33	69	31	73	30	74	23:12
23:18		7		23		36	9	49	17	56	23:18	20	63	30	64	33	68	30	72	29	74	23:18
23:24		5		22		35	8	48	17	55	23:24	19	62	29	64	32	67	30	71	28	73	23:24
23:30		4		21		34	7	48	16	54	23:30	18	61	29	63	31	67	29	70	27	72	23:30
23:36		3		20		34	6	47	15	54	23:36	17	61	28	62	30	66	28	69	27	71	23:36
23:42		2		19		33	5	46	14	53	23:42	17	60	27	62	29	65	27	69	26	70	23:42
23:48		1		18		32	5	46	13	52	23:48	16	59	26	61	29	64	26	68	25	70	23:48
23:54		0		17		31	4	45	12	52	23:54	15	59	25	61	28	64	25	67	24	69	23:54
24:00				16		30	3	44	11	51	24:00	14	58	25	60	27	63	24	66	23	68	24:00
24:06				15		29	2	43	11	50	24:06	13	57	24	59	26	62	23	65	22	67	24:06
24:12				14		29	1	43	10	49	24:12	12	57	23	59	25	61	23	65	21	66	24:12
24:18				13		28	0	42	9	49	24:18	11	56	22	58	25	61	22	64	20	66	24:18
24:24				12		27		41	8	48	24:24	10	56	22	57	24	60	21	63	19	65	24:24
24:30				11		26		41	7	47	24:30	10	55	21	57	23	59	20	62	18	64	24:30
24:36				10		25		40	6	47	24:36	9	54	20	56	22	59	18	61	17	63	24:36
24:42				9		24		39	6	46	24:42	8	54	19	56	22	58	18	61	16	62	24:42
24:48				8		23		39	5	45	24:48	7	53	18	55	21	57	17	60	15	62	24:48
24:54				7		23		38	4	45	24:54	6	52	18	54	20	56	17	59	14	61	24:54
25:00				6		22		37	3	44	25:00	5	52	17	54	19	56	16	58	13	60	25:00
25:06				5		21		37	2	43	25:06	4	51	16	53	18	55	15	58	13	59	25:06
25:12				4		20		36	1	42	25:12	3	50	15	52	18	54	14	57	12	58	25:12
25:18				2		19		35	0	42	25:18	3	50	15	52	17	53	13	56	11	58	25:18
25:24				2		18		34		41	25:24	2	49	14	51	16	53	12	55	10	57	25:24
25:30				1		17		34		40	25:30	1	49	13	51	15	52	11	55	9	56	25:30
25:36				0		17		33		40	25:36	0	48	12	50	15	51	10	54	8	55	25:36
25:42						16		32		39	25:42		47	11	49	14	50	10	53	7	54	25:42
25:48						15		32		38	25:48		47	11	49	13	50	9	52	6	54	25:48
25:54						14		31		38	25:54		46	10	48	12	49	8	51	5	53	25:54
26:00						13		30		37	26:00		45	9	47	11	48	7	51	4	52	26:00
26:06						12		30		36	26:06		45	8	47	11	47	6	50	3	51	26:06
26:12						11		29		35	26:12		44	7	46	10	47	5	49	2	50	26:12
26:18						11		28		35	26:18		43	7	46	9	46	4	48	1	50	26:18
26:24						10		28		34	26:24		43	6	45	8	45	3	47	0	49	26:24
26:30						9		27		33	26:30		42	5	44	7	44	3	47	0	48	26:30
Time	M	F	M	F	M	F	M	F	M	F	Time	M	F	M	F	M	F	M	F	M	F	Time
AGE GROUP	17-21		22-26		27-31		32-36		37-41		AGE GROUP	42-46		47-51		52-56		57-61		62+		AGE GROUP

Scoring standards are used to convert raw scores to point scores after test events are completed. Male point scores are indicated by the M at the top and bottom of the shaded column. Female point scores are indicated by the F at the top and bottom of the unshaded column. To convert raw scores to point scores, find the number of repetitions performed in the left-hand column. Next, move right along that row and locate the intersection of the soldier's appropriate age column. **In all cases, when a time falls between two point values, the lower point value is used.** Record that number in the 2MR points block on the front of the scorecard.

*Figure 14-1 (continued)

SUPERVISION

The APFT must be properly supervised to ensure that its objectives are met. Proper supervision ensures uniformity in the following:

- Scoring the test.
- Training of supervisors and scorers.
- Preparing the test and controlling performance factors.

The goal of the APFT is to get an accurate evaluation of the soldiers' fitness levels. Preparations for administering an accurate APFT include the following:

- Selecting and training supervisors and scorers.
- Briefing and orienting administrators and participants.
- Securing a location for the events.

Commanders must strictly control those factors which influence test performance. They must ensure that events, scoring, clothing, and equipment are uniform. Commanders should plan testing which permits each soldier to perform to his maximal level. They should also ensure the following:

- Soldiers are not tested when fatigued or ill.
- Soldiers do not have tiring duties just before taking the APFT.
- Weather and environmental conditions do not inhibit performance.
- Safety is the first consideration.

Duties of Test Personnel

Testers must be totally familiar with the instructions for each event and trained to administer the tests. Correctly supervising testees and laying out the test area are essential duties. The group administering the test must include the following:

- OIC or NCOIC.
- Event supervisor, scorers, and a demonstrator for each event.
- Support personnel (safety, control, and medical as appropriate). There should be no less than one scorer for each 15 soldiers tested. Twelve to 15 scorers are required when a company-sized unit is tested.

OIC OR NCOIC

The OIC or NCOIC does the following:

- Administers the APFT.
- Procures all necessary equipment and supplies.
- Arranges and lays out the test area.
- Trains the event supervisors, scorers, and demonstrators. (Training video tape No. 21-191 should be used for training those who administer the APFT.)
- Ensures the test is properly administered and the events are explained, demonstrated, and scored according to the test standards in this chapter.
- Reports the results after the test.

EVENT SUPERVISORS

Event supervisors do the following:

- Administer the test events.
- Ensure that necessary equipment is on hand.
- Read the test instructions, and have the events demonstrated.

- Supervise the scoring of events, and ensure that they are done correctly.
- Rule on questions and scoring discrepancies for their event.

SCORERS

Scorers do the following:

- Supervise the performance of testees.
- Enforce the test standards in this chapter.
- Count the number of correctly performed repetitions aloud.
- Record the correct, raw score on each soldier's scorecard, and initial the scorecard block.
- Perform other duties assigned by the OIC or NCOIC.

Scorers must be thoroughly trained to maintain uniform scoring standards. They do not participate in the test.

The goal of the APFT is to get an accurate evaluation of the soldier's fitness levels.

Instructions for Completing DA Form 705, Army Physical Fitness Scorecard, June 1998.

NAME Print soldier's last name, first name and middle initial in NAME block.

SSN Print soldier's social security number in SSN block.

GENDER Print **M** for male or **F** for female in GENDER block.

UNIT Print soldier's unit designation in UNIT block.

DATE Print date the APFT is administered in DATE block.

GRADE Print soldier's grade in GRADE block.

AGE Print soldier's age on the date the APFT is administered in AGE block.

HEIGHT Print soldier's height in HEIGHT block. Height will be rounded to the nearest inch. If the height fraction is less than 1/2 inch, round down to the nearest whole number in inches. If the height fraction is greater than 1/2 inch, round up to the next highest whole number in inches.

WEIGHT Print soldier's weight in WEIGHT block. Weight will be recorded to the nearest pound. If the weight fraction is less than 1/2 pound, round down to the nearest pound. If the weight fraction is 1/2 pound or greater, round up to the nearest pound. Circle **GO** if soldier meets screening table weight IAW AR 600-9. Circle **NO-GO** if soldier exceeds screening table weight IAW AR 600-9.

BODY FAT If soldier exceeds screening table weight, print the soldier's body fat in the BODY FAT block. Percent body fat is recorded from DA Form 5500-R, Body Fat Content Worksheet, Dec 85, for male soldiers and DA Form 5501-R, Body Fat Content Worksheet, Dec 85, for female soldiers. Circle **GO** if soldier meets percent body fat for their age and gender IAW AR 600-9. Circle **NO-GO** if soldier exceeds percent body fat for their age and gender IAW AR 600-9. If soldier does not exceed screening table weight or does not appear to have excessive body fat IAW AR 600-9, print N/A (not applicable) in the BODY FAT block.

PU RAW SCORE The event scorer records the number of correctly performed repetitions of the push-up in the PU RAW SCORE block and prints his or her initials in the INITIALS block.

SU RAW SCORE The event scorer records the number of correctly performed repetitions of the sit-up in the SU RAW SCORE block and prints his or her initials in the INITIALS block.

2MR RAW SCORE The event scorer records the two-mile run time in the 2MR RAW SCORE block. The time is recorded in minutes and seconds. The event scorer then determines the point value for the two-mile run using the scoring standards on the reverse side of the scorecard. The point value is recorded in the 2MR POINTS block and the event scorer prints his or her initials in the INITIALS block. In all cases when a point value falls between two point values, the lower point value is used and recorded. The two-mile run event scorer also determines the point value for push-ups and sit-ups using the scoring standards on the reverse side of the scorecard. The point values are recorded in the appropriate push-up and sit-up POINTS block and the event scorer prints his or her initials in the INITIALS block. The two-mile run event scorer totals the points from the three events and records the total APFT score in the TOTAL POINTS block.

ALTERNATE AEROBIC EVENT The event scorer prints the alternate aerobic event administered (800-yard swim, 6.2-mile-stationary bicycle ergometer, 6.2-mile-bicycle test or 2.5-mile walk) in the ALTERNATE AEROBIC EVENT block. The time the soldier completes the alternate aerobic event is recorded in minutes and seconds in the ALTERNATE AEROBIC EVENT block. The standards for the alternate aerobic event tests are listed in FM 21-20, Chapter 14, Figure 14-9. Scoring for all alternate aerobic events is on a **GO** or **NO-GO** basis. No point values are awarded. Circle **GO** if the soldier completes the alternate aerobic event within the required time or less. Circle **NO-GO** if the soldier fails to complete the alternate aerobic event within the required time. The alternate aerobic event scorer also determines the point value for push-ups and or sit-ups using the scoring standards on the reverse side of the scorecard. The point values are recorded in the appropriate push-up and or sit-up POINTS block and the event scorer prints his or her initials in the 2MR INITIALS block. The alternate aerobic event scorer totals the points from the push-up and or sit-up events and records the total APFT score in the TOTAL POINTS block.

NCOIC/OIC Signature The NCOIC/OIC checks all test scores for accuracy and signs their name in the NCOIC/OIC Signature block.

COMMENTS The event supervisor, event scorer, NCOIC, or OIC may record comments appropriate to the APFT in the COMMENTS block. Appropriate comments may include: weather conditions, injury during APFT and or appeals.

SUPPORT PERSONNEL

Safety and control people should be at the test site, depending on local policy and conditions. Medical personnel may also be there. However, they do not have to be on site to have the APFT conducted. At a minimum, the OIC or NCOIC should have a plan, known to all test personnel, for getting medical help if needed.

Test Site

The test site should be fairly flat and free of debris. It should have the following:
● An area for stretching and warming up.
● A soft, flat, dry area for performing push-ups and sit-ups.

● A flat, 2-mile running course with a solid surface and no more than a three-percent grade. (Commanders must use good judgement; no one is expected to survey terrain.)
● No significant hazards, (for example, traffic, slippery road surfaces, heavy pollution).

When necessary or expedient, a quarter-mile running track can "be used. It can be marked with a series of stakes along the inside edge. When the track is laid out, a horizontal midline 279 feet, 9 3/4 inches long must be marked in the center of a clear area. A 120-foot circle is marked at both ends of this line. The track is formed when the outermost points of the two circles are connected with tangent lines. (See Figure 14-2.)

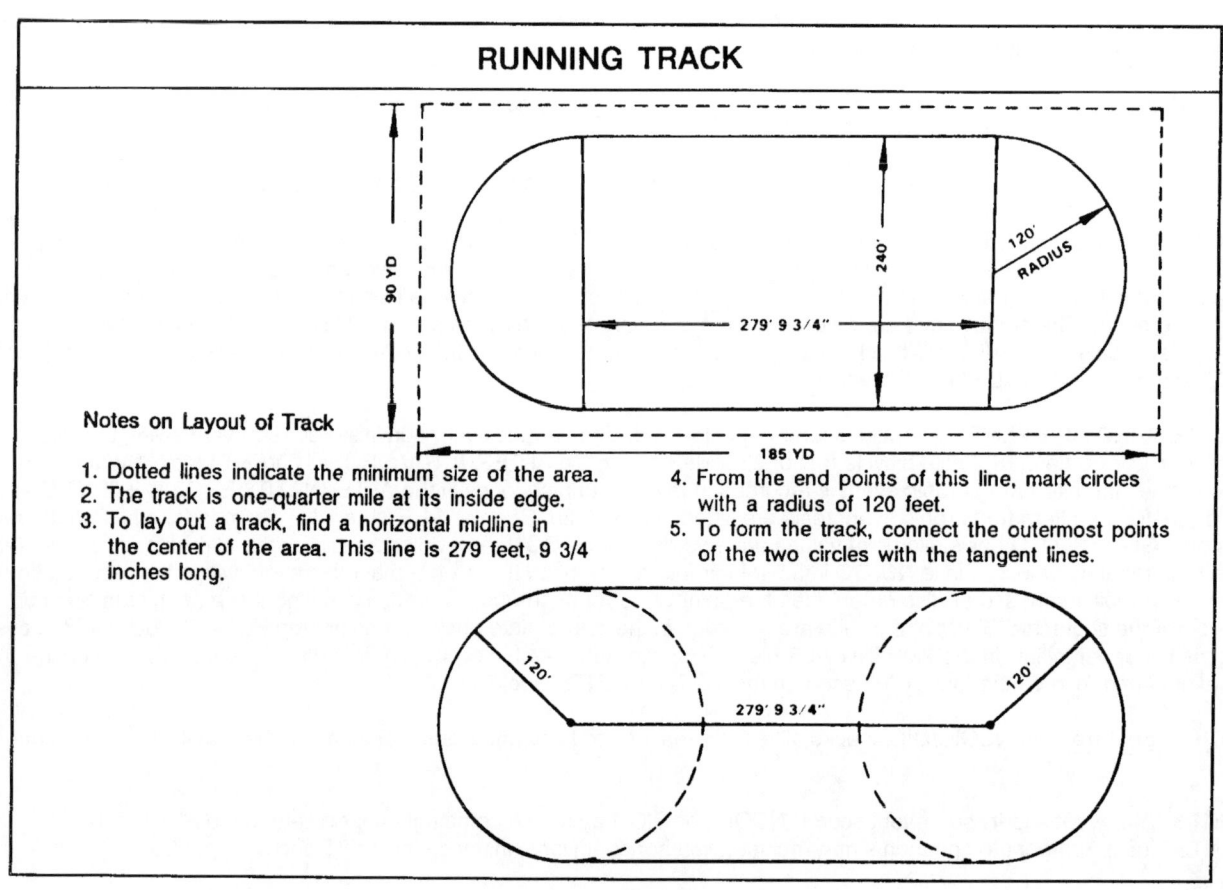

RUNNING TRACK

Notes on Layout of Track

1. Dotted lines indicate the minimum size of the area.
2. The track is one-quarter mile at its inside edge.
3. To lay out a track, find a horizontal midline in the center of the area. This line is 279 feet, 9 3/4 inches long.
4. From the end points of this line, mark circles with a radius of 120 feet.
5. To form the track, connect the outermost points of the two circles with the tangent lines.

Figure 14-2

A 400-meter track may be used in place of the standard quarter-mile (440-yard) track for the 2-mile run, However, one lap run on a 400-meter track is 92 inches shorter than one lap on a 440-yard track. Eight laps on a 400-meter track is 736 inches shorter than eight laps (2 miles) on a 440-yard track. Therefore, soldiers who run the 2-mile event on a 400-meter track must run eight laps plus an additional 61 feet, 4 inches.

Test Procedures

On test day, soldiers are assembled in a common area and briefed by the test OIC or NCOIC about the purpose and organization of the test. The OIC or NCOIC then explains the scorecard, scoring standards, and sequence of events.

The instructions printed here in large type must be read to the soldiers: "YOU ARE ABOUT TO TAKE THE ARMY PHYSICAL FITNESS TEST, A TEST THAT WILL MEASURE YOUR MUSCULAR ENDURANCE AND CARDIORESPIRATORY FITNESS. THE RESULTS OF THIS TEST WILL GIVE YOU AND YOUR COMMANDERS AN INDICATION OF YOUR STATE OF FITNESS AND WILL ACT AS A GUIDE IN DETERMINING YOUR PHYSICAL TRAINING NEEDS. LISTEN CLOSELY TO THE TEST INSTRUCTIONS, AND DO THE BEST YOU CAN ON EACH OF THE EVENTS."

If scorecards have not already been issued, they are handed out at this time. The OIC or NCOIC then says the following: "IN THE APPROPRIATE SPACES, PRINT IN INK THE PERSONAL INFORMATION REQUIRED ON THE SCORECARD." (If scorecards have been issued to the soldiers and filled out before they arrive at the test site, this remark is omitted.)

The OIC or NCOIC pauses briefly to give the soldiers time to check the information. He then says the following: "YOU ARE TO CARRY THIS CARD WITH YOU TO EACH EVENT. BEFORE YOU BEGIN, HAND THE CARD TO THE SCORER. AFTER YOU COMPLETE THE EVENT, THE SCORER WILL RECORD YOUR RAW SCORE, INITIAL THE CARD, AND RETURN IT TO YOU." (At this point, the scoring tables are explained so everyone understands how raw scores are converted to point scores.) Next, the OIC or NCOIC says the following "EACH OF YOU WILL BE ASSIGNED TO A GROUP. STAY WITH YOUR TEST GROUP FOR THE ENTIRE TEST, WHAT ARE YOUR QUESTIONS ABOUT THE TEST AT THIS POINT?"

Groups are organized as required and given final instructions including what to do after the final event. The test is then given.

RETAKING OF EVENTS

Soldiers who start an event incorrectly must be stopped by the scorer before they complete 10 repetitions and told what their errors are. They are then sent to the end of the line to await their turn to retake the event.

A soldier who has problems such as muscle cramps while performing an event may rest if he does not assume an illegal position in the process. If he continues, he receives credit for all correctly done repetitions within the two-minute period. If he does not continue, he gets credit for the number of correct repetitions he has performed up to that time. If he has not done 10 correct repetitions, he is sent to the end of the line to retake that event. He may not retake the event if he has exceeded 10 repetitions. Soldiers who are unable to perform 10 correct repetitions because of low fitness levels may not retake an event.

TEST FAILURES

Soldiers who stop to rest in an authorized rest position continue to receive credit for correct repetitions performed after their rest. Soldiers who rest in an unauthorized rest position will have their performance in that event immediately terminated.

The records of soldiers who fail a record APFT for the first time and those who fail to take the APFT within the required period (AR 350-15, paragraph 11) must be flagged IAW AR 600-8-2 (Reference B).

RETESTING

Soldiers who fail any or all of the events must retake the entire APFT. In case of test failure, commanders may allow soldiers to retake the test as soon as the soldiers and commanders feel they are ready. Soldiers without a medical profile will be retested not-later-than three months following the initial APFT failure in accordance with AR 350-15, paragraph 11.

Test Sequence

The test sequence is the push-up, sit-up, and 2-mile run (or alternate, aerobic event). The order of events cannot be changed. There are no exceptions to this sequence.

Soldiers should be allowed no less than 10 minutes, but ideally no more than 20 minutes, to recover between each event. The OIC or NCOIC determines the time to be allotted between events, as it will depend on the total number of soldiers who are participating in the APFT. If many soldiers are to be tested, staggered starting times should be planned to allow the proper intervals between events. Under no circumstances is the APFT valid if a soldier cannot begin and end all three events in two hours or less.

The following paragraphs describe the equipment, facilities, personnel, instructions, administration, timing techniques, and scorers' duties for the pushup, sit-up, and 2-mile-run events.

PUSH-UPS

Push-ups measure the endurance of the chest, shoulder, and triceps muscles. (See Figure 14-3.)

Equipment

One stopwatch is needed along with one clipboard and pen for each scorer. The event supervisor must have the following the instructions in this chapter on how to conduct the event and one copy of the push-up scoring standards (DA Form 705).

Facilities

There must be at least one test station for every 15 soldiers to be tested. Each station is 6 feet wide and 15 feet deep.

Personnel

One event supervisor must beat the test site and one scorer at each station. The event supervisor may not be the event scorer.

Instructions

The event supervisor must read the following: "THE PUSH-UP EVENT MEASURES THE ENDURANCE OF THE CHEST, SHOULDER, AND TRICEPS MUSCLES. ON THE COMMAND 'GET SET,' ASSUME THE FRONT-LEANING REST POSITION BY PLACING YOUR HANDS WHERE THEY ARE COMFORTABLE FOR YOU. YOUR FEET MAY BE TOGETHER OR UP TO 12 INCHES APART. WHEN VIEWED FROM THE SIDE, YOUR BODY SHOULD FORM A GENERALLY STRAIGHT LINE FROM YOUR SHOULDERS TO YOUR ANKLES.

PUSH-UPS

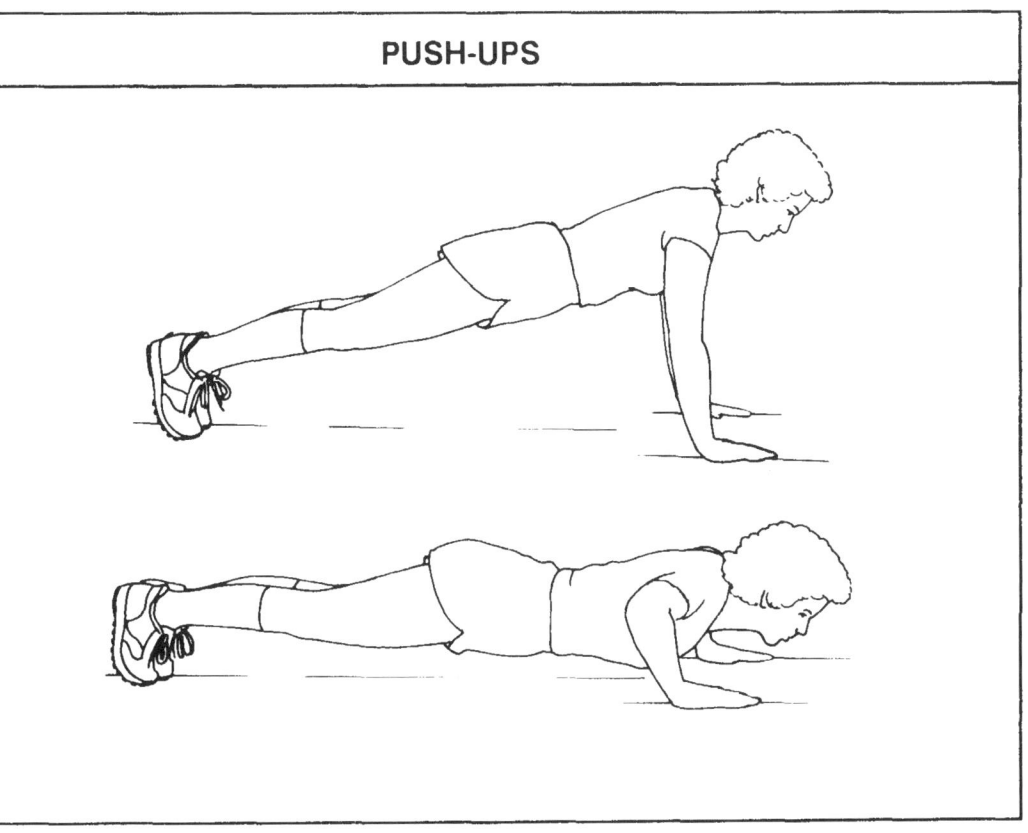

Figure 14-3

ON THE COMMAND 'GO,' BEGIN THE PUSH-UP BY BENDING YOUR ELBOWS AND LOWERING YOUR ENTIRE BODY AS A SINGLE UNIT UNTIL YOUR UPPER ARMS ARE AT LEAST PARALLEL TO THE GROUND. THEN, RETURN TO THE STARTING POSITION BY RAISING YOUR ENTIRE BODY UNTIL YOUR ARMS ARE FULLY EXTENDED. YOUR BODY MUST REMAIN RIGID IN A GENERALLY STRAIGHT LINE AND MOVE AS A UNIT WHILE PERFORMING EACH REPETITION. AT THE END OF EACH REPETITION, THE SCORER WILL STATE THE NUMBER OF REPETITIONS YOU HAVE COMPLETED CORRECTLY. IF YOU FAIL TO KEEP YOUR BODY GENERALLY STRAIGHT, TO LOWER YOUR WHOLE BODY UNTIL YOUR UPPER ARMS ARE AT LEAST PARALLEL TO THE GROUND, OR TO EXTEND YOUR ARMS COMPLETELY, THAT REPETITION WILL NOT COUNT, AND THE SCORER WILL REPEAT THE NUMBER OF THE LAST CORRECTLY PERFORMED REPETITION. IF YOU FAIL TO PERFORM THE FIRST TEN PUSH-UPS CORRECTLY, THE SCORER WILL TELL YOU TO GO TO YOUR KNEES AND WILL EXPLAIN TO YOU WHAT YOUR MISTAKES ARE. YOU WILL THEN BE SENT TO THE END OF THE LINE TO BE RETESTED. AFTER THE FIRST 10 PUSH-UPS HAVE BEEN PERFORMED AND COUNTED, HOWEVER, NO RESTARTS ARE ALLOWED. THE TEST WILL CONTINUE, AND ANY INCORRECTLY PERFORMED PUSH-UPS WILL NOT BE COUNTED. AN ALTERED, FRONT-LEANING REST POSITION IS THE ONLY AUTHORIZED REST POSITION. THAT IS, YOU MAY

SAG IN THE MIDDLE OR FLEX YOUR BACK. WHEN FLEXING YOUR BACK, YOU MAY BEND YOUR KNEES, BUT NOT TO SUCH AN EXTENT THAT YOU ARE SUPPORTING MOST OF YOUR BODY WEIGHT WITH YOUR LEGS. IF THIS OCCURS, YOUR PERFORMANCE WILL BE TERMINATED. YOU MUST RETURN TO, AND PAUSE IN, THE CORRECT STARTING POSITION BEFORE CONTINUING. IF YOU REST ON THE GROUND OR RAISE EITHER HAND OR FOOT FROM THE GROUND, YOUR PERFORMANCE WILL BE TERMINATED. YOU MAY REPOSITION YOUR HANDS AND/OR FEET DURING THE EVENT AS LONG AS THEY REMAIN IN CONTACT WITH THE GROUND AT ALL TIMES. CORRECT PERFORMANCE IS IMPORTANT. YOU WILL HAVE TWO MINUTES IN WHICH TO DO AS MANY PUSH-UPS AS YOU CAN. WATCH THIS DEMONSTRATION." (The exercise is then demonstrated. See Figure 14-4 for a list of points that need to be made during the demonstration.) "WHAT ARE YOUR QUESTIONS?"

Administration

After reading the instructions, the supervisor answers questions. Then he moves the groups to their testing stations. The event supervisor cannot be ready to begin. Successive groups do the event until all soldiers have completed it.

Timing Techniques

The event supervisor is the timer. He calls out the time remaining every 30 seconds and every second for the last 10 seconds of the two minutes. He ends the event after two minutes by the command "Halt!"

Scorers' Duties

Scorers must allow for differences in the body shape and structure of each soldier. The scorer uses each soldier's starting position as a guide throughout the event to evaluate each repetition. The scorer should talk to the soldier before the event begins and have him do a few repetitions as a warm-up and reference to ensure he is doing the exercise correctly.

ADDITIONAL POINTS TO DEMONSTRATE FOR THE PUSH-UP EVENT

The following points must be clarified during the demonstration:

- The soldier's chest may touch the ground (mat or floor) during the push-up as long as the contact does not provide him an advantage. He cannot use the ground to bounce off of or momentarily rest on. However, penalizing a soldier for touching the ground with the chest is unfair. Some soldiers have a large chest or abdomen or are otherwise developed in a way which makes touching the ground unavoidable when they are in the correct down position. Do not count those repetitions in which bouncing off the ground has given the soldier an unfair advantage. Do not count those repetitions in which the long bone of the upper arm does not reach a position parallel to the ground.
- Soldiers may reposition their hands during the push-up event as long as the hands remain in contact with the at all times. The hands can be repositioned either forward, inward, outward, or backward. If a soldier repositions his hands too far backward, the legal front-leaning rest position may be violated.
- If a mat is used, the entire body must be on the mat.

- In the rest position, a soldier may sag in the middle or flex his back in the altered front-leaning rest position; however, he may not readjust his hands backward and/ or bend his knees to such a point that when he bends at the waist and/or knees, he supports most of his body weight with his legs. If this occurs, the soldier's performance in the event will be terminated.
- The feet may not be braced during the push-up event. Test administrators must ensure that a non-slip surface is available.
- Soldiers may do the push-up event on their fists. This may be necessary due to a prior injury. There is no unfair advantage to be gained by doing so.
- Soldiers may not cross their feet while doing the push-up event. This ensures as much standardization as possible and avoids violation of the proper front-leaning rest position, which is the only authorized starting position for this event.
- Soldiers may not take any part of the APFT in bare feet.
- Soldiers should not wear glasses while performing the push-up event.

Figure 14-4

The scorer may either sit or kneel about three feet from the testee's shoulder at a 45-degree angle in front of it. The scorer's head should be about even with the testee's shoulder when the latter is in the front-leaning rest position. Each scorer determines for himself if he will sit or kneel when scoring. He may not lie down or stand while scoring. He counts out loud the number of correct repetitions completed and repeats the number of the last correct push-up if an incorrect one is done. Scorers tell the testees what they do wrong as it occurs during the event. A critique of the performance is done following the test.

When the soldier completes the event, the scorer records the number of correctly performed repetitions, initials the scorecard, and returns it to the soldier.

SIT-UPS

This event measures the endurance of the abdominal and hip-flexor muscles. (See Figure 14-5.)

Equipment

One stopwatch is needed along with one clipboard and pen for each scorer. The event supervisor must have the following: the instructions in this chapter on how to conduct the event and one copy of the sit-up scoring standards (DA Form 705).

Facilities

Each station is 6 feet wide and 15 feet deep. Ensure that no more than 15 soldiers are tested at a station.

Personnel

One event supervisor must be at the test site and one scorer at each station. The event supervisor may not be the event scorer.

Instructions

The event supervisor must read the following: "THE SIT-UP EVENT MEASURES THE ENDURANCE OF THE ABDOMINAL AND HIP-FLEXOR MUSCLES. ON THE

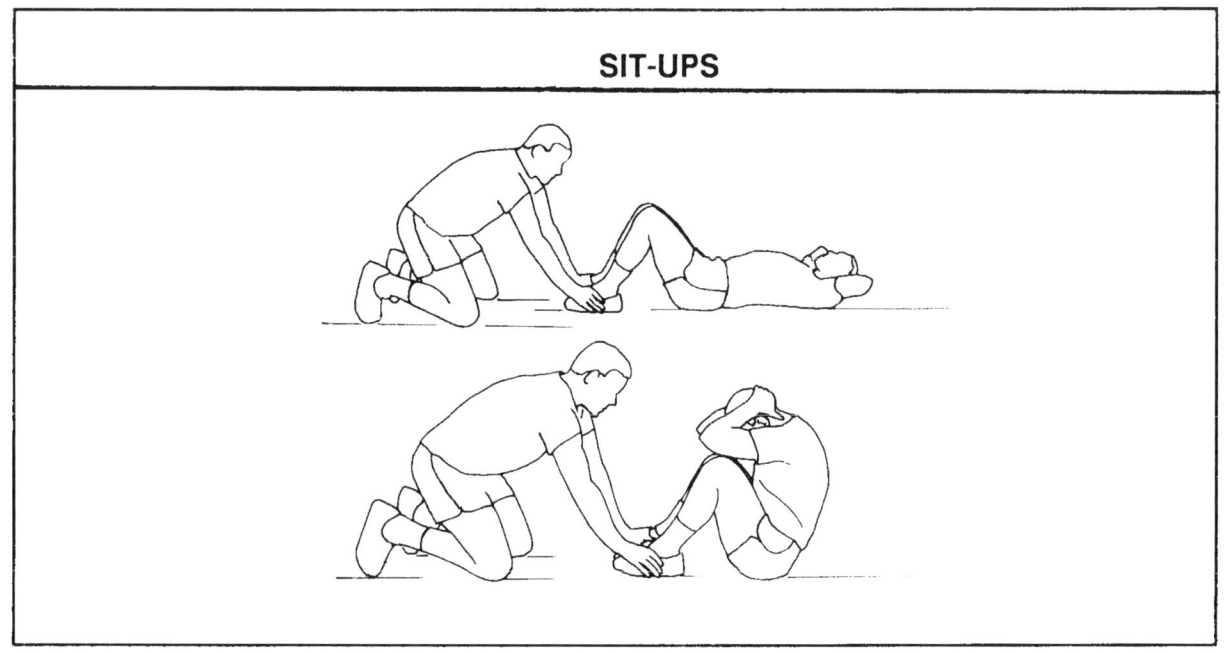

SIT-UPS

Figure 14-5

14-14

COMMAND "GET SET", ASSUME THE STARTING POSITION BY LYING ON YOUR BACK WITH YOUR KNEES BENT AT A 90-DEGREE ANGLE. YOUR FEET MAY BE TOGETHER OR UP TO 12 INCHES APART. ANOTHER PERSON WILL HOLD YOUR ANKLES WITH THE HANDS ONLY. NO OTHER METHOD OF BRACING OR HOLDING THE FEET IS AUTHORIZED. THE HEEL IS THE ONLY PART OF YOUR FOOT THAT MUST STAY IN CONTACT WITH THE GROUND. YOUR FINGERS MUST BE INTERLOCKED BEHIND YOUR HEAD AND THE BACKS OF YOUR HANDS MUST TOUCH THE GROUND. YOUR ARMS AND ELBOWS NEED NOT TOUCH THE GROUND. ON THE COMMAND "GO", BEGIN RAISING YOUR UPPER BODY FORWARD TO, OR BEYOND, THE VERTICAL POSITION. THE VERTICAL POSITION MEANS THAT THE BASE OF YOUR NECK IS ABOVE THE BASE OF YOUR SPINE. AFTER YOU HAVE REACHED OR SURPASSED THE VERTICAL POSITION, LOWER YOUR BODY UNTIL THE BOTTOM OF YOUR SHOULDER BLADES TOUCH THE GROUND. YOUR HEAD, HANDS, ARMS, OR ELBOWS DO NOT HAVE TO TOUCH THE GROUND. AT THE END OF EACH REPETITION, THE SCORER WILL STATE THE NUMBER OF SIT-UPS YOU HAVE CORRECTLY COMPLETED. A REPETITION WILL NOT COUNT IF YOU FAIL TO REACH THE VERTICAL POSITION, FAIL TO KEEP YOUR FINGERS INTERLOCKED BEHIND YOUR HEAD, ARCH OR BOW YOUR BACK AND RAISE YOUR BUTTOCKS OFF THE

ADDITIONAL POINTS TO DEMONSTRATE FOR THE SIT-UP EVENT

The following points must be clarified during the demonstration:

- To minimize stress to the neck, it is recommended that the soldier keep his chin curled downward and touching the top of his chest throughout the performance of the sit-up event.
- From the starting (down) position, or during any phase of the sit-up, the soldier may not use his hands or arms to pull himself up to push off the ground (floor or mat) in order to help himself attain the up position. Any of these procedures can give the violator an unfair advantage. They also violate the intent of the event. The sit-up event will be terminated immediately for those soldiers who, by pushing or pulling, use their arms to assist themselves in attaining the up position.
- If a mat is used, the entire body, including the feet and head, must be on the mat at the start.
- From the starting (down) position, or during any phase of the sit-up, the soldier may not swing his hands or arms in order to help himself attain the up position. If this occurs, that repetition does not count.
- The soldier may wiggle to attain the up position. This gives him no advantage.

- While in the up position, the soldier may not help himself stay in that position by using the elbows or any part of the arms to lock on to or brace against the legs. The elbows can go either inside or outside the knees. However, to push or pull them into the sides or tops of the knees to get extra leverage and rest gives an unfair advantage to that soldier. Therefore, soldiers who use this technique will be warned once for the first violation and immediately terminated if the violation continues or recurs.
- During the performance of the sit-up event, the fingers must be interlocked and behind the head. As long as any of the fingers are overlapping to any degree, the fingers are considered to be interlocked.
- If either foot breaks contact with the ground during a repetition, that repetition will not count. Both heels must stay in contact with the ground (floor or mat) during the performance of the event. The scorer should ensure that the holder has the soldier's feet properly secured. The scorer tells the soldier if his heel(s) is raised from the ground and that the repetition will not count.

Figure 14-6

GROUND TO RAISE YOUR UPPER BODY, OR LET YOUR KNEES EXCEED A 90-DEGREE ANGLE. IF A REPETITION DOES NOT COUNT, THE SCORER WILL RE-PEAT THE NUMBER OF YOUR LAST CORRECTLY PERFORMED SIT-UP. THE UP POSITION IS THE ONLY AUTHORIZED REST POSI-TION. IF YOU STOP AND REST IN THE DOWN (STARTING) POSITION, THE EVENT WILL BE TERMI-NATED. AS LONG AS YOU MAKE A CONTINUOUS PHYSICAL EF-FORT TO SIT UP, THE EVENT WILL NOT BE TERMINATED. YOU MAY NOT USE YOUR HANDS OR ANY OTHER MEANS TO PULL OR PUSH YOURSELF UP TO THE UP (RESTING) POSITION OR TO HOLD YOURSELF IN THE REST POSI-TION. IF YOU DO SO, YOUR PER-FORMANCE IN THE EVENT WILL BE TERMINATED. CORRECT PER-FORMANCE IS IMPORTANT. YOU WILL HAVE TWO MINUTES TO PERFORM AS MANY SIT-UPS AS YOU CAN. WATCH THIS DEMON-STRATION." (The exercise is then demonstrated. See Figure 14-6 for a list of points that need to be made during the demonstration.) "WHAT ARE YOUR QUESTIONS?"

Administration

After reading the instructions, the supervisor answers questions. He then moves the groups to their testing sta-tions. The event supervisor cannot be a scorer. At this point, the testing is ready to begin. Successive groups do the event until all soldiers have com-pleted it.

Timing Techniques

The event supervisor is the timer. He calls out the time remaining every 30 seconds and every second for the last 10 seconds of the two minutes. He ends the event after two minutes by the command "Halt!"

Scorers' Duties

The scorer may either kneel or sit about three feet from the testee's hip. The scorer's head should be about even with the testee's shoulder when the latter is in the vertical (up) position. Each scorer decides for himself whether to sit or kneel down when scoring. He may not lie down or stand while scoring. The scorer counts aloud the number of correctly performed sit-ups and repeats the number of the last correctly performed repetition if an incorrect one is done. Scorers tell the testees what they are doing wrong as it occurs during the event. A critique of his performance is given to each sol-dier after the event. When the soldier completes the event, the scorer records the number of correctly performed sit-ups, initials the scorecard, and returns it to the soldier.

When checking for correct body position, the scorer must be sure that at a 90-degree angle is formed at each knee by the soldier's upper and lower leg. The angle to be measured is not the one formed by the lower leg and the ground. If, while performing the sit-up event, this angle becomes greater than 90 degrees, the scorer should instruct the testee and holder to repo-sition the legs to the proper angle and obtain compliance before allowing the testee's performance to continue. The loss of the proper angle does not terminate the testee's performance in the event. When the soldier comes to the vertical position, the scorer must be sure that the base of the soldier's neck is above or past the base of the spine. A soldier who simply touches his knees with his elbows may not come to a completely vertical position. The scorer must ensure that the holder uses only his hands to brace the exerciser's feet.

TWO-MILE RUN

This event tests cardiorespiratory (aerobic) endurance and the endurance of the leg muscles. (See Figure 14-7.)

Equipment

Two stopwatches for the event supervisor, one clipboard and pen for each scorer, copies of the event's instructions and standards, and numbers for the testees are needed.

Facilities

There must be a level area with no more than a three-degree slope on which a measured course has been marked. An oval-shaped track of known length may be used. If a road course is used, the start and finish and one-mile (half way) point must be clearly marked.

Personnel

One event supervisor and at least one scorer for every 15 runners are required.

Instructions

The event supervisor must read the following: "THE TWO-MILE RUN IS USED TO ASSESS YOUR AEROBIC FITNESS AND YOUR LEG MUSCLES' ENDURANCE. You MUST COMPLETE THE RUN WITHOUT ANY PHYSICAL HELP. AT THE START, ALL SOLDIERS WILL LINE UP BEHIND THE STARTING LINE. ON THE COMMAND 'GO,' THE CLOCK WILL START. YOU WILL BEGIN RUNNING AT YOUR OWN PACE. TO RUN THE REQUIRED TWO MILES, YOU MUST COMPLETE (describe the number of laps, start and finish points, and course layout). YOU ARE BEING TESTED ON YOUR ABILITY TO COMPLETE THE 2-MILE COURSE IN THE SHORTEST TIME POSSIBLE. ALTHOUGH WALKING IS AUTHORIZED, IT IS STRONGLY DISCOURAGED. IF YOU ARE PHYSICALLY HELPED IN ANY WAY (FOR EXAMPLE, PULLED, PUSHED, PICKED UP, AND/OR CARRIED) OR LEAVE THE DESIGNATED RUNNING COURSE FOR ANY

TWO-MILE RUN

Figure 14-7

REASON, YOU WILL BE DISQUALI-FIED. (IT IS LEGAL TO PACE A SOLDIER DURING THE 2-MILE RUN. AS LONG AS THERE IS NO PHYSICAL CONTACT WITH THE PACED SOLDIER AND IT DOES NOT PHYSICALLY HINDER OTHER SOLDIERS TAKING THE TEST, THE PRACTICE OF RUNNING AHEAD OF, ALONG SIDE OF, OR BEHIND THE TESTED SOLDIER, WHILE SERVING AS A PACER, IS PER-MITTED. CHEERING OR CALL-ING OUT THE ELAPSED TIME IS ALSO PERMITTED.) THE NUM-BER ON YOUR CHEST IS FOR IDENTIFICATION. YOU MUST MAKE SURE IT IS VISIBLE AT ALL TIMES. TURN IN YOUR NUMBER WHEN YOU FINISH THE RUN. THEN, GO TO THE AREA DESIG-NATED FOR THE COOL-DOWN AND STRETCH. DO NOT STAY NEAR THE SCORERS OR THE FINISH LINE AS THIS MAY IN-TERFERE WITH THE TESTING. WHAT ARE YOUR QUESTIONS ON THIS EVENT?"

Administration

After reading the instructions, the supervisor answers questions. He then organizes the soldiers into groups of no more than 10. The scorer for each group assigns a number to each soldier in the group. At the same time, the scorer collects the scorecards and records each soldier's number.

Timing Techniques

The event supervisor is the timer. He uses the commands "Get set" and "Go." Two stopwatches are used in case one fails. As the soldiers near the finish line, the event supervisor calls off the time in minutes and seconds (for example, "Fifteen-thirty, fifteen-thirty-one, fifteen-thirty -two," and so on).

Scorers' Duties

The scorers observe those runners in their groups, monitor their laps (if appropriate), and record their times as they cross the finish line. (It is often helpful to record the soldiers' numbers and times on a separate sheet of paper or card. This simplifies the recording of finish times when large groups of soldiers are simultaneously tested.) After all runners have completed the run, the scorers determine the point value for each soldier's run time, record the point values on the scorecards, and enter their initials in the scorers' blocks. In all cases, when a time falls between two point values, the lower point value is used and recorded. For example, if a female soldier, age 17 to 21, runs the two miles in 15 minutes and 19 seconds, the score awarded is 95 points.

At this time, the scorers for the 2-mile run also convert the raw scores for the push-up and sit-up events by using the scoring standards on the back side of the scorecard. They enter those point values on the scorecards and determine the total APFT score for each soldier before giving the score-cards to the test's OIC or NCOIC. After the test scores have been checked, the test's OIC or NCOIC signs all scorecards and returns them to the unit's commander or designated representative.

Test Results

The soldier's fitness performance for each APFT event is determined by converting the raw score for each event to a point score.

Properly interpreted, performance on the APFT shows the following:

- Each soldier's level of physical fitness.
- The entire unit's level of physical fitness.
- Deficiencies in physical fitness.
- Soldiers who need special attention.

(Leaders must develop special programs to improve the performance of soldiers who are below the required standards.)

Commanders should not try to determine the individual's or the unit's strengths and weaknesses in fitness by using only the total scores. A detailed study of the results on each event is more important. For a proper analysis of the unit's performance, event scores should be used. They are corrected for age and sex. Therefore, a female's 80-point push-up score should be considered the same as a male's 80-point push-up score. Using the total point value or raw scores may distort the interpretation.

Scores Above Maximum

Even though some soldiers exceed the maximum score on one or more APFT events, the official, maximum score on the APFT must remain at 300 (100 points per event). Some commanders, however, want to know unofficial point scores to reward soldiers for their extra effort.

Only those soldiers who score 100 points in all three events are eligible to determine their score on an extended scale. To fairly determine the points earned, extra points are awarded at the same rate as points obtained for scores at or below the 100 point level. Each push-up and sit-up beyond the maximum is worth one point as is every six-second decrease in the run time. Take, for example, the following case shown in Figure 14-8. A male soldier performs above the maximum in the 17-21 age group by doing 87 push-ups and 98 sit-ups and by running the two miles in 11 minutes and 12 seconds. His score would be calculated as follows:

CALCULATION FOR AGE 17-21 MALE

PUSH-UPS
Actual	87
Maximum	82
Additional points	+5
Points (official)	100
Points (unofficial)	105

SIT-UPS
Actual	98
Maximum	92
Additional points	+6
Points (official)	100
Points (unofficial)	106

2-MILE RUN
Actual	11:12
Maximum	11:54
	:42
ADDITIONAL POINTS (42 sec/6 = 7)	+7
Points (official)	100
Points (unofficial)	107

Thus, the unofficial total for this soldier in the three events is determined in the following way:

PUSH-UPS	105
SIT-UPS	106
2-MILE RUN	+ 107
UNOFFICIAL TOTAL	318

Figure 14-8

The calculations on the previous page, give the soldier a total score of 318 points. This method lets the commander easily determine the scores for performances that are above the maximum. He may recognize soldiers for their outstanding fitness achievements, not only on the APFT but also for other, unofficial fitness challenges. Using this method ensures that each soldier has an equal chance to be recognized for any of the tested fitness components. Commanders may also establish their own incentive programs and set their own unit's standards (AR 350-15).

Temporary Profiles

A soldier with a temporary profile must take the regular three-event APFT after the profile has expired. (Soldiers with temporary profiles of more than three months may take an alternate test as determined by the commander with input from health-care personnel.) Once the profile is lifted, the soldier must be given twice the time of the profile (but not more than 90 days) to train for the APFT. For example, if the profile period was 7 days, the soldier has 14 days to train for the APFT after the profile period ends. If a normally scheduled APFT occurs during the profile period, the soldier should be given a mandatory make-up date.

Permanent Profiles

A permanently profiled soldier is given a physical training program by the profiling officer using the positive profile form DA 3349 (see Appendix B). The profiling officer gives the unit's commander a list of physical activities that are suitable for the profiled soldier. He also indicates the events and/or alternate aerobic event that the soldier will do on the APFT. This recommendation, made after consultation with the profiled soldier, should address the soldier's abilities and preference and the equipment available. (See DA Form 3349, Physical Profile, referenced in AR 40-501.)

The profiled soldier must perform all the regular APFT events his medical profile permits. Each soldier must earn at least 60 points on the regular events to receive a "go." He must also complete the alternate event in a time equal to or less than the one listed for his age group. For example, a soldier whose profile forbids only running will do the push-up and sit-up events and an alternate aerobic event. He must get at least a minimum passing score on each event to earn a "go" for the test. A soldier whose profile prevents two or more APFT events must complete the 2-mile run or an alternate aerobic event to earn a "go" on the test. Soldiers who cannot do any of the aerobic events due to a profile cannot be tested. Such information will be recorded in their official military record.

The standards for alternate events are listed in Figure 14-9. Scoring for all alternate events is on a go/no go basis. Soldiers who do push-up and sit-up events but who take an alternate aerobic event are not awarded promotion points for APFT performance.

Alternate Events

Alternate APFT events assess the aerobic fitness and muscular endurance of soldiers with permanent medical profiles or long-term (greater than three months) temporary profiles who cannot take the regular, three-event APFT.

The alternate aerobic APFT events are the following:
- 800-yard-swim test.
- 6.2-mile-stationary- bicycle ergometer test with a resistance setting of 2 kiloponds (2 kilograms) or 20 newtons.
- 6.2-mile-bicycle test on a conventional bicycle using one speed.
- 2.5-mile-walk test.

ALTERNATE TEST STANDARDS BY EVENT, SEX, AND AGE												
EVENT	SEX	AGE										
		17-21	22-26	27-31	32-36	37-41	42-46	47-51	52-56	57-61	62+	
800-YARD SWIM	Men	20:00	20:30	21:00	21:30	22:00	22:30	23:00	24:00	24:30	25:00	
	Women	21:00	21:30	22:00	22:30	23:00	23:30	24:00	25:00	25:30	26:00	
6.2-MILE BIKE (Stationary and track)	Men	24:00	24:30	25:00	25:30	26:00	27:00	28:00	30:00	31:00	32:00	
	Women	25:00	25:30	26:00	26:30	27:00	28:00	30:00	32:00	33:00	34:00	
2.5-MILE WALK	Men	34:00	34:30	35:00	35:30	36:00	36:30	37:00	37:30	38:00	38:30	
	Women	37:00	37:30	38:00	38:30	39:00	39:30	40:00	40:30	41:00	41:30	

*Figure 14-9

800-YARD-SWIM TEST

This event is used to assess cardio-respiratory (aerobic) fitness. (See Figure 14-10.)

Equipment

Two stopwatches, one clipboard and pen for each scorer, one copy each of the test instructions and standards, and appropriate safety equipment are needed.

Facilities

A swimming pool at least 25 yards long and 3 feet deep, or an approved facility, is needed.

Personnel

One event supervisor and at least one scorer for every soldier to be tested are required. Appropriate safety, control, and medical personnel must also be present.

Instructions

The event supervisor must read the following statement: "THE 800-YARD SWIM IS USED TO ASSESS YOUR LEVEL OF AEROBIC FITNESS. YOU WILL BEGIN IN THE WATER; NO DIVING IS ALLOWED. AT THE START, YOUR BODY MUST BE IN CONTACT WITH THE WALL OF THE POOL. ON THE COMMAND 'GO,' THE CLOCK WILL START. YOU SHOULD THEN BEGIN SWIMMING AT YOUR OWN PACE, USING ANY STROKE OR COMBINATION OF STROKES YOU WISH. YOU MUST SWIM (tell the number) LAPS TO COMPLETE THIS DISTANCE. YOU MUST TOUCH THE WALL OF THE POOL AT EACH END OF THE POOL AS YOU TURN. ANY TYPE OF TURN IS AUTHORIZED. YOU WILL BE SCORED ON YOUR ABILITY TO COMPLETE THE SWIM IN A TIME EQUAL TO, OR LESS THAN, THAT LISTED FOR YOUR AGE AND SEX. WALKING ON THE BOTTOM TO RECUPERATE IS AUTHORIZED. SWIMMING GOGGLES ARE PERMITTED, BUT NO OTHER EQUIPMENT IS AUTHORIZED. WHAT ARE YOUR QUESTIONS ABOUT THIS EVENT?"

800-YARD SWIM

Figure 14-10

Administration

After reading the instructions, the event supervisor answers only related questions. He assigns one soldier to each lane and tells the soldiers to enter the water. He gives them a short warm-up period to acclimate to the water temperature and loosen up. Above all, the event supervisor must be alert to the safety of the testees throughout the test.

Timing Techniques

The event supervisor is the timer. He uses the commands "Get set" and "Go." Two stopwatches are used in case one fails. As the soldiers near the finish, the event supervisor begins calling off the elapsed time in minutes and seconds (for example, "Nineteen-eleven, nineteen-twelve, nineteen-thirteen," and so on). The time is recorded when each soldier touches the end of the pool on the final lap or crosses a line set as the 800-yard mark.

Scorers' Duties

Scorers must observe the swimmers assigned to the. They must be sure that each swimmer touches the bulkhead at every turn. The scorers record each soldier's time

in the 2-mile-run block on the scorecard and use the comment block to identify the time as an 800-yard-swim time. If the pool length is measured in meters, the scorers convert the exact distance to yards. To convert meters to yards, multiply the number of meters by 39.37 and divide the product by 36; that is, (meters x 39.37)/36 = yards. For example, 400 meters equals 437.4 yards; that is, (400 x 39.37)/36 = 437.4 yards.

6.2-MILE STATIONARY-BICYCLE ERGOMETER TEST

This event is used to assess the soldier's cardiorespiratory and leg-muscle endurance. (See Figure 14-11.)

Equipment

Two stopwatches, one clipboard and pen for each scorer, a copy of the test instructions and standards, and one stationary bicycle ergometer are needed. The ergometers should measure resistance in kiloponds or newtons. The bicycle should be one that can be used for training and testing. Its seat and

6.2 MILE STATIONARY-BICYCLE ERGOMETER TEST

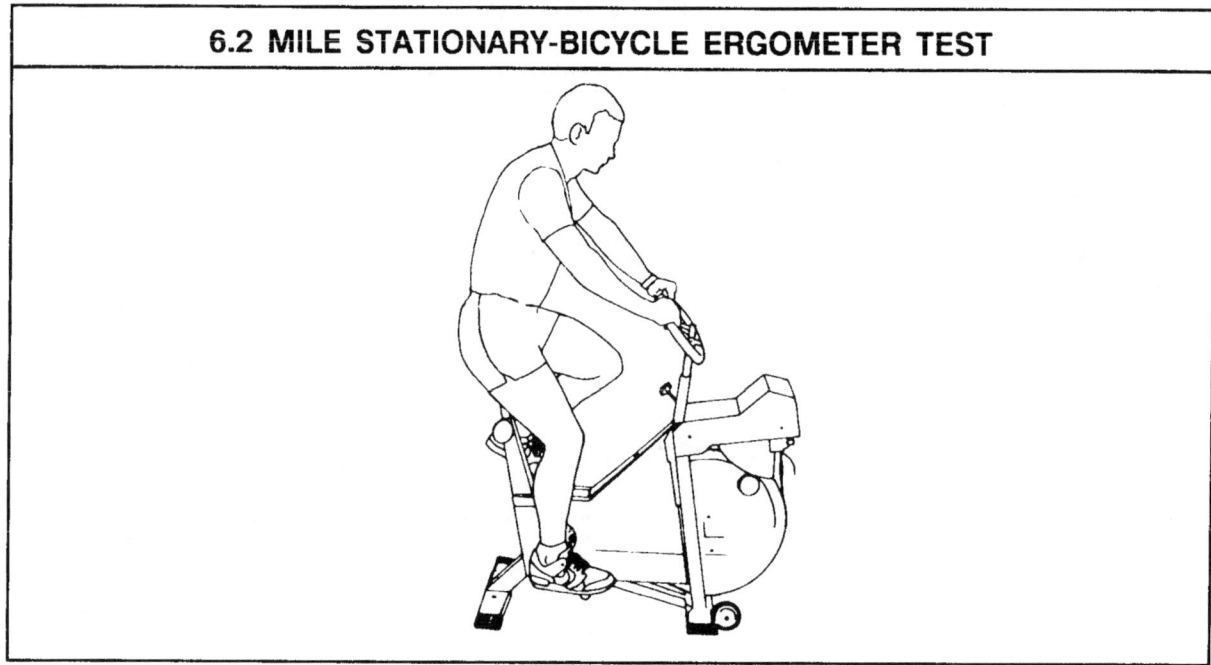

Figure 14-11

handlebars must be adjustable to let the soldier fully extend his legs when pedaling. It should have an adjustable tension setting and an odometer. The resistance is usually set by a tension strap on a weighted pendulum connected to the flywheel. See Appendix D for guidance on using various types of stationary bikes.

Facilities

The test site can be any place where there is an approved bicycle ergometer. This could be the post's fitness facility or the hospital's therapy clinic. Each test station must be two yards wide and four yards deep.

Personnel

One event supervisor and at least one scorer for every three soldiers to be tested are required. Appropriate safety, control, and medical personnel should also be present.

Instructions

The event supervisor must read the following: "THE 6.2-MILE STATIONARY-BICYCLE ERGOMETER EVENT TESTS YOUR CARDIORESPIRATORY FITNESS AND LEG MUSCLE ENDURANCE. THE ERGOMETER'S RESISTANCE MUST BE SET AT TWO KILOPOUNDS (20 NEWTONS). ON THE COMMAND 'GO,' THE CLOCK WILL START, AND YOU WILL BEGIN PEDALING AT YOUR OWN PACE WHILE MAINTAINING THE RESISTANCE INDICATOR AT TWO POUNDS. YOU WILL BE SCORED ON YOUR ABILITY TO COMPLETE 6.2 MILES (10 KILOMETERS), AS SHOWN ON THE ODOMETER, IN A TIME EQUAL TO OR LESS THAN THAT LISTED FOR YOUR AGE AND SEX. WHAT ARE YOUR QUESTIONS ABOUT THIS EVENT?"

Administration

After reading the instructions, the event supervisor answers any related questions. Each soldier is given a short warm-up period and allowed to adjust the seat and handlebar height.

Timing Techniques

The event supervisor is the timer. He uses the commands "Get set" and "Go." Two stopwatches are used in case one fails. As the soldiers pedal the last two-tenths of the test distance, the event supervisor should start calling off the time in minutes and seconds (for example, "Twenty-thirty-one, twenty-thirty-two, twenty-thirty-three," and so on). He calls the time remaining every 30 seconds for the last two minutes of the allowable time and every second during the last ten seconds.

Scorers' Duties

Scorers must ensure that the bicycle ergometer is functioning properly. They must then make sure that the bicycle ergometers' tension settings have been calibrated and are accurate and that the resistance of the ergometers has been set at two kiloponds (20 newtons). The scorers must observe the soldiers throughout the event. From time to time the scorer may need to make small adjustments to the resistance control to ensure that a continuous resistance of exactly 2 kiloponds (20 newtons) is maintained throughout the test. At the end of the test, they record each soldier's time on the scorecard in the 2-mile-run block, initial the appropriate block, and note in the comment block that the time is for a 6.2-mile stationary-bicycle ergometer test.

6.2-MILE BICYCLE TEST

This event is used to assess the soldier's cardiorespiratory and leg-muscle endurance.

Equipment

Two stopwatches, one clipboard and pen for each scorer, a copy of the test instructions and standards, and numbers are needed. Although one-speed bicycles are preferred for this event, multispeed bicycles may be used. If a multispeed bicycle is used, measures must be taken to ensure that only one gear is used throughout the test. (This can usually be done by taping the gear shifters at the setting preferred by the testee.)

Facilities

A relatively flat course with a uniform surface and no obstacles must be used. It must also be clearly marked. Soldiers should not be tested on a quarter-mile track, and they should never be out of the scorers' sight. The course should be completely free of runners and walkers.

Personnel

One event supervisor and at least one scorer for every 10 soldiers are required. Safety, control, and medical personnel should also be present as appropriate.

instructions

The event supervisor must read the following: "THE 6.2-MILE BICYCLE TEST IS USED TO ASSESS YOUR CARDIORESPIRATORY FITNESS AND LEG MUSCLES' ENDURANCE. YOU MUST COMPLETE THE 6.2

MILES WITHOUT ANY PHYSICAL HELP FROM OTHERS. YOU MUST KEEP YOUR BICYCLE IN ONE GEAR OF YOUR CHOOSING FOR THE ENTIRE TEST. CHANGING GEARS IS NOT PERMITTED AND WILL RESULT IN DISQUALIFICA-TION. TO BEGIN, YOU WILL LINE UP BEHIND THE STARTING LINE. ON THE COMMAND 'GO,' THE CLOCK WILL START, AND YOU WILL BEGIN PEDALING AT YOUR OWN PACE. TO COMPLETE THE REQUIRED DISTANCE OF 6.2 MILES, YOU MUST COMPLETE (describe the number of laps, start and finish points, and course layout). YOU WILL BE SCORED ON YOUR ABIL-ITY TO COMPLETE THE DISTANCE-OF 6.2 MILES (10 KILOMETERS) IN A TIME EQUAL TO OR LESS THAN THAT LISTED FOR YOUR AGE AND SEX. IF YOU LEAVE THE DESIGNATED COURSE FOR ANY REASON, YOU WILL BE DISQUALI-FIED. WHAT ARE YOUR QUES-TIONS ABOUT THIS EVENT?"

Administration

After reading the instructions, the event supervisor answers any related questions. He then organizes the soldiers into groups of no more than ten and assigns each group to a scorer. Scorers assign numbers to the soldiers in their groups and record each soldier's number on the appropriate scorecard.

Timing Techniques

The event supervisor is the timer. He uses the commands "Get set" and "Go." Two stopwatches are used in case one fails. As soldiers near the end of the 6.2-mile ride, the event supervisor starts calling off the time in

6.2-MILE BICYCLE TEST

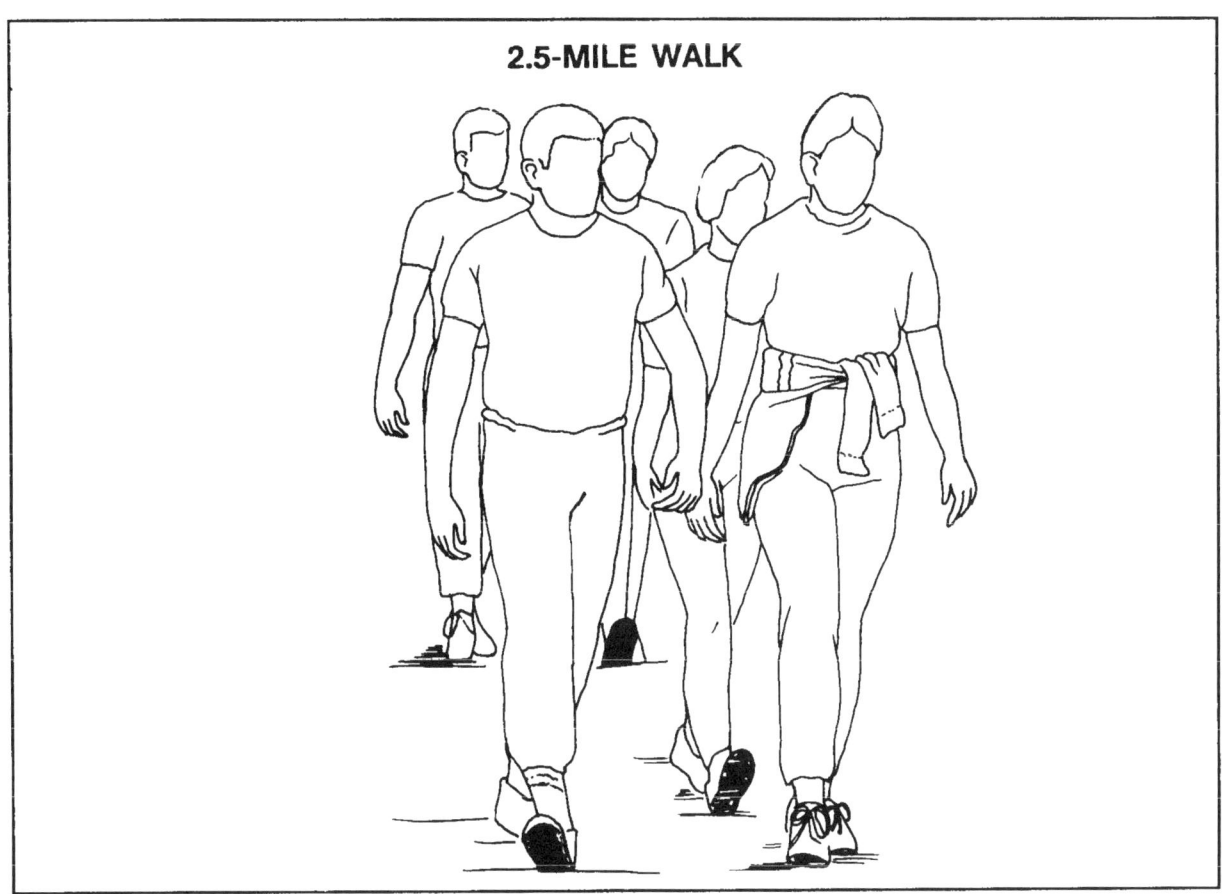

2.5-MILE WALK

minutes and seconds (for example, "Thirty-twenty-one, thirty- twenty-two, thirty -twenty-three," and so on).

Scorers' Duties

When the event is over, scorers record each soldier's time in the 2-mile-run block. They initial the appropriate block and note in the comment block that the time is for a 6.2-mile-bicycle test and whether or not the testee met the required standards for his age and sex.

2.5-MILE WALK

This event serves to assess cardiorespiratory and leg-muscle endurance.

Equipment

Two stopwatches, one clipboard and pen for each scorer, numbers, and copies of the test instructions and standards are needed.

Facilities

This event uses the same course as the 2-mile run.

Personnel

One event supervisor and at least one scorer for every three soldiers to be tested are required. Appropriate safety, control, and medical personnel should be present.

Instructions

The event supervisor must read the following: "THE 2.5-MILE WALK IS USED TO ASSESS YOUR CARDIORESPIRATORY FITNESS AND LEG-MUSCLE ENDURANCE. ON THE COMMAND 'GO,' THE CLOCK WILL START, AND YOU WILL BEGIN WALKING AT YOUR OWN PACE. YOU MUST COMPLETE (describe the number of laps, start and finish points, and course layout). ONE FOOT MUST BE IN CONTACT WITH THE GROUND AT ALL TIMES. IF YOU BREAK INTO A RUNNING STRIDE AT ANY TIME OR HAVE BOTH FEET OFF THE GROUND AT THE SAME TIME, YOUR PERFORMANCE IN THE EVENT WILL BE TERMINATED. YOU WILL BE SCORED ON YOUR ABILITY TO COMPLETE THE 2.5-MILE COURSE IN A TIME EQUAL TO OR LESS THAN THAT LISTED FOR YOUR AGE AND SEX. WHAT ARE YOUR QUESTIONS ABOUT THIS EVENT?"

Administration

After reading the instructions, the event supervisor answers any related questions. He then divides the soldiers into groups of no more than three and assigns each group to a scorer. Each soldier is issued a number which the scorer records on the scorecard.

Timing Techniques

The event supervisor is the timer. He uses the commands "Get set" and "Go." Two stopwatches are used in case one fails. As the soldiers near the end of the 2.5-mile walk, the event supervisor starts calling off the elapsed time in minutes and seconds (for example,"Thirty-three-twenty-two, thirty -three -twenty -three, thirty-three-twenty -four," and so on).

Scorers' Duties

Scorers must observe the soldiers during the entire event and must ensure that the soldiers maintain a walking stride. Soldiers who break into any type of running stride will be terminated from the event and given a "no go." When the event is over, scorers record the time in the 2-mile-run block on the scorecard, initial the appropriate block, and note in the comment block that the time is for a 2.5-mile walk and whether or not the testee received a "go" or "no go."

APPENDIX A

PHYSIOLOGICAL DIFFERENCES BETWEEN THE SEXES

Soldiers vary in their physical makeup. Each body reacts differently to varying degrees of physical stress, and no two bodies react exactly the same way to the same physical stress. For everyone to get the maximum benefit from training, leaders must be aware of these differences and plan the training to provide maximum benefit for everyone. They must also be aware of the physiological differences between men and women. While leaders must require equal efforts of men and women during the training period, they must also realize that women have physiological limitations which generally preclude equal performance. The following paragraphs describe the most important physical and physiological differences between men and women.

SIZE

The average 18- year-old man is 70.2 inches tall and weighs 144.8 pounds, whereas the average woman of the same age is 64.4 inches tall and weighs 126.6 pounds. This difference in size affects the absolute amount of physical work that can be performed by men and women.

MUSCLES

Men have 50 percent greater total muscle mass, based on weight, than do women. A woman who is the same size as her male counterpart is generally only 80 percent as strong. Therefore, men usually have an advantage in strength, speed, and power over women.

FAT

Women carry about 10 percentage points more body fat than do men of the same age. Men accumulate fat primarily in the back, chest, and abdomen; women gain fat in the buttocks, arms, and thighs. Also, because the center of gravity is lower in women than in men, women must overcome more resistance in activities that require movement of the lower body.

BONES

Women have less bone mass than men, but their pelvic structure is wider. This difference gives men an advantage in running efficiency.

HEART SIZE AND RATE

The average woman's heart is 25 percent smaller than the average man's. Thus, the man's heart can pump more blood with each beat. The larger heart size contributes to the slower resting heart rate (five to eight beats a minute slower) in males. This lower rate is evident both at rest and at any given level of submaximal exercise. Thus, for any given work rate, the faster heart rate means that most women will become fatigued sooner than men.

FLEXIBILITY

Women generally are more flexible than men.

LUNGS

The lung capacity of men is 25 to 30 percent greater than that of women. This gives men still another advantage in the processing of oxygen and in doing aerobic work such as running.

RESPONSE TO HEAT

A woman's response to heat stress differs somewhat from a man's. Women sweat less, lose less heat through evaporation, and reach higher body temperatures before sweating starts. Nevertheless, women can adapt to heat stress as well as men. Regardless of gender, soldiers with a higher level of physical fitness generally better tolerate, and adapt more readily to, heat stress than do less fit soldiers.

OTHER FACTORS

Knowing the physiological differences between men and women is just the first step in planning physical training for a unit. Leaders need to understand other factors too.

Women can exercise during menstruation; it is, in fact, encouraged. However, any unusual discomfort, cramps, or pains while menstruating should be medically evaluated.

Pregnant soldiers cannot be required to exercise without a doctor's approval. Generally, pregnant women may exercise until they are close to childbirth if they follow their doctors' instructions. The Army agrees with the position of the American College of obstetricians and Gynecologists regarding exercise and pregnancy. This guidance is available from medical authorities and the U.S. Army Physical Fitness School (USAPFS). The safety and health of the mother and fetus are primary concerns when dealing with exercise programs.

Vigorous activity does not harm women's reproductive organs or cause menstrual problems. Also, physical fitness training need not damage the breasts. Properly fitted and adjusted bras, however, should be worn to avoid potential injury to unsupported breast tissue that may result from prolonged jarring during exercise.

Although female soldiers must sometimes be treated differently from males, women can reach high levels of physical performance. Leaders must use common sense to help both male and female soldiers achieve acceptable levels of fitness. For example, ability-group running alleviates gender-based differences between men and women. Unit runs, however, do not.

APPENDIX B

POSITIVE PROFILE FORM

PHYSICAL PROFILE
For use of this form see AR 40-501, the proponent agency is the Office of The Surgeon General

| 1 MEDICAL CONDITION | 2 | P | U | L | H | E | S |

3 ASSIGNMENT LIMITATIONS ARE AS FOLLOWS

4 THIS PROFILE IS ☐ PERMANENT ☐ TEMPORARY EXPIRATION DATE

5 THE ABOVE STATED MEDICAL CONDITION SHOULD NOT PREVENT THE INDIVIDUAL FROM DOING THE FOLLOWING ACTIVITIES

☐ GROIN STRETCH ☐ THIGH STRETCH ☐ LOWER BACK STRETCH ☐ NECK & SHLDR STRETCH ☐ NECK STRETCH
☐ HIP RAISE ☐ QUADS STRETCH & BAL ☐ SINGLE KNEE TO CHEST ☐ UPPER BACK STRETCH ☐ ANKLE STRETCH
☐ KNEE BENDER ☐ CALF STRETCH ☐ STRAIGHT LEG RAISE ☐ CHEST STRETCH ☐ HIP STRETCH
☐ SIDE STRADDLE HOP ☐ LONG SIT ☐ ELONGATION STRETCH ☐ ONE ARM SIDE STRETCH ☐ UPPER BODY WT TNG
☐ HIGH JUMPER ☐ HAMSTRING STRETCH ☐ TURN AND BOUNCE ☐ TWO ARM SIDE STRETCH ☐ LOWER BODY WT TNG
☐ JOGGING IN PLACE ☐ HAMS & CALF STRETCH ☐ TURN AND BEND ☐ SIDE BENDER ☐ ALL

6 AEROBIC CONDITIONING EXERCISES	7 FUNCTIONAL ACTIVITIES	8 TRAINING HEART RATE FORMULA
☐ WALK AT OWN PACE AND DISTANCE	☐ WEAR BACKPACK (40 LBS)	MALES 220 FEMALES 225
☐ RUN AT OWN PACE AND DISTANCE	☐ WEAR HELMET	MINUS (-) AGE
☐ BICYCLE AT OWN PACE AND DISTANCE	☐ CARRY RIFLE	MINUS (-) RESTING HEART RATE
☐ SWIM AT OWN PACE AND DISTANCE	☐ FIRE RIFLE	TIMES (X) % INTENSITY
☐ WALK OR RUN IN POOL AT OWN PACE	WITH HEARING PROTECTION	PLUS (+) RESTING HEART RATE
	☐ KP/MOPPING/MOWING GRASS	
☐ UNLIMITED WALKING	☐ MARCHING UP TO ___ MILES	50% - EXTREMELY POOR CONDITION
☐ UNLIMITED RUNNING	☐ LIFT UP TO ___ POUNDS	
☐ UNLIMITED BICYCLING	☐ ALL	60% - HEALTHY, SEDENTARY INDIVIDUAL
☐ UNLIMITED SWIMMING		
	PHYSICAL FITNESS TEST	70% - MODERATELY ACTIVE, MAINTENANCE
☐ RUN AT TRAINING HEART RATE FOR ___ MIN	☐ TWO MILE RUN ☐ WALK	
☐ BICYCLE AT TRAINING HEART RATE FOR ___ MIN	☐ PUSH UPS ☐ SWIM	80% - WELL TRAINED PERSON
☐ SWIM AT TRAINING HEART RATE FOR ___ MIN	☐ SIT UPS ☐ BICYCLE	

9 OTHER

| TYPED NAME AND GRADE OF PROFILING OFFICER | SIGNATURE | DATE |
| TYPED NAME AND GRADE OF PROFILING OFFICER | SIGNATURE | |

ACTION BY APPROVING AUTHORITY

| PERMANENT CHANGE OF PROFILE IS | ☐ APPROVED | ☐ NOT APPROVED |
| TYPED NAME, GRADE, & TITLE OF APPROVING AUTHORITY | SIGNATURE | DATE |

ACTION BY UNIT COMMANDER

THIS PERMANENT CHANGE IN THE PHYSICAL PROFILE SERIAL ☐ DOES ☐ DOES NOT REQUIRE A CHANGE IN THE MEMBER'S
☐ MILITARY OCCUPATIONAL SPECIALTY ☐ DUTY ASSIGNMENT BECAUSE

| TYPED NAME AND GRADE OF UNIT COMMANDER | SIGNATURE | DATE |

PATIENT'S IDENTIFICATION (For typed or written entries give Name - Last, first, middle, grade, date, hospital or medical facility)

ISSUING CLINIC AND PHONE NUMBER

DISTRIBUTION
UNIT COMMANDER - ORIGINAL & 1 COPY
HEALTH RECORD JACKET - 1 COPY
CLINIC FILE - 1 COPY
HQDA (DAPC-EPA), 2461 EISENHOWER AVE
ALEXANDRIA, VA 20310-2200 - 1 COPY

DA FORM 3349, MAY 86   REPLACES DA FORM 5302-R (TEST) AND DA FORM 3349 DATED 1 JUN 80.

Figure B-1

APPENDIX C

PHYSICAL FITNESS LOG

Soldiers can use a physical fitness log to record their fitness goals. The log will serve as a diary of how well they achieve them. Fitness goals are determined before the training begins. The results should closely parallel or exceed the unit's goals. While this is not a requirement, the log may also be used by commanders and supervisors as a record of physical fitness training. Figure C-1 shows an example of a physical fitness log that could be reproduced locally.

PHYSICAL FITNESS LOG

Name (Last)	First	MI	Rank	Sex (Circle) M F	Organization

Week	Date	Phase of Training	Exercise Activity No. 1	Reps, Time, Distance, Etc.	Heart Rate Attained	Exercise Activity No. 2	Reps, Time, Distance, Etc.	Heart Rate Attained

SAMPLE

INSTRUCTIONS: 1. Record each workout by entry of date, phase of training, activities performed, and heart rate attained. Also indicate the repetitions, time, or other indication of performance for each activity in the block provided.

2. At the end of each week, extend the last horizontal line to the left to indicate the end of the week, and place the number of the week in the weekly column.

Figure C-1

APPENDIX D

STATIONARY BICYCLE TEST

Only stationary bicycles which can be calibrated and which have mechanically adjustable resistances may be used to test profiled soldiers on the 6.2-mile (1 O-kilometer), alternate APFT event. Therefore, the event supervisor or scorer must be sure that the stationary bicycle can be accurately adjusted to ensure that the soldier pedals against the correct resistance (force) of 2 kiloponds or 20 newtons. If the stationary bicycle cannot be properly calibrated and adjusted, the soldier may end up pedalling against a resistance which is too great or not great enough. In either case, the test would not provide an accurate indication of the soldier's level of cardiorespiratory fitness.

The best type of stationary bicycle for testing has the following features:

• Calibration adjustment.

• Adjustable resistance displayed in kiloponds or newtons.

• Odometer which accurately measures the distance traveled in either miles or tenths of miles or in kilometers and tenths of kilometers.

Examples of stationary bicycles which meet the above criteria are the mechanically braked Bodyguard 990 and Monark 868. Such bicycles can be used to accurately measure a person's rate of work or the total amount of work. They are often called bicycle ergometers.

If the stationary bicycle has an odometer, the soldier must pedal 6.2 miles (10.0 kilometers or 10,000 meters) against a resistance set at 2 kiloponds or 20 newtons. The test is completed when the soldier pedals 6.2 miles (10.0 kilometers). He receives a "Go" if he is below or at the time allotted for his particular age group and gender. Care should be taken to ensure that, when using a stationary bicycle which measures distance in kilometers, the test is ended at 10 kilometers, not 6.2 kilometers.

There are many electrically operated, stationary bicycles (EOSBS) on the market and in gymnasiums on Army installations. Most of them are designed for physical fitness training. Only a limited number of EOSB models are designed to accurately assess a person's energy expenditure during exercise. Such EOSBS are relatively expensive and are generally found in medical and scientific laboratories. Very few, if any, are found in gymnasiums on Army installations.

Because most of the more common training EOSBS were not designed to accurately assess energy expenditure, they should not be used for the alternate, cardiorespiratory APFT event.

For the sake of accuracy and ease of administration, soldiers designated to be tested on either of the two bicycle protocols should be tested using a moving bicycle IAW the guidelines provided elsewhere in this field manuel. If the mechanical y- braked Bodyguard 990 or Monark 868 is used, however, the tester must ensure that the equipment has been properly calibrated prior to each test.

TABLE D-1

MALES

AGE (YEARS)	17-21	22-26	27-31	32-36	37-41	42-46	47-51	52+
TIME ALLOTTED (MINUTES)	24.0	24.5	25.0	25.5	26.0	27.0	28.0	30.0
CALORIES/MIN.	9.8	9.7	9.5	9.3	9.2	8.9	8.6	8.2
CALORIES/HR.	590	580	570	560	550	535	520	490
N.-METERS/SEC. or WATTS	139	136	133	131	128	124	119	111
TOTAL CALORIES EXPENDED	236	237	238	239	240	241	242	245

TABLE D-2

FEMALES

AGE YEARS	17-21	22-26	27-31	32-36	37-41	42-46	47-51	52+
TIME ALLOTTED (MINUTES)	25.0	25.5	26.0	26.5	27.0	28.0	30.0	32.0
CALORIES/MIN.	9.5	9.3	9.2	9.0	8.9	8.6	8.2	7.8
CALORIES/HR.	570	560	550	545	535	520	490	465
N.-METERS/SEC. or WATTS	133	131	128	126	124	120	110	104
TOTAL CALORIES	237.5	238	239	240	240.5	242	245	248

APPENDIX E

SELECTING THE RIGHT RUNNING SHOE

Choosing a running shoe that is suitable for your particular type of foot can help you avoid some common running- related injuries. It can also make running more enjoyable and let you get more mileage out of your shoes.

Shoe manufacturers are aware that, anatomically, feet usually fall into one of three categories. Some people have "floppy" feet that are very "loose- jointed." Because feet like this are too mobile, they "give" when they hit the ground. These people need shoes that are built to control the foot's motion. At the other extreme are people with "rigid" feet. These feet are very tight-jointed and do not yield enough upon impact. To help avoid impact-related injuries, these people need shoes that cushion the impact of running. Finally, the third type, or normal foot, falls somewhere between mobile and rigid. This type of foot can use any running shoe that is stable and properly cushioned. Use the chart at Figure E-1 to help you determine what kind of foot you have. Then, read the information on special features you should look for in a shoe.

When shopping for running shoes, keep the following in mind:
• Expect to spend between $30 and $100 for a pair of good shoes.
• Discuss your foot type, foot problems, and shoe needs with a knowledgeable salesperson.
• Check the PX for available brands and their prices before shopping at other stores.
• Buy a training shoe, not a racing shoe.
• When trying on shoes, wear socks that are as similar as possible to those in which you will run. Also, be sure to try on both shoes.
• Look at more than one model of shoe.
• Choose a pair of shoes that fit both feet well while you are standing.
• Ask if you can try running in the shoes on a non-carpeted surface. This gives you a feel for the shoes.
• Carefully inspect the shoes for defects that might have been missed by quality control. Do the following:
  -Place the shoes on a flat surface and check the heel from behind to see that the heel cup is perpendicular to the sole of the shoe.
  -Feel the seams inside the shoe to determine if they are smooth, even, and well-stitched.
  -Check for loose threads or extra glue spots; they are usually signs of poor construction.

The shoes' ability to protect you from injury decreases as the mileage on them increases. Record the number of miles you run with them on a regular basis, and replace the shoes when they have accumulated 500 to 700 miles even if they show little wear.

FOOT AND SHOE TYPES

HOW TO SELECT THE RIGHT SHOE

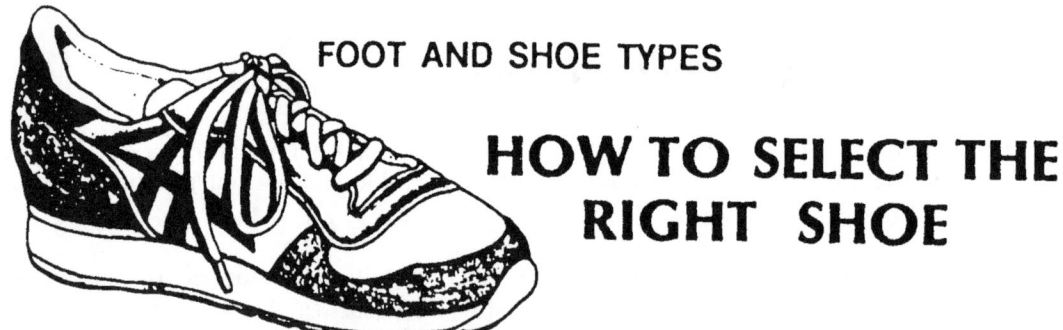

Rigid Foot

Foot tends to stay rigid and does not conform to the ground.

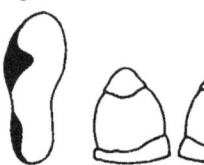 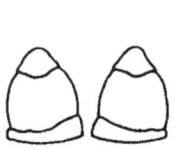

Extreme Wear on Outside of Shoe Sole

Outside Edge of Sole is Broken Down from Rolling Out

High Arch

Typical Injuries

Impact Injuries	Hip Pain
Shin Splints	Heel Pain
Stress Fractures	Ankle Sprains
Knee Pain	

Select a Shoe with these Features

- Maximum Shock Absorbtion and Cushioning
- Dual Density Midsole with the Firmer, Denser Portion on the Outer Edge
- Curved Last
- Flexible Sole
- Elevated Heel
- Avold Flared Heel

Impact Control Shoe

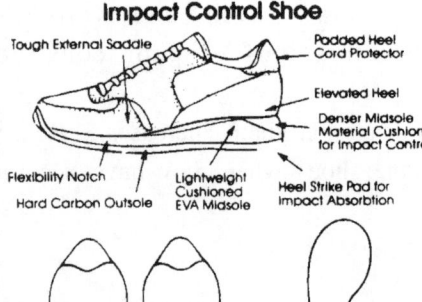

Tough External Saddle

Padded Heel Cord Protector

Elevated Heel

Denser Midsole Material Cushioning for Impact Control

Flexibility Notch

Hard Carbon Outsole

Lightweight Cushioned EVA Midsole

Heel Strike Pad for Impact Absorbtion

Denser Midsole Material

Curver last

Normal Foot

Foot tends to conform to the ground without excess motion.

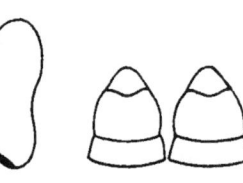 

Normal Wear

Normal Wear

Normal

Select a Shoe with these Features

- Balance of Motion Control and Cushioning
- Flexible Sole
- Durable Outsole Appropriate for the Running Surface

Use this chart to determine the special fit needs you have — then check our selection of shoes.

Floppy Foot

Foot rolls in excessively toward the midline of the body as it bears weight.

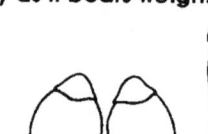 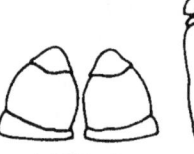

Extreme Wear on Inside and Outside of Shoe Sole

Inside Edge of Sole Is Broken Down from Rolling In

Flat

Typical Injuries

Instability Injuries	Knee Pain (knee cap
Arch Pain	or inside of knee)
Heel Cord Pain	
Shin Pain	

Select a Shoe with these Features

- Dual Density Midsole with the Firmer, Denser Area on the Inside
- External Heel Counter
- Good Arch Support
- Maximum Support
- Straight last

Motion Control Shoe

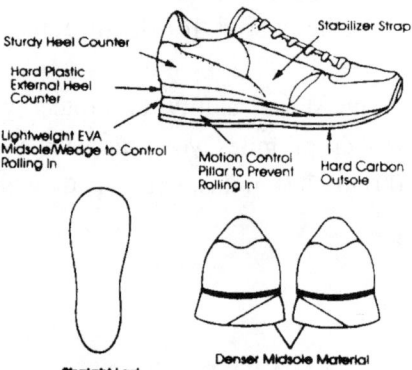

Sturdy Heel Counter

Stabilizer Strap

Hard Plastic External Heel Counter

Lightweight EVA Midsole/Wedge to Control Rolling In

Motion Control Pillar to Prevent Rolling In

Hard Carbon Outsole

Straight Last

Denser Midsole Material

Figure E-1

APPENDIX F

CALCULATION OF $\dot{V}O_2$max

This appendix gives a step-by-step example of how a soldier can calculate $\dot{V}O_2$max using his all-out, 2-mile-run time. This lets interested soldiers compare their fitness levels with others such as athletes whose VO_2max values are published in magazines or journals.

The two equations below convert the 2-mile-run times of males and females to maximum oxygen uptake values. The VO_2max values obtained are shown as the maximum amount of oxygen in milliliters used per kilogram of the person's body weight in one minute during maximum aerobic exercise. $\dot{V}O_2$max values are generally expressed more succinctly as ml O_2/kg x min.

For males, the following equation is used to calculate $\dot{V}O_2$max:

VO_2max = 99.7 - [3.35 x (2-mile-run time in decimal form)].

For females, the following equation is used:

VO_2max = 72.9 - [1.77 x (2-mile-run time in decimal form)].

The example below shows how to use the equation for males. The data is for a 21-year-old male whose all-out, 2-mile-run time is 12 minutes and 36 seconds.

STEP 1. Express the 2-mile-run time as a decimal, and insert it into the equation. When 12 minutes 36 seconds is written as a decimal, it becomes 12.60 minutes. (To determine what fraction of a minute 36 seconds is, divide 36 seconds by the number of seconds in one minute, that is, 60 seconds. Thus, 36/60 = 0.60. This fraction is added to the minute value to give 12.60 as the run time expressed as a decimal.) After putting the decimal form into the equation, the equation should resemble the one below.

VO_2max = 99.7 -[3.35 x (12.60)]

STEP 2. Multiply the decimal form of the 2-mile-run time by 3.35. In this case, we get [3.35 x (12.60)], which equals 42.21. At this point, the equation should resemble the one below.

VO_2max = 99.7 -[42.21].

STEP 3. Subtract the product obtained in Step 2 from 99.7. For our example, we make the following subtraction: 99.7 -[42.21]. This gives a value of 57.49. Thus, the equation should look like the one below.

$\dot{V}O_2$max = 57.49

This calculation reveals that a male whose all-out, 2-mile-run time is 12 minutes 36 seconds will have a $\dot{V}O_2$max of approximately 57.49 ml O_2/kg x min.

To determine how this value or others translates into fitness ratings, refer to Table F-1. It presents information for finding one's level of CR fitness based on $\dot{V}O_2$max. By matching a soldier's value for maximum oxygen uptake with those in the table corresponding to his age group and sex, one gets an adjectival rating (fair, good, superior, etc.).

$\dot{V}O_2max$ AND CR FITNESS CLASSIFICATIONS

CATEGORY	SEX	AGE				
		20-29	30-39	40-49	50-59	60+
SUPERIOR	MALE	54.0+*	52.5+	50.4+	47.1+	45.2+
	FEMALE	46.8+	43.9+	41.0+	36.8+	37.5+
EXCELLENT	MALE	48.2-51.4	46.8-50.4	44.1-48.2	41.0-45.3	38.1-42.5
	FEMALE	41.0-44.2	38.5-41.0	36.3-39.5	32.1-35.2	31.2-35.2
GOOD	MALE	44.2-47.0	42.4-45.3	39.9-43.9	36.7-39.5	33.6-36.7
	FEMALE	36.7-39.5	34.6-37.4	32.3-35.1	29.4-39.9	27.2-30.9
FAIR	MALE	41.0-43.9	38.9-41.6	36.7-39.5	33.8-36.1	30.2-32.4
	FEMALE	33.8-36.1	32.1-33.9	29.5-31.6	26.9-28.7	24.5-26.5
POOR	MALE	37.1-40.3	35.4-38.1	33.0-35.6	30.2-32.5	26.5-29.4
	FEMALE	30.6-32.7	28.7-31.9	26.5-29.4	24.3-26.1	22.8-24.0
VERY POOR	MALE	27.1-36.7	26.5-34.0	24.2-32.3	22.1-29.4	18.3-25.1
	FEMALE	22.6-29.4	22.5-28.0	20.8-25.6	21.1-23.7	17.9-22.1

*$\dot{V}O_2$ max is expressed in ml O_2/kg x min.

Table F-1

Table F-1 lists some values for $\dot{V}O_2max$ along with their associated CR fitness levels. This table was obtained from the Institute for Aerobic Research in Dallas, Texas. These values can be used to classify a soldier's level of CR fitness based on his $\dot{V}O_2max$.

APPENDIX G

PERCEIVED EXERTION

The heart rate has traditionally been used to estimate exercise intensity. However, evidence shows that a person's own perception of the intensity of his exercise can often be just as accurate as the heart rate in gauging his exercise intensity.

The scale in Figure G-1 lets a soldier rate his degree of perceived exertion (PE). This scale consists of numerical ratings for physical exercise followed by their associated descriptive ratings.

PERCEIVED EXERTION (PE) SCALE

NUMERICAL RATING	VERBAL RATING
6	
7	very, very light
8	
9	very light
10	
11	fairly light
12	
13	somewhat hard
14	
15	hard
16	
17	very hard
18	
19	very, very hard
20	

Figure G-1

To judge perceived exertion, estimate how difficult it feels to do the exercise. Do not be concerned with any one single factor such as shortness of breath or work intensity. Instead, try to concentrate on the total inner feeling of exertion.

Multiplying the rating of perceived exertion by 10 roughly approximates the heart rate during exercise. For example, a PE of 14, when multiplied by 10, equals 140.

Most soldiers with THRs between 130 and 170 BPM would exercise between a PE of 13 (somewhat hard) and 17 (very hard).

Although either percent of maximum heart rate or perceived exertion may be used during exercise, the most valid method for calculating THR is percent HRR.

THE MAJOR SKELETAL MUSCLES OF THE HUMAN BODY

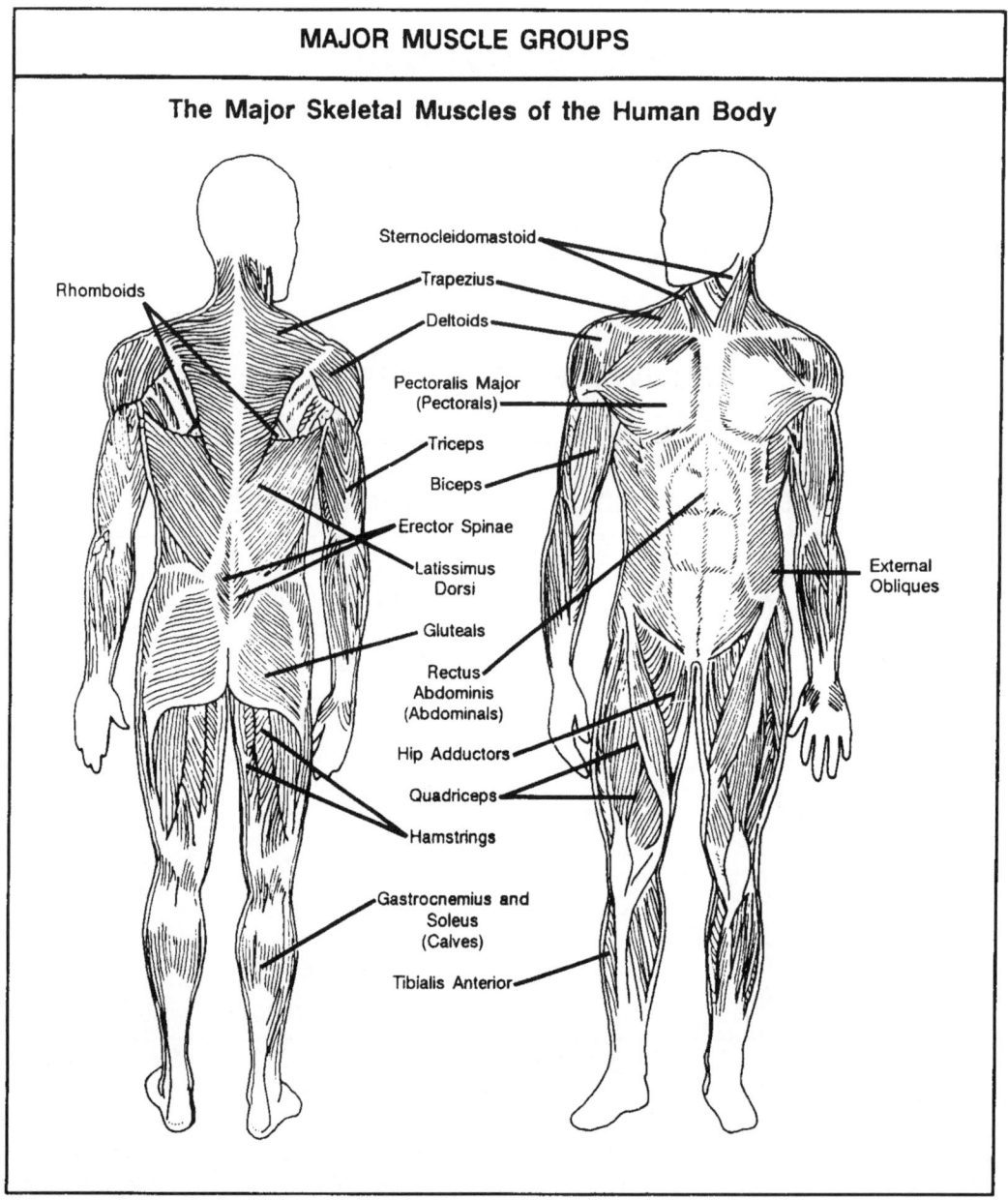

MAJOR MUSCLE GROUPS

The Major Skeletal Muscles of the Human Body

Figure H-1

The iliopoas muscle (a hip flexor) cannot be seen as it lies beneath other muscles. It attaches to the lumbar vertebrae and the femur.

ESSENTIAL ARMY MANUAL DVD

ALL DOCUMENTS ON THE DISC ARE IN THE PDF FORMAT AND IT IS NECESSARY TO HAVE ADOBE ACROBAT INSTALLED IN ORDER TO VIEW THE DOCUMENTS. A LINK IS PROVIDED ON THE DISC TO DOWNLOAD THE ACROBAT READER IF YOU DO NOT ALREADY HAVE IT.

IF YOU ENCOUNTER PROBLEMS READING THE FILES FROM THIS DISC, AND THE ACROBAT READER TROUBLESHOOTING STEPS ARE NOT SUCCESFUL, PLEASE CONTACT US DIRECTLY FOR ASSISTANCE, AND IF NECESSARY A PROMPT REPLACEMENT. DO NOT CONTACT YOUR RETAILER OR RESELLER.

PLEASE E-MAIL US AT: INFO@ARMYMANUALSONLINE.COM

PRIVATELY COMPILED AND PUBLISHED BY
WaterMark, Inc.
NOT THE FEDERAL GOVERNMENT